Selected Titles in This Series

233 **Hanna Nencka, Editor,** Low dimensional topology, 1999

232 **Krzysztof Jarosz, Editor,** Function spaces, 1999

231 **Michael Farber, Wolfgang Lück, and Shmuel Weinberger, Editors,** Tel Aviv topology conference: Rothenberg Festschrift, 1999

230 **Ezra Getzler and Mikhail Kapranov, Editors,** Higher category theory, 1998

229 **Edward L. Green and Birge Huisgen-Zimmermann, Editors,** Trends in the representation theory of finite dimensional algebras, 1998

228 **Liming Ge, Huaxin Lin, Zhong-Jin Ruan, Dianzhou Zhang, and Shuang Zhang, Editors,** Operator algebras and operator theory, 1999

227 **John McCleary, Editor,** Higher homotopy structures in topology and mathematical physics, 1999

226 **Luis A. Caffarelli and Mario Milman, Editors,** Monge Ampère equation: Applications to geometry and optimization, 1999

225 **Ronald C. Mullin and Gary L. Mullen, Editors,** Finite fields: Theory, applications, and algorithms, 1999

224 **Sang Geun Hahn, Hyo Chul Myung, and Efim Zelmanov, Editors,** Recent progress in algebra, 1999

223 **Bernard Chazelle, Jacob E. Goodman, and Richard Pollack, Editors,** Advances in discrete and computational geometry, 1999

222 **Kang-Tae Kim and Steven G. Krantz, Editors,** Complex geometric analysis in Pohang, 1999

221 **J. Robert Dorroh, Giséle Ruiz Goldstein, Jerome A. Goldstein, and Michael Mudi Tom, Editors,** Applied analysis, 1999

220 **Mark Mahowald and Stewart Priddy, Editors,** Homotopy theory via algebraic geometry and group representations, 1998

219 **Marc Henneaux, Joseph Krasil′shchik, and Alexandre Vinogradov, Editors,** Secondary calculus and cohomological physics, 1998

218 **Jan Mandel, Charbel Farhat, and Xiao-Chuan Cai, Editors,** Domain decomposition methods 10, 1998

217 **Eric Carlen, Evans M. Harrell, and Michael Loss, Editors,** Advances in differential equations and mathematical physics, 1998

216 **Akram Aldroubi and EnBing Lin, Editors,** Wavelets, multiwavelets, and their applications, 1998

215 **M. G. Nerurkar, D. P. Dokken, and D. B. Ellis, Editors,** Topological dynamics and applications, 1998

214 **Lewis A. Coburn and Marc A. Rieffel, Editors,** Perspectives on quantization, 1998

213 **Farhad Jafari, Barbara D. MacCluer, Carl C. Cowen, and A. Duane Porter, Editors,** Studies on composition operators, 1998

212 **E. Ramírez de Arellano, N. Salinas, M. V. Shapiro, and N. L. Vasilevski, Editors,** Operator theory for complex and hypercomplex analysis, 1998

211 **Józef Dodziuk and Linda Keen, Editors,** Lipa's legacy: Proceedings from the Bers Colloquium, 1997

210 **V. Kumar Murty and Michel Waldschmidt, Editors,** Number theory, 1998

209 **Steven Cox and Irena Lasiecka, Editors,** Optimization methods in partial differential equations, 1997

208 **Michel L. Lapidus, Lawrence H. Harper, and Adolfo J. Rumbos, Editors,** Harmonic analysis and nonlinear differential equations: A volume in honor of Victor L. Shapiro, 1997

(*Continued in the back of this publication*)

Low Dimensional Topology

CONTEMPORARY MATHEMATICS

233

Low Dimensional Topology

Proceedings of a Conference on
Low Dimensional Topology
January 12–17, 1998
Funchal, Madeira, Portugal

Hanna Nencka
Editor

American Mathematical Society
Providence, Rhode Island

This volume is the record of a conference in Low Dimensional Topology, which took place in Funchal, Madeira, Portugal on January 12–17, 1998.

1991 *Mathematics Subject Classification.* Primary 57Mxx, 57Nxx, 57Pxx, 57Qxx, 57Rxx, 57Sxx, 57Txx.

Library of Congress Cataloging-in-Publication Data

Conference on Low Dimensional Topology (1998 : São João da Madeira, Portugal)
Low dimensional topology : proceedings of a Conference on Low Dimensional Topology, January 12–17, 1998, Funchal, Madeira, Portugal / Hanna Nencka, editor.
p. cm. — (Contemporary mathematics, ISSN 0271-4132 ; 233)
Includes bibliographical references.
ISBN 0-8218-0884-2 (alk. paper)
1. Low dimensional topology Congresses. I. Nencka, Hanna, 1956– . II. Title. III. Series: Contemporary mathematics (American Mathematical Society) ; v. 233.
QA612.14.C63 1998
514′.2—dc21 99-14986
CIP

∞ The paper used in this book is acid-free and falls within the guidelines
established to ensure permanence and durability.
Visit the AMS home page at URL: http://www.ams.org/

10 9 8 7 6 5 4 3 2 1 04 03 02 01 00 99

Contents

PREFACE

This volume of the Contemporary Mathematics series is an outgrowth of the conference on "Low Dimensional Topology" which took place at the University of Madeira (Portugal) in January 1998.

The aim of the conference was to bring together active scientists from America, Asia, and Europe, working in the field of low dimensional topology.

The present collection of contributions, which reflects the many-sided development and actual progress in low dimensional topology, is organized as follows.

The first two papers of this volume are devoted to characterisations of 3-manifolds and PL-manifolds. The first one, by V. Poénaru, considers the following conjecture: *If M^3 is a closed 3-manifold with infinite π_1, then the universal covering space $\widetilde{M^3}$ is simply connected at infinity.* This work is strongly related to Poénaru's programme for the Poincaré conjecture. The second one, by L. Funar, is devoted to the set of cubical decompositions of a compact manifold mod out by some combinatorial moves. It has been shown that there are obstructions for two cubulations of the same PL-manifold. It has been stated that if $\Lambda(n) \subset \mathbb{Z}^{n+1}$ is a sublattice, depending only on the dimension, then the obstruction in $\mathbb{Z}^{n+1}/\Lambda(n)$ is not trivial.

The next two contributions are linked to categories in low dimensional topology. First, I. Bobtcheva and F. Quinn consider a numerical approach to tortile categories, which appears as an intermediate structure in conformal and quantum field theory and low dimensional topology. The objective of the paper is to give a setting for direct calculations in categories derived from representations of root-of-unity deformations of Lie algebras of arbitrary rank (quantum groups) and to describe algorithms used to derive all structure information from a smaller set of data. In the following paper, I. Bobtcheva considers a semisimple stable autonomous tensor category $\mathcal{A}$ over a field K. Associating an element $Q(P) \in K$, invariant under the Andrews-Curtis moves, to any group presentation with finite number of generators, she has shown that this is the same invariant as the one produced by Quinn's algorithm.

The goal of the next article, by A. Kawauchi, is to characterize the infinite cyclic covering quadratic form of a link as a restriction of the Seifert form (extended over rationals) of every connected Seifert surface of the link, giving a generalization of Erle's result. Kawauchi also shows that, for the sphere-slice link, the induced non-singular form of the quadratic form is null-cobordant, the signature of the canonical linking matrix is zero, and the nullity invariants are all even.

The next four papers are devoted to topological quantum field theories, quantum invariants, and mapping class groups. First, H. Murakami considers a polynomial invariant for rational homology 3-spheres, derived from the quantum $SU(2)$-invariants associated with the trivial first cohomology class modulo 2. In the next paper, G. Masbaum gives a simplified proof of the following proposition: *The image of the mapping class* $\alpha^{-1}\beta$ *in the TQFT-representation corresponding to* $SU(2)$ *at level k has infinite order except if k=1,2,3, or 8.* The goal of T. Ohtsuki's paper is to construct quantum Hilbert space for the perturbative $SO(3)$-invariant which is obtained by asymptotic expansion of the quantum $SO(3)$-invariant, and to describe the two structures of the quantum Hilbert space of T^2. The last contribution of this series is by A. Cattaneo, who tells how to associate invariants of 3-dimensional rational homology spheres and knots to certain combinations of trivarient diagrams.

The final part of the volume is devoted to knots, links, braids, and free groups. The first contribution in this series is by H. Morton, who proposes a method for finding the multivariable Alexander polynomial of a link presented as the closure of a braid. He shows that the Alexander polynomial of the closed braid can be derived from the Torres-Fox formula. In the next paper, K. Kobayashi considers four kinds of links, that is link-homotopically trivial links in the strong sense, boundary links, h-split links, and Brunnian links. The relations among them are considered. Some geometrical conditions to become h-split link are shown as well. K. Habiro, T. Kanenobu, and A. Shima construct a Vassiliev-like filtration on the free abelian group generated by the set of ribbon 2-links in 4-space in two equivalent ways. A finite type invariant for a ribbon 2-knot is given, and the notion can be extended to ribbon 2-links. It has been shown that the higher derivatives of the normalized Alexander polynomial of ribbon 2-knots at $t = 1$ are of finite type. S. Kamada is interested in the arrangement of Markov moves for 2-dimensional braids, and he considers quasi-commutativity of 2-dimensional Markov moves. In particular, if two simple 2-dimensional braids represent the same surface link type, then they are related by a Markov moves equivalence through simple 2-dimensional braids. A relation between unknotting numbers of surface links and those of 2-dimensional braids is given. A criterion for signed unknotting number is studied by P. Traczyk in his contribution. Various versions of signed unknotting number are investigated, and tables for prime knots up to nine crossing are given. The goal of the next paper, by H. Nencka, is to give generalisation of Artin's braid relations. A class of relations among the generators of braids groups is proposed which partially solves Xu's problem. The last paper, by J. O'Hara, deals with the asymptotic behavior of energy of polynomial knots. A regularization of "modified electrostatic energy" of charged knots is studied. New energy functionals for smooth knots are defined, and formulae for the asymptotic behaviour of the energy of polynomial knots are given.

The organizers of the conference were particularly pleased to be able to hold the meeting on the island of Madeira, and in receiving financial help from diverse sources which will be named explicitly below.

I wish to express my gratitude to: Francisco Santos (Regional Secretary of Education), Jose Manuel Castanheira da Costa (Rector of the University of Madeira), Ludwig Streit (Director of the Center of Mathematical Science) and Rita Vasconcelos (President of the Mathematics Department), for making the conference possible.

I am particularly grateful to: Valentin Poénaru, Frank Quinn, and Vladimir Turaev for their advice, suggestions, observations, and help, during the conference and the editing of the volume.

I especially thank Philippe Combe for his help and advice in the editing of the book.

I would like to thank Antonio Pires and Gabriel Gomes, my colleagues from the University of Madeira, and Ana Paula Fernandes, our secretary, for their help in the organization of the conference.

The conference was made possible thanks to the support of several organizations. I wish to acknowledge support from Fundaçaõ de Ciéncia e Tecnôlogia, PRAXIS XXI, Pluriannual project, Fundaçaõ Luso-Americana, British Council, Camara Municipal do Funchal, BANIF, and the Secretaria Regional de Turismo e Cultura.

Funchal, January 1999,
Hanna NENCKA

Contemporary Mathematics
Volume **233**, 1999

π_1^∞ and infinite simple homotopy type in dimension 3

Valentin POÉNARU

To the memory of my friend Nico Kuiper, from whom I learned so much.

1. Introduction

In this note we announce several results which, in some way or another, are all related to the following well-known

MAIN CONJECTURE. *"If M^3 is a closed 3-manifold with infinite π_1, then the universal covering space $\widetilde{M^3}$ is simply connected at infinity."*

Since any closed orientable 3-manifold is a connected sum of irreducible factors which are $K(\pi_1, 1)$'s unless they are $S^2 \times S^1$'s or have finite π_1, it would suffice to prove this conjecture for the special case when $M^3 = K(\pi_1 M^3, 1)$. On the other hand, in 1983 M. Davis [Da] showed that for any $n > 3$ there exist closed manifolds $M^n = K(\pi_1 M^n, 1)$ with $\pi_1^\infty \widetilde{M^n} \neq 0$: Here and after, if X is a locally compact space we will write $\pi_1^\infty X = 0$ if X is simply connected at infinity and $\pi_1^\infty X \neq 0$ if it is not. [I remind the reader that a locally compact simply-connected space X is, by definition, simply connected at infinity if for every compact subset $k \subset X$ there is a larger compact set $k \subset K \subset X$ such that any closed loop of $X - K$ is null-homotopic in $X - k$; the classical example of an open simply connected 3-manifold which is not simply connected at infinity is the well-known Whitehead manifold [Wh2], about which more will be said later on.]

The main body of this paper (i.e. this and the next two paragraphs) are supposed to be readable without any special pre-requisites. More technical matters which require at least a superficial knowledge of things like [Po0], [Po1], [Po2], ... have been relegated to the appendices. Actually, the shorter papers [Ga] and [Po12] should be enough to provide adequate background for all of the present work.

Together with the Poincaré Conjecture the "main conjecture" above would give a complete description of all the universal covering spaces of closed 3-manifolds M^3; the (conjectural) statement is that $\widetilde{M^3} = S^3 \quad c(\pi_1 M^3)$, where $e(\pi_1 M^3)$ is the space of ends of $\pi_1 M^3$ (and where it is understood that in the case when there are infinitely many ends, $e(\pi_1 M^3)$ is tamely embedded inside S^3). All this "topological 3-dimensional uniformization theorem" is, of course, a (fairly large?)

1991 *Mathematics Subject Classification.* 57M50, 57M10, 57M30.

chunk of Thurston's larger geometrization conjecture [Th], which would give the "geometrical 3-dimensional uniformization".

In a different vein, let us also notice that to the extent to which the issue of $\pi_1^\infty \widetilde{M^3}$ is a question which only concerns $\pi_1 M^3$, our "main conjecture" is a purely group-theorectical problem, in topological disguise. [It actually makes sense to talk about $\pi_1^\infty G$ for a finitely presented group G.]

Although we will have constantly, at the back of our minds, the main conjecture above, several of our results apply to open simply connected 3-manifolds of the most general type (i.e. not necessarily universal covering spaces). This is, for instance, the case for the next theorem, which sets the stage for the sequel.

THEOREM 0. *"Let V^3 be an open simply-connected 3-manifold. We assume that there is an $n \in Z_+$ such that there is a* **proper** *codimension zero smooth submanifold $X^{n+3} \subset V^3 \times B^n$ (where B^n is the standard smooth n-ball), with the following two properties*

1) X^{n+3} engulfs the zero-crossection of $V^3 \times B^n$

$$V^3 \times 0 \subset X^{n+3} \subset V^3 \times B^n \tag{1}$$

2) X^{n+3} is geometrically simply connected, i.e. it possesses a smooth **proper** *handlebody decomposition without handles of index one.*
Then, $\pi_1^\infty V^3 = 0$."

[One way to explain what geometric simple connectivity means for our X which is noncompact with non-empty boundary, goes as follows. We have a **proper** smooth Morse function $X \xrightarrow{f} R_+$ with all of its critical points contained in $int \quad X$ and such that:

1) All the critical points of f have index $\lambda \neq 1$.
2) Next we consider the Morse function $\partial X \xrightarrow{f|\partial X} R$ and its "**meaningful**" critical points $c \in \partial X$; these are, by definition, those critical points c of $f \quad | \quad \partial X$ for which the inclusion

$$f^{-1}(-\infty, f(c)-\epsilon] \subset f^{-1}(-\infty, f(c)+\epsilon]$$

is **not** a homotopy equivalence. Then, the index of $f \quad | \quad \partial X$ at c is $\lambda \neq 1$.]
Here is an immediate consequence of Theorem 0.

COROLLARY 1. *"Let V^3 be an open simply connected manifold. If there is an $n \in Z_+$ such that $V^3 \times B^n$ is geometrically simply connected, then $\pi_1^\infty V^3 = 0$."*

This is proved in [Po1] and the arguments can be extended so as to prove Theorem 0 too. An even much more general result is given by Louir Funar in [Fu].

Both Theorem 0 and Corrollary 1 become false if V^3 is replaced by a V^d with $d > 4$. In order to explain the special role of dimension three for our present problem, I will start with a very general definition: a locally finite simplicial complex Z will be said to be **weakly** geometrically simply connected if it has an exhaustion by **finite** simplicial subcomplexes which are simply connected. In the context of Corrollary 1 above, the geometric simple connectivity of $V^3 \times B^n$ easily implies weak geometric simple connectivity. After this, the key step is to show that weak geometric simple connectivity descends from $V^3 \times B^n$ to V^3 itself; details can be found in [Po1], [Po12]. But the point is that an analogue of Dehn's Lemma is needed here and hence, the 3-dimensionality of $V^3$is essential for this part of the argument.

Now, a compact simply-connected 3-dimensional manifold has only spheres as boundary components and, via Van Kampen, we get the following implication which also only holds for 3-manifolds:

V^3 is weakly geometrically simply connected$\Rightarrow$ $\pi_1^\infty V^3 = 0$.

This is, in a nutshell, how the proof of Corollary 1 goes. I now open an aside, in order to say a few words concerning a notion which bears some formal resemblence to weak geometric simply connectivity. A smooth n-manifold X^n (which we assume either open or compact bounded) is said to be geometrically simply connected **at long distance** if int X^n has an exhaustion by compact bounded n-manifolds which are geometrically simply connected. So, X^n geometrically simply connected at long distance implies that int X^n is weakly geometrically simply connected, but at least in the critical dimensions $n = 3$, $n = 4$ the converse is far from being clear.

REMARKS.

A) Let X^n be compact bounded and let us denote by $X^n_{\text{small}} = X^n - \partial X^n \times [0, 1)$ a smaller copy of X^n obtained by deleting an open boundary collar. In this context, there is a much more manageable way to express geometric simple connectivity at long distance: one can find a compact geometrically simply connected smooth submanifold Y^n such that $X^n_{\text{small}} \subset Y^n \subset X^n$.

B) In [Po5] it is proved that if Δ^3 is a homotopy 3-ball, then the smooth 4-manifold $\Delta^3 \times I$ is geometrically simply connected at long distance. This is itself a corollary of the following infinitistic

SMOOTH TAMENESS THEOREM. *"For any Δ^3 there is a smooth non-compact 4-manifold with non-empty boundary V^4 such that:*

0) int $V^4 = R^4$, $\partial V^4 = \sum_1^\infty S_i^1 \times$ int D_i^2.

1) We have a diffeomorphism of open 4-manifolds $int((\Delta^3 \times I)\#\infty\#(S^2 \times D^2)) = V^4 + \{$the 2-handles $\sum_1^\infty D_i^2 \times intD_i^2$ which correspond to $\partial V^4\}$.

2) Every finite truncation $V^4 - \sum_{N+1}^\infty S_i^1 \times intD_i^2$ can be smoothly compactified into a standard B^4 "

It is also proved in [Po6] that if $\Delta^3 \times I$ is geometrically simply connected, then $\Delta^3 = B^3$. This is itself a corrollary of the following infinitistic

STRANGE COMPACTIFICATION THEOREM. *"If $\Delta^3 \times I$ is geometrically simply connected, then we can find a smooth pair*

$$LK = (B^4, \sum_1^k S_i^1 \times D_i^2 \subset \partial B^4)$$

such that

1) $(\Delta^3 \times I)\#k\#(S^2 \times D^2) \overset{\text{DIFF}}{=} B^4 + \{$the k handles of index two defined by $LK\}$.

2) Let $Z^4 = B^4\#\infty\#(S^2 \times D^2)$, the connected sum being made far from $\sum_1^k S_i^1 \times D_i^2$ and let

$$V^4(LK) \underset{\text{def}}{=} Z^4 - (\partial Z^4 - \sum_1^k S_i^1 \times intD_i^2)$$

Then, there is a smooth 4-manifold W^4 *such that* $W^3 \underset{\text{def}}{=} \partial W^4$ *is an open connected and simply-connected 3-manifold, endowed with a proper smooth embedding*

$$(V^4(LK), \partial V^4(LK)) \subset (W^4, W^3)$$

inducing a diffeomorphism $intV^4(LK) = intW^4$ *and such that* $\pi_1(W^3 - \partial V^4(LK))$ *is* $F_k =$ *the free group with* k *generators."*

As a first step in trying to connect the two results concerning $\Delta^3 \times I$ stated above, in [Po13], [Po14] we prove the following

THEOREM 2. *"Let* Δ^4 *be a compact simply connected smooth 4-manifold, such that* $\partial\Delta^4$ *is a homology sphere. We also assume that* Δ^4 *is geometrically simply connected at long distance, i.e. that there is a compact smooth geometrically simply connected* X^4 *such that (compare this with (1) above)*

$$\Delta^4_{\text{small}} \subset X^4 \subset \Delta^4.$$

Then, the smooth open 4-manifold

$$int(\Delta^4 \# \infty \# (S^2 \times D^2))$$

is geometrically simply-connected."

If there is a certain similarity between the statements of Theorem 0 and Theorem 2, this is not completely fortuitous. What I want to stress here is only the following fact: both results actually rely heavily on J.H.C. Whitehead's smooth Hauptvermutung Theorem [Wh1], which is true without any dimensional restrictions; for this reason it is essential that we stay in the DIFF category throughout the present paper.

After this long aside, we move back to our main theme.

The results which will be stated in the next paragraphs are part of a general, not yet completely fulfilled program (an early crude version of which can be found in [Po9] and an updated version in the Postscript V at the end of the present paper) for proving the main conjecture by appealing, eventually, to Theorem 0. It turns out, from this viewpoint, that it will be better to replace the open simply-connected V^3 by a related object: we consider an infinite collection of closed, totally discontinuous, tame subsets $K_1, K_2, \ldots$ of V^3. By "tame" it is meant here that each K_i is contained inside a smoothly embedded are $A_i \subset V^3$. Moreover, it will be assumed that $K_i \cap K_j = \emptyset$ (if $i \neq j$) and that any compact set $C \subset V^3$ meets only finitly many K_i's (we will write, symblically, $K_i \to \infty$). So, we will also assume that $A_i \cap A_j = \emptyset$ and $A_i \to \infty$.

We set $F = \sum\limits_1^\infty K_i$ and introduce the following open 3-manifold

$$V^3_h \underset{\text{def}}{=} V^3 - F \tag{2}$$

The precise topology of this V^3_h, which we call, generically, " V^3 with very many holes" depends on the topology of F (and it is of course canonically unique when all the K_i's are Cantor sets), but what ever that might be, it is easy to see that $\pi_1^\infty V^3 = \pi_1^\infty V^3_h$; so in our present context, V^3_h is as good as V^3. I make the following

CONJECTURE (C). *"Let M^3 be a closed 3-manifold with infinite π_1. There exists an $\widetilde{M_h^3}$ (i.e. a specific $\widetilde{M^3}$ with very many holes) and an $n \in Z_+$, for which we can find a* **proper** *submanifold $X^{n+3} \subset \widetilde{M_h^3} \times B^n$, which is geometrically simply connected and which engulfs the 0-section $\widetilde{M_h^3} \times 0$."*

In view of Theorem 0, it is clear that conjecture (C) implies our main conjecture. Postscript V at the end of this note outlines a tentative project for proving conjecture (C).

In order to finish the setting of the stage, let me also add that for M^3's which are such that $\pi_1 M^3$ verifies any of a certain rather large class of "nice geometric conditions" (for the corresponding Cayley graphs), the main conjecture is indeed verified. The list of conditions in question includes Cannon's almost convexity [Can] (i.e. Gromov's hyperbolicity's [Gr] too), Thurston's combability, i.e. automatic groups too [E-C-H-L-P-T] (with some very mild restrictions), Casson's condition [S-G], etc. These kind of results have been proved by the author, by A. Casson, by Bestvina-Mess and by others (see for instance [Po2], [Po3], [Po-T1], [Po-T2], [S-G], [B-M], but this list is far from being exhaustive). It is worthwhile noticing that, with hindsight, the groups appearing in Davis's counterexamples [Da] are very nice indeed (actually they are almost convex or even hyperbolic [Da-S]); this makes the results connecting $\pi_1 M^3$ and $\pi_1^\infty \widetilde{M^3}$ purely 3-dimensional too.

After this setting of the stage, our main results concerning conjecture (C) will be outlined in the next sections.

TERMINOLOGY. *The set-theoretical notion of* **proper inclusion** $X \xrightarrow{i} Y$, *i.e. that for a compact $K \subset Y$ the $i^{-1}K \subset X$ is also compact, obviously plays a key role in the present paper and so it will be written in* **fat** *letters. On the other hand, an inclusion map between manifolds $M \xrightarrow{j} N$ will be called proper (in usual letters) if $j^{-1}(\partial N) = \partial M$, and such a j need not be* **proper** *too.*

I wish to thank Louis Funar, David Gabai, Steve Gersten, Lucien Guillou, John Hubbard, Francois Laudenbach, Curt McMullen, Alexis Marin, Barry Mazur, Larry Siebenmann, John Stallings and Dennis Sullivan for helpful discussions and/or correspondence. I also want to thank the IHES (Bures-sur-Yvette) for its hospitality during part of the time when the results of this paper were obtained. Actually a first draft of the present work was an IHES preprint (M-95-6).

2. UNIVERSAL COVERING SPACES (THEOREM 3).

We consider now an M^3 which is closed and with infinite π_1, like in the main conjecture.

THEOREM 3. *"There exists an $\widetilde{M_h^3}$, such that for any integer $n \geq 3$ there is a* **proper** *condimension one (stratified) smooth submanifold $\Sigma = \Sigma^{n+1} \subset \widetilde{M_h^3} \times S^{n-1} = \partial(\widetilde{M_h^3} \times B^n)$ such that:*

1) *We can find a codimension zero smooth submanifold $X^{n+3} \subset \widetilde{M_h^3} \times B^n - \Sigma$ containing $\widetilde{M_h^3} \times 0$, and which is* **proper** *in $\widetilde{M_h^3} \times B^n - \Sigma$.*
2) *X^{n+3} is geometrically simply connected."*

The proof, which makes heavy use of [Po7], [Po8] and [Po11], will be given elsewhere. The material from [Po7] and [Po8] is briefly reviewed in the Appendices I, II, III to the present paper, while the main result of [Po11] is explained in the next section. We will offer here only some comments.

C) The stratification of the "**bad locus**" Σ is really of the simplest possible type. There are only strata of codimension zero and one (i.e. of dimensions $n+1$ and n respectively), and along any strata of codimension one we have exactly three strata of codimension zero which meet.

D) Here are some other nice extra features, in the context of Theorem 3.

D.1. The bad locus Σ has various $\pi_1 M^3$-invariance properties, the crudest of which can be expressed as follows. There is a $\pi_1 M^3$-invariant **proper** stratified submanifold $\Sigma_1 \subset \widetilde{M^3} \times B^n$ such that $\Sigma = \Sigma_1 \cap (\widetilde{M^3_h} \times B^n)$. [One has actually to use here the "correct" embedding $\widetilde{M^3_h} \subset \widetilde{M^3}$ (see [Po8]), which is not exactly the one suggested by (2).]

D.2. Our X^{n+3} is the regular neighbourhood of a locally finite simplicial complex. (This is neither true for $\widetilde{M^3_h}$, $\widetilde{M^3_h} \times B^n$ nor for $\widetilde{M^3_h} \times B^n - \Sigma$.)

E) But X^{n+3} is only a **proper** submanifold of $\widetilde{M^3_h} \times B^n - \Sigma$ and not necessarily of $\widetilde{M^3_h} \times B^n$ itself. So, our Theorem 3 does not imply conjecture (C) (and hence neither the main conjecture). Actually, without any loss of generality, we can ask that the closure of X^{n+3} in $\widetilde{M^3_h} \times B^n$ be the full $X \cup \Sigma$.

F) At the end of section 1 we had noticed that there are a certain number of theorems ([Po2], [Po3], [S-G], [B-M]...) taking the general form

$$(3) \qquad (\pi_1 M^3 \text{satisfies some nice geometric condition,}$$
$$\text{like for instance almost convexity}) \Rightarrow (\pi_1^\infty \widetilde{M^3} = 0).$$

From the higher, vantage point of Theorem 3, here is how (3) can be understood. Whenever we have a "nice geometric condition" we can apply the machinery which proves Theorem 3 so as to get, additionally, $\Sigma = \emptyset$, and then we can invoke Theorem 0 in order to conclude that $\pi_1^\infty \widetilde{M^3_h} = 0$ (and hence that $\pi_1^\infty \widetilde{M^3} = 0$). But I do not think that when attacking our Conjecture (C) in full generality we can avoid facing the bad locus Σ; we will have to cope with it (see also what is coming next.)

3. OPEN SIMPLY CONNECTED 3-MANIFOLDS, IN GENERAL (THEOREM 4).

The proof of Theorem 3 also relies on ingredients which apply to *all* simply connected open 3-manifolds, even possibly wild ones and which we will partially review now.

I start by reminding the reader what **weak topology** is all about, in a context useful for our present purposes. We will consider, for some fixed $m \in Z_+$, a sequence of smooth m-manifolds and smooth embeddings

$$(4) \qquad Y_1 \subset Y_2 \subset \ldots$$

Here, for $Y_n \overset{i}{\hookrightarrow} Y_{n+1}$ it is only required that i be injective, with nondegenerate tangent map, but no special restrictions are imposed to $i \mid \partial Y_n$, as far as ∂Y_{n+1}

is concerned. We define $\varinjlim Y_n$ as being the set $\overset{\infty}{\underset{1}{\cup}} Y_n$ endowed with the following topology

$$E \subset \overset{\infty}{\underset{1}{\cup}} Y_n \text{ is closed } \leftrightarrow \text{ all } E \cap Y_n\text{'s are closed.} \tag{5}$$

When the ∂Y_n's are non-void this can be a rather weird object. The following example should be useful for understanding our Theorem 4 below.

A TOY MODEL. We work in $R^2 = (x, y)$ and we start with the lower half plane $A = \{y \leq 0\}$. Let $p_1, p_2, \ldots$ be a bounded monotonly increasing sequence of points on ∂A, converging to some p_∞. Let also $I_n \subset \partial A$ be a small arc centered at p_n. We assume the I_n's two-by-two disjoined, and we consider

$$D_n = \{\text{the half-disk of diameter } I_n \text{contained in the upper half-plane}\}.$$

The subsets $Y_n = A \cup \sum_1^n D_i$, $Y_\infty = A \cup \sum_1^\infty D_i$ of R^2 are submanifolds (we can smoothen them in an obvious way). The $\varinjlim Y_n$ is not a manifold but the obvious "identity" map

$$\varinjlim Y_n \xrightarrow{f} Y_\infty$$

is a **continuous bijection**. The inverse map f^{-1} fails to be continuous exactly at p_∞. One should notice that if $q_n \in D_n - I_n$ then, in Y_∞, the sequence $q_1, q_2, \ldots$ is such that $\lim q_n = p_\infty$, while in $\varinjlim Y_n$ the same sequence does not converge at all. Not only is $\varinjlim Y_n$ not a manifold, it isn't even metrizable. This toy-model should be kept in mind for what comes next. We start now with an arbitrary open simple connected 3-manifold V^3.

THEOREM 4. *"There exists a V_h^3 such that for any integer $n \geq 2$ there is a* **proper** *codimension one (stratified) smooth submanifold*

$$\Sigma = \Sigma^{n+1} \subset V_h^3 \times S^{n-1} = \partial(V_h^3 \times B^n) \tag{6}$$

and a sequence of smooth $(n+3)$-dimensional manifolds and smooth embeddings

$$X_1 \subset X_2 \subset \ldots \tag{7}$$

with the following properties.

1) *X_1 is geometrically simply connected. Each inclusion $X_i \subset X_{i+1}$ consists of finitely many Whitehead dilatations and additions of handles of index $\lambda \neq 1$.*
2) *There exist smooth embeddings $X_i \xrightarrow{f_i} V_h^3 \times B^n$ such that the diagram*

$$\begin{array}{ccc} X_i & \xrightarrow{\subset} & X_{i+1} \\ \downarrow f_i & & \downarrow f_{i+1} \\ V_h^3 & \xrightarrow{=} & V_h^3 \end{array}$$

commutes. The map

$$\varinjlim X_i \xrightarrow{f} V_h^3 \times B^n \tag{8}$$

which is induced by the f_i's is a **continuous bijection**.

3) *f^{-1} fails to be continuous exactly along Σ.*

4) $\varinjlim X_i - f^{-1}\Sigma$ *has a natural structure of smooth manifold and the restriction of (8) to this set is a diffeomorphism:*

$$\varinjlim X_i - f^{-1}\Sigma \xrightarrow{f} V_h^3 \times B^n - \Sigma.\text{''} \tag{9}$$

The proof, which makes heavy use of [Po8], is very briefly sketched at the end of Appendix I. It is given with complete details in the handwritten paper [Po11], which hopefully will be typed in a nottoo distant future. But here are some comments.

G) In the special case when $V^3 = \widetilde{M^3}$ and $n \geq 3$, the Σ from Theorem 4 is the same as the Σ from Theorem 3, while the X^{n+3} from Theorem 3 is gotten from X_1, after certain manipulations. Remark D.2 from the last section applies not only to X^{n+3} but also to the X_i's.

H) In the general case, we cannot get rid of the bad locus Σ, since $\Sigma = \emptyset$ would imply that $\pi_1^\infty V^3 = 0$ (see Theorem 0). From this standpoint, Theorem 4 is a best possible result.

I) Generally speaking, the $f_i X_i$'s are not **proper** submanifolds of $V_h^3 \times B^n$.

J) As already said, Theorem 4 applies to any open simply connected 3-manifold, even to a wild one like the classical Whitehead manifold W^3 (see [Wh2]. The standard notation for this venerable object, which is closedly related to the Casson Handles [Ca], [Fr], [Fr-Q], is actually Wh^3, but since we use already the subscript "h", we have changed it here into W^3). At least in this special case of W^3, the game can also be played differently, namely dispensing with the passage (2) from W^3 to W_h^3. We can get something which is exactly like the statement of Theorem 4 with "V_h^3" (or "W_h^3") replaced everywhere by "W^3" (*without* any holes), except that the bad locus Σ is now no longer a proper (stratified) submanifold. Instead, $W^3 \times S^{n-1}$ is endowed with a **lamination** $\mathfrak{L}$ (see [Th]) with smooth $(n+1)$-dimensional leaves, the union of which *is* the bad locus Σ (see (3) and (4) in Theorem 4). So, wanting to keep the bad locus Σ a manifold, rather than a lamination, is one of our reasons for going from $\widetilde{M^3}$ to $\widetilde{M_h^3}$.

The Cantor set C which appears as the transversal structure to the lamination $\mathfrak{L}$ has its own little story. This C is generated by a dynamical feed-back loop which is isomorphic to the one which generates the Julia sets [De] for real quadratic polynomials which are such that the iterates of the unique critical value goes to infinity. So, in some sense, C is a Julia set. All this is explained in detail in [Po-Ta3] (see also Appendix I and Appendix IV below).

K) In the context of (8), $f^{-1}\Sigma$ is also smooth and $f^{-1}\Sigma \xrightarrow{f} \Sigma$ is a diffeomorphism, just like (9).

L) The proof of Theorem 4 uses, in part at least, some techniques borrowed from my Poincaré program (in particular from [Po5]). Also, enough has been said in section 1 of this paper to make it clear that my work concerning the MAIN CONJECTURE (from the beginning of this section) is strongly connected with the program in question. Anyway, in both cases, one stabilizes from dimension three to higher dimensions. But while for the Poincaré Conjecture one has to go exactly to dimension four, for the main conjecture one can stabilize to some arbitrarily high dimension. And as soon as one goes to dimension five or more, the formidable obstruction to COHERENCE, which is not stable (see [Ga]) dies; this should be encouraging, as far as CONJECTURE C is concerned.

APPENDIX I. REPRESENTATIONS OF OPEN SIMPLY-CONNECTED 3-MANIFOLDS

For this appendix, which is a brief review of [Po8], it is assumed that the reader is familiar with section 1 of [Po4] (see also section 1 in David Gabai's review [Ga]). If X is a 2-dimensional simplicial complex and $X \xrightarrow{f} M^3$ is a nondegenerate map into a 3-manifold, a singularity of f is, by definition, a point $x \in X$ in the neighbourhood of which f is not immersive. The ADMISSIBLE SINGULARITIES are exactly the ones described by Figure 1 in [Po4] (sometimes, in Barry Mazur's terminology, such singularities are also called "undrawable"). The main result in [Po4] is the following

THE COLLAPSIBLE PSEUDO-SPINE REPRESENTATION THEOREM. *"Let Σ^3 be any homotopy 3-sphere. Then there exists a nondegenerate simplicial map*

$$K^2 \xrightarrow{f} \Sigma^3 \tag{10}$$

having only admissible singularities, such that
1) *K^2 is a finite simplicial complex which is collapsible.*
2) *The map f can be exhausted by elementary $O(i)$ zipping moves (see Figures 2 to 7 in [Po4]).*
3) *For the open regular neighbourhood of $fK^2 \subset \Sigma^3$ we have a diffeomorphism*

$$N(fK^2) = \Sigma^3 - \{\textit{finite set}\}.\text{"} \tag{11}$$

[Whenever in this paper "regular neighbourhood" is not preceeded by the adjective "open", it should be understood that we talk about the standard transversally compact regular neighbourhoods from simple homotopy theory.]

The proof of Theorem 4 (and of Theorem 3) starts with the attempt to adapt the ideas of the collapsible pseudo-spine representation theorem to open 3-manifolds V^3 with $\pi_1 V^3 = 0$, in lieu of the closed Σ^3. When we do this, a number of changes are forced on us.

I) Instead of a *finite* simplicial K^2 (see (10)) we will have to work with one which is only *locally* finite.

II) Before explaining the second change, I will discuss a bit the notion of collapsibility, in the finite case. Notice one drawback, at least, of this concept: the 2-dimensional squeleton of a collapsible complex is not, generally speaking, collapsible. There is a slightly more flexible concept, which does not have this drawback, and which in many situations is as good as "collapsible". We will say that a 2-dimensional finite complex K^2 is **almost collapsible** if it can be obtained from a collapsible one by adding finitely many 2-cells along their boundaries. It is easy to see that in the context of the theorem above, if we are given an almost-collapsible representation we also get a collapsible one (and conversely, of course, too, that is trivial.) When we go to the context of infinite complexes and open V^3's, instead of asking for K^2 to be collapsible we will rather ask for it to be "**almost arborescent**", in the sense that it is gettable from a point by a sequence of Whitehead dilatations **and** of additions of 2-cells (along their boundaries). [Some explanations are necessary here. The obvious replacement of collapsibility in the infinite case is "**arborescence**" (i.e. being gettable from a point by a sequence of dilatations without, of course, asking for the possibility of collapsing backwards). So if we want a substitute for almost-collapsibility, in the infinite set-up, we are lead to almost-arborescence.] Clearly, a smooth manifold which admits a smooth triangulation with almost-arborescent 2-squeleton is geometrically simply connected. Conversely, a geometrically simply connected manifold always admits smooth triangulations with this property. So, almost collapsibility (in the compact case), respectively almost arborescence (in the non compact case) is very akin to geometric simply connectivity.

III) What is less obvious, a priori, is that we will also be forced to consider spaces $P^2 \subset M^3$ which are no longer simplicial complexes in any reasonable sense, like for instance the subset

$$P^2 = (\text{plane} z = 0) \cup \sum_1^\infty (\text{plane } x = \frac{1}{n}) \subset R^3 = (x, y, z). \tag{12}$$

I will not try to formalize here the notion of "**generalized sub- polyhedra**" of a 3-manifold M^3 (details can be found in [Po8]). It suffices to say that, locally, we start with an infinite but discrete collection of affine planes $R_1^2, R_2^2, \ldots \subset R^3$, and with "reasonable" pieces $L_i \subset R_i^2$. Locally $P^2 = \bigcup_1^\infty L_i$, and the explicit definition is such that **the open regular neighbourhood** $N(P^2) \subset M^3$ **is well defined.**

With all this, the collapsible pseudo-spine representation theorem can be extended as follows, for the open case, with a bit of work,

THEOEREM 5. *"Let V^3 be an open simply connected 3-manifold. Then there exists a non- degenerate map*

$$X^2 \xrightarrow{g} V^3 \tag{13}$$

having only admissible singularities, such that

1) X^2 is a locally finite simplicial complex which is almost arborescent.
2) The mag g can be exhausted by elementary $O(i)$ zipping moves.
3) The image $gX^2 \subset V^3$ is a **generalized sub-polyhedron** *and we have a diffeomorphism of open manifolds.*

$$N(gX^2) = V_h^3, \tag{14}$$

for some V_h^3."

An $X^2 \xrightarrow{g} V^3$ like above will be called an **almost arborescent representation of** V_h^3. Also, by now V_h^3 is in the picture, in lieu of V^3.

Since gX^2 is only a *generalized* subpolyhedron (and as we will soon see we cannot do better than that), the map g is not necessarily simplicial. Actually gX^2 is not a subcomplex because the set of double points of g, denoted by $M_2(g) \subset X^2$, **is not closed**, in general. We will actually show (see Appendix II below) that if $M_2(g)$ is closed then, automatically, we have $\pi_1^\infty V^3 = 0$.

The local structure of $M_2(g)$ can be described as follows. Let $U \subset X^2$ be a smooth sheet of X^2, $U \approx R^2$, which is pre-compact in X^2. Then, after possibly throwing away finitely many double lines and singularities, $M_2(g) \cap U$ breaks up into a finite collection $F_1, F_2, \ldots, F_\alpha$, each F_i being an infinite set of parallel smooth lines (with $\ell \in F_i$ and $\ell' \in F_j$ transversal to each other). Like it is usually done for foliations or for laminations we will consider **tight transversals** $\Lambda \subset U$ to

$M_2(g) \cap U$. By "tight" it is meant that we cannot find a 2-cell $D^2 \subset U$ resting with half of its boundary on Λ and with the other half on a smooth line of $M_2(g) \cap U$ (see Figure 0).

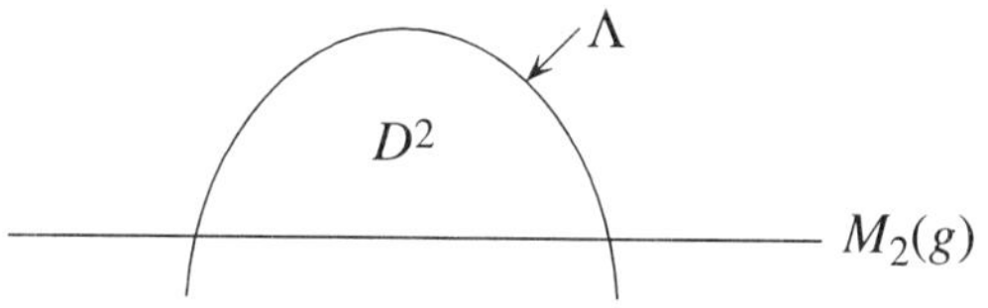

FIGURE 0

We will define the "set of **limit points**"

(15)
$$\lim(M_2(g) \cap \Lambda) \underset{\text{def}}{=} \{\text{the set of points } p_\infty \in \Lambda \text{ such that there is an infinite sequence } p_1, p_2, \ldots \in M_2(g) \cap \Lambda \text{ with } \lim p_n = p_\infty\}.$$

[Notational remark. Our present "lim" is totally unrelated to "$\varinjlim$" from Section 3 above (see for instance (8)).]

In terms of the decomposition

$$M_2(g) \cap U = F_1 \cup F_2 \cup \ldots \cup F_\alpha \cup \{\text{finite residual part}\},$$

in our representations we will always have

$$F_i \cap \lim(F_i \cap \Lambda) = \emptyset, \tag{16}$$

which makes $M_2(g)$ be quite at the opposite pole with respect to laminations; for the latter, instead of (16) we have

$$\lim(\Lambda \cap \mathcal{L}) = \Lambda \cap \mathcal{L}.$$

With quite some extra work, we can sharpen Theorem 5 as follows (and this is really the interesting result for us).

THEOREM 6. (THE FIRST FINITENESS THEOREM). *"For any open simply connected V^3 there is an almost arborescent representation (13) of some V_h^3 (see (14)) such that*

1) For any bounded tight transversal Λ we have

$$\mathit{card}(\lim(M_2(g) \cap \Lambda)) < \infty. \tag{17}$$

2) If we consider the following union over all the tight transversals and afterwards its closure in X^2

$$\mathit{LIMM}_2(g) \underset{\text{def}}{=} \overline{\cup_\Lambda \lim(M_2(g) \cap \Lambda)} \subset X^2, \tag{18}$$

then $g(\mathit{LIMM}_2(g)) \subset gX^2$ is a closed subset. Moreover the inverse image of $g(\mathit{LIMM}_2(g))$ in the open regular neighbourhood $N(gX^2)$ is well-defined, and this inverse image $r^{-1}(g(\mathit{LIMM}_2(g))) \subset N(gX^2)$ is a **proper** *(stratified) submanifold of dimension 2."*

This theorem is proved in [Po8]. Here are some remarks and comments.

M) If $V^3 = \widetilde{M^3}$ (with a closed M^3), then in the context of (17) there is a uniform bound in terms of the length of Λ. (This statement makes sense since, up to quasi-isometry, $\widetilde{M^3}$ carries a canonical riemannian metric.)

N) The bad locus Σ from Theorems 3 and 4 is actually

$$\Sigma = r^{-1}(g(\text{LIM}M_2(g))) \times S^{n-1} \subset V_h^3 \times B^n$$

(see also (14)).

O) In the general case we cannot do better than (17) since we have the following

PROPOSITION 7. *"If for all tight transversals Λ we have* $\lim(M_2(g) \cap \Lambda = \emptyset$ *(which is equivalent to $M_2(g) \subset X^2$ being closed), then $\pi_1^\infty V^3 = 0$."*

One could easily deduce this Proposition 7 from everything said so far, since $M_2(g)$ being closed means that $\text{LIM}M_2(g) = \emptyset$ and hence also $\Sigma = \emptyset$. But there is also a short direct argument which will be given in Appendix II below. This argument should be a good introduction to our whole program for proving the main conjecture.

P) (This continues the Remark J) from Section 3 above.) In going from V^3 to V_h^3 we have vastly increased π_2. Also, the same double point $(x,y) \in X^2 \times X^2$ of g can be joined, inside $X^2 \times X^2$, in many different ways, via double lines, to the singularities. It is this **redundancy** which is the price we have to pay for the finiteness in Theorem6.

The Whitehead manifold W^3 admits also arborescent representations (not just almost arborescent ones), without any kind of redundancy. For such representations $X^2 \xrightarrow{g} W^3$, the $N(gX^2)$ is diffeomorphic to W^3 itself and not just to some W_h^3. But for these particular representations, $M_2(g)$ exhibits very naturally chaotic behaviour. The chaotic dynamics which (for appropriate tight transversals Λ) generates $\Lambda \cap M_2(g)$ and which makes that in this particular situation we can get [Po-T3]

$$\lim(\Lambda \cap M_2(g)) = \text{Julia set}, \tag{19}$$

in lieu of (17), stems from the iterative process which builds up the Whitehead manifold itself (see also Appendix IV below). [As a side comment, Dennis Sullivan once told me that the null-homotopic embedding $S^1 \times D^2 \hookrightarrow S^1 \times D^2$ (see Figure 1) which is the starting point of the Whitehead manifold construction, should be at least as interesting, from the dynamical viewpoint, as the usual "degree two" embedding via which one defines the solenoid.]

The equality (19) is an instance of something which people like John Hubbard, myself and others (see for instance the beginning of Chapter 6 in [Fr-L]) believe to be part of a more general phenomenon, namely a relation between wild low-dimensional differential topology (i.e. V^3's with $\pi_1^\infty V^3 \neq 0$, exotic R^4's, Casson handles which are nonstandard from the DIFF standpoint,...) and dynamical chaos (Julia, Mandelbrot, Hénon,...).

Q) Once we are in possession of Theorems 5 and 6, the idea of the proof of Theorem 4 is to start with a sequence of elementary zipping moves $Y_i \to Y_{i+1}$ (see Figures 2 to 7 in [Po4]), going all the way from $Y_0 = X^2$ to $Y_\infty = gX^2$ (see (13)). Next we smoothen the Y_i's by taking appropriate regular neighbourhoods $N^{n+3}(Y_i)$. (We talk here about usual transversally compact regular neighbourhoods if $i < \infty$, but about the open regular neighbourhood if $i = \infty$.) In order not to have NON-COHERENCE problems (see [Ga], [Po5]) we let $n+3 \geq 5$ and then we get embeddings

$$X_i \underset{\text{def}}{=} N^{n+3}(Y_i) \subset X_{i+1} \underset{\text{def}}{=} N^{n+3}(Y_{i+1})$$

(see (7)).

[As already said, the trouble created by NON-COHERENCE (see [Ga], [Po4], [Po5]) is a purely 4-dimensional phenomenon which is unstable, in the sense that it vanishes after we multiply the 4-dimensional model by any B^n with $n \geq 1$.] Eventually (and perhaps keeping [Po5] at the back of our minds) we try to establish something like $\overset{\infty}{\underset{0}{\cup}} N^{n+3}(Y_i) = N^{n+3}(Y_\infty)$, but the best that can be achieved in this direction is exactly our Theorem 4. Some general position arguments which require $n \geq 3$, rather than just $n \geq 2$, are used in order to get from Theorem 4 to Theorem 3.

APPENDIX II. THE PROOF OF PROPOSITION 7.

For this appendix it is assumed that the reader is familiar with sections 2, 3, 4 of [Po1], or with section 1 in [Po12].

We start with an almost arborescent representation of V_h^3,

$$X^2 \xrightarrow{g} V^3$$

and consider its double point locus $M_2(g) \subset X^2$. Since X^2 is almost arborescent, it admits an exhaustion by finite simplicial complexes which are all simply-connected

(20) $$Y_1 \subset Y_2 \subset \ldots \subset X^2,$$

i.e. it is weakly geometrically simply connected.

In terms of the equivalence relations Ψ and Φ from [Po1] (see also [Po0]), point 2 from Theorem 5 means that $\Psi(g) = \Phi(g)$. [Here $\Phi(g) \subset X^2 \times X^2$ stands for the obvious equivalence relation corresponding to the double points of the map g, while $\Psi(g) \subset \Phi(g)$ is the more subtle "smallest equivalence relation which kills all the singularities" (see [Po1]; more details are also given in [Po0]).]

One can construct a (locally finite) 3-dimensional simplicial complex T^3 mimicking X^2, in the sense that

α) There is a non-degenerate map

(21) $$T^3 \xrightarrow{G} V^3$$

with $GT^3 = N(gX^2) \approx V_h^3$.

β) T^3 has an exhaustion by finite simplicial complexes which are "regular neighbourhoods" of $Y_1, Y_2, \ldots$(and hence are all simply- connected)

(22) $$T_1 \subset T_2 \subset \ldots \subset T^3 = \overset{\infty}{\underset{1}{\cup}} T_i,$$

so T^3 is also weakly geometrically simply connected.

Moreover each $G \mid T_i$ is a bona-fide simplicial map, for a simplicial structure of V^3 which depends on i (actually it gets finer and finer when i goes to infinity; G itself is not simplicial).

γ) We have $\Psi(G) = \Phi(G)$. [In our present context of (T^3, G), the equality $\Psi(g) = \Phi(g)$ means that the double points of the map G can be exhausted by zippings of the most general type, i.e. by folding maps involving two cells of the same dimension with the same image at the target and with a non-void intersection at the source.]

With this, we introduce the notations $\Psi_\infty = \Psi(G), \Phi_\infty = \Phi(G)$ and also

$$\Psi_i = \Psi(G \mid T_i), \Phi_i = \Phi(G \mid T_i) = \Phi_\infty \mid T_i.$$

Notice that $\Psi_i \subset \Psi_\infty \mid T_i$, without necessarily having equality.

LEMMA 7.1. *"For each n there is an $N = N(n) > n$ such that $\Psi_N \mid T_n = \Phi_n$."*

Proof. Since $\Psi_\infty = \Phi_\infty$ this is certainly true for $N = \infty$. By the same compactness argument as in Proposition B, Section 4 of [Po1] there has to exist an $N < \infty$ too.

LEMMA 7.2. *"For any compact subset $K \subset V_h^3$ there is an $n \in Z_+$ such that $GT_n \supset K$ and*

$$(GT_p - GT_n) \cap K = \emptyset \text{ if } p > n."$$

Proof. Because $M_2(g)$ is closed, for any $q \in Z_+$ there exists an $n = n(q) > q$ such that (see (20))

$$(gY_p - gY_n) \cap gY_q = \emptyset \text{ if } p > n.$$

One can deduce our lemma from this last fact. □

Now, for any compact connected $K \subset V_h^3$ choose an n like in Lemma 7.2 and for this n choose an N like in Lemma 7.1. We have a commutative diagram

$$\begin{array}{ccccc} K & \xrightarrow{\subset} & GT_n = T_n/\Phi_n & \xrightarrow{\subset} & T_N/\Psi_N \\ \downarrow{\scriptstyle \cap} & & & & \downarrow{\scriptstyle g_1} \\ V_h^3 & & \xrightarrow{=} & & V_h^3 \end{array}$$

where:

i) g_1 is an immersion (i.e. $\mathrm{Sing}(g_1) = \emptyset$).

ii) $\pi_1(T_N/\Psi_N) = 0$.

[Both i) and ii) are consequences of the general properties of the equivalence relation Ψ (see [Po0], [Po1]).]

iii) As a consequence of Lemma 7.2 we have the Dehn-type condition from [Po1]

$$K \cap M_2(g_1) = \emptyset.$$

One can use then the Dehn-type Lemma from [Po1] and show that $K \subset V_h^3$ can be engulfed by a simply connected compact 3-dimensional submanifold. This implies that $\pi_1^\infty V_h^3 = 0$ and hence $\pi_1^\infty V^3 = 0$

APPENDIX III. THE SECOND FINITENESS THEOREM AND "WHITEHEAD'S NIGHTMARE"

Here we will assume knowledge of the first two sections in [Po12], or Section 3 from [Po2] and/or of Sections 2,3 from [Po3]. We start with a closed 3-manifold M^3, with an infinite π_1, which is represented as the quotient-space of a fundamental domain Δ. Here Δ is a 3-cell the boundary of which is divided into an even set of triangles $\{h_1, h_2, ..., h_{2p}\}$. One is also given a fixed-point free involution j on the set $\{h_1, h_2, ..., h_{2p}\}$ and M^3 is gotten from Δ by glueing each h_i to jh_i just like in the usual representation of compact surfaces. Notice that this construction gives us a prefered system of generators for $\pi_1 M^3$ and hence a norm $\|\dots\|$ on $\pi_1 M^3$. We consider the free monoid G generated by 1 and the h_i's and with this, we introduce the big arborescent (tree-like) non-locally finite object

$$T = \bigcup_{x \in G} x\Delta, \tag{23}$$

where for each x, h_i the h_i-face of $x\Delta$ is glued, in a Cayley-graph manner, to the jh_i-face of $xh_i\Delta$. The obvious tautological surjection $T \to M^3$ can be lifted to a "developing map"

$$T \xrightarrow{F} \widetilde{M}^3 = \bigcup_{g \in \pi_1 M^3} g\Delta, \tag{24}$$

(see, for instance, (3.20) in [Po3]). It is shown in [Po2] that for this particular map F we have $\Psi(F) = \Phi(F)$. Notice also that any compact subset of $\widetilde{M}^3$ is hit **infinitely** many times by the image of F.

Side-Remark. Exactly like the locally finite T^3 from Appendix II, our present nonlocally finite T also admits an exhaustion by finite collapsible unions of fundamental domains

$$T_1 \subset T_2 \subset \dots \subset T.$$

This should of course not be mixed up with (22), but if we define $\Psi_i = \Psi(F \mid T_i), \Phi_i = \Phi(F \mid T_i)$, Lemma 7.1 remains valid in this new context. The way in which the lemma in question is proved does not tell us **anything** about the function $N = N(n)$, except for its mere existence. I think that the growth rate of $N = N(n)$ should be a very interesting piece of information about the group $\pi_1 M^3$. It would be nice to connect it with other known (growth) properties of $\pi_1 M^3$. $\square$

The next result is one of the ingredients for the proof of our Theorem 3.

THEOREM 8. (THE SECOND FINITENESS THEOREM). *"There exists a quotient space T_α of T which gives rise to a commutative diagram*

$$\begin{array}{ccc} T & \xrightarrow{\pi_\alpha} & T_\alpha \\ \downarrow{\scriptstyle F} & & \downarrow{\scriptstyle F_\alpha} \\ \widetilde{M}^3 & \xrightarrow{=} & \widetilde{M}^3 \end{array} \tag{25}$$

such that

1) *π_α is a sequence of folding maps, each involving two cells of the same dimension and same image in $\widetilde{M}^3$, and with at least one point in common at the source.*
2) *For each $x \in T_\alpha$ we have*

$$\text{card } \pi_\alpha^{-1}(x) < \infty. \tag{26}$$

3) *There is a crossection $\widetilde{M}^3 \xrightarrow{\sigma} T_\alpha$ for F_α."*

The proof is given in [Po7]. I will only offer here some comments.

R) This theorem extends, as such to any dimension. In this sense it is a purely group-theoretical fact.

S) For the proof one has to do "geometry" with the large monoid G instead of doing geometry with the group $\pi_1 M^3$, like one did in the context of (3). A priori we have two norms for G, namely the obvious word-length $\mid x \mid$ and also the norm inherited from the $\|\dots\|$ of $\pi_1 M^3$, via the obvious surjective morphism $G \xrightarrow{\chi} \pi_1 M^3$. The norm $\|\chi(x)\|$ is clearly too crude, but $\mid x \mid$ is not good enough either, because T is not locally finite. We can have $\lim |x_n| = \infty$ while at the same time the $x_n\Delta$'s stay all hooked at some fixed place in T, in the sense that

$$x_1\Delta \cap x_2\Delta \cap \dots \neq \emptyset. \tag{27}$$

The good norm on G is gotten by considering all the possible factorizations $x = x'x''$, in G, and taking

$$\||\, x \,\|| \underset{\text{def}}{=} \text{Sup}\|\chi(x')\|. \tag{28}$$

So $\|\chi(x)\| \leq \||x\|| \leq |x|$. Unlike what happens with $\|\chi(x)\|$, we get now the inequality $\||x\|| \leq \||xy\||$ and unlike what happens with $|x|$, when $\lim \||x_n\|| = \infty$ the $x_n\Delta$'s cannot stay hooked at any given spot inside T, they really have to "go to infinity" (see [Po7]).

T) Whenever we have proved a theorem of type (3) (see for instance [Po2], [Po3], [PoTa1], [PoTa2]) the strategy of the proof was always the following. Using the "nice geometric condition on $\pi_1 M^3$" we have managed to construct, for any compact set $K \subset \widetilde{M^3}$, a nondegenerate map analogous to (24), having as source an arborescent union of fundamental domains (possibly a subset of T)

$$\Theta \xrightarrow{H} \widetilde{M^3}, \tag{29}$$

(with Θ, H depending of K), such that $\Psi(H) = \Phi(H)$, and which has the property that the image of H hits the given K only **finitely** many times. It can be shown (see, for instance, [Po3]) that if such a construction can be performed for an arbitrary K, then $\pi_1^\infty \widetilde{M^3} = 0$; the argument is really the same as the one given in Appendix II above. But in order to show that $\Psi = \Phi$ we always had to rely on the following fact, which should be compared with our Theorem 8.

PROPOSITION 9. *(see [Po3]) "Let X be union of fundamental domains and $X \xrightarrow{g} \widetilde{M^3}$ a surjective nondegerate map. It is assumed that any pair of fundamental domains in X are connected by a finite chain of fundamental domains where two consecutive ones have a codimension one face in common. We have $\Psi(g) = \Phi(g)$ if and only if $X/\Psi(g) \to \widetilde{M^3}$ houses a section $X/\Psi(g) \underset{s}{\leftarrow} \widetilde{M^3}$."*

In the context of [Po2], [Po3],... and of (29) the hardest thing was always to have such a section s, *without* the map H hitting K infinitely many times. I will try to analyze here the exact nature of the difficulty we encounter, when no special restrictions are imposed to $\pi_1 M^3$. Let us introduce the notation

$$B_n \underset{\text{def}}{=} \bigcup_{\|g\| \leq n} g\Delta \subset \widetilde{M^3}$$

and assume that we have already constructed, inductively, a first approximation of the desired (29), namely a map with an arborescent source

$$\Theta_n \xrightarrow{H_n} \widetilde{M^3} \tag{30}$$

which carries a partial section $B_n \xrightarrow{s_n} \Theta_n/\Psi(H_n)$ and which hits K only finitely many times. For simplicity's sake we will assume that n is sufficiently large so that we have already $K \subset H_n\Theta_n$.

We want to extend Θ_n to a larger arborescent union of fundamental domains

$$\begin{array}{ccc} \Theta_n & \xrightarrow{\subset} & \Theta_{n+1} \\ \downarrow H_n & & \downarrow H_{n+1} \\ \widetilde{M^3} & \xrightarrow{=} & \widetilde{M^3} \end{array} \tag{31}$$

such that s_n extends to a section $B_{n+1} \xrightarrow{s_{n+1}} \Theta n + 1/\Psi(H_{n+1})$. There are various recepees for achieving this and for the sake of clarity I will fix one of them. But all the other scenarios are eventually equivalent to the one I will outline now.

I start by introducing the notation

$$\Theta_n/\Psi(H_n) \underset{r_n}{\longleftarrow} \Theta_n \supset r_n^{-1}(s_n B_n), \tag{32}$$

where r_n is the obvious quotient-space projection.

Let $\overline{\Delta_1}, \overline{\Delta_2}, \ldots$ be the fundamental domains of $B_{n+1} - B_n \subset \widetilde{M^3}$. For each $\overline{\Delta_\alpha}$ there is a fundamental domain $\Delta_\alpha \subset B_n$, touching ∂B_n, such that $\overline{\Delta_\alpha} \cap \Delta_\alpha$ consists of a common face δ_α. With this, we start by adding to Θ_n, for each occurance of a $\Delta_\alpha(i) \subset r_n^{-1}(s_n B_n)$ living above Δ_α, a $\overline{\Delta_\alpha}(i)$ glued to $\Delta_\alpha(i)$ along $\delta_\alpha(i)$. This leads to a first arborescent extension

$$\Theta_n' = \Theta_n \underset{\delta}{\cup} \sum_{\alpha,i} \overline{\Delta_\alpha}(i) \xrightarrow{H_n'} \widetilde{M^3}. \tag{33}$$

[At the price of making first connected packages out of the $\overline{\Delta}$'s and then glueing them to the approriate Δ's, we can pretend here that the δ's are 2-dimensional.]

The problem now is that the equivalence relation $\Psi(H_n')$ does not necessarily force the identification of pairs like $\overline{\Delta_\alpha}(i), \overline{\Delta_\alpha}(j) (i \neq j)$, and thus prevents $\Theta_n'/\Psi(H_n')$ from carrying the desired section s_{n+1}. We can correct this, as follows. In $\Theta_n'/\Psi(H_n)$ we consider a path ℓ connecting $\overline{\Delta_\alpha}(i)$ to $\overline{\Delta_\alpha}(j)$, and what we want is to add, arborescently, enough material to Θ_n' so as to get an extension $\Theta_{n+1} \xrightarrow{H_{n+1}} \widetilde{M^3}$ for which the equivalence relation $\Psi(H_{n+1})$ **forces the path ℓ to close.**

There is a more or less canonical recipee for performing this. Start by considering ℓ as a closed path in the Cayley graph Γ of $\pi_1 M^3$, with a preferred vertex corresponding to $\overline{\Delta_\alpha}(i) = \overline{\Delta_\alpha}(j)$. If we add to Γ all the 2-cells corresonding to the relators (so as to get the universal covering space of a certain "2-spine" of M^3) then we get a 2-disk D^2 of boundary ℓ with a cell-decomposition into relator-cells. Choose a collapse of D^2 onto a tree containing the arc ℓ disconnected at the preferred vertex $\overline{\Delta_\alpha}(i) = \overline{\Delta_\alpha}(j)$. This collapse gives, essentially, a recipee for adding, arborescently, fundamental domain to Θ_n' so as to get the desired Θ_{n+1}, for which $\Theta_{n+1}/\Psi(H_{n+1})$ houses s_{n+1}.

But now another problem has to be faced. We might well have new contacts with K, coming from occurances which I will denote symbolically by

$$H_{n+1}D^2 \cap K \neq \emptyset. \tag{34}$$

If such a thing will happen for **an infinity** of partial section $s_{\ell>n}$, piling infinitely much stuff on toop of K, then $H\Theta$ will hit K infinitely many times and we will be in trouble, unless we can repair this "**first trouble**", to begin with. So let us take care of the (34), for H_{n+1}, first. It so happens that we can afford a slight slackening of the arborescence condition, via which double points $(x, y) \in \Theta_{n+1} \times \Theta_{n+1}$ of H_{n+1} corresponding to (34) could be avoided:

α) Join x to y in Θ_{n+1} by an arc ℓ_1. [For certain technical reasons, this operation might be preceeded by a finite family of preliminaty zippings, provided that some **properness** condition is fulfilled when we iterate the steps leading to $s_{n+1}, s_{n+2}, \ldots$]

β) Glue x to y, add to the corresponding quotient of Θ_{n+1} a 2-disk D_1^2 which cobounds the loop ℓ_1 and then map everything into $\widetilde{M^3}$ (extending the already existing map from Θ_{n+1})

$$(\Theta_{n+1}/(x = y)) \underset{\ell_1}{\cup} D_1^2 \overset{H_{n+1}^1}{\longrightarrow} \widetilde{M^3}. \tag{35}$$

One should really think here of the source of H_{n+1}^1 as being realized as a union of fundamental domains.

As far as s_{n+1} is concerned everything is still fine, but we can have now a repetition of the (34) trouble, namely

$$H_{n+1}^1 D_1^2 \cap K \neq \emptyset. \tag{36}$$

We can certainly try to take care of this, just as we did before, with a new pair (ℓ_2, D_2^2),... But what we cannot afford, with **any** reasonably useful extension of arborescence, is an **infinite** cascade

$$H_{n+1}^j D_j^2 \cap K = \emptyset, \text{for } j = 1, 2, \ldots, \tag{37}$$

which I call our "**second trouble**". I call the "**Whitehead nightmare**" the infinite trap which is opened by the combination of our two troubles. The usefulness of the "nice geometric conditions" in (3) is exactly to avoid the Whitehead nightmare; by now it should be pretty obvious why almost convexity [Can] appears completely naturally as a first candidate for the job. On the other hand, whenever the infinite trap of the Whitehead nightmare is, in principle at least, unescapable, our bad locus $\sum$ is unavoidable too. The reader might have also noticed that the first trouble is not unrelated with the construction of the Whitehead manifold itself [Wh2], while the second trouble is clearly reminiscent of the Casson Handles [Ca], [Fr], [Fr-Q]. As a personal touch I might add that my own work on the 3-dimensional Poincaré Conjecture made me already familiar with this infinite trap, quite a long time ago; [Po10] is an early written trace of this.

U) Closely related to the Whitehead nightmare, is the following fact. Consider a sequence of fundamental domains of T (see (23))

$$x\Delta, \quad xh_{i_1}\Delta, \quad xh_{i_1}h_{i_2}\Delta, \ldots$$

which from the standpoint of T "advances forwardly", in the sense that we obviously have

(38) $$| x |<| xh_{i_1} |<| xh_{i_1}h_{i_2} |< ...$$

When we project down to $\pi_1 M^3$ (and/or to $\widetilde{M}^3$), the sequence

(39) $$\chi(x), \chi(xh_{i_1}), \chi(xh_{i_1}h_{i_2}), ...$$

is, generally speaking, recurrent, in the sense that for an arbitrary $R \in Z_+$ it might have infinitely many terms with $\| \dots \| < R$.

In contrast with this, for the $||| \dots |||$ norm we find that

(40) $$|||x||| \leq |||xh_{i_1}||| \leq |||xh_{i-1}h_{i_2}||| \leq \dots .$$

The offshot of this is that in the context of (the proof of) Theorem 8 the Whitehead nightmare is absent. The section $\widetilde{M}^3 \xrightarrow{s} T_\alpha$ (see Theorem 8) can then be gotten, more or less, like in the proof of (3), but without imposing now any special geometrical conditions to π_1.

APPENDIX IV. MORE ON THE WHITEHEAD MANIFOLD AND JULIA SETS

I will briefly describe what one sees when one tries to apply, directly, to the Whitehead manifold W^3 [Wh2] the mechanism of the collapsible pseudo-spine representation theorem for homotopy 3-spheres $\sum^3$ (see [Po4] and Appendix I above). Remember that W^3 is obtained by iterating the embedding described by Figure 1 of one copy of $S^1 \times D^2$, call it T_n into the interior of a second one, call it T_{n+1}. With this sequence of embeddings

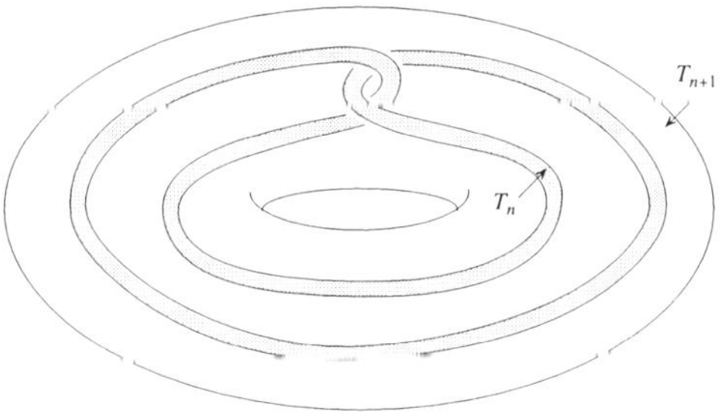

FIGURE 1

(41) $$T_0 \subset T_1 \subset ... \subset T_n \subset T_{n+1} \subset ...$$

we have $W^3 \underset{\text{def}}{=} \overset{\infty}{\underset{0}{\cup}} T_n$. It is possible to find *arborescent* representations $X^2 \xrightarrow{g} W^3$ for W^3 (see Appendix I above). If one does things in the most economical way (see

[Po-Ta3]) then g has no triple points and also the set of double points $M_2(g) \subset X^2$ has no cycles. The locally finite simplicial complex X^2 is a union of disks

$$X^2 = D_0 \cup D_1 \cup D_n \cup D_{n+1} \cup ... \tag{42}$$

the D_n from (42) corresponding to the T_n from (41). We consider a sufficiently long tight transversal to $M_2(g)$.

$$\Lambda \subset D_n. \tag{43}$$

The set $\Lambda \cap M_2(g)$ has a partition

$$\Lambda \cap M_2(g) = F_1 + F_2 + ... \tag{44}$$

where F_p stands for those points $x \in \Lambda \subset D_n$ which are such that there is also a point $y \in D_{n+p}$ with $g(x) = g(y)$.

From the viewpoint of the linear order on the arc Λ, the relative positions of the F_p's presents quite a peculiar recursive pattern which I will describe in the following graphical manner. The tight transversal Λ is the vertical arc $[a_{\text{down}}, a_{\text{up}}]$ from Figure 2. Generically, the points of $F_p \subset \Lambda$ will be labelled by the numeral p. In Figure 2 a second **folded** copy of Λ, denoted by Λ_0, has also been drawn. With this, here is the

DYNAMICAL RULE FOR CONSTRUCTING (44). The set F_1 has exactly two points which we draw on Λ_0 at the same level as the endpoints of Λ. No points p have been yet drawn on Λ, at this initial stage of the game and no points $p > 1$ on Λ_0 either. Next we apply the following

Rule I. *"Whenever some p has been drawn on Λ_0 (respectively on Λ) it has to be drawn on Λ (respectively on Λ_0) too, with a similar location with respect to the points labelled by $p-1, p-2, ...$"*

So the points 1 from Λ_0 have also to be drawn on Λ. Now here comes a second rule.

Rule II. *" Whenever a p has been drawn on Λ, then the two points where the horizontal lines through p cuts Λ_0 are $(p+1)$'s."*

This rule II creates four points 2 on Λ_0 (see Figure 2) and then rule I forces us to draw such points on Λ too. Then eight points 3 appear on Λ_0, via rule II, and by now it should be clear how the process continues indefinitely until exactly the **same** infinite pattern has been drawn on both Λ and Λ_0, in agreement with our rule I.

It is very easy to see that if f_λ is a real quadratic polynomial with $\lim_{n=\infty} f_\lambda^n(c) = \infty$ where c is the unique critical point and if a set $E \subset R$ of cardinality two is correctly choosen, then our dynamical rule above is exactly the same as the one which builds up the disjoined union

$$E \cup f_\lambda^{-1} E \cup f_\lambda^{-2} E \cup ... \subset R,$$

with $f_\lambda^{-n} E$ corresponding to F_{n-1} from (44). Our parameter λ lives outside the Mandelbrot set (which accounts for the relative simple-mindedness of our little dynamical process above). In term of the usual graphical display of the Mandelbrot set M (see for instance Plate 5 in [De]) our λ's are located on the half-infinite horizontal line to the left of M and, up to a quasiconformal homeomorphism, the Julia set [De]

$$\mathcal{J}_\lambda = \lim(E \cup f_\lambda^{-1} E \cup f_\lambda^{-2} E \cup \ldots) \subset R$$

is essentially unique.

If in its (improper) isotopy class the map $X^2 \xrightarrow{g} W^3$ is chosen correctly, metrically speaking, then we get exactly (19), or more preceisely

$$\lim(\Lambda \cap M_2(g)) = \mathcal{J}_\lambda.$$

POSTSCRIPT V. A tentative project for proving Conjecture (C) (and hence the MAIN CONJECTURE.)

The present section is of a highly speculative character and so it does not have the same mathematical status as the rest of this paper. On the other hand what we give here is only a bird's eye view of a lot of material and work in progress, and this might obscure the connection with the previous theorems. For instance, although the bad locus $\sum$ will hardly be mentioned explicitely in the following discussion, it is actually ever present, when we look at things in greater detail. In some form or other, most of the things said before in the present paper, are supposed to be used in the project which I will outline now.

The whole conjectural approach depends on some notion of "**structure**" which one can introduce, to begin with, on open simply connected 3-manifolds V^3. Some hints of what such structures might be will be given below. It will be more useful, for the time being, to look at some much more mundane "structures" and to try to extract some "axiomatic" properties. Here are two examples: a foliation $\mathcal{F}$ on a manifold M and a triangulation T of some space X. As far as we are concerned here, the common features of these two examples are the following items.

i) The existence of an "**underlying space**" $|\ldots|$ which, in our present examples is, quite trivially, $|\mathcal{F}| = M$, respectively $|T| = X$.
ii) In the presence of appropriate maps having M or X (i.e. $|\ldots|$ itself in our two examples) as a source, respectively as a target, the structures can be pushed forward, respectively pulled back.

Notice that in both cases, what we are actually given is a certain amount of 2-dimensional material piling up inside M or X.

With this, we are ready to start.

STEP I. We will fix once and for all a "level" $n \in Z_+$ with $n \geq 3$, like we did in the context of Theorem 3. With this, for any open simply connected 3-manifold V^3, a whole class of "structures", denoted generically by $S_u V^3$ will be introduced. In the end of this section we will give a simple example of what such a structure might be, in the compact context. Essentialy, in the case of an open simply connected V^3, an $S_u V^3$ structure will be an almost arborescent representation $X^2 \xrightarrow{g} V^3$, with certain additional features. It will, anyway, be amenable to push forward or pull-back operations, based on V^3. The subscipt "u" in $S_u V^3$ (which will distinguish it from $S_b V^3$, about which more will be said later) stands for "usual". But $|S_u V^3|$ will not be V^3; it will actually be a more exotic, higher-dimensional object. The idea is this: as already said, $gX^2 \subset V^3$ does not possess a transversally compact regular neighbourhood, only the open regular neighbourhood $N(gX^2) \underset{\text{DIFF}}{=} V_h^3$ is

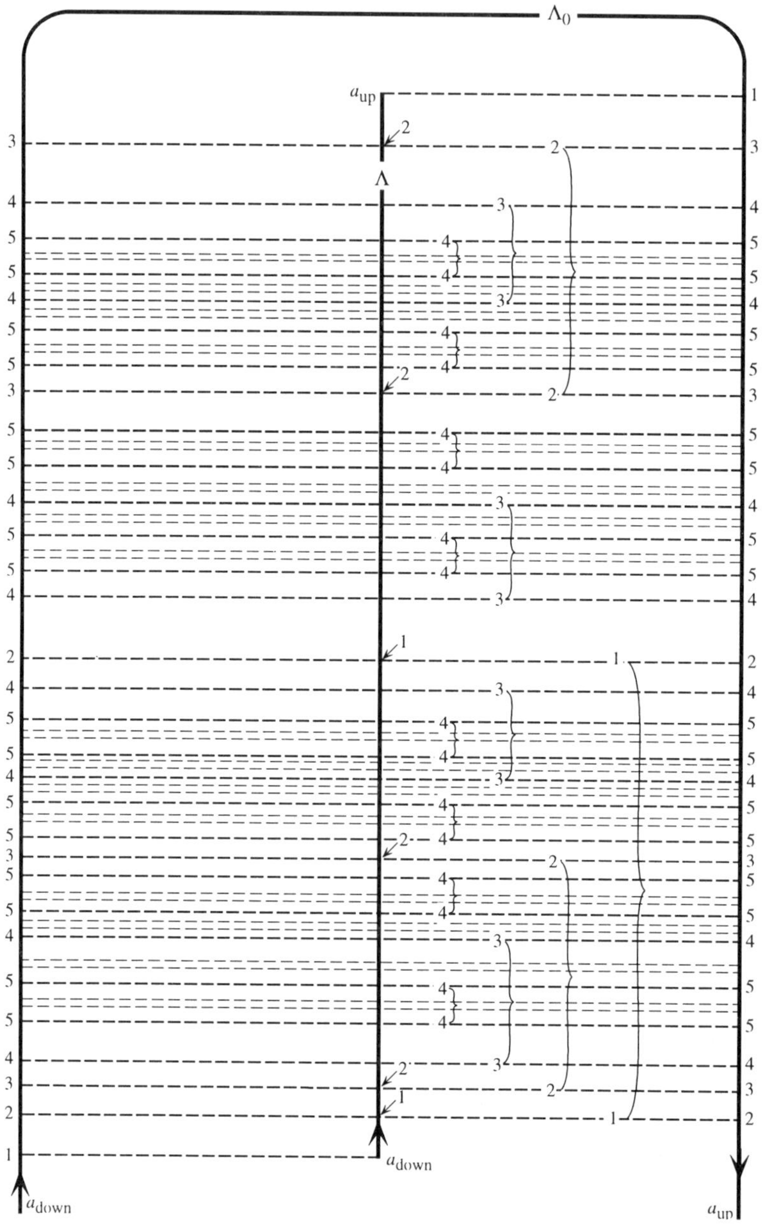

FIGURE 2

well-defined. Roughly speaking, we will have

$$|S_u V^3| = \{ \text{ a certain 3-dimensional thickening of } gX^2, \text{ which is "as much as possible transversally compact"} \} \times B^n.$$

The reason for the B^n factor is to allow us, later on, the comparison with $|S_b V^3|$. Anyway our $|S_u V^3|$ will be a **proper** submanifold of codimension zero of $V_h^3 \times B^n$, which is such that

$$V_h^3 \times 0 \subset |S_u V^3| \subset V_h^3 \times B^n; \tag{45}$$

this should be compared to (1). For a given V^3 and $S_u V^3$, we will also have a second, more "bizarre" structure $S_b V^3$, where $|S_b V^3|$ is again a smooth noncompact $(n+3)$-dimensional manifold, not necessarily the same as the $|S_u V^3|$ which appears in (45). A priori, $|S_b V^3|$ has no direct relation with $V_h^3 \times B^n$, unlike $|S_u V^3|$. As far as our "structures" will be concerned, we will always move inside the DIFF category.

In the end of the section a simple example of what S_b might be will be given, in the compact context, where quite naturally we will find that $|S_u \dots|$ and $|S_b \dots|$ coincide. But in our context of an open simply connected V^3, an $S_b V^3$ will be a commutative diagram

$$\begin{array}{ccc} X^2 & \xrightarrow{i} & Y^2 \\ \big\downarrow g & & \big\downarrow \bar{g} \\ V^3 & \xrightarrow{=} & V^3 \end{array}$$

where:

a) $X^2 \xrightarrow{g} V^3$ is, again, an almost arborescent representation, with some extra features.

b) Y^2 is a (not locally finite) 2-dimensional object obtained from X^2 by dilatations and additions of 2-cells.

c) In an apropriate sense, the map $Y^2 \xrightarrow{\bar{g}} V^3$ "imitates" the exhaustion of g by $O(i)$ zipping moves. We will have

$$|S_b V^3| = \{\text{an apropriate } (n+3)\text{-dimensional thickening of } Y^2\}.$$

Remarks. So, now the underlying space $|\dots|$ of the structure is by no means the 3-manifold on which the structure is defined. *But pushing forward or pulling back will still correspond to maps involving the 3-manifold in question*, and not the "underlying spaces" $|\dots|$ themselves. The important point is that via such operations the structures S_u, S_b might now be defined on 3-manifolds which, in principle at least, can be neither simply-connected, nor open, nor both. Step II will make use exactly of this flexibility.

The precise construction of S_b is such that, for an open simply-connected V^3, the manifold $|S_b V^3|$ will be geometrically simply-connected, a property which, generally speaking will not be shared by $|S_u V^3|$; but $|S_b V^3|$ itself does not verify the analogue of (45) either. Here is an immediate consequence of the things just said and of theorem 0.

COROLLARY 10. *"If*

$$|S_u V^3| \underset{\text{DIFF}}{=} |S_b V^3|, \tag{46}$$

then $\pi_1^\infty V^3 = 0$."

STEP II. We consider now a closed 3-manifold M^3 like in our main conjecture and move from the general V^3 to the universal covering space $\widetilde{M^3} \xrightarrow{\pi} M^3$, with its full symmetry

$$\pi_1 M^3 \times \widetilde{M^3} \to \widetilde{M^3}. \tag{47}$$

I hope I am in possession of all the ingredients necessary for a complete proof of the following

STATEMENT 11. *"1) On $\widetilde{M^3}$ we can construct a structure $S_u\widetilde{M^3}$ which is $\pi_1 M^3$-invariant. The corresponding system of holes h (see (45)) is $\pi_1 M^3$-invariant too.*

2) This being granted, the push-forward (= direct image) $\pi_*(S_u\widetilde{M^3})$ is a well-defined structure on M^3, which will be denoted by $S_u M^3$. If we consider the map

$$\widetilde{M^3} \times B^n \xrightarrow{\pi\times \mathrm{id}} M^3 \times B^n,$$

then $|S_u M^3| \underset{\mathrm{DIFF}}{=} (\pi \times id)|S_u\widetilde{M^3}|$. *Moreover, if we introduce*

$$M^3_{\pi(h)} \underset{\mathrm{def}}{=} \pi\widetilde{M^3_h} = M^3 - \{a \text{ certain closed, tame, totally discontinuous subset}\},$$

then $|S_u M^3|$ is a **proper** *codimension zero submanifold of $M^3_{\pi(h)} \times B^n$, having the property that*

$$M^3_{\pi(h)} \times 0 \subset |S_u M^3| \subset M^3_{\pi(h)} \times B^n, \tag{48}$$

which should be compared to (45).

3) Under the conditions above, the $S_b\widetilde{M^3}$ attached to $S_u\widetilde{M^3}$ can also be pushed forward to a structure $S_b M^3 \underset{\mathrm{def}}{=} \pi_* S_b\widetilde{M^3}$ on M^3."

Both $|S_u M^3|$ and $|S_b M^3|$ are now *non-simply connected* non compact smooth $(n+3)$-manifolds.

The next statement is essentially formal.

PROPOSITION 12. *"1) Using the covering map $\widetilde{M^3} \xrightarrow{\pi} M^3$ we can pull the structures $S_u M^3$ and $S_b M^3$ back to $\widetilde{M^3}$ and get*

$$\pi^* S_u M^3 = S_u\widetilde{M^3}, \quad \pi^* S_b M^3 = S_b\widetilde{M^3}. \tag{49}$$

2) When we go to the universal covering spaces for the underlying spaces downstairs we simply get

$$|S_u M^3|^\sim = |S_u\widetilde{M^3}|, |S_b M^3|^\sim = |S_b\widetilde{M^3}|." \tag{50}$$

The whole point of the present Step II is that M^3 is **compact**, and hence quite trivially one has $\pi_1^\infty M^3 = 0$ or, more accurately, for the open manifold $M^3_{\pi(h)}$ we have $\pi_1^\infty M^3_{\pi(h)} = 0$. [Of course $\pi_1 M^3 = \pi_1 M^3_{\pi(h)}$ is nontrivial, but we are talking here about π_1^∞, and not about π_1.]

With all this, we can move now to the last step, which is my present conjectual strategy for getting around the "Whitehead nightmare" discussed in Appendix III above.

CONJECTURAL STEP III. Notice, to begin with that our Corollary 10 takes the general form of an implication

$$\{|S_u(\dots)| \underset{\text{DIFF}}{=} |S_b(\dots)|\} \Rightarrow \{\pi_1^\infty(\dots) = 0\}. \tag{51}$$

The conjectual guess is that enough of the implication **converse** to (51) is true (and there is some "empirical" evidence for this) so that from the simple connectivity at infinity of $M^3_{\pi(h)}$ (and/or of M^3) we could draw the conclusion that

$$|S_u M^3| \underset{\text{DIFF}}{=} |S_b M^3|. \tag{52}$$

[The general idea is that (52) becomes true after an infinity of deletion of handles of codimension two has been performed on both sides and that, afterwards, one should be able to recover the whole (52) from this, using the $\pi_1^\infty = 0$ condition.]

Assuming that all this is so, we could then use (50) and get the equality

$$|S_u \widetilde{M^3}| \underset{\text{DIFF}}{=} |S_b \widetilde{M^3}|. \tag{53}$$

Now, for the simply connected $\widetilde{M^3}$ we know that $|S_b\widetilde{M^3}|$ is geometrically simply connected, and if we plug all this information into (45), specialized to the case $V^3 = \widetilde{M^3}$, we would get (if everything works well), the proof of the CONJECTURE (C) and hence the proof of our MAIN CONJECTURE too (see also Corollary 10 above).

Some Comments.

V) We know on "philosophical grounds" that any proof of the main conjecture has to involve both of the following two ingredients: dimension three and free action with compact fundamental domain. In our tentative approach, the first ingredient occurs in step I (and III) while the second ingredient certainly occurs in step II (and also III).

W) I will give now the simplest and most trivial example of structures S_u/S_b. No holes h will be involved here. I consider a compact 2-dimensional space K^2 which is the union of the two rectangles C, D from Figure 3, with C and D glued together along the arcs $[x,\sigma_1], [y,\sigma_2]$, so that $\pi_1 K^2 = Z$. With this particular K^2 we consider a non-degenerate map $K^2 \xrightarrow{f} R^3$ which is injective when restricted to C, D
individually and has fC and fD transversal to each other. Figures 3, 4 are supposed to give a complete description of this map. The points σ_1, σ_2 are singularities of f, in the sense that in their neighbourhood f is not immersive. (Readers which are familiar with [Po4] and/or with [Ga] will have recognized the definition of an 0(3) move, by now. Clearly, also, our singularities are of the "undrawable" kind.)

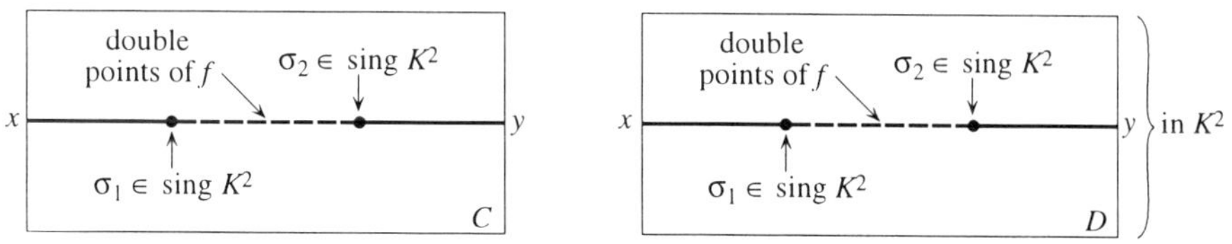

FIGURE 3

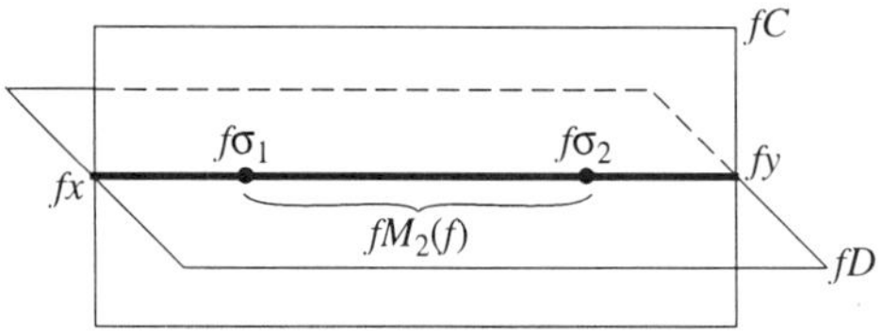

FIGURE 4

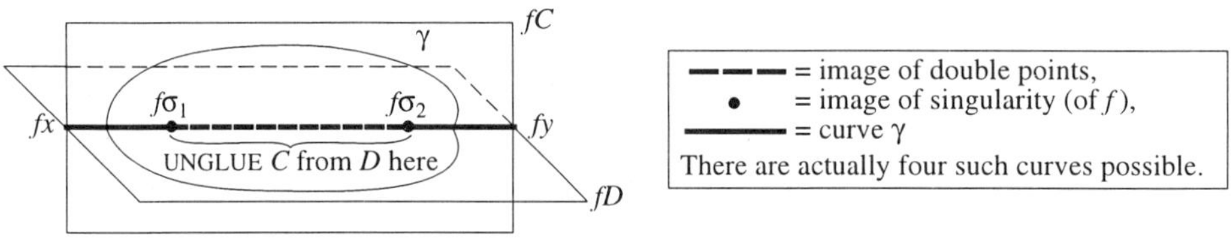

FIGURE 5

Let $U \subset R^3$ be a regular neighbourhood of $fK^2 \subset R^3$ with $\partial U \cap fK^2 = \partial(fK^2)$. This $K^2 \xrightarrow{f} U$ *is* an $S_u U$ structure, with $|S_u U| \underset{\text{def}}{=} \{$ regular neighbourhood of fK^2 in dimension $(n+3)\} = U \times B^n$.

Now we move to $S_b U$. On the "surface" of $fK^2 \subset R^3$ we draw the curve γ from Figure 5 and then we add to fK^2 an embedded 2-cell $\delta \subset R^3$ like in Figure 6.

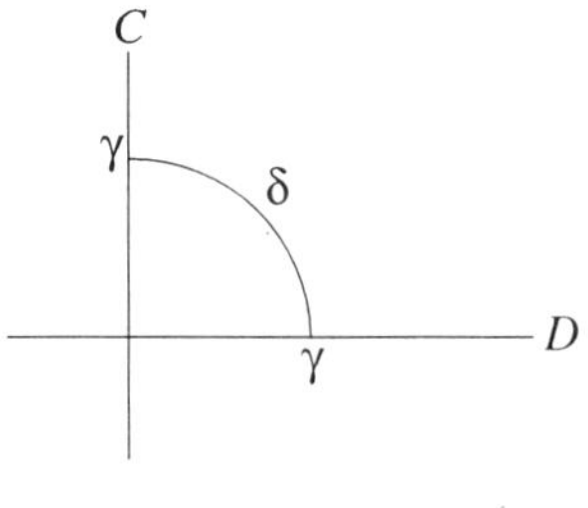

FIGURE 6

Next we move to $R^3 \times [-1,1] \supset R^3 \times 0 = R^3$ and (without touching γ) we unglue $fC \subset fK^2 \cup \delta$ from $fD \subset fK^2 \cup \delta$ along $fM_2(f)$ by pushing the brach D into the fourth dimension, in the direction $t = +1$ (and, for the sake of symmetry C in the direction $t = -1$). This gives us a 2-dimensional space P^2 of which $fK^2 \cup \delta$ is a quotient-space, and which is embedded inside $R^3 \times [-1,+1]$. The obvious map $P^2 \to U$ *is* our structure $S_b U$ and $|S_b U| \underset{\text{def}}{=} \{$the regular neighbourhood of P^2 in $R^3 \times [-1,1]\} \times B^{n-1}$. Trivially, of course, $|S_u U| \underset{\text{DIFF}}{=} |S_b U|$, in this case. [Any question of possible NON-COHERENCE (see [Po4], [Po5], [Po6], [Ga]) for our 0(3)-move has been avoided by chosing n high enough.]

For an open simply-connected manifold V^3, in defining $S_u V^3$ the role of the $K^2 \xrightarrow{f} U$ from our simple example, is essentially played by the almost-arborescent representation $X^2 \xrightarrow{g} V^3$ of V_h^3 from Theorem 6. What is far from being trivial is to devise the correct way of playing something like the game $S_u U/S_b U$ with this rather exotic map $X^2 \xrightarrow{g} V^3$, in lieu of the simple minded $K^2 \xrightarrow{f} U$. All the techniques of proof for Theorems 3, 4, 5, 6, in particular the very detailed constructions from [Po8] (and [Po11]) are necessary at this point. The mutual inter-relation between $V_h^3 \times B^n$ and $\varinjlim X_i$ from our Theorem 4 is a primitive version of the more sophisticated inter-relation between $|S_u V^3|$ and $|S_b V^3|$. The big progress is that the very pathological space $\varinjlim X_i$ has been replaced now by the smooth manifold $|S_b V^3|$.

In the same vein I should add that when the bad locus $\sum$ is $\emptyset$, then it can be shown by "local" arguments that $|S_u V^3|$ is diffeomorphic to $|S_b V^3|$. In the presence of a bad locus $\sum \neq \emptyset$ the situation is considerably trickier, any kind of "local" augument breaks down, and if and when the equality $|S_u V^3| \underset{\text{DIFF}}{=} |S_b V^3|$ might still be valid, establishing it requires a "global" proof.

This is as much as I can say about S_u/S_b in the present format.

References

[B-M] M. Bestvina and G. Mess, *The boundary of negatively curved groups*, Preprint (1990).

[Can] J.W Cannon, *Almost convex groups*, Geom. Dedicata **22** (1987), 197-210.

[Ca] A. Casson, *A la recherche de la Topologie perdue (ed. L. Guillou and A. Marin)*, Birkhäuser, 1986, pp. 201-244..

[Da] M.W. Davis, *Groups generated by reflexions and aspherical manifolds not covered by Euclidean Spaces*, Ann. of Math. **117** (1983), 293-324.

[Da-S] M. Davis and M. Shapiro, *Coxeter groups are almost convex*, Preprint (1991).

[De] R. Devaney, *An introduction to Chaotic Dynamical Systems*, Addison-Wesley, 1989.

[E-C-et al] D.B.A. Epstein, J.W. Cannon, D.F. Holt, S.V.F. Levy, M.S. Patterson and W. Thurston, *Word processing in groups*, James and Bartlett, 1992.

[Fr] M.H. Freedman, *The topology of four-dimensional manifolds*, J. Diff. Geom. **17** (1982), 357-453.

[Fr-L] M.H. Freedman and F. Luo, *Marker Lectures in Math*, Princeton, 1989.

[Fr-Q] M.H. Freedman and F. Quinn, *The Topology of 4-Manifolds* (1990), Princeton Univ. Press.

[Fu] L.Funar, *Infinite simple homotopy and the simple connectivity at infinity of open 3-manifolds*, Preprint, Univ. of Grenoble (1996).

[Ga] D. Gabai, *in the volume "Geometry, Topology and Physics", Raoul Gott (edited by S.T. Yau)*, International Press, 1994, pp. 139-166.

[Gr] M. Gromov, *Hyperbolic groups*, MSRI Publ **8** (1987), 72-263.

[Po0] V. Poénaru, *On the equivalence relation forced by the singularities of a nondegenerate simplicial map*, Duke Math. J. **63** (1991), 421-429.

[Po1] V. Poénaru, *Killing Handles of Index one stably and* π_1^∞, Duke Math. J. **63** (1991), 431-447.

[Po2] V. Poénaru, *Almost convex groups, Lipschitz Combings and* π_1^∞ *for Universal Covering spaces of closed 3-manifolds*, J. Diff. Geom. **35** (1992), 103-130.

[Po3] V. Poénaru, *Geometry "à la Gromov" for the fundamental group of a closed 3-manifold* M^3 *and the simple connectivity at infinity of* $\widetilde{M^3}$, Topology **33** (1994), 181-196.

[Po4] V. Poénaru, *The collapsible pseudo-spine representation theorem*, Topology **31** (1992), 625-636.

[Po5] V. Poénaru, *Processus infinis et conjecture de Poincareé en dimension trois, IV: Le théorème de non sauvagerie lisse (The smooth tameness theorem)*, Part A, Orsay Preprint, 92-83 (1992); Part B,Orsay Preprint, 95-33 (1995).

[Po6] V. Poénaru, *The strange compactification theorem*, Part A, IHES M-95-15 (1995).

[Po7] V. Poénaru, *A general finiteness theorem in group theory*, Orsay Preprint, 92-33 (1992), 19-33.

[Po8] V. Poénaru, *Representations of open simply-connected 3-manifolds; a finiteness result*, Orsay Preprint (1992), 69-92.

[Po9] V. Poénaru,, *Letter to John Stallings*, May (1991).

[Po10] V. Poénaru, *A remark on simply-connected 3-manifolds*, BAMS **88** (1974), 1203-1204.

[Po11] V. Poénaru, *The Handles of index one of the product of an open simply-connected 3-manifold with a high dimensional ball*, manuscript.

[Po12] V. Poénaru, *Three lectures on higher-dimensional topological methods in three-dimensional topology*, Suppl-Rend. Palermo (1997), 203-217.

[Po13] V. Poénaru, *Geometric simple connectivity in smooth four-dimensional topology : an outline*, Orsay Preprint (1998).

[Po14] V. Poénaru, *Geometric simple connectivity in smooth four-dimensional topology*, manuscript.

[Po-T1] V. Poénaru and C. Tanasi, *Hausdorff combing of groups ...*, Annali Sc. Norm. Sup. Pisa **20** (1993), 387-414.

[Po-T2] V. Poénaru and C. Tanasi, *k-almost convex groups and* $\pi_1^{\infty}\widetilde{M^3}$, Geom. Dedicata **48** (1993), 57-81.

[Po-T3] V. Poénaru and C. Tanasi,, *Representations of the Whitehead manifold* Wh^3 *and Julia sets*, Ann. de Toulouse **V.4.N.3** (1995 pazges 655-694).

[S-G] J. Stallings and S.M. Gersten, *Casson's idea about 3-manifolds whose universal cover is* R^3, Preprint (1991).

[Th] W.P.Thurston, *Three dimensional manifolds, Kleinian groups and hyperbolic geometry*, BAMS **6** (1982), 357-381.

[Wh1] J.H.C. Whitehead, *On* C^1*-complexes*, Ann. of Math. **41** (1940), 804-824.

[Wh2] J.H.C. Whitehead, *A certain open manifold whose groups is unity*, Q. J. of Math. **6** (1935), 268-279.

UNIVERSITY PARIS-SUD, MATHIMATIQUE BATIMENT 425, F-91405 ORSAY ADEX FRANCE
E-mail address: poe@topo.math.u-psud.fr

Contemporary Mathematics
Volume **233**, 1999

Cubulations mod bubble moves

Louis Funar

ABSTRACT. The aim of this paper is to consider the set of cubical decompositions of a compact manifold mod out by some combinatorial moves (bubble) analogous to the bistellar moves earlier considered by Pachner. We prove that, in general there are obstructions for two cubulations of the same PL-manifold to be related by bubble moves, answering negatively a question of Habegger.

1. Introduction

In the twenties Alexander proved that two triangulations of a polyhedron (or equivalently of two PL-homeomorphic manifolds) are related by a set of combinatorial moves, called stellar moves. After seventy years Alexander's moves were refined to a set of finite local moves on the triangulations of manifolds which were used to prove that certain state-sums associated to a triangulation provide topological invariants of 3-manifolds, the so-called Turaev-Viro invariants. The new moves are the bistellar moves and Pachner ([**27**]) has proved that they relate any two triangulations of a polyhedron, settling a long standing conjecture in combinatorial topology. Basically such a move in dimension n corresponds to replace a ball B by another ball B', where B and B' are complementary balls, unions of simplexes in the boundary of the standard $(n+1)$-simplex. For a nice exposition of Pachner's result and various extensions, see [**22**].

On the other hand Alexander's theorem becomes trivial in the context of some more general cell decompositions ("cellulation régulière") considered by Siebenmann [**28**] where the analogous moves are called "bisections". Here the cells are convex subsets in some Euclidean space, with an arbitrary number of vertices.

The Turaev-Viro invariants carry less information than the Reshetikhin-Turaev invariants, which are defined using Dehn surgery presentations of the manifolds instead of triangulations. Actually the latter have a strong 4-dimensional flavor, as explained by the theory of shadows developed by Turaev (see [**31**]). This motivates the study of state-sums based on cubulations, as an alternative way to get intrinsic invariants possibly containing more information (e.g. the phase factor). In order to apply the state-sum machinery to these decompositions we need an analogue of the Alexander's or Pachner's theorems. Specifically, N.Habegger asked (see problem 5.13 from R.Kirby's list ([**18**])) the following:

PROBLEM 1. *Suppose M and N are PL-homeomorphic cubulated n-manifolds. Are they related by the following set of moves: excise B and replace it by B', where B and B' are complementary balls (union of n-cubes) in the boundary of the standard $(n+1)$-cube?*

1991 *Mathematics Subject Classification.* Primary 57 Q 15, 57 R 42, Secondary 51 M 20, 57 Q 35.

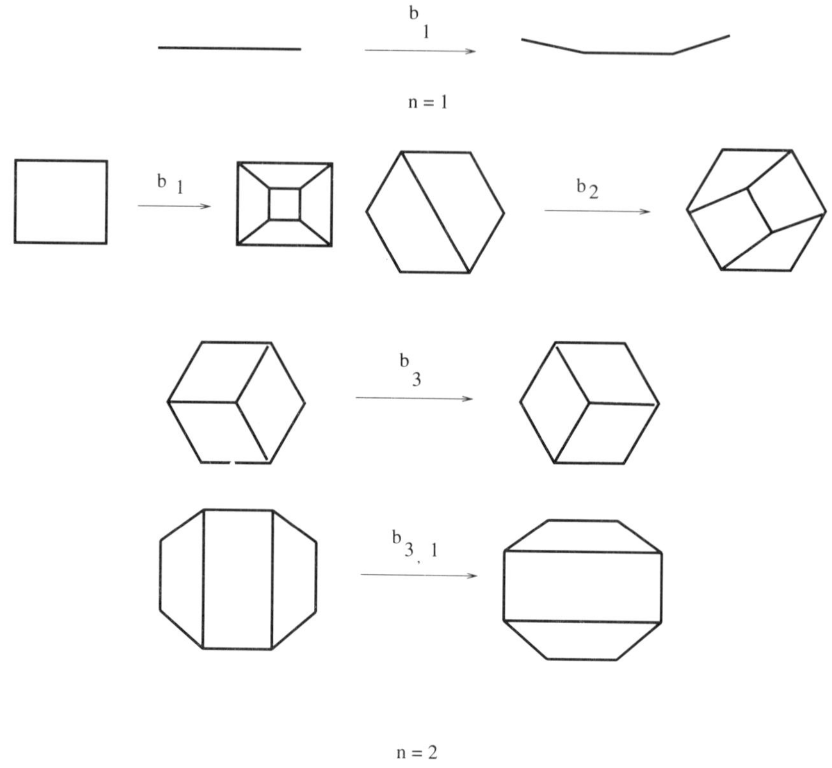

FIGURE 1. Bubble moves for $n = 1$ and $n = 2$

These moves will be called *bubble* moves in the sequel. The subfamily of those bubble moves for which at least one of B and B' does not contain parallel (when viewed in the $n+1$-cube) faces form the *np-bubble* moves. There are $n+1$ distinct np-bubble moves b_k, $k = 1, 2, ..., n+1$ and their inverses, b_k replacing B which is the union of exactly k cubes by its complementary. For $n = 2$ there is one bubble moves which is not a np-bubble (see picture 1).

Set $C(M)$ for the the set of cubulations of a closed manifold M (without boundary), $CBB(M)$ for the equivalence classes of cubulations mod np-bubble moves and $CB(M)$ for the equivalence classes of cubulations mod bubble moves.

For $n = 1$ the move b_1 divides an edge into three smaller edges and it follows that: $CB(\bigsqcup_n S^1) = CBB(\bigsqcup_n S^1) = (\mathbf{Z}/2\mathbf{Z})^n$, where n is the number of components. Thus there are non-trivial obstructions for two cubulations be np-bubble (bubble) equivalent. Assume from now on, that the manifolds considered are connected unless the contrary will be specified.

In the following section we will describe similar obstructions in higher dimensions. The set of all transforms of a given f-vector $\mathbf{f}$ under np-bubble/bubble moves is $\mathbf{f} + \Lambda(n)$, where $\Lambda(n) \subset \mathbf{Z}^{n+1}$ is a sublattice, depending only on the dimension. Therefore the class of $\mathbf{f} \in \mathbf{Z}^{n+1}/\Lambda(n)$ is an invariant taking values in a finite Abelian group. Our first result states that

THEOREM 2. *The obstruction in* $\mathbf{Z}^{n+1}/\Lambda(n)$ *is not trivial.*

We cannot expect the obstruction map $f : CB(M) \longrightarrow \mathbf{Z}^{n+1}/\Lambda(n)$ be injective in high dimensions.

Acknowledgements: Part of this work was done during the author's visit at Columbia University, whose support and hospitality are gratefully acknowledged. I'm thankful to E.Babson, J.Birman, C.Blanchet, R.Casali, C.Chan, N.Habegger, L.Guillou, T.Kashiwabara, A.Marin, S.Matveev, V.Poénaru for helpful discussions, suggestions and improvements.

2. Other statements and comments

2.1. Elementary obstructions. We want to identify obstructions, similar to that for the 1-dimensional case, in higher dimensions: set for $x \in C(M)$, and M of dimension n, $f_i(x) = card\{i\text{-dimensional cubes in } x\}$. Putting all these together we get a map $f : C(M) \longrightarrow \mathbf{Z}^{n+1}$ whose components are f_i, and it is usually called the f-vector in the theory of polytopes. Notice that once we start transforming a cubication of f-vector $\mathbf{f}$ by using all np-bubble/bubble moves we obtain a set of possible f-vectors having the form $\mathbf{f} + \Lambda(n) \subset \mathbf{Z}^{n+1}$, where $\Lambda(n)$ is a lattice. Therefore the first obstruction we encounter is the class of $\mathbf{f} \in \mathbf{Z}^{n+1}/\Lambda(n)$. The latter is a finite Abelian group and we will see that it is non-trivial.

PROPOSITION 2.1. *There exist some natural numbers $a_i(n) \in \mathbf{Z}_+$ such that:*

1. *All the $a_i(n)$ are non-trivial, and divisible by 2.*
2. *The map f induces a well-defined map $fb : CBB(M) \longrightarrow \prod_{i=0}^{n} \mathbf{Z}/a_i(n)\mathbf{Z}$ by*
$$fb(x) = (f_i(x)(mod\ a_i(n))_{i=0,1,\ldots,n}.$$
3. *The greatest such numbers $a_i(n)$ verify*
$$a_n(n) = 2, a_{n-1}(n) = 2n, a_{n-2}(n) = 2, a_0(n) = 2, a_1(n) = 3 + (-1)^n, (n > 2)$$

We computed the vector $\mathbf{a}(n)$ whose components are $a_i(n)$ for small dimensions: $\mathbf{a}(2) = (2, 4, 2)$, and $\mathbf{a}(3) = (2, 2, 6, 2)$. After some messy computations we obtain $\mathbf{a}(4) = (2, 4, 2, 8, 2)$, $\mathbf{a}(5) = (2, 2, 4, 2, 10, 2)$, and $a_2(6) = 6$. We denote by $fb^{(2)}$ the values of fb reduced mod $(2, 2, 2, ..., 2, 2n, 2)$.

Next we can extend this result on obstructions to bubble moves:

PROPOSITION 2.2. *There exist some natural numbers $\tilde{a}_i(n) \in \mathbf{Z}_+$ such that:*

1. *All the $\tilde{a}_i(n)$ are non-trivial, and divisible by 2.*
2. *The map f induces a well-defined map $fb : CB(M) \longrightarrow \prod_{i=0}^{n} \mathbf{Z}/\tilde{a}_i(n)\mathbf{Z}$ by*
$$fb(x) = (f_i(x)(mod\ \tilde{a}_i(n))_{i=0,1,\ldots,n}.$$
3. *The greatest such numbers $\tilde{a}_i(n)$ verify*
$$\tilde{a}_n(n) = 2, \tilde{a}_{n-1}(n) = 2n, \tilde{a}_{n-2}(n) = 2, \tilde{a}_0(n) = 2, \tilde{a}_1(n) = 3 + (-1)^n, (n > 2)$$

The two sequences a_j and $\tilde{a}_j$ are not identical since we have:
$$\tilde{a}_2(6) = 2 \neq a_2(6),\ \tilde{a}_2(4) = 2 \neq a_2(4).$$

Notice that $f_{n-1}(x) = nf_n(x)$ for all x. This means that the image of fb_{n-1} is $\mathbf{Z}/2\mathbf{Z} \cong \{0, n\} \subset \mathbf{Z}/2n\mathbf{Z}$. By the way the component fb_{n-1} is determined by fb_n. Notice that for $n = 2$ the image of fb is determined by f_0, and for $n = 3$ by f_0 and f_1.

A natural problem now is to know the images $fb(CB(M))$, $fbb(CBB(M))$, or at least their mod 2 reduction. This will give a hint about how much these invariants are powerful. We are far from having a complete answer now, and this problem is more difficult than it seems at first glance. There are some partial results for the case of the mod 2 reductions $fb^{(2)}(CB(M))$ and $fbb^{(2)}(CBB(M))$. Actually this is equivalent to characterize those f-vectors mod 2 which can be realized by cubulations of the manifold M. Obviously there are constraints for the existence of a simplicial polyhedra with a given f-vector and fixed topological type. For convex simplicial polytopes we have for instance the McMullen conditions (conjectured by McMullen in [**23**] and proved in [**4, 5, 29**]; the reader may consult also other proofs and results in [**23, 3, 24, 25**]). The complete characterization of the f-vectors of simplicial polytopes (and PL-spheres) was obtained by Stanley in [**30**]. The analogous problem of the realization of f-vectors by cubical polytopes has also been addressed in some recent papers, for example [**7, 2, 16, 17**] and references therein. The Dehn-Sommerville equations have a counterpart for cubical polytopes as in [**15**]. The lower bound conjecture and the upper bound conjecture have counterparts in the cubical case. The new feature is that, unlike in the simplicial case, there are parity restrictions on the f-vectors. This was firstly observed in [**7**]. Remark that it is exactly these restrictions in which we are interested. We have, as a simple application of the Dehn-Sommerville equation, a first constraint on the range of the mod 2 image:

PROPOSITION 2.3. *The rank of the affine module $fb^{(2)}(CB(M))$ is at most* $\left[\frac{n+1}{2}\right]$.

The relationship between cubical PL n-spheres and the immersions was described in the following beautiful result of Babson and Chan (see [**2**]):

PROPOSITION 2.4. *Let $\varphi : M \longrightarrow S^n$ a codimension 1 normal crossing immersion. Then there exists a PL cubical n-sphere K, such that*

$$f_i(K) = \chi(X_i(M, \varphi))(mod\ 2),$$

where χ denotes the Euler characteristics, and $X_i(M, \varphi) = \{x \in S^n;\ \ card\, \varphi^{-1}(x) = i\}$.

As a consequence, there exists a PL cubical n-sphere K with given $f_i(mod\ 2)$ if and only if there exists a codimension 1 immersion (M, φ) in S^n, for which the Euler characteristic of the multiple point loci $X_i(M, \varphi)$ of degree i equals f_i mod 2.

Remark that this result extends immediately to other varieties than the spheres. We have only to consider immersions $\varphi : M \longrightarrow N$ such that the image $\varphi(M)$ is a spine of N, which means that $N - \varphi(M)$ is an union of balls.

There is a wide literature on immersions, and especially on the following function θ_n, considered first by Freedman ([**13**]), where $\theta_n(\varphi)$ is the number of multiple n-points mod 2. The beginning of this theory was the result of Banchoff [**1**] saying that the number of normally triple points of a closed surface immersed in $\mathbf{R}^3$ is congruent mod 2 with its Euler characteristic. The function θ_n is easily seen to be well-defined as a function on the Abelian group B_n of bordism classes of immersions of $(n-1)$-manifolds in S^n. We have therefore an induced homomorphism:

$$\theta_n : B_n \longrightarrow \mathbf{Z}/2\mathbf{Z}.$$

Remark that the question on whether θ_n is surjective (i.e. nontrivial) is equivalent to find the image of $fb^{(2)}_{n-1}(S^n)$. From the results concerning the function θ_n obtained in [**13, 10, 11, 12, 20, 21, 8, 9**] we deduce that the f-vectors of a n-sphere have the following properties (see also [**2**]):

1. For $n = 2$ we have $f_0 = f_2(mod\ 2)$ and $f_1 = 0(mod\ 2)$ and thus $fb^{(2)}(CB(S^2)) = fb^{(2)}(CBB(S^2)) = \mathbf{Z}/2\mathbf{Z}$.
2. For $n = 3$, $f_0 = f_1 = 0(mod\ 2)$, $f_2 = f_3(mod\ 2)$. From the existence of Boy's immersion $j : \mathbf{R}P^2 \longrightarrow S^2$, with a single degree 3 intersection we find that there exists a PL 3-sphere with an odd number of facets. Therefore $fb^{(2)}(CB(S^3)) = fb^{(2)}(CBB(S^3)) = \mathbf{Z}/2\mathbf{Z}$.
3. The problem of characterizing the image $fb^{(2)}_{n-1}(S^n)$ is reduced to a homotopy problem. Namely, the image is $\mathbf{Z}/2\mathbf{Z}$ if and only if
   (a) either n is $1, 3, 4$ or 7.
   (b) or else $n = 2^a - 2$, with $a \in \mathbf{Z}_+$, and there exists a framed n-manifold with Kervaire invariant 1. The latter is known to be true for $n = 2, 6, 14, 30, 62$.
4. If we consider only the class of edge orientable cubulations in the sense of [**16**] the problem of characterizing the image $fb^{(2)}_{n-1}(S^n)$ is also reduced to a homotopy problem, which is completely solved. In fact the condition of edge orientability is equivalent to ask that the associated manifold M immersed in S^n be orientable. Thus we have to consider only the restriction of the map θ_n at the subgroup of oriented bordism classes of immersions, as originally considered by Freedman [**13**]. Away from the trivial cases $n = 1, 2$ the only case when the restriction of θ_n remains surjective in the orientable context is $n = 4$. Thus $f_{n-1} = 0(mod\ 2)$ if $n \neq 1, 2, 4$.

Thus, at this time only a finite number of n is known, for which the last component of $fb^{(2)}$ is nontrivial. Anyway the remarks from above show that the invariants fb and fbb are interesting and nontrivial. Notice that fb does not determine the class of the f-vector in $\mathbf{Z}^{n+1}/\Lambda(n)$, since the lattice $\Lambda(n)$ is not a product, in general. We have for instance:

PROPOSITION 2.5. *The complete f-obstruction for 3-manifolds is specified by $(f_0, f_1) \in \mathbf{Z}/2\mathbf{Z}$ and the additional $f_0 + f_1 \in \mathbf{Z}/4\mathbf{Z}$.*

2.2. The 2-dimensional case. In the 2-dimensional case the image of the immersion associated to a cubulation C is an union of immersed circles K_i.

PROPOSITION 2.6. *The collection of homotopy classes of the circles K_i is np-bubble invariant.*

PROOF. The local pictures of the moves b_1, b_2, b_3, on the boundary of the 3-ball are relative homotopy equivalences. Since the moves have their supports in small disks (part of the cubulated surface) we are done. □

Notice that we have a collection, and not a set, since some elements can be multiple. It follows that $CBB(M)$ is infinite provided that the genus of M is at least 1. This generalizes also in higher dimensions for manifolds with nontrivial topology. This makes the np-bubble equivalence too rigid: for manifolds with nontrivial topology the homothetic cubulations M and λM will be np-bubble nonequivalent. Thus the np-bubble equivalence is interesting only for spheres. We say that a two dimensional cubulation is *simple* if the associated circles K_i are individually embedded in the respective surface. A cubulation is called now *semi-simple* if each image circle $\varphi(K_i)$ has an even number of double points, which form cancelling pairs, i.e. pairs of double points connected by two distinct and disjoint arcs bounding a disk. We have the following characterization of $CB(S^2)$:

PROPOSITION 2.7. *The np-bubble moves act transitively on the set of simple cubulations of* S^2. *The orbit of the cubulation* ∂C^{n+1} *is the set of semi-simple cubulations. The map* $fb^{(2)}$ *is an isomorphism between* $CB(S^2)$ *and* $\mathbf{Z}/2\mathbf{Z}$.

PROOF. The moves b_2, b_3 act like the second and the third Reidemester moves in plane, and b_1 creates (its inverse destroys) a circle. There is a complication due to the presence of some other arcs in the diagrams, which remain untouched by the moves. A combinatorial analysis shows that all immersed circles can be reduced to some standard circles similar to the framed versions of the unknot, with exactly one curl for each component. The move $b_{3,1}$ can add several components and then creating new cancelling pairs which can be annihilated by np-bubbles. The details are given in [**14**]. □

2.3. Around bordisms. Let now consider the set $\mathcal{I}(M)$ (respectively $\mathcal{I}^+(M)$ in the orientable case) of bordisms of codimension 1 nc (i.e. normal crossings) immersions into a differentiable manifold M. Two nc-immersions $f_i : N_i \longrightarrow M$ of the $n-1$-dimensional manifolds N_i are bordant if there exists a proper nc-immersion $f : N \longrightarrow M \times [0,1]$ from some cobordism N between N_1 and N_2 such that the restriction of f to N_i is isotopic to f_i. Notice that the transversality allows us to get rid of the nc-assumption.

We define now a *marked* cubulation as a cubulation C of the manifold M, endowed with a PL-homeomorphism of the subjacent space $|C|$, $|C| \longrightarrow M$ up to an isotopy. We ask for simplicity the compatibility of the PL and DIFF structures on the image, and M to be a DIFF manifold. The set of marked combinatorial cubulations is denoted by $\widetilde{CB}(M)$. We can associate to each marked cubulation C of the DIFF manifold M (compatible with the DIFF structure) a codimension 1 nc-immersion $\varphi_C : N_C \longrightarrow M$. We associate to each cube the set of n section hyperplanes, each of them splitting it into two equal halves. The union of all these section hyperplanes form the image of a normal crossings codimension 1-immersion after a suitable smoothing. If the cubulation C is edge-orientable (following [**16**]), and M is oriented then N_C is an oriented manifold. We denote therefore by $\widetilde{CB}^+(M)$ and $CB^+(M)$ the associated objects in the oriented case. We have then the following result:

PROPOSITION 2.8. *There are applications* $I : \widetilde{CB}(M) \longrightarrow \mathcal{I}(M)$, *and in the oriented case* $\widetilde{CB}^+(M) \longrightarrow \mathcal{I}^+(M)$, *given by* $C \to \varphi_C$.

The map I is injective for the 2-sphere (see below).

PROOF. We consider the local picture of a bubble move, viewed in the boundary of the $n+1$-cube. The set of sections which make the immersion on the boundary are intersections of the section hyperplanes of the $n+1$-cube with the faces. The n-balls B and B' (interchanged by the bubble move) form the boundary of the sphere S^n. We infer between B and B' a short cylinder which identifies the $n+1$-ball with a cobordism between B and B'. When adding to this picture the hyperplane sections (trivially extended over the cylinder) we find a bordism (with normal crossings) between the immersions φ_B and $\varphi_{B'}$ into the n-ball. Then a sequence of bubble moves induces a composition of bordisms relating the initial and final immersions. □

A partial result about the converse direction was obtained in [**2**]. We think that $\widetilde{CB}(M)$ depends only on the homotopy type of M and the functor $\widetilde{CB}$ which associates to M the set $\widetilde{CB}(M)$ is (homotopically) representable, i.e. there exists a space $\mathcal{CB}$ such that $\widetilde{CB}(M) = [M, \mathcal{CB}]$.

Notice that $CB(M) = \widetilde{CB}(M)/\mathcal{M}(M)$, where $\mathcal{M}(M)$ is the mapping class group of M, i.e. the group of diffeomorphisms of M up to an isotopy. However the description of $\mathcal{I}(M)$ is not a simple task. The classical Pontryagin-Thom construction (see e.g. [**32**]) implies that we have a homotopical description as follows: $\mathcal{I}(M) = [M_c, \Omega^\infty S^\infty \mathbf{RP}^\infty]$, and $\mathcal{I}^+(M) = [M_c, \Omega^\infty S^\infty S^1]$, where M_c is the one point compactification, Ω is the loop space, S the reduced suspension and the brackets denote the homotopy classes of maps. Moreover $\mathcal{I}^+(M) = \pi^1(M_c)$ can be identified then with the first cohomotopy group $\pi^1(M_c)$, which is however hard to compute. There are some cases when the cohomotopy groups could in principle be computed, for instance in the case of spheres. This gives $\mathcal{I}(S^n) = \pi_n^s(\mathbf{RP}^\infty)$, where $\pi_n^s(\mathbf{RP}^\infty)$ is the n-th stable homotopy group, and $\mathcal{I}^+(S^n) = \pi_n^s(S^1) = \pi_{n-1}^s$. For instance we can use $\pi_1^s(\mathbf{RP}^\infty) = \pi_2^s(\mathbf{RP}^\infty) = \mathbf{Z}/2\mathbf{Z}$, and $\pi_3^s(\mathbf{RP}^\infty) = \mathbf{Z}/8\mathbf{Z}$, and few values of the stable stems are tabulated below:

n	0	1	2	3	4	5	6	7	8
π_n^s	$\mathbf{Z}$	$\mathbf{Z}/2$	$\mathbf{Z}/2$	$\mathbf{Z}/24$	0	0	$\mathbf{Z}/2$	$\mathbf{Z}/240$	$\mathbf{Z}/2 \oplus \mathbf{Z}/2$

Eventually let us introduce the bordism set of cubulations $\mathcal{C}(M)$ of the manifold M: two cubulations C_1 and C_2 are bordant if there exists a cubulation C of $M \times [0,1]$ whose restrictions on the boundaries are C_i. Notice that this definition has a strong combinatorial flavor, since the topology of the cobordant cubulation is fixed. Observe that the identity induces a map $CB(M) \longrightarrow \mathcal{C}(M)$,and we would like to know whether there is an inverse arrow. This has some similitudes with Wall's theorem about the existence of formal deformations between simple homotopy equivalent CW complexes of dimension n, by passing throughout $n+1$-dimensional cells ($n \neq 2$). Remark that any two cubulations become bordant after suitable subdivisions. Alternatively let us look at cubulations of the sphere S^n which are bubble equivalent to the standard one. We can view the bubble moves as the result of gluing and deleting $n+1$ cubes (after a suitable thickening) to the given cubulation. It follows that any such cubulations bounds, i.e. it is the boundary of a cubulation of the $n+1$-ball. For instance if $n=1$ we should have a polygon with an even number of vertices. It is possible that for $n = 2,3$ the converse is also true: the boundary of a ball cubulation is bubble equivalent to a trivial one. We don't expect the same phenomenon for $n \geq 4$. let put it another way: it is known that there exist non-shellable triangulations of the ball for $n \geq 3$. We define a *shuffling* of a cubulation (triangulation) as being a sequence of moves where alternate shellings (adding iteratively cells which intersect the previous stage union into balls) and inverse shellings (deleting iteratively cells which intersect the previous stage boundary into balls). Then Pachner theorem says that all triangulations can be shuffled. We saw that an obstruction for shuffling is that the cubulation bounds. We expect however that cubulations of the $n+1$-ball which cannot be shuffled actually exist. A good candidate is the double of a cubulation whose image by I is the generator of the third stable stem. The idea is the following: If all bounding cubulation can be shuffled then $CB(S^n)$ is a product of $\mathbf{Z}/2\mathbf{Z}$'s. We consider a cubulation x of the sphere. It is proved in [**14**] that the connected sum of cubulations makes $CB(S^n)$ a monoid. The connected sum is obtained by removing one cell from each cubulation and then joining them by a standardly cubulated cylinder as a piping tube. Moreover it is sufficient to show that $x\sharp x$ the connected sum with itself is bubble equivalent to a trivial one. But it is clear that $x\sharp x$ is a cubulated sphere which bounds a cubulated ball, namely $(x -$ one cell $) \times [0,1]$. Then we will have a bubble equivalent cubulation, meaning that any element is of order 2. However if there is a cubulation x such that the image $I(x)$ in the bordism group $I^+(S^4) = \mathbf{Z}/24\mathbf{Z}$ is a generator then, x cannot have order 2.

3. Proofs

3.1. The proof of Proposition 2.1. At the beginning we make some notations: D_k^+ is the union of k cubes (of dimension n) which are the faces of a $(n+1)$-dimensional cube, and no two of them are parallel faces, for $k = 1, 2, ..., n+1$. The complementary union of $2(n+1) - k$ cubes is denoted D_k^-. So a np-bubble move b_k replaces D_k^+ by D_k^-. Now $f_p(D_k^+)$ is the number of p-dimensional cubes in D_k^+. The number of interior p-cubes is $f_p(int(D_k^+)) = f_p(D_k^+) - f_p(\partial D_k^+)$. Notice that this is a notational convention because it does not count the number of p-cubes in the interior of D_k^+ but the number of open such p-cubes sitting in the interior of D_k^+.

In order to find the corresponding $a_i(n)$ we have to compute the numbers $f_p(b_k(x)) - f_p(x) = f_p(D_k^+) - f_p(D_k^-) = a(k,n,p)$; then $a_i(n) = gcd\{f_p(b_k(x)) - f_p(x), k = 1, 2, ..., n+1\}$.

We use the method of generating functions: set $F_X(T) = \sum_{p=0}^n f_p(X)T$. It is well-known that $f_p(C_n) = \begin{pmatrix} n \\ p \end{pmatrix} 2^{n-p}$ for the n-cube C_n so that $F_{C_n}(T) = (2+T)^n$.

Let e_i, $i = 1, ..n+1$ be the vectors spanning C_{n+1} and L_i be the n-cube spanned by e_1, e_2,..., e_{i-1}, e_{i+1},..., e_{n+1}, where e_i is omitted. Then a combinatorial model for D_k^+ is $\bigsqcup_{i=1}^k L_i$. The inclusion-exclusion principle states that:

$$\begin{aligned} f_p(\bigsqcup_{i=1}^k L_i) &= \sum_{i=1}^k f_p(L_i) - \sum_{i<j}^k f_p(L_i \cap L_j) + ... \\ &+(-1)^{l+1} \sum_{i_1<i_2<...<i_l}^k f_p(L_{i_1} \cap L_{i_2} \cap ... \cap L_{i_l}) + ... \\ &+(-1)^{k+1} f_p(L_1 \cap L_2 \cap ... \cap L_k). \end{aligned}$$

We observe that $L_{i_1} \cap L_{i_2} \cap ... \cap L_{i_l}$ is combinatorially the cube C_{n+1-l} and we derive

$$f_p(\bigsqcup_{i=1}^k L_i) = \sum_{i=1}^k (-1)^{i+1} \begin{pmatrix} n-i+1 \\ p \end{pmatrix} \begin{pmatrix} k \\ i \end{pmatrix} 2^{n+1-i-p}.$$

It follows that, at the level of generating functions, we have:

$$\begin{aligned} F_{D_k^+}(T) &= \sum_{i=1}^{k}(-1)^{i+1}\left(\begin{array}{c} k \\ i \end{array}\right)_{F_{C_{n+1-i}}(T)=\sum_{i=1}^{k}(-1)^{i+1}}\left(\begin{array}{c} k \\ i \end{array}\right)_{(2+T)^{n+1-i}=} \\ &(2+T)^{n+1-i}\left((2+T)^k-\sum_{i=1}^{k}(-1)^i\left(\begin{array}{c} k \\ i \end{array}\right)_{(2+T)^{k-i}}\right) = \\ &(2+T)^{n+1}-(2+T)^{n+1-k}(1+T)^k. \end{aligned}$$

Therefore the generating function counting the interior cubes in D_k^- is simply

$$F_{int(D_k^-)}(T) = (2+T)^{n+1-k}(1+T)^k - T^{n+1} - kT^n.$$

In fact the total number of p-cubes in D_k^+ and D_k^- is $f_p(C_{n+1})$, but there are no $(n+1)$-dimensional faces and also the n-dimensional cubes are not interior in D_k^-, so k of them have to be removed from the total.

It remains to compute the number of interior p-cubes in D_k^+: all of them come as intersections $L_{i_1}\cap L_{i_2}\cap\ldots\cap L_{i_{n-p+1}}$ with the additional condition that $n-p+1\geq 2$. It follows that

$$f_p(int(D_k^+)) = \begin{cases} \left(\begin{array}{c} k \\ n-p+1 \end{array}\right) & \text{if} \quad n-1\geq p\geq n-k+1 \\ 0 & \text{elsewhere} \end{cases}$$

The generating function is therefore

$$F_{int(D_k^+)}(T) = \sum_{p=n-k+1}^{n-1}\left(\begin{array}{c} k \\ n-p+1 \end{array}\right)_{T^p=T^{n-k+1}[(1+T)^k-kT^{k-1}-T^k].}$$

We find that the series associated to the jumps of the f-vector is

$$-F_{int(D_k^+)}(T) + F_{int(D_k^-)}(T) = \left((2+T)^{n+1-k} - T^{n+1-k}\right)(T+1)^k.$$

so that

$$\sum_{p=0}^{n} a(k,n,p)T^p = \left((2+T)^{n+1-k} - T^{n+1-k}\right)(T+1)^k$$

As an immediate corollary we derive that all $a(k,n,p)$ are divisible by 2 since $(2+T)^{n+1-k} - T^{n+1-k}$ has even coefficients. This prove the first two claims. Developing the terms we obtain by a simple computation $a(k,n,n) = 2(n+1-k)$, $a(k,n,n-1) = 2n(n+1-k)$, and

$$a(k,n,n-2) = \frac{n+1-k}{3}\left(3n(n-1)+(n-k)(n-1-k)\right).$$

Also

$$a(k,n,0) = \begin{cases} 2^{n-k+1}, & if \quad k<n+1 \\ 0 & elsewhere \end{cases}$$

$$a(k,n,1) = \begin{cases} 2^{n-k}(n+1+k), & if \quad k\leq n-1 \\ 2n, & if \quad k=n \\ 0 & elsewhere \end{cases}$$

$$a(k,n,2)=\begin{cases} 2^{n-k-2}\left((n+1-k)(n+k)+4k(k-1)\right), & if \quad k<n-1 \\ 2n(n-1), & if \quad k=n-1 \\ n(n-1), & if \quad k=n \\ 0 & elsewhere \end{cases}$$

A tedious verification ends the proof.

3.2. The proof of Proposition 2.2. Set more generally $D_{k,r}^{+}$ for the union of k cubes (of dimension n) which are the faces of a $(n+1)$-dimensional cube, and exactly $2r$ of them arise in pairs of parallel faces, for $k=1,2,...,n+1$, and $2r\leq k$. The complementary set is denoted by $D_{k,r}^{-}$. Denote by $b_{k,r}$ the bubble move which replaces $D_{k,r}^{+}$ by $D_{k,r}^{-}$.

LEMMA 3.1. *The set of all combinatorially distinct unions of k cubes, $k\leq n+1$, of dimension n, which are faces of the $n+1$-dimensional cube, and are topologically disks, as well as their complementary sets, is exactly the set of all $D_{k,r}^{+}$ with $2r\leq k$. Therefore the bubble moves are the $b_{k,r}$ and their inverses.*

PROOF. Let γ_n be the adjacency graph of the cube ∂C^{n+1}. Then γ_{n+1} is the suspension $S(\gamma_n)$: this means that there are two new vertices which are connected with all of γ_n vertices but there is no edge between them. In fact it suffices to see that after excising two opposite faces of ∂C^{n+1} we obtain $\partial C^{n}\times[0,1]$, and the latter has the adjacency graph γ_n. Alternatively, this shows that γ_n is the complementary of the pairing graph P_n as subgraph of the complete graph $K_{2(n+1)}$. The pairing graph P_n has $2(n+1)$ vertices which are connected by $n+1$ edges, in pairs. Now we can use a recurrence argument. If $k\leq n$ and Γ is the adjacency graph of the union D of k cubes, then there exists a pair of opposite faces such that none of them is contained in D. Thus D is contained in $\partial C^{n}\times[0,1]$. We can write then $D=\tilde{D}\times[0,1]$, where $\tilde{D}\subset\partial C^{n}$ is the union of k cubes one dimension lower. First $\tilde{D}$ is a ball because it has trivial homotopy. The complementary of D inside $\partial C^{n}\times[0,1]$ should be also a ball, because it is the same as $\partial C^{n+1}-D$ without two cells touching the boundary (which are shelled of). Thus the complementary of $\tilde{D}$ in ∂C^{n} must be a ball and we can apply the recurrence hypothesis, in order to find the shape of D.

If $k=n+1$ there are two cases: either there is again one pair of opposite faces outside D (and then $D\subset\partial C^{n}\times[0,1]$, and the argument from above works), or else no such pair exists. Therefore D contains from each pair of opposite faces exactly one. In this case there is only one such D, up to the cubical symmetries, namely the $n+1$ coordinate hyperplanes intersection with the standard cubes. □

We have the following:

LEMMA 3.2. *For all k,r,p we have the identity:*

$$f_p(D_{k,r}^{+})-f_p(D_{k,r}^{-})=f_p(D_k^{+})-f_p(D_k^{-})(\ mod\ 2).$$

PROOF. Set $k=2r+q$. We denote the faces of C_{n+1} by L_i^{ε}, where $L_i^0=L_i$ and L_i^1 is the opposite face. Then $D_{k,r}$ is combinatorially equivalent to some

$$D_{k,r}\cong\bigsqcup_{i=1}^{r}L_i^0\bigsqcup_{i=1}^{r}L_i^1\bigsqcup_{i=r+1}^{r+q}L_i^{\varepsilon_i},$$

for some ε_i. But now the choice of the ε_i is irrelevant because there is an isometry of C_{n+1} which transforms an ε_j into $1-\varepsilon_j$ and preserve the others. In terms of coordinates x_i this is defined by $x_i \longrightarrow x_i$ for $i \neq j$, and $x_j \longrightarrow x_j$. Therefore we can choose a combinatorial model for $D_{k,r}$ with all $\varepsilon_i = 0$.

The only difference with the previous case is that

$$\bigcap_{s=1}^{m} L_{i_s}^{\varepsilon_s} = \begin{cases} C_{n+1-m} & \text{if there are no } s_1, s_2 \text{ such that } i_{s_1} = i_{s_2}, \varepsilon_{s_1} \neq \varepsilon_{s_2} \\ \emptyset & \text{otherwise} \end{cases}$$

Therefore, with respect to the computation we made previously we have to take into account that some of intersections are void. The void intersections correspond to the combinations of h-tuples $L_i^0, ..., L_i^1, ...$ (the remaining $h-2$ faces being arbitrary) which has been counted as cubes C_{n+1-h} previously. In the calculation of $f_p(\bigsqcup_{i=1}^{r} L_i^0 \bigsqcup_{i=1}^{r} L_i^1 \bigsqcup_{i=r+1}^{r+q} L_i^{\varepsilon_i})$ we have to see how each term arising in the inclusion-exclusion principle, namely $X_h = \sum_{L_{i_s}^{\varepsilon_s} \in A} f_p(\cap_{s=1}^{h} L_{i_s}^{\varepsilon_s})$, has changed. Here A is the set of faces of $D_{k,r}$.

For $h = 1$ there are no intersections so that this factor is conserved. For $h = 2$ all the factors $L_i^0 \cap L_i^1$ have been counted before, but now their contribution is zero. There are r such factors which implies that:

$$X_2 = \binom{n-1}{p} 2^{n-p-1} \left(\binom{k}{2} - r \right)$$

For $h = 3$ the now vanishing combinations are $r(k-2)$ since a couple can be chosen among the r pairs and the third can be chosen in $k-2$ ways. It follows that :

$$X_3 = \binom{n-2}{p} 2^{n-p-2} \left(\binom{k}{3} - r(k-2) \right)$$

We continue with $h = 4$: we have $r \binom{k-2}{2}$ possibilities to get at least one couple of parallel faces but the couples $L_i^0, L_i^1, L_j^1, L_j^0$ are counted two times. Applying again the inclusion-exclusion principle we derive that

$$X_4 = \binom{n-3}{p} 2^{n-p-3} \left(\binom{k}{4} - r \binom{k-2}{2} + \binom{r}{2} \right)$$

This generalizes easily by induction to:

$$X_s = \binom{n+1-s}{p} 2^{n+1-p-s} \left(\sum_{j=1}^{2j \leq s} (-1)^j \binom{k-2j}{s-2j} \binom{r}{j} \right)$$

We are able now to write:

$$f_p(D_{k,r}^+) = f_p(D_k^+) + \sum_{s=2}^{n-p+1} (-1)^s \binom{n+1-s}{p} 2^{n+1-p-s} \left(\sum_{j=1}^{2j \leq s} (-1)^j \binom{k-2j}{s-2j} \binom{r}{j} \right)$$

Now we need to know how the number of interior p-cubes $f_p(int(D_{k,r}^+))$ has been changed. Of course these interior p-cubes are always coming as intersections

$$\bigcap_{s=1; L_{i_s}^{\varepsilon_s} \in A}^{n-p+1} L_{i_s}^{\varepsilon_s}$$

Again some of these intersections are void because parallel faces are allowed to be in A. But the inclusion-exclusion principle gives (in the non-trivial case $n-1 \geq$

$p \geq n-k+1)$:

$$f_p(int(D^+_{k,r})) = f_p(int(D^+_{k,0})) \sum_{j=1}^{2j \leq n-p+1} (-1)^j \binom{k-2j}{n-p+1-2j} \binom{r}{j}$$

It suffices now to see that the difference $f_p(D^+_{k,r}) - f_p(int(D^+_{k,r}))$ has the same parity for all r. In the previous formula for $f_p(D^+_{k,r})$ only the term for $s = n-p+1$ has not a coefficient divisible by 2. But for $s = n-p+1$ the contributing term in $f_p(D^+_{k,r})$ is exactly the same as the total contributing term in $f_p(int(D^+_{k,r}))$ and they cancel each other.

Remark that in fact the greater common divisors of $f_p(D^+_{k,r}) - f_p(D^-_{k,r})$ are the analogous $\tilde{a}_i(n)$ of $a_i(n)$. It is clear than $\tilde{a}_i(n)$ are divisors of $a_i(n)$. However the explicit computations are more difficult. We omit the annoying details for checking that $\tilde{a}_0(n), \tilde{a}_1(n), \tilde{a}_{n-1}(n)$ are exactly those claimed. This ends the proof of the Proposition 2.2. □

3.3. The proof of Proposition 2.3. Let P^{n+1}_c denotes the family of convex cubical polytopes. Some of the cubulations of the sphere corresponds to the boundaries ∂P^{n+1}_c of elements from P^{n+1}_c. The last component of the f-vector of elements in P^{n+1}_c is trivial. Consider the affine $\mathbf{Z}$-submodule (or $\mathbf{Z}$-submodule coset) A^{n+1}_c generated by all the f-vectors of elements of P^{n+1}_c , viewed in $\mathbf{Z}^{n+1}$.

LEMMA 3.3. *(Dehn-Sommerville Equations) The affine space $A^{n+1}_c \otimes \mathbf{Q} \subset \mathbf{Q}^{n+1}$ is of dimension $\left[\frac{n+1}{2}\right]$. A set of defining equation is obtained from the Euler-Poincaré equation*

$$\sum_{i=0}^{n} (-1)^i f_i(x) = 1 - (-1)^{d+1}$$

together with

$$\sum_{j=k}^{n} (-1)^j \binom{j}{k} 2^{j-k} f_j = (-1)^{d-1} f_k, k=0,1,2,\ldots,n-1.$$

Equivalently we have the Euler-Poincaré equation and the set of independent equations

$$\sum_{j=k}^{n} (-1)^j \binom{j}{k} f_j = 0, k \equiv n+1 (mod\ 2), 1 \leq k \leq n-1.$$

PROOF. See [**15**], p. 156-159. □

In the general case of an arbitrary manifold M, not necessary S^n, we have to replace the Euler-Poincaré equation by the corresponding:

LEMMA 3.4. *(Dehn-Sommerville Equations for a manifold) The affine space $A^{n+1}_c(M) \otimes \mathbf{Q} \subset \mathbf{Q}^{n+1}$ is of dimension at most $\left[\frac{n+1}{2}\right]$. A set of equations which define a flat containing $A^{n+1}_c(M) \otimes \mathbf{Q}$ is obtained from the Euler-Poincaré equation*

$$\sum_{i=0}^{n} (-1)^i f_i(x) = \chi(M)$$

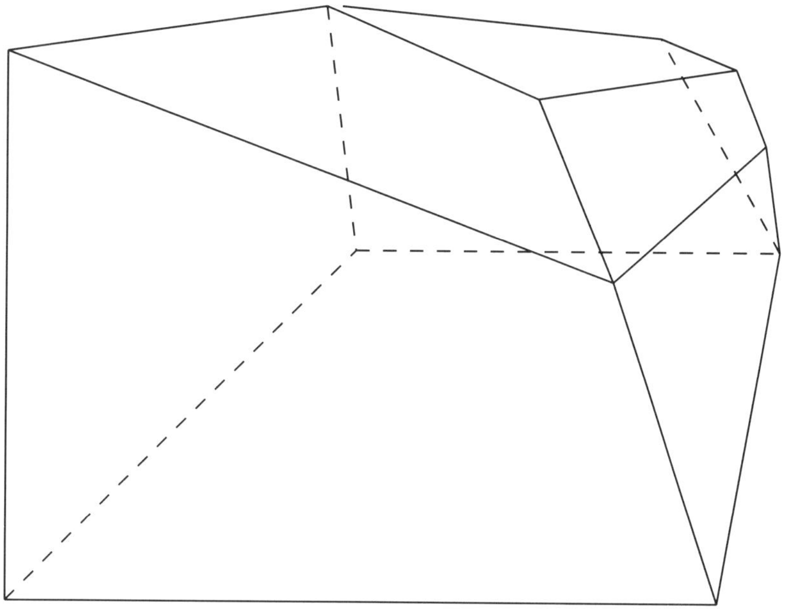

FIGURE 2. A cubulation with odd f_2

together with

$$\sum_{j=k}^{n}(-1)^j\binom{j}{k}2^{j-k}f_j=(-1)^{d-1}f_k, k=0,1,2,\ldots,n-1.$$

Equivalently, we have the Euler-Poincaré equation and the set of independent equations

$$\sum_{j=k}^{n}(-1)^j\binom{j}{k}f_j=0, k\equiv n+1(mod\ 2), 1\le k\le n-1.$$

This was proved by Klee for simplicial complexes in [**19**], see also [**15**], p. 152. The case of cubulations is similar, and we omit the proof.

As a corollary we derive that the affine $\mathbf{Z}/2\mathbf{Z}$-submodule of $(\mathbf{Z}/2\mathbf{Z})^{n+1}$ generated by $fb^{(2)}(CB(M))$ has rank at most $\left[\frac{n+1}{2}\right]$. In fact, the system of independent equations written above has the determinant 1 mod 2.

In the case $M = S^n$ the affine $\mathbf{Z}/2\mathbf{Z}$-submodule is a $\mathbf{Z}/2\mathbf{Z}$-submodule because the only hyperplane which was not incident to the origin was the (Euler-Poincaré)-hyperplane, but $1 + (-1)^d \equiv 0(mod\ 2)$, so that also this hyperplane pass through the origin, when tensorizing with $\mathbf{Z}/2\mathbf{Z}$.

There exist cubulations of the sphere S^2 having an odd number of faces (or vertices), because immersed circles with an odd number of double points are easy to construct. An example of a convex cubical polytope with the f-vector (11,18,9) is drawn in picture 2. This proves that the image is in fact $\mathbf{Z}/2\mathbf{Z}$.

3.4. The complete f-obstruction. The complete f-obstruction is the class of the f-vector in the finite Abelian group $\mathbf{Z}^{n+1}/\Lambda(n)$. In general $\Lambda(n)$ is not a product in $\mathbf{Z}^{n+1}$ so that the projections fb do not contain all the information. In other words we saw that $\mathbf{f}(b_k(x)) = \mathbf{f}(x) + a(k,n)$, where $a(k,n) = (a(k,n,p)_{p=0,\ldots,n})$.

Therefore, starting from the f-vector α we will obtain β after some np-bubble moves if and only if the system of linear equations

$$\sum_{k=1}^{n+1} x_k a(k,n) = \beta - \alpha = b,$$

has integer solutions $x_k \in \mathbf{Z}$. Here x_k is the algebraic number of times the move b_k has been used (an inverse move b_k^{-1} counts as -1). Such a linear system has integer solutions only if $gcd(a(k,n,p), k = 1, n+1)$ divides b_p, for each $p = 0, 1, ..., n$. These obstructions on $\alpha - \beta$ are exactly those described in the first part, namely $f_i(mod\ a_i(n))$. . From 2.2 our the linear system is under-determined, of rank less than $\left[\frac{n+1}{2}\right]$. The complete obstruction is the set of compatibility relations making this system having solutions.

In the case of our system modeling the transformations of the f-vector by bubble/np-bubble moves, for $n = 1, 2$, we do not find any new obstructions. Now the result of proposition 2.6 can be deduced immediately:

PROOF. For $n = 3$ there is one new obstruction. There are at most two independent f_i, let choose for instance f_0 and f_1. Then the np-bubble moves b_i have the matrix $a(*, *)$:

$$\begin{pmatrix} 0 & 2 & 4 & 8 \\ 0 & 6 & 12 & 20 \end{pmatrix}.$$

This gives, away from the f_0, f_1 (mod 2) the that $(f_0 + f_1)(mod\ 4)$ is an invariant.

In the case we have to do with the bubble moves, there are two more moves to add to the previous list, namely $b_{3,1}$ and $b_{4,1}$, and the matrix is:

$$\begin{pmatrix} 0 & 2 & 4 & 8 & 0 & 0 \\ 0 & 6 & 12 & 20 & 0 & 4 \end{pmatrix}.$$

Thus $(f_0 + f_1)(mod\ 4)$ is also an invariant for bubble moves. □

Let observe that the standard cubulation of S^3 (the boundary of the 4-cube) has trivial $(f_0 + f_1)(mod\ 4)$. On the other hand, $CB(S^3)$ is a group whose elements are of order two, hence $2(f_0 + f_1) = 0(mod\ 4)$ for any cubulation of the sphere. This implies that $f_0 = f_1(mod\ 2)$, which agrees with the results of [**2**] for the polytopes. Remark however that the obstructions $f_0(mod\ 2)$, $f_1(mod\ 2)$ and $(f_0 + f_1)(mod\ 4)$. are defined on $CB(M)$ for all 3-manifolds M, not only for the sphere.

References

[1] T.Banchoff, *Triple points and surgery of immersed surfaces* Proc.Amer.Math.Soc., **46**(1974), 407–413.

[2] E.K.Babson and C.S.Chan, *Counting faces of cubical spheres modulo two*, Preprint, 1997.

[3] D.W. Barnette, *The minimum number of vertices of a simple polytope*, Israel J.Math., **10**(1971), 121-125.

[4] D.W. Barnette, *A proof of the lower bound conjecture for convex polytopes* Pacific J.Math., **46**(1973), 349-354.

[5] L.J.Billera and C.W.Lee, *A proof of the sufficiency of McMullen's conditions for f-vectors of simplicial convex polytopes* J.Combinatorial Theory Ser.A , **31**(1981), 237-255.

[6] R.H.Bing, *Some aspects of the topology of 3-manifolds related to the Poincaré conjecture*, Lectures on Modern Mathematics, (T.L.Saaty, Editor), 93-128, John Wiley and Sons, New York, 1964.

[7] G. Blind and R. Blind, *Gaps in the number of faces of cubical polytopes*, *Discrete and Comp. Geometry*, **11** (1994), 351–356.

[8] J. Scott Carter, *A further generalization of Boy's surface*, *Houston J.Math.*, **12**(1986), 11–31.

[9] J. Scott Carter, *Surgery on codimension one immersions*, Trans.Amer.Math.Soc., **298**(1986), 83–101.

[10] P.J. Eccles, *Codimension 1 immersions and the Kervaire invariant one problem* Math.Proc.Cambridge Phil.Soc.,**90**(1981), 483–493.

[11] P.J. Eccles, *Multiple points of codimension 1 immersions*, in Topology Symposium, Siegen 1979, U.Koschorke and W.Neumann Eds., Lecture Notes in Math. no.788, 23–38, 1980.

[12] P.J. Eccles *Multiple points of codimension one immersions of oriented manifolds*, Math.Proc.Cambridge Phil.Soc., **87**(1980), 213–220.

[13] M. Freedman, *Quadruple points of 3-manifolds in S^4*, Comment.Math.Helvetici, **53**(1978), 385–394.

[14] L.Funar, *Cubulations of compact manifolds and a problem of Habegger*, preprint IF, Grenoble, 1997. (available at `http://www-fourier.ujf-grenoble.fr/~funar`)

[15] R. Grünbaum, *Convex polytopes.* Pure and Appl. Math. , Interscience Publ. , John Wiley and Sons, London-New York - Sydney, vol. 16, 1967.

[16] G. Hetyei, *On the Stanley ring of a cubical complex*, Discrete and Comp. Geometry, **14**(1995), 305–330.

[17] W. Jockusch *The lower and upper bound problems for cubical polytopes*, Discrete and Comp. Geometry,**9**(1993), 159–163.

[18] R. Kirby, *Problems in low-dimensional topology*, in "Geometric Topology", Georgia International Topology Conference, (W.H.Kazez, Editor), AMS-IP Studies in Advanced Math., **2**, part 2, 35-473, 1995.

[19] V. Klee, *A combinatorial analogue of Poincare's duality theorem*, Canadian J.Math., **16**(1964), 517-531.

[20] U. Koschorke and B.Sanderson, *Selfintersections and higher Hopf invariants*, Topology, **17**(1978), 283–290.

[21] J. Lannes, *Sur les immersions de Boy*, in Algebraic Topology, Aarhus, 1982, Lecture Notes in Math., no.1051, 263–270, 1984.

[22] R. Lickorish, *Piecewise linear manifolds and Cerf theory*, in "Geometric Topology", Georgia International Topology Conference, (W.H.Kazez, Editor), AMS-IP Studies in Advanced Math., **2**, part 1, 375-387, 1995.

[23] P. McMullen, *On the number of faces of a simplicial polytope* Israel J.Math.,**9**(1971), 559-570.

[24] P. McMullen, *The maximum number of faces of a convex polytope*, Mathematika, **17**(1970), 179-184.

[25] P. McMullen, *On simple polytopes*, Inventiones Math., **113**(1993), 419-444.

[26] U. Pachner, *Shellings of simplicial balls and p.l. manifolds with boundary*, Discrete Math., **81**(1990), 37-47.

[27] U. Pachner, *Homeomorphic manifolds are equivalent by elementary shellings*, European J.Combinatorics, **12**(1991), 129-145.

[28] L. Siebenmann, *Les bisections expliquent le théorème de Reidemester-Singer*, Preprint Orsay, 81T16, 1981.

[29] R. Stanley, *The number of faces of a simplicial convex polytope*, Adv. Math., **35**(1980), 236-238.

[30] R. Stanley, *The number of faces of a simplicial polytopes and spheres*, in Discrete Geometry and Convexity, J.E.Goodman et. all. Eds., Annals of the N.Y.Academy of Sci., 440(1985), 212-223.

[31] V. Turaev, *Quantum invariants of links and 3-valent graphs in 3-manifolds*, de Gruyter Studies in Mathematics, vol.18, 1994.

[32] P. Vogel, *Cobordismes d'immersions* Ann.Ecole Normale Sup. **7**(1974), 317-358.

INSTITUT FOURIER, BP 74, UNIV.GRENOBLE I, MATHÉMATIQUES, 38402 SAINT-MARTIN-D'HÈRES CEDEX, FRANCE

E-mail address: `funar@fourier.ujf-grenoble.fr`

Contemporary Mathematics
Volume **233**, 1999

NUMERICAL PRESENTATIONS OF TORTILE CATEGORIES

Ivelina Bobtcheva
Frank Quinn

The objective is to compute with objects and morphisms in a tortile category as explicitly as one would work with elements of a group. Such explicit access is needed for numerical calculations in topological quantum field theories. This paper provides algorithms that describe the whole structure of a category from data files of special cases, somewhat analogous to the use of a compressed multiplication table to describe a group.

0. Introduction

A tortile category is an additive category with, among other things, a product which is associative and commutative up to natural isomorphism. The categories considered here are finite semisimple, which is to say every object is a finite sum of copies of finitely many simple objects. The semisimple structure gives a way to construct bases for modules of morphisms. A numerical presentation of the category consists of matrices describing the various structures in terms of these bases. These include the "fusion rule algebra" and "$6j$ symbols" of the physics literature.

These categories appear as intermediate structures in conformal and quantum field theory, and low dimensional topology. The simplest cases, representations of the rank 1 algebra $sl(2)$ and its deformations, have been well explored, see eg. [CFS, Kassel, Turaev]. The rank 2 case is also somewhat accessible [Kuperberg]. In higher-rank cases some special calculations can be done using general principles, but generic examples require numerical access to the category structure. Data for this requires direct calculation in a model for the category. The first author has developed programs for direct calculation in categories derived from representations of mod p or root-of-unity deformations ("quantum groups") of Lie algebras of arbitrary rank, [Bobtcheva]. The first objective of the present paper is to give a setting, or "user interface," for the output of such calculations. Full data sets for this are usually enormous, and our second objective is to describe algorithms used to derive all structure information from a significantly smaller set of "primitive" data.

Computer implementations of these algorithms, and data sets for categories derived from Lie algebras, are posted on the Quantum Topology Project web page [QTP].

1991 *Mathematics Subject Classification.* Primary 1804, Secondary 18D10, 17B37, 57N10.
Partially supported by the National Science Foundation

Contents

The first part of the paper gives abstract definitions of the structure matrices, and specifies names used in implementations. It also describes construction of bases for general morphism modules from fixed choices for morphisms from one simple object into the product of two other simples. A graphic notation is used to visualize the bases and morphisms. This information is sufficient for practical use of the presentations.

The second part begins the reductions, with conventions concerning choices of bases for morphisms from one simple into the product of two others. These conventions give closed-form descriptions of some simple structure functions, and simplify relations among others.

The third section gives a further reduction, by describing how the hexagon relation and the simplest case of the pentagon can be used to deduce some associativity functions from others. This gives a roughly twelve-fold reduction in the size of the data sets, and the algorithms can be used at run time.

The fourth section shows how the full pentagon axiom can be used to obtain further reductions in the data. The reductions obtained are substantial, but the algorithms are recursive and in current form are not suitable for use at run time. In practice these are used to "decompress" data to obtain an input file for the earlier algorithms.

We compare the amount of data required in the various reductions. The focus is on the matrices describing associativity isomorphisms, since they dominate the data. In general there is such a matrix associated to each sequence of four simple objects. A "type 0" category file simply stores all of these, so if there are n simple objects a type 0 category file would have n^4 associativity matrices. The basis choices of section 2 essentially give closed form expressions for associativities that involve the unit simple element. A "type 1" file stores all others, so requires $(n-1)^4$ matrices. For small n this can be helpful, but this is still a 4th degree polynomial with leading coefficient 1, so for large n there is little improvement. A "type 2" file makes use of the permutation reductions of section 3. The number of matrices required is still 4th degree, but the leading coefficient is 1/12. Thus for large n there is a roughly twelve-fold reduction in data. Finally a "type 3" file uses the pentagon reduction described in section 4. This requires only some of the associativities in which one of the elements is in a "generating set." The number of these needed depends on the structure of products in the category, but it is a third degree polynomial with leading coefficient often substantially less than 1/5. Thus for large n there is a savings of at least $n/7$ over type 2 files. Some actual numbers (with file type across the top, and number of simple elements on the left) are:

	0	1	2	3
5	625	256	36	14
10	10^4	$6.5 \cdot 10^3$	621	129
15	$5 \cdot 10^5$	$3.8 \cdot 10^4$	$3.4 \cdot 10^3$	469
20	$1.6 \cdot 10^5$	$1.3 \cdot 10^5$	$1.1 \cdot 10^4$	$1.2 \cdot 10^3$

Here for type 3 we have used figures for the "worst case" coming from representations of $SL(2)$; most others are significantly better. We see for instance that for 20 simple elements the use of a type 3 file gives a compression by a factor of over 100.

There are only very weak theoretical limit to improvements on this. As a system of equations the axioms for a tortile category are hugely overdetermined, so solutions are probably specified by very little data. In some cases (eg. center-1 categories from quantum deformations of Lie algebras of type A over the complex numbers, [Kazhdan-Wenzl]) these categories are determined simply by the semiring of isomorphism classes of objects (a.k.a. fusion rule algebra, dimension function, based Grothendieck ring). This is not true in general [FK, Quinn]. In any case algorithmic reconstruction is quite a different matter. The most abstractly effective algorithms are too recursive for use at run time, so in practice they are used to expand to eg. a type 2 file before use.

Some of the proofs are collected in the final section.

1. The presentations

Denote the coefficient ring by k. In many treatments this is taken to be the complex numbers, or at least a field. It will be important not to make this assumption. For instance categories obtained from representations of deformations of Lie algebras ("quantum groups") are usually taken over the complex numbers, but at roots of unity Lusztig [Lusztig] has shown that the categories can be defined with coefficients the cyclotomic field. Much of the construction can be done over cyclotomic integers, and it seems likely there will eventually be categories in this context. Such reductions simplify computation, and may permit important abstract constructions, e.g. reduction modulo primes.

The product in the category is denoted by $\diamond$, or simply by juxtaposition: $A \diamond B = AB$. The unit object for the product will be denoted by "1". Notations for the unit, commutativity, associativity, twist morphisms, forms and coforms are:

$$\begin{aligned}
\rho_A &: A \longrightarrow 1A \\
\sigma_{A,B} &: AB \longrightarrow BA \\
\alpha_{A,B,C} &: (AB)C \longrightarrow A(BC) \\
\theta_A &: A \longrightarrow A \\
\lambda_A &: A\,\overline{A} \longrightarrow 1 \\
\Lambda_A &: 1 \longrightarrow \overline{A}\,A
\end{aligned}$$

Here the "$\overline{A}$" in the form and coform notations stands for the duality involution on objects in the category. The axioms are standard, cf. [MacLane, Shum, Yetter] except that particular choices of forms and coforms have been subsituted for a duality functor. Properties of these choices equivalent to naturality, etc. of the functor are reviewed in Details 1a. We also normalize the forms to satisfy a symmetry condition.

1.1 Semi-simplicity. The category under consideration is an additive category over k. This means the morphism sets $\hom(A, B)$ have k-module structures that are natural with respect to composition. Further, there is a sum operation on objects that induces direct sums of k-modules on the morphism level.

We assume that the category is finite semisimple in the following sense:

(1) there is a finite set of "simple" objects so that every other object is a finite sum of copies of these;

(2) if a and b are simple then they are either isomorphic or 0 is the only morphism between them; and
(3) the endomorphism ring of each simple object is naturally isomorphic to k as a k-algebra.

The last condition is the strong one. The categories of Lie type (derived from mod p or p^{th}-root deformations of Lie algebras) satisfy this condition. However representations of finite groups usually have simples with more complicated endomorphism algebras.

The finite semisimple condition implies that the category has a very simple additive structure: it is a sum of copies of the category of finitely generated free k-modules, one copy for each simple object. If the simples are denoted by $\{a_i\}$, $1 \leq i \leq n$ then a specific equivalence of categories is given by

$$A \mapsto (\hom(a_1, A), \ldots, \hom(a_n, A)).$$

A consequence of this is that all morphism sets are free k-modules.

1.2 Structure functions. We fix representatives of the simple objects. Specifically S is a set of simple objects which contains exactly one in each isomorphism class. The structure in the category is described in terms of effect on morphisms from simples. So for example the commutativity isomorphism $\sigma_{a,b}\colon ab \longrightarrow ba$ is described by the induced homomorphisms $\hom(x, ab) \longrightarrow \hom(x, ba)$ for $x \in S$.

There is an ordering chosen on S, with the unit 1 as minimal element. In practice the simples are referred to by number, so $a \in S$ means a is an integer $1 \leq a \leq \#S$.

Bases (generators as free k-modules) are given for the module $\hom(a, bc)$, for all $a, b, c \in S$. These induce bases for morphisms into iterated products; this is discussed in 1.3. The structure functions give matrices describing induced homomorphisms in terms of these bases. The functions are (for standard simples a, b, $\ldots$):

$$\begin{aligned}
\texttt{spl} &= \#S, \text{ the number of simple objects} \\
\texttt{bar}[a]&\colon S \longrightarrow S, \text{ the duality involution} \\
\texttt{tw}[a] &= \theta_a \in \hom(a,a) = k \\
\texttt{rank}[a] &= \lambda_a \Lambda_{\overline{a}} \in \hom(1,1) = k \\
\texttt{sgn}[a] &\in k \text{ such that } \lambda_a = \texttt{sgn}[a]\texttt{tw}[a]\lambda_{\overline{a}}\sigma_{a,\overline{a}} \\
\texttt{dim}[a,b,c] &= \dim_k(\hom(a,bc)) \\
\texttt{com}[a,b,c]&\colon \hom(a,bc) \longrightarrow \hom(a,cb), \text{ induced by } \sigma_{b,c} \\
\texttt{dual}[a,b,c]&\colon \hom(a,bc) \otimes \hom(\overline{a}, \overline{c}\,\overline{b}) \longrightarrow k, \text{ the duality pairing} \\
& \quad f \otimes g \mapsto \lambda_{bc}(f \diamond g)\Lambda_{\overline{a}} \\
\texttt{asoc}[a,b,c,d]&\colon \hom(a,(bc)d) \longrightarrow \hom(a, b(cd)) \text{ induced by } \alpha_{b,c,d}
\end{aligned}$$

The function "sgn" is 1 unless a is self-dual, and satisfies $\texttt{sgn}[a]^2 = 1$, see Details 1a. In these terms we can be more explicit about "numerical presentations." A numerical presentation of a category is a data file, together with algorithms for reconstructing the structure functions from the information in the data file.

Bases for morphisms into the triple products used in "asoc" are described below in 1.3. In our computer implementations elements in hom sets are thought of as column vectors and the functions act by matrix multiplication with the argument on the *right*. So if $\texttt{f}\colon \hom_1 \to \hom_2$ is a function and $x \in \hom_1$, then $\texttt{f}(x) = \texttt{f} \cdot x$. The duality pairing is represented by a matrix "`dual`" so that the value on $x \otimes y$ is $\texttt{Transpose}[x] \cdot \texttt{dual} \cdot y$.

1.3 Bases for morphisms. The basic unit for description of morphisms into products are the modules $\hom(a, bc)$, with a, b, c simple objects. Reduction to this case uses the following:

Lemma. *Suppose A, B, C are objects, then the natural map*

$$\oplus_{x \in S} \hom(A, x\, C) \otimes_k \hom(x, B) \longrightarrow \hom(A, BC)$$

given by $f \otimes g \mapsto (g \diamond \mathrm{id}) \circ f$, and where the sum is taken over simple objects x, is an isomorphism.

Similarly the other factor can be decomposed:

$$\hom(A, BC) \simeq \oplus_y \hom(A, By) \otimes_k \hom(y, C).$$

Proof. This expression is additive in B. Since B is isomorphic to a sum of copies of simple objects, the additivity reduces the lemma to the case where B is simple. In that case $\hom(x, B) = 0$ unless $x \simeq B$, and then $\hom(x, B) = k$. The factors in the sum on the left side of the expression are thus all zero except for $\hom(A, xC) \otimes_k k$ when $x \simeq B$, and this does go by isomorphism to $\hom(A, BC)$.

We note the strong simplicity hypothesis (endomorphism algebras isomorphic to k) is used here. In general $\hom(A, xC)$ and $\hom(x, B)$ are modules over the endomorphisms of x, and the tensor products in the lemma must be taken over this algebra.

1.4 Bases in iterated products. If bases have been chosen for all modules $\hom(a, bc)$ with a, b, c simple then this lemma can be iterated to give generating sets for morphisms into iterated products. These are not immediately bases: bases are *ordered* generating sets and there is some ambiguity in the ordering. In practice choosing and changing orderings are major activities.

For instance if a, b, etc. are all simple, then

$$\hom(a, (bc)d) = \oplus_x \hom(a, xd) \otimes \hom(x, bc).$$

The sum is over $x \in S$, so terms in the sum are ordered by the ordering on S. Fix x and denote the standard bases by $f_i \in \hom(a, xd)$ and $g_j \in \hom(x, bc)$. Then products $f_i \otimes g_j$ generate $\hom(a, xd) \otimes \hom(x, bc)$. We can order them by lexicographic order on (i, j), or on (j, i). If we think of $f_i \otimes g_j$ as doubly indexed (a "matrix" of objects) then one total order is obtained by flattening the matrix, the other by flattening its transpose.

The orderings used in the standard associativity matrices have the $\hom(x, -)$ terms second. In the notation of the previous paragraph the order on $f_i \otimes g_j$ is lexicographic in (i, j). In a more elaborate product, for instance

$$\hom(a, (bc)(de)) = \oplus_{x,y} \hom(a, xy) \otimes \hom(x, bc) \otimes \hom(y, de)$$

the choices are more general: priority must be established among the indices (x, y) of the sum, as well as among the three indices of each individual tensor product.

1.5 Graphic notation. Labeled graph notations for specifying morphisms of products are helpful and widely used (cf. [Shum, Yetter]. Here we describe the version used in this paper. A sample manipulation is given in Details 1.5a: the `dual` matrices are derived from associativities.

First a direction on the page is specified, to determine direction of morphisms and order of compositions. Then a graph is given, with edges labeled by objects (usually simple), and the vertices labeled by morphisms. For example Figure 1.5.1 is typical of the trees used to specify bases into iterated products, in this case of a into $b \diamond c \diamond d \diamond e$.

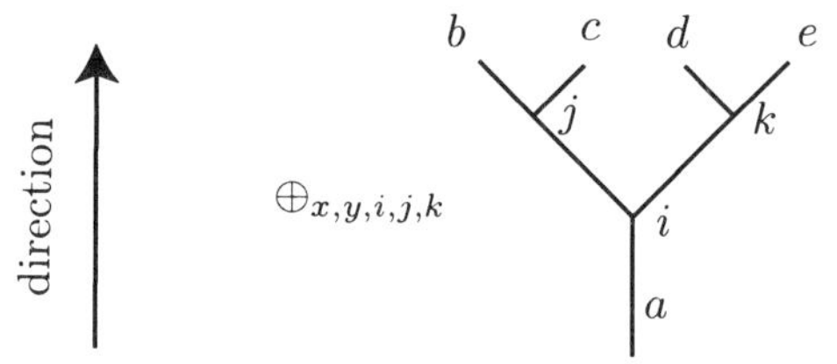

Figure 1.5.1

We describe how to interpret this. This example was oriented with morphism direction from bottom to top so that the important information (the range b, c, d, e) will be displayed at the top. We caution that other authors sometimes fix the composition direction to be top to bottom, and in these terms the figure is upside-down. Next, objects are specified by slices perpendicular to the specified direction and missing the vertices: take the product of the objects labeling the edges, reading from left to right. Finally, morphisms are specified by slices passing through vertices: take products (again left to right) of morphisms labeling the vertices, or identity morphisms when the slice passes through edges away from a vertex. In Figure 1.5.1 the first slices (starting at the bottom) hit only the edge a. Moving up past the first vertex, slices passes through two edges indicating the product $x \diamond y$ (reading left to right). The vertex represents a morphism $a \to x \diamond y$, specifically the ith element in the basis chosen for $\hom(a, xy)$. The next vertex indicates a morphism on the x factor into $b \diamond c$, multiplied by the identity on y. The third vertex indicates the product of identities with a morphism $y \to d \diamond e$. At the top a slice reads off $b \diamond c \diamond d \diamond e$. More precisely, the order of appearence of the factors also specifies an association, namely $(b \diamond c) \diamond (d \diamond e)$. All together we get a morphism $a \to (b \diamond c) \diamond (d \diamond e)$. The corresponding algebraic expression is $(\mathrm{id}_{b \diamond c} \diamond k)(j \diamond \mathrm{id}_y)(i)$. Finally the notation $\oplus_{(x,y,i,j,k)}$ indicates that we take all such compositions, as x, y vary over the simple objects and (i, j, i) vary over bases of the indicated hom sets. According to Section 1.4 this gives a basis (including an ordering) for $\hom(a, (bc)(de))$.

There are three further conventions used in this notation. First, associations and associativity morphisms are not indicated. Associations of input and output are either specified in algebraic notation in the text, or understood to be whatever takes the least rearrangement to fit the morphisms. Inbetween input and output we simply put in whatever associativities are needed, and note that the result is well-defined by coherence properties of monoidal categories [MacLane]. This simplifies the notation, but it must be used with care: associativities are by far the largest and most complicated part of a numerical presentation, and it is important not to miss them in invisible notations. See the derivation of the duality morphisms from associativities in Details 1.5a for an example. The second convention is that commutativity isomorphisms are denoted by an overcrossing, rather than by a labeled vertex. Finally we tend to omit edges labeled by the identity object. These are understood to be absorbed in some other edge via the unit isomorphism, and this does not need to be explicitly displayed since the result is well-defined. Figure 1.5.2 illustrates these conventions.

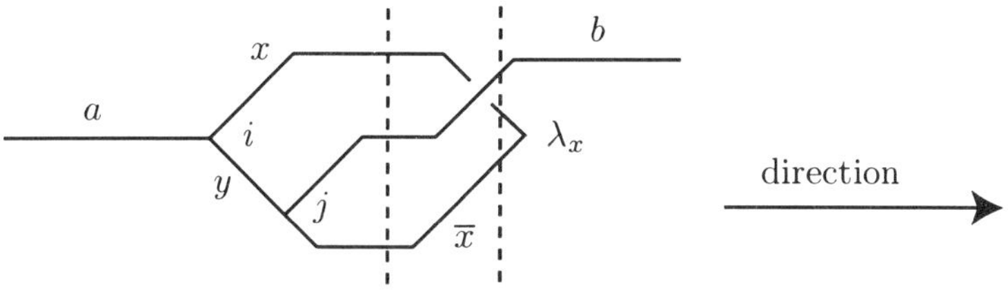

Figure 1.5.2

The morphism direction in this figure is left to right, since it fits better on the page. Products are read off going top to bottom along vertical slices. The first two vertices indicate basis morphisms $a \to x \diamond y \to x \diamond (b \diamond \overline{x})$, as in the previous figure. The third vertex is an overcrossing, indicating the commutativity $\sigma_{x,b}\colon x \diamond b \to b \diamond x$. But an invisible associativity $x \diamond (b \diamond \overline{x}) \simeq (x \diamond b) \diamond \overline{x}$ is needed between the second vertex and the overcrossing (indicated here by a dotted line) in order for the composition to be defined. The final vertex indicates composition with the form $\lambda\colon x \diamond \overline{x} \to 1$. Another invisible association is required for this composition to make sense. Finally we have omitted an edge labeled "1" coming out of the λ_x vertex. This edge would go from the λ vertex to a vertex on the b edge, and the new vertex should be labeled with the unit morphism $\rho_b^{-1}\colon b \diamond 1 \to b$.

Relations among functions correspond to modification of diagrams. For example in Figure 1.5.3 composition with an `asoc` function corresponds to moving a branch in the graph. In detail, the left-hand diagram describes the composition $\oplus_x \hom(a, xd) \otimes \hom(x, bc) \to \hom(a, (bc)d)$. On the right we have $\oplus_x \hom(a, bx) \otimes \hom(x, cd) \to \hom(a, b(cd))$. The left function is obtained from the right by composition with $\mathtt{asoc}[a, b, c, d]$, and an obvious "invisible reassociation." See the derivation of `dual` in Details 1.5a for a sample calculation.

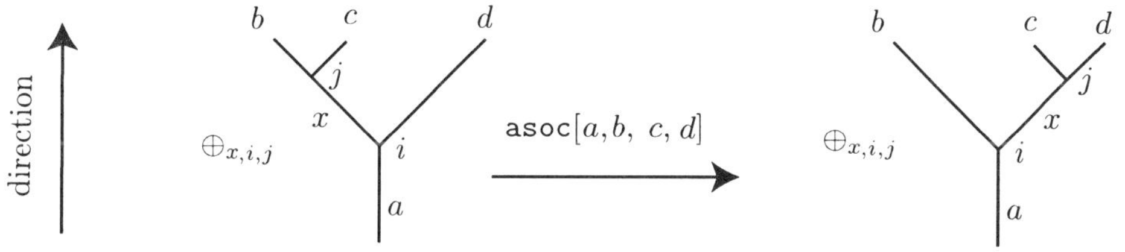

Figure 1.5.3

1.6 Representations of the braid groups. Representations of the framed braid group are one of the standard consequences of a tortile category. Here we show how to use the structure functions in a numerical presentation to give the matrices generating these representations.

To describe the braid group on n strands start with a 2-sphere with n holes, or in other words a planar surface with n boundary components. The braid group is the group of isotopy classes of orientation-preserving diffeomorphisms of this surface. To get framed braids choose parameterizations of the boundary components by S^1, and require that the diffeomorphisms preserve these parameterizations.

A description that looks more like a braid is obtained by considering n monotone arcs going from one end of $S^2 \times I$ to the other. "Monotone" means any slice $S^2 \times \{t\}$ meets each arc in a single point. The monotonicity condition implies that there is a monotone flow that has the given arcs as flow arcs. The flow gives a diffeomorphism between the endsof $S^2 \times I$, and this gives the connection to diffeomorphisms of S^2. Monotone arcs give an attractive and convenient picture, but it is good to remember that the real activity concerns the complement of these arcs.

The group is generated by diffeomorphisms that interchange two adjacent holes, by a twist along a circle enclosing them, or equivalently by crossing of two strands of the braid. To describe a representation of the group we need a vector space, and matrices giving the action of these generators. The vector space we use is the space of morphisms of the unit into n-fold products. We use the "left" association of the product, so for example the 5-fold product would be $(((ab)c)d)e$. To interchange two holes first associate to get the corresponding objects paired together, apply the commutativity isomorphism, and then associate back. For instance interchanging the third and fourth of a 5-fold product is given by

$$(((ab)c)d)e \xrightarrow{\alpha} ((ab)(cd))e \xrightarrow{\sigma} ((ab)(dc))e \xrightarrow{\alpha^{-1}} (((ab)d)c)e.$$

A basis for $\hom(1, (((ab)c)d)e)$ is described in the graphic notation by labeled trees, see Figure 1.6.1. The full space on which the group acts is obtained by also summing over all (a, b, c, d, e).

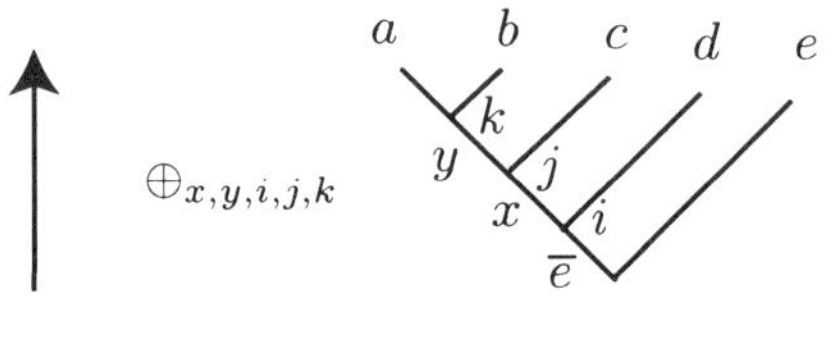

Figure 1.6.1

We describe the connection between these trees and braid diagrams. The trees lie in the plane and thereby in 3-space. A thickening is a 3-disk; a thickening relative to the ends of the graph gives a 3-disk with tree ends on the boundary, see Figure 1.6.2. The thickening is reasonably canonically isomorphic to the standard disk, so we can think of the trees as all being in D^3, with endpoints in S^2. Manipulations of trees correspond to mutations inside D^3, and movements of the endpoints on S^2. The moving endpoints trace out the braid.

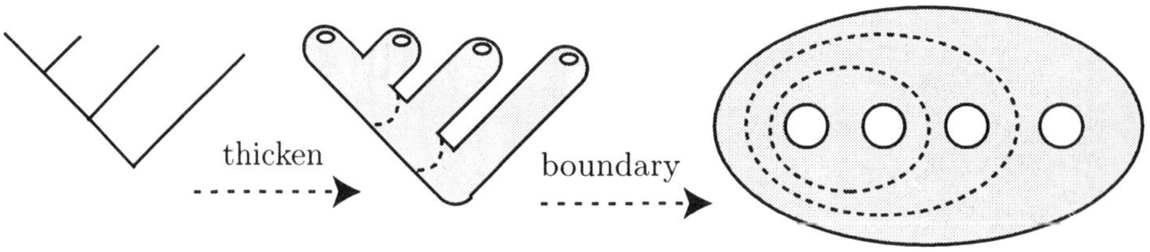

Figure 1.6.2

The figure also illustrates how the structure of the graph as a triadic tree determines a decomposition of the thickening, and thereby a decomposition of the boundary punctured sphere into planar surfaces with three boundary components ("pairs of pants"). Therefore labels on the tree can be thought of as labels on this decomposition, giving a description of the basis in terms of labeled pants. Such descriptions occur frequently in the literature, but trees are easier to work with and more fundamental.

We describe how to obtain an explicit matrix corresponding to crossing the $n+1$ strand over the n strand. Let the labels in Figure 1.6.3 establish notation for the basis.

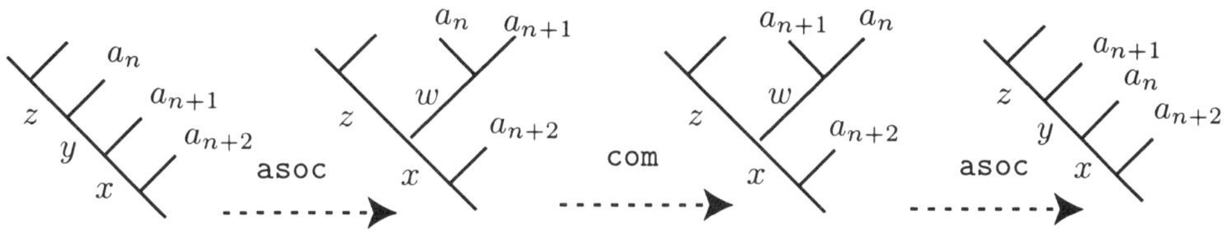

Figure 1.6.3

Reorder the basis to the form $(\dots, y, i, j, \dots)$. In this ordering the operation is a sum of matrices of the form $\mathrm{id} \otimes M[x, z, a_n, a_{n+1}] \otimes \mathrm{id}$, where M acts on the (y, i, j) factors. Figure 1.6.3 shows that $M[x, z, a_n, a_{n+1}]$ is given by

$$\texttt{asoc}[x, z, a_{n+1}, a_n]^{-1} \cdot (\oplus_w \mathrm{id}_{\texttt{dim}[x,z,w]} \otimes \texttt{com}[w, a_n, a_{n+1}]) \cdot \texttt{asoc}[x, z, a_n, a_{n+1}].$$

All the representations (any number of strands) can be reconstructed from these matrices and the dimension function. We remark that the hexagon identity can be used to give another expression for M, as a composition of two `com` and one `asoc` function.

2. Choice of bases

Suppose (a, b, c) are simple objects, and (x, y, z) is some permutation of this set. Then there are isomorphisms $\hom(\overline{a}, bc) \simeq \hom(\overline{x}, yz)$. In this section we choose a basis for $\hom(\overline{a}, bc)$ with $a \le b \le c$, and then get bases for permutations by using canonical choices of isomorphisms. These choices simplify `com` and a few of the associativity morphisms. If the set of duals $\{\overline{a}, \overline{b}, \overline{c}\}$ is different from $\{a, b, c\}$ then duality gives a basis for one from the other. This simplifies `dual`.

The canonical isomorphisms to other permutations are constructed from isomorphisms

$$\texttt{cycle}[a, b, c] \colon \hom(\overline{a}, bc) \longrightarrow \hom(\overline{b}, ca)$$
$$\texttt{flip}[a, b, c] \colon \hom(\overline{a}, bc) \longrightarrow \hom(\overline{a}, cb)$$

defined below. The algorithm is: if (x, y, z) is an even permutation of the standard ordering (a, b, c) then apply `cycle` an appropriate number of times to get an isomorphism $\hom(\overline{a}, bc) \longrightarrow \hom(\overline{x}, yz)$. If (x, y, z) is not obtained from an even permutation, apply `flip` to go from (a, b, c) to (a, c, b), and then cycle to get (x, y, z).

In this section we define `cycle` and `flip`, then work out the consequences of these basis choices for the standard functions.

2.1 Morphisms from the unit. If a and b are simple, then $\hom(1, ab)$ is trivial unless $b = \overline{a}$, and in that case the simplicity hypothesis 1.1(3) implies that it is 1-dimensional with basis the coform Λ_b. The first basis conventions are:

Basis conventions.

(1) for each a take the coform $\Lambda_a\colon 1 \longrightarrow \overline{a} \diamond a$ as the basis for $\hom(1, \overline{a}a)$, and
(2) take the standard unit isomorphisms as bases for $\hom(a, a1)$ and $\hom(a, 1a)$.

Note that the coforms are required to satisfy the symmetry condition $\Lambda_{\overline{a}} = \mathtt{sgn}[a](\theta_a \diamond \mathrm{id}) \circ \sigma \circ \Lambda_a$ (see Details 1a). Since $\mathtt{com}$ is the function induced by σ, and θ_a is multiplication by $\mathtt{tw}[a]$, this gives

$$\mathtt{com}[1, a, \overline{a}] = \mathtt{sgn}[a]\mathtt{tw}[a]^{-1}.$$

It follows from the second part of the convention that $\mathtt{com}[a, b, c]$ and $\mathtt{asoc}[a, b, c, d]$ are identities if b, c or d is the unit.

This convention is also used to put hom sets in a more symmetric form, using the basis conventions for iterated products given in 1.4. According to 1.4 composition gives an isomorphism

$$\oplus_x \hom(1, ax) \otimes \hom(x, bc) \xrightarrow{\simeq} \hom(1, a(bc)).$$

But the first factor on the left is nonzero only when $x = \overline{a}$, and then it is 1-dimensional with a standard basis, so this degenerates to an isomorphism

$$\hom(\overline{a}, bc) \simeq \hom(1, a(bc))$$

which by convention preserves basis (ie. is represented by the identity matrix). Explicitly this isomorphism is given by $f \mapsto (f \diamond \mathrm{id})\Lambda_{\overline{a}}$. There is similarly a basis-preserving isomorphism $\hom(\overline{a}, bc) \longrightarrow \hom(1, (bc)a)$.

2.2 The $\mathtt{cycle}$ and $\mathtt{flip}$ functions. Define

$$\mathtt{cycle}[a, b, c]\colon \hom(\overline{a}, bc) \longrightarrow \hom(\overline{b}, ca)$$

to be $\mathtt{sgn}[b]$ times the composition

$$\hom(\overline{a}, bc) \xrightarrow{\simeq} \hom(1, (bc)a) \xrightarrow{\mathtt{asoc}} \hom(1, b(ca)) \xrightarrow{\simeq} \hom(\overline{b}, ca).$$

As a matrix it is given by $\mathtt{sgn}[b]\mathtt{asoc}[1, b, c, a]$.

Define $\mathtt{flip}[a, b, c]\colon \hom(\overline{a}, bc) \longrightarrow \hom(\overline{a}, cb)$ by

$$\mathtt{flip}[a, b, c] = \mathtt{sgn}[a]\mathtt{tw}[a]^{-1}\mathtt{com}[\overline{a}, b, c]$$

Lemma.

(1) $\mathtt{cycle}^3 = \mathrm{id}$, *or more precisely,*

$$\mathtt{cycle}[c, a, b] \cdot \mathtt{cycle}[b, c, a] \cdot \mathtt{cycle}[a, b, c] = \mathrm{id};$$

(2) $\mathtt{flip} \cdot \mathtt{cycle}^2 = \mathtt{cycle} \cdot \mathtt{flip}$, *or more precisely*

$$\mathtt{flip}[c, a, b] \cdot \mathtt{cycle}[b, c, a] \cdot \mathtt{cycle}[a, b, c] = \mathtt{cycle}[a, c, b] \cdot \mathtt{flip}[a, b, c];$$

and

(3)
$$\mathtt{flip}[a, c, b] \cdot \mathtt{flip}[a, b, c] = (\mathtt{tw}[a]\mathtt{tw}[b]\mathtt{tw}[c])^{-1}.$$

The second identity comes from the hexagon, and the third follows immediately from the balance relation. The proof is given in Details 2.2a.

2.3 Bases and simple functions. As indicated above, we suppose that bases are given for $\hom(\overline{a}, bc)$ when $a \leq b \leq c$. Bases for other permutations of the set $\{a, b, c\}$ are obtained by:

Basis conventions, continued.

(1) If $a \leq b \leq c$ are not all equal, then the cycle morphisms

$$\hom(\overline{a}, bc) \xrightarrow{\mathtt{cycle}[a,b,c]} \hom(\overline{b}, c, a) \xrightarrow{\mathtt{cycle}[b,c,a]} \hom(\overline{c}, ab)$$

preserve bases, and

(2) if $a \leq b \leq c$ are distinct and different from 1 then

$$\hom(\overline{a}, bc) \xrightarrow{\mathtt{flip}[a,b,c]} \hom(\overline{a}, cb) \xrightarrow{\mathtt{cycle}[a,c,b]} \hom(\overline{c}, ba) \xrightarrow{\mathtt{cycle}[c,b,a]} \hom(\overline{b}, ac)$$

preserve bases.

We caution that the second convention may not be satisfied when $a = 1$ since the choices in 2.1 give $\mathtt{flip}[1, b, \overline{b}] = \mathtt{sgn}[b]\mathtt{tw}[b]^{-1}$.

These choices give simple expressions for the functions used to define `cycle` and `flip`:

Proposition.

(1) *If a, b, c are not all equal, then* $\mathtt{asoc}[1, a, b, c] = \mathtt{sgn}[a]\mathrm{id}$*;*

(2) *if $\{x, y, z\} = \{a, b, b\}$ as sets, then*

$$\mathtt{com}[\overline{x}, y, z] = \mathtt{sgn}[x]\mathtt{sgn}[a]\mathtt{tw}[x]\mathtt{tw}[a]^{-1}\mathtt{com}[\overline{a}, b, b];$$

and

(3) *if a, b, c are distinct and different from 1 then*

$$\mathtt{com}[\overline{a}, b, c] = \begin{cases} \mathtt{sgn}[a]\mathtt{tw}[a] & \textit{for even permutations} \\ \mathtt{sgn}[a]\mathtt{tw}[b]^{-1}\mathtt{tw}[c]^{-1} & \textit{for odd permutations} \end{cases}$$

In (3) "even permutations" means (a, b, c) is obtained by an even permutation of the standard ordering of the triple. The primitive cases that still have to be computed are $\mathtt{asoc}[1, a, a, a]$ and $\mathtt{com}[a, b, b]$.

Proof. If a, b, c are not all equal, then by choice $\mathtt{cycle} = \mathrm{id}$ for two of the cyclic permutations of $\{a, b, c\}$. The identity $\mathtt{cycle}^3 = \mathrm{id}$ of 2.2 shows the remaining function is also the identity. This proves the first case of the proposition.

Since $\mathtt{cycle} = \mathrm{id}$, the second identity of lemma 2.2 reduces to $\mathtt{flip}[c, a, b] = \mathtt{flip}[a, b, c]$. Subsituting the definition of `flip` gives

$$\mathtt{sgn}[c]\mathtt{tw}[c]^{-1}\mathtt{com}[\overline{c}, a, b] = \mathtt{sgn}[a]\mathtt{tw}[a]^{-1}\mathtt{com}[\overline{a}, b, c].$$

Setting $b = c$ or $a = b$ gives (2) of the proposition.

If a, b, c are distinct, ordered, and different from 1, then by choice $\mathtt{flip}[a, b, c]$ is the identity. The formula in the previous paragraph gives $\mathtt{com}[\overline{x}, y, z] = \mathtt{sgn}[x]\mathtt{tw}[x]$ for x, y, z any even permutation of a, b, c. The balance relation gives the odd permutations from this.

2.4 Duality. "Duality" here refers to the bilinear pairing

$$\hom(a, bc) \otimes \hom(\overline{a}, \overline{c}\,\overline{b}) \longrightarrow k$$

defined by $f \otimes g \mapsto \lambda_b(\mathrm{id} \diamond \lambda_c \diamond \mathrm{id})(f \diamond g)\Lambda_{\overline{a}}$. More precisely, it refers to the adjoint of this:

$$\hom(a, bc) \longrightarrow (\hom(\overline{a}, \overline{c}\,\overline{b}))^*$$

where $(-)^*$ indicates the the hom-dual over the coefficient ring, $\hom_k(M, k)$. We add to the basis conventions of 2.1 and 2.3:

Basis conventions, concluded.

(1) If $1 < a \le b \le c$ and (a, b, c) is smaller than $(\overline{a}, \overline{b}, \overline{c})$ after sorting, then the signed duality

$$\hom(\overline{a}, bc) \xrightarrow{\mathtt{sgn}[a]\mathtt{dual}[\overline{a},bc]} (\hom(a, \overline{b}\,\overline{c}))^*$$

preserves bases.

The convention is not used in following proposition, but together they imply that in non-self-dual cases all dual functions are scalar multiples of identity matrices, see 2.7.

Proposition. *Suppose (a, b, c) is ordered, and (x, y, z) is a permutation of this set. Then $\mathtt{sgn}[x]\mathtt{dual}[\overline{x}, y, z]$ is*

(1) $\mathtt{sgn}[x]\mathtt{rank}[b]$ *if $a = 1$ and $\overline{b} = c$;*
(2) $\mathtt{sgn}[a]\mathtt{dual}[\overline{a}, b, c]$ *if $a > 1$ and either (x, y, z) is an even permutation of (a, b, c), or $(\overline{a}, \overline{b}, \overline{c})$ is an even permutation of the standard order, and*
(3) $\mathtt{sgn}[a](\mathtt{tw}[a]\mathtt{tw}[b]\mathtt{tw}[c])^{-1}\mathtt{dual}[\overline{a}, b, c]$ *otherwise.*

We can usually arrange the last case to be rare. The *adjacent dual* condition is that if a and $\overline{a}$ are different then they are adjacent in the ordering on the simple objects. This condition implies that if $a > b$ then $\overline{a} > \overline{b}$ or $a = \overline{b}$. It follows from this that $(\overline{a}, \overline{b}, \overline{c})$ is an odd permutation of the standard order if and only if a, b, c are all distinct and two are dual.

The proof uses material from the next several sections, and is given in Details 2.4a.

2.5 Duality and products. The first step is to describe bar, forms, and coforms for products. The dual of an iterated product is defined to be the product in reverse order, with bar applied to the factors. For example $\overline{A((BC)D)} = (\overline{D}(\overline{C}\,\overline{B}))\overline{A}$. The form λ for a product is defined by: re-associate to group the innermost dual pair, apply the form to eliminate this pair, and repeat. For instance the form for $(AB)C$ is defined by

$$((AB)C)(\overline{C}(\overline{B}\,\overline{A})) \xrightarrow{\alpha} ((AB)(C\overline{C}))(\overline{B}\,\overline{A}) \xrightarrow{\mathrm{id}\diamond\lambda_C\diamond\mathrm{id}} (AB)(\overline{B}\,\overline{A})$$
$$\xrightarrow{\alpha} (A(B\overline{B}))\overline{A} \xrightarrow{\mathrm{id}\diamond\lambda_B\diamond\mathrm{id}} A\overline{A} \xrightarrow{\lambda_A} 1.$$

It is not necessary to be explicit about the associations needed for this because the coherence principle for monoidal categories [MacLane] asserts that the final result is well-defined.

Coforms for products are defined dually.

The first consequence of this is a decomposition for duality pairings of iterated products. In stating this we extend the notation. $\mathtt{dual}[A, B]$ is used to denote the duality pairing for arbitrary objects A, B in the category. In these terms the earlier notation with three simple arguments is $\mathtt{dual}[a, b, c] = \mathtt{dual}[a, bc]$.

Lemma. *In terms of the decompositions of Lemma 1.3 as sums over simple objects*

$$\hom(A, BC) \simeq \oplus_x \hom(A, xC) \otimes \hom(x, B) \text{ and}$$

$$\hom(\overline{A}, \overline{C}\,\overline{B}) \simeq \oplus_y \hom(\overline{A}, \overline{C}\,y) \otimes \hom(y, \overline{B})$$

the pairing

$$\mathtt{dual}[A, BC]\colon \hom(A, BC) \otimes \hom(\overline{A}, \overline{C}\,\overline{B}) \to k$$

is 0 on summands with $y \neq \overline{x}$, and is $\mathtt{rank}[x]^{-1}\mathtt{dual}[A, xC] \otimes \mathtt{dual}[x, B]$ when $y = \overline{x}$.

In particular

$$\mathtt{dual}[a, (bc)d] = \oplus_x \mathtt{rank}[x]^{-1}\, \mathtt{dual}[a, x, d] \otimes \mathtt{dual}[x, b, c].$$

The C term can be decomposed similarly to give

$$\mathtt{dual}[a, b(cd)] = \oplus_y \mathtt{rank}[y]^{-1}\, \mathtt{dual}[a, b, y] \otimes \mathtt{dual}[y, c, d].$$

The unit source cases ($a = 1$) reduce to the single terms $x = \overline{d}$ or $y = \overline{b}$. Since $\mathtt{dual}[a, \overline{d}, d] = \mathtt{rank}[d]$, we get $\mathtt{dual}[1, (bc)d] = \mathtt{dual}[\overline{d}, b, c]$, and similarly $\mathtt{dual}[1, b(cd)] = \mathtt{dual}[\overline{b}, c, d]$.

The proof of the lemma is given in Details 2.5a.

2.6 Duality and other structure. The relations are: for duality,

$$\mathtt{dual}[a, b, c] = \mathtt{Transpose}[\mathtt{dual}[\overline{a}, \overline{b}, \overline{c}]]\mathtt{sgn}[a]\mathtt{sgn}[b]\mathtt{sgn}[c]. \tag{2.6.1}$$

For commutivity, the diagram

$$\begin{array}{ccc}
\hom(a, bc) \otimes \hom(\overline{a}, \overline{b}\,\overline{c}) & \xrightarrow{\mathrm{id}\otimes\mathtt{com}[\overline{a}, \overline{b}, \overline{c}]} & \hom(a, bc) \otimes \hom(\overline{a}, \overline{c}\,b) \\
\Big\downarrow{\scriptstyle \mathtt{com}[a,b,c]\otimes\mathrm{id}} & & \Big\downarrow{\scriptstyle \mathtt{dual}[a,b,c]} \\
\hom(a, cb) \otimes \hom(\overline{a}, \overline{b}\,\overline{c}) & \xrightarrow{\mathtt{dual}[a,c,b]} & k
\end{array}$$

commutes. Recalling that $\mathtt{dual}$ acts on the first variable by multiplication of the transpose on the left, we get

$$\mathtt{dual}[a, b, c] \cdot \mathtt{com}[\overline{a}, \overline{b}, \overline{c}] = \mathtt{Transpose}[\mathtt{com}[a, b, c]] \cdot \mathtt{dual}[a, c, b]. \tag{2.6.2}$$

For associativity we have the similar diagram

$$\begin{array}{ccc}
\hom(a, (bc)d) \otimes \hom(\overline{a}, (\overline{d}\,\overline{c})\,b) & \xrightarrow{\mathrm{id}\otimes\mathtt{asoc}[\overline{a}, \overline{d}, \overline{c}, \overline{b}]} & \hom(a, (bc)d) \otimes \hom(\overline{a}, \overline{d}(\overline{c}\,\overline{b})) \\
\Big\downarrow{\scriptstyle \mathtt{asoc}[a,b,c,d]\otimes\mathrm{id}} & & \Big\downarrow{\scriptstyle \mathtt{dual}[a,(bc)d]} \\
\hom(a, b(cd)) \otimes \hom(\overline{a}, (\overline{d}\,\overline{c})\,\overline{b}) & \xrightarrow{\mathtt{dual}[a,b(cd)]} & k
\end{array}$$

This gives

(2.6.3) $\texttt{dual}[a,(bc)d]\cdot\texttt{asoc}[\,\overline{a},\,\overline{d},\,\overline{c},\,\overline{b}] = \texttt{Transpose}[\texttt{asoc}[a,b,c,d]]\cdot\texttt{dual}[a,b(cd)]$.

The dual relation results from the symmetry of forms and coforms. The com formula uses naturality and product structure of σ. The associativity statement is an immediate consequence of the definition of the pairing. Proofs are straightforward, so are omitted.

The following gives a relation among the primitive com cases of Proposition 2.3.

2.7 Lemma. *Assuming the* com *and* dual *basis conventions, if a, b are greater than 1 and $\{a,b\} \neq \{\,\overline{a},\,\overline{b}\}$ then*

$$\texttt{com}[\,\overline{a},b,b] = \texttt{Transpose}[\texttt{com}[a,\,\overline{b},\,\overline{b}]].$$

Proof. One of $\{a,b,b\}$ and $\{\,\overline{a},\,\overline{b},\,\overline{b}\}$ is larger after sorting, so by convention the signed duality function for the standard order is the identity. According to the duality relation (2.6.1) the dual function for some permutation of the dual set is also the identity. Since there is a repetition any permutation of either set is even, so Proposition 2.4 implies that the dual functions are all the identity. The lemma now follows from the com-dual relation (2.6.2).

In a similar way associativity functions are determined when the arguments are not self-dual (as sets). The relationship is not nearly so tidy, however, and we will not use it.

2.8 Symmetry. We have used cycle, flip, and dual to define bases when they go between different spaces. When they go from a space to itself they are primitive and must be calculated. However in these cases they can be thought of as symmetries, and frequently bases can be chosen to put them into standard forms.

For example, $\hom[\,\overline{a},aa]$ is taken to itself by both $\texttt{flip}[a,a,a]$ and $\texttt{cycle}[a,a,a]$. Lemma 2.2 shows that they define a "projective" action of the permutation group. In detail, the permutation group on three elements has a presentation

$$\langle c,f \mid c^3=1, f^2=1, fc^2=cf\rangle$$

where c takes $(1,2,3)$ to $(2,3,1)$, and f takes $(1,2,3)$ to $(1,3,2)$. Lemma 2.2 shows cycle and flip satisfy the first and third relation, and the second is "projectively" true in the sense that f^2 is multiplication by a scalar (specifically, $\texttt{tw}[a]^{-3}$). If $\texttt{tw}[a]$ has a square root then we can normalize flip to get a genuine representation of the permutation group, but in any case this representation usually decomposes into irreducible factors. We can choose a basis for $\hom(\,\overline{a},aa)$ that puts the representation into a canonical form. This gives a further compression of the data: instead of matrices $\texttt{com}[\,\overline{a},a,a]$ and $\texttt{asoc}[1,a,a,a]$ we need only store the number of irreducible factors of each type.

Similarly, if $\{a,b,c\}$ and $\{\,\overline{a},\,\overline{b},\,\overline{c}\}$ are equal as sets then duality defines a bilinear pairing on $\hom(\,\overline{a},bc)$. Frequently a basis can be chosen so this pairing has a special form. And when there is both a symmetry and a pairing they can frequently be standardized simultaneously.

These refinements have little immediate practical benefit because the reductions achieved are so small compared to the storage problem for associativities. They may well be useful in studying endomorphisms of categories, or determining when two categories are isomorphic, because they reduce the "free variables" in the problem.

3. Permutation reduction of associativity

In this section we "solve" relations among associativity functions to express all of them in terms of a basic collection. This uses the hexagon axiom and the simplest case of the pentagon, and does not depend on choice of bases as does section 2.

The result is that asoc for any permutation of arguments (a, b, c, d) can be derived from the values for a few standard choices. The "easy" relations reflection (3.2), and rotation (3.3) give one associativity from another one, and commutativity and cyclic symmetries. There are at most three distinct classes of permutations equivalent under these moves. The full hexagon (3.4) relates these three classes, so if representatives for two are known it gives one in the third.

3.1 The data and outline of the algorithm. The data needed is: suppose (a, b, c, d) is ordered and either $a > 1$ or $b = c = d$. Then we save

(1) $\mathtt{asoc}[\overline{a}, b, c, d]$; and

(2) if a, b, c, d are all distinct, then $\mathtt{asoc}[\overline{a}, c, d, b]$.

Now suppose w, x, y, z are arbitrary, and let a, b, c, d be $\overline{w}, x, y, z$ sorted. If $a = 1$ then $\mathtt{asoc}[w, x, y, z]$ is a cycle morphism (and is the identity unless the other element are all equal). If all elements are greater than 1 then $\mathtt{asoc}[w, x, y, z]$ can be obtained by reflections and rotations from one of $\mathtt{asoc}[\overline{a}, b, c, d]$, $\mathtt{asoc}[\overline{a}, c, d, b]$, or $\mathtt{asoc}[\overline{a}, b, d, c]$. The first case is always stored. If the elements are distinct then the first two are stored, and the third can be obtained from these using the full hexagon. Finally if they are not distinct then either the second or third case is equivalent to the first under reflection and rotation. Again, two are known, and the third can be obtained from the hexagon.

There are further reductions that are not used because the savings do not justify the complexity of the algorithms. For instance, we only need the data above if $(\overline{a}, b, c, d)$ is not larger than its dual after sorting. When it is larger the associativity is determined by the dual-associativity relation 2.6.3.

3.2 Reflection. This relation is obtained by joining two hexagons. The full statement of the axiom asserts that the diagram in 3.4 also commutes if inverses of σ functions are used: $\sigma^{-1}_{cd,b}$ instead of $\sigma_{b,cd}$ in the top row, etc. We use this "inverse hexagon" with the functions themselves rather than the inverses. The resulting diagram still "commutes" in the sense that the two ways of getting from one vertex to another by composing functions or inverses give the same results.

There is a three-term segment of the standard hexagon containing two associativities that matches a three-term segment of an inverse hexagon. Glueing the two diagrams along these segments gives a relation that involves only two associativities:

$$\begin{array}{ccccccc}
(bc)d & & & \xrightarrow{\alpha} & & & b(cd) \\
\downarrow{\scriptstyle \sigma\diamond \mathrm{id}} & & & & & & {\scriptstyle \sigma}\downarrow \\
(cb)d & \xrightarrow{\alpha} & c(bd) & \xrightarrow{\mathrm{id}\diamond\sigma} & c(db) & \xleftarrow{\alpha} & (cd)b \\
\downarrow{\scriptstyle \sigma} & & & & & & \downarrow{\scriptstyle \sigma\diamond \mathrm{id}} \\
d(cb) & & & \xleftarrow{\alpha} & & & (dc)b
\end{array}$$

Mapping $\overline{a}$ into the outer square gives a relation

$$\mathtt{com1} \cdot \mathtt{asoc}[\overline{a}, b, c, d] = \mathtt{asoc}[\overline{a}, d, c, b]^{-1} \cdot \mathtt{com2}$$

where com1 and com2 are given by sums of tensor products of commutativity functions, see Details 3.2a. This relation is called "reflection" because it reflects the second and fourth arguments across the third.

3.3 Rotation. This relation is obtained by mapping the unit object into the pentagon. Write the pentagon as

$$\begin{array}{ccc} ((ab)c)d & \xrightarrow{\alpha_{a,b,c}\diamond \mathrm{id}} & (a(bc))d \\ \downarrow \alpha & & \downarrow \alpha \\ (ab)(cd) & & \\ \downarrow \alpha & & \downarrow \\ a(b(cd)) & \xleftarrow{\mathrm{id}\diamond\alpha} & a((bc)d) \end{array}$$

Consider morphisms of 1 into this. Using the standard isomorphisms $\hom(1, xA) \simeq \hom(\overline{x}, A) \simeq \hom(1, Ax)$ gives

$$\begin{array}{ccc} \hom(\overline{d}, (ab)c) & \xrightarrow[\mathtt{asoc}[\overline{d},a,b,c]]{} & \hom(\overline{d}, a(bc)) \\ \downarrow \mathtt{asoc1} & & \downarrow \mathtt{asoc2} \\ \hom(\overline{a}, b(cd)) & \xleftarrow{\mathtt{asoc}[\overline{a},b,c,d]} & \hom(\overline{a}, (bc)d) \end{array}$$

or in other words a relation

$$\mathtt{asoc}[\overline{a}, b, c, d] = \mathtt{asoc1} \cdot \mathtt{asoc}[\overline{d}, a, b, c]^{-1} \cdot \mathtt{asoc2}^{-1}.$$

The "asoc2" term is given by

$$\mathtt{asoc2} = \oplus_x \mathtt{asoc}[1, a, x, d] \otimes \mathrm{id}_{\mathtt{dim}[x,b,c]}.$$

The 'asoc1" term is more complicated since it involves a change of ordering. It is

$$\left(\oplus_y \mathtt{asoc}[1, a, b, y] \otimes \mathrm{id}_{\mathtt{dim}[y,c,d]}\right) \cdot P \cdot \left(\oplus_x \mathtt{asoc}[1, x, c, d] \otimes \mathrm{id}_{\mathtt{dim}[x,a,b]}\right).$$

The y terms in the first factor correspond to $\overline{x}$ terms in the second. P is the permutation of bases corresponding to changing from the ordering determined by (y, i, j) to $(\mathtt{bar}[y], j, i)$. Note that the $\mathtt{asoc}[1, \ldots]$ terms are cycle functions (2.2), and are frequently identities up to sign.

3.4 The full hexagon. If b, c, d are objects then the hexagon axiom asserts that the diagram

$$\begin{array}{ccccc} (bc)d & \xrightarrow{\alpha} & b(cd) & \xrightarrow{\sigma_{b,cd}} & (cd)b \\ \downarrow \sigma_{b,c}\diamond\mathrm{id} & & & & \downarrow \alpha \\ (cb)d & \xrightarrow{\alpha} & c(bd) & \xrightarrow{\mathrm{id}\diamond\sigma_{b,d}} & c(db) \end{array}$$

commutes. Mapping an object $\overline{a}$ into this gives an equation

$$\mathtt{asoc}[\overline{a}, c, d, b] \cdot \mathtt{com3} \cdot \mathtt{asoc}[\overline{a}, b, c, d] =$$
$$\mathtt{com4} \cdot \mathtt{asoc}[\overline{a}, c, b, d] \cdot \mathtt{com5}$$

where the terms com3, com4, com5 are induced by the σ terms indicated in the diagram, and are described explicitly in terms of standard bases and functions in Details 3.4a.

4. Pentagon reductions

The basic idea is that if a is a summand of the product rs, then sometimes we can find $\mathtt{asoc}[\overline{a}, b, c, d]$ from associativities involving r, s, b, c, d, and summands of products of these. The pentagon (see 3.3) gives a rather complicated relation among such functions. We restrict to "linear" cases that can be solved for one in terms of others that we can assume are known. Most of the relations are higher-order polynomials we cannot yet use effectively.

4.1. The relation. The pentagon is an identity involving five morphisms, each a sum of associativities. We organize these as follows: consider morphisms of $\overline{r}$ into the pentagon for s, b, c, d. The five morphisms are represented by

(1) $(rs), b, c, d$
(2) $r, (sb), c, d$
(3) $r, s, (bc), d$
(4) $r, s, b, (cd)$
(5) $(dr), s, b, c$

where for instance $r, (sb), c, d$ indicates a sum over summands x of (sb), of functions $\mathtt{asoc}[\overline{r}, x, c, d]$. The list is not in the order they appear going around the pentagon. For consistency we will take (1) to be the target, to be solved for in terms of (2–5).

4.2 Reduction to associations with a generating object. A *generating set* is a set of simple objects containing the unit, and so that any object is a summand of products of copies of the given ones. In the basic examples derived from representations of Lie algebras the "fundamental representations" give generating sets. These sets are small (one to three objects, plus the unit) and the representations have relatively small dimension. The small dimension gives a substantial advantage to computations using these representations.

Note that any collection (containing the unit) is a generating set for a monoidal subcategory, namely the one with simple objects appearing as summands of products of the given set. Many interesting examples have no nontrivial monoidal subcategories so *any* nonunit object will form a generating set. There is quite a bit of variation in the speed with which different objects generate the category, however.

Proposition. *Fix a generating set. Then any associativity can be derived via pentagons from associativies in which one object is in the generating set.*

Proof. We outline the algorithm. Suppose G is the generating set, and suppose $\mathtt{asoc}[\overline{w}, x, y, z]$ is known when one of w, x, y, z are in G. Suppose r, s are objects in the generating set and consider the pentagon for r, s, b, c, d as above, b, c, d arbitrary. The functions (2–5) all involve either r or s (or both) so are known. Solving for (1) gives a, b, c, d for any summand a of rs. Denote the set of such summands by G^2. The conclusion is: we can obtain all associativities involving at least one element of G^2.

Now repeat: assume r, s are in G^2 and solve for summands of rs. This gives associativities with one element in G^4. Iteration eventually gives all associativities, since we assumed G is a generating set.

4.3 Associations with two generating objects. We would like to continue the process of 4.2, and reduce to cases involving two generating objects. This is only partially successful: a restriction on the ordering implies we sometimes get only one of the orbits of the reflection-rotation operations of section 3. Recall that there are three such orbits if the objects are distinct, and two if there is a duplication. The full hexagon of 3.4 gives the third orbit if two are known, so we still need one orbit as initial data.

Proposition. *Suppose G is a generating set, then all associativities can be derived from ones satisfying the conditions of 3.1 and additionally either*

(i) *two of the objects are in G, or*

(ii) *one objects is in G and for some $k \geq 1$, one is in $G^{k+1} - G^k$ and the remaining two are* **not** *in G^{k+1}, and are adjacent in the cyclic sense.*

Pairs of elements of (a, b, c, d) are "adjacent in the cyclic sense" if they are adjacent or are the two ends a, d. Note any pair of elements from the set are adjacent in one or the other of the standard orbit representatives of 3.1.

Proof. According to (i), any associativity involving two elements of G is known. Suppose r, s, t are in G and consider the pentagon relation using r, s, b, t, c. Derived lists (2–5) of 4.1 all have two elements of G, so solving for (1) gives a, b, t, c with $t \in G$ and $a \in G^2$. If $a \in G$ this is not new. If one of b, c is also in G^2 we can interchange it with a and find new r, s, t to give this list in the same way. This gives a representative of a second reflection-rotation orbit, and from these two we obtain all orbits. However we are stuck if $a \in G^2 - G$ and neither of b, c is in G^2: we only get the orbit where a, t alternate with b, c. And unfortunately this is not the main orbit in the two-orbit case $b = c$. Therefore in these cases (case (ii) of the Proposition) we must assume another orbit is already known.

We now know all cases involving one element of G and one in G^2. Consider the pentagon for (r, s, b, t, c), where $r \in G^2$ and $s, t \in G$. Cases (2–5) of 4.1 are known, so (1) gives a, b, t, c with a a summand of rs, ie., an element of G^3. Again we are missing the orbit in which b, c are adjacent if $a \in G^3 - G^2$ and $b, c \notin G^3$. Again if this is given in the data then we get all associativities with one element in G and one in G^3.

Continuing in this fashion eventually gives all cases with one element in G and the others arbitrary, because G is a generating set. Proposition 4.2 then shows how to get all other cases.

5. Details

Details 1a *Axioms.* The hexagon and pentagon axioms are given in 3.4 and 3.3 respectively. The balance relation is

$$\theta_{A \diamond B} = (\theta_A \diamond \theta_B)\sigma_{A,B}\sigma_{B,A}.$$

The duality axiom is described in terms of specific data rather than as a functor. S denotes a set of chosen representatives of the simple objects, so $a \in S$ refers to a specific object, not just an isomorphism class.

(1) there is an involution $a \mapsto \overline{a}$ on S so that $\overline{1} = 1$. This is also denoted by "`bar`", so we write `bar`: $S \to S$, and $\overline{a} = \mathtt{bar}(a)$;

(2) there are "bilinear forms" with dual coforms: morphisms $\lambda_a : a \diamond \overline{a} \to 1$ and $\Lambda_a : 1 \to \overline{a} \diamond a$ satisfying the identities:

$$a \xrightarrow{\mathrm{id} \diamond \Lambda_a} a(\diamond \overline{a} \diamond a) \xrightarrow{\text{associate}} (a \diamond \overline{a}) \diamond a \xrightarrow{\lambda_a \diamond \mathrm{id}} a$$

and

$$\overline{a} \xrightarrow{\Lambda_a \diamond \mathrm{id}} (\overline{a} \diamond a) \diamond \overline{a} \xrightarrow{\text{associate}} \overline{a}(\diamond a \diamond \overline{a}) \xrightarrow{\mathrm{id} \diamond \lambda_a} \overline{a}$$

are identity maps; and

(3) these are symmetric in the sense that there is a function sgn satisfying $\mathtt{sgn}[a]^2 = 1$ and if a is not self-dual then $\mathtt{sgn}[a] = 1$, and

$$\overline{a} \diamond a \xrightarrow{\text{twist}\diamond\text{id}} \overline{a} \diamond a \xrightarrow{\text{commute}} a \diamond \overline{a} \xrightarrow{\lambda_a} 1$$

and

$$1 \xrightarrow{\Lambda_a} \overline{a} \diamond a \xrightarrow{\text{commute}} a \diamond \overline{a} \xrightarrow{\text{twist}\diamond\text{id}} a \diamond \overline{a}$$

are $\mathtt{sgn}[a]\lambda_{\overline{a}}$ and $\mathtt{sgn}[a]\Lambda_{\overline{a}}$ respectively.

The symmetry condition results from a normalization, and is not a restriction on the category. If a is not self-dual fix λ_a, Λ_a and use the symmetry relation with $\mathtt{sgn}[a] = 1$ to define $\lambda_{\overline{a}}$, $\Lambda_{\overline{a}}$. A calculation with the balance relation shows symmetry also holds for $\overline{a}$. If a is self-dual then use the condition to define $\mathtt{sgn}[a]$. The same calculation with the balance relation shows $\mathtt{sgn}[a]^2 = 1$. Note that in a field $\mathtt{sgn}[a]$ must be ± 1. In physics applications the symmetric ($\mathtt{sgn}[a] = 1$) objects are called "fermions", the antisymmetric elements "bosons."

Details 1.5a *Evaluation of the duality morphism.* The duality pairing is defined by

$$\mathtt{dual}[a, b, c](i \otimes j) = \lambda_b(\mathrm{id}_b \diamond \lambda_c \diamond \mathrm{id}_{\overline{b}})(i \diamond j)\Lambda_{\overline{a}}.$$

The pairing is represented graphically on the left of Figure D1.5a.1. Note there is an invisible reassociation in both the diagram and the algebraic formula. The output of $(i \diamond j)\Lambda_{\overline{a}}$ is naturally associated $(* \diamond *) \diamond (* \diamond *)$, and must be reassociated to $((* \diamond (* \diamond *)) \diamond *)$ to apply $\mathrm{id}_b \otimes \lambda_c \mathrm{id}_{\overline{b}}$. In the diagram this occurs on a horizontal line just below the λ_c vertex.

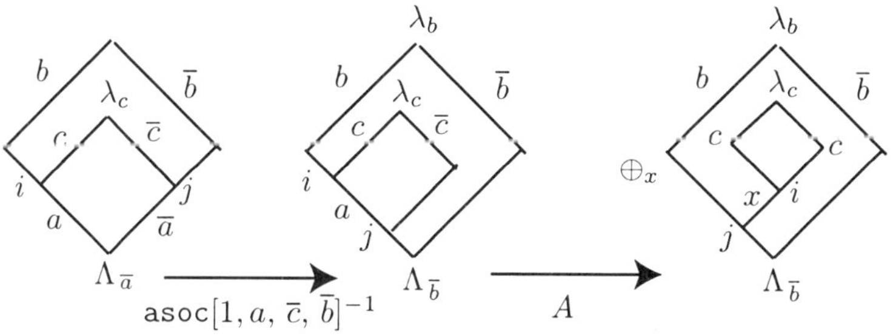

Figure D1.5a.1

The middle diagram in the figure defines a function $\hom(a, bc) \otimes \hom(b, a\,\overline{c}) \to k$. The duality function is obtained by composing this with $\mathrm{id} \otimes (\mathtt{asoc}[1, a, \overline{c}, \overline{b}])^{-1}$. Recall that dual is a matrix whose action on $x \otimes y$ is $\mathtt{Transpose}[x] \cdot \mathtt{dual} \cdot y$. Represent the middle function similarly by a matrix M, then

$$\mathtt{dual} = M \cdot \mathtt{asoc}[1, a, \overline{c}, \overline{b}]^{-1}.$$

Note that in 2.2 this associativity is denoted by $\mathtt{sgn}[a]\mathtt{cycle}[\overline{c}, a, \overline{b}]$, and by convention 2.3 cycle is the identity matrix unless $\overline{c} = a = \overline{b}$.

The middle diagram is obtained by composing the right one with the $w = a$ part of $\mathtt{asoc}[b, b, c, \overline{c}]\colon \oplus_w \hom(b, w\,\overline{c}) \otimes \hom(w, bc) \to \oplus_x \hom(b, bx) \otimes \hom(x, c\,\overline{c})$. To analyse this first note that λ_c is applied to the $\hom(x, c\,\overline{c})$ factor, giving morphisms $x \to 1$. These are 0 unless $x = 1$, in which case the space is 1-dimensional with specified generator $\Lambda_{\overline{c}}$. Since the ordering of the basis in the range of asoc is lexicographic in (x, i, j), the $x = 0$ subspace is spanned by the first basis vector. Thus the part of the asoc matrix that concerns us is a segment of the first column. The next step is to rearrange this into a matrix A that will eventually give M. The specifications for M are that it acts on $i \otimes j \in \hom(a, bc) \otimes \hom(b, a\,\overline{c})$ by $i \cdot M \cdot j$. The asoc matrix acts on vectors with basis ordered by (in this notation) (a, j, i), and the first column takes this to an element of the coefficient ring k. Thus the rows of A are segments of the first column of $\mathtt{asoc}[b, b, c, \overline{c}]$, of length $\mathtt{dim}[a, b, c]$, starting at $\sum_{x \leq a} \mathtt{dim}[x, b, c]^2 + 1$.

The end of the calculation is shown in Figure D1.5a.2.

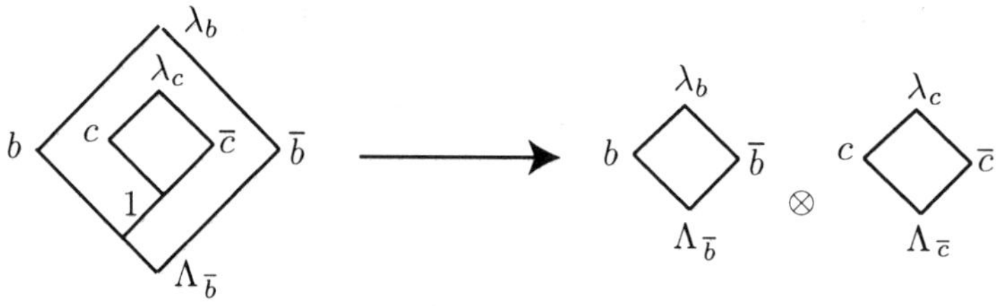

Figure D1.5a.2

We saw that the only relevant x in Figure D1.5a.1 is $x = 1$, and 1 trivially commutes and associates with everything, so the diagram separates into two pieces. This is shown on the right, where edges with unit labels have been omitted. This diagram encodes multiplication by the product $\mathtt{rank}[b]\,\mathtt{rank}[c]$. Putting this all together gives

$$\mathtt{dual}[a, b, c] = \mathtt{rank}[b]\,\mathtt{rank}[c]\,(A \cdot \mathtt{asoc}[1, a, \overline{c}, \overline{b}]^{-1}),$$

with A described above.

Details 2.2a *Proof of lemma 2.2.* The following diagram commutes:

$$\begin{array}{ccccc}
(bc)a & \xrightarrow{\alpha} & b(ca) & \xrightarrow{\sigma} & (ca)b \\
\downarrow{\scriptstyle \sigma\diamond \mathrm{id}_a} & & & & \downarrow{\scriptstyle \alpha} \\
(cb)a & \xrightarrow[\alpha]{} & c(ba) & \xrightarrow{\mathrm{id}_c \diamond \sigma} & c(ab) \\
\downarrow{\scriptstyle \sigma\diamond \mathrm{id}_a} & & \downarrow{\scriptstyle \sigma} & & \downarrow{\scriptstyle \sigma} \\
(bc)a & & (ba)c & \xrightarrow{\sigma \diamond \mathrm{id}_c} & (ab)c \\
\downarrow{\scriptstyle \alpha} & & \downarrow{\scriptstyle \alpha} & & \downarrow{\scriptstyle \alpha} \\
b(ca) & \xrightarrow{\mathrm{id}_a \diamond \sigma} & b(ac) & & \\
\uparrow{\scriptstyle \alpha} & & & & \downarrow \\
(bc)a & \longrightarrow & & \xrightarrow{\sigma} & a(bc)
\end{array}$$

The upper and middle left regions commute by the hexagon axiom. The lower region is the "inverse" hexagon, where inverses of the σ are used. The right middle square commutes by naturality of σ.

Now consider the diagram of hom sets obtained by mapping the unit into this. According to 2.1 the σ terms are multiplication by scalars. The composition across the top and down the right side is therefore $\mathtt{tw}[b]^{-1}\mathtt{tw}[c]^{-1}\mathtt{cycle}^3$. Going down the left side and across the bottom gives $\mathtt{tw}[a]^{-1}\alpha^{-1}\alpha(\sigma^2 \diamond \mathrm{id})$. Cancelling $\alpha^{-1}\alpha$ and using the balance relation on the σ^2 term gives $\mathtt{tw}[b]^{-1}\mathtt{tw}[c]^{-1}$. Since the diagram commutes the two compositions are equal, and $\mathtt{cycle}^3 = \mathrm{id}$.

For the second identity in the lemma, map the unit into the standard hexagon, and apply 2.1 to the σ term. This gives

$$\mathtt{tw}[b]^{-1}\mathtt{cycle}[c,a,b]\mathtt{cycle}[a,b,c] = \mathtt{com}[\overline{c},b,a]\mathtt{cycle}[a,c,b]\mathtt{com}[\overline{a},b,c].$$

Multiplying on the right by $\mathtt{tw}[a]^{-1}\mathtt{com}[\overline{c},b,a]^{-1} = \mathtt{tw}[c]^{-1}\mathtt{tw}[b]\mathtt{com}[\overline{c},a,b]$ gives the identity.

The third identity follows immediately from the balance relation.

Details 2.4a *Proof of Proposition 2.4.* First set $a = 1$ in the $\mathtt{dual}-\mathtt{asoc}$ relation 2.6.3 to get

$$\mathtt{dual}[\overline{d},b,c]\cdot\mathtt{asoc}[1,\,\overline{d},\,\overline{c},\,\overline{b}] = \mathtt{Transpose}[\mathtt{asoc}[1,b,c,d]]\cdot\mathtt{dual}[\overline{b},c,d].$$

Here we have used the identity $\mathtt{dual}[1,AB] = \mathtt{dual}[\overline{A},B] = \mathtt{dual}[\overline{B},A]$ from 2.5. If b,c,d are not all equal then Proposition 2.3(1) gives $\mathtt{sgn}[d]\mathtt{dual}[\overline{d},b,c] = \mathtt{sgn}[b]\mathtt{dual}[\overline{b},c,d]$. This shows the signed dual is invariant under cyclic permutation of the arguments. This completes the proof if a,b,c are not all distinct, since in that case every permutation can also be obtained by an even permutation.

The case $a = 1$, $b \neq c$ is easily checked directly, so we can assume a,b,c are distinct and greater than 1.

The $\mathtt{dual}-\mathtt{com}$ relation 2.6.2 is

$$\mathtt{dual}[\overline{a},b,c]\cdot\mathtt{com}[a,\,\overline{b},\,\overline{c}] = \mathtt{Transpose}[\mathtt{com}[\overline{a},b,c]]\cdot\mathtt{dual}[\overline{a},c,b].$$

Proposition 2.3(3) implies that the $\mathtt{com}$ functions are multiplication by scalars. If a,b,c and $\overline{a},\,\overline{b},\,\overline{c}$ are obtained by the same parity permutations from the standard order then the scalars are equal, and cancel to leave $\mathtt{dual}[\overline{a},b,c] = \mathtt{dual}[\overline{a},c,b]$. This implies the signed dual is invariant under an odd permutations of the argument, so is constant under any permutation. If the parities are different then the formulas of 2.3 give the final identity of 2.4.

Details 2.5a *Proof of Lemma 2.5.* The decompositions are given by $f\otimes g \in \hom(A,xC)\otimes\hom(x,B) \mapsto (g\diamond\mathrm{id})f$, and similarly for $r\otimes s$ in the dual. The dual pairing on $(f\otimes g)\otimes(r\otimes s)$ is

$$\lambda_{BC}(((g\diamond\mathrm{id})f)\diamond(\mathrm{id}\diamond s)r)\Lambda_{\overline{A}}.$$

Subsituting the definitions of form and coform we can write this as the composition of

$$(\mathrm{id}\diamond\lambda_C\diamond\mathrm{id})(f\diamond r)\Lambda_{\overline{A}}\colon 1 \longrightarrow A\overline{A} \longrightarrow xC\,\overline{C}y \longrightarrow x\,y, \text{ and}$$

$$\lambda_B(g \diamond s)\colon x\, y \longrightarrow B\,\overline{B} \longrightarrow 1.$$

A morphism $1 \longrightarrow x\,y$ for x, y simple is nonzero only if $y = \overline{x}$, and then it must be a multiple of $\Lambda_{\overline{x}}$. Suppose the composition obtained from $f \otimes r$ is $\alpha\Lambda_{\overline{x}}$ for some α. Composing this with the $g \otimes s$ term gives, by definition, $\alpha\,\mathtt{dual}[x, B](g \otimes s)$. It remains to evaluate α. Composing the definition with λ_x gives

$$\alpha\lambda_x\Lambda_{\overline{x}} = \lambda_x(\mathrm{id} \diamond \lambda_C \diamond \mathrm{id})(f \diamond r)\Lambda_{\overline{A}} = \lambda_{xC}(f \diamond r)\Lambda_{\overline{A}} = \mathtt{dual}[A, xc](f \otimes r).$$

Recall $\mathtt{rank}[x] = \lambda_x\Lambda_{\overline{x}}$. Solving for α and subsituting in the first expression gives the expression of the Lemma.

Details 3.2a *The* com *functions used in reflections.* The function "com1" of 3.2 is induced by $(\sigma_{c,d} \diamond \mathrm{id})(\sigma_{b,cd})$, so in standard coordinates

$$\begin{array}{ccc}
\hom(\overline{a}, b(cd)) & \longrightarrow & \hom(\overline{a}, (dc)b) \\
\downarrow \simeq & & \downarrow \simeq \\
\oplus_x \hom(\overline{a}, bx) \otimes \hom(x, cd) & \longrightarrow & \oplus_x \hom(\overline{a}, xb) \otimes \hom(x, dc)
\end{array}$$

it is given by

$$\mathtt{com1} = \oplus_x \mathtt{com}[\overline{a}, b, x] \otimes \mathtt{com}[x, c, d].$$

This is a composition of com3 and com4 terms, as used in 3.4.

Similarly com2 has standard coordinates

$$\begin{array}{ccc}
\hom(\overline{a}, (bc)d) & \longrightarrow & \hom(\overline{a}, d(cb)) \\
\downarrow \simeq & & \downarrow \simeq \\
\oplus_x \hom(\overline{a}, xd) \otimes \hom(x, bc) & \longrightarrow & \oplus_x \hom(\overline{a}, dx) \otimes \hom(x, cb)
\end{array}$$

in which it is given by:

$$\mathtt{com2} = \oplus_x \mathtt{com}[\overline{a}, x, d] \otimes \mathtt{com}[x, b, c].$$

In comparing with 3.4 this is a composition of com3 and com5-type expressions.

Details 3.4a *The* com *functions used in the full hexagon.* The first com term in 3.4 is

$$\begin{array}{ccc}
\hom(\overline{a}, b(cd)) & \xrightarrow{\sigma_{b,cd}} & \hom(\overline{a}, (cd)b) \\
\downarrow \simeq & & \downarrow \simeq \\
\oplus_x \hom(\overline{a}, bx) \otimes \hom(x, cd) & \xrightarrow{\mathtt{com3}} & \oplus_x \hom(\overline{a}, xb) \otimes \hom(x, cd)
\end{array}$$

so is given by $\mathtt{com3} = \oplus_x \mathtt{com}[\overline{a}, b, x] \otimes \mathrm{id}$. The second com term is

$$\begin{array}{ccc}
\hom(\overline{a}, c(bd)) & \xrightarrow{\mathrm{id} \diamond \sigma_{b,d}} & \hom(\overline{a}, c(db)) \\
\downarrow \simeq & & \downarrow \simeq \\
\oplus_x \hom(\overline{a}, cx) \otimes \hom(x, bd) & \xrightarrow{\mathtt{com4}} & \oplus_x \hom(\overline{a}, cx) \otimes \hom(x, db)
\end{array}$$

so is given by $\texttt{com4} = \oplus_x \mathrm{id} \otimes \texttt{com}[x, b, d]$. The third term is similar to the second, differing mainly in the position of the variable x:

$$\begin{array}{ccc} \hom(\overline{a}, (bc)d) & \xrightarrow{\sigma_{b,c}\diamond \mathrm{id}} & \hom(\overline{a}, (cb)d) \\ \downarrow \simeq & & \downarrow \simeq \\ \oplus_x \hom(\overline{a}, xd) \otimes \hom(x, bc) & \xrightarrow{\texttt{com5}} & \oplus_x \hom(\overline{a}, xd) \otimes \hom(x, cb) \end{array}$$

so is given by $\texttt{com5} = \oplus_x \mathrm{id} \otimes \texttt{com}[x, b, c]$.

References

[Bobtcheva] Ivelina Bobtcheva, *Dissertation, Virginia Tech 1996.*

[CFS] J. S. Carter, D. E. Flath and M. Saito, *The classical and quantum 6i-symbols*, Math. Notes 43, Princeton University Press, 1996.

[FK] J. Frölich and T. Kerler, *Quantum groups, quantum categories, and quantum field theories*, Lecture Notes in Mathematics, vol. 1542, Springer-Verlag, 1993.

[Kassel] Christian Kassel, *Quantum Groups*, Springer-Verlag Graduate Text in Mathematics, vol. 155, 1995.

[Kazhdan-Wenzl] D. Kazhdan and H. Wenzl, *Reconstructing monoidal categories*, Advances in Soviet Math. **16** (1993), 111–136.

[Kuperberg] G. Kuperberg, *Spiders for rank 2 Lie algebras.*, Comm. Math. Phys. **180** (1996), 109–151.

[Lusztig] G. Lusztig, *Introduction to quantum groups*, Progress in Mathematics, vol. 110, Birkhäuser, 1993.

[MacLane] Saunders MacLane, *Categories for the working mathematician*, Graduate texts in mathematics, vol. 5, Springer-Verlag, 1971.

[QTP] Quantum Topology Project, http://www.math.vt.edu/quantum_topology.

[Quinn] F. Quinn, *Group-categories and their field theories* (to appear).

[Turaev] Vladimir Turaev, *Quantum invariants of knots and 3-manifolds*, W. de Gruyter, Berlin, 1994.

[Shum] M. C. Shum, *Tortile tensor categories*, J. Pure Appl. Algebra **93** (1994), 57–110.

[Yetter] D. N. Yetter, *State-sum invariants of 3-manifolds associated to Artinian semisimple tortile categories*, Topology Appl. **58** (1994), 47–80.

E-mail address: quinn@math.vt.edu, demeio@anvax1.cineca.it

Contemporary Mathematics
Volume **233**, 1999

On Quinn's Invariants of 2-dimensional CW-complexes

Ivelina Bobtcheva

ABSTRACT. Given a semisimple stable autonomous tensor category $\mathcal{A}$ over a field K, to any group presentation with finite number of generators we associate an element $Q(P) \in K$ invariant under the Andrews-Curtis moves. We show that in fact, this is the same invariant as the one produced by the algorithm introduced by Frank Quinn in [**8**]. The new definition allows us to present a relatively simple proof of the invariance and to evaluate $Q(P)$ for some presentations. On the basis of some numerical calculations over different Gelfand-Kazhdan categories, we make a conjecture which allows us to relate the value of $Q(P)$ for two different classes of presentations.

1. Introduction

The original formulation of the Andrews-Curtis conjecture [**1**] has the following algebraic form. Suppose that $< x_1, x_2, \dots, x_n | \; R_1, R_2, \dots, R_n >$ is a presentation of the trivial group. Then this can be reduced to the empty presentation by a finite sequence of operations of the following types:

(i) The places of R_1 and R_s are interchanged.
(ii) R_1 is replaced with gR_1g^{-1}, where g is any element in the group.
(iii) R_1 is replaced with R_1^{-1}.
(iv) R_1 is replaced with R_1R_2.
(v) Adding of an additional generator y and an additional relator y.
(vi) The reverse of (v).

We will refer to these six operations as AC-moves. The Andrews-Curtis conjecture is a particular case of the more general deformation conjecture stating that a simple homotopy equivalence of 2-complexes can be obtained by a deformation through 2-complexes (2-deformation [1]). It is known that if two complexes X and Y are simple homotopy equivalent, then for some k, there exists a 2-deformation from the one point union of X with k copies of S^2 to the one point union of Y with k copies of S^2.

An algebraic formulation of this deformation conjecture comes from the following result [**10**]: two 2-complexes are simple homotopy equivalent if and only

1991 *Mathematics Subject Classification.* Primary 57M20; Secondary 57M05.
Partially supported by grant from the US National Science Fondation.

[1] In part of the literature [**5**] this is actually called a 3-deformation, since it can be achieved through expansions and collapses of disks of dimension at most three.

if the corresponding group presentations have the same difference # relation $-$# generators, and they can be reduced to the same group presentation via the moves (i)-(vi) above plus the additional move

(vii) $< x_1, x_2, \ldots, x_n | R_1, R_2, \ldots, R_m >$ is replaced with $< x_1, x_2, \ldots, x_n | Q_1, Q_2, \ldots, Q_m >$ if there are words $w_{i,j}$, $v_{i,j}$ for $1 \leq i \leq m$, $1 \leq j \leq k(i)$, and indices $r(i,j)$, $s(i,j)$ such that for all i,

$$R_i(Q_i)^{-1} = \sqcap_{j=1}^{k(i)} R_{i,j} Q_{i,j} R_{i,j}^{-1} Q_{i,j}^{-1},$$

where $R_{i,j} = w_{i,j} R^{\pm 1}_{r(i,j)} w^{-1}_{i,j}$ and $Q_{i,j} = v_{i,j} Q^{\pm 1}_{s(i,j)} v^{-1}_{i,j}$. Then, in its general form, the Andrews-Curtis conjecture states that two presentations which have the same difference # relation $-$ # generators and are equivalent through the moves (i)-(vii), are actually equivalent through the moves (i)-(vi).

In [**8**] an algorithm is described for computing topological quantum field theories on $1+1$-complexes associated to stable subcategories of the autonomous tensor categories of Gelfand and Kazhdan over the finite field Z_p. These two dimensional topological theories have the property that the invariant corresponding to S^2 is 0, and therefore give the opportunity of looking for counterexamples of the Andrews-Curtis conjecture. The invariants have been approached numerically [**8, 2, 9**] where every CW-complex is being sliced in standard elementary slices. Then the invariant of the complex is obtained by composing the morphisms corresponding to each slice. Up to now no counterexamples have been detected, and the analytical value of the invariant even for simple complexes wasn't known.

The goal of this work is to present a "global" definition of the invariant, in sense that, given a group presentation P with a finite number of generators, we associate to it an element $Q(P) \in K$, which is invariant under the AC-moves. There is an explicit formula for $Q(P)$ in terms of some kind of trace (partial trace) of a given morphism in the underlying semisimple stable autonomous tensor category $\mathcal{A}$.

The paper is structured in the following way. In section 2 some basic definitions and properties of semisimple stable autonomous tensor categories are presented. In section 3 the diagrammatic notation , which will be the main tool of manipulating the expressions is described . Section 4 is concerned with the definition and properties of partial traces in the category. The invariant $Q(P)$ is defined in section 5, and the proof of the invariance under the AC-moves is presented in section 6. Section 7 reviews the algorithm introduced in [**8**] and shows that it produces the same invariant. In the last section we list some conjectures on the dimension functions in the Gelfand-Kazhdan categories and on the basis of one of them show that if P' is a presentation obtained from P by adding a generator y and a relator $xyx^{-1}y^{-1}$ (where x is a generator of P), and if P'' is obtained from P by simply adding a relator x, then $Q(P') = NQ(P'')$, where N is the number of the simple objects in the category.

I would like to thank Frank Quinn for introducing me into the subject, and patiently explaining to me many basic concepts and ideas. His help and encouragement have been decisive for the completion of this work.

2. General category statements

2.1. $\mathcal{A}$ is a tensor category if it is supplied with a bifunctor $\diamond : \mathcal{A} \times \mathcal{A} \to \mathcal{A}$ and an identity object $\mathbf{1}$, such that for any $A, B, C \in \mathcal{A}$ there are natural isomorphisms

$$\alpha_{A,B,C} : (A \diamond B) \diamond C \to A \diamond (B \diamond C) \quad \text{(associativity morphisms)},$$
$$\eta_A : \mathbf{1} \diamond A \to A, \quad {}_A\eta : A \diamond \mathbf{1} \to A,$$
$$\gamma = \gamma_{A,B} : A \diamond B \to B \diamond A \quad \text{(commutativity morphisms)}.$$

Moreover, $\gamma_{AB}\gamma_{BA} = id_{AB}$. These isomorphisms satisfy a number of axioms as listed in [**7**].

In order to simplify the notation, in the future we will often write AB instead of $A \diamond B$.

2.2. Let $\mathcal{P}$ be the free $(\otimes, I)$-algebra on a single symbol O. The number of O's in an element T in $\mathcal{P}$ will be called its length. Let now $\mathcal{A}$ be a tensor category,. Following [**6**], we define a category $\mathcal{P} \circ \mathcal{A}$, with objects $T[\underline{B}] \equiv T[B_1, B_2, \ldots, B_n]$, where T is an object of length n in $\mathcal{P}$, and $\underline{B} = (B_1, B_2, \ldots, B_n)$ is a sequence of objects in $\mathcal{A}$. The morphisms have the form $\pi[f_1, f_2, \ldots, f_n] : T[B_1, B_2, \ldots, B_n] \to S[A_1, A_2, \ldots, A_n]$, where π is a bijection from the O's in T to the O's in S, i.e. an element in the symmetric group $\mathbf{S}_n$, and $f_i : B_{\pi^{-1}(i)} \to A_i$ are maps in $\mathcal{A}$. We use the following convention for the special maps:

$$T[f_1, f_2, \ldots, f_n] \equiv \iota(n)[f_1, f_2, \ldots, f_n] : T[B_1, B_2, \ldots, B_n] \to T[A_1, A_2, \ldots, A_n],$$
$$\pi \equiv \pi[id_{A_1}, id_{A_2}, \ldots, id_{A_n}] : T[A_{\pi(1)}, A_{\pi(2)}, \ldots, A_{\pi(n)}] \to S[A_1, A_2, \ldots, A_n],$$

where $f_i : B_i \to A_i$, and $\iota(n)$ denotes the identity element $\mathbf{S_n}$. We will refer to the objects in $\mathcal{P} \circ \mathcal{A}$ as bracketings in $\mathcal{A}$, and to the maps of type $\pi = \pi[id_{A_1}, id_{A_2}, \ldots, id_{A_n}]$ as permutations.

2.3. $\mathcal{P} \circ \mathcal{A}$ possesses a structure of a tensor category with identity object $I[\,]$, and product $\otimes$ defined in the following way:

$$S[B_1, B_2, \ldots, B_n] \otimes T[B_{n+1}, B_2, \ldots, B_m] = (S \otimes T)(B_1, B_2, \ldots, B_{n+m}).$$

The associativity and commutativity morphisms are defined to be the corresponding permutations. Then the coherence theorem of MacLane [**7**] is equivalent to the assertion that there is a strict tensor functor $\mathcal{P} \circ \mathcal{A} \to \mathcal{A}$. We write $T(B_1, B_2, \ldots, B_n)$ and $\pi(f_1, f_2, \ldots, f_n)$ for the images of $T[B_1, B_2, \ldots, B_n]$ and $\pi[f_1, f_2, \ldots, f_n]$ under this functor; for example $((O \otimes O) \otimes I)(a, b) = (a \diamond b) \diamond \mathbf{1}$. The images of the permutations will be called again permutations, and we will denote them with the corresponding cyclic decomposition. For example, $(1,2)(3,5,4) : T(A, B, C, D, G) \to S(B, A, D, G, C)$.

In the case when $B_i = B$ for any i, we will use the convention $T(B^k) = T(B_1, B_2, \ldots, B_k)$. Then the permutations define an action of $\mathbf{S}_k$ on $T(B^k)$, and we denote the corresponding representation of $\mathbf{S}_k$ as $\rho[T(B^k)]$.

2.4. Let $P\mathcal{A}$ be the category with the same objects as $\mathcal{P} \circ \mathcal{A}$, but has different set of morphisms: for any morphism $T(\underline{B}) \to R(\underline{B})$ in $\mathcal{A}$ there is a morphism $T[\underline{B}] \to R[\underline{C}]$, which will be denoted in the same way. The composition is obvious. Then there are functors $\mathcal{P} \circ \mathcal{A} \to P\mathcal{A}$ and $P\mathcal{A} \to \mathcal{A}$, whose compositions is exactly the functor $\mathcal{P} \circ \mathcal{A} \to \mathcal{A}$ given by the coherence result of MacLane.

2.5. An *autonomous tensor* category in the terminology of **[11]** is a tensor category $\mathcal{A}$ in which to every object A is assigned an object $A^* \in \mathcal{A}$ and morphisms $\Lambda_A : \mathbf{1} \to A^*A$ (coform) and $\lambda_A : AA^* \to \mathbf{1}$ (form) such that the compositions

$$A^* \xrightarrow{\eta_{A^*}^{-1}} \mathbf{1}\, A^* \xrightarrow{\Lambda_A \diamond id_{A^*}} (A^*A)A^* \xrightarrow{\alpha} A^*(AA^*) \xrightarrow{id_{A^*} \diamond \lambda_A} A^*\, \mathbf{1} \xrightarrow{\eta_{A^*}} A^*,$$
$$A \xrightarrow{A\eta^{-1}} A\, \mathbf{1} \xrightarrow{id_A \diamond \Lambda_A} A(A^*A) \xrightarrow{\alpha^{-1}} (AA^*)A \xrightarrow{\lambda_A \diamond id_A} \mathbf{1}\, A \xrightarrow{\eta_A} A$$

act as identities. In any such category there is a canonical isomorphism $w_A : A^{**} \to A$, for which $\gamma_{A^*A} \circ \Lambda_A = (w_A \diamond id) \circ \Lambda_{A^*}$ We call an autonomous tensor category $\mathcal{A}$ *stable* if $(A^*)^* = A$ and $w_A = id_A$.

2.6. A category $\mathcal{A}$ is called *semisimple* if it is abelian over a field K, and if there is a finite subset of objects Σ in $\mathcal{A}$, such that every other objects is isomorphic to a direct sum of objects in Σ, and for any $a, b \in \Sigma$, $hom_{\mathcal{A}}(a, b) \simeq \begin{cases} 0, & \text{if } a \neq b, \\ K, & \text{if } a = b. \end{cases}$ In general, we will use small latin letters $a, b, c, \ldots$ to denote objects in Σ, and capital letters to indicate an arbitrary object in the category. We also use the notation $F(A, B) = hom_{\mathcal{A}}(A, B)$.

The fact that an object A is isomorphic to a direct sum of objects in Σ means that for any $a \in \Sigma$ there exists a basis $\{\epsilon_i(a, A) : a \to A\}_i$ of $F(a, A)$ and a basis $\{\epsilon_i(a, A)^* : A \to a\}_i$ of $F(A, a)$, which are dual in sense that $\epsilon_i(a, A)^* \circ \epsilon_j(a, A) = \delta_{i,j} id_a$. Moreover, $\sum_{a,j} \epsilon_j(a, A) \circ \epsilon_j(a, A)^* = id_A$. Then, given the objects A, B and C, the map

$$\begin{aligned} \nabla : \quad \oplus_{b \in \Sigma} F(a, A \diamond (b \diamond C)) \otimes F(b, B) &\to \quad F(a, A \diamond (B \diamond C)), \\ \psi \otimes \phi &\to \quad (id_A \diamond (\phi \diamond id_C)) \circ \psi \end{aligned}$$

is an isomorphism. In fact, the inverse of ∇ is given by the map

$$\begin{aligned} \nabla^{-1} : F(a, A \diamond (B \diamond C)) &\to \quad \oplus_{b \in \Sigma} F(a, A \diamond (b \diamond C)) \otimes F(b, B), \\ \phi &\to \quad \oplus_{b \in \Sigma, i}[(id_A \diamond (\epsilon_i(b, B)^* \diamond id_C)) \circ \phi] \otimes \epsilon_i(b, B). \end{aligned}$$

In a semisimple tensor category for any simple object b, the representation $\rho[T(a^k)]$ from 2.3, induces representations $\rho_b[T(a^k)]$ of $\mathbf{S_k}$ on the space $F(b, T(a^k))$.

2.7. The concrete categories, used in the numerical computations of quantum invariants of 2-dimensional CW-complexes, are the ones studied in **[4]**. These are semisimple tensor categories, defined over the finite field $K = Z_p$, and are constructed as a quotient of the category of some modular finite-dimensional modules of a semisimple algebraic group G, in the case when p is greater then the Coxeter number of the corresponding Lie-algebra. If $\mathcal{A}$ is such Gelfand-Kazhdan category, its simple objects correspond to the simple modules with highest weights inside the fundamental alcove. We remind that *involution* on a set S is a bijection of S into itself whose square is equal to the identity. Then on the set of simple objects Σ there is an involution denoted by $a \to a^*$, such that a^* is isomorphic to the dual module of a. A form and coform can be introduced, and this makes $\mathcal{A}$ an autonomous tensor category. But this category is not always stable, i.e. the condition $\gamma_{a,a^*} \circ \Lambda_{a^*} = \Lambda_a : \mathbf{1} \to a^* \diamond a.$ is not always satisfied. The simplest examples come from the categories of modular representations of $sl(2)$. In this case if a is an odd highest weight module in Σ, $\gamma_{a,a^*} \circ \Lambda_{a^*} = -\Lambda_a$. But the subcategory generated by the even highest weights modules is stable.

The general situation is not very different from the one of $sl(2)$, and is described, for example, in **[2]**. It is shown that there is always a subcategory of $\mathcal{A}$ which is stable, for example the one generated by the modules with highest weights in the root lattice. In general, there are number of sublattices of the weight lattice which produce subcategories with the desired property.

3. Diagrammatic notation

The diagrammatic notation used here is based on the coherence results in [**6**, **11**, **3**]. The particular case of autonomous tensor categories is studied in [**6**], and the results there are extended to tortile categories in [**11**, **3**]. The version we use here is the one described in [**12**], specified to the case of stable autonomous tensor category.

The idea behind the diagrammatic notation is that to each composition of $\mathcal{A}$-morphisms there corresponds a labeled link diagram with coupons. Then there are rules for transforming these diagrams, and the statement is that if by such transformations the diagrams corresponding to $\psi_1 : A \to B$ and $\psi_2 : A \to B$ can be reduced to the same diagram, then $\psi_1 = \psi_2$. In this sense, a morphism in $\mathcal{A}$ is determined by its diagram.

PROPOSITION 3.1. *Suppose that the group* $Z_2 \times Z_2 =< \theta_1, \theta_{-1} | \, \theta_1^2, \theta_{-1}^2 >$ *acts on a set* S. *Let* S_i, $i = \pm 1$, *be the set of fixed points of the subgroup generated by* θ_i, *and suppose that* $S_1 \cap S_{-1} = \emptyset$. *Then if* $\tilde{S} = S \setminus (S_1 \cup S_{-1})$ *and* $\mathcal{O}$ *is any orbit, we have that either* $\mathcal{O} \setminus \tilde{S}$ *is a two elements set, or* $\mathcal{O} \subset \tilde{S}$.

The proof is straightforward.

3.2. The *coupon category* $\mathbf{C}\mathcal{A}$ associated to a stable autonomous tensor category $(\mathcal{A}, \alpha, \gamma, \eta, \lambda, \Lambda, *)$ is defined as a quotient of another category $\tilde{\mathbf{C}\mathcal{A}}$. The objects of $\tilde{\mathbf{C}\mathcal{A}}$ are finite sequences of objects in $\mathcal{A}$. Below, we will use the notation $\underline{A}$ both for the sequence $(A_1, A_2, \dots, A_l)$ of length $|\underline{A}| = l$, and for the set of objects $\{A_i\}_{i=1}^l$ in $\mathcal{A}$. In particular, $\underline{A} \sqcup \underline{B}$ denotes the disjoint union of the two sets, and $\underline{A} \setminus \underline{B}$ denotes the complement of $\underline{B}$ in $\underline{A}$.

We introduce a $*$ operation on the objects in $\tilde{\mathbf{C}\mathcal{A}}$: $\underline{A} = (A_1, A_2, \dots, A_l) \to \underline{A}^* = (A_l^*, A_{l-1}^*, \dots, A_1^*)$. Then a morphism $(F, \theta) : \underline{X} \to \underline{Y}$ in $\tilde{\mathbf{C}\mathcal{A}}$ is given by a finite set $F = \{f^i : T^i[\underline{A}^i] \to R^i[\underline{B}^i]\}_i$ of morphisms in $P\mathcal{A}$, and an involution θ on the set $S_F = X \sqcup Y^* \sqcup (\sqcup_i(A^{i*} \sqcup B^i))$ such that $\theta(C)^* = C$ and θ is free of fixed points. The elements of $\underline{A}^i$ and $\underline{B}^i$ are called correspondingly the lower and the upper ends of f^i. The identity morphisms in $\tilde{\mathbf{C}\mathcal{A}}$, $id : S \to S$ are given by $(\{\}, *)$.

Let now $(F, \theta) : \underline{X} \to \underline{Y}$ and $(G, \eta) : \underline{Y} \to \underline{Z}$ be two morphisms in $\tilde{\mathbf{C}\mathcal{A}}$. To define their composition, we would like to take $F \sqcup G$ with the involution map obtained by identifying the two copies of $\underline{Y}$, but in this way one obtains an involution, which may contain fixed points. To avoid this, identity morphisms are introduced as follows.

Let $S' = S_F \setminus \underline{Y}^*$ and $S'' = S_G^* \setminus \underline{Y}^*$. We extend θ to an involution on $S = S' \sqcup \underline{Y}^* \sqcup S''$ by defining θ to be the identity on the elements in S''. In a similar way we define an involution η' on S as $\eta'(x) = \begin{cases} \eta(x^*)^* & \text{if } x \in \underline{Y}^* \sqcup S'' \\ x & \text{otherwise} \end{cases}$.

Then θ and η' define an action of $Z_2 \times Z_2$ on S, which satisfy the requirements of proposition 3.1, where $\tilde{S} = \underline{Y}^*$, $S_1 = S'$, and $S_{-1} = S''$. Then, according to the proposition, the orbits under this action are divided in two - the ones which are entirely contained in $\underline{Y}^*$, and ones which have exactly two elements outside this set. Let $\mathbf{E}$ be the subset of orbits which are entirely contained in $\underline{Y}^*$. If $\mathcal{O} \in \mathbf{E}$, let

$A_{\mathcal{O}}$ be the smallest element in the orbit (with respect to the ordering which comes with $\underline{Y}^*$), and let $\mathbf{T} = \{id_{A_{\mathcal{O}}}\}_{\mathcal{O}\in\mathbf{E}}$. Then composition of (F,θ) and (G,η) is given by $(F \sqcup G \sqcup T, \xi)$, where ξ is an involution on the set $S' \sqcup (S'')^* \sqcup (\sqcup_{\mathcal{O}\in\mathbf{E}}\{A^*_{\mathcal{O}}, A_{\mathcal{O}}\})$ defined in the following way. Let $\imath : S' \sqcup (S'')^* \overset{id\sqcup *}{\longrightarrow} S' \sqcup S'' \subset S$. Then if A is an element in $S' \sqcup (S'')^*$, we define $\xi(A) = \imath^{-1}(B)$, where B is the unique element in $S' \sqcup S''$ which belongs to the same $Z_2 \times Z_2$-orbit as $\imath(A)$. Moreover, $\xi(A_{\mathcal{O}}) = A^*_{\mathcal{O}}$.

3.3. The coupon category $\mathbf{C}\mathcal{A}$ is defined to be the quotient of $\tilde{\mathbf{C}\mathcal{A}}$ under the following relations. Let (F,θ) be a morphism in $\tilde{\mathbf{C}\mathcal{A}}$. Then the following replacements between morphisms in F can be made:

(a) $id_A : O[A] \to O[A]$, whose ends are not mapped into each other, can be removed. Then θ changes to θ' where $\theta'(X)$ is equal to $\theta(X)$ if X is not mapped into some of the ends of id_A, otherwise $\theta'(\theta(A)) = \theta(A^*)$.

(b) If $f : T[\underline{A}] \to Q[\underline{B}]$ and $g : Q[\underline{C}] \to R[\underline{D}]$ are such that $|\underline{B}| = |\underline{C}|$ and the upper ends of f are mapped into the lower ends of g, i.e. $\theta(B_i) = C_i^*$, then f, g can be replaced with $f \circ g$.

(c) $f : T[\underline{A}] \to Q[\underline{B}] \longleftrightarrow f' = \iota(l)^{-1} \circ f \circ \iota(k) : T'[\underline{A}] \to Q'[\underline{B}]$, where $k = |\underline{A}|$, $l = |\underline{B}|$, and $\iota(k) : T'(\underline{A}) \to T(\underline{A})$ and $\iota(l) : Q'(\underline{B}) \to Q(\underline{B})$ are the identity permutations as defined in 2.2.

(d) $\gamma_{A,B} : (O \otimes O)[A,B] \to (O \otimes O)[B,A] \longleftrightarrow id : (O \otimes O)[A,B] \to (O \otimes O)[A,B]$.

(e) $\{f : T[\underline{A}] \to T'[\underline{B}], g : Q[\underline{C}] \to Q'[\underline{D}]\} \longleftrightarrow g \diamond f : T[\underline{A}] \otimes Q[\underline{B}] \to T'[\underline{C}] \otimes Q'[\underline{D}]$.

(f) $\eta_A : (O \otimes I)[A] \to O[A] \longleftrightarrow id_A : O[A] \to O[A]$, and the analogous statement for ${}_A\eta$.

(g) $\lambda_A : (O \otimes O)[A, A^*] \to I[\{\}] \longleftrightarrow id_{A^*} : O[A^*] \to O[A^*]$.

(h) $\Lambda_A : I[\{\}] \to (O \otimes O)[A^*, A] \longleftrightarrow id_A : O[A] \to O[A]$.

3.4. The category $\mathbf{C}\mathcal{A}$ is a stable autonomous tensor category. The product $\underline{A} \otimes \underline{B}$ is simply the sequence obtained by putting the elements of $\underline{B}$ after the elements of $\underline{A}$, the identity object is the empty set, and the associativities, $\eta_{\underline{A}}$ and ${}_{\underline{A}}\eta$ are the identity morphisms. γ, λ and Λ are given by $(\{\}, *)$. Moreover, there is a strict tensor functor $P\mathcal{A} \to \mathbf{C}\mathcal{A}$ which maps $T[\underline{B}]$ into $\underline{B}$ and $f : T[\underline{B}] \to R[\underline{C}]$ into $(\{f\}, *) : \underline{B} \to \underline{C}$. Then the coherence result which underlines the use of the diagrams is that if $\psi, \phi : T(\underline{B}) \to R(\underline{C})$ are two morphisms in $\mathcal{A}$ which have the same image in $\mathbf{C}\mathcal{A}$, then $\psi = \phi$. In this way, given $T, R \in \mathcal{P}$, every morphism $(F,\theta) : \underline{B} \to \underline{C}$ determines a unique morphism $T(\underline{B}) \to R(\underline{C})$ in $\mathcal{A}$, which also will be denoted as (F,θ).

3.5. A morphism $(F,\theta) : \underline{X} \to \underline{Y}$ in $\mathbf{C}\mathcal{A}$ can be described by a diagram in the following way. First we draw a set of $|\underline{X}| + |\underline{Y}|$ points labeled by the corresponding elements in $\underline{X}$ and $\underline{Y}$, and for each morphism $f : T[\underline{A}] \to R[\underline{B}]$ we draw a rectangle (coupon) labeled by f with $|\underline{A}|$ lines labeled by A_i's, attached to its lower side, and $|\underline{B}|$ lines labeled by B_j's, attached to its upper side. Then we connect with lines every two ends of coupons and labeled points which are mapped into each other by the involution θ. In this way one looses any information about the elements in $\mathcal{P}$ associated with the morphism of a coupon except their length, but note that according to 3.3 (c) this is the only essential one. An example for the type of diagram obtained in this way is presented in figure 1. Then the equivalence relations in 3.3 imply that one can perform the local diagram moves represented on figure

2. Note that if in the diagram there is a line which starts and finishes in a coupon and is labeled by the identity object **1**, this line can be removed.

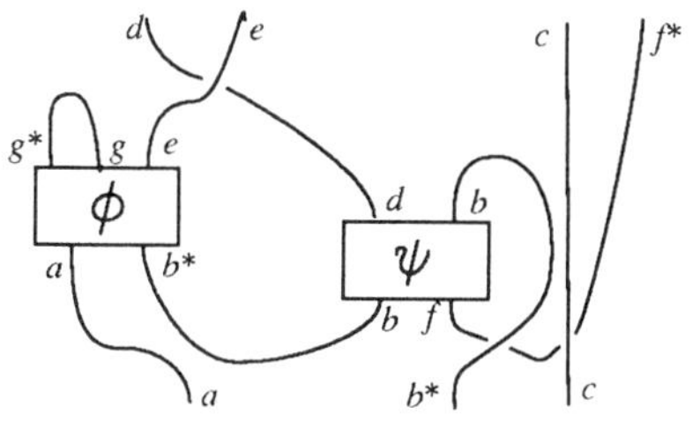

FIGURE 1. General diagram.

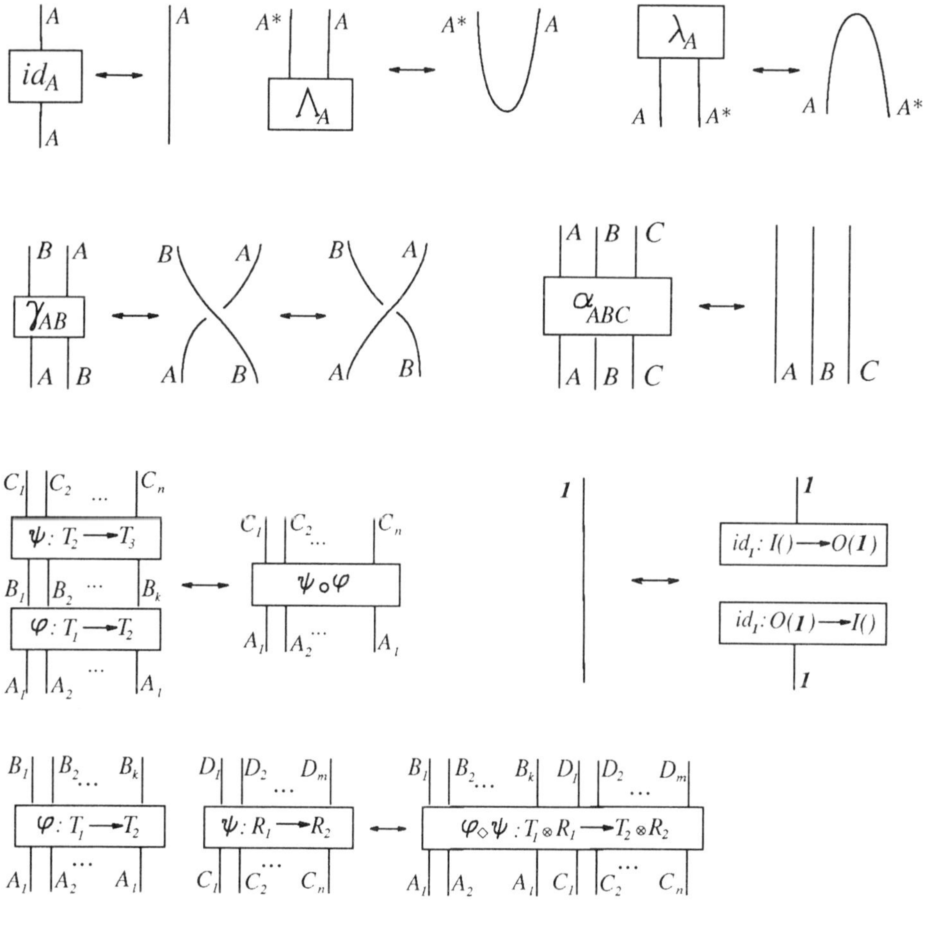

FIGURE 2. Diagram presentation of the relation 3.3

3.6. We introduce a special notation for the basic elements of the homomorphism spaces. If $\{\epsilon_i\}_i$ is a basis of $F(b, T(A_1, A_2, \dots, A_l))$ and $\{\epsilon_i^*\}_i$ is its dual basis of $F(T(A_1, A_2, \dots, A_l), b)$ (see 2.6), we represent these basic elements in the diagrams as it is shown in figure 3 (a) and (b). In the special and most often used case of a basis for $F(a, bc)$, where $a, b, c \in \Sigma$, the notation will be simplified by replacing the coupon with a vertex labeled by i. Figure 4 represents the defining identities of the dual basis. From the discussion in 2.6 it follows that if $\{\zeta_k\}_k$ is a basis for $F(a, bC)$ and $\{\epsilon_i\}_i$ is a basis for $F(b, AB)$, then as one varies b, i, k, the morphisms $(\epsilon_i \diamond id_C) \circ \zeta_k$ form a basis of $F(a, (AB)C)$ (figure 3 (c)).

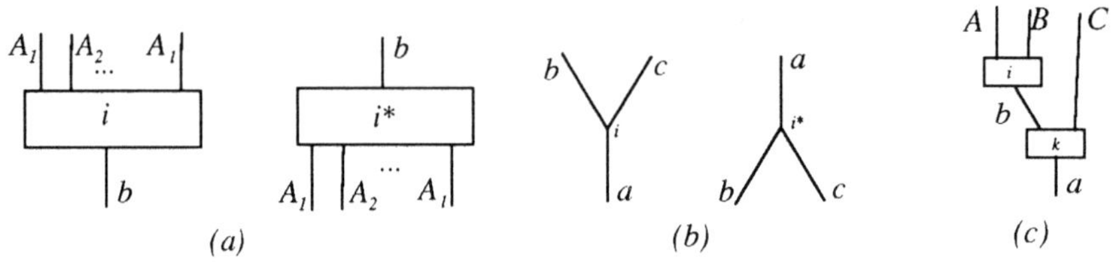

FIGURE 3. Diagram presentation of basic elements.

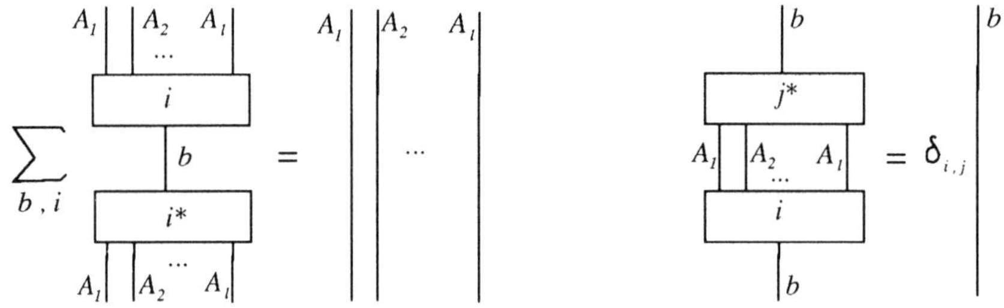

FIGURE 4. Diagram presentation of the defining identities for the dual basis.

3.7. Given a morphism φ of $F(A, B)$, we define

$$\hat{\varphi} = \eta_{A^*}(\lambda_{B^*} \diamond id_{A^*})\alpha^{-1}_{B^*,B,A^*}(id_{B^*} \diamond (\varphi \diamond id_{A^*}))(id_{B^*} \diamond \Lambda_{A^*})({}_{B^*}\eta)^{-1} \in F(B^*, A^*).$$

The corresponding diagram is presented in figure 5. It is easy to see that if $\{\epsilon_i\}_i$ is a basis for $F(A, B)$, then $\{\hat{\epsilon}_i\}_i$ is a basis for $F(B^*, A^*)$ with dual $\{\hat{\epsilon_i^*}\}_i$.

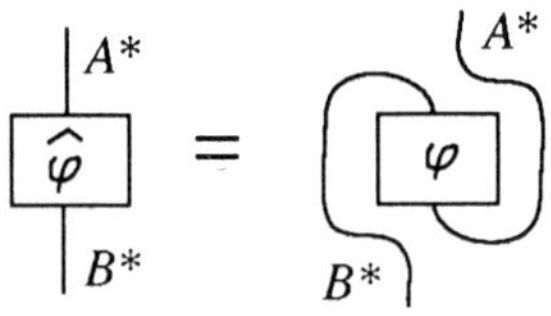

FIGURE 5. Diagram presentation of $\hat{\varphi}$.

3.8. If $\varphi : A \to A$ is a morphism in $\mathcal{A}$, define the rank of φ to be the morphism $r(\varphi) = \lambda_{A^*} \circ (id_{A^*} \diamond \varphi) \circ \Lambda_A$, and $r(A) \equiv r(id_A)$. We think of $r(\varphi)$ as an element in the field via the standard isomorphism $F(\mathbf{1}, \mathbf{1}) \sim K$. The diagrammatic notation for the rank of a morphism is presented in figure 6 (a). As an easy application of the diagrams one can see that

(a) $r(A) = r(A^*)$;
(b) $r(T(A_1, A_2, \dots, A_l)) = r(A_1)r(A_2)\dots r(A_l)$;

(a) allows as to use the simplified notation on figure 6 (b) for a rank of an object. The identity in figure 6 (c) follows from the fact that for any simple object b, $F(b,b) \sim K$.

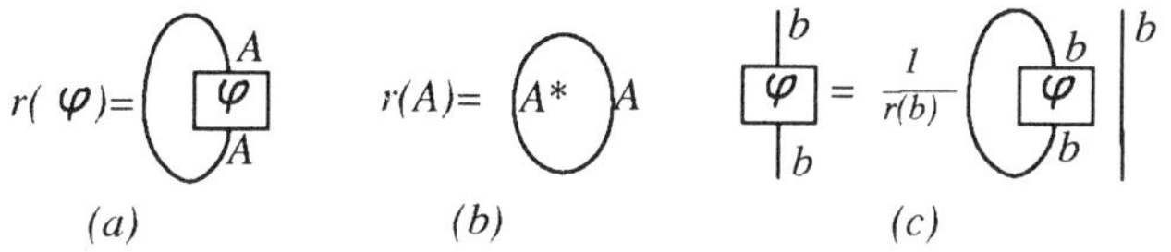

FIGURE 6. Rank of a morphism and rank of an object.

Let $\{\epsilon_i(a, bc)\}_i$ be a basis for $F(a, bc)$, and $\{\epsilon_i(a, bc)^*\}_i$ be the corresponding dual basis. Then the morphisms $\bar{\zeta}_i(c, b^*a) = \eta_c(\lambda_{b^*} \diamond id_c)(id_{b^*} \diamond \epsilon_i)$ and $\zeta_i(c, b^*a) = (id_{b^*} \diamond \epsilon_i^*)(\Lambda_b \diamond id_c)\eta_c{}^{-1}$ represent sets of elements in $F(b^*a, c)$ and $F(c, b^*a)$ correspondingly.

PROPOSITION 3.9. *$\{\zeta_i(c, b^*a)\}_i$ forms a basis for $F(c, b^*a)$ with a dual basis given by $\zeta_i(c, b^*a)^* = \frac{r(c)}{r(a)}\bar{\zeta}_i(c, b^*a)$. In particular, $\sum_{c,i} \frac{r(c)}{r(a)}(\zeta_i(c, b^*a) \circ \bar{\zeta}_i(c, b^*a)) = id_{b^*a}$.*

The proof is presented in figure 7.

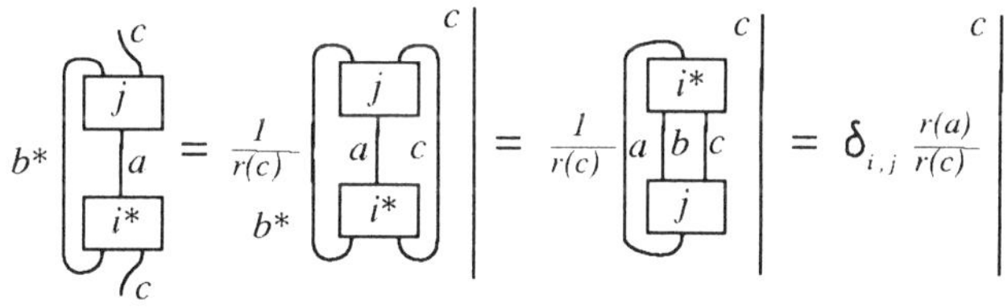

FIGURE 7. Proof of proposition 3.9.

3.10. To make it easier in the future to construct diagrams and to study parts of them, we introduce the notion of an *open morphism*, correspondingly *open diagrams*. An open morphism in $\mathbf{C}\mathcal{A}$ is a morphism $(F, \theta) : \underline{A} = (A_1, A_2, \dots, A_l) \to \underline{B} = (B_1, B_2, \dots, B_k)$ where some subset $In(F)$ of the lower ends is labeled as *in*-ends, and some subset $Out(F)$ of the upper ends is labeled as *out*-ends. We refer to the set of *in*- and *out*- ends as *free* ends.

If $(F, \theta) : \underline{A} \to \underline{B}$ and $(G, \xi) : \underline{C} \to \underline{D}$ are two open morphisms, and $S \subset Out(F) \cap In(G)$, we define the product of the two morphisms over S to be the open morphism obtained by mapping the *out*-ends of (F, θ) labeled by S into the *in*-ends of (G, ξ) labeled by S as shown on figure 8 (a). To be more precise, let $\underline{B}' = \underline{B} \setminus S$ and $\underline{C}' = \underline{C} \setminus S$. Then $(F, \theta) \circ_S (G, \xi) : \underline{A} \otimes \underline{C}' \to \underline{B}' \otimes \underline{D}$ with $In(G \circ_S F) = In(F) \sqcup (In(G) \setminus S)$ and $Out(G \circ_S F) = (Out(F) \setminus S) \sqcup Out(G)$ is the following composition in $\mathbf{C}\mathcal{A}$:

$$(F, \theta) \circ_S (G, \xi) = (id_{\underline{B}'} \otimes (G, \xi)) \circ (\{\}, \epsilon) \circ ((F, \theta) \otimes id_{\underline{C}'}),$$

where $\epsilon : \underline{B} \sqcup \underline{C}' \to (\underline{B}' \sqcup \underline{C})^*$ is the involution map sending $S \subset \underline{B} \to (S \subset \underline{C})^*$.

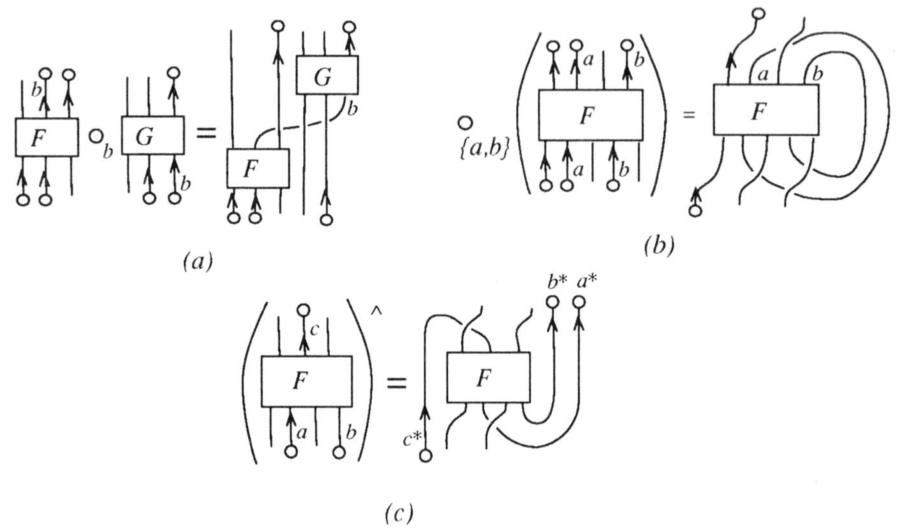

FIGURE 8. Product, closure and reverse of open morphisms.

In the diagrams we indicate the *in*-ends with a circle and an arrow which comes out of it, and the *out*-ends with a circle and an arrow which goes into it. We will think of an open diagram as a sub-diagram of a bigger one, obtained by cutting out in the places of the free ends. For this reason, often the free ends will be left hanging without putting them in a specific place among the lower or the upper ends of the diagram, indicating in this way that the only important thing is how these will glue with the free ends of other open diagrams in the final "closed" one.

3.11. Given an open morphism $(F,\theta) : \underline{A} \to \underline{B}$ we define the closure and the reverse of (F,θ). Given $S \subset In(F) \cap Out(F)$, the closure of (F,θ) over S is defined to be the open morphisms $\circ_S(F,\theta) = (F,\theta')$, where θ' is the extension of θ obtained by mapping into each other $S \subset \underline{A}$ and $S^* \subset \underline{B}^*$ as shown in figure 8 (b). By definition $In(\circ_S(F)) = In(F) \setminus S$ and $Out(\circ_S(F)) = Out(F) \setminus S$.

The reverse of (F,θ) is defined to be a morphism $(F,\theta)\hat{} = (\hat{F},\hat{\theta}) : Out(F)^* \otimes (\underline{A} \setminus In(F)) \to (\underline{B} \setminus Out(F)) \otimes In(F)^*$ with $In(\hat{F}) = Out(F)^*$ and $Out(\hat{F}) = In(F)^*$. $(F,\theta)\hat{}$ is obtained from (F,θ) by exchanging the places of the *in*- and *out*- ends through composition with the corresponding form and coform in $\mathbf{C}\mathcal{A}$ as it is shown in figure 8 (c).

4. Partial traces

4.1. Given a homomorphism $\phi : T(A_1, A_2, \ldots, A_l) \to T(A_1, A_2, \ldots, A_l)$, $b \in \Sigma$ and a basis $\{\epsilon_i\}_i$ of $F(b, A_k)$, the partial trace of ϕ as A_k goes to b is defined to be the morphism

$$\mathtt{Tr}^{\epsilon_i}_{A_k \to b}(\phi) = \sum_i (id \diamond \epsilon_i^* \diamond id) \circ \phi \circ (id \diamond \epsilon_i \diamond id) :$$

$$T(A_1, \ldots, A_{k-1}, b, A_{k+1}, \ldots, A_l) \to T(A_1, \ldots, A_{k-1}, b, A_{k+1}, \ldots, A_l),$$

where $\{\epsilon_i^*\}_i$ is the dual basis of $F(A_k, b)$, and $(id \diamond \psi \diamond id)$ stands for $T(id_{A_1}, \ldots, id_{A_{k-1}}, \psi, id_{A_{k+1}}, \ldots, id_{A_l})$. The corresponding diagram is presented in figure 9 (a).

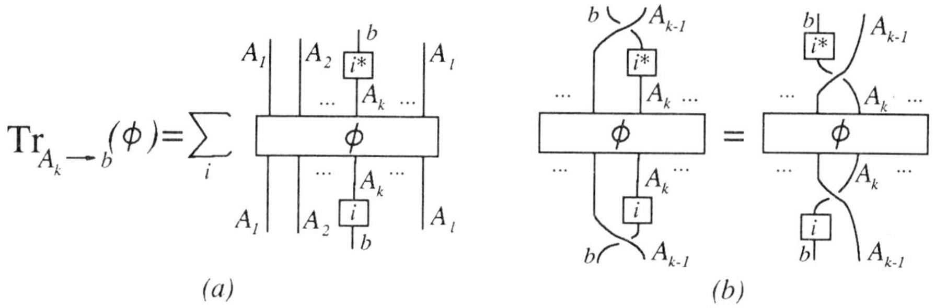

FIGURE 9. Partial trace.

4.2. Let ϕ be as above.

(a) If $\psi : A_k \to A'_k$ is an isomorphism and ϵ_i, ϵ'_i are bases of $F(b, A_k)$ and $F(b, A'_k)$ correspondingly, then

$$\mathtt{Tr}^{\epsilon_i}_{A_k \to b}(\phi) = \mathtt{Tr}^{\epsilon'_i}_{A'_k \to b}((id \diamond \psi \diamond id) \circ \phi \circ (id \diamond \psi^{-1} \diamond id));$$

(b) If $k \neq m$, and $a, b \in \Sigma$, then

$$\mathtt{Tr}_{A_m \to a}(\mathtt{Tr}_{A_k \to b}(\phi)) = \mathtt{Tr}_{A_k \to b}(\mathtt{Tr}_{A_m \to a}(\phi)) \equiv \mathtt{Tr}_{\{A_k \to b, A_m \to a\}}(\phi);$$

(c) If $A_k = R(B_1, B_2, \dots, B_m)$, i.e. for some T',

$$T(A_1, A_2, \dots, A_l) = T'(A_1, A_2, \dots, A_{k-1}, B_1, \dots, B_m, A_{k+1}, \dots, A_l),$$

then

$$\mathtt{Tr}_{A_k \to b}(\phi) = \sum_{a \in \Sigma} \mathtt{Tr}_{R(B_1, \dots, B_{s-1}, a, B_{s+1}, \dots, B_m) \to b}(\mathtt{Tr}_{B_s \to a}(\phi));$$

(d) If π is a permutation in $\mathbf{S}_l$, then $\pi \circ \mathtt{Tr}_{A_k \to b}(\phi) \circ \pi^{-1} = \mathtt{Tr}_{A_{\pi(k)} \to b}(\pi \circ \phi \circ \pi^{-1})$.

(a) follows from the semisimplicity of the category in the following way:

$$\sum_i (id \diamond (\epsilon_i'^* \circ \psi) \diamond id) \circ \phi \circ (id \diamond (\psi^{-1} \circ \epsilon'_i) \diamond id) =$$

$$= \sum_{i,j} (id \diamond (\epsilon_i'^* \circ \psi \circ \epsilon_j) \diamond id) \circ (id \diamond \epsilon_j^* \diamond id) \circ \phi \circ (id \diamond (\psi^{-1} \circ \epsilon'_i) \diamond id) =$$

$$= \sum_{i,j} (id \diamond \epsilon_j^* \diamond id) \circ \phi \circ (id \diamond (\psi^{-1} \circ \epsilon'_i) \diamond id) \circ (id \diamond (\epsilon_i'^* \circ \psi \circ \epsilon_j) \diamond id) =$$

$$= \sum_j (id \diamond \epsilon_j^* \diamond id) \circ \phi \circ (id \diamond \epsilon_j \diamond id).$$

(b) follows directly from the definition. To prove (c), we observe that from 2.6 follows that if $\{\tau_i\}_i$ is a basis for $F(b, R(B_1, \dots, B_{s-1}, a, B_{s+1}, \dots, B_m))$, and $\{\epsilon_j\}_j$ is a basis for $F(a, B_s)$, then $\{R(id_{B_1}, \dots, id_{B_{s-1}}, \epsilon_j, id_{B_{s+1}} \cdots, id_{B_m}) \circ \tau_i\}_{a,i,j}$ is a basis for $F(b, A_k)$. The use of this basis in the evaluation of the partial trace leads to the expression in (c). The statement in (d) is straightforward: every permutation is a product of ones of the form $(k-1, k)$, and for those is shown in figure 9 (b).

As a consequence of (a) above, the indication of the basis in the notation of the partial trace will be omitted.

5. Definition of the invariant

To simplify the notation, we make a standard choice of bracketing. When T is of the form $(\ldots((O \otimes O) \otimes O) \cdots \otimes O) \otimes O$, we will write $T(B^k)$ as simply B^k, and $T(B_1, B_2, \ldots, B_k)$ as $B_1 \diamond B_2 \diamond \cdots \diamond B_k$. In particular, $A(BC)B^3C = (((A,(BC))((B,B),B))C$.

5.1. We start by introducing some definitions. Given any $b \in \Sigma$ define

$$b^l \equiv \begin{cases} b^l \text{ if } l > 0, \\ (b^*)^{|l|} \text{ if } l < 0 \end{cases}$$

Then if $l > 0$, let $f(b,l) = id_{b^{l+1}} : (b, b, \ldots, b) \to (b, b, \ldots, b)$ be the open morphism in $\mathbf{C}\mathcal{A}$, where $In(f(b,l))$ is the first copy of b in the lower ends and $Out(f(b,l))$ is the last copy of b among the upper ends. For $l < 0$, we define $f(b,l) = f(b^*, |l|)^{\vee} : (b, b^*, \ldots, b^*) \to (b^*, \ldots, b^*, b)$. The diagrams corresponding to $f(b,l)$ in both cases are presented in figures 10 (a) and (b). Note that with this definition, for any integer l, $f(b,l)^{\vee} = f(b^*, -l)$.

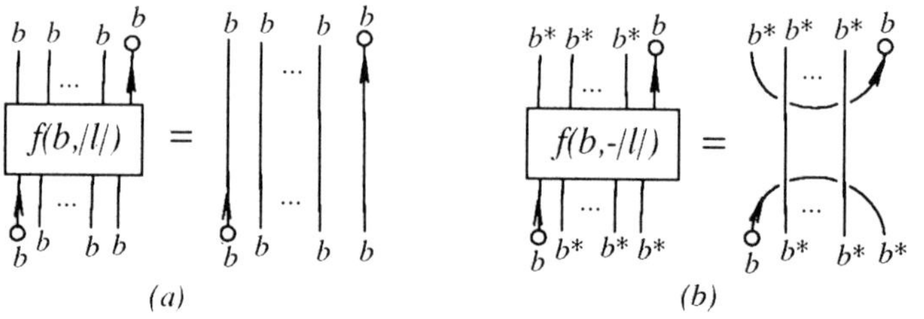

FIGURE 10. Diagram presentation of $f(b,l)$.

5.2. Suppose we are given a nonzero integer l and $b_1, b_2, a_i \in \Sigma$, $1 \le i \le |l|$. Let $\chi : b_1 b_1^l b_2 b_2^l \to (b_1 b_2)(\diamond_i (b_1 b_2)_i)$ and $\chi' : b_1^l b_1 b_2^l b_2 \to (b_1 b_2)(\diamond_i (b_1 b_2)_i)$ be the corresponding permutations, where

$$(b_1 b_2)_i = \begin{cases} b_1 b_2 \text{ if } l > 0, \\ b_1^* b_2^* \text{ if } l < 0 \end{cases} .$$

Then $\mathtt{Tr}_{\{(b_1b_2)_i \to a_i\}_i}(\chi' \circ (f(b_1,l) \otimes f(b_2,l)) \circ \chi^{-1}) : b_1 b_2 (\diamond_i a_i) \to b_1 b_2 (\diamond_i a_i)$ is zero unless a_i are not all the same, in which case is closely related to $f(a_1, l)$. To be able to give the precise statement, let $\{\epsilon_k(a_i, b_1 b_2) : a_i \to b_1 b_2\}_k$ be a basis for $F(a_i, b_1 b_2)$ and $\{\epsilon_k(a_i, b_1 b_2))^*\}_k$ be its dual basis. We will denote with the same symbols the open morphism in $\mathbf{C}\mathcal{A}$: $(a_i) \to (b_1, b_2)$ $((b_1, b_2) \to (a_i))$ which have a unique coupon labeled by $\epsilon_k(a_i, b_1 b_2)$ (correspondingly $(\epsilon_k(a_i, b_1 b_2))^*$), and all lower ends are *in*, and all upper ends are *out*. Let $f^\epsilon(a_i, l) = \sum_k \epsilon_k(a_i, b_1 b_2)^* \circ_{a_i} f(a_i, l) \circ_{a_i} \epsilon_k(a_i, b_1 b_2)$: $b_1 b_2 a_1^l \to a_1^l b_1 b_2$. Then if $l > 0$ we have

$$\mathtt{Tr}_{\{(b_1b_2)_i \to a_i\}_i}(\chi' \circ (f(b_1,l) \otimes f(b_2,l)) \circ \chi^{-1}) = \sqcap_{i=1}^{l-1} \delta_{a_i, a_{i+1}} \sum_k \pi \circ f^\epsilon(a_1, l).$$

Here $\pi : a_1^l b_1 b_2 \to b_1 b_2 a_1^l$ is the corresponding permutation. The proof is shown in figure 11.

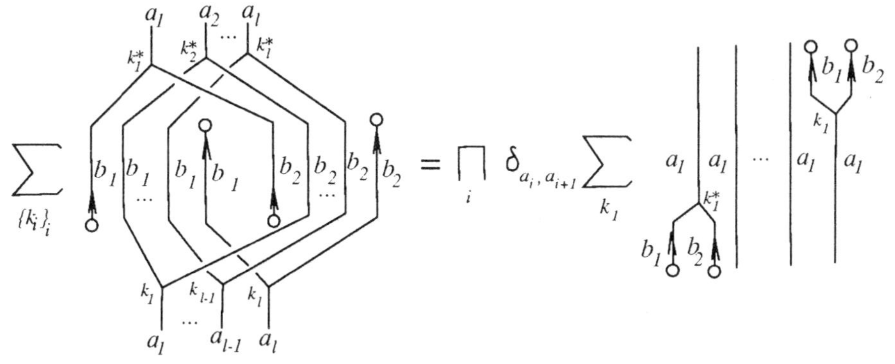

FIGURE 11. Proof of the property in 5.2 for positive l.

When $l < 0$, the morphisms $\{\zeta_k(a_i^*, b_1^* b_2^*) = \gamma_{b_2^*, b_1^*} \circ \hat{\epsilon}_k(a_i, b_1 b_2)\}_k$ form a basis for $F(a_i^*, b_1^* b_2^*)$. Then taking the partial trace with respect to this basis, also in this case, we obtain that

$$\mathtt{Tr}_{\{(b_1 b_2)_i \to a_i^*\}_i}(\chi' \circ (f(b_1, l) \otimes f(b_2, l)) \circ \chi^{-1}) = \sqcap_{i=1}^{|l|-1} \delta_{a_i, a_{i+1}} \sum_k \pi \circ f^{\epsilon}(a_1, l).$$

The proof is shown in figure 12.

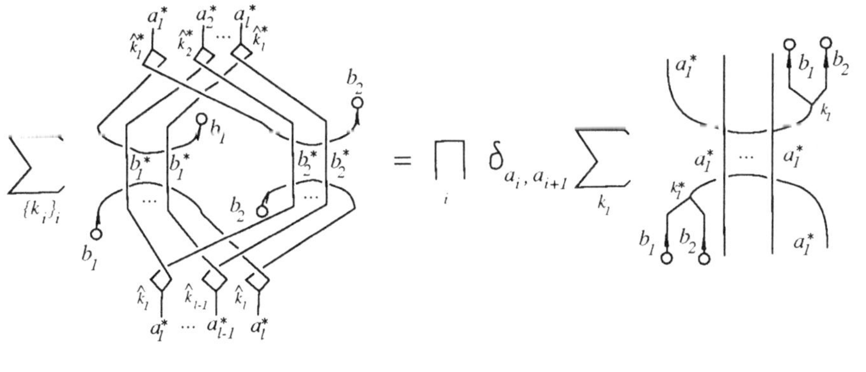

FIGURE 12. Proof of the property in 5.2 for negative l.

5.3. First we will define the invariant for the case of a presentation with one relation $P =< x_1, x_2, \ldots, x_n | R >$, where

$$R = x_{i_1}^{l_1} x_{i_2}^{l_2} \ldots x_{i_k}^{l_k}.$$

To the relation R and to any $b \in \Sigma$ we associate an element $R(b) = b^{l_1} b^{l_2} \ldots b^{l_k} \in \mathcal{A}$ and a morphism $[R, b] : R(b) \to R(b)$ defined in the following way:

$$[R, b] = \circ_b (r(b) \, f(b, l_1) \circ_b f(b, l_2) \circ_b \ldots \circ_b f(b, l_k)).$$

In other words, the *out*-end of $f(b, l_i)$ is connected with the *in*-end of $f(b, l_{i+1})$ and at the end the morphism is closed by connecting the *out*-end of $f(b, l_k)$ with the *in*-end of $f(b, l_1)$. The result is multiplied by the rank of b. The open morphism without the final closure will be denoted with $[R, b]^o$. We will refer to the factors b^l

in $R(l)$ as b-factors. As examples, the diagrams corresponding to $[x^2y^{-2}x^{-1}y, b]^o$ and $[x^2y^{-2}x^{-1}y, b]$ are presented in figure 13 (a) and (b).

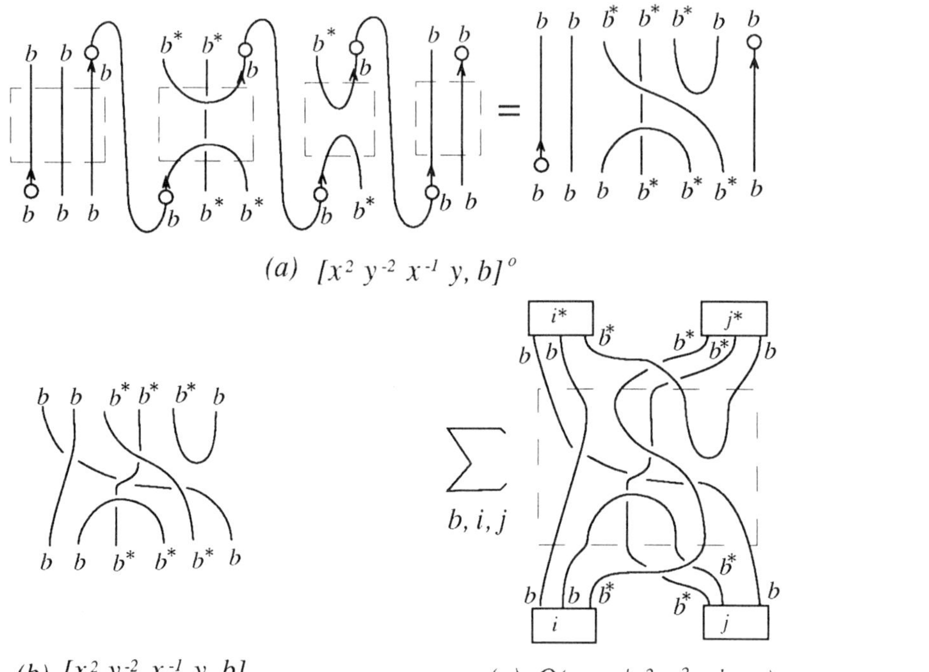

(a) $[x^2\, y^{-2}\, x^{-1}\, y, b]^o$

(b) $[x^2\, y^{-2}\, x^{-1}\, y, b]$

(c) $Q(<x,y/x^2 y^{-2} x^{-1} y>)$

FIGURE 13. Diagram presentation of $Q(< x, y|x^2y^{-2}x^{-1}yx >)$.

5.4. Now to the relation R, to $b \in \Sigma$, and to any generator x_k, we associate an element $g_k(R, b) \in \mathcal{A}$, defined as the product of all b-factors in $R(b)$ corresponding to the generator x_k : $g_k(R, b) = \diamond_{r|x_{i_r}=x_k} b^{l_r}$. Let $\kappa(R) : R(b) \to g_1(R, b)g_2(R, b)\ldots g_n(R, b)$ be the permutation of factors. Then the invariant of the presentation P is defined to be:

$$Q(P) = \sum_{b\in\Sigma} \mathtt{Tr}_{\{g_k(R,b)\to\mathbf{1}\}_k}(\kappa(R) \circ [R, b] \circ \kappa(R)^{-1}).$$

The diagram corresponding to $Q(< x, y|x^2y^{-2}x^{-1}y >)$ is presented in figure 13 (c). The evaluation of the invariant for some simple examples is presented on figures 14 and the results are collected below.

(a) $Q(< x|x^k >) = \sum_{b\in\Sigma} r(b)\mathtt{Trace}[\rho_{\mathbf{1}}[b^k](1, 2, 3, \ldots, k)]$. In particular, $Q(< x|x^2 >) = \sum_{b\in\Sigma|b=b^*} r(b)$;

(b) The invariant for the standard presentation of the fundamental group of a surface of genus n is $Q(< x_1, x_2, \ldots, x_{2n}| \sqcap_{j=1}^{n-1} x_{2j}x_{2j+1}x_{2j}^{-1}x_{2j+1}^{-1} >) = \sum_{b\in\Sigma} r(b)^{-2(n-1)}$;

(c) $Q(< x, y|xyxy^{-1}x^{-1}y^{-1} >) = \sum_{b\in\Sigma|b=b^*} \mathtt{Trace}[\rho_{\mathbf{1}}[b^3](1, 2, 3)]$;

(d) $Q(< x, y|xyx^{-1}y >) = |\{b \in \Sigma|b = b^*\}|$.

Here $\mathtt{Trace}$ denotes the usual trace of the corresponding element in the representation of the symmetric group.

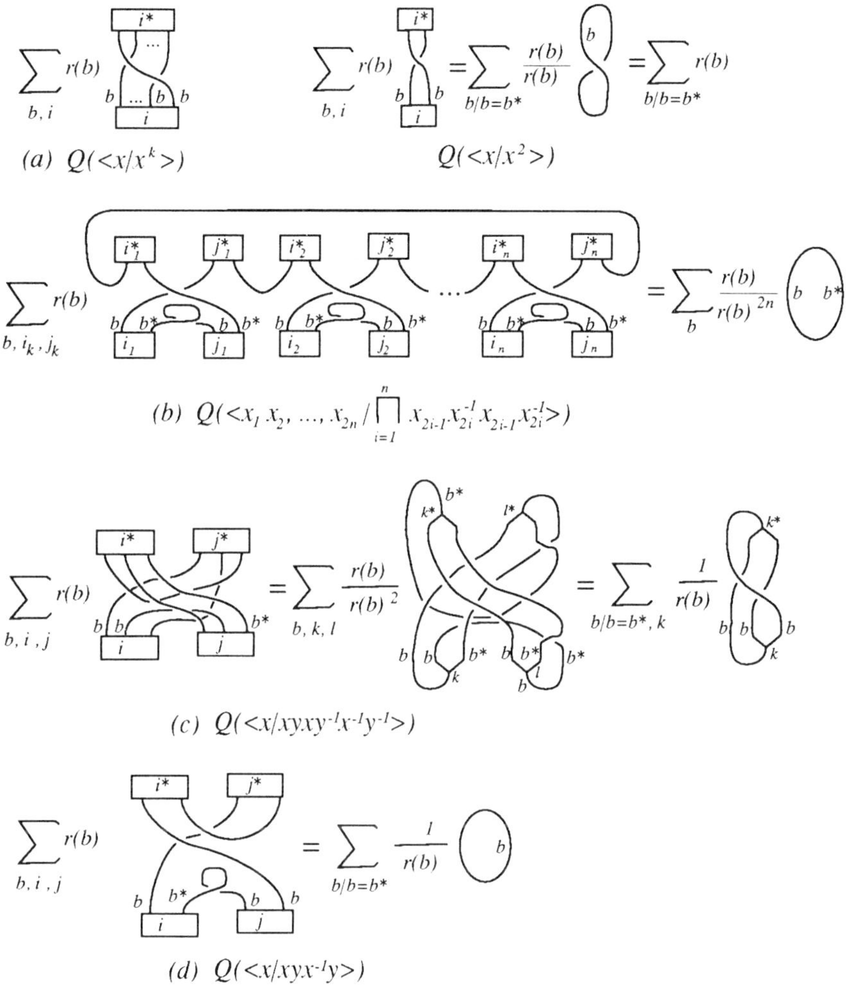

FIGURE 14. Some examples.

5.5. Now we extend the definition to the case of presentations with more then one relation. Let $P =< x_1, x_2, \dots, x_n | R_1, R_2, \dots, R_m >$ be such a presentation, and let $\underline{b} = (b_1, b_2 \dots b_m) \in \Sigma^{\times m}$. Then define

$Rel(\underline{b}) = \diamond_{j=1}^{m} R(b_j)$,

$[P, \underline{b}] = [R_1, b_1] \diamond [R_2, b_2] \diamond \dots \diamond [R_m, b_m] : Rel(\underline{b}) \to Rel(\underline{b})$,

$G_k(\underline{b}) = \diamond_{j=1}^{m} g_k(R_1, b_1) g_k(R_2, b_2) \dots g_k(R_m, b_m)$,

$\xi(P) : Rel(\underline{b}) \to \diamond_{k=1}^{n} G_k(\underline{b})$ to be the corresponding permutation of b-factors.

Note that the permutation $\xi(P)$ groups the b-factors in $Rel(\underline{b})$ according to the generator to which they correspond. Then the invariant of the presentation P is defined to be

$$Q(P) = \sum_{\underline{b} \in \Sigma^{\times m}} \mathtt{Tr}_{\{G_k(\underline{b}) \to \mathbf{1}\}_k} (\xi(P) \circ [P, \underline{b}] \circ \xi(P)^{-1}).$$

As example the diagram corresponding to $Q(<x,y|x^2y^2,xyx^{-1}y^{-1}>)$ is shown in figure 15.

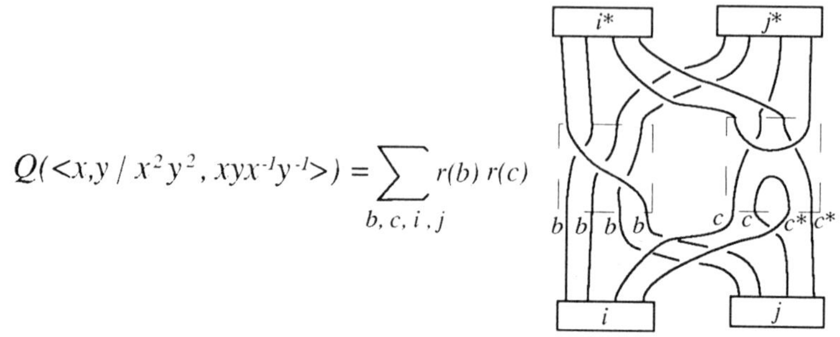

FIGURE 15. Example with two relations.

THEOREM 5.6. *$Q(P)$ is invariant under the AC-moves (i)-(vi).*

The proof is contained in the next section.

6. Invariance under the AC - moves

We consider each move separately. The presentation obtained after performing the move will be denoted with P' and the corresponding products of b-factors will be denoted with $Rel'(\underline{b})$ and $G'_k(\underline{b})$.

6.1. The first move consists in exchanging the places of R_1 and R_s. Let

$$\underline{b}' = (b_s, b_2, b_3, \dots, b_{s-1}, b_1, b_{s+1}, \dots, b_m)$$

and let $\pi_k : G_k(\underline{b}) \to G'_k(\underline{b}')$ be the permutation which exchanges the places of $g_k(R_1, b_1)$ and $g_k(R_s, b_s)$. Then $\pi = \pi_1 \diamond \pi_2 \diamond \dots \diamond \pi_n : \diamond_{k=1}^n G_k(\underline{b}) \to \diamond_{k=1}^n G'_k(\underline{b}')$. As morphisms in $\mathbf{C}\mathcal{A}$ we have that

$$\pi \circ \xi(P) \circ [P,\underline{b}] \circ \xi(P)^{-1} \circ \pi^{-1} = \xi(P') \circ [P',\underline{b}'] \circ \xi(P')^{-1}.$$

This fact is illustrated in figure 16. Then according to 4.2 (a) we have

$$\begin{aligned} Q(P') &= \sum_{\underline{b}' \in \Sigma^{\times m}} \mathtt{Tr}_{\{G'_k(\underline{b}') \to \mathbf{1}\}_k} (\pi \circ \xi(P) \circ [P,\underline{b}] \circ \xi(P)^{-1} \circ \pi^{-1}) = \\ &= \sum_{\underline{b}' \in \Sigma^{\times m}} \mathtt{Tr}_{\{G_k(\underline{b}) \to \mathbf{1}\}_k} (\xi(P) \circ [P,\underline{b}] \circ \xi(P)^{-1}) = Q(P). \end{aligned}$$

This shows the invariance under the first move.

6.2. The second move conjugates R_1 with an arbitrary element in the group. Obviously, it is enough to show the invariance under conjugation with one of the generators $R_1 \to R'_1 = x_l R_1 x_l^{-1}$. Then G_k changes only when $k = l$, and $G'_l(\underline{b}) = b_1 g_l(R_1, b_1) b_1^* g_l(R_2, b_2) \dots g_l(R_m, b_m)$. Let

$$\pi : (\diamond_{k<l} G_k(\underline{b})) G'_l(\underline{b}) (\diamond_{k>l} G_k(\underline{b})) \to (\diamond_{k<l} G_k(\underline{b})) b_1 b_1^* G_l(\underline{b}) (\diamond_{k>l} G_k(\underline{b}))$$

be the corresponding permutation of factors. The goal now is to show that

$$E = \mathtt{Tr}_{G'_l(\underline{b}) \to \mathbf{1}} (\xi(P') \circ [P',\underline{b}] \circ \xi(P')^{-1}) = \mathtt{Tr}_{G_l(\underline{b}) \to \mathbf{1}} (\xi(P) \circ [P,\underline{b}] \circ \xi(P)^{-1}),$$

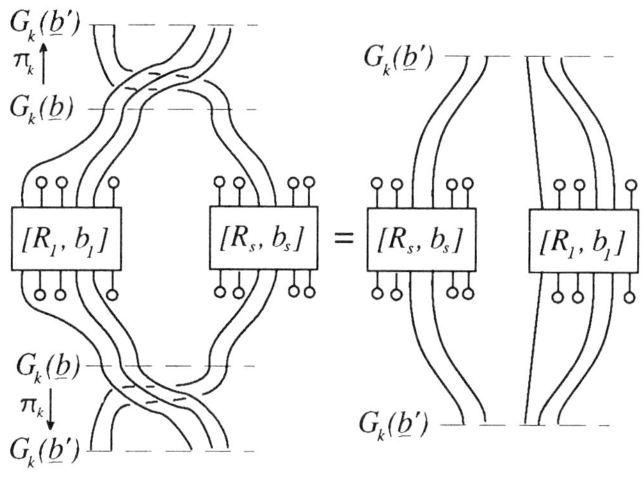

FIGURE 16. Invariance under move (i).

which according to 4.2 (b) would imply the statement. By using properties 4.2 (a) and (c) of the partial trace we obtain

$$\begin{aligned} E &= \mathtt{Tr}_{b_1 b_1^* G_l(\underline{b}) \to \mathbf{1}}(\pi \circ \xi(P') \circ [P', \underline{b}] \circ \xi(P')^{-1} \circ \pi^{-1}) = \\ &= \sum_{a \in \Sigma} \mathtt{Tr}_{a G_l(\underline{b}) \to \mathbf{1}}[\mathtt{Tr}_{b_1 b_1^* \to a}(\pi \circ \xi(P') \circ [P', \underline{b}] \circ \xi(P')^{-1} \circ \pi^{-1})]. \end{aligned}$$

We remind that $[P', \underline{b}] = \diamond_{i=1}^{m}[R_i', b_i]$, but the partial trace $\mathtt{Tr}_{b_1 b_1^* \to a}$ actually involves only the morphism $[R_1', b_1]$, in sense that $[R_1', b_1]$ maps $b_1 R_1(b_1) b_1^*$ into itself, as shown on figure 17 (a), and the effect of the permutation π is to move the b-factors b_1 and b_1^* together, so that the partial trace can be taken. Then figure 17 (b) proves that

$$\mathtt{Tr}_{b_1 b_1^* \to a}(\pi \circ \xi(P') \circ [P', \underline{b}] \circ \xi(P')^{-1} \circ \pi^{-1}) = \delta_{a,\mathbf{1}}\, \nu \circ (id_{\mathbf{1}} \diamond (\xi(P) \circ [P, \underline{b}] \circ \xi(P)^{-1})) \circ \nu^{-1}.$$

Here $\nu : \mathbf{1}\, G_1(\underline{b}) G_2(\underline{b}) \ldots G_2(\underline{b}) \to G_1(\underline{b}) G_2(\underline{b}) \ldots G_{l-1}(\underline{b})\, \mathbf{1}\, G_l(\underline{b}) \ldots G_n(\underline{b})$ is the corresponding permutation. The factor $\mathbf{1}$ can be removed from $\mathtt{Tr}_{\mathbf{1} G_l(\underline{b}) \to \mathbf{1}}$ by conjugating the argument with $\eta_{G_l(\underline{b})}$ and using 4.2 (a). This completes the proof of the invariance under (ii).

6.3. The third move consists of replacing R_1 with R_1^{-1}. First we observe that if $R = x_{i_1}^{l_1} x_{i_2}^{l_2} \ldots x_{i_s}^{l_s}$ is a relation, then

$$\begin{aligned} [R^{-1}, b] &= r(b) \circ_b (f(b, -l_s) \circ_b f(b, -l_{s-1}) \circ_b \ldots \circ_b f(b, -l_1)) = \\ &= r(b^*) \circ_b (f(b^*, l_s)^{\check{}} \circ_b f(b^*, l_{s-1})^{\check{}} \circ_b \ldots \circ_b f(b^*, l_1)^{\check{}}\,) = \\ &= \pi(R) \circ [R, b^*] \circ \pi(R)^{-1}, \end{aligned}$$

where $\pi(R) : b^{-l_1} b^{-l_2} \ldots b^{-l_s} \to b^{-l_s} b^{-l_{s-1}} \ldots b^{-l_1}$ is the permutation of b-factors. The last equality is illustrated in figure 18. In the case we want to study, $\pi(R_1)$ induces a permutation $\pi = \pi(R_1) \diamond id : R_1(b_1^*) \diamond (\diamond_{i=2}^{m} R_i(b_i)) \to R_1^{-1}(b_1) \diamond (\diamond_{i=2}^{m} R_i(b_i))$. Then from the discussion above it follows that

$$[P', \underline{b}] = \pi \circ [P, \underline{b}] \circ \pi^{-1},$$

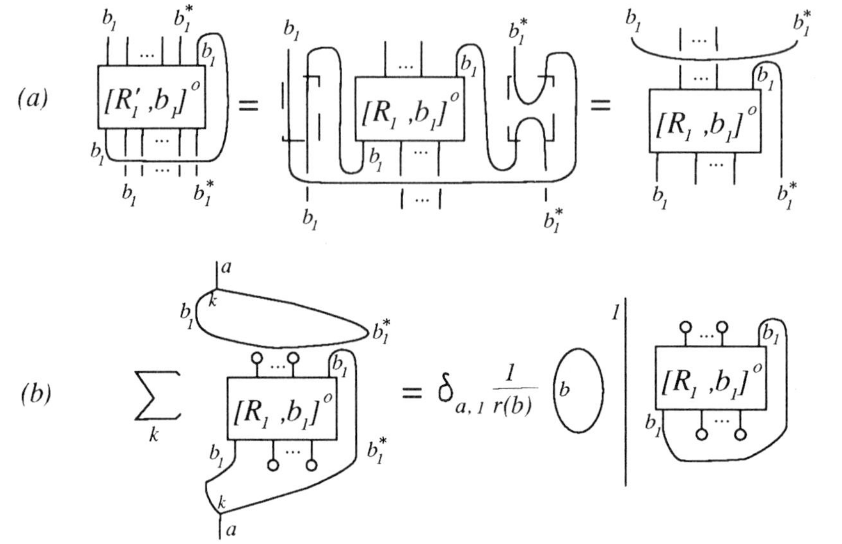

FIGURE 17. Invariance under move (ii).

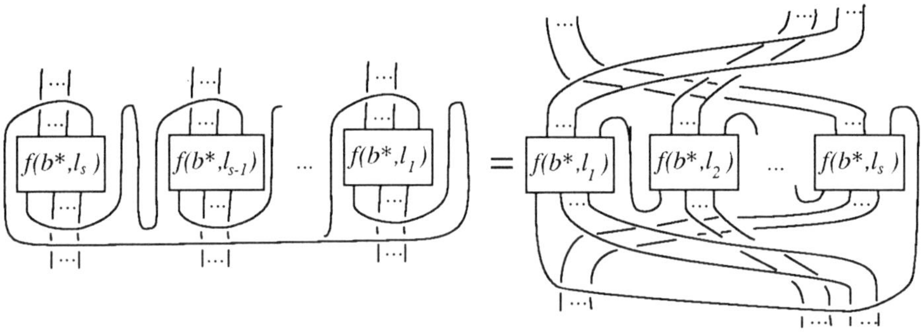

FIGURE 18. Invariance under move (iii).

where $\hat{\underline{b}} = (b_1^*, b_2, \dots, b_m)$. Moreover, $G'_k(\underline{b})$ can be obtained from $G_k(\hat{\underline{b}})$ by reversing the order of the b-factors of the form $b_1^{l_j}$. Let the corresponding permutation be $\chi_k : G'_k(\underline{b}) \to G_k(\hat{\underline{b}})$, and let $\chi = \chi_1 \diamond \chi_2 \diamond \dots \diamond \chi_m : \diamond_{k=1}^n G'_k(\underline{b}) \to \diamond_{k=1}^n G_k(\hat{\underline{b}})$. Then from property 4.2 we obtain

$$\begin{aligned} Q(P') &= \sum_{\underline{b}\in\Sigma^{\times m}} \mathtt{Tr}_{\{G'_k(\underline{b})\to\mathbf{1}\}_k}(\xi(P')\circ[P',\underline{b}]\circ\xi(P')^{-1}) = \\ &= \sum_{\hat{\underline{b}}\in\Sigma^{\times m}} \mathtt{Tr}_{\{G_k(\hat{\underline{b}})\to\mathbf{1}\}_k}(\chi\circ\xi(P')\circ\pi\circ[P,\hat{\underline{b}}]\circ\pi^{-1}\circ\xi(P')^{-1}\circ\chi^{-1}) = \\ &= \sum_{\hat{\underline{b}}\in\Sigma^{\times m}} \mathtt{Tr}_{\{G_k(\hat{\underline{b}})\to\mathbf{1}\}_k}(\xi(P)\circ[P,\hat{\underline{b}}]\circ\xi(P)^{-1}) = Q(P). \end{aligned}$$

This completes the proof of the invariance under move (iii).

6.4. The forth move consists of replacing R_1 with R_1R_2. Figure 19 presents the corresponding change in $[P,\underline{b}]$. Let $R_2 = x_{i_1}^{l_1}x_{i_2}^{l_2}\dots x_{i_s}^{l_s}$, and

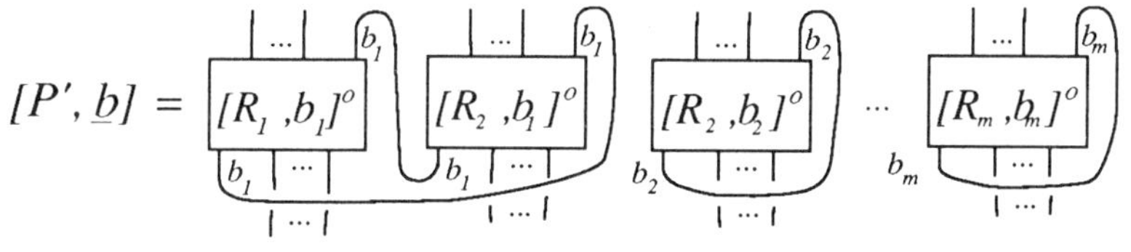

FIGURE 19. Change in $[P,\underline{b}]$ under move (iv).

$\chi : R_1(b_1)R_2(b_1)R_2(b_2) \to R_1(b_1)\diamond_{i=1}^{s}\diamond_{j=1}^{|l_i|}(b_1b_2)_{i,j}$ be the corresponding permutation, where

$$(b_1b_2)_{i,j} = \begin{cases} b_1b_2 \text{ if } l_i > 0 \\ b_1^*b_2^* \text{ if } l_i < 0 \end{cases}.$$

Then $\chi \circ ([R_1R_2,b_1]\diamond[R_2,b_2])\circ\chi^{-1}$ maps $R_1(b_1)\diamond_{i=1}^{s}\diamond_{j=1}^{|l_i|}(b_1b_2)_{i,j}$ into itself. As a first step we prove that

$$W = \sum_{b_2\in\Sigma} \mathtt{Tr}_{\{(b_1b_2)_{i,j}\to a_{i,j}\}_{i,j}}(\chi\circ([R_1R_2,b_1]\diamond[R_2,b_2])\circ\chi^{-1}) =$$

$$\sqcap_{i=1}^{s-1}(\delta_{a_{i,1},a_{i+1,1}}\sqcap_{j=1}^{|l_i|-1}\delta_{a_{i,j},a_{i,j+1}})([R_1,b_1]\diamond[R_2,a_{1,1}])$$

Let $a_i \equiv a_{i,1}$, $1\le i\le s$. Recalling 5.1, where the partial trace involving $f(b_1,l)\otimes f(b_2,l)$ was evaluated, and we have that $W = \circ_{b_1}([R_1,b_1]^o\circ_{b_1}G)$, where

$$G = (\sqcap_{i=1}^{s-1}\sqcap_{j=1}^{|l_i|-1}\delta_{a_{i,j},a_{i,j+1}})\sum_{b_2\in\Sigma}\circ_{b_2}(f^\epsilon(a_1,l_1)\circ_{\{b_1,b_2\}}f^\epsilon(a_2,l_2)\circ_{\{b_1,b_2\}}\dots\circ_{\{b_1,b_2\}}f^\epsilon(a_s,l_s)).$$

Then the statement for W follows from the fact that $\epsilon_k(a_i,b_1b_2)^*\circ\epsilon_l(a_{i+1},b_1b_2) = \delta_{a_i,a_{i+1}}\delta_{k,l}\,id_{a_i}$ and from proposition 3.9 as it is shown in figure 20.

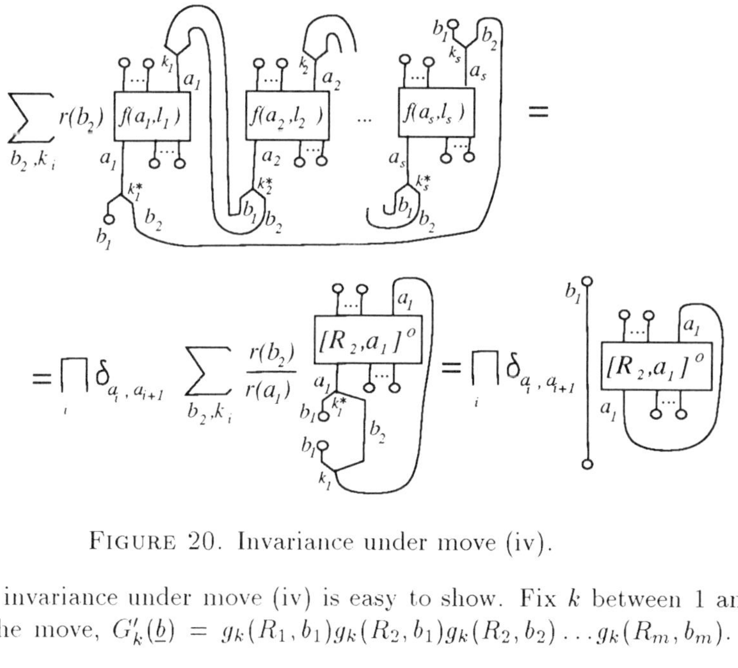

FIGURE 20. Invariance under move (iv).

Now the invariance under move (iv) is easy to show. Fix k between 1 and n. Then after the move, $G_k'(\underline{b}) = g_k(R_1,b_1)g_k(R_2,b_1)g_k(R_2,b_2)\dots g_k(R_m,b_m)$. Let

$G_k^{>2}(\underline{b}) = \diamond_{i=3}^{s} g_k(R_i, b_i)$ and $g_k(R_2, b) = \diamond_{r=1}^{s_k} b^{l_{i_r}}$. Then define $\tau_k : G'_k(\underline{b}) \to g_k(R_1, b_1) \diamond_{r=1}^{s_k} \diamond_{j=1}^{|l_{i_r}|}(b_1 b_2)_{i_r,j} G_k^{>2}(\underline{b})$ to be the corresponding permutation, and let $\tau = \tau_1 \diamond \tau_2 \diamond \ldots \diamond \tau_m : \diamond_{k=1}^{n} G'_k(\underline{b}) \to \diamond_{k=1}^{n} g_k(R_1, b_1) \diamond_{r=1}^{s_k} \diamond_{j=1}^{|l_{i_r}|}(b_1 b_2)_{i_r,j} G_k^{>2}(\underline{b})$. Then from 4.2 (a) and (c) we obtain,

$$\sum_{\underline{b} \in \Sigma^{\times m}} \mathrm{Tr}_{G'_k(\underline{b}) \to \mathbf{1}}(\xi(P') \circ [P', \underline{b}] \circ \xi(P')^{-1}) =$$

$$= \sum_{\underline{b} \in \Sigma^{\times m}} \mathrm{Tr}_{g_k(R_1,b_1) \diamond_{r=1}^{s_k} \diamond_{j=1}^{|l_{i_r}|}(b_1 b_2)_{i_r,j} G_k^{>2}(\underline{b}) \to \mathbf{1}}(\tau \circ \xi(P') \circ [P', \underline{b}] \circ \xi(P')^{-1} \circ \tau^{-1}) =$$

$$= \sum_{\underline{b} \in \Sigma^{\times m}} \sum_{\underline{a}_r \in \Sigma^{\times s_k}} \mathrm{Tr}_{g_k(R_1,b_1) \diamond_{r=1}^{s_k} \diamond_{j=1}^{|l_{i_r}|} a_{r,j} G_k^{>2}(\underline{b}) \to \mathbf{1}} \mathrm{Tr}_{\{(b_1 b_2)_{i_r,j} \to a_{r,j}\}_{r,j}}$$
$$(\tau \circ \xi(P') \circ [P', \underline{b}] \circ \xi(P')^{-1} \circ \tau^{-1}).$$

Now we observe that

$$\tau \circ \xi(P') \circ [P', \underline{b}] \circ \xi(P')^{-1} \circ \tau^{-1} = \xi(P) \circ ((\chi \circ ([R_1 R_2, b_1] \diamond [R_2, b_2]) \circ \chi^{-1}) \diamond_{i=3}^{m} [R_i, b_i]) \circ \xi(P)^{-1}.$$

Here the use of $\xi(P)$ is somewhat abusive. By definition $\xi(P)$ is the permutation of b-factors $\diamond_{j=1}^{m} R(b_j) \to \diamond_{k=1}^{n} G_k(\underline{b})$, while on the r.h.s. above the same notation is used to indicate the same permutation between expressions where each factors $b_2^{l_i}$ has been replaced with $\diamond_{j=1}^{l_i}(b_1 b_2)_{i,j}$. Then the statement follows from 4.2 (d).

6.5. The invariance under the moves (v) and (vi) is straightforward: according to the definition, these moves change the value of the invariant by multiplication or division by

$$\sum_{b \in \Sigma} r(b) \mathrm{Tr}_{b \to \mathbf{1}} id_b = 1.$$

7. The definition of Quinn's invariant

7.1. In this section we show that the invariant defined above is actually the one produced by the algorithm described in [**8**]. First we introduce some morphisms which are being used in the algorithm.

(i) Given any $a, b, c \in \Sigma$ define $cycl(a, b, c) : F(a, bc) \to F(b^*, ca^*)$ as
$$cycl(a, b, c)(\phi) : \quad b^* \to b^* \mathbf{1} \xrightarrow{\Lambda_{a^*}} b^*(aa^*) \xrightarrow{id_{b^*} \diamond (\phi \diamond id_{a^*})}$$
$$b^*((bc)a^*) \to b^* b(ca^*) \xrightarrow{\lambda_{b^*}} \mathbf{1}(ca^*) \to ca^*.$$
$cycl(a, b, c)$ is actually an isomorphisms with inverse $cycl(c^*, a^*, b) \circ cycl(b^*, c, a^*)$.

(ii) Given any $a, b \in \Sigma$ and $A \in \Sigma$ define
$$m_b : F(a, A) \longrightarrow F(\mathbf{1}, (bb^*)A), \quad \psi \to (\Lambda_{b^*} \diamond \psi) \circ \eta_a{}^{-1}$$
$$d_b : F(\mathbf{1}, (bb^*)A) \longrightarrow F(a, A), \quad \phi \to \eta_A \circ (\lambda_b \diamond id_A) \circ \phi.$$

The corresponding diagrams are presented in figures 21.

7.2. The invariant of a presentation $P = (x_1, x_2, \ldots, x_n | R_1, R_2, \ldots, R_m)$ as defined in [**8**] is a map $Q(P) : F(\mathbf{1}, \mathbf{1}) \to F(\mathbf{1}, \mathbf{1})$, i.e. an element in $F(\mathbf{1}, \mathbf{1}) \simeq K$. This map is obtained as a composition of the morphisms which are listed below. Here we will refer to the space $V(k) = \oplus_{y_j \in \Sigma} F(\mathbf{1}, y_1 y_1^* y_2 y_2^* \ldots y_k y_k^*)$ as the *state space of k generators*, and to the summand $F(\mathbf{1}, \mathbf{1}\mathbf{1} \ldots \mathbf{1}\mathbf{1})$ of $V(k)$ as the trivial summand of this state space.

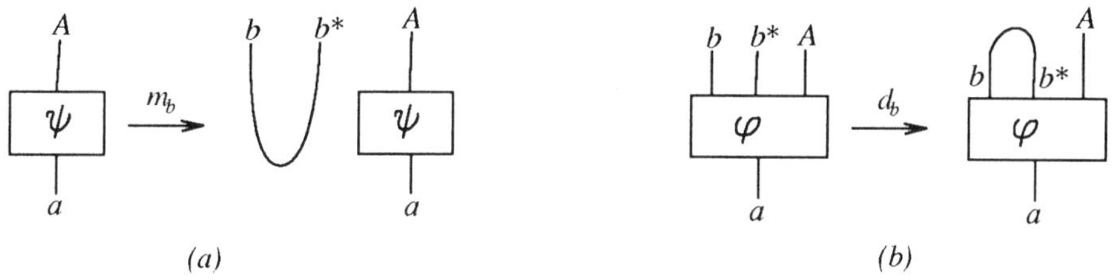

FIGURE 21. The morphisms m_b and d_b.

(i) **Beginning presentation.** This morphism maps $F(\mathbf{1},\mathbf{1})$ into $V(n)$ by embedding $F(\mathbf{1},\mathbf{1})$ into the trivial summand.

$$\begin{aligned} K = \quad & F(\mathbf{1},\mathbf{1}) \to \quad \oplus_{y_j\in\Sigma} F(\mathbf{1}, y_1 y_1^* y_2 y_2^* \dots y_n y_n^*), \\ & 1 \to \quad 1 \in F(\mathbf{1}, \mathbf{1}\,\mathbf{1} \dots \mathbf{1}\,\mathbf{1}). \end{aligned}$$

(ii) **Beginning relation.** Starting a relation leads to the appearance of an additional generator, and therefore maps $V(n)$ into $V(n+1)$ by using the map m_b:

$$\oplus_{y_j\in\Sigma} F(\mathbf{1}, y_1 y_1^* y_2 y_2^* \dots y_n y_n^*) \overset{\oplus_b r(b)^2 m_{b\bullet}}{\longrightarrow} \oplus_{b,y_j\in\Sigma} F(\mathbf{1}, b^* b y_1 y_1^* y_2 y_2^* \dots y_n y_n^*).$$

(iii) **The morphism corresponding to a factor $(y_i)^s$ in a relation.** The main ingredient in it is the *circulator* which is the following map of $\oplus_{x\in\Sigma} F(a, bxx^*)$ in itself.

$$\begin{aligned} CR(a,b): \quad & \oplus_{x\in\Sigma} F(a, bxx^*) \overset{(1,3,2)}{\longrightarrow} \oplus_{x\in\Sigma} F(a, x(x^* b)) \overset{\nabla^{-1}}{\to} \\ & \oplus_{x,z\in\Sigma} F(a, xz^*) \otimes F(z^*, x^* b) \overset{id\circ cycl}{\longrightarrow} \oplus_{x,z\in\Sigma} F(a, xz^*) \otimes F(x, bz) \overset{\nabla}{\to} \\ & \oplus_{z\in\Sigma} F(a, bzz^*). \end{aligned}$$

Here ∇ is as in 2.6. The diagram describing $CR(a,b)$ is presented in figure 22. We note that the circulator is actually an isomorphism and a specific expression for its inverse will be provided later. Then a factor $(y_i)^s$ in a

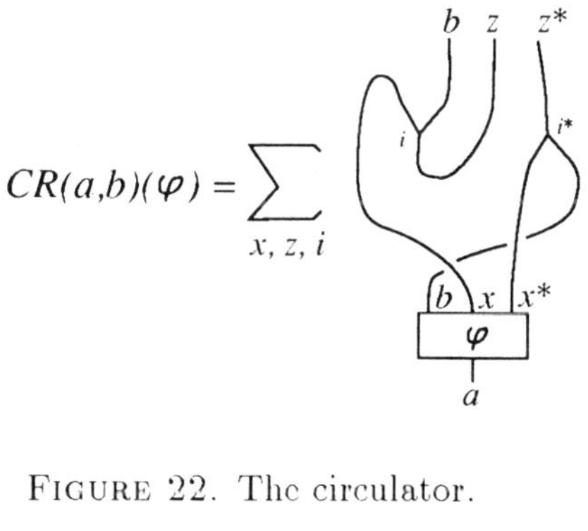

FIGURE 22. The circulator.

relation corresponds to the following composition of morphisms. First, in the state space of the $n+1$ generators, $y_i y_i^*$ is moved next to b. Then the part corresponding to $b y_i y_i^*$ is separated from the rest by using the isomorphism

∇^{-1}, and the circulator CR is applied s times on it. Then following the inverse steps one goes back to the original form of the state space.

$$\oplus_{b,y_j\in\Sigma}F(\mathbf{1},b^*by_1y_1^*y_2y_2^*\dots y_ny_n^*)\to$$

$$\oplus_{b,y_j\in\Sigma}F(\mathbf{1},b^*(by_iy_i^*)y_1y_1^*\dots\hat{y_iy^*}_i\dots y_ny_n^*)\stackrel{\nabla^{-1}}{\to}$$

$$\oplus_{y_j\in\Sigma,j\neq i}\oplus_{a,b\in\Sigma}F(\mathbf{1},b^*ay_1y_1^*\dots\hat{y_iy^*}_i\dots y_ny_n^*)\otimes(\oplus_{y_i\in\Sigma}F(a,by_iy_i^*))\stackrel{id\otimes(CR(b,a))^s}{\longrightarrow}$$

$$\oplus_{y_j\in\Sigma,j\neq i}\oplus_{a,b\in\Sigma}F(\mathbf{1},b^*ay_1y_1^*\dots\hat{y_iy^*}_i\dots y_ny_n^*)\otimes(\oplus_{y_i\in\Sigma}F(a,by_iy_i^*))\stackrel{\nabla}{\to}$$

$$\oplus_{b,y_j\in\Sigma}F(\mathbf{1},b^*(by_iy_i^*)y_1y_1^*\dots\hat{y_iy^*}_i\dots y_ny_n^*)\to\oplus_{b,y_j\in\Sigma}F(\mathbf{1},b^*by_1y_1^*y_2y_2^*\dots y_ny_n^*),$$

where $\hat{y}$ indicates that the corresponding term is missing.

(iv) **Ending relation.** This is a map from $V(n+1)$ generators back into $V(n)$ proportional to the one induced by d_b.

$$\oplus_{b,y_j\in\Sigma}F(\mathbf{1},b^*by_1y_1^*y_2y_2^*\dots y_ny_n^*)\stackrel{\oplus_b r(b)^{-1}d_b}{\longrightarrow}\oplus_{y_j\in\Sigma}F(\mathbf{1},y_1y_1^*y_2y_2^*\dots y_ny_n^*).$$

(v) **Ending presentation.** This is a map from $V(n)$ generators back to the ring $V(0)$, which is injective on the trivial summand on the state space and sends all other summands of $V(n)$ to 0.

$$\oplus_{y_j\in\Sigma}F(\mathbf{1},y_1y_1^*y_2y_2^*\dots y_ny_n^*)\to F(\mathbf{1},\mathbf{1}),\ 1\in F(\mathbf{1},\mathbf{1}\mathbf{1}\dots\mathbf{1}\mathbf{1})\to 1.$$

THEOREM 7.3. *Given a group presentation P, the invariant of P, produced by the above algorithm, is exactly $Q(P)$.*

The key is to understand the powers of the circulator. Given $b\in\Sigma$ and a nonzero integer l, we use the definitions of b^l and $f(b,l)$ made in 5.1. Let $x,w\in\Sigma$, $\{\epsilon_i\}_i$ be a basis for $F(w^*,x^*b^l)$ and let $\{\epsilon_i^*\}_i$ be its dual. We define morphisms $\phi(l,b,x,w,i): bxx^*\to x^*(bb^l)w$ and $\psi(l,b,x,w,i): x^*(b^lb)w\to bww^*$ in the following way:

$$\phi(l,b,x,w,i):\quad bxx^*\to bxx^*\mathbf{1}\stackrel{(1,2,3)\circ\Lambda_w}{\longrightarrow}(x^*bx)(w^*w)\stackrel{id\circ(\epsilon_i\circ id_w)}{\longrightarrow}$$
$$\to(x^*bx)(x^*b^lw)\to x^*(b(xx^*)b^l)w\stackrel{\lambda_x}{\longrightarrow}x^*(bb^l)w,$$
$$\psi(l,b,x,w,i):\quad x^*(b^lb)w\to(x^*b^l)bw\stackrel{\epsilon_i^*}{\longrightarrow}w^*bw\stackrel{(1,3,2)}{\longrightarrow}bww^*.$$

PROPOSITION 7.4. *Given $f\in F(a,bxx^*)$, $CR(b,a)^l(f)=\oplus_{w\in\Sigma}cr(l,b,x,w)\circ f$, where $cr(l,b,x,w)=\sum_i\psi(l,b,x,w,i)\circ(id_{x^*}\circ f(b,l)\circ id_w)\circ\phi(l,b,x,w,i)$.*

The corresponding diagram is shown in figure 23. Note that $cr(l,b,x,w)$ can actually be written as the composition of a partial trace and another morphism, and therefore it is independent on the particular choice of the basis $\{\epsilon_i\}_i$. The fact that the circulator $CR(a,b): F(a,bxx^*)\to F(a,bww^*)$ acts as $cr(1,b,x,w)$ follows from the comparison of the diagrams in figure 22 and 23 (a). Then in figure 24 we prove that $cr(-1,b,x,w)$ is a right inverse of the circulator. In a similar fashion one can see that it is a left inverse as well. Then, by assuming that for $l>1$, the $(l-1)$-th power of the circulator acts as $cr(l-1,b,x,w)$, figure 25 proves that the statement is also true for the l-th power. The inductive proof for the negative powers of the circulator goes in a similar fashion.

Now theorem 7.3 follows from the following statement. Let P be a presentation with n generators, $R=x_{i_1}^{l_1}x_{i_2}^{l_2}\dots x_{i_s}^{l_s}$ be a relation in P, and given $\underline{y},\underline{z}\in\Sigma^{\times n}$ and $b\in\Sigma$, let $\{\epsilon_j^i(z_i^*,y_i^*y_i(R,b))\}_j$, $i=1\div n$, be a set of bases for the homomorphisms

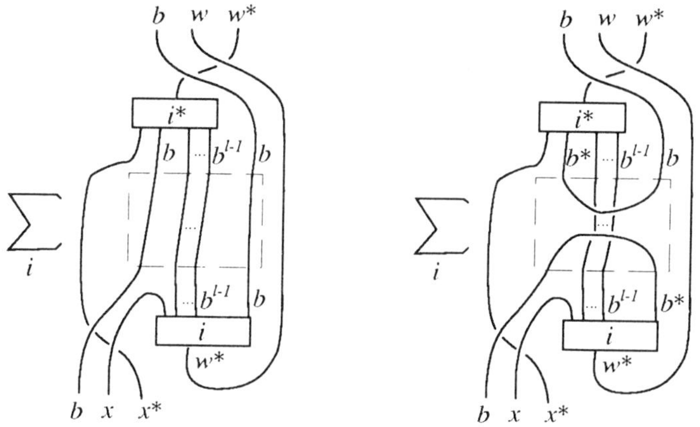

(a) c(/l/,b,x,w) (b) c(-/l/,b,x,w)

FIGURE 23. The morphism $c(l, b, x, w)$.

spaces $F(z_i^*, y_i^* g_i(R, b))$. Then the map $V(n+1) \to V(n+1)$, which the relation R induces according to the algorithm in 7.2, is given by $\varphi \to \oplus_{\underline{y},\underline{z},j} rl(\underline{y}, \underline{z}, j) \circ \varphi$, where $rl(\underline{y}, \underline{z}, j)$ is the following composition of maps:

$$b^* b y_1 y_1^* y_2 y_2^* \dots y_n y_n^* \to b^*(\diamond_i y_i^*) b(\diamond_i y_i) \overset{id \diamond (\diamond_i \Lambda_{z_i})}{\longrightarrow}$$

$$b^*(\diamond_i y_i^*) b(\diamond_i y_i)(\diamond_i z_i^* z_i) \overset{id \diamond (\diamond_i (\epsilon_j^! \diamond id_{z_i}))}{\longrightarrow}$$

$$b^*(\diamond_i y_i^*) b(\diamond_i y_i)(\diamond_i y_i^* g_i(R, b) z_i) \to b^*(\diamond_i y_i^*) b(\diamond_i y_i y_i^*)(\diamond_i g_i(R, b))(\diamond_i z_i) \overset{id \diamond (\diamond_i \lambda_{y_i}) \diamond id}{\longrightarrow}$$

$$b^*(\diamond_i y_i^*) b(\diamond_i g_i(R, b))(\diamond_i z_i) \overset{id \diamond \phi(R) \diamond id}{\longrightarrow} b^*(\diamond_i y_i^*)(\diamond_i g_i(R, b)) b(\diamond_i z_i) \to$$

$$b^*(\diamond_i y_i^* g_i(R, b)) b(\diamond_i z_i) \overset{id \diamond (\diamond_i \epsilon_j^{!*}) \diamond id}{\longrightarrow} b^*(\diamond_i z_i^*) b(\diamond_i z_i) \to b^* b z_1 z_1^* z_2 z_2^* \dots z_n z_n^*,$$

where $\phi(R) = (id_b \diamond \kappa(R))[R, b]^o(\kappa(R)^{-1} \diamond id_b) : b(\diamond_i g_i(R, b)) \to (\diamond_i g_i(R, b))b$ (see 5.4). The corresponding diagram is presented in figure 26, where for simplicity we have restricted ourselves to the case of two generators. The statement is proved by induction over s. For $s = 1$ it is reduced to proposition 7.4. Assume it is true for R, and let $R' = Rx_k^l$ for some k and l. Then

$$g_i(R', b) = \begin{cases} g_k(R, b) b^l & \text{if } i = k, \\ g_i(R, b) & \text{otherwise} \end{cases}, \quad \text{and} \quad [R', b] = \circ_b([R, b]^o \circ_b f(b, l)).$$

Then the statement follows from figure 27.

8. Conjectures and consequences

As we said, this work was inspired by an extensive numerical study of the invariants defined in [8]. The numerical project has been carried out by Frank Quinn, the author, and Luoqi Zhang, and the autonomous tensor categories used in it are subcategories of the Gelfand-Kazhdan categories, as described in 2.7. We remind that such a subcategory is determined by a type and rank of simple Lie algebra (A, B, C, D, E or F), a prime number p, and an invariant sublattice of the weight

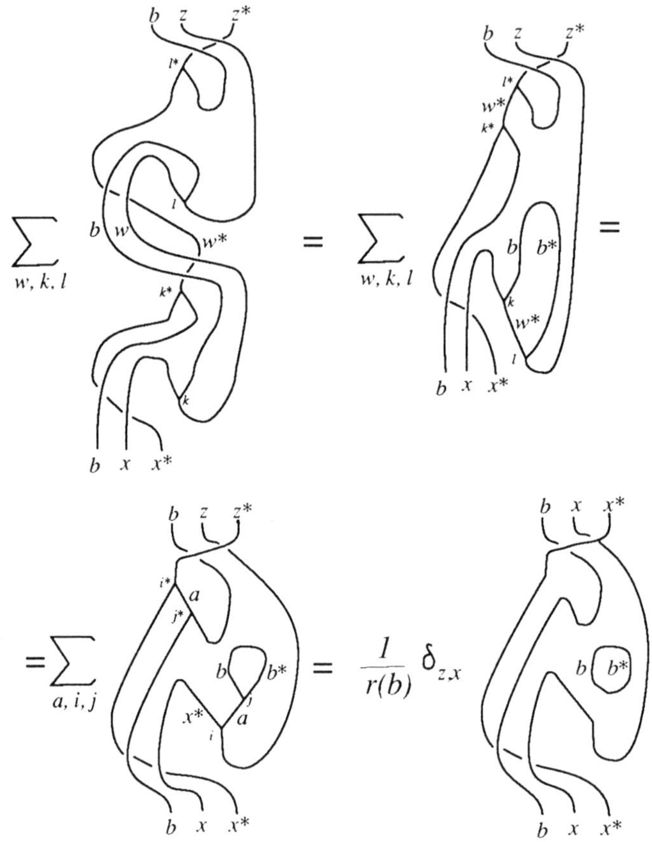

FIGURE 24. Proof that $c(-1, b, x, w)$ represents the inverse of $c(1, b, x, w)$.

lattice (containing the root lattice). But actually many of the categories corresponding to different algebras are equivalent. Conjecturally, any two autonomous tensor categories with sets of simple objects Σ_1 and Σ_2, such that there is a bijection $\mu : \Sigma_1 \to \Sigma_2$, and $dim(F(a, bc)) = dim(F(\mu(a), \mu(b)\mu(c)))$, are equivalent. In other words, the dimensions of the homomorphism spaces determine the category. Let $dim(a, bc)$ denote the dimension of $F(a, bc)$ and we will refer to those as dimension functions. On web page http://www.math.vt.edu/quantum_topology has been collected a list with the values of the dimension functions for some categories corresponding to algebras type A, B, C, D, small primes, and either the full weight lattice or the root lattice. Let L be the order of the abelian group obtained as a quotient of the lattice over which the category is defined, modulo the root lattice. Then all numerically generated examples satisfy the following conjecture:

CONJECTURE 8.1. *In the case of stable Gelfand-Kazhdan categories, for any* $b \in \Sigma$

(a) $dim(a, b^{Lp}) = 0 \pmod p$, *if* $a \neq \mathbf{1}$;

(b) $r(b)^{-1} \sum_{c\in\Sigma} dim(b, cc^*) = \begin{cases} |\Sigma| \text{ if } b \text{ belongs to the root lattice,} \\ 0 \text{ otherwise} \end{cases} \pmod p$.

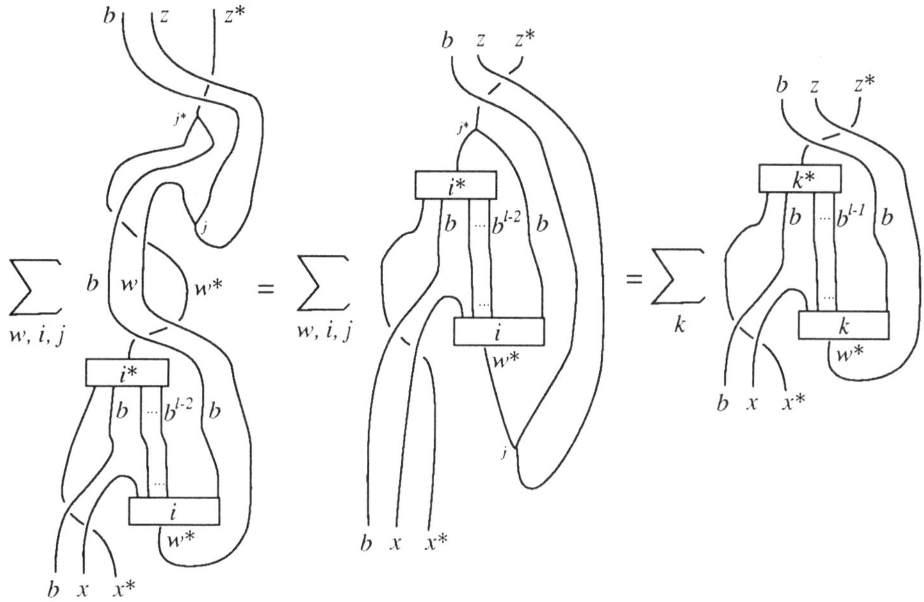

FIGURE 25. Inductive step in the proof of proposition 7.4.

We make few comments on this. From 3.8 and (a) above we obtain

$$dim(\mathbf{1}, b^{Lp}) = \sum_{a\in\Sigma} dim(a, b^{Lp})r(a) = r(b^{Lp}) = r(b)^{Lp} = r(b)^{L}, \quad (mod\ p).$$

In connection with (b) we comment that the statement can be proved easily for algebra type A_1. Moreover, the 0 part can be derived from a conjecture of Frank Quinn, according to which the category over an arbitrary lattice is a product of the category over the root lattice and a finite group category. If this is true, it would imply that the only interesting categories are the ones coming from the root lattice, since a finite group category brings to a classical invariant. For categories over the root lattice, the conjecture 8.1 brings to the following result.

COROLLARY 8.2. *Let $P = < x_1, x_2, \ldots, x_n | R_1, R_2, \ldots, R_m >$ be a group presentation. If we are working with a Gelfand-Kazhdan category over the root lattice, and 8.1 (b) is true, then $Q(P') = |\Sigma|\, Q(P'')$, where*

$$P' = < x_1, x_2, \ldots, x_n, y | R_1, R_2, \ldots, R_m, x_k y x_k^{-1} y^{-1} > \quad and$$
$$P'' = < x_1, x_2, \ldots, x_n | R_1, R_2, \ldots, R_m, x_k > .$$

Let $\underline{b} \in \Sigma^{\times m}$, and $a, c \in \Sigma$. If $(\underline{b}, c)$ denotes the sequence $(b_1, b_2, \ldots, b_m, c)$ we have that

$$[P', (\underline{b}, c)] = [P, \underline{b}] \diamond [x_k y x_k^{-1} y^{-1}, c] : Rel(\underline{b})ccc^*c^* \to Rel(\underline{b})ccc^*c^*,$$

$$G'_l(\underline{b}, c) = \begin{cases} G_k(\underline{b})(cc^*) \text{ if } l = k, \\ G_l(\underline{b}) \text{ otherwise,} \end{cases} \qquad \text{and} \qquad G'_y(\underline{b}, c) = cc^*,$$

$$\xi(P') : Rel(\underline{b})ccc^*c^* \overset{\xi(P)\diamond(2,3)}{\longrightarrow} (\diamond_l G_l(\underline{b}))cc^*cc^* \to$$
$$\to (\diamond_{l<k} G_l(\underline{b}))G_k(\underline{b})(cc^*)(\diamond_{l>k} G_l(\underline{b}))cc^*.$$

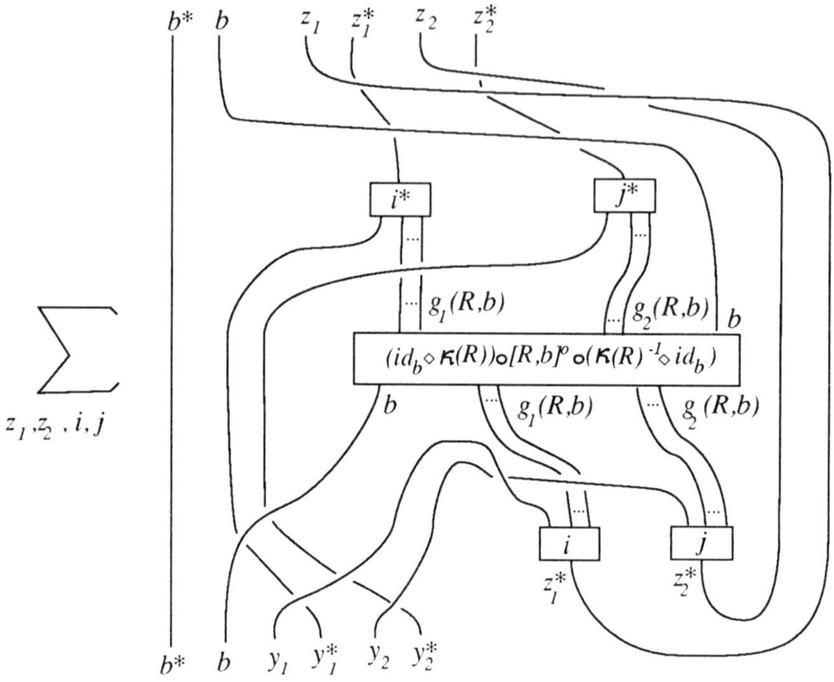

FIGURE 26. The morphisms $V(n+1) \to V(n+1)$ induced by a relation R (case $n=2$).

Moreover,

$$[P'', (\underline{b}, a)] = r(a)[P, \underline{b}] \diamond id_a : Rel(\underline{b})a \to Rel(\underline{b})a,$$

$$G_l''(\underline{b}, a) = \begin{cases} G_k(\underline{b})a \text{ if } l = k, \\ G_l(\underline{b}) \text{ otherwise,} \end{cases}$$

$$\xi(P'') = Rel(\underline{b})a \overset{\xi(P)\diamond id_a}{\longrightarrow} (\diamond_l G_l(\underline{b}))a \to (\diamond_{l<k} G_l(\underline{b}))G_k(\underline{b})a(\diamond_{l>k} G_l(\underline{b})),$$

where we have used the notations in 5.5. Then by using 4.2 (c) and (d) we obtain the following for the part of the partial trace involving $G_k'(\underline{b}, c)$ and $G_y'(\underline{b}, c)$:

$$\begin{aligned} W &= \sum_{c \in \Sigma} \mathtt{Tr}_{\{G_k'(\underline{b},c)\to \mathbf{1},\ G_y'(\underline{b},c)\to \mathbf{1}\}}(\xi(P') \circ ([P, \underline{b}] \diamond [x_k y x_k^{-1} y^{-1}, c]) \circ \xi(P')^{-1}) = \\ &= \sum_{c, a \in \Sigma} \mathtt{Tr}_{G_k(\underline{b})a \to \mathbf{1}}(\xi(P'') \circ ([P, \underline{b}] \diamond K) \circ \xi(P'')^{-1}), \end{aligned}$$

where $K = \mathtt{Tr}_{\{cc^* \to a, cc^* \to \mathbf{1}\}}((2,3) \circ [x_k y x_k^{-1} y^{-1}, c] \circ (2,3))$. Figure 28 shows that in fact, $K = dim(a, cc^*) id_a \diamond id_{\mathbf{1}}$. Hence we obtain :

$$\begin{aligned} W &= \sum_{c, a \in \Sigma} dim(a, cc^*) \mathtt{Tr}_{G_k(\underline{b})a \to \mathbf{1}}(\xi(P'') \circ ([P, \underline{b}] \diamond id_a) \circ \xi(P'')^{-1}) = \\ &= |\Sigma| \sum_{a \in \Sigma} \mathtt{Tr}_{G_k(\underline{b})a \to \mathbf{1}}(\xi(P'') \circ [P'', (\underline{b}, a)] \circ \xi(P'')^{-1}), \end{aligned}$$

which is exactly the part of the partial trace in the definition of $Q(P'')$ involving $G_k''(\underline{b}, a)$.

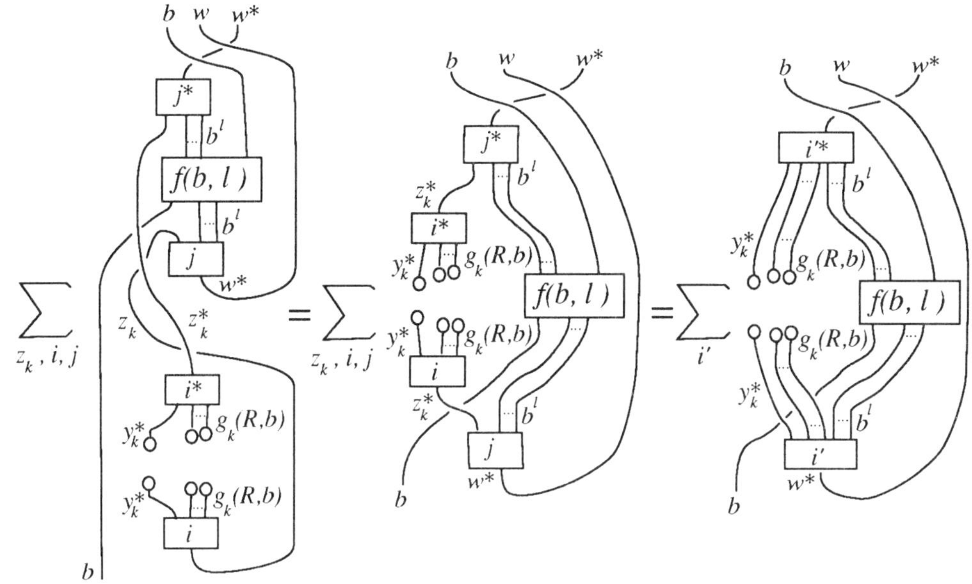

FIGURE 27. Inductive step in the proof of theorem 7.3.

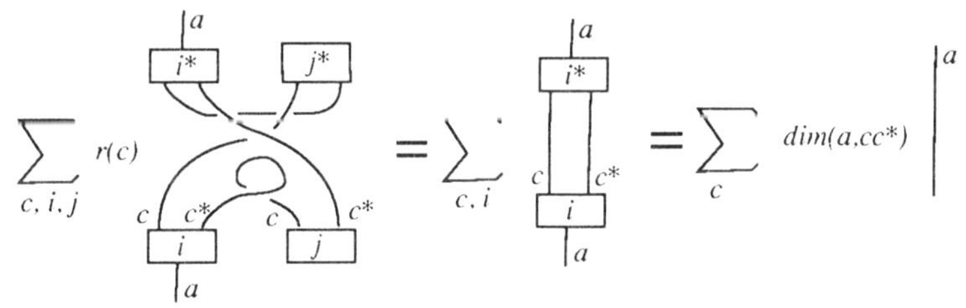

FIGURE 28. Proof of corollary 8.2.

Another conjecture we would like to discuss concerns the order of the circulator. It is observed in [**8**], that for all numerically generated categories, the order of the circulator is in fact Lp. On the basis of this and the expression for the powers of the circulator 7.4 we may state the following

CONJECTURE 8.3. *Let $\sigma^{Lp+1}(b) = (1, 2, 3, \ldots, Lp+1) : bb^{Lp} \to bb^{Lp}$. Then in the case of stable Gelfand-Kazhdan categories,*

$$\mathtt{Tr}_{b^{Lp}\to w}\sigma^{Lp+1}(b) = \begin{cases} 0 & \text{if} \quad w \neq \mathbf{1}, \\ id_b \diamond id_{\mathbf{1}} & \text{otherwise.} \end{cases}$$

The corresponding diagram is presented in figure 29.

The conjecture about the finite order of the circulator implies that given a presentation, the exponents of the generators in the relations matter only mod Lp, in particular it is enough to look at presentations P with only positive exponents. In this last case, the morphism $[P, \underline{b}]$ has a particularly simple form - it is a product

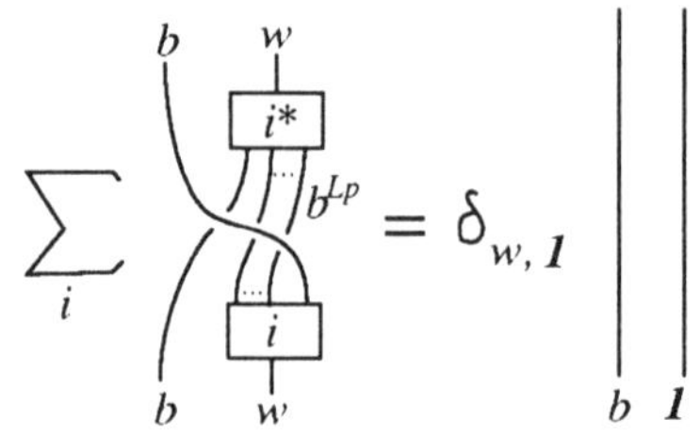

FIGURE 29. Conjecture 8.3.

of cyclic permutations:

$$[P, \underline{b}] = (1, 2, 3, \ldots, e_1) \diamond (1, 2, 3, \ldots, e_2) \diamond \cdots \diamond (1, 2, 3, \ldots, e_m),$$

where e_i is the total exponent of the relation R_i. This fact doesn't seem to help though with evaluating the invariant, at least until some more information is available for the symmetry group representations $\rho[b^k]$.

In conclusion we point out that clearly the main question is how the invariant behaves under the move (vii) in section 1, i.e. under simple homotopy equivalence. The numerical examples haven't exhibited any nontrivial behavior, but using the formalism in this paper to write down the change in the value of the invariant under (vii), brings to expressions which look quite complicated and we don't know how to analyze. We suspect that if the behavior of the computed invariant under this move is actually trivial, this is a consequence of some specific properties of the Gelfand - Kazhdan categories being used (like the ones listed in conjecture 8.1), rather then a consequence of the general setup of the theory.

References

[1] J.Andrews, M.Curtis, *Free groups and handlebodies*, Proc.AMS **16** (1965),192-195.

[2] I.Bobtcheva, *Numerical generation of semisimple tortile categories coming from quantum groups*, PhD thesis, Virginia Polytechnic Institute and State University, Blacksburg, VA, August 1996.

[3] P.Freyd and D.Yetter, *Brided compact closed categories with applications in law-dimensional topology*, Adv. in Math. **77** (1989),156-182.

[4] S.Gelfand and D.Kazhdan, *Examples of tensor categories*, Invent.Math. **109** (1992),595-617.

[5] C.Hog-Angeloni, W.Metzler and A.Sieradski, *Two-dimensional homotopy and combinatorial group theory*, Cambridge University Press, London Mathematical Society Lecture Notes Series, vol.**197**, 1993.

[6] G.Kelly and M.Laplaza, *Coherence for compact closed categories*, J.Pure and Applied Math. **19** (1980),193-213.

[7] S.MacLane, *Natural associativities and commutativities*, Rice Univ. Studies **49** (1963),28-46.

[8] F.Quinn, *Lectures on Axiomatic Topological Quantum Field Theory*, LAS/Park City Mathematical Series, vol.**1**, 1995.

[9] I.Bobtcheva and F.Quinn, *Numerical presentations of tortile categories*, internal report.

[10] F.Quinn, *Handlebodies and 2-complexes*, Lecture Notes in Mathematics **1167**, Geometry and Topology, Proceedings, University of Maryland, 1983-1984, ,245-259.

[11] Mei Chee Shum, *Tortile tensor categories*, Journal of Pure and Applied Algebra **93** Berlin Heidelberg New York (1994), 57-110.

[12] D.Yetter, *State-sum invariants of 3-manifolds associated to artinian semisimple tortile categories*, Topology and its Applications **58** (1994),47-80.

VIRGINIA POLYTECHNIC INSTITUTE AND STATE UNIVERSITY BLACKSBURG, VA, 24061
E-mail address: demeio@anvax1.cineca.it

Contemporary Mathematics
Volume **233**, 1999

The Quadratic Form of a Link

Akio Kawauchi

Abstract. We characterize the (infinite cyclic covering) quadratic form of a link as a restriction of the Seifert form (extended over the rationals) of every connected Seifert surface of the link, giving a generalization of Erle's result for a knot. The signature and nullity invariants are also discussed in comparison with those of the Seifert form. As an application, the quadratic form of a sphere-slice link (= slice link in the ordinary sense) is studied.

1. Introduction. *Throughout this paper, links will mean oriented links in the 3-sphere* S^3. The quadratic form of a knot was introduced by J. W. Milnor [18] using the Milnor duality on infinite cyclic covering. In [4], this quadratic form was naturally generalized to the case of a general link by establishing a generalization of the Milnor duality. In §2, we explain the quadratic form as well as the Seifert form associated with a Seifert surface. The quadratic form of a link is in general a singular form, and in §3 we introduce four kinds of nullity invariants together with other related invariants. In the case of knot, D. Erle [1] clarified a relationship between the quadratic form and the Seifert form. For a general link, however, this direct generalization is impossible because the signature invariants of the quadratic form and the Seifert form are in general distinct (cf. §5). In §4, we show that *the quadratic form of a general link can be characterized in terms of a restriction of the Seifert form extended over the rationals* Q, thus obtaining a generalization of Erle's result. In §5, we establish an exact relation between the signature invariants of the quadratic form and the Seifert form for a general link, and give the proof in §6. As applications, in §7 we review a signature estimation of the 4-dimensional genus of a link, more or less familiar to the experts except that we relate the signature invariants to the invariants of infinite cyclic covering (and work in the TOP category), and in §8 we show for a sphere-slice link that *the induced non-singular form of the quadratic form is null-cobordant, the signature of the canonical linking matrix is zero, and the four kinds of nullity invariants are all even.* In the final section (§9), some remaining problems are discussed.

2. The quadratic form. Let L be an $r(\geq 1)$-component link. Let E be the compact exterior $\mathrm{cl}(S^3 - N(L))$ for a regular neighborhood $N(L)$ of L in S^3. Let

---

1991 *Mathematics Subject Classification.* Primary 57M25.

Key words and phrases. Link, Quadratic form, Seifert form, Sphere-slice link.

$\gamma : \pi_1(E) \to Z$ be the epimorphism sending every (oriented) meridian of L to 1. Let $p : \tilde{E} \to E$ be the covering associated with the kernel subgroup of γ which is called the *infinite cyclic covering* of the link L. The homology groups $H_*(\tilde{E}; Z)$, $H_*(\partial\tilde{E}; Z)$, and $H_*(\tilde{E}, \partial\tilde{E}; Z)$ form finitely generated Λ-modules for the integral group ring $\Lambda = Z[Z] = Z[t, t^{-1}]$. The non-trivial part of the homology long exact sequence of the pair $(\tilde{E}, \partial\tilde{E})$ is only the following part:

$$0 \to H_2(\tilde{E}; Z) \xrightarrow{j_*} H_2(\tilde{E}, \partial\tilde{E}; Z) \xrightarrow{\partial} H_1(\partial\tilde{E}; Z) \xrightarrow{i_*} H_1(\tilde{E}; Z)$$
$$\xrightarrow{j_*} H_1(\tilde{E}, \partial\tilde{E}; Z) \xrightarrow{\partial} H_0(\partial\tilde{E}; Z) \xrightarrow{i_*} H_0(\tilde{E}; Z) \to 0.$$

Throughout this paper, we use the notation $\lambda_\varepsilon = t - \varepsilon$ for $\varepsilon = \pm 1$. Then we have the following calculation results (cf. [9]):

$$H_0(\tilde{E}; Z) \cong \Lambda/(\lambda_1), H_0(\partial\tilde{E}; Z) \cong H_1(\partial\tilde{E}; Z) \cong [\Lambda/(\lambda_1)]^r, \text{ and}$$
$$H_2(\tilde{E}; Z) \cong \Lambda^s, H_2(\tilde{E}, \partial\tilde{E}; Z) \cong \Lambda^s \oplus \Lambda/(\lambda_1) \quad \text{for some} \quad s \le r - 1.$$

More generally, we consider the infinite cyclic connected covering $p : \tilde{E}' \to E'$ of a (possibly bounded) compact connected oriented n-dimensional manifold E' associated with an indivisible element $\gamma \in H^1(E'; Z)$. Then we still have $H_{n-1}(\tilde{E}', \partial\tilde{E}'; Z) \cong \Lambda^s \oplus \Lambda/(\lambda_1)$ for some s by [9] and we can take a unique element μ' of the Λ-cyclic direct summand $\Lambda/(\lambda_1)$ so that $p_*(\mu') = \gamma \cap [E']$ in $H_{n-1}(E', \partial E'; Z)$. This element $\mu' \in H_{n-1}(\tilde{E}', \partial\tilde{E}'; Z)$ is called the *fundamental class* of the covering $p : \tilde{E}' \to E'$ (see [9,11]). For example, let F be a connected oriented Seifert surface of the link L. Then we see that any connected lift of the proper surface $F_E = F \cap E (\cong F) \subset E$ to $\tilde{E}$ represents the fundamental class $\mu \in H_2(\tilde{E}, \partial\tilde{E}; Z)$ of the covering $p : \tilde{E} \to E$. For a Λ-module H, let TH be the torsion part of H and $BH = H/TH$. Let $H_Q = H \otimes_Z Q$, forming a Λ_Q-module. We have $BH_Q = (BH)_Q = H_Q/TH_Q$ for the Λ_Q-torsion part $TH_Q = (TH)_Q$ of the Λ_Q-module H_Q, a natural Λ_Q-isomorphism $\mathrm{Hom}_Z[H, Q] \cong \mathrm{Hom}_Q[H_Q, Q]$, and a natural short exact sequence

$$0 \to \mathrm{Hom}_Z[BH, Q] \to \mathrm{Hom}_Z[H, Q] \to \mathrm{Hom}_Z[TH, Q] \to 0.$$

If H is a finitely generated Λ-module, then TH and BH are finitely generated over Λ, so that the Λ_Q-module TH_Q splits into finitely many cyclic torsion Λ_Q-modules and hence is (non-canonically) Λ_Q-isomorphic to $\mathrm{Hom}_Z[TH, Q]$, and the Λ_Q-module BH_Q is free of finite rank. The homology $H_*(\tilde{E}', \tilde{E}''; Z)$ for the preimage $\tilde{E}'' = p^{-1}(E'')$ of a compact submanifold $E'' \subset E'$ forms a finitely generated Λ-module and the Q-homology $H_*(\tilde{E}', \tilde{E}''; Q) = H_*(\tilde{E}', \tilde{E}''; Z)_Q$ forms a finitely generated Λ_Q-module. We use the following notations:

$$T^*(\tilde{E}', \tilde{E}''; Q) = \mathrm{Hom}_Z[TH_*(\tilde{E}', \tilde{E}''; Z), Q] = \mathrm{Hom}_Q[TH_*(\tilde{E}', \tilde{E}''; Q), Q],$$
$$B^*(\tilde{E}', \tilde{E}''; Q) = \mathrm{Hom}_Z[BH_*(\tilde{E}', \tilde{E}''; Z), Q] = \mathrm{Hom}_Q[BH_*(\tilde{E}', \tilde{E}''; Q), Q].$$

Since the Q-cohomology $H^*(\tilde{E}', \tilde{E}''; Q)$ is identified with $\mathrm{Hom}_Z[H_*(\tilde{E}', \tilde{E}''; Z), Q]$, we obtain the following short exact sequence:

$$0 \to B^*(\tilde{E}', \tilde{E}''; Q) \to H^*(\tilde{E}', \tilde{E}''; Q) \to T^*(\tilde{E}', \tilde{E}''; Q) \to 0.$$

The following duality theorem is given in [4] (see also [7]), which generalizes the Milnor duality in [18]:

DUALITY THEOREM 2.1. *(1) The cup product pairing*

$$H^*(\tilde{E}';Q) \times H^{n-1-*}(\tilde{E}',\partial\tilde{E}';Q) \xrightarrow{\cup} H^{n-1}(\tilde{E}',\partial\tilde{E}';Q)$$

induces a t-isometric non-singular pairing

$$T^*(\tilde{E}';Q) \times T^{n-1-*}(\tilde{E}',\partial\tilde{E}';Q) \xrightarrow{\cup} T^{n-1}(\tilde{E}',\partial\tilde{E}';Q).$$

(2) The cap product homomorphisms

$$\cap\mu : H^*(\tilde{E}',\partial\tilde{E}';Q) \to H_{n-1-*}(\tilde{E}';Q) \quad \text{and}$$
$$\cap\mu : H^*(\tilde{E}';Q) \to H_{n-1-*}(\tilde{E}',\partial\tilde{E}';Q)$$

induce t-anti Q-isomorphisms

$$\cap\mu : T^*(\tilde{E}',\partial\tilde{E}';Q) \cong TH_{n-1-*}(\tilde{E}';Q) \quad \text{and}$$
$$\cap\mu : T^*(\tilde{E}';Q) \cong TH_{n-1-*}(\tilde{E}',\partial\tilde{E}';Q).$$

When $\partial E' \neq \emptyset$, we note that the inclusion $j : \tilde{E}' \subset (\tilde{E}',\partial\tilde{E}')$ induces the following commutative square:

$$\begin{array}{ccc} H^*(\tilde{E}',\partial\tilde{E}';Q) & \longrightarrow & T^*(\tilde{E}',\partial\tilde{E}';Q) \\ {\scriptstyle j^*}\downarrow & & {\scriptstyle j^*}\downarrow \\ H^*(\tilde{E}';Q) & \longrightarrow & T^*(\tilde{E}';Q). \end{array}$$

Hence from the naturality of cup product and Duality Theorem 2.1(1), we see that the cup product pairing

$$H^*(\tilde{E}',\partial\tilde{E}';Q) \times H^{n-1-*}(\tilde{E}',\partial\tilde{E}';Q) \xrightarrow{\cup} H^{n-1}(\tilde{E}',\partial\tilde{E}';Q)$$

induces a pairing

$$T^*(\tilde{E}',\partial\tilde{E}';Q) \times T^{n-1-*}(\tilde{E}',\partial\tilde{E}';Q) \xrightarrow{\cup} T^{n-1}(\tilde{E}',\partial\tilde{E}';Q).$$

By a *quadratic form* on a finitely generated torsion Λ_Q-module T, we mean a symmetric t-isometric (possibly singular) Q-bilinear form $b : T \times T \to Q$.

The following definition of the quadratic form of a link in [4] is a natural generalization of the quadratic form of a knot by J. W. Milnor [18]:

DEFINITION 2.2. The *quadratic form* of a link L is the pairing

$$b_L : T^1(\tilde{E},\partial\tilde{E};Q) \times T^1(\tilde{E},\partial\tilde{E};Q) \to Q$$

defined by $b(x,y) = < x \cup (t - t^{-1})y, \mu >$ for all $x, y \in T^1(\tilde{E},\partial\tilde{E};Q)$.

The symmetry $b(x,y) = b(y,x)$ and the t-isometry $b(tx,ty) = b(x,y)$ for all $x, y \in T^1(\tilde{E},\partial\tilde{E};Q)$ are easily checked. This quadratic form b_L is in general a singular form, whose nullity is also an interesting invariant discussed in the next section. For a connected Seifert surface F of a link L, the *Seifert form (associated with F)* is a bilinear form

$$\phi_F : H_1(F;Z) \times H_1(F;Z) \to Z$$

defined by the identity $\phi_F(x,y) = \mathrm{Link}(c_x^+, c_y)$ for $x = \{c_x\}$ and $y = \{c_y\}$ in $H_1(F;Z)$, where c_x^+ denotes a cycle in S^3 obtained by translating the cycle c_x off in the positive normal direction and Link denotes the linking number in S^3. The matrix representing ϕ_F under a Z-basis for $H_1(F;Z)$ is called a *Seifert matrix (associated with F)*. The S-equivalence class of a Seifert matrix is well-known to be a link invariant (see F. Trotter[24] for a knot and [12] for a link). Since the Seifert form ϕ_F is not symmetric for every F with positive genus, we often consider the symmetric bilinear form

$$b_F : H_1(F;Z) \times H_1(F;Z) \to Z$$

defined by the identity $b_F(x,y) = \phi_F(x,y) + \phi_F(y,x)$ for $x,y \in H_1(F;Z)$. We call this form b_F the *Seifert quadratic form* (on F). The link invariance of the S-equivalence class of a Seifert matrix implies that the Seifert quadratic form b_F is an invariant of the link L up to orthogonal additions of hyperbolic plane forms and Z-isomorphisms, although b_F itself is not a link invariant. The quadratic form b_D of a link diagram D of L discussed by R. H. Kyle [15] is related to this Seifert quadratic form b_F in the associated torsion linking forms. Our result of this paper will mean that the quadratic form b_L is Q-isomorphic to a restriction of the Seifert quadratic form b_F extended over Q for every connected Seifert surface F of L.

3. The null space of the quadratic form. We denote by $T_N(L;Q)$ the null space of the quadratic form b_L which is a Λ_Q-submodule of $T^1(\tilde{E},\partial\tilde{E};Q)$, and $n(L) = \dim_Q T_N(L;Q)$ which is called the *nullity of b_L*. We denote the induced non-singular form of b_L by

$$\bar{b}_L : \bar{T}^1(\tilde{E},\partial\tilde{E};Q) \times \bar{T}^1(\tilde{E},\partial\tilde{E};Q) \to Q$$

with $\bar{T}^1(\tilde{E},\partial\tilde{E};Q) = T^1(\tilde{E},\partial\tilde{E};Q)/T_N(L;Q)$. Let $T_\delta(L;Q)$ be the image of the (coboundary) homomorphism $\delta : T^0(\partial\tilde{E};Q) \to T^1(\tilde{E},\partial\tilde{E};Q)$ which is Q-dual to the boundary operator $\partial : TH_1(\tilde{E},\partial\tilde{E};Z) \to TH_0(\partial\tilde{E};Z)(= H_0(\partial\tilde{E};Z))$. Let $\delta(L) = \dim_Q T_\delta(L;Q)$. Since the sequence

$$T^0(\partial E;Q) \xrightarrow{\delta} T^1(\tilde{E},\partial\tilde{E};Q) \xrightarrow{j^*} T^1(\tilde{E};Q)$$

is exact, we see from Duality Theorem 2.2 that the cup product $\cup : T^1(\tilde{E},\partial\tilde{E};Q) \times T^1(\tilde{E},\partial\tilde{E};Q) \to T^2(\tilde{E},\partial\tilde{E};Q)$ induces a non-singular pairing (denoted by the same symbol)

$$\cup : \hat{T}^1(\tilde{E},\partial\tilde{E};Q) \times \hat{T}^1(\tilde{E},\partial\tilde{E};Q) \to T^2(\tilde{E},\partial\tilde{E};Q)$$

with $\hat{T}^1(\tilde{E},\partial\tilde{E};Q) = T^1(\tilde{E},\partial\tilde{E};Q)/T_\delta(L;Q)$. We also note that the duality isomorphism $\cap\mu : T^1(\tilde{E},\partial\tilde{E};Q) \cong TH_1(\tilde{E};Q)$ induces an isomorphism

$$\cap\mu : \hat{T}^1(\tilde{E},\partial\tilde{E};Q) \cong \hat{T}H_1(\tilde{E};Q)$$

for the cokernel $\hat{T}H_1(\tilde{E};Q)$ of the natural homomorphism $i_* : TH_1(\partial\tilde{E};Q) \to TH_1(\tilde{E};Q)$. The induced quadratic form of b_L on $\hat{T}^1(\tilde{E},\partial\tilde{E};Q)$ is denoted by

$$\hat{b}_L : \hat{T}^1(\tilde{E},\partial\tilde{E};Q) \times \hat{T}^1(\tilde{E},\partial\tilde{E};Q) \to Q.$$

The form $\hat{b}_L$ has a natural geometric interpretation. To see it, let M be the Dehn surgery manifold of S^3 along the link L with Seifert surface framing. The proper surface $F_E = F \cap E(\cong F) \subset E$ naturally extends to a closed connected surface F_M in M. Let $U = \mathrm{cl}(M - E)$ which is the disjoint union of r solid tori. The infinite cyclic covering $\tilde{E} \to E$ extends to a covering $(\tilde{M},\tilde{U}) \to (M,U)$. Then we define the quadratic form

$$b_M : T^1(\tilde{M};Q) \times T^1(\tilde{M};Q) \to Q$$

in a similar way to the definition of b_L, namely by $b_M(x,y) = < x \cup (t-t^{-1})y, \mu_M >$ for $x, y \in T^1(\tilde{M}; Q)$, where μ_M denotes the fundamental class of the covering $\tilde{M} \to M$ represented by a connected lift of F_M to $\tilde{M}$. Then we have the following lemma:

LEMMA 3.1. *The induced quadratic form* $\hat{b}_L : \hat{T}^1(\tilde{E}, \partial\tilde{E}; Q) \times \hat{T}^1(\tilde{E}, \partial\tilde{E}; Q) \to Q$ *is* Λ_Q*-isomorphic to the quadratic form* $b_M : T^1(\tilde{M}; Q) \times T^1(\tilde{M}; Q) \to Q$.

PROOF. Since there is an excision isomorphism $H_*(\tilde{E}, \partial\tilde{E}; Z) \cong H_*(\tilde{M}, \tilde{U}; Z)$, the quadratic form b_L is Λ_Q-isomorphic to the quadratic form

$$b_{M,U} : T^1(\tilde{M}, \tilde{U}; Q) \times T^1(\tilde{M}, \tilde{U}; Q) \to Q$$

defined similarly with the class $\mu_{M,U} \in H_2(\tilde{M}, \tilde{U}; Z)$ corresponding to μ by the excision isomorphism. Since each connected component of $\tilde{U}$ is contractible, we have a natural isomorphism $H_2(\tilde{M}; Z) \to H_2(\tilde{M}, \tilde{U}; Z)$ sending μ_M to $\mu_{M,U}$ and an exact sequence $H^0(\tilde{U}; Q) \xrightarrow{\delta} H^1(\tilde{M}, \tilde{U}; Q) \xrightarrow{j^*} H^1(\tilde{M}; Q) \to 0$, which induces an exact sequence

$$T^0(\tilde{U}; Q) \xrightarrow{\delta} T^1(\tilde{M}, \tilde{U}; Q) \xrightarrow{j^*} T^1(\tilde{M}; Q) \to 0.$$

Let $T_\delta(\tilde{M}, \tilde{U}; Q)$ be the image of $\delta : T^0(\tilde{U}; Q) \to T^1(\tilde{M}, \tilde{U}; Q)$, which is Λ_Q-isomorphic to $T_\delta(L; Q)$ under the isomorphism $T^1(\tilde{M}, \tilde{U}; Q) \cong T^1(\tilde{E}, \partial\tilde{E}; Q)$. Then the quadratic form $\hat{b}_{M,U}$ on $\hat{T}^1(\tilde{M}, \tilde{U}; Q) = T^1(\tilde{M}, \tilde{U}; Q)/T_\delta(\tilde{M}, \tilde{U}; Q)$ induced from $b_{M,U}$ is Λ_Q-isomorphic to $\hat{b}_L$. Using further a natural isomorphism $j^* : B^1(\tilde{M}, \tilde{U}; Q) \to B^1(\tilde{M}; Q)$ and the naturality of cup product, we see that $\hat{b}_{M,U}$ is Λ_Q-isomorphic to the quadratic form $b_M : T^1(\tilde{M}; Q) \times T^1(\tilde{M}; Q) \to Q$. □

By a dual argument to the proof of Lemma 3.1, we have a natural exact sequence

$$TH_1(\partial\tilde{E}; Q) \xrightarrow{i_*} TH_1(\tilde{E}; Q) \to TH_1(\tilde{M}; Q) \to 0,$$

so that $\hat{T}H_1(\tilde{E}; Q) \cong TH_1(\tilde{M}; Q)$. For a Laurent polynomial $f \in \Lambda_Q$, we denote the Λ_Q-submodule $\{x \in H | fx = 0\}$ of a Λ_Q-module H by $\mathrm{Ker}(f; H)$. We note that $\mathrm{Ker}(\lambda_\varepsilon; \hat{T}^1(\tilde{E}, \partial\tilde{E}; Q))$ for $\varepsilon = \pm 1$ is a null space of the quadratic form $\hat{b}_L$. For $\varepsilon = \pm 1$, we shall use the notation $H^{(\varepsilon)} = H/\mathrm{Ker}(\lambda_\varepsilon; H)$. Then we note that $\hat{b}_L$ induces a quadratic form

$$\hat{b}_L^{(1)} : \hat{T}^1(\tilde{E}, \partial\tilde{E}; Q)^{(1)} \times \hat{T}^1(\tilde{E}, \partial\tilde{E}; Q)^{(1)} \to Q.$$

We shall characterize in §4 this quadratic form $\hat{b}_L^{(1)}$ in terms of a restriction of the Seifert form ϕ_F associated with every connected Seifert surface F of the link L. Let

$$\hat{\kappa}_\varepsilon(L) = \dim_Q \mathrm{Ker}(\lambda_\varepsilon; \hat{T}^1(\tilde{E}, \partial\tilde{E}; Q)) = \dim_Q \mathrm{Ker}(\lambda_\varepsilon; T^1(\tilde{M}; Q)).$$

It is convenient to introduce further numerical link invariants in order to understand these nullity invariants. Let $\kappa_\varepsilon(L) = \dim_Q \mathrm{Ker}(\lambda_\varepsilon; TH_1(\tilde{E}; Q))$. Let $\beta(L) = \mathrm{rank}_{\Lambda_Q} BH_1(\tilde{E}; Q)$. Let $n_S(L)$ be the nullity of the Seifert quadratic form $b_F : H_1(F; Z) \times H_1(F; Z) \to Z$. Let V be the *canonical linking matrix* (i.e., the linking matrix associated with the Seifert surface framing) of an ordered link $L = \cup_{i=1}^r L_i$, meaning a symmetric matrix such that the (i,j)-entry $v_{ij} = \mathrm{Link}(L_i, L_j)$ for $i \neq j$ and the (i,i)-entry v_{ii} is determined by the identity $\sum_{j=1}^r v_{ij} = 0$. Let $\mathrm{r}_V(L)$ be the rank of V. Then we have the following lemma:

LEMMA 3.2. *For an r-component link L, we have the following identities:*
1. $n(L) = \delta(L) + \hat{\kappa}_1(L) + \hat{\kappa}_{-1}(L)$.
2. $\delta(L) = \hat{\kappa}_1(L) + \mathrm{r}_V(L) = \kappa_1(L)$, $\hat{\kappa}_{-1}(L) = \kappa_{-1}(L)$.
3. $\beta(L) + \kappa_1(L) = r - 1$.
4. $\beta(L) + \kappa_{-1}(L) = n_S(L)$.

PROOF. 1 is direct from the definitions of b_L and b_M. Before showing 2, we show 3. The Q-vector space $H_1(\tilde{E};Q)/\lambda_1 H_1(\tilde{E};Q)$ has Q-dimension $r-1$ by the Wang exact sequence and also has Q-dimension $\beta(L) + \kappa_1(L)$ by taking a Λ_Q-cyclic splitting of $H_1(\tilde{E};Q)$, from which 3 is obtained. For 2, we note that $\dim_Q H_1(M;Q) = r - \mathrm{r}_V(L)$. By a similar way to 3, we have $\hat{\beta}(L) + \hat{\kappa}_1(L) = (r - \mathrm{r}_V(L)) - 1$. By 3 and a Λ_Q-isomorphism $BH_1(\tilde{E};Q) \cong BH_1(\tilde{M};Q)$, we have $\kappa_1(L) = \hat{\kappa}_1(L) + \mathrm{r}_V(L)$. We show that $\delta(L) = \kappa_1(L)$. First, we note that $\delta(L)$ is the Q-dimension of the image of the boundary homomorphism $\partial : TH_1(\tilde{E},\partial\tilde{E};Q) \to \tilde{H}_0(\partial\tilde{E};Q) \cong [\Lambda_Q/\lambda_1]^{r-1}$ which we denote by $T\partial(L;Q)$. The homology exact sequence of the pair $(\tilde{E},\partial\tilde{E})$ induces a Λ_Q-exact sequence

$$0 \to BH_1(\tilde{E};Q) \xrightarrow{j_*} BH_1(\tilde{E},\partial\tilde{E};Q) \to \tilde{H}_0(\partial\tilde{E};Q)/T\partial(L;Q) \to 0.$$

Let B be the image of the monomorphism j_* in this sequence. We show that $B = \lambda_1 BH_1(\tilde{E},\partial\tilde{E};Q)$. Then $\delta(L) = \kappa_1(L)$ is obtained from 1), since $BH_1(\tilde{E},\partial\tilde{E};Q)$ is Λ_Q-free of rank $\beta(L)$ and $\tilde{H}_0(\partial E;Q)/T\partial(L;Q) \cong [\Lambda_Q/(\lambda_1)]^{r-1-\delta(L)}$. Since $\lambda_1(\tilde{H}_0(\partial E;Q)/T\partial(L;Q)) = 0$, we have $B \supset \lambda_1 BH_1(\tilde{E},\partial\tilde{E};Q)$. We consider the non-singular Λ_Q-intersection form

$$\mathrm{Int}_{\Lambda_Q} : BH_1(\tilde{E},\partial\tilde{E};Q) \times BH_2(\tilde{E};Q) \to \Lambda_Q.$$

Since E is in S^3, the usual intersection form $H_1(E;Q) \times H_2(E;Q) \to Q$ is the zero-form which implies that $\mathrm{Int}_{\Lambda_Q}(j_*(x),y)(1) = 0$ for all $x \in BH_1(\tilde{E};Q)$ and all $y \in BH_2(\tilde{E};Q)$. By the non-singularity of Int_{Λ_Q}, we have $B \subset \lambda_1 BH_1(\tilde{E},\partial\tilde{E};Q)$, and $B = \lambda_1 BH_1(\tilde{E},\partial\tilde{E};Q)$, showing $\delta(L) = \kappa_1(L)$. Since $H_1(\tilde{E};Q)$ and $H_1(\tilde{M};Q)$ are Λ_Q-isomorphic except the λ_1-primary components, $\hat{\kappa}_{-1}(L) = \kappa_{-1}(L)$ is obtained, showing 2. The Q-vector space $H_1(\tilde{E};Q)/\lambda_{-1}H_1(\tilde{E};Q)$ is seen to have Q-dimension $\beta(L)+\kappa_{-1}(L)$ by taking a Λ_Q-cyclic splitting of $H_1(\tilde{E};Q)$. Since the matrix $tS - S'$ for any Seifert matrix S of L is a Λ_Q-presentation matrix for $H_1(\tilde{E};Q)$, the matrix $-(S+S')$ is a Q-presentation matrix for $H_1(\tilde{E};Q)/\lambda_{-1}H_1(\tilde{E};Q)$, obtaining 4. □

We consider two kinds of Laurent polynomial invariants of a link L, namely, the zeroth and $\beta(L)$-th characteristic polynomials $A_L(t)$ and $A_L^T(t)$ of the Λ-module $H_1(\tilde{E};Z)$ which are respectively called the *Alexander polynomial* and the *torsion Alexander polynomial* of L. It is well-known that $A_L(t)$ and $A_L^T(t)$ are invariants of L up to units of Λ, and $A_L^T(t)$ is the zeroth characteristic polynomial of $TH_1(\tilde{E};Z)$. In particular, we have always $A_L^T(t) \neq 0$. When L has r components, we can see directly that $0 \le \beta(L) \le r-1$ and $A_L(t) = 0 \Leftrightarrow \beta(L) > 0$. (By definition, if $\beta(L) = 0$, then $A_L(t)$ and $A_L^T(t)$ are identical.) Let Λ^+ be the subring of the quotient field $Q(\Lambda)$ of Λ such that $\Lambda \subset \Lambda^+$ and the set of units of Λ^+ is equal to the set of Laurent polynomials $f \in \Lambda$ with $f(1) = \pm 1$. For a Λ-module H, we denote $H \otimes_\Lambda \Lambda^+$ by H^+. The following lemma is observed in [5]:

LEMMA 3.3. *The Λ^+-module $H_1(\tilde{E};Z)^+$ is an invariant of link concordance (= link cobordism) of L up to Λ^+-isomorphisms. In particular, $\beta(L)$ and the torsion Alexander polynomial $A_L^T(t)$ are link concordance invariants of L up to unit multiplications of Λ^+.*

The following corollary is direct from Lemmas 3.2 and 3.3:

COROLLARY 3.4. *The nullity invariants $n(L)$, $\delta(L)$, $\hat{\kappa}_1(L)$, $\hat{\kappa}_{-1}(L)$, and $n_S(L)$ are all link concordance invariants.*

When L has r components, it is also known that $A_L^T(1) = 0 \Leftrightarrow \beta(L) < r-1$ and $A_L^T(1) = \pm 1 \Leftrightarrow \beta(L) = r-1$ (cf. [5]). Hence we see from Lemma 3.2 the following corollary:

COROLLARY 3.5. *For every r-component link L, the quadratic form b_L is non-singular if and only if $\beta(L) = r-1$.*

For example, if L is concordant to a homology boundary link, then the quadratic form b_L is non-singular.

4. Characterizing the quadratic form in terms of a restriction of Seifert form. For simplicity, we identify the Seifert surface F with the proper surface F_E in E in considering the Seifert form $\phi_F : H_1(F;Z) \times H_1(F;Z) \to Z$. We consider the kernel $K(F;Q)$ of the natural homomorphism $I_* : H_1(F;Q) \to H_1(E;Q)$. In terms of a lifting homomorphism $\tilde{I}_* : H_1(F;Q) \to H_1(\tilde{E};Q)$, we have $K(F;Q) = \tilde{I}_*^{-1}(\lambda_1 H_1(\tilde{E};Q))$ by the Wang exact sequence. Let

$$K^T(F;Q) = \tilde{I}_*^{-1}(\lambda_1 TH_1(\tilde{E};Q)) \quad \text{and} \quad K^0(F;Q) = \mathrm{Ker}\tilde{I}_*.$$

Similarly, let $K(F_M;Q)$ be the kernel of the homomorphism $(I_M)_* : H_1(F_M;Q) \to H_1(M;Q)$, namely $K(F_M;Q) = (\tilde{I}_M)_*^{-1}(\lambda_1 H_1(\tilde{M};Q))$. Let

$$K^T(F_M;Q) = (\tilde{I}_M)_*^{-1}(\lambda_1 TH_1(\tilde{M};Q)) \quad \text{and} \quad K^0(F_M;Q) = \mathrm{Ker}(\tilde{I}_M)_*.$$

Let $K^T(\partial F;Q)$ be the Q-submodule of $K^T(F;Q)$ consisting of an element represented by a rational 1-cycle in ∂F.

LEMMA 4.1. *The natural homomorphism $k_*^T : K^T(F;Q) \to K^T(F_M;Q)$ is onto and* $\mathrm{Ker}k_*^T = K^T(\partial F;Q)$ *and* $(k_*^T)^{-1}(K^0(F_M;Q)) = K^0(F;Q) + K^T(\partial F;Q)$.

PROOF. We consider the following commutative diagram with natural horizontal sequences exact:

$$\begin{array}{ccccccc}
H_2(F_M,F;Q) & \xrightarrow{\partial} & H_1(F;Q) & \xrightarrow{k_*} & H_1(F_M;Q) & \longrightarrow & 0 \\
\downarrow{\scriptstyle (\tilde{I}_U)_*} & & \downarrow{\scriptstyle \tilde{I}_*} & & \downarrow{\scriptstyle (\tilde{I}_M)_*} & & \\
H_2(\tilde{M},\tilde{E};Q) & \xrightarrow{\tilde{\partial}} & H_1(\tilde{E};Q) & \xrightarrow{\tilde{k}_*} & H_1(\tilde{M};Q) & \longrightarrow & 0.
\end{array}$$

For any element $x \in K^T(F_M;Q)$, we have an element $x' \in H_1(F;Q)$ and an element $y \in TH_1(\tilde{M};Q)$ such that $k_*(x') = x$ and $(\tilde{I}_M)_*(x) = \lambda_1 y$. Since $H_2(\tilde{M},\tilde{E};Q) \cong H_*(\tilde{U},\partial\tilde{U};Q)$ is a torsion Λ_Q-module, we have an element $y' \in TH_1(\tilde{E};Q)$ with $\tilde{k}_*(y') = y$. Then $\tilde{k}_*(\tilde{I}_*(x') - \lambda_1 y') = 0$. Since $(I_U)_*$ is a Q-isomorphism, we have an element $w \in H_2(F_M,F;Q)$ such that $\tilde{I}_*(\partial(w)) = \tilde{I}_*(x') - \lambda_1 y'$. Then $\tilde{I}_*(x' - \partial(w)) =$

$\lambda_1 y'$ and $k_*^T(x' - \partial(w)) = k_*(x') = x$. Hence $k_*^T(K^T(F;Q)) = K^T(F_M;Q)$. By the exact sequence of the pair (F_M, F), we have $\mathrm{Ker} k_* = K^T(\partial F;Q) = \mathrm{Ker} k_*^T$. Since $\tilde{k}_* : H_1(\tilde{E};Q) \to H_1(\tilde{M};Q)$ induces a Λ_Q-epimorphism $\lambda_1 TH_1(\tilde{E};Q) \to \lambda_1 TH_1(\tilde{M};Q)$ whose kernel is generated by some rational 1-cycles in ∂F, we have $(k_*^T)^{-1}(K^0(F_M;Q)) = K^0(F;Q) + K^T(\partial F;Q)$. □

LEMMA 4.2. *There is an orthogonal splitting*

$$K^T(F;Q) = S(F;Q) \perp (K^0(F;Q) + K^T(\partial F;Q))$$

with respect to the Q-intersection form Int_F *on* $H_1(F;Q)$.

PROOF. By [7], we have a Q-monomorphism $\pi : TH_1(\tilde{M};Q) \to H_1(F_M;Q)$ such that the composite

$$TH_1(\tilde{M};Q) \xrightarrow{\pi} H_1(F_M;Q) \xrightarrow{(\tilde{I}_M)_*} H_1(\tilde{M};Q)$$

is the inclusion map $TH_1(\tilde{M};Q) \subset H_1(\tilde{M};Q)$. Let $TH_1(F_M;Q)$ be the image of $TH_1(\tilde{M};Q)$ under π, and $H_1^T(F_M;Q)$ be the preimage of $TH_1(\tilde{M};Q)$ under $(\tilde{I}_M)_*$. We have

$$H_1^T(F;Q) = TH_1(F_M;Q) \oplus K^0(F_M;Q).$$

We can regard $TH_1(F_M;Q)$ as a Λ_Q-module by the identity $\pi(tx) = t\pi(x)$ for all $x \in TH_1(\tilde{M};Q)$. Then we have Λ_Q-isomorphisms $\pi : TH_1(\tilde{M};Q) \cong TH_1(F_M;Q)$ and $(\tilde{I}_M)_* : TH_1(F_M;Q) \cong TH_1(\tilde{M};Q)$ which are mutually the inverse maps. In particular,

$$K^T(F_M;Q) = \lambda_1 TH_1(F_M;Q) \oplus K^0(F_M;Q).$$

From [7;Lemma 2.1], we see that $TH_1(F_M;Q) \perp K^0(F_M;Q)$ with respect to Int_{F_M}. By Lemma 4.1, we have a Q-isomorphism $K^T(F;Q)/\mathrm{Ker} k_*^T \cong K^T(F_M;Q)$ induced from k_*^T. Since $\mathrm{Ker} k_*^T = K^T(\partial F;Q)$, $\mathrm{Ker} k_*^T$ is a null space on Int_F. Further, using $(k_*^T)^{-1}(K^0(F_M;Q)) = K^0(F;Q) + K^T(\partial F;Q)$, we can find an orthogonal splitting $S(F;Q) \perp (K^0(F;Q) + K^T(\partial F;Q))$ of $K^T(F;Q)$ with $k_*^T(S(F;Q)) = \lambda_1 TH_1(F_M;Q)$. □

In Lemma 4.2, we note that the orthogonal summand $S(F;Q)$ does not naturally determined from $K^T(F;Q)$ in general, although $\dim_Q S(F;Q)$ is uniquely determined. Let

$$\phi_{S(F;Q)} = \phi_F|_{S(F;Q) \times S(F;Q)} : S(F;Q) \times S(F;Q) \to Q$$

be the restricted Seifert form of the Seifert form ϕ_F extend over Q. The following lemma is important to our argument:

LEMMA 4.3. *Let* $K^T(F;Q) = S(F;Q) \perp (K^0(F;Q) + K^T(\partial F;Q))$ *be any orthogonal splitting with respect to* Int_F. *Then the restricted Seifert form* $\phi_{S(F;Q)}$: $S(F;Q) \times S(F;Q) \to Q$ *is a non-singular bilinear form and a link invariant of* L *up to Q-isomorphisms.*

PROOF. By Lemma 4.1, the Q-isomorphism $K^T(F;Q)/\mathrm{Ker}k_*^T \cong K^T(F_M;Q)$ gives an orthogonal splitting $K^T(F_M;Q) = S(F_M;Q) \perp K^0(F_M;Q)$ with respect to Int_{F_M}. We write $K^0(F_M;Q) = K^0(F_M;Q)' \oplus O(F_M;Q)$ where $O(F_M;Q)$ is the null space and $K^0(F_M;Q)'$ is a non-singular part with respect to the restriction of Int_{F_M} to $K^0(F_M;Q)$. From [7;Lemmas 3.1, 3.2], we see that the orthogonal splitting

$$K^T(F_M;Q) = \lambda_1 TH_1(F_M;Q) \oplus K^0(F_M;Q)' \oplus O(F_M;Q)$$

with respect to Int_{F_M} also gives an orthogonal splitting with $O(F_M;Q)$ a null space with respect to the Seifert form

$$\phi_{F_M} : K(F_M;Q) \times K(F_M;Q) \to Q$$

associated with the Q-linking number in M. Since $S(F_M;Q)$ and $\lambda_1 TH_1(F_M;Q)$ coincide in $K^T(F_M;Q)/O(F_M;Q)$ and $\dim_Q S(F_M;Q) = \dim_Q \lambda_1 TH_1(F_M;Q)$, we see that the restriction of the Seifert form ϕ_{F_M} to $\lambda_1 TH_1(F_M;Q)$ is Q-isomorphic to the restriction of the Seifert form ϕ_{F_M} to $S(F_M;Q)$, and hence Q-isomorphic to the restriction of ϕ_F to $S(F;Q)$, because $\mathrm{Ker}k_*^T = K^T(\partial F;Q)$ is seen to be a null space of the restriction of ϕ_F to $K^T(F;Q)$ (by considering ϕ_{F_M} in M).

Let $\lambda_1 x$ and $\lambda_1 y$ be any elements in $\lambda_1 TH_1(F_M;Q)$. Then we see from [7;3.3] that

$$\phi_{F_M}(\lambda_1 x, \lambda_1 y) = -\mathrm{Int}_{F_M}(x, \lambda_1 y).$$

By [7;Lemma 1.5], the restriction of Int_{F_M} to $TH_1(F_M;Q)$ is a non-singular skew-symmetric t-isometric form which is Λ_Q-isomorphic to the form

$$* : TH_1(\tilde{M};Q) \times TH_1(\tilde{M};Q) \to Q$$

defined by using Duality Theorem 2.1 as the identity $x * y = < u_x \cup u_y, \mu_M >$ for u_x and u_y in $T^1(\tilde{M};Q)$ with $x = u_x \cap \mu_M$ and $y = u_y \cap \mu_M$. Thus, by the Λ_Q-isomorphisms $(\tilde{I}_M)_* : TH_1(F_M;Q) \cong TH_1(\tilde{M};Q)$ and $\lambda_1 : TH_1(\tilde{M};Q)^{(1)} \cong \lambda_1 TH_1(\tilde{M};Q)$, the restriction of the Seifert form ϕ_{F_M} to $\lambda_1 TH_1(F_M;Q)$ is Λ_Q-isomorphic to the non-singular form

$$*' : \lambda_1 TH_1(\tilde{M};Q) \times \lambda_1 TH_1(\tilde{M};Q) \to Q,$$

which is well defined by the identity $\lambda_1 x *' \lambda_1 y = -x * \lambda_1 y$. This last form is uniquely determined from the link L itself. □

Since the non-singular form $*' : \lambda_1 TH_1(\tilde{M};Q) \times \lambda_1 TH_1(\tilde{M};Q) \to Q$ is t-isometric, the proof of Lemma 4.3 also shows the existence of "t-isometric Seifert form". When L is a knot, every Seifert matrix S is S-equivalent to a non-degenerate matrix S_* (possibly empty) by H. F. Trotter [24]. Then we can see that the matrix S_* represents a restricted Seifert form $\phi_{S(F;Q)} : S(F;Q) \times S(F;Q) \to Q$, implying that the following theorem is a generalization of D. Erle's result:

THEOREM 4.4. *For every orthogonal splitting*

$$K^T(F;Q) = S(F;Q) \perp (K^0(F;Q) + K^T(\partial F;Q))$$

with respect to Int_F, *let* S_T *be a matrix representing the restricted Seifert form*

$$\phi_{S(F;Q)} : S(F;Q) \times S(F;Q) \to Q$$

under a Q-basis of $S(F;Q)$. Then the quadratic form

$$\hat{b}_L^{(1)} : \hat{T}^1(\tilde{E}, \partial\tilde{E}; Q)^{(1)} \times \hat{T}^1(\tilde{E}, \partial\tilde{E}; Q)^{(1)} \to Q$$

and the isometry

$$t : \hat{T}^1(\tilde{E}, \partial\tilde{E}; Q)^{(1)} \to \hat{T}^1(\tilde{E}, \partial\tilde{E}; Q)^{(1)}$$

are represented by $S_T + S'_T$ and $(S_T)^{-1}S'_T$ under the Q-basis of $\hat{T}^1(\tilde{E}, \partial\tilde{E}; Q)^{(1)}$ specified from the given Q-basis of $S(F;Q)$ by the composite Q-isomorphism

$$\hat{T}^1(\tilde{E}, \partial\tilde{E}; Q)^{(1)} \overset{(\cap\mu)}{\cong} \hat{T}H_1(\tilde{E}; Q)^{(1)} \overset{\lambda_1}{\cong} \lambda_1 \hat{T}H_1(\tilde{E}; Q) \overset{(\tilde{I}_*)^{-1}}{\cong} S(F;Q).$$

PROOF. Since the duality isomorphism $\cap\mu_M : T^1(\tilde{M};Q) \cong T_1(\tilde{M};Q)$ is a t-anti Q-isomorphism, we see from Lemma 3.1 that the quadratic form $\hat{b}_L^{(1)}$ is t-anti Q-isomorphic to the quadratic form

$$b_{F_M}^{(1)} : TH_1(F_M;Q)^{(1)} \times TH_1(F_M;Q)^{(1)} \to Q$$

defined by $b_{F_M}^{(1)}(x,y) = \mathrm{Int}_{F_M}(x, (t^{-1} - t)y)$. Then by the proof of Lemma 4.3 we have

$$\begin{aligned} b_{F_M}^{(1)}(x,y) &= \mathrm{Int}_{F_M}(x, (t^{-1} - 1)y) + \mathrm{Int}_{F_M}(x, (1-t)y) \\ &= \mathrm{Int}_{F_M}(y, (1-t)x) + \mathrm{Int}_{F_M}(x, (1-t)y) \\ &= \phi_{F_M}((1-t)y, (1-t)x) + \phi_{F_M}((1-t)x, (1-t)y). \quad (4.4.1) \end{aligned}$$

Thus, the quadratic form $b_{F_M}^{(1)}$ is Λ_Q-isomorphic to the quadratic form

$$\hat{b}_{F_M} : \lambda_1 TH_1(F_M;Q) \times \lambda_1 TH_1(F_M;Q) \to Q$$

defined by writing the sum (4.4.1) as $\hat{b}_{F_M}(\lambda_1 x, \lambda_1 y)$ (which may be regarded as a restriction of the Seifert quadratic form b_F extended over Q) under the Λ_Q-isomorphism $\lambda_1 : TH_1(F_M;Q)^{(1)} \cong \lambda_1 TH_1(F_M;Q)$. Next, by Lemma 4.3 we note that the restricted Seifert form

$$\phi_{\lambda_1 TH_1(F_M;Q)} : \lambda_1 TH_1(F_M;Q) \times \lambda_1 TH_1(F_M;Q) \to Q$$

is non-singular. Let

$$\varphi''\varphi' : \lambda_1 TH_1(F_M;Q) \cong \lambda_1 TH_1(F_M;Q)$$

be the composite of the Q-isomorphisms

$$\varphi' : \lambda_1 TH_1(F_M;Q) \to \mathrm{Hom}_Q[\lambda_1 TH_1(F_M;Q), Q]$$

and

$$\varphi'' : \mathrm{Hom}_Q[\lambda_1 TH_1(F_M;Q), Q] \to \lambda_1 TH_1(F_M;Q)$$

which send $\lambda_1 x$ to the homomorphism associated with $\phi_{\lambda_1 TH_1(F_M;Q)}(\lambda_1 x, \bullet)$ and the homomorphism associated with $\phi_{\lambda_1 TH_1(F_M;Q)}(\bullet, \lambda_1 y)$ to $\lambda_1 y$, respectively.

Suppose $\phi_{F_M}(\lambda_1 x, \lambda_1 z) = \phi_{F_M}(\lambda_1 z, \lambda_1 y)$ for all $z \in TH_1(F_M;Q)$. By the identities $\phi_{F_M}(\lambda_1 x, \lambda_1 z) = \mathrm{Int}_{F_M}(t^{-1}\lambda_1 x, z)$ and $\phi_{F_M}(\lambda_1 z, \lambda_1 y) = \mathrm{Int}_{F_M}(\lambda_1 y, z)$, we have

$$\mathrm{Int}_{F_M}(t^{-1}\lambda_1 x, z) = \mathrm{Int}_{F_M}(\lambda_1 y, z)$$

for all $z \in TH_1(F_M;Q)$. By the non-singularity of the restriction of Int_{F_M} to $TH_1(F_M;Q)$, we have $t^{-1}\lambda_1 x = \lambda_1 y$, implying that $\varphi''\varphi' = t^{-1}$ on $\lambda_1 TH_1(F_M;Q)$. Let S_T be the matrix representing

$$\phi_{F_M} : \lambda_1 TH_1(F_M;Q) \times \lambda_1 TH_1(F_M;Q) \to Q$$

with respect to the Q-basis of $\lambda_1 TH_1(F_M;Q)$ specified by the given basis of $S(F;Q)$ under the composite isomorphism

$$S(F;Q) \overset{\tilde{I}_*}{\cong} \lambda_1 TH_1(\tilde{M};Q) \overset{(\tilde{I}_M)_*^{-1}}{\cong} \lambda_1 TH_1(F_M;Q)$$

preserves ϕ_F and ϕ_{F_M}. The matrix $S_T + S_T'$ represents $\hat{b}_{F_M}$ on the specified Q-basis of $\lambda_1 TH_1(F_M;Q)$. Taking the dual Q-basis of $\mathrm{Hom}_Q[\lambda_1 TH_1(F_M;Q),Q]$, we see from the identity $t^{-1} = \varphi''\varphi'$ that the isometry t^{-1} of $\hat{b}_{F_M}$ is represented by $(S_T)^{-1}S_T'$. Noting that the isometry t of the quadratic form $\hat{b}_L^{(1)}$ on $\hat{T}^1(\tilde{E},\partial\tilde{E};Q)^{(1)}$ corresponds to the isometry t^{-1} of the quadratic form $\hat{b}_{F_M}$ on $\lambda_1 TH_1(F_M;Q)$, we obtained the desired result. □

5. The signature invariants. We can consider two kinds of signature invariants; namely, the signature $\sigma(L)$ of the quadratic form b_L and the signature $s(L)$ of the Seifert quadratic form $b_F : H_1(F;Z) \times H_1(F;Z) \to Z$ for a connected Seifert surface F of L, which is called the *Musasugi signature* of L after [19]. First of all, we observe that the signature invariants $\sigma(L)$ and $s(L)$ are distinct for a general link L:

EXAMPLE 5.1. Let L be a positive Hopf link. Then the matrix (-1) is a Seifert matrix of L and we have $s(L) = \mathrm{sign}(-2) = -1$. On the other hand, since $E \cong S^1 \times S^1 \times [0,1]$, we have $\tilde{E} \cong S^1 \times [0,1] \times R$. Hence we have

$$j_* = 0 : H_1(\tilde{E};Z) = \Lambda/(\lambda_1) \to H_1(\tilde{E},\partial\tilde{E};Z) = \Lambda/(\lambda_1),$$

so that $j^* = 0 : T^1(\tilde{E},\partial\tilde{E};Q) \to T^1(\tilde{E};Q)$ and $\hat{T}^1(\tilde{E},\partial\tilde{E};Q) = 0$. This implies that b_L is a zero form. Thus, we have

$$\sigma(L) = 0 \neq -1 = s(L).$$

By a similar consideration, if L is an r-component link which is any $(r-1)$-fold connected sum of the positive Hopf link, then we see that $\sigma(L) = 0$ and $s(L) = 1-r$. Thus, for such a link L the quadratic form b_L is not Q-isomorphic to the Seifert quadratic form b_F extended over Q in any sense.

We further consider the local signature invariants of the quadratic form b_L by extending the coefficient field Q to the real numbers R:

$$b_L^R : T^1(\tilde{E},\partial\tilde{E};R) \times T^1(\tilde{E},\partial\tilde{E};R) \to R$$

where $T^1(\tilde{E},\partial\tilde{E};R)$ is a (finitely generated torsion) Λ_R-module with $\Lambda_R = \Lambda \otimes_Z R$. Let $\lambda_a = t^2 - 2at + 1$ for every $a \in (-1,1)$. Recall that $\lambda_\varepsilon = t - \varepsilon$ for $\varepsilon = \pm 1$. For a Λ_R-module H and $a \in [-1,1]$, let H_{λ_a} denote the λ_a-primary component of H. We denote by $\sigma_a(L)$ the signature of the quadratic form

$$b_L^{\lambda_a} : T^1(\tilde{E},\partial\tilde{E};R)_{\lambda_a} \times T^1(\tilde{E},\partial\tilde{E};R)_{\lambda_a} \to R$$

defined by b_L^R. Then $\sigma_a(L)$ is zero except a finite number of $a \in [-1,1]$ and an even integer for all $a \in (-1,1)$. Further, we have

$$\sigma(L) = \sum_{x\in[-1,1]} \sigma_x(L)$$

(cf. [6], [18]). Let

$$\sigma_{[a,1]}(L) = \sum_{x\in[a,1]} \sigma_x(L) \quad \text{for each} \quad a \in [-1,1] \quad \text{and}$$
$$\sigma_{(a,1]}(L) = \sigma_{[a,1]}(L) - \sigma_a(L) \quad \text{for each} \quad a \in [-1,1).$$

Clearly, we have $\sigma_{[-1,1]}(L) = \sigma(L)$. For every $x \in (-1,1)$, the complex number $x + \sqrt{1-x^2}\sqrt{-1}$ with norm one is denoted by ω_x. For a Seifert matrix S of a link L, we consider the signature of the Hermitian matrix $(1-\omega_x)S + (1-\bar{\omega}_x)S'$ which is a step function on x (cf. [7]). This kind of signature on Seifert matrix was first considered by A. G. Tristram [23].

NOTATION 5.2.

$$\begin{aligned}
s_{[a,1]}(L) &= \lim_{x\to a-0} \operatorname{sign}((1-\omega_x)S + (1-\bar{\omega}_x)S') \quad \text{for each} \quad a \in (-1,1],\\
s_{[-1,1]}(L) &= s(L),\\
s_{(a,1]}(L) &= \lim_{x\to a+0} \operatorname{sign}((1-\omega_x)S + (1-\bar{\omega}_x)S') \quad \text{for each} \quad a \in [-1,1),\\
s_a(L) &= s_{[a,1]}(L) - s_{(a,1]}(L) \quad \text{for each} \quad a \in (-1,1),\\
s_1(L) &= s_{[1,1]}(L),\\
s_{-1}(L) &= s_{[-1,1]}(L) - s_{(-1,1]}(L).
\end{aligned}$$

We denote by $v(L)$ the signature of the canonical linking matrix V of L (introduced in §3). By convention, $v(L) = 0$ when L is a knot. The following identity connects the Murasugi-Tristram signatures and the infinite cyclic covering signatures:

THEOREM 5.3. *For every $a \in [-1,1]$, we have the identity:*

$$\sigma_{[a,1]}(L) = s_{[a,1]}(L) - v(L).$$

When L is a knot, this identity is given by D. Erle [1] and T. Matumoto [17]. By definition, we have the following corollary:

COROLLARY 5.4.

$$\begin{aligned}
\sigma_1(L) &= s_1(L) - v(L),\\
\sigma_a(L) &= s_a(L) \quad \textit{for all} \quad a \in [-1,1).
\end{aligned}$$

Since $|v(L)| \leq r-1$ for any r-component link L, we also have the following corollary:

COROLLARY 5.5. *For every r-component link L and every $a \in [-1,1]$, we have*

$$|\sigma_{[a,1]}(L) \quad s_{[a,1]}(L)| \leq r-1.$$

The following example shows that this estimation is best possible.

EXAMPLE 5.6. Let L be a positive Hopf link. The linking matrix V of L is $\begin{pmatrix} -1 & 1 \\ 1 & -1 \end{pmatrix}$. Directly, we have $v(L) = -1$ and $s(L) - v(L) = -1 - (-1) = 0$. On the other hand, b_L is a zero form by Example 5.1. Hence we have $\sigma_a(L) = 0$ for all $a \in [-1, 1]$ and hence $s_1(L) = -1$ and $s_a(L) = 0$ for all $a \in [-1, 1)$. If L is an r-component link which is any $(r-1)$-fold connected sum of the positive Hopf link, then we have $\sigma_a(L) = 0$ for all $a \in [-1, 1]$ and hence $v(L) = s_1(L) = 1 - r$ and $s_a(L) = 0$ for all $a \in [-1, 1)$.

It is well-known that the signature invariants $\sigma_{[a,1]}(L)$, $v(L)$, and $s_{[a,1]}(L)$ are all link concordance invariants (cf. [6]). For $\varepsilon = \pm 1$, let $a^T_\varepsilon(L) = \dim_R TH_1(\tilde{E}; R)_{\lambda_\varepsilon}$, which is also a link concordance invariant by Lemma 3.3. We have the following congruences:

LEMMA 5.7.
1. $s_1(L) \equiv \kappa_1(L) \equiv a^T_1(L) \pmod 2$ *and* $\sigma_1(L) \equiv \hat{\kappa}_1(L) \pmod 2$.
2. $s_{-1}(L) = \sigma_{-1}(L) \equiv \hat{\kappa}_{-1}(L) = \kappa_{-1}(L) \pmod 2$ *and* $a^T_{-1}(L) \equiv 0 \pmod 2$.

PROOF. Since the restriction of the cup product

$$\cup : T^1(\tilde{M}; Q) \times T^1(\tilde{M}; Q) \to T^2(\tilde{M}; Q) \cong Q$$

to $T^1(\tilde{M}; Q)_{\lambda_\varepsilon}$ is non-singular and skew-symmetric, it follows from Lemma 3.1 that $a^T_1(L) - \delta(L)$, $a^T_{-1}(L)$, $\sigma_1(L) + \hat{\kappa}_1(L)$, and $\sigma_{-1}(L) + \hat{\kappa}_{-1}(L)$ are even. Using that $v(L) \equiv \mathrm{r}_V(L) \pmod 2$, we obtain the desired congruences from Lemma 3.2 and Corollary 5.4. □.

6. Proof of identifying the signature invariants.

In this section, we show Theorem 5.3. For the proof, we use the signature theorems on infinite cyclic coverings of a 4-manifold which is done in [8]. From the quadratic form b_M, we can define the signature invariants $\sigma(M)$, $\sigma_a(M)$ $(a \in [-1, 1])$, $\sigma_{[a,1]}(M)$, and $\sigma_{(a,1]}(M)$ by the method which we have done for b_L in §5, although they are identical to $\sigma(L)$, $\sigma_a(L)$ $(a \in [-1, 1])$, $\sigma_{[a,1]}(L)$, and $\sigma_{(a,1]}(L)$, respectively, by Lemma 3.1. Further, if $\tilde{Y}$ is an infinite cyclic covering of a compact oriented 4-manifold Y, then the signature invariant $s_{[a,1]}(\tilde{Y})$ $(a \in (-1, 1])$ of $\tilde{Y}$ is defined as follows (cf. [8]): Namely,

$$s_{[a,1]}(\tilde{Y}) = \lim_{x \to a-0} \mathrm{sign} S(\omega_x)$$

for a t-Hermitian $Q(\Lambda)$-matrix $S(t)$ representing the $Q(\Lambda)$-extension form of the Λ-intersection form

$$\mathrm{Int}_\Lambda : H_2(\tilde{Y}; Z) \times H_2(\tilde{Y}; Z) \to \Lambda.$$

We use the following lemma:

LEMMA 6.1. *Let F be a (possibly disconnected) compact oriented surface embedded properly and locally-flatly in the 4-disk D^4. Let $L = \partial F$ be the link in $S^3 = \partial D^4$. Let D^4_2 be the double covering space over D^4 branched along F. Let $Y = \mathrm{cl}(D^4 - F \times D^2)$ for a tubular neighborhood $F \times D^2$ of F in D^4, and $\tilde{Y} \to Y$ the infinite cyclic covering space over Y extending the covering $p : \tilde{E} \to E$. Then we have $s(L) = \mathrm{sign} D^4_2$ and $s_{[a,1]}(L) = s_{[a,1]}(\tilde{Y})$ for every $a \in (-1, 1]$.*

PROOF. By the covering version of the Novikov addition theorem for signatures (cf. [8;Lemmas 2.2,2.3]), we can assume that F is a proper surface obtained from a connected Seifert surface for L by pushing the interior into $\mathrm{int}D^4$. Then the desired identities are obtained from an analogous argument of L. H. Kauffman [3] (see [8;Lemmas 3.1,3.3]). $\square$

PROOF OF THEOREM 5.3. We construct a 4-manifold X from the 4-disk D^4 by attaching the 2-handles $h_j(\cong D^2\times D^2)$ $(j=1,2,\dots,r)$ along L with Seifert surface framing. We regard X as a union $D^4\cup X'$ with $X'=S^3\times[0,1]\cup(\cup_{j=1}^r h_j)$ by regarding D^4 as $D^4\cup S^3\times[0,1]$. For a connected Seifert surface F of L, let F_D be a proper surface in D^4 obtained from F by pushing $\mathrm{int}F$ into $\mathrm{int}D^4$. Let $F_X=F_D\cup D_X$ be a closed surface in X where $D_X=L\times[0,1]\cup(\cup_{j=1}^r D(h_j))$ with $D(h_j)$ the core disk $(\cong D^2\times 0)$ of the 2-handle h_j. By construction, the tubular neighborhood $N(F_X)$ of F_X in X is diffeomorphic to $F_X\times D^2$. Let W be a 4-manifold obtained from X by replacing $F_X\times D^2$ with $V\times S^1$ where V is a handlebody with $\partial V=F_X$. We note $\partial X=\partial W=M$. The signature theorem of [8] says that

$$\sigma_{[a,1]}(M)=s_{[a,1]}(\tilde{W})-\mathrm{sign}(W)\quad\text{for every}\quad a\in(-1,1].$$

Let $Y=\mathrm{cl}(X-N(F_X))$. Then we have $Y=Y_1\cup Y_2$ where $Y_1=\mathrm{cl}(D^4-N(\hat{F}))\cup E\times[0,1]\cong Y_0$ with $Y_0=\mathrm{cl}(D^4-N(\hat{F}))$, and $Y_2=\cup_{j=1}^r\mathrm{cl}(h_j-N(\dot{\mathrm{h}}_j))$ which is a disjoint union of r copies of $S^1\times D^3$. Let $Y_{1,2}=Y_1\cap Y_2$ which is a disjoint union of r copies of $S^1\times S^1\times[0,1]$. Let $(\tilde{Y},\tilde{Y}_0,\tilde{Y}_1,\tilde{Y}_2,\tilde{Y}_{1,2})\to(Y,Y_0,Y_1,Y_2,Y_{1,2})$ be the covering induced from the covering $\tilde{W}\to W$. Since the lift of $V\times S^1$ under the covering is $V\times R$ whose integral homology is Λ-torsion, we have a natural isomorphism $H_2(\tilde{Y};Z)\otimes_\Lambda Q(\Lambda)\cong H_2(\tilde{W};Z)\otimes_\Lambda Q(\Lambda)$. Since $H_*(\tilde{Y}_2;Z)$ and $H_*(\tilde{Y}_{1,2};Z)$ are torsion Λ-modules, we also have a natural composite isomorphism

$$H_2(\tilde{Y}_0;Z)\otimes_\Lambda Q(\Lambda)\cong H_2(\tilde{Y}_1;Z)\otimes_\Lambda Q(\Lambda)\cong H_2(\tilde{Y};Z)\otimes_\Lambda Q(\Lambda).$$

By these observations and Lemma 6.1, we have $s_{[a,1]}(\tilde{W})=s_{[a,1]}(\tilde{Y}_0)=s_{[a,1]}(L)$ for all $a\in(-1,1]$. On the other hand, by the Novikov addition theorem on signatures, we have $\mathrm{sign}(W)=\mathrm{sign}(Y)=\mathrm{sign}(X)$, for $\mathrm{sign}(V\times S^1)=\mathrm{sign}(N(F_X))=0$. By construction, we have $\mathrm{sign}(X)=v(L)$, so that $\mathrm{sign}(W)=v(L)$. Hence, we have

$$\sigma_{[a,1]}(L)=\sigma_{[a,1]}(M)=s_{[a,1]}(L)-v(L)\quad\text{for every}\quad a\in(-1,1].$$

To obtain the result for $a=-1$, we consider, instead of the infinite cyclic covering, the associated double covering $(W_2,X_2,D_2^4,X_2',(S^3\times[0,1])_2)$ of $(W,X,D^4,X',S^3\times[0,1])$ branched along $(\emptyset,F_X,\hat{F},D_X,L\times[0,1])$. In this case, the signature theorem of [8] says that

$$\sigma(M)=\mathrm{sign}(W_2)-2\mathrm{sign}(W).$$

By the Novikov addition theorem on signatures, we have the following:

$$\mathrm{sign}(W_2)=\mathrm{sign}(X_2)=\mathrm{sign}(D_2^4)+\mathrm{sign}(X_2')\ \text{ and }\ \mathrm{sign}(W)=\mathrm{sign}(X)=v(L).$$

By Lemma 6.1, we have $\mathrm{sign}(D_2^4)=s(L)$. Thus, to complete the proof, it suffices to show that $\mathrm{sign}(X_2')=v(L)$. Let $L_j(j=1,2,\dots,r)$ be the components of L. Let F_j be a Seifert surface of L_j in S^3 which meets $L-L_j$ transversely. We identify S^3 with $S^3\times 1\subset S^3\times[0,1]\subset X'$. Let $F_j^{(2)}$ and $\dot{\mathrm{h}}_j^{(2)}$ be the lifts of F_j and $\dot{\mathrm{h}}_j$ under the branched double covering $X'_2 \to X'$, respectively. Since the lifting part of $\mathrm{int}F_j$ is a double branched covering space of $\mathrm{int}F_j$ branched along $F_j \cap (L - L_j)$ and in particular, it is an oriented surface, we see that the 2-cycles $c_j = -F_j^{(2)} + 2\dot{\mathrm{h}}_j$ $(j = 1, 2, \ldots, r)$ in X'_2 have the intersection numbers $\mathrm{Int}(c_j, c_k) = 2\mathrm{Link}(L_j, L_k)$ for all j, k (where the self-linking number is taken with respect to the Seifert surface framing of L). We consider the following natural exact sequence:

$$0 \to H_2((S^3 \times [0,1])_2; Q) \to H_2(X'_2; Q) \to H_2(X'_2, (S^3 \times [0,1])_2; Q) \to 0.$$

Since by excision the cycles $c_j (j = 1, 2, \ldots, r)$ form a Q-basis of $H_2(X'_2, (S^3 \times [0,1])_2; Q)$, we see that $H_2(X'_2; Q)$ has a Q-basis consisting of the cycles $c_j (j = 1, 2, \ldots, r)$ and the cycles in $\partial X'_2$ forming a Q-basis of $H_2((S^3 \times [0,1])_2; Q)$. Since any cycles of $\partial X'_2$ do not contribute the signature of X'_2, we obtain $v(L) = \mathrm{sign}(X'_2)$. □

7. Estimating the 4-genus of a link. The *4-genus* of a link L, denoted by $g^*(L)$ is the minimum of the genus $g(F)$ of a compact connected oriented surface F embedded properly and locally-flatly in D^4 which bounds the link L in S^3 in the TOP category, where the locally-flatness of F is understood that F admits a normal disk bundle in D^4. For every link L, we have the following estimate:

PROPOSITION 7.1.

$$\begin{aligned} |s(L)| + \kappa_{-1}(L) &\leq 2g^*(L) + \kappa_1(L) \quad \textit{and} \\ |s_{[a,1]}(L)| &\leq 2g^*(L) + \kappa_1(L) \quad \textit{for every} \quad a \in (-1, 1]. \end{aligned}$$

REMARK 7.2. If L' is obtained from L by a surgery (= hyperbolic transformation) along a band, then we have

$$\begin{aligned} |s(L') - s(L)| + |n_S(L') - n_S(L)| &= 1 \quad \text{and} \\ |s_{[a,1]}(L') - s_{[a,1]}(L)| + |\beta(L') - \beta(L)| &\leq 1 \quad \text{for every} \quad a \in (-1, 1]. \end{aligned}$$

This inequality or equality is called the *Murasugi-Tristram inequality*. The inequalities of Theorem 7.1 can be deduced when we combine it with Lemma 3.2 and work in the smooth or PL category (cf. [19], [23]).

A *weakly slice link* is a link L in S^3 with $g^*(L) = 0$.

COROLLARY 7.3. *If L is a weakly slice link, then we have the following inequalities:*

$$\begin{aligned} |s(L)| + \kappa_{-1}(L) &\leq \kappa_1(L) \quad \textit{and} \\ |s_{[a,1]}(L)| &\leq \kappa_1(L) \quad \textit{for every} \quad a \in (-1, 1]. \end{aligned}$$

REMARK 7.4. The Alexander polynomial $A_L(t)$ of a link L does not contribute to the determination of an $r(> 1)$-component weakly slice link. In fact, examining the construction of F. Hosokawa in [2], we see that the Alexander polynomial $A_L(t)$ of every $r(> 1)$-component link L is realizable by an r-component weakly slice link.

PROOF OF PROPOSITION 7.1. Let $g = g^*(L)$. Let $Y_0 = D^4 - \mathrm{int}(F \times D^2)$ for a tubular neighborhood $F \times D^2$ of a compact connected oriented proper surface F of genus g in D^4 such that $\partial F = L$. Then $H_q(Y_0; Z)$ is isomorphic to Z (if $q = 0, 1$), Z^{2g+r-1} (if $q = 2$), and 0 (if $q \neq 0, 1, 2$). By Lemma 6.1, we have $s_{[a,1]}(\tilde{Y}_0) = s_{[a,1]}(L)$ for $a \in (-1, 1]$. Further, $H_2(Y_0, \partial Y_0; Z) \cong Z^{2g+r-1}$. By the Wang exact sequence, we have

$$\dim_{Q(\Lambda)} H_2(\tilde{Y}_0, \partial\tilde{Y}_0; Z) \otimes_\Lambda Q(\Lambda) \leq 2g + r - 1.$$

Since $\dim_{Q(\Lambda)} H_1(\tilde{Y}_0; Z) \otimes_\Lambda Q(\Lambda) = 0$ and $\dim_{Q(\Lambda)} H_1(\partial\tilde{Y}_0; Z) \otimes_\Lambda Q(\Lambda) = \beta(L)$, we have $|s_{[a,1]}(\tilde{Y}_0)| \leq 2g + r - 1 - \beta(L) = 2g + \kappa_1(L)$ by Lemma 3.2. By Lemma 6.1, we have $|s_{[a,1]}(L)| \leq 2g + \kappa_1(L)$ for $a \in (-1, 1]$. Let D^4_2 be the double covering space of D^4 branched along F. Then $\mathrm{sign}(D^4_2) = s(L)$ by Lemma 6.1. The Z-ranks of $H_2(D^4_2; Z)$ and $H_1(\partial D^4_2; Z)$ are respectively $2g + r - 1$ and $n_S(L)$. Using $H_1(D^4_2; Q) = 0$, we see from Lemma 3.2 that

$$|s(L)| \leq 2g + r - 1 - n_S(L) = 2g + \kappa_1(L) - \kappa_{-1}(L). \quad \square$$

8. The quadratic form of a sphere-slice link. A *strongly slice link* in S^3 is a link which is the boundary of the union of mutually disjoint disks embedded properly and locally-flatly in D^4 (in the TOP category). A *sphere-slice link* in S^3 is a link which is obtained as the cross-section of a locally flat 2-sphere in the 4-sphere S^4 by the equatorial 3-sphere S^3 (in the TOP category). A two-component sphere-slice link is strongly slice. A sphere-slice link is known by E. Ogasa [20] and S. Satoh [21] to be proper, i.e., to have the canonical linking matrix whose diagonal entries are even. Further, S. Satoh [21] classified the canonical linking matrices of all sphere-slice links with at most 5 components.

DEFINITION 8.1 (J. Levine[16]). Two non-singular quadratic forms $b_i : T_i \times T_i \to Q (i = 1, 2)$ are *cobordant* if there is a Λ_Q-submodule T_0 of the direct sum $T_1 \oplus T_2$ which is self-orthogonal with respect to the orthogonal sum $b_1 \perp - b_2$. Further, b_1 is *null-cobordant* if $T_2 = 0$.

We show the following theorem:

THEOREM 8.2. *If L is a sphere-slice link, then $v(L) = 0$ and the induced non-singular quadratic form*

$$\bar{b}_L : \bar{T}^1(\tilde{E}, \partial\tilde{E}; Q) \times \bar{T}^1(\tilde{E}, \partial\tilde{E}; Q) \to Q$$

of the quadratic form $b_L : T^1(\tilde{E}, \partial\tilde{E}; Q) \times T^1(\tilde{E}, \partial\tilde{E}; Q) \to Q$ is null-cobordant. In particular, $\sigma_a(L) = 0$ for all $a \in [-1, 1]$ and the nullity invariants $\delta(L)$, $\hat{\kappa}_1(L)$, $\hat{\kappa}_{-1}(L)$, and $n(L)$ are all even. Further, the torsion Alexander polynomial $A^T_L(t)$ has the form $\pm t^m f(t) f(t^{-1})$ for an integer m and a Laurent polynomial $f \in \Lambda$.

By this theorem and Lemma 3.2, we have also $n_S(L) \equiv \beta(L) \equiv r-1 \pmod 2$ for every r-component sphere-slice link L, and in particular, the Alexander polynomial $A_L(t)$ must be zero for every even-component sphere-slice link L.

REMARK 8.3. If L is a strongly slice link, then we have $\beta(L) = r - 1$ and $A^T_L(1) = \pm 1$, so that $f(1) = \pm 1$ in Theorem 8.2. In particular, $n(L) = \delta(L) = \kappa_{\pm 1}(L) = 0$ (cf. Corollary 3.5).

PROOF OF THEOREM 8.2. Let $D^4_\pm$ be the 4-disks obtained from S^4 by splitting along the equatorial 3-sphere S^3. Let $L = S^3 \cap S^2$ and $F_\pm = D^4_\pm \cap S^2$ for a locally flat 2-sphere S^2 in S^4. Let $X_- = D^4_- \cup F_+ \times D^2$ for a tubular neighborhood $F_+ \times D^2$ of F_+ in D^4_+. Let ℓ_i $(i = 1, 2, \ldots, s)$ be mutually disjoint, simple loops in $\mathrm{int}(F_+)$ such that every component of $F_+ - \cup_{i=1}^s \ell_i$ contains just one component of L. Let $\ell_i \times I$ be a bi-collar of ℓ_i in F_+ with $I = [-1, 1]$ and $\ell_i = \ell_i \times 0$. Let $D_i (i = 1, 2, \ldots, s)$ be the s copies of the 2-disk. Let $X'_- = X_- \cup_{i=1}^s D_i \times I \times S^1$ by identifying $\ell_i \times I \times (\partial D^2)$ with $(\partial D_i) \times I \times S^1$ factorwise for $i = 1, 2, \ldots, s$. Then we note that $\partial X'_- = M$. We consider the following natural exact sequence:

$$0 \to H_2(X_-; Z) \to H_2(X'_-; Z) \to H_2(X'_-, X_-; Z) \to 0.$$

By excision, $H_2(X'_-, X_-; Z) \cong \oplus_{i=1}^s H_2((D_i, \ell_i) \times S^1; Z)$ is a free abelian group with $[D_i \times 1] (i = 1, 2, \ldots, s)$ a basis. Cutting S^2 by each ℓ_i and then attaching a disk extending $D_i \times 1$ and then slight deforming, we can find mutually disjoint 2-spheres $S_i (i = 1, 2, \ldots, s)$ with trivial normal bundles in $\mathrm{int}(X'_-)$ such that the natural epimorphism $H_2(X'_-; Z) \to H_2(X'_-, X_-; Z)$ sends $[S_i]$ to $[D_i \times 1]$ for all i. Then the intersection number $\mathrm{Int}_{X'_-}([S_i], [S_j]) = 0$ for all i and j. Since X_- is in S^4, we have $\mathrm{Int}_{X'_-}(x, y) = 0$ for all x, y in the image of $H_2(X_-; Z) \to H_2(X'_-; Z)$. Thus, we can write $H_2(X'_-; Z) = A + B$ for some abelian groups A and B with $\mathrm{Int}_{X'_-}(A, A) = \mathrm{Int}_{X'_-}(B, B) = 0$. This implies that $\mathrm{sign}(X'_-) = 0$. Let $X'_0 = \mathrm{cl}(X'_- - S^2 \times D^2)$. Let $(\tilde{X}'_0, \tilde{M}) \to (X'_0, M)$ be the infinite cyclic covering extending the infinite cyclic covering $\tilde{E} \to E$. Then for every $a \in (-1, 1]$, we have

$$\sigma_{[a,1]}(L) = \sigma_{[a,1]}(M) = s_{[a,1]}(\tilde{X}'_0) - \mathrm{sign} X'_0.$$

Writing $Y_- = \mathrm{cl}(D^4_- - F_- \times D^2)$ for a tubular neighborhood $F_- \times D^2$ of F_- in D^4_-, we have $s_{[a,1]}(\tilde{X}'_0) = s_{[a,1]}(\tilde{Y}_-) = s_{[a,1]}(L)$ by Lemma 6.1. Since $\mathrm{sign} X'_0 = \mathrm{sign} X'_- = 0$, we see that $\sigma_{[a,1]}(L) = s_{[a,1]}(L)$ and by Theorem 2.4 $v(L) = s_{[a,1]}(L) - \sigma_{[a,1]}(L) = 0$. Let $M_- = \partial Y_-$. We consider the sequence:

$$TH_2(\tilde{Y}_-, \tilde{M}_-; Z) \xrightarrow{\partial} TH_1(\tilde{M}_-; Z) \xrightarrow{i_*} TH_1(\tilde{Y}_-; Z). \tag{8.2.1}$$

Let $W = S^4 - \mathrm{int} N(S^2)$ for the 2-sphere S^2 giving the sphere-slice of L, and $\tilde{W}$ the infinite cyclic covering space of W. Since $BH_*(\tilde{W}; Z) = 0$, the Λ-intersection number on $BH_2(\tilde{Y}_-; Z)$ is trivial and hence by the zeroth duality in [9] the natural map $BH_2(\tilde{Y}_-; Z) \to BH_2(\tilde{Y}_-, \tilde{M}_-; Z)$ is trivial. Thus, the sequence (8.2.1) is exact. For a finitely generated torsion Λ_Q-module T, let T' be the (unique) submodule of T given by the splitting $T = T_{\lambda_1} \oplus T' \oplus T_{\lambda_{-1}}$ of T. We have $T^1(\tilde{E}, \partial\tilde{E}; Q)' = T^1(\tilde{M}; Q)' - T^1(\tilde{M}_-; Q)'$ on which the quadratic forms b_L, b_M, and b_{M_-} are non-singular and identical. By the exactness of (8.2.1), the following sequence

$$T^1(\tilde{Y}_-; Q)' \xrightarrow{i^*} T^1(\tilde{M}_-; Q)' \xrightarrow{\delta} T^2(\tilde{Y}_-, \tilde{M}_-; Q)'$$

is exact. This means that the restriction of b_{M_-} to $T^1(\tilde{M}_-; Q)'$, namely the restriction of $\bar{b}_L$ to $\bar{T}^1(\tilde{E}, \partial\tilde{E}; Q)'$ is null-cobordant. In particular, we have $\sigma_a(L) = 0$ for all $a \in (-1, 1)$. Let $T_{\lambda_\varepsilon} = \bar{T}^1(\tilde{E}, \partial\tilde{E} : Q)_{\lambda_\varepsilon}$, and b_{λ_ε} the restriction of b_L to T_{λ_ε}. To complete the proof, it suffices to show that $(b_{\lambda_\varepsilon}, T_{\lambda_\varepsilon})$ is null-cobordant for $\varepsilon = \pm 1$. By J. Levine [16, Lemma 12], $(b_{\lambda_\varepsilon}, T_{\lambda_\varepsilon})$ is null-cobordant or cobordant to a quadratic form $(\tilde{b}_{\lambda_\varepsilon}, \tilde{T}_{\lambda_\varepsilon})$ such that $\lambda_\varepsilon \tilde{T}_{\lambda_\varepsilon} = 0$. If $\mathrm{sign}(\tilde{b}_{\lambda_\varepsilon}) = \sigma_\varepsilon(L) = 0$, then the form $(\tilde{b}_{\lambda_\varepsilon}, \tilde{T}_{\lambda_\varepsilon})$ and hence $(b_{\lambda_\varepsilon}, T_{\lambda_\varepsilon})$ are null-cobordant. We show that $\sigma_\varepsilon(L) = 0$ for $\varepsilon = \pm 1$. Since the Λ_Q-intersection form Int_{Λ_Q} on $H_2(\tilde{Y}_-; Q)$ is a zero form, we have $s_1(L) = s_1(\tilde{Y}_-) = 0$ by Lemma 6.1. Using $v(L) = 0$, we have $\sigma_1(L) = 0$. Let $(D^4_-)_2$ be the double covering space of D^4_- branched along F_-. Since $(D^4_-)_2$ is embedded in the double covering space S^4_2 of S^4 branched along S^2 which is a rational homology 4-sphere, we see from Lemma 6.1 that $\sigma(L) + v(L) = s(L) = \mathrm{sign}(D^4_-)_2 = 0$, so that $\sigma_{-1}(L) = 0$. The results on the nullity invariants now follow from Lemmas 3.2 and 5.7. Since $\tilde{b}_L$ is null-cobordant and by Lemma 5.7 $a^T_{\pm 1}(L)$ are even, $A^T_L(t)$ must have a form $\pm t^m b f'(t) f'(t^{-1})$ for a positive integer b and a Laurent polynomial $f' \in \Lambda$ which are coprime in Λ. Using $\tilde{H}_*(\tilde{W}; Z)^+ = 0$, we see that the exact sequence (8.2.1) induces an exact sequence:

$$(8.2.2) \qquad 0 \to TH_2(\tilde{Y}_-, \tilde{M}_-; Z)^+ \xrightarrow{\partial^+} TH_1(\tilde{M}_-; Z)^+ \xrightarrow{i^+_*} TH_1(\tilde{Y}_-; Z)^+ \to 0.$$

If $h(t)$ is the zeroth characteristic polynomial of $TH_1(\tilde{Y}_-; Z)$, then we see from the first duality (=Blanchfield duality) of [9] that the zeroth characteristic polynomial of $TH_2(\tilde{Y}_-, \tilde{M}_-; Z)$ is given by $h(t^{-1})$. If $A'(t)$ is the zeroth characteristic polynomial of $TH_1(\tilde{M}_-; Z)$, then the exact sequence (8.2.2) implies that $A'(t)$ is equal to $h(t)h(t^{-1})$ up to unit multiplications of Λ^+. Using that $\lambda_1 H_*(\tilde{M}_-, \tilde{E}; Z) = 0$, we see also that $A^T_L(t)$ is equal to $A'(t)$ up to multiplications of λ_1 and units of Λ and hence equal to $h(t)h(t^{-1})$ up to multiplications of λ_1 and units of Λ^+. Since Λ is a unique factorization domain, the positive integer b must be a square $(b')^2$, and we have $A^T_L(t) = \pm t^m f(t) f(t^{-1})$ by taking $b' f'(t)$ as $f(t)$. □

9. Some remaining problems. Let $\mathcal{S}$ be the set of integral square matrices. For $S \in \mathcal{S}$, let H_S be the Λ-module with $tS - S'$ as a Λ-presentation matrix. Given a Seifert matrix $S \in \mathcal{S}$ on a Z-basis of the free abelian group

$$K(F; Z) = \mathrm{Ker}(I_* : H_1(F; Z) \to H_1(E; Z)),$$

we have a practical method of constructing from S the matrix S_T in Theorem 4.4. This is because every rational square matrix is rationally S-equivalent to a block sum of a non-singular matrix (possibly the empty matrix) and a zero matrix. Further, this non-singular matrix is unique up to rational congruences. This result has been observed by P. W. Keef[14]. Then we can take this non-singular matrix as S_T which will be seen from the arguments of §4 and [7] using F_M. On the other hand, there exist lots of $S \in \mathcal{S}$ which are not (integrally) S-equivalent to a block sum of any non-degenerate matrix (i.e., any matrix with non-zero determinant) and any zero matrix. To see this, we note that if $S \in \mathcal{S}$ is (integrally) S-equivalent to a block sum of a non-degenerate matrix and a zero matrix, then the torsion-free part BH_S of H_S is Λ-free. Since in [10] we have constructed lots of links L such that $BH_1(\tilde{E}; Z)$ is not Λ-free, every integral Seifert matrix S of such a link L cannot be S-equivalent to a block sum of any non-degenerate matrix and any zero matrix. From this viewpoint, the following problem may be interesting as a generalization of (integral) S-equivalence:

PROBLEM 9.1. Find operations on the set $\mathcal{S}$ such that
1. the torsion part TH_S of the Λ-module H_S for every $S \in \mathcal{S}$ is unchanged under these operations, and

2. every $S \in \mathcal{S}$ is transformed into a block sum of a non-degenerate matrix and a zero matrix.

Another problem is a characterization problem of the torsion Alexander polynomial $A_L^T(t)$ of a link L (cf.§3). The polynomial $A_L^T(t)$ of a general link L with $\delta(L) > 0$ is characterized by the form $(t-1)^{\delta(L)} f$ (up to units of Λ) for an $f \in \Lambda$ such that $f(t) = f(t^{-1})$ and span(f) is even. When $\delta(L) = 0$, we need further the condition that $f(1) = \pm 1$. Using Lemma 3.2, we can derive this characterization from the well-known characterization of the usual Alexander polynomial due to H. Seifert [22] and F. Hosokawa [2] by adding a suitable trivial link as a split summand for realization. The polynomial $A_L^T(t)$ of a weakly slice link L with $\delta(L) > 0$ is characterized by the same form by Remark 7.4. The polynomial $A_L^T(t)$ of a weakly slice link L with $\delta(L) = 0$ is characterized by the form $f'(t)f'(t^{-1})$ (up to units of Λ) for an $f' \in \Lambda$ with $f'(1) = \pm 1$ as well as the polynomial $A_L^T(t)$ of a strongly slice link L. This result can be shown by a method similar to the proof of Theorem 8.2 as a standard application of the quadratic form b_L of the link L. As the remaining case, we have the following problem:

PROBLEM 9.2. Characterize the torsion Alexander polynomial $A_L^T(t)$ of sphere-slice link L.

In Theorem 8.2, some necessary condition is stated.

References

[1] D. Erle, *Die quadratische Form eines Knotens und Satz über Knotenmanigfaltigkeiten*, J. Reine Angew. Math. **236** (1969), 174-218.

[2] F. Hosokawa, *On ∇-polynomials of links*, Osaka Math. J. **10** (1958), 273-282.

[3] L. H. Kauffman, *Branched coverings, open books and knot periodicity*, Topology **13** (1974), 143-160.

[4] A. Kawauchi, *On quadratic forms of 3-manifolds*, Invent. Math. **43** (1977), 177-198.

[5] A. Kawauchi, *On the Alexander polynomials of cobordant links*, Osaka J. Math. **15** (1978), 151-159.

[6] A. Kawauchi, *On a 4-manifold homology equivalent to a bouquet of surfaces*, Trans. Amer. Math. Soc. **262** (1980), 95-112.

[7] A. Kawauchi, *The signature invariants of infinite cyclic coverings of closed odd dimensional manifolds*, In: Algebraic and Topological Theories-to the memory of Dr. T. Miyata, pp.52-85, Kinokuniya Co. Ltd., 1985.

[8] A. Kawauchi, *On the signature invariants of infinite cyclic coverings of even dimensional manifolds*, Advanced Studies in Pure Math. **9** (1986), 177-188.

[9] A. Kawauchi, *Three dualities on the integral homology of infinite cyclic coverings of manifolds*, Osaka J. Math. **23** (1986), 633-651.

[10] A. Kawauchi, *On the integral homology of infinite cyclic coverings of links*, Kobe J. Math. **4** (1987), 31-41.

[11] A. Kawauchi, *On the fundamental class of an infinite cyclic covering*, Kobe J. Math. **15** (1998), 103-114.

[12] A. Kawauchi, *A survey of knot theory*, Birkhäuser, 1996.

[13] A. Kawauchi, *The quadratic form of a link and a Seifert matrix,*, In: The 5th Korea-Japan School of Knots and Links, Proc. Applied Math. Workshop **8** (1997), 119-129.

[14] P. W. Keef, *On the S-equivalence of some general sets of matrices*, Rocky Mountain J. Math. **13** (1983), 541-551.

[15] R. H. Kyle, *Branched covering spaces and the quadratic forms of links*, Ann. of Math. **59** (1954), 539-548.

[16] J. Levine, *Invariants of knot cobordism*, Invent. Math. **8** (1969), 98-110.

[17] T. Matumoto, *On the signature invariants of a non-singular complex sesqui-linear form*, J. Math. Soc. Japan **29** (1977), 67-72.

[18] J. W. Milnor, *Infinite cyclic coverings*, In: Conference on the Topology of Manifolds, pp. 115-133, Prindle, Weber & Schmidt. 1968.

[19] K. Murasugi, *On a certain numerical invariant of link types*, Trans. Amer. Math. Soc. **117** (1965), 387-422.

[20] E. Ogasa, *Some properties of ordinary sense slice 1-links: some answers to the problem (26) of Fox*, Proc. Amer. Math. Soc. (to appear).

[21] S. Satoh, *On sphere-slice links with at most five components*, J. Knot Theory Ramifications **7** (1998), 217-230.

[22] H. Seifert, *Über das Geschlecht von Knoten*, Math. Ann. **110** (1934), 571-592.

[23] A. G. Tristram, *Some cobordism invariants for links*, Proc. Cambridge Phil. Soc. **66** (1969), 251-264.

[24] H. F. Trotter, *Homology of group systems with applications to knot theory*, Ann. of Math. **76** (1962), 464-498.

Department of Mathematics, Osaka City University, Osaka 558, Japan
E-mail address: kawauchi@sci.osaka-cu.ac.jp

Contemporary Mathematics
Volume **233**, 1999

Quantum $SU(2)$-invariants of three-manifolds associated with the trivial first cohomology class modulo two

Hitoshi Murakami

After E. Witten proposed invariants associated with a compact Lie group and an integer (level) [**36**] using path integral, many mathematically rigorous definitions follow ([**28, 16, 3, 11**]for $SU(2)$,[**12, 37**] for $SU(n)$ and [**35, 32**] for other classical cases). We denote by $\tau_r^{SU(2)}(M)$ the $SU(2)$-invariants associated with the r-th root of unity $q = \exp(2\pi\sqrt{-1}/r)$ (r equals the level plus 2 for $SU(2)$). If r is odd, then $\tau_r^{SU(2)}(M)$ splits into the product of the $SO(3)$-invariant $\tau_r^{SO(3)}(M)$ and $\tau_3^{SU(2)}(M)$ (or its complex conjugate) [**33, 9**].

The author proved that if r is an odd prime and M is a $\mathbf{Z}/r\mathbf{Z}$-homology three-sphere, then $\tau_r^{SO(3)}(M)$ belongs to $\mathbf{Z}[q]$ with $q = \exp(2\pi\sqrt{-1}/r)$ [**20**] (for a simpler proof see [**18**] and for the $PSU(N)$ case see [**19, 31**]). This implies that we can write

$$\tau_r^{SO(3)}(M) = a_{r,0} + a_{r,1}(q-1) + a_{r,2}(q-1)^2 + \dots$$

for some $a_{r,n} \in \mathbf{Z}$. Though these coefficients are well defined only modulo r and for $n \le r-2$ [**25**, Introduction], T. Ohtsuki showed that they have well-defined $r \to \infty$ limits. He actually defined a formal power series invariant

$$\tau(M) = \sum_{n=0}^{\infty} \lambda_n(M)(t-1)^n$$

with $\lambda_n(M) \in \mathbf{Z}[1/2, 1/3, \dots, 1/m]$ with $m = \max\{2n+1, |H_1(M;\mathbf{Z})|\}$ for an oriented rational homology three-sphere M such that (roughly speaking) $\tau_r^{SO(3)}(M) = \left(\frac{|H_1(M;\mathbf{Z})|}{r}\right)\tau(M) \mod r$ [**25**]. It is known [**21**] that $\lambda_0(M) = 1/|H_1(M;\mathbf{Z})|$ and $\lambda_1(M) = 6\lambda(M)/|H_1(M;\mathbf{Z})|$, where $\lambda(M)$ is the Casson-Walker invariant with A. Casson's normalization [**1, 34**].

1991 *Mathematics Subject Classification.* Primary 57N10, 57M25; Secondary 81R50.

Partially supported by Waseda University Grant for Special Research Projects (No. 97A-342) and Grant-in-Aid for Scientific Research (C) (No. 09640135), the Ministry of Education, Science, Sports and Culture.

There are another series of invariants which refines the quantum $SU(2)$-invariants. For an odd integer $p > 1$ and $\alpha \in H^1(M; \mathbf{Z}/2\mathbf{Z})$, an invariant $\tau_{2p}^{SU(2)}(M, \alpha)$ is defined and satisfies

$$\tau_{2p}^{SU(2)}(M) = \sum_{\alpha \in H^1(M;\mathbf{Z}/2\mathbf{Z})} \tau_{2p}^{SU(2)}(M, \alpha)$$

[**33, 9, 2**]. In this paper we will study the case where α is trivial and show a similar result to [**20, 21**].

Let p be an odd prime, M a $\mathbf{Z}/p\mathbf{Z}$-homology three-sphere and $|H_1(M;\mathbf{Z})|$ the number of element in $H_1(M;\mathbf{Z})$. Put $\xi = \exp\left(\frac{2\pi\sqrt{-1}}{p}\right)$ and denote $\mathbf{Z}[1/2]$ by $\mathbf{Z}_{(2)}$. We define $T : \mathbf{Z}_{(2)}[\xi] \to \mathbf{Z}/p\mathbf{Z}$ by $T(\sum_{j=0}^{r-2} a_j/2^{k_j}\xi^j) = \sum_{j=0}^{r-2} \overline{2}^{k_j} a_j$ (see [**9**]). Here $\overline{x} \in \mathbf{Z}/p\mathbf{Z}$ is the inverse of x, i.e., $x\overline{x} = 1 \in \mathbf{Z}/p\mathbf{Z}$. Then we will show

THEOREM 1.

(a)
$$\tau_{2p}^{SU(2)}(M, 0) \in \mathbf{Z}_{(2)}[\xi],$$

(b)
$$T\left(\tau_{2p}^{SU(2)}(M, 0)\right) = \left(\frac{|H_1(M;\mathbf{Z})|}{p}\right),$$

$$\left(\textit{Therefore } \frac{\tau_{2p}^{SU(2)}(M,0) - \left(\frac{|H_1(M;\mathbf{Z})|}{p}\right)}{\xi - 1} \in \mathbf{Z}_{(2)}[\xi].\right)$$

and

(c)
$$T\left(\frac{\tau_{2p}^{SU(2)}(M,0) - \left(\frac{|H_1(M;\mathbf{Z})}{p}\right)}{\xi - 1}\right) \equiv 3\lambda(M) \mod p,$$

where $\lambda(M)$ is the Casson-Walker invariant [**1, 34**] *with A. Casson's normalization (half of K. Walker's normalization) and $\left(\frac{x}{p}\right)$ is the Legendre symbol (1 if x is a quadratic residue modulo p and -1 otherwise).*

ACKNOWLEDGMENT. The author thanks K. Kodama for his computer program KNOT (available at ftp://ftp.math.s.kobe-u.ac.jp/pub/knot/), which was used for experimental calculation for Lemma 4.1. He also thanks organizers, especially Hanna Nencka, of the Low Dimension Topology Conference, Funchal, Portugal for their hospitality. Thanks are also due to J. Andersen, S. Garoufalidis, and G. Masbaum for useful comments.

1. Preliminaries

We put $r = 2p$ with p an odd prime, $p = 4u + \varepsilon$ with $\varepsilon = \pm 1$. We also put $\xi = \exp\frac{2\pi\sqrt{-1}}{p}$, $q = \exp\frac{2\pi\sqrt{-1}}{r}$, $s = \exp\frac{2\pi\sqrt{-1}}{2r}$. So $\xi = q^2 = s^4$. Note that $q = -\xi^{-2\varepsilon u} = -\xi^{\overline{2}}$ and $s = \varepsilon\sqrt{-1}\xi^{-\varepsilon u} = \varepsilon\sqrt{-1}\xi^{\overline{4}}$. We put $[k] = (s^k - s^{-k})/(s - s^{-1})$. In particular, $[2] = s + s^{-1} = \varepsilon\sqrt{-1}(\xi^{\overline{4}} - \xi^{-\overline{4}})$.

Bold faces are used for multi-indices with m entries. For example $\mathbf{k} = (k_1, k_2, \ldots, k_m)$. We use the following conventions. For multi-indices $\mathbf{j}$ and $\mathbf{k}$,

$$\mathbf{j} + \mathbf{k} = (j_1 + k_1, j_2 + k_2, \ldots, j_m + k_m),$$

$$x^{\mathbf{k}} = \prod_{i=1}^{m} x^{k_i},$$
$$|\mathbf{k}| = k_1 + k_2 + \cdots + k_m,$$

By $\mathbf{j} \geq \mathbf{k}$ we mean $j_i \geq k_i$ for every i and by $\mathbf{j} > \mathbf{k}$ we mean $j_i \geq k_i$ for every i but $\mathbf{j} \neq \mathbf{k}$. For a scalar x, we put $\mathbf{x} = (x, x, \dots, x)$.

The Conway polynomial $\nabla(L; z)$ of a link L [**4**, **7**] is defined by

$$\nabla(\text{trivial knot}; z) = 1$$

and

$$\nabla(L_+; z) - \nabla(L_-; z) = -z\nabla(L_0; z)$$

for the usual skein triple (L_+, L_-, L_0).

L_+ L_- L_0

We denote by $\varphi(L)$ the coefficient of $z^{\sharp(L)+1}$, where $\sharp(L)$ is the number of components of L.

The Jones polynomial $V(L; t)$ [**5**] is defined by

$$V(\text{trivial knot}; t) = 1$$

and

$$t^{-1}V(L_+; t) - tV(L_-; t) = (t^{1/2} - t^{-1/2})V(L_0; t).$$

We will often use a version $\tilde{V}(L; t)$ of $V(L; t)$ defined as follows.

$$\tilde{V}(\text{trivial knot}; t) = 1$$

and

$$t\tilde{V}(L_+; t) - t^{-1}\tilde{V}(L_-; t) = (t^{1/2} - t^{-1/2})\tilde{V}(L_0; t).$$

We put $\tilde{V}(\emptyset; t) = 1/(t^{1/2} + t^{-1/2})$ by convention. Then it is not hard to see that $\tilde{V}(L; t) = (-1)^{\sharp(L)+1}V(L; t^{-1})$.

For an m-component framed link $\mathcal{L} = \mathcal{L}_1 \cup \mathcal{L}_2 \cup \cdots \cup \mathcal{L}_m$, we denote by $w(\mathcal{L}_i)$ the writhe of the ith component. It is defined to be the sum of the signs of the self-crossings of $\mathcal{L}_i$. In this paper, we use the black-board framing for framed link diagrams and so $w(\mathcal{L}_i)$ coincides with the framing of $\mathcal{L}_i$. We denote by $\mathcal{L}^{\mathbf{n}}$ the $\mathbf{n}$-parallel of $\mathcal{L}$, i.e., we take n_i parallels of $\mathcal{L}_i$ on the plane for each $\mathcal{L}_i$.

A framed link is called algebraically split if the linking number of each pair of components vanishes. The following is (a special case of) a result of M. Sakuma on the Jones polynomial of $\mathbf{2}$-parallel links.

THEOREM 2. [**29**, Theorems 1 and 2] *Let $\mathcal{L}$ be an m-component algebraically split framed link. Then $V(|\mathcal{L}^{\mathbf{2}}|; t)$ is divisible by $(t+1)^{m-1}$, where $|\mathcal{L}^{\mathbf{2}}|$ is the underlying link of the framed link $\mathcal{L}^{\mathbf{2}}$. Moreover*

$$\left.\frac{V(|\mathcal{L}^{\mathbf{2}}|)}{(-t^{1/2} - t^{-1/2})^{m-1}}\right|_{t=-\sqrt{-1}} = \sqrt{-1}^{\,m-2\sum_{i=1}^{m} w(\mathcal{L}_i)} 2^m \prod_{i=1}^{m} w(\mathcal{L}_i).$$

If $w(L_i) = 0$ for every i, then $V(|\mathcal{L}^2|)$ is divisible by $(-t^{1/2} - t^{-1/2})^m$. Since $|\mathcal{L}^2|$ has even number of components, $V(|\mathcal{L}^2|; t)$ is a Laurent polynomial of odd powers of $\sqrt{t}$. So as a corollary, we have

COROLLARY 1.1. *If $\mathcal{L}$ is an m-component algebraically split framed link with every framing zero. Then*

$$V(|\mathcal{L}^2|; t) = (t+1)^m \sqrt{t} W(|\mathcal{L}|; t)$$

with $W(|\mathcal{L}|; t)$ a Laurent polynomial in t.

Substituting $q = -\xi^2$ with t and $s = \varepsilon\sqrt{-1}\xi^4$ with $\sqrt{t}$, we can write $\tilde{V}(|\mathcal{L}^2|; q)$ as $(\xi - 1)^m \varepsilon\sqrt{-1} \sum_{d=0}^{p-2} w_d(|\mathcal{L}|)(\xi - 1)^d$ with $w_d(|\mathcal{L}|) \in \mathbf{Z}$. Moreover we have the following lemma which is proved in Section4

LEMMA 1.2. *Let $\mathcal{L}$ be an algebraically split m-component link with zero framing. Then*

$$\tilde{V}(|\mathcal{L}^2|; q) = (\xi - 1)^m \varepsilon\sqrt{-1} \times \sum_{d=0}^{p-2} w_d(|\mathcal{L}|)(\xi - 1)^d$$

with

$$w_0(|\mathcal{L}|) = -2^m \times 6\varphi(|\mathcal{L}|) + \frac{1}{2}\delta_{m,1},$$

where $\varphi(|\mathcal{L}|)$ is the z^2-coefficient of the Conway polynomial of $|\mathcal{L}|$.

We follow R. Kirby and P. Melvin to define the quantum $SU(2)$-invariants associated with the trivial first cohomology class.

DEFINITION 1.3. [9, Theorem 8.32] Let M be the 3-manifold obtained from S^3 by Dehn surgery along an m-component framed link $\mathcal{L}$. Then

$$\tau_{2p}^{SU(2)}(M, 0) = \alpha(\mathcal{L})\Sigma(\mathcal{L}),$$

with

$$\alpha(\mathcal{L}) = \left(\sqrt{\frac{1}{p}} \sin\frac{\pi}{2p}\right)^m \exp\left(-\frac{3(p-1)}{8p} \times 2\pi\sqrt{-1}\right)^{\sigma(\mathcal{L})}$$

and

$$\Sigma(\mathcal{L}) = \sum_{\mathbf{l}=0}^{(\mathbf{p}-\mathbf{1})/2} 2^{|T_{\mathbf{l}}|}[2\mathbf{l}+\mathbf{1}] J_{\mathcal{L},2\mathbf{l}+\mathbf{1}}.$$

Here $\sigma(\mathcal{L})$ is the signature of $\mathcal{L}$ (the signature of the linking matrix of $\mathcal{L}$; diagonals are framings and off-diagonals are linking numbers), $J_{\mathcal{L},2\mathbf{l}+\mathbf{1}}$ is the colored Jones polynomial defined in [9, Definition 3.25], and $T_{\mathbf{l}} = (T_{l_1}, T_{l_2}, \ldots, T_{l_m})$ with T_{l_i} is 1 or 0 according to whether $l_i < (p-1)/2$ or $l_i = (p-1)/2$.

2. Proof of Theorem 1

In this section we give the proof of Theorem 1. We will prove the following proposition instead of them.

PROPOSITION 2.1. *Suppose that a $\mathbf{Z}/p\mathbf{Z}$-homology three-sphere M is obtained from S^3 by Dehn surgery along an m-component, algebraically split framed link $\mathcal{L}$. Then the conclusion of Theorem 1 holds.*

The proof of Theorem 1 now follows from Ohtsuki's diagonalizing lemma [**25**, Lemma 2.4] (see [**25**, Proof of Theorem 1.1 assuming Proposition 2.3, p.248]). So we will assume that $\mathcal{L}$ is algebraically split with framing $\mathbf{f} = (f_1, f_2, \ldots, f_m)$. Since M is a $\mathbf{Z}/p\mathbf{Z}$-homology sphere we also assume that $(f_i, p) = 1$ for $1 \le i \le m$.

The following three lemmas are proved in the next section.

LEMMA 2.2.

$$\alpha(\mathcal{L}) = \frac{(\xi^{\bar{4}} + \xi^{-\bar{4}})^m}{(2G_{0,1}(\xi))^m}(-1)^{u\sigma(\mathcal{L})+\sigma_-(\mathcal{L})(\varepsilon-1)/2}\xi^{3\times\bar{8}\varepsilon\sigma(\mathcal{L})},$$

where $G_{0,1}(\xi) = \sum_{k=0}^{p-1}\xi^{k^2}$ *is a Gaussian sum and* $\sigma_-(\mathcal{L})$ *is the number of negatives in* $\mathbf{f}$.

Note that $G_{0,1}(\xi)$ is of the form $u(\xi-1)^{(p-1)/2}$ with u invertible in $\mathbf{Z}[\xi]$. (See [**20**].)

LEMMA 2.3.

$$\Sigma(\mathcal{L}) = [2]\sum_{\mathbf{n}=0}^{(\mathbf{p}-\mathbf{1})/2}(-1)^{\mathbf{n}}\tilde{V}(|\mathcal{L}_{\mathbf{0}}^{2\mathbf{n}}|)S_{\mathbf{n}}(\mathbf{f}),$$

where $\mathcal{L}_{\mathbf{0}}$ *is the* $\mathbf{0}$*-framed link obtained from* $\mathcal{L}$, *and* $S_{\mathbf{n}}(\mathbf{f}) = (S_{n_1}(f_1), S_{n_2}(f_2), \ldots, S_{n_m}(f_m))$ *with*

$$S_n(f) = \sum_{k=n}^{p-n-1}[2k+1]\xi^{(k^2+k)f/2}(-1)^k\binom{k+n}{k-n}.$$

LEMMA 2.4. $S_n(f)$ *is divisible by* $(\xi-1)^{(p-1)/2-n}$ *in* $\mathbf{Z}[\xi]$ *and*

$$T\left((\xi-1)^n\frac{S_n(f)}{G_{0,1}(\xi)}\right) \equiv \begin{cases} \left(\dfrac{2f}{p}\right) & \textit{if } n=0 \\ -\overline{2f}\left(\dfrac{2f}{p}\right) & \textit{if } n=1 \end{cases} \mod p.$$

We put

$$\Sigma_0(\mathcal{L}) = [2]\tilde{V}(|\mathcal{L}_{\mathbf{0}}^{0}|)S_{\mathbf{0}}(\mathbf{f}) = S_{\mathbf{0}}(\mathbf{f}),$$

$$\Sigma_1(\mathcal{L}) = \sum_{0<\mathbf{n}\le\mathbf{1}}[2](-1)^{\mathbf{n}}\tilde{V}(|\mathcal{L}_{\mathbf{0}}^{2\mathbf{n}}|)S_{\mathbf{n}}(\mathbf{f}),$$

$$\Sigma_2(\mathcal{L}) = \sum_{\mathbf{n}\not\le 1}[2](-1)^{\mathbf{n}}\tilde{V}(|\mathcal{L}_{\mathbf{0}}^{2\mathbf{n}}|)S_{\mathbf{n}}(\mathbf{f}),$$

so that $\Sigma(\mathcal{L}) = \Sigma_0(\mathcal{L}) + \Sigma_1(\mathcal{L}) + \Sigma_2(\mathcal{L})$.

Now the proof of Proposition 2.1 follows from the following five claims assuming Lemmas 2.2–2.4 (see [**21**, Proof of Proposition 2.1]).

CLAIM 2.5. $\alpha(\mathcal{L})\Sigma_n(\mathcal{L}) \in \mathbf{Z}_{(2)}[\xi]$ *for every* n.

CLAIM 2.6.

$$T(\alpha(\mathcal{L})\Sigma_0(\mathcal{L})) \equiv \left(\frac{|H_1(M;\mathbf{Z})|}{p}\right) \mod p$$

and

$$T\left(\frac{\alpha(\mathcal{L})\Sigma_0(\mathcal{L})-\left(\frac{|H_1(M;\mathbf{Z})|}{p}\right)}{\xi-1}\right)$$
$$\equiv \overline{8}\left(\frac{|H_1(M;\mathbf{Z})|}{p}\right)\left\{3\sigma(\mathcal{L})-\sum_{l=1}^{m}(f_l+\overline{f_l})\right\} \quad \bmod p.$$

CLAIM 2.7. *$\alpha(\mathcal{L})\Sigma_1(\mathcal{L})$ is divisible by $\xi-1$ in $\mathbf{Z}_{(2)}[\xi]$ and moreover*

$$T\left(\frac{\alpha(\mathcal{L})\Sigma_1(\mathcal{L})}{\xi-1}\right)\equiv\left(\frac{|H_1(M;\mathbf{Z})|}{p}\right)\left\{-\overline{8}\sum_{l=1}^{m}\overline{f_l}+3\sum_{\tilde{\mathcal{L}}\in\mathcal{L}}\varphi(\ell)\prod_{\mathcal{K}_l\in\tilde{\mathcal{L}}}\overline{f_l}.\right\} \quad \bmod p$$

CLAIM 2.8. *$\alpha(\mathcal{L})\Sigma_2(\mathcal{L})$ is divisible by $(\xi-1)^2$ in $\mathbf{Z}_{(2)}[\xi]$.*

PROOF OF CLAIM 2.5. If $n=0$, the claim follows from Lemma 2.4 since $G_{0,1}(\xi)$ is divisible by $(\xi-1)^{(p-1)/2}$. (We can also show that it is in $\mathbf{Z}[\xi]$.) If $n>0$, we see that $\Sigma_n(\mathcal{L})$ is divisible by $(\xi-1)^{(p-1)/2-n+1}$ (note that $[2]=\varepsilon\sqrt{-1}(\xi^{\overline{4}}-\xi^{-\overline{4}})$) and so $\alpha(\mathcal{L})\Sigma_n(\mathcal{L})\in\mathbf{Z}_{(2)}[\xi]$ (moreover it is divisible by $\xi-1$). □

PROOF OF CLAIM 2.6. By a calculation using [**20**, Lemma 6.2], we have

$$S_0(f)=\frac{2\left(\frac{2f}{p}\right)\xi^{-\overline{8}(f+\overline{f})}G_{0,1}}{\xi^{\overline{4}}+\xi^{-\overline{4}}}.$$

From Lemmas 2.2 and 2.4, we have

$$\alpha(\mathcal{L})\Sigma_0(\mathcal{L})=(-1)^{u\sigma(\mathcal{L})+\sigma_-(\mathcal{L})(\varepsilon-1)/2}\xi^{\overline{8}(3\varepsilon\sigma(\mathcal{L})-\sum_{l=1}^{m}(f_l+\overline{f_l}))}\prod_{l=1}^{m}\left(\frac{2f_l}{p}\right). \tag{2.1}$$

Therefore

$$T\left(\alpha(\mathcal{L})\Sigma_0(\mathcal{L})\right)=(-1)^{u\sigma(\mathcal{L})+\sigma_-(\mathcal{L})(\varepsilon-1)/2}\prod_{l=1}^{m}\left(\frac{2f_l}{p}\right).$$

Since $\left(\frac{2}{p}\right)=(-1)^u$ and $\left(\frac{-1}{p}\right)=(-1)^{(p-1)/2}$, we have

$$\begin{aligned}T\left(\alpha(\mathcal{L})\Sigma_0(\mathcal{L})\right)&=(-1)^{u\sigma(\mathcal{L})+\sigma_-(\mathcal{L})(\varepsilon-1)/2+mu+\sigma_-(\mathcal{L})(p-1)/2}\left(\frac{\prod_{l=1}^{m}|f_l|}{p}\right)\\&=\left(\frac{\prod_{l=1}^{m}|f_l|}{p}\right)\\&=\left(\frac{|H_1(M;\mathbf{Z})|}{p}\right).\end{aligned}$$

To prove the second statement, we define a derivative $D:\mathbf{Z}_{(2)}[\xi]\to(\mathbf{Z}/p\mathbf{Z})[\xi]$ by $D(\sum_j a_j/2^{k_j}\xi^j)=\sum_j\overline{2}^{k_j}ja_j\xi^{j-1}$. Then it is easy to check that for any $x\in\mathbf{Z}_{(2)}[\xi]$, $x-T(x)$ is divisible by $\xi-1$ and $T\left(\frac{x-T(x)}{\xi-1}\right)\equiv T(D(x)) \bmod p$. (See [**21**, Lemma 3.3].)

Therefore we will show that

$$T(D(\alpha(\mathcal{L})\Sigma_0(\mathcal{L}))) \equiv \overline{8}\left(\frac{|H_1(M;\mathbf{Z})|}{p}\right)\left\{3\sigma(\mathcal{L}) - \sum_{l=1}^{m}(f_l + \overline{f_l})\right\} \mod p,$$

which follows from Equation (2.1) and the proof of the first statement. □

PROOF OF CLAIM 2.7. First note that the summation over $\mathbf{0} < \mathbf{n} \leq \mathbf{1}$ can be regarded as the summation over all the nonempty sublinks of L. Therefore

$$\Sigma_1(\mathcal{L}) = [2] \sum_{\emptyset \neq \tilde{\mathcal{L}} \subset \mathcal{L}} (-1)^{\sharp(\tilde{\mathcal{L}})}\tilde{V}(|\tilde{\mathcal{L}_0}^2|) \prod_{\mathcal{K}_l \in \tilde{\mathcal{L}}} S_1(f_l) \prod_{\mathcal{K}_l \notin \tilde{\mathcal{L}}} S_0(f_l)$$

and

$$\begin{aligned}
&\alpha(\mathcal{L})\Sigma_1(\mathcal{L}) \\
&= (-1)^{u\sigma(\mathcal{L})+\sigma_-(\mathcal{L})(\varepsilon-1)/2}\xi^{3\times\overline{8}\varepsilon\sigma(\mathcal{L})}\left(\frac{\xi^{\overline{4}}+\xi^{-\overline{4}}}{2}\right)^m \\
&\quad \times [2]\sum_{\tilde{\mathcal{L}} \subset \mathcal{L}}(-1)^{\sharp(\tilde{\mathcal{L}_0})}\tilde{V}(|\tilde{\mathcal{L}_0}^2|)\prod_{\mathcal{K}_l \in \tilde{\mathcal{L}}}\frac{S_1(f_l)}{G_{0,1}(\xi)}\prod_{\mathcal{K}_l \notin \tilde{\mathcal{L}}}\frac{S_0(f_l)}{G_{0,1}(\xi)}.
\end{aligned}$$

From Lemma 1.2, this becomes

$$\begin{aligned}
&(-1)^{u\sigma(\mathcal{L})+\sigma_-(\mathcal{L})(\varepsilon-1)/2}\xi^{3\times\overline{8}\varepsilon\sigma(\mathcal{L})}\left(\frac{\xi^{\overline{4}}+\xi^{-\overline{4}}}{2}\right)^m \\
&\times \varepsilon\sqrt{-1}(\xi^{\overline{4}}-\xi^{-\overline{4}})\sum_{\tilde{\mathcal{L}} \subset \mathcal{L}}(-1)^{\sharp(\tilde{\mathcal{L}})}\varepsilon\sqrt{-1}\sum_{d=0}^{p-1}w_d(|\tilde{\mathcal{L}_0}|)(\xi-1)^d \\
&\times \prod_{\mathcal{K}_l \in \tilde{\mathcal{L}}}\frac{S_1(f_l)(\xi-1)}{G_{0,1}(\xi)}\prod_{\mathcal{K}_l \notin \tilde{\mathcal{L}}}\frac{S_0(f_l)}{G_{0,1}(\xi)}.
\end{aligned}$$

Since $T(\xi^{\overline{4}} - \xi^{-\overline{4}}) \equiv 0$ and $T\left(\frac{\xi^{\overline{4}}-\xi^{-\overline{4}}}{\xi-1}\right) \equiv \overline{2} \mod p$, we have from Lemma 2.4 and the proof of Claim 2.6 that $\alpha(\mathcal{L})\Sigma_1(\mathcal{L})$ is divisible by $\xi - 1$ and

$$\begin{aligned}
T\left(\frac{\alpha(\mathcal{L})\Sigma_1(\mathcal{L})}{\xi-1}\right) &\equiv -\overline{2}(-1)^{u\sigma(\mathcal{L})+\sigma_-(\mathcal{L})(\varepsilon-1)/2}\sum_{\tilde{\mathcal{L}} \subset \mathcal{L}}(-1)^{\sharp(\tilde{\mathcal{L}})}w_0(|\tilde{\mathcal{L}_0}|) \\
&\quad \times \prod_{\mathcal{K}_l \in \tilde{\mathcal{L}}}(-1)\left(\frac{2f_l}{p}\right)\overline{2f_l}\prod_{\mathcal{K}_l \notin \tilde{\mathcal{L}}}\left(\frac{2f_l}{p}\right) \\
&\equiv -\overline{2}\left(\frac{|H_1(M;\mathbf{Z})|}{p}\right)\sum_{\tilde{\mathcal{L}} \subset \mathcal{L}}\left\{2^{\sharp(\tilde{\mathcal{L}})}\times(-6)\varphi(|\tilde{\mathcal{L}}|)+\overline{2}\delta_{\sharp(\tilde{\mathcal{L}}),1}\right\}\prod_{\mathcal{K}_l \in \tilde{\mathcal{L}}}\overline{2f_l} \\
&\equiv \left(\frac{|H_1(M;\mathbf{Z})|}{p}\right)\left\{3\sum_{\tilde{\mathcal{L}} \subset \mathcal{L}}\varphi(|\tilde{\mathcal{L}}|)\prod_{\mathcal{K}_l \in \tilde{\mathcal{L}}}\overline{f_l} - \overline{8}\sum_{l=1}^{m}\overline{f_l}\right\} \mod p,
\end{aligned}$$

completing the proof. □

PROOF OF CLAIM 2.8. As noted in the proof of Claim 2.5, we see that $\alpha(\mathcal{L})\Sigma_2(\mathcal{L})$ is divisible by $(\xi - 1)$. So it is sufficient to prove that $T\left(\frac{\alpha(\mathcal{L})\Sigma_2(\mathcal{L})}{\xi-1}\right) \equiv 0 \mod p$. Since each term is a multiple of $T(\tilde{V}(|\mathcal{L}_0^{2\mathbf{n}}|))$, we only need to show that $T(\tilde{V}(|\mathcal{L}_0^{\mathbf{n}}|)) \equiv$

0. But it equals $T(w_0(|\mathcal{L}_0^{\mathbf{n}}|))$ which vanishes if some n_l is greater than 1. (This is because in such a case $|\mathcal{L}_0^{\mathbf{n}}|$ has two components parallel to each other with zero framing and so has vanishing Conway polynomial.) □

3. Proof of Lemmas 2.2-2.4

In this section we will give proofs of lemmas used in the previous section.

PROOF OF LEMMA 2.2. First note that $G_{0,1}(\xi) = (\sqrt{-1})^{(1-\varepsilon)/2}\sqrt{p}$. Therefore

$$\sqrt{\frac{1}{p}}\sin\frac{\pi}{2p} = \frac{(\sqrt{-1})^{(1-\varepsilon)/2}(s-s^{-1})}{2\sqrt{-1}G_{0,1}(\xi)} = \frac{\varepsilon\sqrt{-1}^{(1-\varepsilon)/2}(\xi^{\bar{4}}+\xi^{-\bar{4}})}{2G_{0,1}(\xi)}.$$

Now since

$$\exp\left(-\frac{3(p-1)}{8p}\times 2\pi\sqrt{-1}\right) = s^{-3(p-1)/2} = (\varepsilon\sqrt{-1})^{-3(p-1)/2}\xi^{-3\times\bar{4}(p-1)/2},$$

we have

$$\alpha(\mathcal{L}) = \left(\frac{\varepsilon\sqrt{-1}^{(1-\varepsilon)/2}(\xi^{\bar{4}}+\xi^{-\bar{4}})}{2G_{0,1}(\xi)}\right)^m (\varepsilon\sqrt{-1})^{-3\sigma(\mathcal{L})(p-1)/2}\xi^{-3\times\bar{4}\sigma(\mathcal{L})(p-1)/2}$$
$$= \left(\frac{\xi^{\bar{4}}+\xi^{-\bar{4}}}{2G_{0,1}(\xi)}\right)^m (-1)^{u\sigma(\mathcal{L})+\sigma_-(\mathcal{L})(\varepsilon-1)/2}\xi^{3\times\bar{8}\sigma(\mathcal{L})}$$

□

PROOF OF LEMMA 2.3. From [9], we have

$$\begin{aligned}\Sigma(\mathcal{L}) &= \sum_{\mathbf{l}=0}^{(\mathbf{p}-1)/2} 2^{|T_{\mathbf{l}}|}[2\mathbf{l}+1]J_{L,2\mathbf{l}+1} \\ &= \sum_{\mathbf{l}=0}^{(\mathbf{p}-1)/2} 2^{|T_{\mathbf{l}}|}[2\mathbf{l}+1]s^{\mathbf{f}((2\mathbf{n}+1)^2-1)/2}J_{L_0,2\mathbf{l}+1} \\ &= \sum_{\mathbf{l}=0}^{(\mathbf{p}-1)/2} 2^{|T_{\mathbf{l}}|}[2\mathbf{l}+1]s^{\mathbf{f}(2\mathbf{l}^2+2\mathbf{l})}J_{L_0,2\mathbf{l}+1} \\ &= \sum_{\mathbf{l}=0}^{(\mathbf{p}-1)/2} 2^{|T_{\mathbf{l}}|}[2\mathbf{l}+1]\xi^{\mathbf{f}(\mathbf{l}^2+\mathbf{l})/2}[2]\sum_{\mathbf{j}=0}^{\mathbf{l}}(-1)^{\mathbf{j}}\binom{2\mathbf{l}-\mathbf{j}}{\mathbf{j}}\tilde{V}(L_0^{2\mathbf{l}-2\mathbf{j}}).\end{aligned}$$

Now we put $\mathbf{n}=\mathbf{l}-\mathbf{j}$ and replace the summation $\sum_{\mathbf{l}=0}^{(\mathbf{p}-1)/2}\sum_{\mathbf{j}=0}^{\mathbf{l}}$ with $\sum_{\mathbf{n}=0}^{(\mathbf{p}-1)/2}\sum_{\mathbf{l}=\mathbf{n}}^{(\mathbf{p}-1)/2}$.

Then

$$\begin{aligned}\Sigma(\mathcal{L}) &= \sum_{\mathbf{n}=0}^{(\mathbf{p}-1)/2}\sum_{\mathbf{l}=\mathbf{n}}^{(\mathbf{p}-1)/2} 2^{|T_{\mathbf{l}}|}[2\mathbf{l}+1]\xi^{\mathbf{f}(\mathbf{l}^2+\mathbf{l})/2}[2](-1)^{\mathbf{l}-\mathbf{n}}\binom{\mathbf{l}+\mathbf{n}}{\mathbf{l}-\mathbf{n}}\tilde{V}(L_0^{2\mathbf{n}}) \\ &= [2]\sum_{\mathbf{n}=0}^{(\mathbf{p}-1)/2}(-1)^{\mathbf{n}}\tilde{V}(L_0^{2\mathbf{n}})\sum_{\mathbf{l}=\mathbf{n}}^{(\mathbf{p}-1)/2}(-1)^{\mathbf{l}}2^{|T_{\mathbf{l}}|}[2\mathbf{l}+1]\xi^{\mathbf{f}(\mathbf{l}^2+\mathbf{l})/2}\binom{\mathbf{l}+\mathbf{n}}{\mathbf{l}-\mathbf{n}}.\end{aligned}$$

Now we calculate

$$\begin{aligned}
&\sum_{l=n}^{(p-1)/2} (-1)^l 2^{|T_{2l+1}|}[2l+1]\xi^{f(l^2+l)/2}\binom{l+n}{l-n} \\
&= \sum_{l=n}^{(p-3)/2} (-1)^l 2[2l+1]\xi^{f(l^2+l)/2}\binom{l+n}{l-n} \\
&\quad + (-1)^{(p-1)/2}[p]\xi^{f((p-1)/2)^2+(p-1)/2)/2}\binom{(p-1)+n}{(p-1)-n}.
\end{aligned}$$

Since $[2l+1]$, $\xi^{(l+1)l/2}$ and $\binom{l+n}{l-n}$ does not alter if we replace l with $p-1-l$, we see the right hand side of the formula above coincides with

$$S_n(f) = \sum_{l=n}^{p-1-n} (-1)^l [2l+1]\xi^{f(l^2+l)/2}\binom{l+n}{l-n},$$

completing the proof. □

Put $G_{l,a}(\xi) = \sum_{k=0}^{p-1} k^l \xi^{ak^2}$. We use the following lemma first proved by T. Ohtsuki and improved by X.-S. Lin and X. Wang.

LEMMA 3.1. (Corollary to **[22**, Lemmas 8.1, 8.2 and 8.7], **[17**, Lemma 6.2]) *If l is odd, then $G_{l,a}(\xi)$ is divisible by $(\xi-1)^{p-1}$.*

If l is even and $l \le p-1$, then $G_{l,a}(\xi)$ is divisible by $(\xi-1)^{(p-1)/2-l/2}$. Moreover

$$T\left((\xi-1)^{l/2}\frac{G_{l,a}(\xi)}{G_{0,1}}\right) \equiv \begin{cases} \left(\dfrac{a}{p}\right) & \text{if } l=0 \\ -\overline{2a}\left(\dfrac{a}{p}\right) & \text{if } l=2 \end{cases} \mod p.$$

Using this lemma we have the proof of Lemma 2.4.

PROOF OF LEMMA 2.4. Note that since $\binom{k+n}{k-n}$ vanishes modulo p for $0 \le k \le n-1$ and for $p-n \le k \le p-1$, we have

$$\begin{aligned}
S_n(f) &\equiv \sum_{k=0}^{p-1}[2k+1]\xi^{f(k^2+k)/2}(-1)^k\binom{k+n}{k-n} \\
&\equiv \frac{1}{\xi^{\overline{4}}+\xi^{-\overline{4}}}\sum_{k=0}^{p-1}\left\{\xi^{\overline{2}fk^2+\overline{2}(f-1)k-\overline{4}}\binom{k+n}{k-n} + \xi^{\overline{2}fk^2+\overline{2}(f+1)k+\overline{4}}\binom{k+n}{k-n}\right\} \\
&\mod p.
\end{aligned}$$

Now since $\binom{k+n}{k-n}$ is a polynomial in k of degree $2n$, $S_n(f)$ is divisible by $(\xi-1)^{(p-1)/2-n}$ from Lemma 3.1. Moreover since the coefficient of k^{2n} in $\binom{k+n}{k-n}$ is 1, the conclusion follows immediately. □

4. Jones polynomial of a two-parallel link

In this section we show the following result about the Kauffman bracket of a two-parallel link with respect to the zero framing.

Let $\langle \mathcal{L} \rangle \in \mathbf{Z}[A, A^{-1}]$ be the Kauffman bracket of a framed link $\mathcal{L}$ defined by the following formulae.

$$\langle \mathcal{L} \sqcup \mathcal{O} \rangle = (-A^2 - A^{-2}) \langle \mathcal{L} \rangle,$$

and

$$\langle \mathcal{O} \rangle = 1,$$

where $\mathcal{L} \bigsqcup \mathcal{O}$ denotes the disjoint union of $\mathcal{L}$ and the trivial knot with zero framing $\mathcal{O}$.

We also use L.H. Kauffman's convention for two-parallels. Let $\langle\langle \mathcal{L} \rangle\rangle$ denote the two-parallel of a framed link $\mathcal{L}$. We take the two-parallel with respect to the framing which are presented by the blackboard framing. (The writhe of a component equals its framing.)

LEMMA 4.1. *Let $\mathcal{L}$ be an m-component, algebraically split link with zero framing. Then $\langle\langle \mathcal{L} \rangle\rangle$ is divisible by d^m, where $d = -A^2 - A^{-2}$. Moreover*

$$\left. \frac{\langle\langle \mathcal{L} \rangle\rangle}{d^m} \right|_{A=\zeta} = (-4\sqrt{-1})^m (-6\sqrt{-1}\varphi(L)) + \delta_{m,1},$$

where $\zeta = \exp(-\pi\sqrt{-1}/4)$, $\delta_{m,1}$ is 1 or 0 according to whether $m = 1$ or $m > 1$, and $\varphi(L)$ is the coefficient of z^{m+1} of the Conway polynomial of L, the underlying link of $\mathcal{L}$.

PROOF. We use Kauffman's switching formula [**6**]:

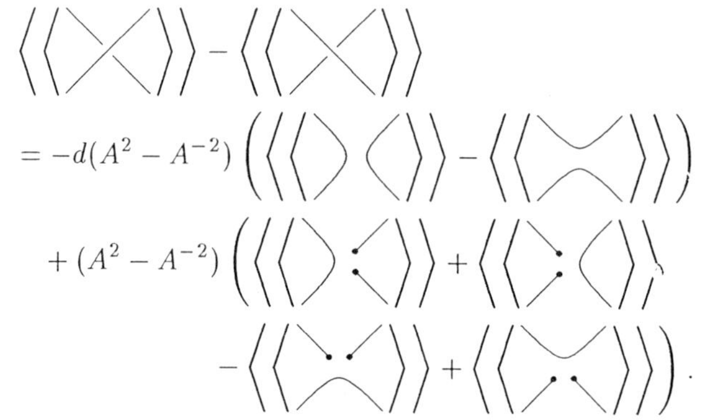

Here 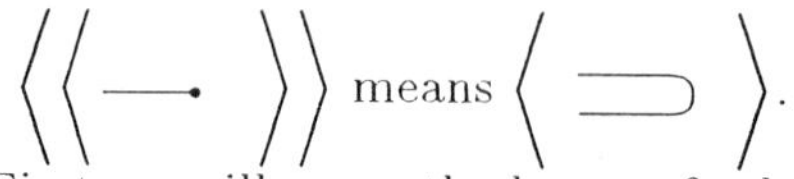.

First we will prove the lemma for knots.

Since

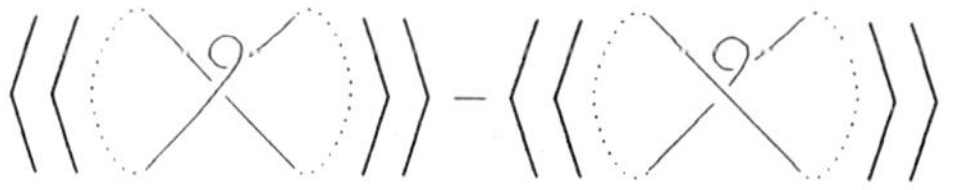

$= -d(A^2 - A^{-2})$ 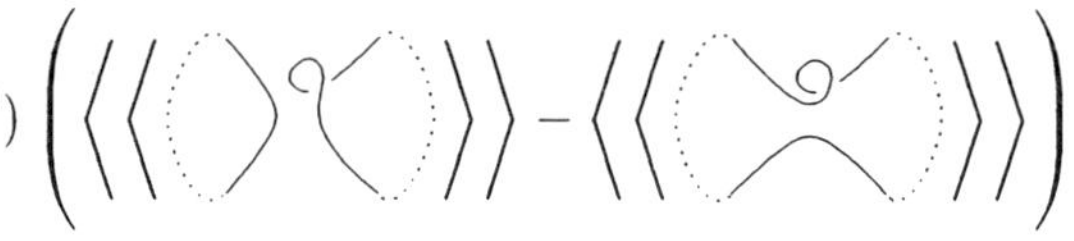

$+ (A^2 - A^{-2})$ 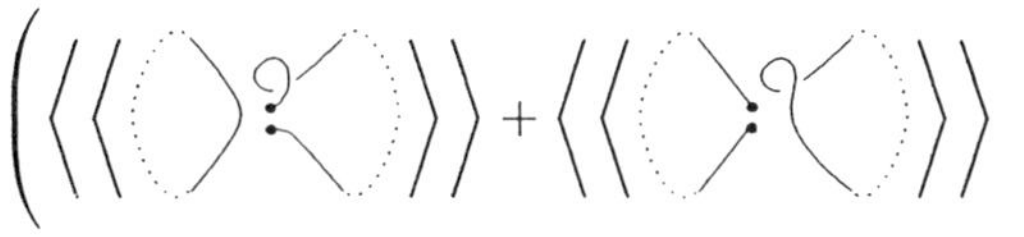

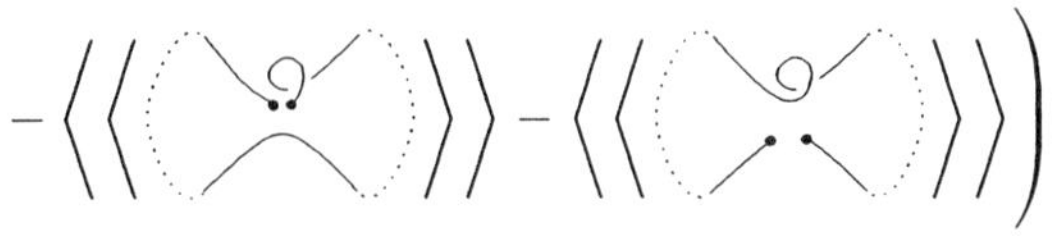

$= d(A^2 - A^{-2})$ 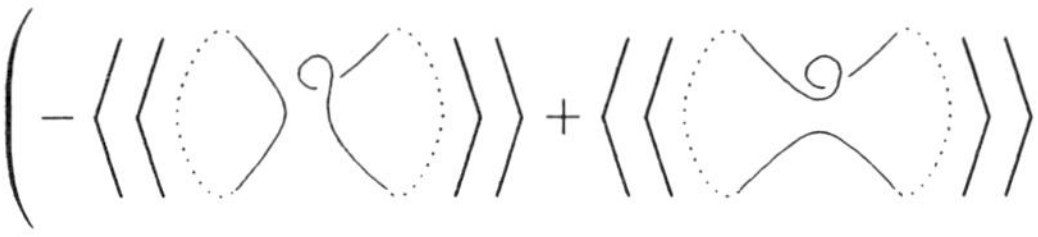

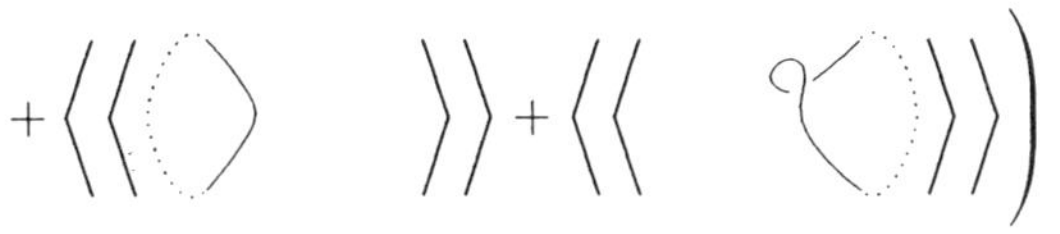

$- 2(A^2 - A^{-2})$

and

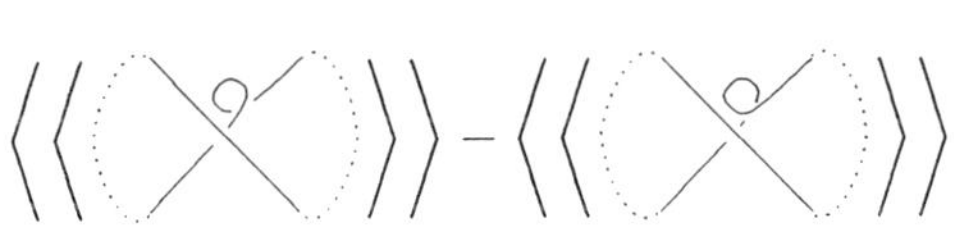

$= -d(A^2 - A^{-2})$ 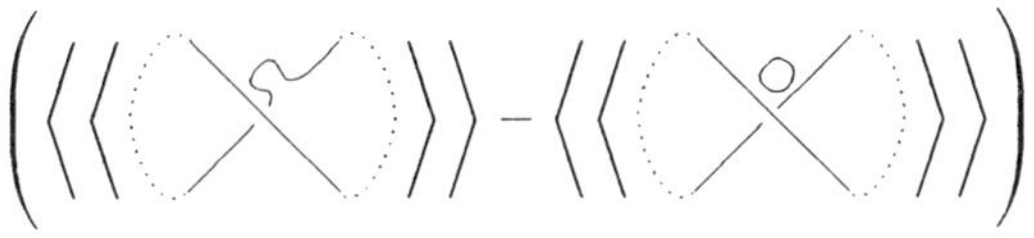

$+ (A^2 - A^{-2})$ 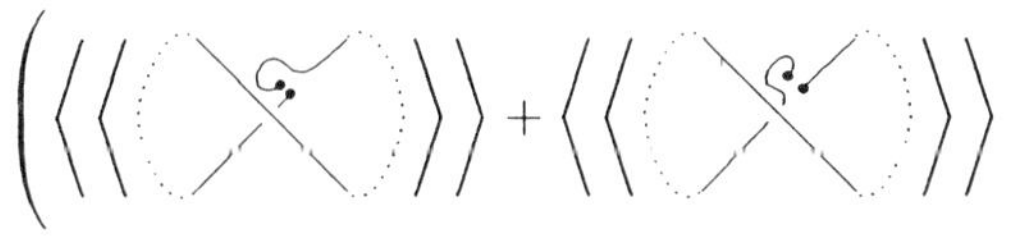

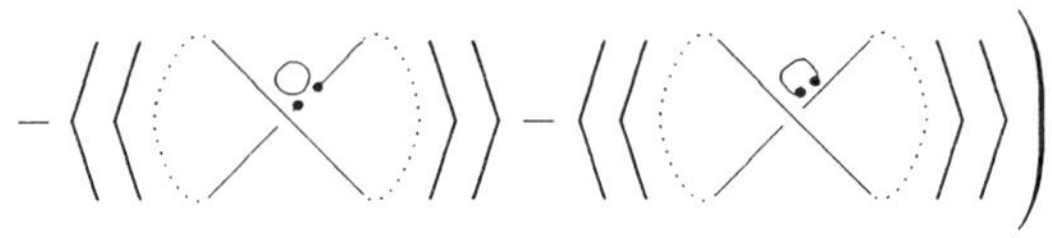

$= d(d^2 - 2)(A^2 - A^{-2})$ 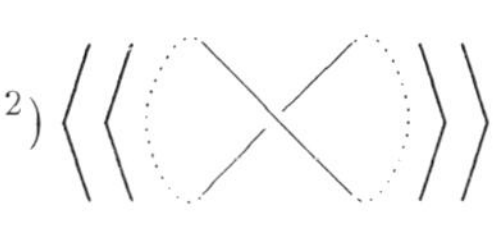

$- (d^2 - 2)(A^2 - A^{-2}),$

we have

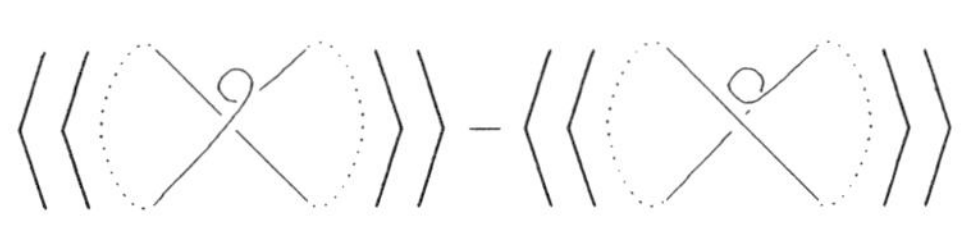

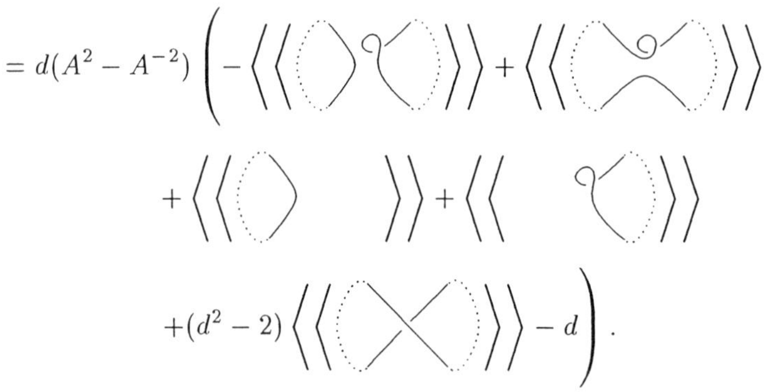

Here we assume that the two framed knots appeared in the right hand side have zero framing. Since $\langle\langle \mathcal{O} \rangle\rangle = d$, it follows that for any 0-framed knot $\mathcal{K}$, $\langle\langle \mathcal{K} \rangle\rangle$ is divisible by d and

(4.1)

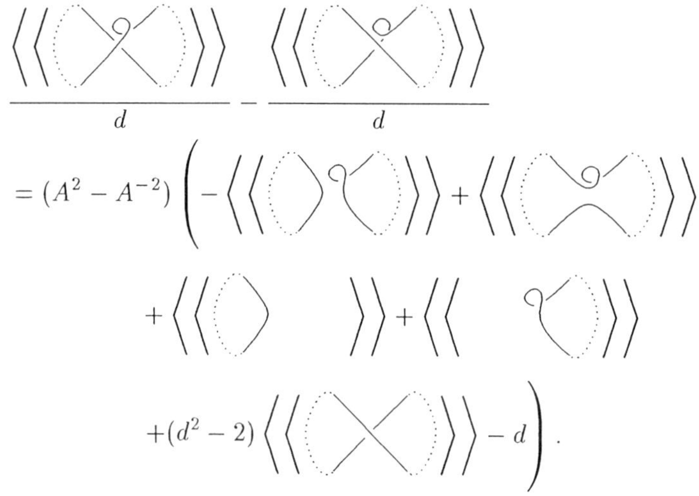

Let α (β, respectively) be the arc connecting the left-top and the left-bottom (right-top and right-bottom, respectively) and orient them in this way. (They are indicated by dotted lines in figures.) Denote by $w(\alpha)$ ($w(\beta)$) the writhe (sum of the signs of the self-crossings) of α (β, respectively) and by $\mathrm{lk}(\alpha, \beta)$ the linking number (half of the sum of the signs of crossings between) of α and β. Since the writhe of

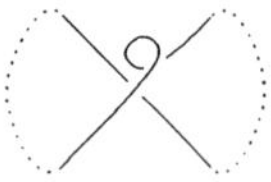 is zero,

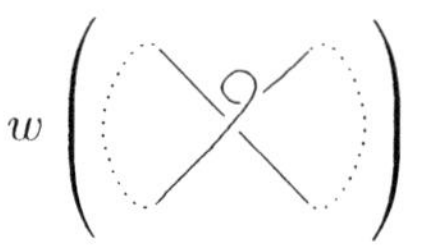 $= w(\alpha) + w(\beta) + 2\mathrm{lk}(\alpha, \beta) = 0.$

Therefore

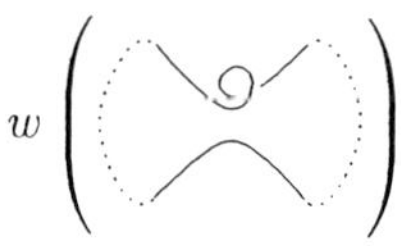 $= w(\alpha) + w(\beta) - 2\mathrm{lk}(\alpha, \beta) - 1 = -4\mathrm{lk}(\alpha, \beta) - 1$

w 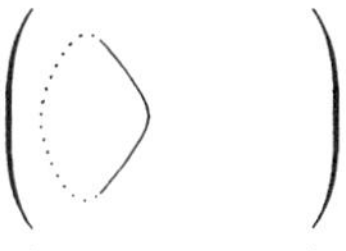 $= w(\alpha)$

w 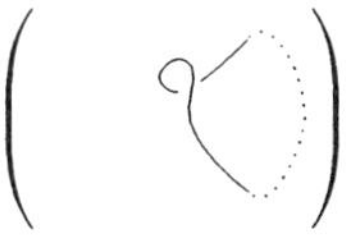 $= w(\beta) - 1$

w 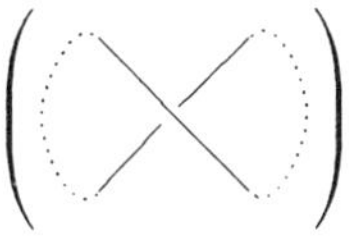 $= -1.$

From [**29**, Theorem 2], we have

$$\begin{aligned}\langle\langle \mathcal{K} \rangle\rangle|_{A=\zeta} &= (-A^3)^{4w(\mathcal{K})} V(K; A^{-4})\Big|_{A=\zeta} \\ &= (-1)^{w(\mathcal{K})} 2\sqrt{-1}^{1-2w(\mathcal{K})} w(\mathcal{K}) \\ &= 2\sqrt{-1}w(\mathcal{K})\end{aligned}$$

since the writhe of the two-parallel of $\mathcal{K}$ is $4w(\mathcal{K})$ and $\langle \mathcal{K} \rangle = V(K; A^{-4})$. Here K is the underlying knot of a framed knot $\mathcal{K}$. Therefore we have

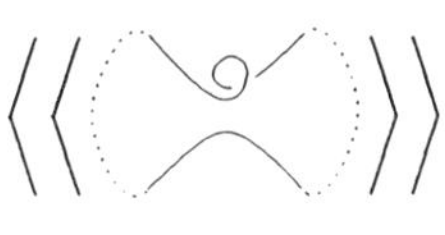 $\Big|_{A=\zeta} = 2\sqrt{-1}\,(-4\mathrm{lk}(\alpha, \beta) - 1)$

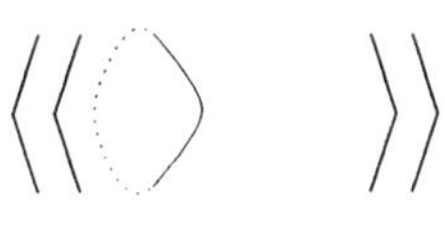 $\Big|_{A=\zeta} = 2\sqrt{-1}w(\alpha)$

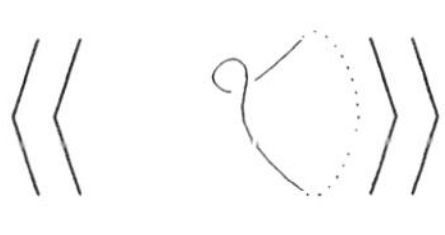 $\Big|_{A=\zeta} = 2\sqrt{-1}\,(w(\beta) - 1)$

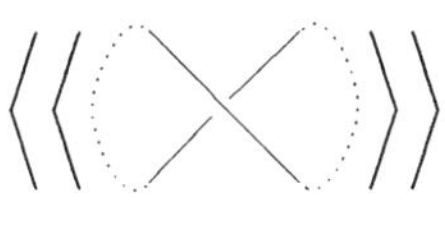 $\Big|_{A=\zeta} = -2\sqrt{-1}.$

From (4.1), we have

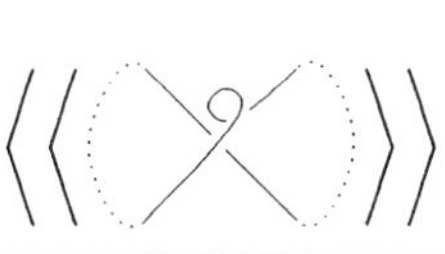 $\Big/ d \,\Big|_{A=\zeta} -$ 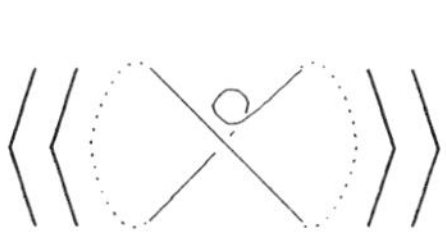 $\Big/ d \,\Big|_{A=\zeta}$

$$\begin{aligned}&= 4\,(-4\mathrm{lk}(\alpha, \beta) - 1 + w(\alpha) + w(\beta) - 1 + 2) \\ &= -24\mathrm{lk}(\alpha, \beta)\end{aligned}$$

since 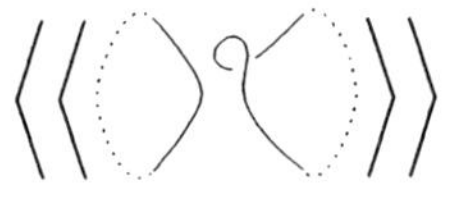 is divisible by d from [**29**, Theorem 2] again. Comparing this with the formula (see for example [**8**])

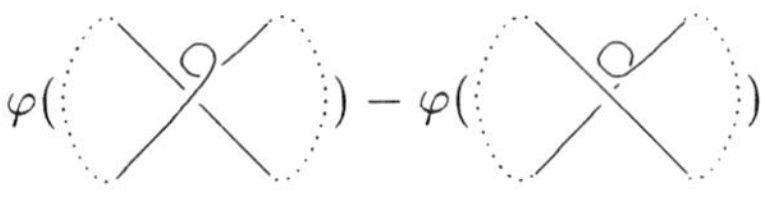

$$= \mathrm{lk}(\alpha, \beta)$$

and

$$\left.\frac{\langle\langle \mathcal{O} \rangle\rangle}{d}\right|_{A=\zeta} = 1,$$

we have

$$\left.\frac{\langle\langle \mathcal{K}_0 \rangle\rangle}{d}\right|_{A=\zeta} = -24\varphi(K) + 1$$

if $\mathcal{K}_0$ is a 0-framed knot with underlying knot K.

Next we will prove the lemma when $m > 1$. We will proceed using the double crossing change technique used in [**20**]. In the following figure, we will look at a pair of small discs and other parts of link diagrams are assumed to be identical. The connectivity of arcs are indicated by dotted lines. We have

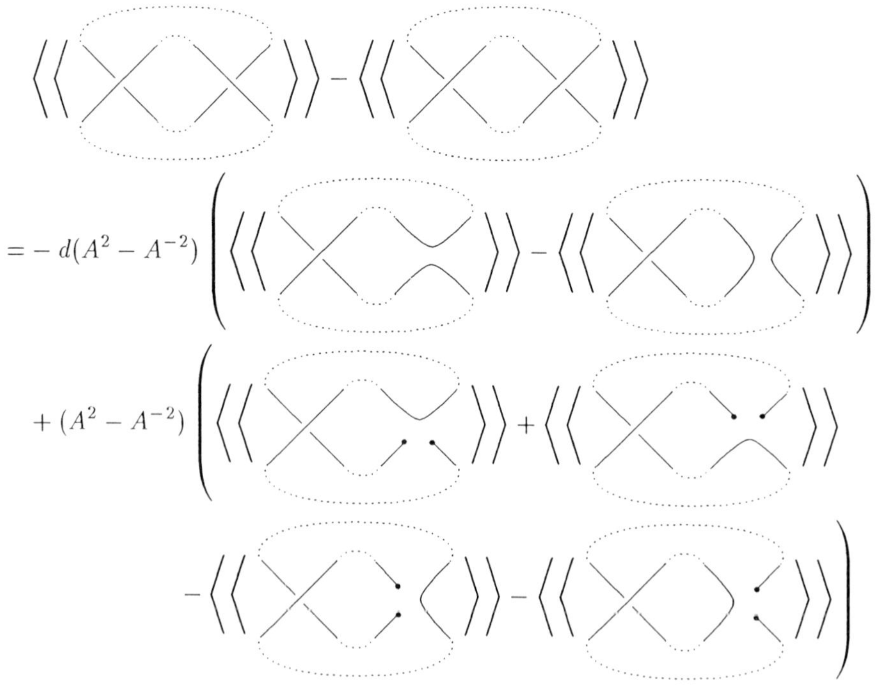

$= -d(A^2 - A^{-2})$ 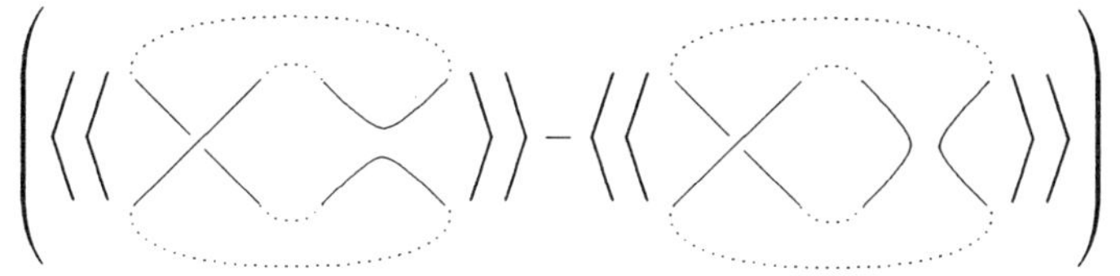

and

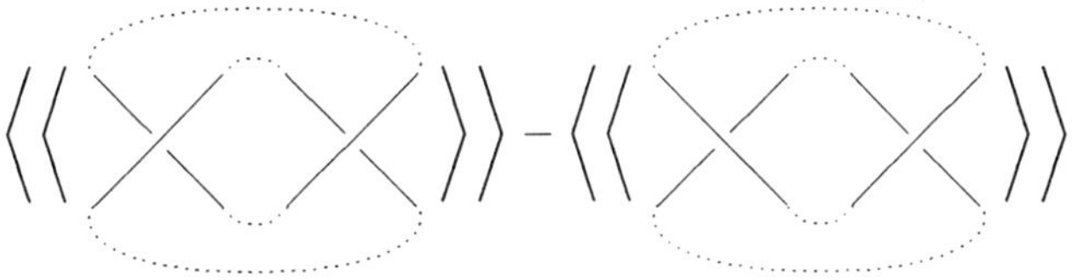

$= -d(A^2 - A^{-2})$ 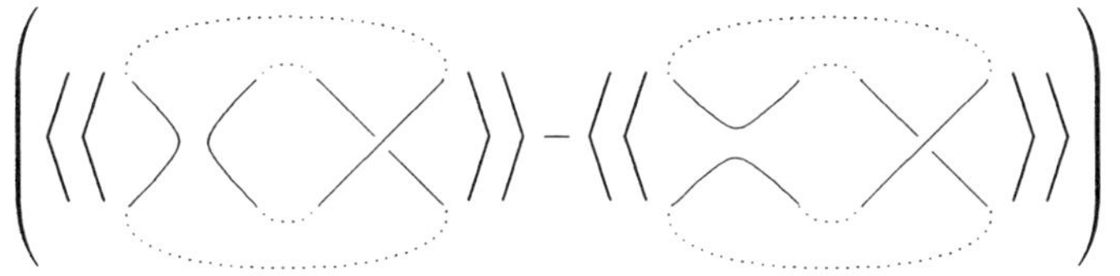

$+ (A^2 - A^{-2})$ 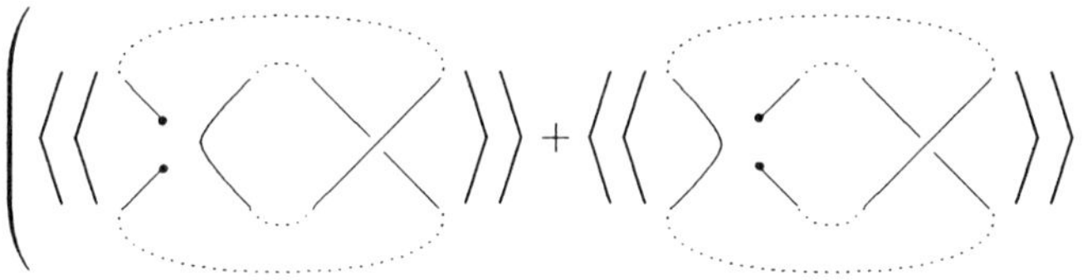

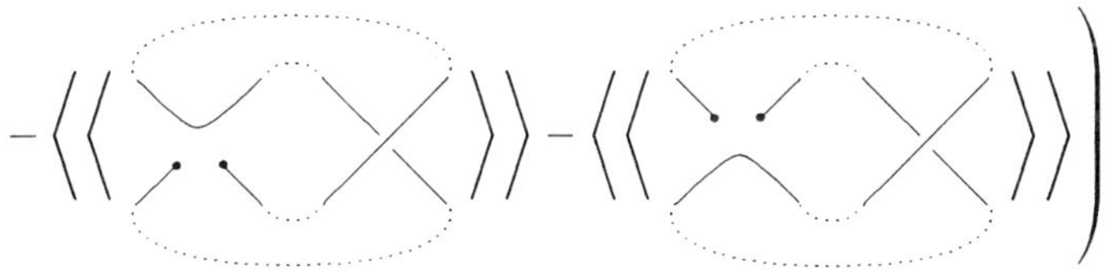

$= -d(A^2 - A^{-2})$ 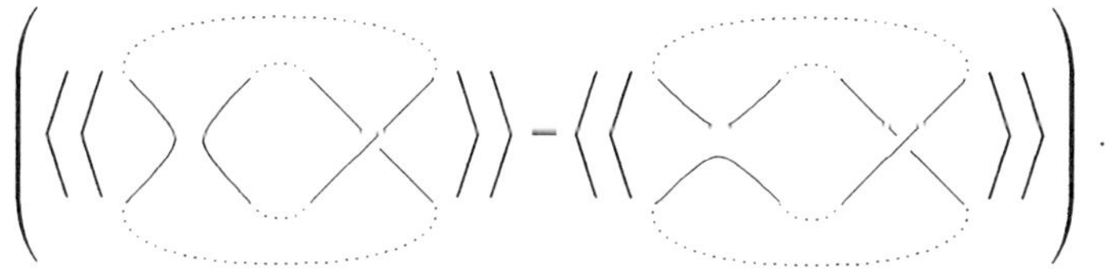

Therefore

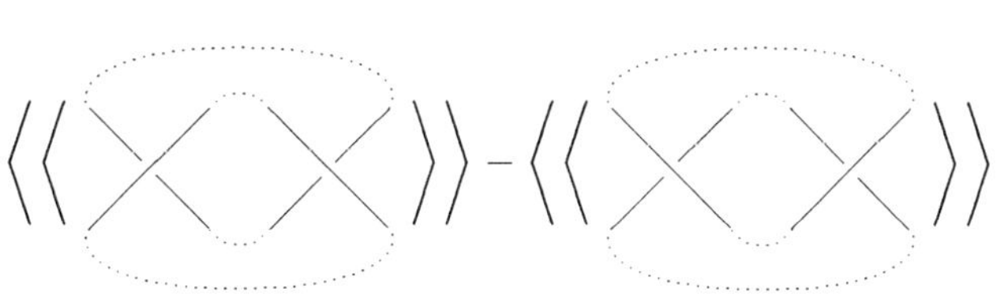

$= -d(A^2 - A^{-2})$ 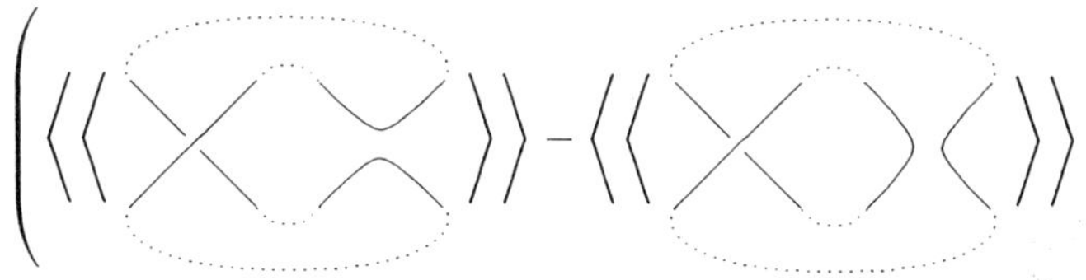

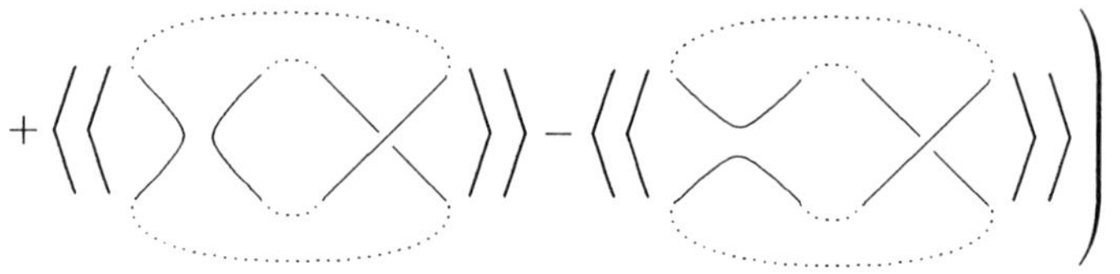

and so

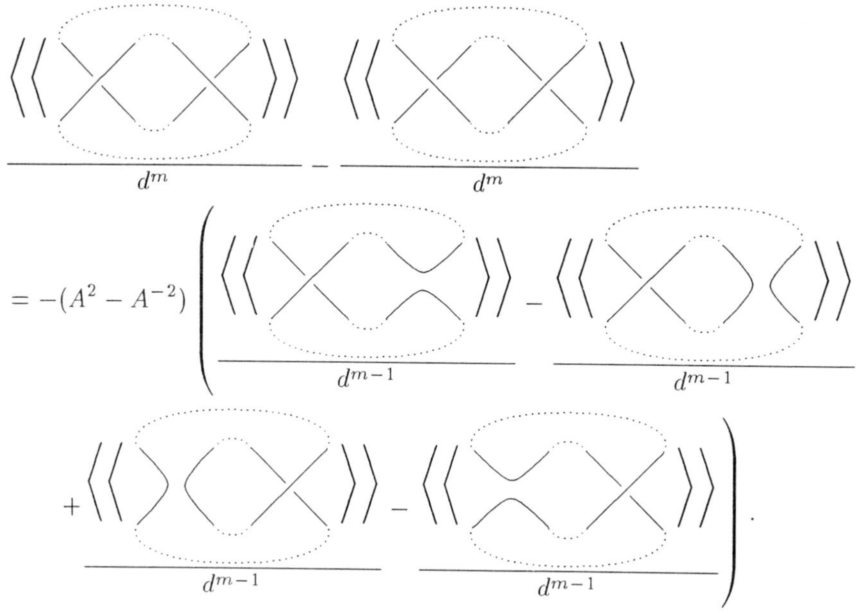

Therefore by induction on the number that is needed to change a link into the trivial link by double crossing changes, we conclude that $\langle\langle\mathcal{L}\rangle\rangle$ is divisible by $d^{\sharp(\mathcal{L})}$. Moreover we have

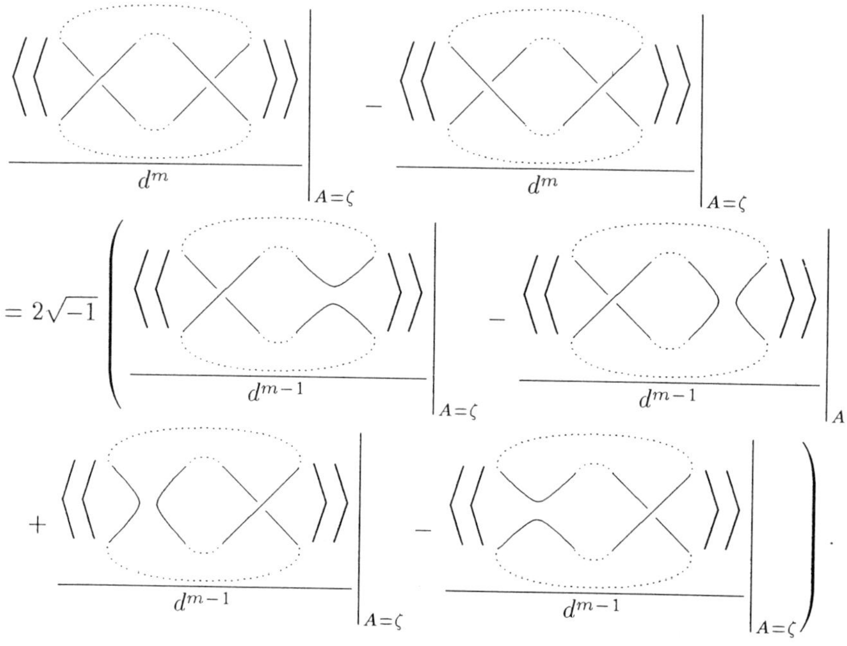

By the way from the definition of the Conway polynomial, we have

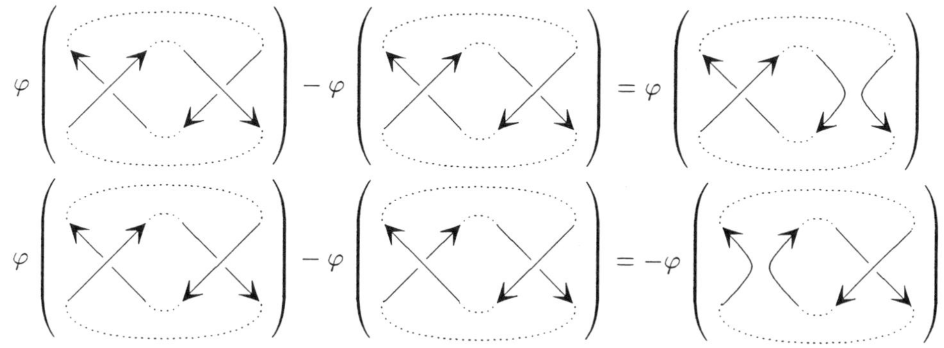

and

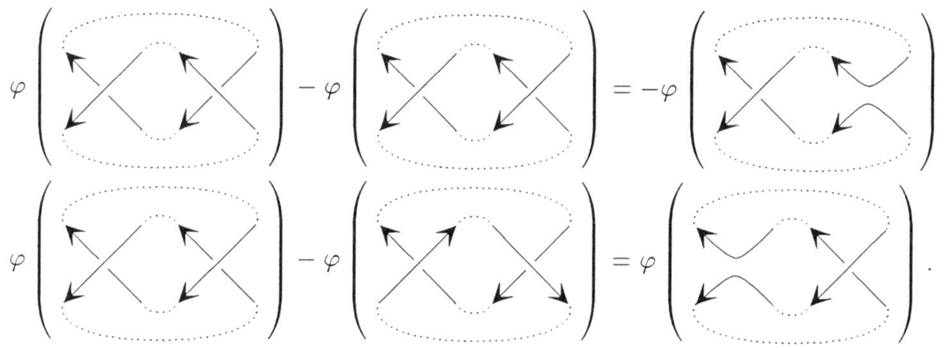

Since $\varphi(L)$ does not depend on the orientation if L is algebraically split [**20**, Corollary 3.13], we have

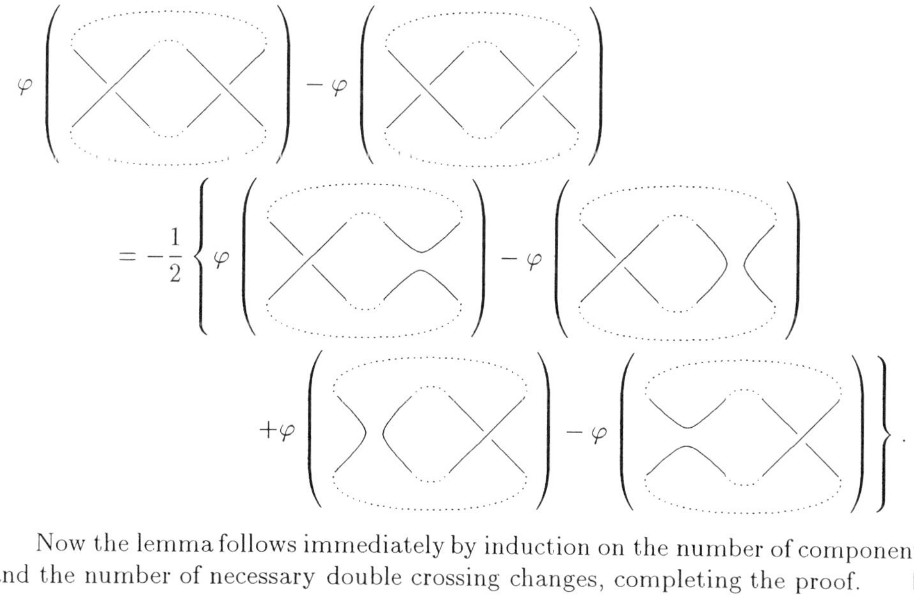

Now the lemma follows immediately by induction on the number of components and the number of necessary double crossing changes, completing the proof. □

Now since $\tilde{V}(L) = (-1)^{m+1} \langle L \rangle \Big|_{A=t^{-1/4}}$, we have Lemma 1.2 immediately.

5. Remarks, conjectures and problems

In this final section we collect some remarks, conjectures and problems.

REMARK 5.1. We first note that R. Kirby, P. Melvin and X. Zhang gives a complete topological interpretation for $\tau_6^{SU(2)}(M,\alpha)$ for any $\alpha \in H^1(M;\mathbf{Z}/2\mathbf{Z})$ [**10**, Theorem]. In particular, if M is a $\mathbf{Z}/3\mathbf{Z}$ homology 3-sphere then $\tau_6^{SU(2)}(M,0) = \sqrt{-1}^{-w(M)}$ with $w(M) \in \mathbf{Z}/4\mathbf{Z}$ the mod 3 Witt invariant of M defined there. It is not hard to show that this coincides with $\left(\dfrac{|H_1(M;\mathbf{Z})|}{3}\right)$. Note also since $p=3$, our theorem says nothing about the Casson-Walker invariant in this case.

REMARK 5.2. In the main theorem we only prove that $\tau_{2p}^{SU(2)}(M,0) \in \mathbf{Z}_{(2)}[\xi]$. The author does not know whether it is in $\mathbf{Z}[\xi]$.

In the conference the author stated that we could obtain a formal power series invariant which is an '$r \to \infty$ limit' of $\tau_{2p}^{SU(2)}(M,0)$ but the proof was erroneous. So we state it as the following conjecture.

CONJECTURE 5.3. *There exist a series of invariants* $\lambda_n^{SU(2)}(M,0) \in \mathbf{Q}\setminus\mathbf{Z}[1/p]$ *for* $n = 0,1,2,\dots$ *such that*

$$\tau_{2p}^{SU(2)}(M,0) = \left(\frac{|H_1(M;\mathbf{Z})|}{p}\right)\sum_{k=0}^{N} \underline{\lambda_0^{SU(2)}(M,0)}(\xi-1)^k + u(\xi-1)^{N+1}$$

for some u *and* N. *Here* $\underline{a/b} \in \mathbf{Z}/p\mathbf{Z}$ *with* $(b,p)=1$ *is defined to be* $\bar{b}\times a$.

Moreover $\lambda_0^{SU(2)}(M,0) = 1$ *and* $\lambda_1^{SU(2)}(M,0) = 3\lambda(M)$.

REMARK 5.4. C. Sato showed that this holds if M is a rational homology Seifert fibered space [**30**]. (He also shows that $\tau_{2p}^{SU(2)}(M,0) \in \mathbf{Z}[\xi]$ in this case.)

REMARK 5.5. S. Garoufalidis and G. Masbaum conjectured at the conference that these two are same up to a constant, i.e. the following formula holds.

$$|H_1(M;\mathbf{Z})|\tau(M;t) = \tau^{SU(2)}(M,0;t^2),$$

where $\tau^{SU(2)}(M,0;t)$ is the formal power series $\sum_{n=0}^{\infty}\lambda_n^{SU(2)}(M,0)(t-1)^n$ and $\tau(M;t)$ is T. Ohtsuki's $SO(3)$ formal power series invariant. But this is not true. In fact, Sato calculated $\tau_{2p}^{SU(2)}$ for rational homology Seifert fibered space in [**30**] and gave a relation between $\lambda_1^{SU(2)}(M,0)$ and $\lambda_1^{SU(2)}(M,0)$, which differs from the $SO(3)$ case.

But we do not know whether these two are independent or not.

PROBLEM 5.6. Is $\lambda_n^{SU(2)}(M,0)$ independent of $\lambda_n(M)$?

Note that in the definitions of both $\tau_r^{SO(3)}(M)$ and $\tau_{2p}^{SU(2)}(M,0)$, we take the sum of the Jones polynomials of odd colors (see [**9**]).

T. Ohtsuki introduced a notion of finite type for integral homology three-spheres [**24**].

PROBLEM 5.7. Is $\lambda_n^{SU(2)}(M,0)$ of finite type? The answer is yes for $n=0$ and 1. Note that λ_n is of finite type for every n [**13**]. This is equivalent to the following as pointed out by J. Andersen.

PROBLEM 5.8. Can $\lambda_n^{SU(2)}(M,0)$ be recovered from the universal perturbative invariant [**15**, **14**]? Note again that one can recover λ_n from it. [**26**] (see also [**23**, **27**]).

PROBLEM 5.9. Can we prove integrality of $\tau_{2p}^{SU(2)}(M,0)$ using the technique described in [**18**]? Note that G. Masbaum and H. Wenzl [**19**] proved integrality of the quantum $PSU(N)$-invariants (see also [**31**]).

References

1. S. Akbulut and J.D. McCarthy, *Casson's Invariant for Oriented Homology 3-spheres - An Exposition*, Mathematical Notes, vol. 36, Princeton University Press, Princeton, 1990.
2. C. Blanchet, *Invariants on three-manifolds with spin structure*, Comment. Math. Helv. **67** (1992), 406–427.
3. C. Blanchet, N. Habegger, G. Masbaum, and P. Vogel, *Three-manifold invariants derived from the Kauffman bracket*, Topology **31** (1992), 685–699.
4. J.H. Conway, *An enumeration of knots and links, and some of their algebraic properties*, Computational Problems in Abstract Algebra (J. Leech, ed.), Pergamon, 1970, pp. 329–358.
5. V.F.R. Jones, *A polynomial invariant for knots via von Neumann algebras*, Bull. Amer. Math. Soc. (N.S.) (1985), 103–111.
6. L.H. Kauffman, *State models for knot polynomials*, preprint.
7. ______, *The Conway polynomial*, Topology **20** (1981), 101–108.
8. ______, *Formal Knot Theory*, Mathematical Notes, vol. 30, Princeton University Press, Princeton, 1983.
9. R. Kirby and P. Melvin, *The 3-manifold invariants of Witten and Reshetikhin–Turaev for $sl(2,\mathbb{C})$*, Invent. Math. **105** (1991), 473–545.
10. R. Kirby, P. Melvin, and X. Zhang, *Quantum invariants at the sixth root of unity*, Comm. Math. Phys. **151** (1993), 607–617.
11. T. Kohno, *Topological invariants for 3-manifolds using representations of mapping class group I*, Topology **31** (1992), 203–230.
12. T. Kohno and T. Takata, *Symmetry of Witten's 3-manifold invariants for $sl(n,\mathbb{C})$*, J. Knot Theory Ramifications **2** (1993), 149–169.
13. A. Kricker and B. Spence, *Ohtsuki's invariants are of finite type*, J. Knot Theory Ramifications **6** (1997), no. 4, 583–597.
14. T.T.Q. Le, J. Murakami, and T. Ohtsuki, *An invariant of 3-manifolds derived from the universal Vassiliev-Kontsevich invariant*, preprint, 1995.
15. T.T.Q. Le, J. Murakami, and T. Ohtsuki, *On a universal perturbative invariant of 3-manifolds*, Topology **37** (1998), no. 3, 539–574.
16. W.B.R. Lickorish, *The skein method for three-manifold invariants*, J. Knot Theory Ramifications **2** (1993), 171–194.
17. X.-S. Lin and Z. Wang, *On Ohtsuki's invariants of integral homology 3-spheres, I*, preprint, 1996 (second version, q-alg/9509009).
18. G. Masbaum and J.D. Roberts, *A simple proof of integrality of quantum invariants at prime roots of unity*, Math. Proc. Cambridge Philos. Soc. **121** (1997), 443–454.
19. G. Masbaum and H. Wenzl, *Integral modular categories and integrality of quantum invariants at roots of unity of prime order*, preprint, March 1997, Université Paris VII and University of California, San Diego.

20. H. Murakami, *Quantum $SU(2)$-invariants dominate Casson's $SU(2)$-invariant*, Math. Proc. Cambridge Philos. Soc. **115** (1993), 253–281.
21. ———, *Quantum $SO(3)$-invariants dominate the $SU(2)$-invariant of Casson and Walker*, Math. Proc. Cambridge Philos. Soc. **117** (1995), 237–249.
22. T. Ohtsuki, *A polynomial invariant of integral homology 3-spheres*, Math. Proc. Cambridge Philos. Soc. **117** (1995), 83–112.
23. ———, *Combinatorial quantum method in 3-dimensional topology*, Lecture Notes of the workshop at Oiwake Seminar House, Waseda University, Karuizawa, September, 1996.
24. ———, *Finite type invariants of integral homology 3-spheres*, J. Knot Theory Ramifications **5** (1996), 101–115.
25. ———, *A polynomial invariant of rational homology 3-spheres*, Invent. Math. **123** (1996), 241–257.
26. ———, *The perturbative $SO(3)$ invariant of rational homology 3-spheres recovers from the universal perturbative invariant*, preprint, Tokyo Institute of Technology, 1997.
27. ———, *Ryoshi-fuhenryo wo meguru 3-jigen isou-kikagaku no kumiawase-teki houhou*, Infinite Analysis Lecture Notes, No. 15, June 1997, Japanese version of [**23**] tranlated by M. Shiomi.
28. N.Yu. Reshetikhin and V.G. Turaev, *Ribbon graphs and their invariants derived from quantum groups*, Comm. Math. Phys. **127** (1990), 1–26.
29. M. Sakuma, *An evaluation of the Jones polynomial of a parallel link*, Math. Proc. Cambridge Philos. Soc. **104** (1988), 105–113.
30. C. Sato, *A polynomial invariant of seifert fibered rational homology spheres derived from the quantum $su(2)$-invariants associated with the trivial cohomology class modulo two*, preprint, 1998.
31. T. Takata and Y. Yokota, *The $PSU(N)$ invariants of 3-manifolds are polynomials*, preprint, Kyushu Univ., 1996.
32. V. Turaev and H. Wenzl, *Quantum invariants of 3-manifolds associated with classical simple Lie algebras*, Internat. J. Math. **4** (1993), 323–358.
33. V.G. Turaev, *State sum models in low-dimensional topology*, Proceedings of the International Congress of Mathematicians, (Kyoto, 1990), Vol. I, Math. Soc. Japan, Tokyo, 1991, pp. 689–698.
34. K. Walker, *An Extension of Casson's Invariant*, Annals of Mathematics Studies, vol. 126, Princeton University Press, Princeton, 1992.
35. H. Wenzl, *Braids and invariants of 3-manifolds*, Invent. Math. **114** (1993), 235–275.
36. E. Witten, *Quantum field theory and the Jones polynomial*, Comm. Math. Phys. **121** (1989), 351–399.
37. Y. Yokota, *Skeins and quantum $SU(N)$ invariants of 3-manifolds*, Math. Ann. **307** (1997), 109–138.

Department of Mathematics, School of Science and Engineering, Waseda University, Ohkubo, Shinjuku, TOKYO 169, JAPAN
E-mail address: `hitoshi@uguisu.co.jp`

Contemporary Mathematics
Volume **233**, 1999

An element of infinite order in TQFT-representations of mapping class groups

Gregor Masbaum

Quantum invariants of 3-manifolds and the associated 'Topological Quantum Field Theories' (TQFT) give rise to interesting finite-dimensional representations of mapping class groups of surfaces. Recently, there has been some interest in the question of whether these representations have finite image. For the $SU(2)$-theory, Gilmer [G] has shown that this is the case in genus 1, while Funar [F] has shown that the image of the mapping class group in genus $g > 1$ is an infinite group, except for the representations corresponding to a few small levels.

To do so, Funar observed that it is enough to prove this for the mapping class group of a 4-punctured sphere. In this case, his proof that the image cannot be a finite group uses the classification of finite subgroups of $SO(3)$, following an argument used by V. Jones to show the generic infiniteness of braid group representations into Hecke algebras. Although this is a nice argument, it does not actually show that there is a mapping class whose image in the TQFT-representations has infinite order.

In this note we will complete Funar's result, and at the same time simplify his proof, by exhibiting such a mapping class explicitly. This class has infinite order in the TQFT-representations for $SU(2)$ at level k except if $k = 1, 2, 4$, or 8 (in other words, except if the order $r = k+2$ of the root of unity q is 3, 4, 6, or 10). Although the basic idea is the same as Funar's, the proof here is by a simple Kauffman bracket computation which is perhaps easier to check than the considerations in [F].

Let Σ be a surface of genus $g > 1$, and let α and β be Dehn twists along simple closed curves a and b meeting transversely in exactly two points but with opposite signs. Note that a regular neighborhood of $a \cup b$ in Σ is a four-holed sphere, S, say. Let us assume that none of the four boundary circles of S is null-homologous in $\Sigma - S$. Such a pair of simple closed curves always exists if $g > 1$.

PROPOSITION. *The image of the mapping class $\alpha^{-1}\beta$ in the TQFT-representation corresponding to $SU(2)$ at level k has infinite order except if $k = 1, 2, 4$, or 8.*

NOTE. As the TQFT-representations of the mapping class group are only projective-linear, the above statement means more precisely that no power of the mapping class $\alpha^{-1}\beta$ acts as a scalar matrix. Also note that every Dehn twist has

---

1991 *Mathematics Subject Classification.* 57M99,57R99,20F38.

finite order in the TQFT-representations, so that a product of at least two twists is needed to get a matrix of infinite order.

The proof is very simple and goes as follows. We will take the point of view of the TQFT-functors V_p constructed in [BHMV] from the Kauffman bracket, with Kauffman's skein variable A being a primitive root of unity of order $2p$. The $SU(2)$-theory at level k corresponds to the case $p = 2k + 4$.

Consider the four-holed sphere S which is a regular neighborhood of $a \cup b$. We think of the curves a and b as the equator and a meridian of that sphere, respectively. Since (by assumption) none of its boundary circles is homologically trivial in $\Sigma - S$, it follows from the 'fusion rules' (as shown in theorem 1.14 of [BHMV]) that the vector space $V_p(\Sigma)$ contains a direct summand $V \otimes W$ where $V = V_p(S^2, (1, 1, 1, 1))$ is the vector space associated to a four-punctured sphere with all punctures (corresponding to the boundary circles of our 4-holed sphere S) colored by 1 (corresponding to the fundamental representation of $SU(2)$). We will compute a matrix for the mapping class $\alpha^{-1}\beta$ acting on V, and then show it has infinite order, proving the proposition.

The vector space V is 2-dimensional, and is isomorphic to the skein module of banded $(2, 2)$-tangles in a 3-ball, modulo the usual Kauffman skein relations. Since the mapping classes α and β extend to the 3-ball, their action on V is just the natural action on tangles (see [BHMV, MR]). A basis of V is given by the tangles

$v_0 =$  $h_0 =$ 

The mapping classes α and β are given by a 'horizontal' resp. a 'vertical' full twist, so that α leaves fixed v_0 and β leaves fixed h_0. The skein elements v_2 (resp. h_2) determined by the following graphs

$v_2 =$ 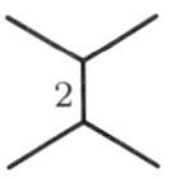

$h_2 =$ 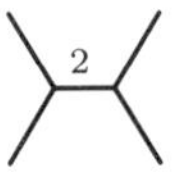

are also eigenvectors for α (resp. β), with eigenvalue A^8 in both cases (with the notations of [BHMV], this is μ_2 where $\mu_i = (-1)^i A^{i^2+2i}$). Here the 2 beneath an edge means cabling by a Jones-Wenzl idempotent f_2. Thus one has $v_2 = h_0 - \delta^{-1} v_0$ and $h_2 = v_0 - \delta^{-1} h_0$, where $\delta = -A^2 - A^{-2}$. (Indeed, f_2 is given by $f_2 = h_0 - \delta^{-1} v_0$ where h_0 is viewed as the identity $(2, 2)$-tangle.) From this, it is easy to work out the matrix for the base change from (v_0, v_2) to (h_0, h_2). One finds that $\alpha^{-1}\beta$ in the basis (v_0, v_2) is given by the product of matrices

$$\begin{pmatrix} 1 & 0 \\ 0 & A^{-8} \end{pmatrix} \begin{pmatrix} \delta^{-1} & 1-\delta^{-2} \\ 1 & -\delta^{-1} \end{pmatrix} \begin{pmatrix} 1 & 0 \\ 0 & A^{8} \end{pmatrix} \begin{pmatrix} \delta^{-1} & 1-\delta^{-2} \\ 1 & -\delta^{-1} \end{pmatrix}$$

$$= \begin{pmatrix} A^8 + \delta^{-2}(1 - A^8) & \star \\ \star & A^{-8} + \delta^{-2}(1 - A^{-8}) \end{pmatrix}$$

This matrix has determinant 1 and its trace, t, say, is easily computed to be

$$t = 2 - q - q^{-1} + q^2 + q^{-2},$$

where we have put $q = A^4$. If a power of this matrix is a scalar multiple of the identity matrix, then the scalar must be ± 1, and the eigenvalues λ, λ^{-1} of the matrix must be roots of unity. Hence for every embedding $\mathbb{Q}(q) \to \mathbb{C}$, we must have

$$|t| = |\lambda + \lambda^{-1}| \leq 2.$$

But if the order $r = k + 2$ of q is ≥ 5 and different from 6 or 10, then there is a primitive r-th root of unity $q \in \mathbb{C}$ such that $Re(q^2) > Re(q)$ and hence $|t| > 2$. (Just take $q = e^{2i\pi l/r}$, where $l = 1 + r/2$ if $r \equiv 0$ (4) and $r \geq 8$, $l = (1 + r)/2$ if $r \equiv 1$ (2) and $r \geq 5$, and $l = 2 + r/2$ if $r \equiv 2$ (4) and $r \geq 14$.) Thus, $\alpha^{-1}\beta$ has infinite order in the TQFT-representation except if the level is $k = 1, 2, 4$, or 8, as asserted.

REMARKS. (i) If $k = 1, 2, 4$, or 8, then the matrix has order $1, 4, 4, 6$, respectively.

(ii) The same argument shows that $\alpha^{-1}\beta$ has infinite order in the TQFT-representation in the $SO(3)$-case (corresponding to V_p with p odd) under the same conditions on the order of $q = A^4$. (See [BHMV], thm. 4.14.)

(iii) Funar [F] conjectured that a class similar to $\alpha^{-1}\beta$ but made out of half-twists in place of full twists has infinite order. He also has an argument (which I haven't checked) in the case of level 8 (i.e., $r = 10$) but he needs the genus to be at least 3.

(iv) The projective-linear representation of the mapping class group may also be thought of as a linear representation of a central extension of the mapping class group by $\mathbb{Z}$. Although this extension is non-trivial, there are 'preferred' lifts of Dehn twists to this extension (see [BHMV, MR]). The matrices for α and β described above correspond to the 'geometric' lifts considered in [MR].

ACKNOWLEDGMENTS. I would like to thank J. E. Andersen and B. van Geemen for asking me this question.

References

[BHMV] C. BLANCHET, N. HABEGGER, G. MASBAUM, P. VOGEL. Topological Quantum Field Theories derived from the Kauffman bracket. *Topology* **34** (1995), 883-927.

[G] P. GILMER. On the WRT representations of mapping class groups. *Proc. AMS* (to appear) q-alg/9701042

[F] L. FUNAR. On the TQFT representations of the mapping class groups. *Pacific J. Math.* (to appear) math.GT/9804017

[MR] G. MASBAUM, J. D. ROBERTS. On central extensions of mapping class groups. *Math. Ann.* **302** (1995) 131-150.

[RT] N. YU. RESHETIKHIN, V. G. TURAEV. Invariants of 3-manifolds via link polynomials and quantum groups, *Invent. Math.* **103** (1991), 547-597.

[W] E. WITTEN. Quantum field theory and the Jones polynomial. *Commun. Math. Phys* **121** (1989) 351-98.

INSTITUT DE MATHÉMATIQUES DE JUSSIEU (UMR 7586 CNRS), CASE 7012, UNIVERSITÉ PARIS 7 - DENIS DIDEROT, 75251 PARIS CEDEX 05, FRANCE

E-mail address: masbaum@math.jussieu.fr

Contemporary Mathematics
Volume **233**, 1999

The perturbative $SO(3)$ invariant of homology circles

Tomotada Ohtsuki

ABSTRACT. We describe two structures of quantum Hilbert space of a torus in topological quantum field theory for the perturbative $SO(3)$ invariant. To describe the action of a central extension of mapping class group, which is one of the two structures, we introduce the notion of extended 3-manifold and construct central extensions of the mapping class groups of closed surfaces by using the notion.

The notion of topological quantum field theory (TQFT) was introduced by Atiyah *et al.* [**4, 3**], see also [**6**], [**29**, Chapter III] and [**27**]. The theory consists of axioms for functors of the category of closed surfaces and 3-cobordisms to the category of vector spaces and linear maps between them. The image of a closed surface Σ by a functor of TQFT is denoted by $\mathcal{H}_\Sigma$, which is called *quantum Hilbert space* of Σ. We denote by $Z(M)$ the image of a 3-cobordism M.

For a functor of TQFT, we have two structures of $\mathcal{H}_\Sigma$. One is the action of the mapping class group $\mathcal{G}$ of the surface Σ on the space $\mathcal{H}_\Sigma$; by the action a homeomorphism f of Σ takes $Z(M)$ to $Z(M \cup_\Sigma C_f)$, where M is a 3-cobordism with $\partial M = \Sigma$ and C_f is the mapping cylinder of f. The other structure is the inner product in $\mathcal{H}_\Sigma$, induced by the pairing

$$< Z(M_1), Z(M_2) >= Z\big(M_1 \cup_\Sigma (-M_2)\big) \in \mathbb{C}, \tag{0.1}$$

where M_1 and M_2 are 3-cobordisms with common boundary Σ, the 3-cobordism $-M_2$ is M_2 with opposite orientation and the right hand side belongs $\mathcal{H}_{\text{empty surface}}$ which is equal to $\mathbb{C}$ by an axiom of TQFT.

Quantum invariants of 3-manifolds, defined by using surgery presentations of 3-manifolds as in [**22**], induce functors satisfying the axioms of TQFT based on surgery presentations of 3-cobordisms, though the functors satisfy the composition law of TQFT only up to anomaly, see for example [**28, 29, 8**]. By restricting each of such functors, we obtain a projective representation of the mapping class group $\mathcal{G}$ on $\mathcal{H}_\Sigma$; we also have a representation of $\tilde{\mathcal{G}}$ for a suitable central extension $\tilde{\mathcal{G}}$ of $\mathcal{G}$.

Our aim of this paper is to construct quantum Hilbert space $\mathcal{H}_{T^2}$ for the perturbative $SO(3)$ invariant, which is obtained by asymptotic expansion (in the sense of [**20**]) of the quantum $SO(3)$ invariant, and to describe the two structure of the

---

1991 *Mathematics Subject Classification.* Primary 57M25, 57M.

This research is supported in part by Grand-in-Aid for Scientific Research, The Ministry of Education, Science, Sports and Culture.

space $\mathcal{H}_{T^2}$. Note that, since the perturbative $SO(3)$ invariant is defined only for rational homology 3-spheres, we can consider the formula (0.1) under homological assumption of 3-cobordisms M_1 and M_2 such that $M_1 \cup_\Sigma (-M_2)$ becomes a rational homology 3-sphere. Hence TQFT for the perturbative $SO(3)$ invariant will be constructed as restricted TQFT with homological assumption on the formula (0.1).

In Section 1 we give a formulation of central extensions of mapping class groups of closed surfaces by using the notion of classes of surgery presentations of 3-cobordisms. In Section 2 we show relations between the quantum $SO(3)$ invariant and the perturbative $SO(3)$ invariant of links and 3-manifolds. Using the relations we give the two structures of quantum Hilbert space of a torus for the perturbative $SO(3)$ invariant in Section 3. He also thanks the referee for careful comments.

The author would like to thank Jun Murakami, Thang Le and Nariya Kawazumi for valuable comments and suggestions, Hitoshi Moriyoshi for letting the author know his work on geometric construction of central extensions. He is also grateful to Justin Roberts, Thomas Kerler, Stephen Sawin and Vladimir Turaev for many helpful and technical comments on surgery presentations for TQFT in an early version of this draft.

1. Extended 3-manifolds and central extensions of mapping class groups of closed surfaces

1.1. Surgery presentations of 3-manifolds. When a 3-cobordism M has empty boundary, we call a such framed link L a *surgery presentation* of M that M is obtained from S^3 by integral surgery along L. In this subsection we give a formulation of surgery presentations of 3-cobordisms with non-empty boundaries. A geometric technique of surgery presentations of 3-cobordisms was introduced in [**22**, Section 4]; see also [**29**, Chapter IV] and [**7**] for detailed formulations of the presentations. For similar approaches for surgery presentations of 3-cobordisms, see [**14, 27**].

A *chain graph* is a trivalent graph obtained by adding arcs to a line linearly in order as shown in Figure 1 where two chain graphs are depicted. We fix orientation and framing of a chain graph as in Figure 1. We consider a standard embedding of a chain graph Γ in S^3, *i.e.*, an embedding whose image is included in a plane. A closed surface Σ is often identified with the boundary of a tubular neighborhood $N(\Gamma)$ of the embedded Γ. For two chain graphs Γ_1 and Γ_2 shown in Figure 1, we put $\Sigma_i = \partial N(\Gamma_i)$ $(i = 1, 2)$. Fixing an orientation of S^3, the orientation of $S^3 - N(\Gamma_i)$ induces an orientation of Σ_i. We regard Σ_2 as $-\Sigma_1$ as follows. We obtain Γ_2 as the mirror image of Γ_1, ignoring the orientation of Γ_i. We have orientation reversing homeomorphism between $N(\Gamma_1)$ and $N(\Gamma_2)$, which induces identification between $-\Sigma_1$ and Σ_2.

Let M be a compact 3-manifold with boundary such that each component of the boundary is identified with the boundary of a tubular neighborhood of a chain graph. Let $L \cup G$ be a union of a framed link and an embedding of the disjoint union of chain graphs Γ in S^3 such that L and G does not intersect and that M is obtained from $S^3 - N(G)$ by surgery along L and that the identification of ∂M and $\partial N(\Gamma)$[1] is given by $\partial M = \partial N(G)$. We call such $L \cup G$ a *surgery presentation* of M.

[1] $N(\Gamma)$ denotes a tubular neighborhood of a standard embedding of Γ in S^3. We often identify $N(\Gamma)$ and $N(G)$ in a natural way.

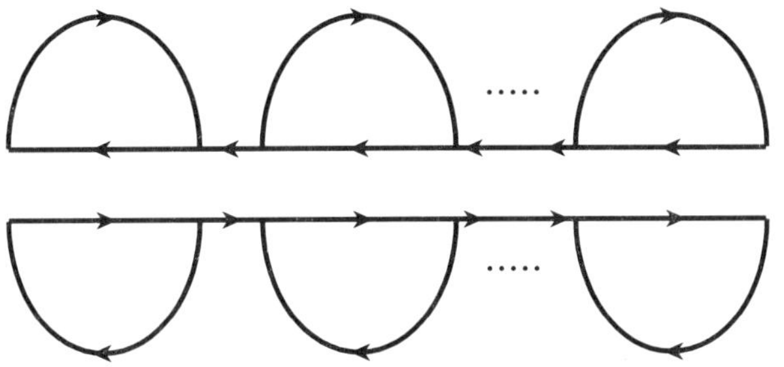

FIGURE 1. A chain graph Γ_1 (the upper one) and its mirror image Γ_2 with opposite orientation (the lower one). These graphs are framed by the blackboard framing.

Let M and M' be 3-manifolds with boundaries Σ and $-\Sigma$ respectively. When we give their surgery presentations $L \cup G$ and $L' \cup G'$, the two surgery presentations determine a surgery presentation of $M \cup_\Sigma M'$ as the following lemma, which is essentially due to [**22**, Lemma 4.4], see also [**29**, Section IV.2] and [**18**].

LEMMA 1.1. *The framed link* $L \cup (G \hat{\cup} G') \cup L'$ *becomes a surgery presentation of* $M \cup_\Sigma M'$ *where* $\hat{\cup}$ *denotes the operation of gluing* G *and* G' *along horizontal lines and removing the horizontal line; see Figure 2 for its pictorial definition.*

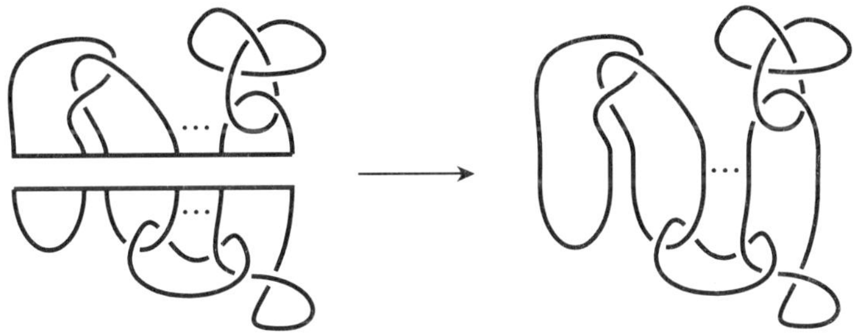

FIGURE 2. The framed link $L \cup (G \hat{\cup} G') \cup L'$ obtained from $L \cup G$ and $L' \cup G'$

REMARK 1.2. When we glue two components of the boundary of the same 3-manifold, the new surgery presentation is obtained as in Figure 3. The detailed arguments are discussed in [**29**, Section IV.2]; see also [**7**, **18**].

1.2. Extended 3-manifolds. It is well known that two surgery presentations of the same closed 3-manifold are related by Kirby moves KI and KII [**9**]. By using restricted Kirby moves, the notion of "extended 3-manifold" [2] was introduced for closed 3-manifolds by Roberts in [**23**], where the notion is called "structure". In this subsection we extend the notion to that for 3-manifolds with boundaries.

We define Kirby moves KI and KII for a surgery presentation $L \cup G$ of a 3-cobordism as follows. *KI move* is a move adding or removing the trivial knot with ± 1 framing to or from $L \cup G$. *KII move* is a handle slide move taking a segment of

[2] We use the terminology "extended 3-manifold" in a different meaning of [**29**].

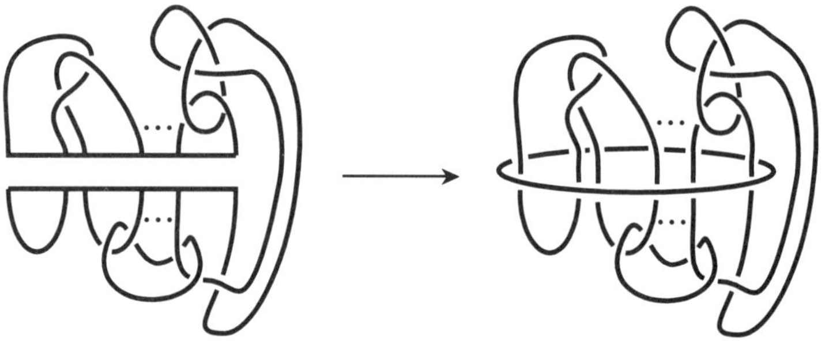

FIGURE 3. The framed link obtained from $L \cup G \cup G'$

$L \cup G$ sliding over a component of L such that the segment is not included in the component.

The following lemma is essentially due to [**22**, Section 7.3], which is an extension of Kirby's theorem to our surgery presentations of 3-cobordisms.

LEMMA 1.3. *Let $L \cup G$ and $L' \cup G'$ be surgery presentations of two 3-cobordisms M and M' respectively, where G and G' are embeddings of the same chain graph Γ. Then M and M' are homeomorphic preserving identifications of the boundaries with $N(\Gamma)$ if and only if $L \cup G$ can be obtained from $L' \cup G'$ by a series of KI, KII moves and isotopy of S^3.*

Following [**23**] we introduce a restricted KI move, say $\widehat{\mathrm{KI}}$ move, such that *$\widehat{KI}$ move* is a move adding or removing the split union of the trivial knot with $+1$ framing and that with -1 framing. Among the surgery presentations of a 3-manifold (possibly with boundary) we consider the equivalence relation generated by $\widehat{\mathrm{KI}}$ and KII moves. An *extended 3-manifold* is a 3-manifold with a class of surgery presentations of the 3-manifold with respect to the equivalence relation. When gluing two extended 3-manifolds along their common boundaries we obtain the extended 3-manifold by composing two surgery presentations by Lemma 1.1.

REMARK 1.4. There is a slightly different approach[3] to give an alternative definition of the equivalence relation among surgery presentations of 3-cobordisms by Kirby moves in the following way. We consider handle slide moves, say KII$'$ moves, not allowing slides of a segment of a chain graph G over any component of a link, unlike our KII moves. The notion of extended 3-manifolds are alternatively defined by using the equivalence relation generated by $\widehat{\mathrm{KI}}$, KII$'$ and $\mathcal{O}_3$ moves for links in a fixed 3-manifold with boundary, for example a handlebody, where the move $\mathcal{O}_3$ is defined in [**24**].

1.3. The central extensions of mapping class groups derived from extended mapping cylinders. By identifying homeomorphisms of a closed surface with mapping cylinders of the homeomorphisms, we obtain surgery presentations of the mapping class group $\mathcal{G}$ of the closed surface; this technique is due to [**22**, Section 4.6], see also [**29**, Section IV.5] and [**14**]. In this section, based on the technique, we construct a central extension of $\mathcal{G}$ using extended mapping cylinders. Moreover we show the central extension is isomorphic to the one given in Appendix. For a concrete presentation of the central extension, see [**17**].

[3]This approach was suggested by Justin Roberts.

Let $\tilde{\mathcal{G}}$ be the set of extended mapping cylinders of homeomorphisms of Σ. There is a natural projection $\tilde{\mathcal{G}} \to \mathcal{G}$, obtained by forgetting classes of surgery presentations.

THEOREM 1.5. *As extensions of $\mathcal{G}$, the extension $\tilde{\mathcal{G}}$ is isomorphic to the central extension $\tilde{\mathcal{G}}'$ derived from "signature cocycle"; see Appendix for definition of the central extension $\tilde{\mathcal{G}}'$. In particular the element of $H^2(\mathcal{G};\mathbb{Z})$ which determines the central extension $\tilde{\mathcal{G}}$ is equal to four times a generator of $H^2(\mathcal{G};\mathbb{Z})$.*

PROOF. We give a map $\tilde{\mathcal{G}}' \to \tilde{\mathcal{G}}$ as follows. Let (C_f, λ) be an element in $\tilde{\mathcal{G}}'$, and choose $L \cup G \cup G'$ representing λ. The framed link, say L', obtained from $L \cup G \cup G'$ as shown in Figure 3 is a surgery presentation of the mapping torus T_f of f. Hence we obtain T_f as the boundary of the 4-manifold, say W, obtained from a 4-ball B^4 by attaching 2-handles along $L' \subset \partial B^4$. We put the image of (C_f, λ) in $\tilde{\mathcal{G}}$ to be $(T_f, [W])$.

We show that the map $\tilde{\mathcal{G}}' \to \tilde{\mathcal{G}}$ given above is a homomorphism as follows. Let $a_1 = (C_{f_1}, \lambda_1)$ and $a_2 = (C_{f_2}, \lambda_2)$ be two elements in $\tilde{\mathcal{G}}'$, and let $(T_{f_1}, [W_1])$ and $(T_{f_2}, [W_2])$ be their images in $\tilde{\mathcal{G}}$. We choose $L_i \cup G_i \cup G'_i$ representing λ_i $(i = 1, 2)$. The product $a_1 a_2$ is $C_{f_1 f_2}$ with the class of surgery presentations including

$$G_1 \cup L_1 \cup (G'_1 \hat{\cup} G_2) \cup L_2 \cup G'_2. \tag{1.1}$$

Its image in $\tilde{\mathcal{G}}$ is $(T_{f_1 f_2}, [W])$ where W is the 4-manifold obtained from (1.1) as above. Hence W is alternatively obtained from W_1 and W_2 by taking connected sum along a fiber Σ of Σ-bundles T_{f_1} and T_{f_2}. Further this is homeomorphic to $W_1 \cup_{T_{f_1}} Z \cup_{T_{f_2}} W_2$ defined in Appendix. This implies that the map $\tilde{\mathcal{G}}' \to \tilde{\mathcal{G}}$ defined above is a homomorphism.

By their definition the homomorphism is compatible with natural projections of them to $\mathcal{G}$. By Theorem A.1 we obtain the theorem. □

1.4. The case of genus one. We consider the case of the mapping class group of homeomorphisms of a torus in more detail in this subsection. In this case the mapping class group can be identified with $SL(2;\mathbb{Z})$, which is also identified with the set of mapping cylinders of homeomorphisms of a torus to itself. Let $\widetilde{SL(2;\mathbb{Z})}$ be the set of extended mapping cylinders of homeomorphisms of a torus to itself. As shown in the previous subsection, $\widetilde{SL(2;\mathbb{Z})}$ is a central extension of $SL(2;\mathbb{Z})$.

By using continued fraction expansion any element of $\widetilde{SL(2;\mathbb{Z})}$ can be expressed by a link of the shape in Figure 4; for example see [**10**] for such expression based on continued fraction expansion. We denote the link in Figure 4 by $\alpha = (n_1, n_2, \cdots, n_k)$. The projection of $\widetilde{SL(2;\mathbb{Z})}$ to $SL(2;\mathbb{Z})$ takes α to

$$A_\alpha = -ST^{-n_1}ST^{-n_2}S\cdots ST^{-n_k}S,$$

where $S = \begin{pmatrix} 0 & -1 \\ 1 & 0 \end{pmatrix}$ and $T = \begin{pmatrix} 1 & 1 \\ 0 & 1 \end{pmatrix}$ are standard generators of $SL(2;\mathbb{Z})$. We also put $|\alpha| = \sum_i n_i$.

LEMMA 1.6. *Let φ be the map of $\widetilde{SL(2;\mathbb{Z})}$ to $\mathbb{Z} \times SL(2;\mathbb{Z})$ which takes α to $(|\alpha|, A_\alpha)$. Then φ is an injective homomorphism whose image is a subgroup of $\mathbb{Z} \times SL(2;\mathbb{Z})$ of index 3.*

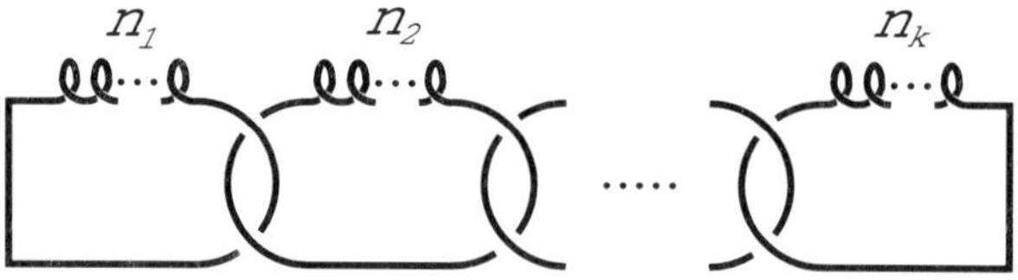

FIGURE 4. Any element of $\widetilde{SL(2;\mathbb{Z})}$ can be expressed by a link of this shape. We denote this link by $\alpha = (n_1, n_2, \cdots, n_k)$.

PROOF. We denote by $\check{A}_\alpha$ the linking matrix of the link in Figure 4, *i.e.*, we put

$$\check{A}_\alpha = \begin{pmatrix} n_1 & 1 & & & 0 \\ 1 & n_2 & 1 & & \\ & 1 & \cdots & \cdots & \\ & & \cdots & \cdots & 1 \\ 0 & & & 1 & n_k \end{pmatrix}$$

Kirby and Melvin [**10**] showed the equation $|\alpha| + \phi(A_\alpha) = 3 \operatorname{sign}\check{A}_\alpha$, where ϕ denotes Radmacher's ϕ function [**21**] and $\operatorname{sign}\check{A}_\alpha$ denotes the signature of $\check{A}_\alpha$. Hence the image of φ is included in the following set

$$\{(m, A) \in \mathbb{Z} \times SL(2;\mathbb{Z}) \mid m + \phi(A) \in 3\mathbb{Z}\}. \tag{1.2}$$

Further, if $\varphi((n_1, \cdots, n_k)) = (m, A)$, then we have $\varphi((n_1, \cdots, n_k, 1, 1, 1)) = (m + 3, A)$ and $\varphi((n_1, \cdots, n_k, 0, 0, -1, -1, -1)) = (m - 3, A)$. Hence the image of φ becomes equal to the set (1.2). Since the set (1.2) is equal to the image of a group by a homomorphism, it becomes a subgroup of $\mathbb{Z} \times SL(2;\mathbb{Z})$; in particular the subgroup is of index 3 by its definition (1.2). □

2. Relations between quantum invariants and perturbative invariants

2.1. The perturbative sl_2 invariant of links. Let $L \cup K$ be an algebraically split framed link in S^3 such that for simplicity we assume that each of L and K consists of one component and their framings are f and 0 respectively. Let V_k be the k-dimensional irreducible representation of the Lie algebra sl_2. We denote by $Q^{sl_2;V_j,V_i}(L \cup K) \in \mathbb{Z}[q^{\pm 1/2}]$ the quantum $(sl_2; V_j, V_i)$ invariant of $L \cup K$. There exists

$$P^{sl_2}(L \cup K) \in q^{fj^2/4} \cdot \mathbb{Q}[q^{\pm 1/2}, q^{\pm j/2}, q^{\pm i/2}][[q^{1/2} - q^{-1/2}]],$$

called the *perturbative sl_2 invariant* of $L \cup K$, which satisfies

$$P^{sl_2}(L \cup K)^{(\leq N)}\big|_{\text{fix } i,j} \equiv Q^{sl_2,V_j,V_i}(L \cup K) \qquad \bmod (q-1)^{N+1} \tag{2.1}$$

for any i and j, where the upper index "$(\leq N)$" in the left hand side means cutting off the part of degree $> N$ as a power series of $q^{1/2} - q^{-1/2}$. The existence of the perturbative sl_2 invariant can be shown by using the universal sl_2 invariant[4]; see also [**25**] for the existence of the perturbative sl_2 invariant.

2.2. The perturbative $SO(3)$ invariant of rational homology 3-spheres. Let M be the rational homology 3-sphere obtained by integral surgery along L and r an odd prime. Then the quantum $SO(3)$ invariant $\tau_r^{SO(3),V_i}(M,K)$ is defined by

$$\tau_r^{SO(3),V_i}(M,K) = Q_r^{sl_2;\omega,V_i}(L \cup K) \tag{2.2}$$

where we put

$$\omega = c \sum_{\substack{0<j<r \\ j \text{ is odd}}} [j]_\zeta V_j, \quad c = -\frac{\zeta^{1/2}-\zeta^{-1/2}}{\sum_{i\in\mathbb{Z}/r\mathbb{Z}} \zeta^{i^2}}, \quad [j]_\zeta = \frac{\zeta^{j/2}-\zeta^{-j/2}}{\zeta^{1/2}-\zeta^{-1/2}}.$$

There exists $\tau^{SO(3)}(M,K) \in \mathbb{Q}[q^{\pm 1/2}, q^{\pm i/2}][[q^{1/2}-q^{-1/2}]]$, called the *perturbative* $SO(3)$ *invariant* of (M,K), which satisfies the following formula[5]

(2.3)

$$\tau^{SO(3)}(M,K)^{(\leq N)}\Big|_{\substack{\text{fix } i \\ q=\zeta}} \equiv \left(\frac{|H_1(M;\mathbb{Z})|}{r}\right) \tau_r^{SO(3),V_i}(M,K) \qquad \text{mod } (\zeta-1)^{N+1},$$

for any i, where $\left(\frac{\cdot}{r}\right)$ denotes the Legendre symbol and $|H_1(M;\mathbb{Z})|$ denotes the order of the homology group $H_1(M;\mathbb{Z})$. For the existence of the perturbative invariant see [**20, 26, 12**].

3. The structures of $\mathcal{H}_{T^2}$ for the perturbative $SO(3)$ invariant

A *homology circle* is a compact 3-manifold with torus boundary which have the same rational homology group as S^1. Let N be a homology circle with fixed basis of $H_1(\partial N;\mathbb{Z}) \cong \mathbb{Z}\oplus\mathbb{Z}$, called "meridian" and "framing". We often express N as $N = M - N(K)$, which is the 3-manifold obtained from M by removing a tubular neighborhood of K, where M is a rational homology 3-sphere including a framed knot K such that the meridian and the framing of ∂N coincide with the meridian and the framing of K. In this section we investigate the structure of $\mathcal{H}_{T^2}$ for the quantum $SO(3)$ invariant and the perturbative $SO(3)$ invariant by computing the invariants of homology circles.

3.1. The case of the quantum $SO(3)$ invariant. Before considering the perturbative case, we review the structures of $\mathcal{H}_{T^2}$ for the quantum $SO(3)$ invariant. In this case we put $\mathcal{H}_{T^2} = \mathbb{C}^{(r-1)/2}$ and define $\tau_r^{SO(3)}(N)$ for a homology circle N by

$$\tau_r^{SO(3)}(N) = \left\{\tau_r^{SO(3),V_i}(M,K)\right\}_{i=1,3,5,\cdots,r-2} \in \mathcal{H}_{\partial N}.$$

[4]For an algebraically-split link L with l components, we have the universal sl_2 invariant of L which belongs to $\left(U_q(sl_2)/I\right)^{\otimes l}$, where I is the vector subspace of the quantum group $U_q(sl_2)$ spanned by $ab - ba$ for any $a, b \in U_q(sl_2)$; see [**11, 19**] for the definition of the universal sl_2 invariant. The quantum (sl_2, V_i) invariant is obtained from the universal sl_2 invariant through the trace Tr_{V_i}. By applying a "universal" trace $\mathrm{Tr}_{V_\star}$ to the universal sl_2 invariant, we obtain the perturbative sl_2 invariant.

[5]To be precise, before the formula (2.3), we should show that the prime factors of the denominators of the coefficients of $\tau^{SO(3)}(M,K)^{(\leq N)}$ are less than r, for any N which is relatively small to r; see [**20**]. Note that the formula (2.3) is available only for such N.

We consider the action of $\alpha = (n_1, n_2) \in \widetilde{SL(2;\mathbb{Z})}$ on ∂N. Let $K_1 \cup K_2$ be Hopf link of framing (n_1, n_2). By Lemma 1.1, α takes (M, K) to $(M', K') = (M_{K\#K_1}, K_2)$, where $M_{K\#K_1}$ denotes the 3-manifold obtained by surgery along the connected sum $K\#K_1$. Hence we have

$$\begin{aligned}\tau_r^{SO(3),V_j}(M', K') &= \tau_r^{SO(3),\omega,V_j}(M, K\#K_1, K_2)\\ &= c \sum_{\substack{0<i<r\\ i \text{ is odd}}} \tau_r^{SO(3),V_i}(M, K) \times \zeta^{\frac{i^2-1}{4}n_1 + \frac{j^2-1}{4}n_2}[ij]_\zeta,\end{aligned}$$

where the last term in the second line implies the quantum (sl_2, V_i, V_j) invariant of the Hopf link $K_1 \cup K_2$. Hence the action of $\alpha \in \widetilde{SL(2;\mathbb{Z})}$ on $\mathcal{H}_{T^2}$ takes $\{x_i\}_{i=1,3,\cdots,r-2}$ to $\{y_j\}_{j=1,3,\cdots,r-2}$ where $y_j = c\sum_i x_i \zeta^{\frac{i^2-1}{4}n_1 + \frac{j^2-1}{4}n_2}[ij]_\zeta$. Further it follows that α takes

$$\left(\frac{\mu}{r}\right)\zeta^{\frac{\nu}{\mu}\frac{i^2}{4}}\left[\frac{ki}{\mu}\right]_\zeta \qquad \text{to} \qquad \left(\frac{\mu'}{r}\right)\zeta^{\frac{\nu'}{\mu'}\frac{j^2}{4} - \frac{|\alpha|}{4} + \frac{k^2}{4\mu\mu'}}\left[\frac{kj}{\mu'}\right]_\zeta \tag{3.1}$$

where $\alpha = (n_1, n_2)$ and $\begin{pmatrix}\mu'\\ \nu'\end{pmatrix} = A_\alpha \begin{pmatrix}\mu\\ \nu\end{pmatrix}$. By iterating similar calculation, we can show that (3.1) also holds for any $\alpha \in \widetilde{SL(2;\mathbb{Z})}$.

3.2. The case of the perturbative $SO(3)$ invariant. For each pair of coprime integers μ and ν, we define quantum Hilbert space $\mathcal{H}_{T^2,\mu,\nu}$ for the perturbative $SO(3)$ invariant to be the vector space spanned by $q^{\frac{\nu}{\mu}\frac{i^2}{4}}\left[\frac{ki}{|\mu|}\right]$, $(k = 1, 2, 3, \cdots)$ over the ring $\mathbb{Q}[q^{\pm 1/2}][[q^{1/2} - q^{-1/2}]]$, where $[m] = \frac{q^{m/2} - q^{-m/2}}{q^{1/2} - q^{-1/2}}$; note that i is a variable in this space.

Let N be a homology circle with fixed meridian and framing in $H_1(\partial N)$. Let μ and ν be the images of meridian and framing by $H_1(\partial N) \to H_1(N) = \mathbb{Q}$. Since the perturbative $SO(3)$ invariant of (M, K) can be expressed as a linear sum of $q^{\frac{\nu}{\mu}\frac{i^2}{4}}\left[\frac{ki}{|\mu|}\right]$ (see **[26]**), we define $\tau^{SO(3)}(N)$ to be $\tau^{SO(3)}(M, K) \in \mathcal{H}_{T^2,\mu,\nu}$.

By applying the relations (2.1), (2.2) and (2.3) to (3.1), we obtain (1) of the following theorem, where we obtain (2) by similar calculation.

THEOREM 3.1. *The two structures of quantum Hilbert space $\mathcal{H}_{T^2}$ for the perturbative $SO(3)$ invariant are given as follows.*

(1) The action of $\alpha \in \widetilde{SL(2;\mathbb{Z})}$ on $\bigoplus_{\mu,\nu} \mathcal{H}_{T^2,\mu,\nu}$ takes

$$q^{\frac{\nu}{\mu}\frac{i^2}{4}}\left[\frac{ki}{|\mu|}\right] \qquad \text{to} \qquad q^{\frac{\nu'}{\mu'}\frac{j^2}{4} - \frac{|\alpha|}{4} + \frac{k^2}{4\mu\mu'}}\left[\frac{kj}{|\mu'|}\right],$$

where $\begin{pmatrix}\mu'\\ \nu'\end{pmatrix} = A_\alpha \begin{pmatrix}\mu\\ \nu\end{pmatrix}$.

(2) Put $H = \mu_2\nu_1 + \mu_1\nu_2$. If $H \neq 0$, then the inner product in $\mathcal{H}_{T^2}$ is given by

$$\left(q^{\frac{\nu_1}{\mu_1}\frac{i^2}{4}}\left[\frac{k_1 i}{|\mu_1|}\right], q^{\frac{\nu_2}{\mu_2}\frac{i^2}{4}}\left[\frac{k_2 i}{|\mu_2|}\right]\right) = q^{\frac{1}{4H}\left(\frac{\mu_2}{\mu_1}k_1^2 - \frac{\mu_1}{\mu_2}k_2^2\right)}\left[\frac{k_1 k_2}{|H|}\right].$$

EXAMPLE 3.2. By gluing two solid tori by $\begin{pmatrix} a & b \\ c & d \end{pmatrix} \in SL(2;\mathbb{Z})$, we obtain the lens space[6] $L(c,d)$. Let $\alpha \in \widetilde{SL(2;\mathbb{Z})}$ be a lift of $\begin{pmatrix} a & b \\ c & d \end{pmatrix}$. We denote by $\lambda_1(M)$ the first coefficient of $\tau^{SO(3)}(M)$. By Theorem 3.1 we have

$$\lambda_1\big(L(c,d)\big) = \frac{3}{4}\mathrm{sign}\check{A}_\alpha - \frac{|\alpha|}{4} - \frac{a}{4c} - \frac{d}{4c}. \tag{3.2}$$

On the other hand we have

$$\lambda_1\big(L(c,d)\big) = 3\lambda\big(L(c,d)\big) = -3s(d,c), \tag{3.3}$$

where $\lambda(M)$ denotes Casson invariant of M, the first equality is derived from the relation $\lambda_1(M) = 3\lambda(M)$ (see [**20**]) and the second equality is derived from the fact that Casson invariant of a lens space is equal to Dedekind sum. By (3.2) and (3.3) we obtain an alternative construction of Dedekind sum by a representation of $\widetilde{SL(2;\mathbb{Z})} \subset \mathbb{Z} \times SL(2;\mathbb{Z})$.

References

[1] Atiyah, M.F., *The logarithm of the Dedekind η-function*, Math. Ann. **278** (1987) 335–380.

[2] ———, *On framings of 3-manifolds*, Topology **29** (1990) 1–7.

[3] ———, *Topological quantum field theories*, Publ. Math. IHES **68** (1989) 175–186.

[4] Atiyah, M.F., Hitchin, N., Lawrence, R., Segal, G., Oxford seminar on Jones-Witten theory (1988).

[5] Atiyah, M.F., Singer, I.M., *The index of elliptic operators: III*, Ann. of Math. **87** 546–604.

[6] Blanchet, C., Habegger, N., Masbaum, G., Vogel, P., *Topological quantum field theories derived from the Kauffman bracket*, Topology **34** (1995) 883–927.

[7] Kerler, T., *Bridged links and tangle presentations of cobordism categories*, to appear in Adv. Math.

[8] ———, *On the connectivity of cobordisms and half-projective TQFT's*, preprint, 1996.

[9] Kirby, R.C., *A calculus for framed links in S^3*, Invent. Math. **65** (1978) 35–56.

[10] Kirby, R.C., Melvin, P., *Dedekind sums, μ-invariants and the signature cocycle*, Math. Ann. **299** (1994) 231–267.

[11] Lawrence, R.J., *A universal link invariant using quantum groups*, in Differential geometric methods in theoretical physics (Chester, 1988) 55-63.

[12] Le, T.T.Q., *On perturbative $PSU(n)$ invariants of rational homology 3-spheres*, preprint, math.GT/9802032.

[13] Masbaum, G., Roberts, J., *On central extensions of mapping class groups*, Math. Ann. **302** (1995) 131–150.

[6] For example, in the case of the matrix $\begin{pmatrix} \pm 1 & b \\ 0 & \pm 1 \end{pmatrix}$, the lens space becomes $L(0,\pm 1)$ which is equal to $S^2 \times S^1$. In this case H in Theorem 3.1 (2) becomes 0 and we can not apply Theorem 3.1 (2). In general the assumption $H \neq 0$ in Theorem 3.1 (2) implies the condition that $M_1 \cup_\Sigma (-M_2)$ in the formula (0.1) is a rational homology 3-sphere, *i.e.*, the condition that its perturbative $SO(3)$ invariant can be defined.

[14] Matveev, S., Polyak, M., *On a tangle presentation of the mapping class groups of surfaces*, Contemp. Math. **164** (1994) 219–229.

[15] Meyer, W., *Die Signatur von Flächenbündeln*, Math. Ann. **201** (1973) 239–264.

[16] Moriyoshi, H., *Geometric construction of central group extensions for the mapping class groups*, in preparation.

[17] Murakami, J., *Representation of mapping class groups via the universal perturbative invariant*, Proceedings of Knots 96, ed. by S. Suzuki, World Scientific, 573–586 (1997).

[18] Murakami, J., Ohtsuki, T., *Topological Quantum Field Theory for the universal quantum invariant*, Commun. Math. Phys. **188** (1997) 501–520.

[19] Ohtsuki, T., *Invariants of 3-manifolds derived from universal invariants of framed links*, Math. Proc. Camb. Phil. Soc. **117** (1995) 259–273.

[20] ———, *A polynomial invariant of rational homology 3-spheres*, Invent. Math. **123** (1996) 241–257.

[21] Rademacher, H., Grosswald, E., Dedekind sum. (Carus Math. Monogr., **16**) Math. Assoc. Am. 1972.

[22] Reshetikhin, N., Turaev, V.G., *Invariants of* 3*-manifolds via link polynomials and quantum groups*, Invent. Math. **103** (1991) 547–597.

[23] Roberts, J., *Skeins and mapping class groups*, Math. Proc. Camb. Phil. Soc. **115** (1994) 53–77.

[24] ———, *Kirby calculus in manifolds with boundary*, preprint, 1996.

[25] Rozansky, L., it Higher order terms in the Melvin-Morton expansion of the colored Jones polynomial, preprint, q-alg/9601009.

[26] ———, *On finite type invariants of links and rational homology spheres derived from the Jones polynomial and Witten-Reshetikhin-Turaev invariant*, preprint, q-alg/9511025.

[27] Sawin, S., *Links, quantum groups and TQFTs*, Bull. Amer. Math. Soc. **33** (1996) 413–445.

[28] Turaev, V.G., *Quantum representations of modular groups and Maslov indices*, Topology '90 (ed. by B. Apanasov et al.), 381–390. De Gruyter, Berlin, New York, 1992.

[29] ———, Quantum invariants of knots and 3-manifolds, de Gruyter Studies in Math. **18**, de Gruyter, 1994.

Appendix A. Central extensions of mapping class groups of closed surfaces derived from signature cocycle

In this section we review a geometric construction of central extensions of mapping class groups by signature cocycle due to [**2**]; for detailed formulation we follow [**16**].

Let M be a closed 3-manifold. We consider the equivalence relation among 4-manifolds with boundary M such that two 4-manifolds W and W' with $\partial W = \partial W' = M$ are *equivalent* if $\operatorname{sign}\big(W \cup_M (-W')\big) = 0$, where "sign" denotes the signature of a closed 4-manifold. Such classes of bounding 4-manifolds play a role of framings of the 3-manifold M.

Let $\mathcal{G}$ be the mapping class group of a closed surface Σ. We obtain a central extension of $\mathcal{G}$ using classes of bounding 4-manifolds as follows. For $f \in \mathcal{G}$ we denote by T_f the mapping torus of f, *i.e.*, the surface bundle over S^1 with a fixed base point whose monodromy is f. We put $\tilde{\mathcal{G}}'$ to be the set of mapping tori with classes of bounding 4-manifolds. We define a product in $\tilde{\mathcal{G}}'$ as follows. Let (T_f, α) and (T_g, β) be two elements in $\tilde{\mathcal{G}}'$, and choose 4-manifolds W_1 and W_2 representing the classes α and β respectively. Let X be the surface obtained from 2-sphere by removing three open disks. Let $Z_{f,g}$ be a surface bundle over X with a fiber Σ such that the monodromies of the bundle along the three components of the boundary of X are equal to f, g and fg. We define the product of (T_f, α) and (T_g, β) to be (T_{fg}, γ) where γ is the class including $W_1 \cup_{T_f} Z_{f,g} \cup_{T_g} W_2$. Note that γ is well defined by additivity of the signature; see [**5**, Proposition 7.1] for a proof of the additivity. There is a natural homomorphism $\tilde{\mathcal{G}}' \to \mathcal{G}$ obtained by forgetting classes of bounding 4-manifolds.

In the above construction of $\tilde{\mathcal{G}}'$, $\mathrm{sign}(Z_{f,g})$ determines the extension $\tilde{\mathcal{G}}'$; in fact $\mathrm{sign}(Z_{f,g})$ induces a 2-cocycle of $\mathcal{G}$, which is what we call *signature cocycle.*

THEOREM A.1 ([**15, 1**], see also [**13**, Remark 3.11] and [**16**]). *$\tilde{\mathcal{G}}'$ is a central extension of $\mathcal{G}$, i.e., there is an exact sequence*

$$0 \longrightarrow \mathbb{Z} \longrightarrow \tilde{\mathcal{G}}' \longrightarrow \mathcal{G} \longrightarrow 1$$

such that the image of $\mathbb{Z}$ is central in $\tilde{\mathcal{G}}'$. Moreover the element of $H^2(\mathcal{G};\mathbb{Z})$ which determines the extension is the class including signature cocycle, which is equal to four times a generator of $H^2(\mathcal{G};\mathbb{Z})$.

DEPARTMENT OF MATHEMATICAL AND COMPUTING SCIENCES, TOKYO INSTITUTE OF TECHNOLOGY, OH-OKAYAMA, MEGURO-KU, TOKYO 152-8552, JAPAN
E-mail address: tomotada@is.titech.ac.jp

Contemporary Mathematics
Volume **233**, 1999

Configuration Space Integrals and Invariants for 3-Manifolds and Knots

Alberto S. Cattaneo

1. Introduction

In this paper we give a brief description of the way proposed in [**4**] of associating invariants of both 3-dimensional rational homology spheres (r.h.s.) and knots in r.h.s.'s to certain combinations of trivalent diagrams. In addition, we discuss the relation between this construction and Kontsevich's proposal [**8**].

The same diagrams appear in the LMO invariant [**9**] for 3-manifolds, and it would be very interesting to know if there exists any relationship between the two approaches in the case of r.h.s.'s.

The reason for restricting here to r.h.s.'s is quite technical, as will be clear in Sect. 4. Until that point, without any loss of generality we can assume M to be any connected, compact, closed, oriented 3-manifold.

Our construction yields the invariants in terms of integrals over a suitable compactification of the configuration space of points on M. More precisely, the number of points corresponds to the number of vertices in the trivalent diagram, and the integrand is obtained by associating to each edge in the diagram a certain 2-form that represents the integral kernel of an "inverse of the exterior derivative d."

One reason for constructing invariants in terms of "d^{-1}" comes from perturbative Witten–Chern–Simons theory [**11**]. (More precisely, one should invert the covariant derivative with respect to a flat connection; so the present construction is related to the trivial-connection contribution.)

Another reason, which is perhaps more transparent to topologists, relies on the definition of the linking number of two curves in $\mathbb{R}^3$ as the intersection number of one curve with any surface cobounding the other. Thus, linking number may be defined in terms of an inverse of the boundary operator ∂. If one wants to represent the linking number with an integral formula (viz., Gauss's formula), then one must consider the Poincaré duals of curves and apply "d^{-1}."

The exterior derivative d is neither injective nor surjective. Thus, to invert it, one must restrict it to the complement of its kernel and invert it on its image.

1991 *Mathematics Subject Classification.* Primary 57M99.
Supported by MURST and partially by INFN.

Notice that one needs an explicit choice of the complement of the kernel, and this introduces an element of arbitrariness in the construction. Actually, our main task will be to prove that the invariants we define are really independent of this arbitrary choice.

A general way of defining the inverse of d is by introducing a *parametrix*, i.e., a linear operator on $\Omega^*(M)$ that decreases by one the form degree and satisfies the following equation:

$$d \circ P + P \circ d = I - S, \tag{1}$$

where I is the identity operator and S is a suitable smooth operator such that (1) has a solution.

The definition of the parametrix is far from unique. For the choice of S is to a large extent arbitrary and, moreover, given a solution P, we get another solution of the form $P + d \circ Q - Q \circ d$ for any linear operator Q that decreases by two the form degree. These ambiguities reflect the ambiguities in defining an inverse of d.

REMARK 1.1. One possible choice for the parametrix is the Riemannian parametrix P_g, which is based on the choice of a Riemannian metric g:

$$P_g = d^* \circ (\Delta + \pi_{\mathrm{Harm}})^{-1}, \qquad \alpha \in \Omega^*(M). \tag{2}$$

where $*$ is the Hodge-$*$ operator, $\Delta = d^*d + dd^*$ is the Laplace operator, and π_{Harm} is the projection to harmonic forms. P_g satisfies (1) with $S = \pi_{\mathrm{Harm}}$.

The main point is now to define an integral kernel (of course not unique) for P: i.e., we want to represent P as a convolution. In the language of differential topology, convolution can be written as

$$P\alpha = -\pi_{2*}(\hat{\eta}\ \pi_1^*\alpha), \qquad \alpha \in \Omega^*(M),$$

where π_1 and π_2 are the projections from $M \times M$ to either copy of M, and $\hat{\eta}$ is the integral kernel for P (wedges will be understood throughout). By dimensional reasons it is clear that $\hat{\eta}$ must be a 2-form on $M \times M$. Because of the identity operator in (1), it is also clear that $\hat{\eta}$ cannot be a smooth form. This is clearly a problem since we want to use $\hat{\eta}$ to define smooth invariants for the manifold M. The solution consists in replacing $M \times M$ with a suitable compactification $C_2(M)$ of the configuration space.

The ambiguities in the definition of the parametrix imply that the form $\hat{\eta}$ is not unique. Thus, the main point will be to prove that the invariants are independent of the choices involved in the construction of $\hat{\eta}$. The plan of the construction is then the following:

1. One constructs a form $\hat{\eta}$ to represent the integral kernel of a parametrix with a prescription on its behavior on the boundary of $C_2(M)$.
2. One introduces the unit interval I as a space of parameters to take care of the arbitrary choices involved in this construction.
3. One associates to each trivalent diagram with n vertices a function on I by integrating pullbacks of the form $\hat{\eta}$ over a suitable compactification $C_n(M)$ (to be defined in Sect. 2) of the configuration space of n points on M. An invariant will then be a constant function on I.
4. If the integrand form is closed, then the differential of the corresponding function on I is determined only by contributions on the boundary of $C_n(M)$. One shows then that if M is a r.h.s., it is possible to associate a closed non-trivial integrand to each trivalent diagram.

5. Because of the prescribed behavior of $\hat{\eta}$ on the boundary, one can cancel the boundary contributions by summing up appropriate combinations of diagrams.

We discuss points 1. and 2. in Sect. 3 and points 3., 4. and 5. in Sect. 4.

More concisely, the final statement is that to certain combinations (cocycles) of trivalent diagrams it is possible to associate a well-defined element of $H^{3n}(C_n(M), \partial C_n(M))$, where M is a r.h.s. The invariant for M is then obtained by comparing this element with the unit generator.

As we will see, the form $\hat{\eta}$ is not closed, and this accounts for the complications in point 4. This is due to the fact that the cohomology of M is not trivial. There are however cases when one can obtain a closed form; viz.:

1. One can introduce a flat bundle E over M and consider the relevant covariant derivative instead of the exterior derivative. If the bundle is non trivial, it may happen that the complex $H^*(M; E)$ is acyclic. In this case, one can construct a covariantly closed form to represent the inverse of the covariant derivative.
2. If M is a r.h.s., one can remove one point, thus obtaining a rational homology disc and, consequently, a closed form $\hat{\eta}$.

Case 1. was studied by Axelrod and Singer [**3**] in the Riemannian framework of eq. (2) (with Δ the covariant Laplace operator and, by hypothesis, $\pi_{\mathrm{Harm}} = 0$). For a more general treatment of this case, s. [**5**]. Notice, however, that this approach does not apply in general: e.g., it does not even work for S^3 whose only flat connection is trivial.

Case 2. was proposed by Kontsevich [**8**]. A realization of this proposal for the simplest invariant—known as the Θ-invariant—was then studied by Taubes [**10**]. He also proved his version of the Θ-invariant to be trivial on integral homology spheres. This rules out any relationship with the LMO invariants which predict the Θ-invariant on integral homology spheres to be the Casson invariant.

In Sect. 5 we will apply Kontsevich's proposal to the invariants constructed in [**4**] (and Sect. 4), and will compare our result with Taubes's.

REMARK 1.2. The above construction can be used to define invariants for knots in a r.h.s. as well, s. [**4**] (and Sect. 4).

Also in this case the construction is simplified if one gets a closed form $\hat{\eta}$. This happens, e.g., when $M = \mathbb{R}^3$ [**6, 1**], or when $M = \Sigma \times [0, 1]$ with Σ a connected, compact, closed, oriented 2-manifold [**2**].

Another possibility is an approach à la Kontsevich when M is a r.h.s., s. Sect. 5.

2. A compactification of configuration spaces

In this section we assume M to be a d-dimensional connected, compact, closed, oriented manifold.

The (open) configuration space $C_n^0(M)$ of n points in M is obtained by removing all diagonals from the Cartesian product M^n.

A C^∞-compactification for these spaces was proposed in [**3**] (generalizing the algebraic compactification of [**7**]). This compactification is obtained by taking the closure

$$C_n(M) \doteq \overline{C_n^0(M)} \subset M^n \times \prod_{\substack{S \subset \{1,2,\dots,n\} \\ |S| \geq 2}} Bl(M^S, \Delta_S),$$

where $Bl(M^S, \Delta_S)$ denotes the differential-geometric blowup obtained by replacing the principal diagonal Δ_S in M^S with its unit normal bundle $N(\Delta_S)/\mathbb{R}^+$. Notice that Δ_S is diffeomorphic to M and that

$$N(\Delta_S) \simeq TM^{\oplus S}/\text{global translations}.$$

Thus, the boundary of $Bl(M^S, \Delta_S)$ is a bundle over M associated to the tangent bundle.

Because of all the blowups, the spaces $C_n(M)$ turn out to be manifolds with corners ($C_2(M)$ is simply a manifold with boundary). The codimension-one components of the boundary of $C_n(M)$ are labeled by subsets of $\{1, \ldots, n\}$. By permuting the factors, we can always put ourselves in the case when this subset is $\{1, \ldots, k\}$, $2 \leq k \leq n$. We will denote by $\mathcal{S}_{n,k} \subset \partial C_n(M)$ a face of this kind. Then we have the following functorial description of $\mathcal{S}_{n,k}$:

$$\begin{array}{ccc} \mathcal{S}_{n,k} \simeq (\pi_1)^{-1}\widehat{C}_k(TM) & \longrightarrow & \widehat{C}_k(TM) \\ \downarrow & & \downarrow \\ C_{n-k+1}(M) & \xrightarrow[\pi_1]{} & M \end{array} \tag{3}$$

Here π_1 is the projection onto the first copy of M (i.e., where the first k points have collapsed) and $\widehat{C}_k(TM)$ is a bundle associated to the tangent bundle of M whose fiber $\widehat{C}_k(\mathbb{R}^d)$, $d = \dim M$, is obtained from $(\mathbb{R}^d)^k/G$—G being the group of global translations and scalings—by blowing up all diagonals. More precisely,

$$\widehat{C}_k(\mathbb{R}^d) = \overline{C^0_k(\mathbb{R}^d)/G} \subset \left((\mathbb{R}^d)^k \times \prod_{\substack{S\subset\{1,2,\ldots,k\} \\ |S|\geq 2}} Bl((\mathbb{R}^d)^S, \Delta_S) \right) /G.$$

Notice that each $\widehat{C}_k(\mathbb{R}^d)$ is a compact manifold with corners. In the simplest case we have $\widehat{C}_2(\mathbb{R}^d) = S^{d-1}$.

The diagonal action of $SO(d)$ on $(\mathbb{R}^d)^k$ descends to $\widehat{C}_k(\mathbb{R}^d)$. If we choose a Riemannian metric on M, we can write

$$\widehat{C}_k(TM) = OM \times_{SO(d)} \widehat{C}_k(\mathbb{R}^d),$$

where OM is the orthonormal frame bundle of TM. In particular, when $n = 2$, we have

$$C_2(M) = Bl(M \times M, \Delta),$$

and $\partial C_2(M) \simeq S(TM) = OM \times_{SO(d)} S^{d-1}$.

3. The construction of a parametrix

In this section we assume that M is any connected, compact, closed, oriented 3-manifold.

We start considering the following commutative diagram:

$$\begin{array}{ccc} \partial C_2(M) & \xrightarrow{\iota^\partial} & C_2(M) \\ {\scriptstyle\pi^\partial}\downarrow & & \downarrow{\scriptstyle\pi} \\ \Delta & \xrightarrow[\iota^\Delta]{} & M \times M \end{array}$$

Then we define the involution T that exchanges the factors in $M \times M$. By abuse of notation, we will denote by T also the corresponding involution on $C_2(M)$ and on its boundary $\partial C_2(M)$. On the latter T acts as the antipodal map on the fiber crossed with the identity on the base. We will denote by $H^*_\pm$ the + and − eigenspaces of T in the cohomology of any of the above spaces.

We will denote by $\chi_\Delta \in \Omega^3(M \times M)$ a representative of the Poincaré dual of the diagonal Δ. Since $[\chi_\Delta] \in H^3_-(M \times M)$, there is really no loss of generality in choosing an odd representative.

On the sphere bundle $\partial C_2(M) \to \Delta$, one can introduce a global angular form η; i.e., a form $\eta \in \Omega^2(\partial C_2(M))$ with the following properties:

1. the restriction of η to each fiber is a generator of the cohomology of the fiber;
2. $d\eta = -\pi^{\partial *}e$, where e is a representative of the Euler class.

Since M is 3-dimensional, the Euler class is trivial. Moreover, since $H^2_+(S^2) = 0$, we may choose the global angular form to be odd. Since T acts as the identity on the base, 2. is then replaced by $d\eta = 0$. We have the following

PROPOSITION 3.1. *Given an odd global angular form η and an odd representative χ_Δ of the Poincaré dual of the diagonal Δ in $M \times M$, there exists a form $\hat{\eta} \in \Omega^2(C_2(M))$ with the following properties:*

(4a) $$d\hat{\eta} = \pi^* \chi_\Delta,$$

(4b) $$\iota_\partial^* \hat{\eta} = -\eta,$$

(4c) $$T^* \hat{\eta} = -\hat{\eta}.$$

This is a simple generalization of the analogous proposition in [**4**] for the case when M is a r.h.s.

PROOF. Let U be a tubular neighborhood of Δ in $M \times M$, and $\tilde{U} = \pi^{-1}U$ its preimage in $C_2(M)$. Then $\tilde{U}$ has the structure of $\partial C_2(M) \times [0,1]$. Let us denote by $\partial_0 \tilde{U} = \partial C_2(M)$ and by $\partial_1 \tilde{U} = \pi^{-1}\partial U$ the two boundary components of $\tilde{U}$.

Let ρ be a function on $\tilde{U}$ which is constant and equal to -1 in a neighborhood of $\partial_0 \tilde{U}$ and is constant and equal to 0 in a neighborhood of $\partial_1 \tilde{U}$. Moreover, assume that ρ is even under the action of T.

Let $p : \tilde{U} \to \partial C_2(M)$ be the natural projection. Then consider the form $\tilde{\eta} = \rho\, p^* \eta$. Since η is a global angular form, $d\tilde{\eta} = d\rho\, p^* \eta$ is a representative of the Thom class of the normal bundle of Δ. Therefore, if we extend $\tilde{\eta}$ by zero on the whole of $C_2(M)$, we have that $d\tilde{\eta}$ is the pullback of a representative of the Poincaré dual of the diagonal.

This might not be our choice χ_Δ. Anyhow, $d\tilde{\eta} = \pi^*(\chi_\Delta + d\alpha)$, and it is not difficult to check that one can choose $\alpha \in \Omega^2_-(M \times M)$. So we set $\hat{\eta} = \tilde{\eta} - \pi^* \alpha$, and it is an immediate check that properties (4) hold. □

Notice that, as is clear from the proof, the definition of $\hat{\eta}$ is not unique, even for fixed η and χ_Δ.

With such a form $\hat{\eta}$ we can finally define a parametrix. In fact, let us denote by ρ_1 and ρ_2 the projections from $M \times M$ to each factor, and by π_1 and π_2 the corresponding projections from $C_2(M)$. By defining the push-forward as acting from the left, we have the following

PROPOSITION 3.2. *If $\hat{\eta} \in \Omega^2(M)$ satisfies (4a) and (4b), then*

$$P\alpha = -\pi_{2*}(\hat{\eta}\,\pi_1^*\alpha), \qquad \alpha \in \Omega^*(M),$$

is a parametrix with $S = -\rho_{2*}(\chi_\Delta\,\rho_1^*\alpha)$.

The proof is a simple exercise of fiber integration in the case when the fiber has a boundary (s. [**4**]).

REMARK 3.3. Property (4c) is not needed in the definition of the parametrix, but is natural and simplifies the writing of the invariants. Notice incidentally that the *propagator* in perturbative Witten–Chern–Simons theory is odd under the action of T, being related to the *expectation value* of a connection 1-form placed at two different points.

3.1. The global angular form. To end with the construction, we have to specify a choice for the global angular form.

Following [**3**] and [**4**], we pick up a Riemannian metric g. Then $\partial C_2(M) \simeq OM \times_{SO(3)} S^2$. Let

$$p : OM \times S^2 \to OM \times_{SO(3)} S^2$$

be the natural projection, and let θ be a connection form on OM (i.e., a metric connection). By abuse of notation, we will write θ also for its pullback to $OM \times S^2$. if it is a polynomial in θ and $d\theta$—with coefficients in $\Omega^*(S^2)$—and it is basic (i.e., $\bar{\eta} = p^*\eta$). For a given θ, the equivariant global angular form is unique and its θ-independent part is the $SO(3)$-invariant unit volume form on S^2, which we will denote by ω throughout. See [**4**] for an explicit construction.

CONDITION 3.4. We assume that the restriction of $\hat{\eta}$ to the boundary is such that its pullback to $OM \times S^2$ is the equivariant global angular form.

We denote by ω_{ij} the pullbacks of ω by the projections $\widehat{C}_n(\mathbb{R}^3) \to \widehat{C}_2(\mathbb{R}^3) = S^2$. Then a useful property proved in [**8**] (s. also [**4**]) is expressed by the following

LEMMA 3.5 (Kontsevich). *If x_i, $1 \leq i \leq n$, is any coordinate in $\widehat{C}_n(\mathbb{R}^3)$, then, for any indices j and k ($j \neq i$, $k \neq i$)*

$$\int_{x_i} \omega_{ij}\,\omega_{ik} = 0.$$

3.2. Parametrizing the choices. We have constructed an integral kernel $\hat{\eta}$ for a parametrix depending on the following choices: a metric g, a compatible connection form θ, a representative χ_Δ of the Poincaré dual of Δ (and a function ρ and a 2-form α as in the proof of Prop. 3.1).

To take care of these choices, we introduce the parameter space $I = [0,1]$. Then, denoting by σ the inclusion $M \times M \hookrightarrow M \times M \times I$, we take $\chi_\Delta \in \Omega^3(M \times M \times I)$ such that $d\chi_\Delta = 0$ and $\sigma^*\chi_\Delta$ is a representative of the Poincaré dual of Δ. (We treat similarly ρ and α.)

As for g and θ, we operate as follows. We take a block-diagonal metric g on $M \times I$ (i.e., a metric such that $g_{(m,t)}(v,w) = 0$ for all $v \in T_mM$ and $w \in T_tI$), and consider the orthonormal frame bundle $\widetilde{OM}$ of $TM \to M \times I$. Then we choose a connection form θ on $\widetilde{OM}$ and define the equivariant global angular form on $\widetilde{OM} \times S^2$.

Using the projections $C_n(M) \times I \to C_2(M) \times I$, we can pull back the form $\hat{\eta}$ in $n(n-1)/2$ different ways which we denote by $\hat{\eta}_{ij}$. if it is a product of pullbacks of $\hat{\eta}$. Each special form can be graphically associated to a diagram, each edge representing a pullback of $\hat{\eta}$.

Let $\mathcal{S}_{n,k}$ denote, as in the previous section, the face in $\partial C_n(M)$ corresponding to the collapse of the first k points (we consider only this case since all other codimension-one faces can be reduced to this one by simply applying a permutation of the factors). Let $\pi^{\mathcal{S}}$ denote the induced projection $\mathcal{S}_{n,k} \times I \to C_{n-k+1}(M) \times I$, and let $\pi_1 : C_{n-k+1}(M) \times I \to M \times I$ denote the projection on the first point (i.e., where the first k points have collapsed). Then we have the following

LEMMA 3.6 (Axelrod and Singer). *If $\alpha \in \Omega^*(\mathcal{S}_{n,k})$ is the restriction of a special form, then*

$$\pi^{\mathcal{S}}_* \alpha = \beta\, \pi_1^* \gamma,$$

where β is special and γ is either a constant or a multiple of the first Pontrjagin form p_1 associated to θ.

For the proof, s. [**3**] or [**4**]. Using the same notations, we also have the following

COROLLARY 3.7. *γ (and hence $\pi_1^*\gamma$) is a constant in the case when no parameter space is introduced.*

4. An invariant for rational homology spheres

In this section we assume M to be a 3-dimensional r.h.s. We then choose a representative v of the unit generator of $H^3(M)$, so we can take, as the Poincaré dual of the diagonal in $M \times M$,

$$\chi_\Delta = v_2 - v_1, \tag{5}$$

where $v_i = \rho_i^* v$, and ρ_i, $i = 1, 2$, is the projection to the i-th factor.

We now define our form $\hat{\eta}$ as in the preceding section—i.e., satisfying (4) and Condition 3.4—with χ_Δ as in (5).

Next we consider the three projections $\pi_{ij} : C_3(M) \to C_2(M)$, and write $\hat{\eta}_{ij} = \pi_{ij}^* \hat{\eta}$. Then (5) implies $d\hat{\eta}_{ij} = v_j - v_i$. Thus, we can define the following non-trivial closed form in $\Omega^2(C_3(M))$:

$$\hat{\eta}_{123} \doteq \hat{\eta}_{12} + \hat{\eta}_{23} + \hat{\eta}_{31}. \tag{6}$$

Notice that on any configuration space $C_n(M)$, $n > 2$, we can analogously define closed forms $\hat{\eta}_{ijk}$ for any triple of distinct indices i, j, k.

Now consider graphs with numbered vertices, and set equivalent to zero all graphs with an edge connecting a vertex to itself. We have then an induced orientation of the edges (viz., each edge is oriented from the lower to the higher end-point).

To each trivalent graph Γ of the above type we can associate the following number:

$$A_\Gamma(M) \doteq \int_{C_{n+1}(M)} v_0 \prod_{(ij)\in E(\Gamma)} \hat{\eta}_{ij0}, \tag{7}$$

where n is the number of vertices and $E(\Gamma)$ is the set of ordered edges. The point labeled by 0 is an extra point and not a vertex of Γ. We extend A_Γ to combinations of graphs by linearity.

We are interested in the dependence of A_Γ on the choices in the construction of $\hat{\eta}$. So we introduce a parameter space I as in the preceding section and consider A_Γ as a function on I. (As for v, we take it in $\Omega^3(M\times I)$ and such that it is closed and its restriction to M is a representative of the unit generator of $H^3(M)$.) Then we consider the differential of this function. Since the integrand form is closed, this differential is given only by boundary terms. These are dealt with by using Lemmata 3.5 and 3.6.

REMARK 4.1. A_Γ can be defined also if M is not a r.h.s. However, in this case the integrand form is not closed. So in differentiating A_Γ we also have a bulk contribution which we do not know how to deal with.

We now define a coboundary operator δ that acts on graphs by contracting each edge one at a time, with a sign given by the parity of the higher end-point. In [**4**], it is shown that δ is a coboundary operator.

We call a *cocycle* a δ-closed combination of graphs. We say that it is connected (trivalent) if all its terms are connected (trivalent) graphs.

Finally, we consider the Chern–Simons integral,

$$\mathrm{CS}(M,f) = -\frac{1}{8\pi^2}\int_M f^*\,\mathrm{Tr}\left(\theta\,d\theta + \frac{2}{3}\,\theta^3\right),$$

where f is a section of OM and θ is the same connection form as in the construction of $\hat{\eta}$. In [**4**], the following was proved:

THEOREM 4.2. *If Γ is a connected, trivalent cocycle, then there exists a constant $\phi(\Gamma)$ such that*

$$I_\Gamma(M,f) = A_\Gamma(M) + \phi(\Gamma)\,\mathrm{CS}(M,f)$$

is an invariant for the framed rational homology 3-sphere M.

REMARK 4.3. Instead of defining the equivariant global angular form, one could repeat the previous construction by choosing a trivialization of $S(TM)$ and by defining the global angular form to be the (pullback of the) $SO(3)$-invariant unit volume form ω on S^2 (as suggested in [**8**]).

The equivariant treatment shows, cf. Thm. 4.2, that under a change of framing the invariants A_Γ behave as multiples of the Chern–Simons integral. In particular, they are invariant under a homotopic change of framing.

All graphs in a cocycle have the same number n of vertices. If the cocycle is trivalent, this number is even. In this case, one can define

$$\mathrm{ord}\,\Gamma = \frac{n}{2}. \tag{8}$$

The constant $\phi(\Gamma)$ depends only on Γ and not on M. One can show [**3, 4**] that $\phi(\Gamma)=0$ if $\mathrm{ord}\,\Gamma$ is even. Moreover, in [**4**] it is shown that $\phi(\Theta)=1/4$, with

$$A_\Theta = \int_{C_3(M)} v_0\,\hat{\eta}^3_{012}. \tag{9}$$

This allows for the definition of the following unframed invariants (for $\Gamma\neq\Theta$):

$$J_\Gamma(M) = A_\Gamma(M) - 4\,\phi(\Gamma)\,A_\Theta(M).$$

A computation in [**4**] shows that $J_\Gamma(S^3) = J_\Gamma(SO(3)) = 0$ if $\mathrm{ord}\,\Gamma$ is odd.

4.1. Knot invariants. In [**4**], invariants for knots in a r.h.s. are studied.

If K is an imbedding $S^1 \hookrightarrow M$, one has induced imbeddings $\widetilde{C}_n(S^1) \hookrightarrow C_n(M)$, where $\widetilde{C}_n(S^1)$ is the connected component of $C_n(S^1)$ defined by an ordering of the points on S^1. The configuration space $\widetilde{C}^K_{n,t}(M)$ of n points on the knot and t points in M is then defined by pulling back the bundle $C_{n+t}(M) \to C_n(M)$.

All the forms introduced before can be pulled back to $\widetilde{C}^K_{n,t}$, and by abuse of notation we will keep calling them with the same names. One should keep in mind, however, that the pulled-back forms depend on the imbedding K. This understood, one defines the self-linking number

$$\operatorname{sln}(K,M) \doteq \int_{\widetilde{C}^K_{2,0}(M)} \hat{\eta}_{12}.$$

Now consider graphs with a distinguished loop, which represents the knot. We call *external* the vertices and the edges on this loop, and *internal* all the others. To a trivalent graph we can then associate the number

$$A_\Gamma(K,M) \doteq \int_{\widetilde{C}^K_{n,t+1}(M)} v_0 \prod_{(ij)\in I(\Gamma)} \hat{\eta}_{ij0}, \tag{10}$$

where n and t are the numbers of external and internal vertices in Γ, and $I(\Gamma)$ is the set of internal edges. Again we extend (10) to combinations of graphs by linearity.

Next we define a coboundary operator δ as before with the only difference that now δ does not contract internal edges connecting two external vertices. Graph combinations killed by δ are called cocycles. (An explicit computation of these cocycles is presented in [**1**].)

We name *prime* a graph which is connected after removing any pair of external edges (in [**4**] a graph of this kind was called connected). A cocycle will be called prime (trivalent) if all its terms are prime (trivalent). In [**4**], the following was proved:

THEOREM 4.4. *If K is a knot in the rational homology 3-sphere M, and Γ is a prime, trivalent cocycle, then there exists a constant $\mu(\Gamma)$ such that*

$$I_\Gamma(K,M) = A_\Gamma(K,M) + \mu(\Gamma)\operatorname{sln}(K,M)$$

is a knot invariant. Moreover, $\mu(\Gamma) = 0$ if $\operatorname{ord}\Gamma$ is even.

5. Relationship with Kontsevich's proposal

As we recalled in the Introduction, in [**8**] Kontsevich proposed a different way of constructing invariants for r.h.s.'s. His proposal differs from our construction since *i*) one point is removed from M in order to make its rational homology trivial, and *ii*) the global angular form on the boundary of the configuration space is defined via a trivialization of TM. Let us consider part *i*) of the proposal first.

We introduce the compactified configuration space $C_n(M, x_\infty)$ of n points on $M\backslash x_\infty$ (where x_∞ is an arbitrary base point) as the fiber of $C_{n+1}(M) \to M$ over the point x_∞ in the last copy of M; viz.:

$$\begin{array}{ccc} C_n(M,x_\infty) & \xrightarrow{\ p\ } & C_{n+1}(M) \\ \downarrow & & \downarrow{\scriptstyle \pi_{n+1}} \\ x_\infty & \longrightarrow & M \end{array}$$

Consider now the projections $\pi_{ij} : C_{n+1}(M) \to C_2(M)$, $i < j \leq n+1$, and set $p_{ij} = \pi_{ij} \circ p$ for $i < j \leq n$ and $p_{i\infty} = \pi_{i,n+1} \circ p$ for $i \leq n$. Then we will denote by $\hat{\eta}_{ij}$ and $\hat{\eta}_{i\infty}$ the pullbacks of $\hat{\eta}$ to $C_n(M, x_\infty)$. Accordingly, we will define $\hat{\eta}_{ijk}$ and $\hat{\eta}_{ij\infty}$. On the boundary faces we will have the pullbacks of global angular forms η_{ij} and $\eta_{i\infty}$. Observe that $\eta_{i\infty} = \omega_{i\infty}$, with the notations of Lemma 3.5.

We also have projections $C_n(M, x_\infty) \to C_n(M)$. The forms $\hat{\eta}_{ij}$ we have written above can also be seen as the pullbacks of the forms with the same name on $C_n(M)$.

In particular, $C_1(M, x_\infty)$ is just M blown up at x_∞, and $\partial C_1(M, x_\infty) = S^2$. If we pull v back by the projection $\tau : C_1(M, x_\infty) \to M$, we get an exact form $\tau^* v = dw$, where the two-form w restricted to the boundary must be a representative of the unit generator of $H^2(S^2)$. We may choose this representative to be ω since, by Thm. 4.2, the explicit choice of v does not affect the invariant. We have then the following

THEOREM 5.1. *If Γ is a connected, trivalent cocycle and M is a rational homology 3-sphere, then*

$$A_\Gamma(M) = A'_\Gamma(M) + B_\Gamma,$$

with

$$A'_\Gamma(M) = \int_{C_n(M,x_\infty)} \prod_{(ij)\in E(\Gamma)} \hat{\eta}_{ij\infty},$$

$$B_\Gamma = \int_{\widehat{C}_{n+2}(\mathbb{R}^3)} \omega_{0\infty} \prod_{(ij)\in E(\Gamma)} \omega_{ij0}.$$

Moreover, if $\operatorname{ord}\Gamma$ *is odd,* B_Γ *vanishes.*

Notice that B_Γ does not depend on M or on any arbitrary choice. Thus, even if it should not vanish, it would just represent a constant shift in the invariant. As a consequence, Thm. 4.2 holds with $A_\Gamma(M)$ replaced by $A'_\Gamma(M)$.

PROOF. We can pull back the integrand form in (7) to $C_{n+1}(M, x_\infty)$ and integrate it over there. Using the fact that the pullback of v_0 is exact, by Stokes's theorem we can rewrite (7) as

$$A_\Gamma(M) = \int_{\partial C_{n+1}(M,x_\infty)} w_0 \prod_{(ij)\in E(\Gamma)} \hat{\eta}_{ij0}.$$

The codimension-one faces in $\partial C_{n+1}(M, x_\infty)$ are labeled by subsets of $\{0, 1, \ldots, n, \infty\}$. Denote by $\mathcal{S}$ any of these subsets.

Assume now that the cardinality of $\mathcal{S}' = \mathcal{S} \cap \{1, \ldots, n\}$ is k. Since points in $\mathcal{S}'$ label vertices in the graph and the graph is trivalent, we have the relation

$$3k = 2e + e_0,$$

where e denotes the number of edges connecting points in $\mathcal{S}'$, and e_0 denotes the number of edges with exactly one end-point in $\mathcal{S}'$. Now we have four cases, according to

$$\mathcal{S} \backslash \mathcal{S}' = \begin{cases} \{0\} & \text{(a)} \\ \emptyset & \text{(b)} \\ \{0, \infty\} & \text{(c)} \\ \{\infty\} & \text{(d)} \end{cases}$$

The cardinality r of $\mathcal{S}$ is then $k+1$, k, $k+2$ and $k+1$ respectively. The boundary face labeled by $\mathcal{S}$ is a bundle over $C_{n+2-r}(M, x_\infty)$ with projection $\pi^{\mathcal{S}}$ and fiber $\widehat{C}_r(\mathbb{R}^3)$. So the fiber dimension is $3k-1$, $3k-4$, $3k+2$ and $3k-1$ respectively.

We now write the integrand form α restricted to this boundary as $(\pi^{\mathcal{S}*}\beta)\,\alpha'$, where β is the product of the pullbacks of $\hat{\eta}$ corresponding to edges with at least one end-point not in $\mathcal{S}$, times w_0 in cases (a), (b) and (d).

In cases (b) and (d), the term $\hat{\eta}_{ij0}$ contributes to α' only if both i and j are in $\mathcal{S}'$. In case (a) and (c), also terms with either i or j in $\mathcal{S}'$ contribute. Moreover, w_0 contributes to α' only in case (c). As a consequence, the degree of α' will be: (a) $2e+2e_0$, (b) $2e$, (c) $2e+2e_0+2$, (d) $2e$.

By using all the above results, we see that the degree of γ in Corollary 3.7 is e_0+1, $4-e_0$, e_0 and $1-e_0$ respectively. Since γ must be a constant zero-form, we see that the contribution of the face $\mathcal{S}$ vanishes unless we are in case (b) with $e_0=4$, in case (c) with $e_0=0$, or in case (d) with $e_0=1$. Notice, moreover, that we can replace $\hat{\eta}$ by ω in α'. Thus, in the last case above we conclude that the contribution vanishes by Lemma 3.5. The first case is taken care of by the fact that Γ is a cocycle.

We are then left with case (c) and $e_0=0$. Since Γ is connected, there are only two possibilities: 1) only point 0 has collapsed at x_∞, 2) all points have collapsed at x_∞. In case 1), $\alpha' = \omega_{0\infty}$ and the fiber is S^2. After this trivial integration we get $A'_\Gamma(M)$. Case 2) yields B_Γ.

To prove that B_Γ vanishes if $\operatorname{ord}\Gamma = n/2$ is odd, consider the involution $x_i \to -x_i$, $i=0,1,\dots,n,\infty$. All the pullbacks of ω change signs. Since the number of edges is $3n/2$, the integrand form gets the sign $(-1)^{3n/2+1}$. On the other hand, since $\widehat{C}_{n+2}(\mathbb{R}^3)$ is S^{3n+2} with some submanifolds blown up, under the involution the orientation gets the sign $(-1)^{n+1}$. □

In the particular case of the Θ-invariant (9), we have

$$A'_\Theta(M) = \int_{C_2(M,x_\infty)} \hat{\eta}^3_{12\infty}.$$

By our construction, $\hat{\eta}_{12\infty}$ is a closed form on $C_2(M, x_\infty)$ which reduces to the global angular form when restricted to the faces (1∞), (2∞) and (12).

As observed in remark 4.3, one can also modify the construction by choosing a trivialization of $S(TM)$ at the very beginning, and this corresponds to part *ii*) of Kontsevich's proposal. Invariance under homotopic changes of framing is then guaranteed (while under non-homotopic changes, the invariant behaves as $-1/4\,\mathrm{CS}$). In this case, we have the additional property that $\hat{\eta}^2_{12\infty}$ vanishes close to the faces (1∞), (2∞) and (12). However, close to (12∞) neither $\hat{\eta}^2_{12\infty}$ nor $\hat{\eta}^3_{12\infty}$ vanish.

This is to be compared with Taubes's invariant

$$\widetilde{A}_\Theta(M) \doteq \int_{C_2(M,x_\infty)} \omega^3,$$

where ω is a 2-form on $C_2(M, x_\infty)$ with the following properties:

1. ω restricted to the faces (1∞), (2∞) and (12) is a global angular form;
2. ω^2 vanishes not only close to (1∞), (2∞) and (12) but also close to (12∞).

The latter property is achieved only by choosing what Taubes names a *singular framing* for $T(M\backslash x_\infty)$. As a consequence, ω^2 (and hence ω^3) is a form with compact support, and Taubes's Θ-invariant can actually be defined as an integral over the uncompactified configuration space $C_2^0(M, x_\infty)$. Moreover, property 2. is crucial in Taubes's proof that his invariant is trivial on integral homology spheres.

Now the main question is if there is any relationship between the two different ways, A'_Θ and $\widetilde{A}_\Theta$, of realizing Kontsevich's proposal for the Θ-invariant.

5.1. The case of knots. Let us consider an imbedding K of S^1 in the interior of $M\backslash x_\infty$. This induces imbeddings $\widetilde{C}_n(S^1) \hookrightarrow C_n(M, x_\infty)$. By pulling back the bundles $C_{n+t}(M, x_\infty) \to C_n(M, x_\infty)$, we then obtain the configuration spaces $\mathcal{C}_{n,t}^K(M, x_\infty)$. We have the following

THEOREM 5.2. *If Γ is a prime, trivalent cocycle and K is a knot in the rational homology 3-sphere M, then*

$$A_\Gamma(K, M) = \int_{\widetilde{C}_{n,t}^K(M, x_\infty)} \prod_{(ij)\in I(\Gamma)} \hat{\eta}_{ij\infty},$$

$$\mathrm{sln}(K, M) = \int_{\widetilde{C}_{2,0}^K(M, x_\infty)} \hat{\eta}_{12}.$$

In particular, if $M = S^3$ and we choose the Euclidean metric on $\mathbb{R}^3 = S^3\backslash x_\infty$, we recover Bott and Taubes's result [**6**]. As a consequence, the anomaly coefficients $\mu(\Gamma)$ are the same in the two cases.

PROOF. We work as in the proof of Thm. 5.1. The only difference is that we must distinguish between the cases when the collapse is at a point on K or otherwise.

Notice that, since x_∞ does not belong to the image of K, there is no such term as B_Γ. By the same reason, when we consider a collapse at a point on K, we only have points in $\{0, 1, \ldots, n+t\}$. If 0 is involved, the term vanishes since w_0 is basic and is a 2-form. If 0 is not involved, reasoning as in the proof of Thm. 5.1 and applying Corollary 3.7 shows that the term vanishes unless $e_0 = 2$. But this is taken care of by the fact that Γ is a cocycle. □

ACKNOWLEDGMENT. I thank J. E. Andersen, R. Bott, P. Cotta-Ramusino, N. Habegger, R. Longoni, G. Masbaum and P. Vogel for very useful discussions.

For partial support at Madeira conference on "Low Dimensional Topology," January 1998, I acknowledge the Center of Mathematical Sciences (CCM) PRAXIS XXI project. I especially thank H. Nencka for her efforts in organizing the conference.

I am finally thankful to the Institut de Mathematiques de Jussieu and to the University of Nantes for warm hospitality and financial support.

References

[1] D. Altschuler and L. Freidel, "Vassiliev Knot Invariants and Chern–Simons Perturbation Theory to All Orders," Commun. Math. Phys. **187** (1997), 261–287.

[2] J. E. Andersen and J. Mattes, "Configuration Space Integrals and Universal Vassiliev Invariants over Closed Surfaces," q-alg/9704019.

[3] S. Axelrod and I. M. Singer, "Chern–Simons Perturbation Theory," in *Proceedings of the XXth DGM Conference*, edited by S. Catto and A. Rocha (World Scientific, Singapore, 1992), pp. 3–45; "Chern–Simons Perturbation Theory. II," J. Diff. Geom. **39** (1994), 173–213.

[4] R. Bott and A. S. Cattaneo, "Integral Invariants of 3-Manifolds," J. Diff. Geom. **48** (1998), 91–133.

[5] ———, "Integral Invariants of 3-Manifolds. II," math/9802062, to appear in J. Diff. Geom.

[6] R. Bott and C. Taubes, "On the Self-Linking of Knots," J. Math. Phys. **35** (1994), 5247–5287.
[7] W. Fulton and R. MacPherson, "A Compactification of Configuration Spaces," Ann. Math. **139** (1994), 183–225.
[8] M. Kontsevich, "Feynman Diagrams and Low-Dimensional Topology," First European Congress of Mathematics, Paris 1992, Volume II, *Progress in Mathematics* **120** (Birkhäuser, 1994), 120.
[9] T. Q. T. Le, J. Murakami and T. Ohtsuki, "On a Universal Perturbative Invariant of 3-Manifolds," Topology **37** (1998), 539–574.
[10] C. Taubes, "Homology Cobordism and the Simplest Perturbative Chern–Simons 3-Manifold Invariant," in *Geometry, Topology, and Physics for Raoul Bott*, edited by S.-T. Yau (International Press, Cambridge, 1994), pp. 429–538.
[11] E. Witten, "Quantum Field Theory and the Jones Polynomial," Commun. Math. Phys. **121** (1989), 351–399.

Dipartimento di Matematica "F. Enriques" — Università degli Studi di Milano — via Saldini, 50 — I-20133 Milano — Italy
E-mail address: cattaneo@elanor.mat.unimi.it

Contemporary Mathematics
Volume **233**, 1999

The Multivariable Alexander Polynomial for a Closed Braid

H. R. Morton

ABSTRACT. A simple multivariable version of the reduced Burau matrix is constructed for any braid. It is shown how the multivariable Alexander polynomial for the closure of the braid can be found directly from this matrix.

1. Introduction

It has been known for some time that the Alexander polynomial of a closed braid $\hat{\beta}$ can be found from the Burau matrix β, [**1**]. This relation was extended in [**3**] to present the 2-variable Alexander polynomial $\Delta_{\hat{\beta}\cup A}(t,x)$ of the link consisting of the closed braid $\hat{\beta}$ together with its axis A as the characteristic polynomial, $\det(I - x\overline{B}_\beta(t))$, of the reduced $(n-1)\times(n-1)$ Burau matrix $\overline{B}_\beta(t)$ of the braid β. The Alexander polynomial of $\hat{\beta}$ can be recovered by applying the Torres-Fox formula to the 2-variable polynomial to get the equation

$$\frac{\Delta_{\hat{\beta}}(t)}{1-t} = \frac{\Delta_{\hat{\beta}\cup A}(t,1)}{1-t^n}.$$

In this paper I give a similar method for finding the multivariable Alexander polynomial of a link L presented as the closure of a braid β. The main ingredient is a readily constructed multivariable version of the reduced Burau matrices. Other versions of 'coloured' Burau matrices have been developed, for example by Penne, [**4**], which can be interpreted as determining linear presentations of suitably extended versions of the braid group.

The most useful feature of the matrices which are used here is that they are extremely simple to remember and they give an immediate and very straightforward construction of the Alexander polynomial of a closed braid and axis, leading at once to the polynomial of the closed braid. For a pure braid the resulting matrix is conjugate to a reduced version of the Gassner matrix; the construction given here has the advantage that it applies to any braid which presents the link, and does not require the braid to be rewritten in any special form.

An implementation of this calculation by a Maple procedure, which returns the multivariable Alexander polynomial of $\hat{\beta}$ given the braid β, was made in early

1991 *Mathematics Subject Classification.* Primary 57M25.

This paper was prepared during the low-dimensional topology program at MSRI, Berkeley in 1996-97. I am grateful to MSRI for their support.

by suppressing the axis and applying the Torres-Fox formula. Put $x = 1$ and note that $\varphi(A) = t_1 t_2 \cdots t_n$ in the complement of L, to get

$$D_L = \frac{\det(I - \overline{B}_\beta(t_1, \dots, t_n))}{1 - t_1 t_2 \cdots t_n},$$

making any identifications $t_{\pi(i)} = t_i$ required where strings of β belong to the same component of L. In the case when L has one component this gives the well-known formula $\Delta_L(t) = \dfrac{\det(I - \overline{B}_\beta(t))}{1 - t^n}(1 - t)$ quoted earlier.

PROOF OF THEOREM 1. Apply Fox's free differential calculus to a presentation of the fundamental group of the closed braid and axis, very much as in [**1**] or [**3**]. The heart of the proof lies in relating the fundamental group of the n-punctured disk spanning the axis A which meets $\hat{\beta}$ at the bottom of β to the corresponding group for the disk at the top of β. Write $x_1, \dots, x_n$ for the generators of the group at the bottom represented by meridian loops around the punctures, and $X_1, \dots, X_n$ for the meridian loops around the punctures at the top. It is well known that $X_1, \dots, X_n$ can be expressed in terms of $x_1, \dots, x_n$ as $X_i = F_\beta(x_i)$, where $F_\beta : F_n \to F_n$ is an automorphism of the free group F_n determined by the braid β. Furthermore the map $\beta \mapsto F_\beta$ from the braid group B_n to $\operatorname{Aut} F_n$ is itself a group homomorphism. Then F_β is the composite of elementary automorphisms corresponding to the elementary braids making up β, which are given explicitly by $F_{\sigma_i}(x_i) = x_{i+1}$, $F_{\sigma_i}(x_{i+1}) = x_{i+1} x_i x_{i+1}^{-1}$ and $F_{\sigma_i}(x_j) = x_j$, $j \neq i, i+1$.

The group of the link $\hat{\beta} \cup A$ is presented by a generator x, arising from a meridian of A, and generators $x_1, \dots, x_n$ as above, with n relations $F_\beta(x_i) = x^{-1} x_i x$. The reduced Burau matrix shows up most naturally by using a different system of generators $g_1, \dots, g_n$ for F_n, defined recursively by $g_1 = x_1$, $g_{i+1} = x_{i+1} g_i$. The element g_i can be represented by a loop in the disk which encircles the first i punctures as indicated in figure 2. The automorphism F_{σ_i} then satisfies $F_{\sigma_i}(g_i) = F_{\sigma_i}(x_i g_{i-1}) = x_{i+1} g_{i-1} = g_{i+1} g_i^{-1} g_{i-1}$ and $F_{\sigma_i}(g_j) = g_j$, $j \neq i$.

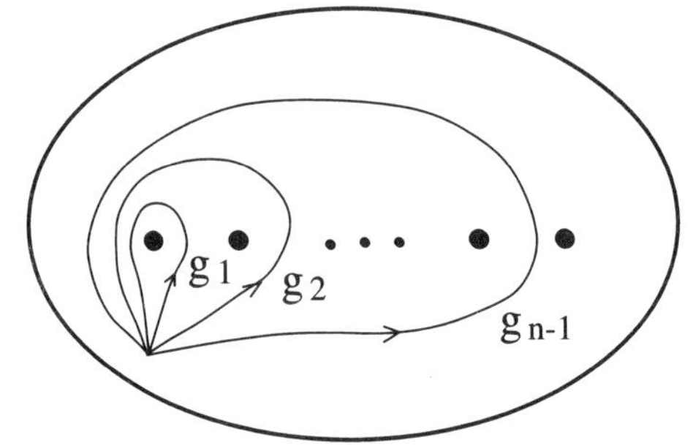

FIGURE 2. The generators in the punctured disk.

The standard method for finding the Alexander invariant of a link (or knot) L evaluated in $\mathbf{Z}[H]$ using a homomorphism $\varphi : G_L \to H$ starting from a presentation of G_L by $n+1$ generators $\{g_j\}$ and n relations $\{r_i\}$ is the following. Calculate the $n \times (n+1)$ matrix $\dfrac{\partial r_i}{\partial g_j}$ of free derivatives, and evaluate the entries in $\mathbf{Z}[H]$ by applying φ. Then delete one column corresponding to a generator c, say, with $\varphi(c) \neq 1$, and divide the determinant of the remaining matrix by $1 - \varphi(c)$. The result is the evaluation of the Alexander invariant, that is, the multi-variable Alexander polynomial when L has more than one component, or $\Delta_L(t)/(1-t)$ when L is a knot. Where a relation is written in the form $r = s$ the entries $\frac{\partial(r-s)}{\partial g_j}$ work equally well in the matrix.

The presentation of G_L by generators $g_1, \dots, g_n$ and x above will give an immediate calculation of the Alexander invariant, once we know the entries $\frac{\partial F_\beta(g_i)}{\partial g_j}$.

The chain rule for free derivatives [**1**] shows that if we write $\beta = \beta_1\beta_2$ and set $G_i = F_{\beta_2}(g_i)$ then $F_\beta(g_i) = F_{\beta_1}(G_i)$ and

$$\frac{\partial F_\beta(g_i)}{\partial g_j} = \sum_{k=1}^{n} \frac{\partial F_{\beta_1}(G_i)}{\partial G_k}\frac{\partial G_k}{\partial g_j}.$$

We can then compute the matrix $\frac{\partial F_\beta(g_i)}{\partial g_j}$ as the product of matrices $\frac{\partial F_\gamma(G_k)}{\partial G_j}$, where γ runs through the elementary braids $\sigma_i^{\pm 1}$ in β.

Now when $\gamma = \sigma_i$ we have

$$\begin{aligned} F_\gamma(G_i) &= G_{i+1}G_i^{-1}G_{i-1} \\ F_\gamma(G_j) &= G_j,\ j \neq i. \end{aligned}$$

Then

$$\begin{aligned} \frac{\partial F_\gamma(G_i)}{\partial G_{i-1}} &= G_{i+1}G_i^{-1} \\ \frac{\partial F_\gamma(G_i)}{\partial G_i} &= -G_{i+1}G_i^{-1} \\ \frac{\partial F_\gamma(G_i)}{\partial G_{i+1}} &= 1 \\ \frac{\partial F_\gamma(G_k)}{\partial G_j} &= \delta_{j\,k},\ \text{otherwise.} \end{aligned}$$

Apply φ to the matrix to get

$$\varphi\left(\frac{\partial F_\gamma(G_k)}{\partial G_j}\right) = \begin{pmatrix} 1 & & & & & \\ & \ddots & & & & \\ & & 1 & & & \\ & & a & -a & 1 & \\ & & & & 1 & \\ & & & & & \ddots \\ & & & & & & 1 \end{pmatrix},$$

where $a = \varphi(G_{i+1}G_i^{-1})$. This is the value of φ on the meridian at the undercrossing in the current elementary braid $\gamma = \sigma_i$. Similarly, when $\gamma = \sigma_i^{-1}$ we get the matrix

$$\begin{pmatrix} 1 & & & & & & \\ & \ddots & & & & & \\ & & 1 & & & & \\ & & 1 & -a^{-1} & a^{-1} & & \\ & & & & 1 & & \\ & & & & & \ddots & \\ & & & & & & 1 \end{pmatrix}$$

with $a = \varphi(G_iG_{i-1}^{-1})$ which is again the value of φ on the meridian of the undercrossing string.

The generator g_n is clearly unchanged by any automorphism F_β, so the last row in each of the matrices of partial derivatives is $(0 \cdots 0\ 1)$. The leading $(n-1)\times(n-1)$ submatrix for the partial derivatives of $F_\beta(g_i)$ is then the product of the leading submatrices from the constituent elementary braids, and so is the reduced coloured Burau matrix of β introduced earlier, with $t_j = \varphi(x_j)$. Thus

$$\varphi\left(\frac{\partial F_\beta(g_i)}{\partial g_j}\right) = \begin{pmatrix} \overline{B}_\beta(t_1, \dots, t_n) & \mathbf{v} \\ \mathbf{0} & 1 \end{pmatrix} = \tilde{B}_\beta(t_1, \dots, t_n), \text{ say,}$$

for some column $\mathbf{v}$. The matrix of free derivatives arising from the presentation of the group by generators $g_1, \dots, g_n$ and x given above is the $n \times (n+1)$ matrix whose first n columns, corresponding to the generators $g_1, \dots, g_n$, are $\varphi(\frac{\partial}{\partial g_j}(F_\beta(g_i) - x^{-1}g_ix))$. These columns form the matrix $\tilde{B}_\beta(t_1, \dots, t_n) - x^{-1}I_n$. They arise from the full matrix of free derivatives by deleting the column corresponding to the generator x, so the Alexander invariant for the link is the determinant of this matrix divided by $1-x$. Now clearly $\det(\tilde{B}_\beta(t_1, \dots, t_n) - x^{-1}I_n) = x^{-n}\det(x\overline{B}_\beta(t_1, \dots, t_n) - I)(1-x)$. Then $\det(I - x\overline{B}_\beta(t_1, \dots, t_n))$, making the identifications of variables $t_j = t_{\pi(j)}$ forced by matching the strings at the top and bottom, is the Alexander polynomial of the closed braid and its axis, up to a power of x. □

The Alexander polynomial of the closed braid itself can then be found from the Torres-Fox formula as shown above.

References

[1] J.S.Birman. 'Braids, links and mapping-class groups'. Ann. Math. Studies, 82. Princeton Univ. Press, (1974).

[2] R.H.Crowell and R.H.Fox. 'Introduction to knot theory'. Ginn and Co. Boston (1963); Springer Graduate Texts 57, (1977).

[3] H.R.Morton. Exchangeable braids In 'Low Dimensional Topology'. London Mathematical Lecture Notes, **95**, Edited R.A. Fenn. Cambrige University press (1985), 86–105.

[4] R.Penne. 'Multi-variable Burau matrices and labeled line configurations'. J.Knot Theory Ramif. 4 (1995), 235-262.

Department of Mathematical Sciences, University of Liverpool, Liverpool L69 3BX, England.

E-mail address: `morton@liv.ac.uk`

Contemporary Mathematics
Volume **233**, 1999

Boundary links and h-split links

Kazuaki Kobayashi

1. Introduction

Many kinds of links are introduced in the litterature, among them, an interesting class is given by links which verify the splittness property side. The geometrically split link has the strongest splittness property but it is a "product" of knots. In this paper we shall consider the following four kinds of links:

1. link-homotopically trivial links in the strong sense (ss l-ht links) ,
2. boundary links,
3. h-split (homotopically split) links,
4. Brunnian links.

Previous three links (1,2,3) are homotopically trivial in nature but they have homotopically distinct properties as strength. We investigate the relations among them with the aid of Brunnian property which means every proper sublink trivial. The literature concerning h-trivial in the strong sense and the boundary links is very large ([H1], [S1], [S2]). In particular, the boundary links have been studied from the point view of the cobordism ([C1], [C2], [C-O], [K1], [O1], [O2], [O3]). However, the invariants studied in those papers are all vanish on the boundary links. That is, they cannot distinguish the h-split links from the boundary links. There are few investigations about the h-split links because of lack of invariants we think. So we mainly consider the h-split links in the second half of this paper and give frame observations for them. We show some geometrical conditions to become h-split link and give a characterization of Seifert matrices for h-split links defined originally for boundary links by ([K1]). In this paper all links are tame, and the spanning surface of knots or links are orientable (or oriented).

2. Definitions for links with respect to the splitness property side

Let $L = (K_1,\ K_2, \cdots,\ K_n)$ be an n-component link in a 3-sphere $\mathbf{S}^3$ (or an Euclidean 3-space $\mathbf{R}^3$).

DEFINITION 1. *Two n-component links L and L' are link-homotopic (abr. homotopic) if there is a map $H : n\mathbf{S}^1 \times [0,1] \longrightarrow \mathbf{S}^3$ such that*
$\mathbf{S}^1 \times \{0\}) = L$,
$H(n\mathbf{S}^1 \times \{1\}) = L'$,

1991 *Mathematics Subject Classification.* 57M25, 57M12.

and

$H(n\mathbf{S}^1 \times \{t\} \times \{i\}) \cap H(n\mathbf{S}^1 \times \{t\} \times \{j\}) = \phi$

for all $t \in [0,1]$ *and for all* $1 \leq i \neq j \leq n$ *where* $n\mathbf{S}^1$ *means the disjoint union of* n *copies of* $\mathbf{S}^1$.

If L' *is the trivial link, we say that* L *is link-homotopically (abr. homotopically) trivial.*

DEFINITION 2. *A link* L *is homotopically trivial (abr. h-trivial) in the strong sense if for any* i $[K_i] = 1$ *in* $\pi_1(\mathbf{S}^3 - (K_1, \cdots, \check{K}_i, \cdots, K_n))$ *where* $[K_i]$ *is the homotopy class of* K_i.

EXAMPLE 1. *(Figure 1.)* $[K_1] = 1 \in \pi_1(\mathbf{S}^3 - K_2)$, $[K_2] \neq 1 \in \pi_1(\mathbf{S}^3 - K_1)$. *So* L *is h-trivial in the usual sense but not h-trivial in the strong sense.*

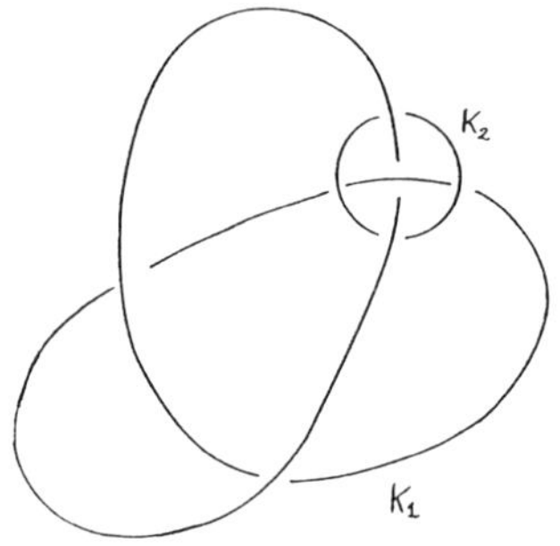

Figure 1.

DEFINITION 3. L *is a boundary link if there are orientable disjoint spanning surfaces* F_i $(i = 1, 2, \cdots, n)$, *that is, (1)* $\partial F_i = K_i$ *and (2)* $F_i \cap F_j = \phi$ $(i \neq j)$.

DEFINITION 4. L *is homotopically split (h-split) if there are maps* $\phi_i : D^2 \longrightarrow \mathbf{S}^3$ $(i = 1, 2, \cdots, n)$ *such that (1)* $\phi_i|_{\partial D^2}$ *is an embedding (2)* $\phi_i(\partial D^2) = K_i$ *and (3)* $\phi_i(D^2) \cap \phi_j(D^2) = \phi$ $(i \neq j)$.

We may assume that all singularities of ϕ_i *are clasp type or ribbon type (see Figure 2).*

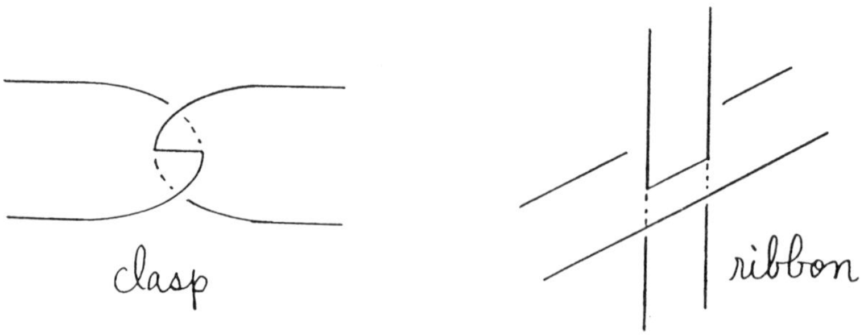

Figure 2.

DEFINITION 5. L *is a Brunnian link if every proper sublink of* L *is the trivial link.*

3. Properties and examples

PROPOSITION 1. *If* L *is an h-split link,* L *is a boundary link.*

Proof. Since L is an h-split link, there are singular disks $\phi_i : D^2 \longrightarrow \mathbf{S}^3$ ($i = 1, 2, \cdots, n$) such that (1) $\phi_i|_{\partial D^2}$ is an embedding (2) $\phi_i(\partial D^2) = K_i$ and (3) $\phi_i(D^2) \cap \phi_j(D^2) = \phi$ ($i \neq j$). Performing surgery on singular disks as Figure 3, we get disjoint orientable spanning surfaces F_i ($i = 1, 2, \cdots, n$) for K_i. For this reason L is a boundary link. □

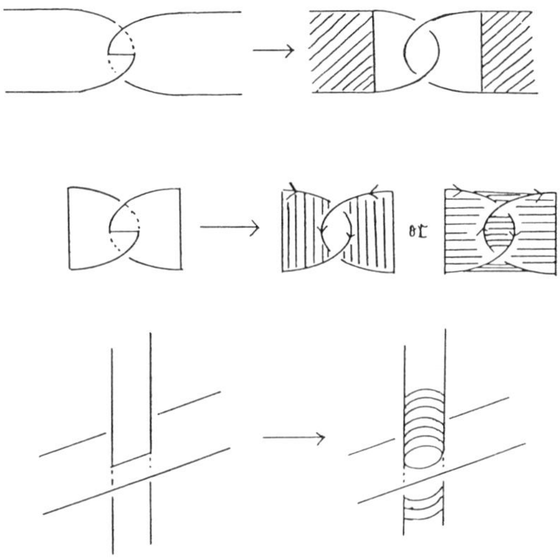

Figure 3.

EXAMPLE 2. *(Figure 4.) L is a h-split link and also a boundary link.*

Figure 4.

EXAMPLE 3. *(Figurc 5.) L is a Brunnian, boundary link and is also h-trivial in the strong sense.*

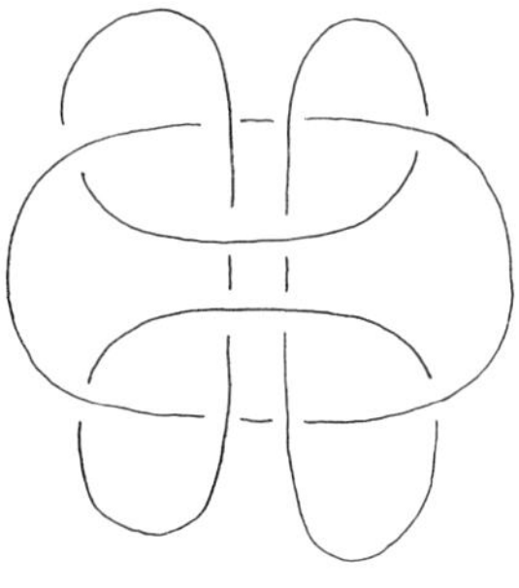

Figure 5.

PROPOSITION 2. *Let $L = (K_1, K_2)$ be a boundary link and K_1 be a trivial knot. If $p : (M_k(K_1), \tilde{K}_1) \longrightarrow (\mathbf{S}^3, K_1)$ is a k-fold cyclic branched covering space branched* *along K_1, then for a k-component link $p^{-1}(K_2) = (\tilde{K}_{21}, \cdots, \tilde{K}_{2k})$ is also a boundary link.*

Remark. Since K_1 is a trivial knot, $M_k(K_1)$ is homeomorphic to the 3-sphere.

Proof. Since L is a boundary link, there exist disjoint spanning surfaces F_1, F_2 of K_1, K_2 respectively. To construct a k-fold cyclic branched covering space $M_k(K_1)$, we cut out $\mathbf{S}^3$ along F_1 and take k copies of this compact manifold. Nextly, we paste the boundaries of these compact manifolds properly. Since $F_1 \cap F_2 = \phi$, F_2 can be lifted to $\tilde{F}_{2i}$ which is bounded by each component $\tilde{K}_{2i}$ of $p^{-1}(K_2)$, that is, $\tilde{F}_{2i} \cap \tilde{F}_{2j} = \phi$, then $p^{-1}(K_2)$ is a boundary link.

EXAMPLE 4. *(Figure 6.) Whitehead link $L = (K_1, K_2)$, L is Brunnian and h-trivial in the strong sense but not boundary link by Proposition 2. The reason is that it takes a 2-fold branched covering space $M_2(K_1)$, which is branched along K_1. The lifting of K_2 is a 2-component link, $p^{-1}(K_2) = (\tilde{K}_{21}, \tilde{K}_{22})$ with $lk(\tilde{K}_{21}, \tilde{K}_{22}) = -2$.*

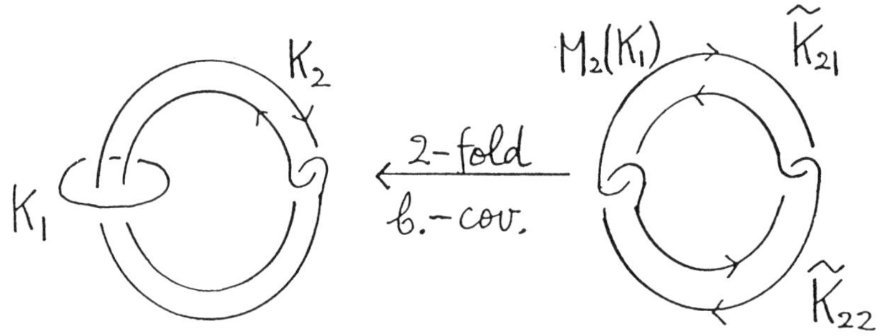

Figure 6.

EXAMPLE 5. *(Figure 7.) $L = (K_1, K_2)$ is a boundary link but not a h-trivial in the strong sense. K_1 is a trivial knot and K_2 is a trefoil knot, so $[K_2] = 1$ in $\pi_1(\mathbf{S}^3 - IntN(K_1))$ but $[K_1] \neq 1$ in $\pi_1(\mathbf{S}^3 - IntN(K_2))$. So, L is not h-split.*

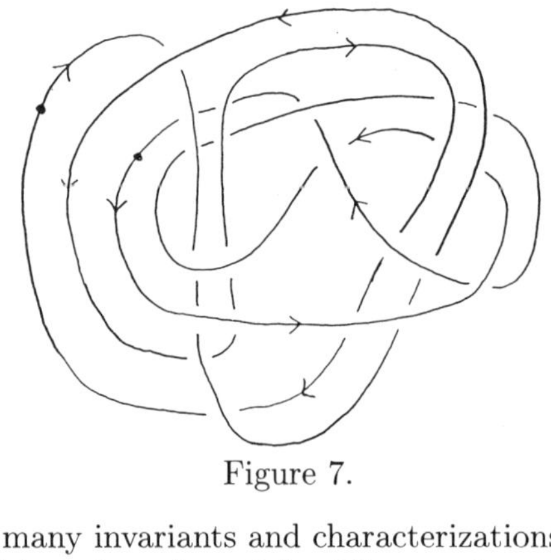

Figure 7.

FACT. There are many invariants and characterizations for boundary links. Let $L = (K_1, \cdots, K_n)$ be a boundary link . Then we have the following:

1. for the k-th elementary ideal $\mathcal{E}_k(L), \mathcal{E}_0(L) \subset \cdots \subset \mathcal{E}_{n-1}(L) = 0$
2. for the Seifert matrix M, the nullity of $M + M^T$ is greater than or equal to $n - 1$.
3. the Conway polynomial $\nabla_L(z) = 0$.

4. if and only if there is a homomorphism ϕ of $\pi_1(\mathbf{S}^3 - L)$ onto the free group of rank n, $F(n) =< x_1, ..., x_n | - >$ such that $\phi(m_i) = x_i$ for i-th meridian m_i of L ([S1]).
5. "higher order linking" is defined ([C1]).
6. There are many results related to cobordism ([C2], [C-O], [Ha], [K1], [O1], [O2], [O3]).

However, these are all invariants to distinguish boundary links from other kinds of links and vanish on the boundary links. So, they cannot be used to distinguish h-split links from boundary links.

PROPOSITION 3. *Let $L = (K_1, K_2)$ be a Brunnian, boundary link, and F_i ($i = 1, 2$) be disjoint spanning surface for K_i, then for any simple closed curve c_1 on F_1 (resp. c_2 on F_2), $lk(c_1, K_2) = 0$ (resp. $lk(c_2, K_1) = 0$).*

Proof. If there is a simple closed curve c_1 on F_1 with $lk(c_1, K_2) \neq 0$, then K_2 is not null homologous in $\mathbf{S}^3 - c_1$. But $F_1 \cap F_2 = \phi$ and so $K_2 \sim *$ in $\mathbf{S}^3 - F_1 \subset \mathbf{S}^3 - c_1$. It is a contradiction. □

PROPOSITION 4. *Let $L = \{K_1, K_2, \cdots, K_n\}$ be an h-split link and $\phi : D^2 \longrightarrow \mathbf{S}^3$ ($i = 1, 2, \cdots, n$) be disjoint spanning singular disks for K_i. Then for any simple closed curve c_i on $\phi_i(D^2)$, $[K_j] = 1 \in \pi_1(\mathbf{S}^3 - (K_1, \cdots, K_i, c_i, K_{i+1}, \cdots, K_j, \cdots, K_n))$*

Proof.Note that $\phi_j(D^2) \cap \coprod_{i \neq j} \phi_i(D^2) = \phi$ □

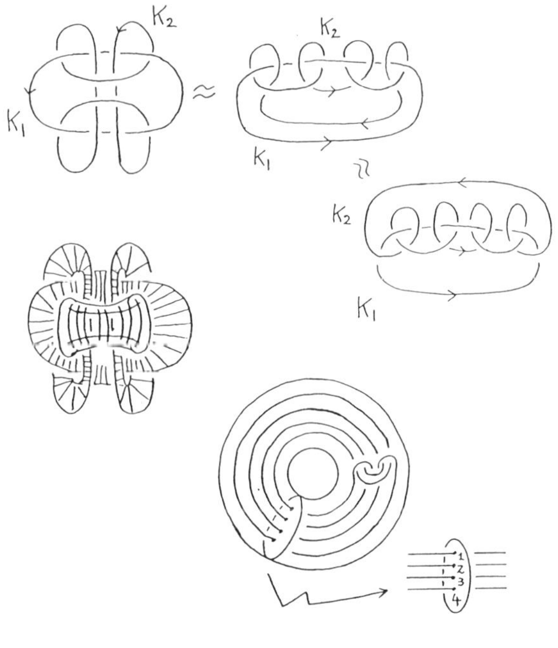

Figure 8.

EXAMPLE 6. *(Figure 8.) $L = (K_1, K_2)$ is a boundary link, but not h-split.*
L is a boundary link by Figure 8.
Now we will prove that L is not h-split. Firstly, we deform L by ambient isotopy as Figure 8. $V = Cl(\mathbf{S}^3 - IntN(K_1))$ is a solid torus and it contains K_2. K_2 is contractible in $\mathbf{S}^3 - IntN(K_1)$. We will show that for any spanning singular disk

$\phi_2(D^2)$ of K_2 in $\mathbf{S}^3 - IntN(K_1)$, there is a simple closed curve c_2 on $\phi_2(D^2)$ such that $lk(c_2, K_1) \neq 0$. However,it contradicts Proposition 4. Nextly, we may assume that the singularities of ϕ_2 are all clasp types and so $Sing(\phi_2) = \{simple\ arcs\}$. Then there exists a meridian disk D_v of V such that $D_v \cap Sing(\phi_2) = \phi$. Remark, (Fig. 8) that, $D_v \cap K_2$ are 4 points , (we note them as "1, 2, 3, 4" (Fig. 8)). So, $D_v \cap \phi_2(D^2)$ is a set of 2 arcs joining pairs of points in $\{1, 2, 3, 4\}$:

1. *joining 1 and 2; 3 and 4.*
2. *joining 1 and 4; 2 and 3.*
3. *joining 1 and 3; 2 and 4.*

For all cases (1),(2) and (3), there exists a simple closed curve c_2 such that c_2 is not contractible in V. □

PROPOSITION 5. *Let L be a boundary link and F_i $(i = 1, \cdots, n)$ be disjoint spanning surfaces for K_i. If there is a link $L_0 = (K_1^0, \cdots, K_n^0)$ such that each regular neighborhood $N(K_i^0)$ contains F_i, $N(K_i^0) \supset F_i$ and $N(K_i^0) \cap N(K_j^0) = \phi$ for any i, j $(i \neq j)$, then L is h-split.*

Proof. Since $K_i = \partial F_i \subset F_i \subset N(K_i^0)$, $K_i \sim *$ in $N(K_i^0)$. On the other hand $\pi_1(N(K_i^0)) \cong \mathbf{Z} \cong H_1(N(K_i^0))$. So $K_i \simeq *$ in $N(K_i^0)$ and hence L is h-split. □

EXAMPLE 7. *(Figure 9.) This boundary link is also h-split.*

Figure 9.

4. System of companion handlebodies of L

DEFINITION 6. *If $L = \{K_1, \cdots, K_n\}$ is a boundary (resp. a h-split) link, then there are handlebodies $V_i (i = 1, 2, \cdots, n)$ of genus g_i such that*

1. *$K_i \subset V_i$ $(i = 1, 2, \cdots, n)$*
2. *$K_i \sim *$ (resp. $K_i \simeq *$) in V_i $(i = 1, 2, \cdots, n)$ and*
3. *$V_i \cap V_j = \phi$ $(i \neq j)$*

As we are considering only geometrically non-split link, we may assume $g_i \geq 1$ $(i = 1, 2, \cdots n)$. (For example, since L is a boundary (resp. a h-split) link, there are disjoint spanning surface F_i (resp. disjoint spanning singular disks $\phi_i(D^2)$)$(i = 1, 2, \cdots, n)$,so we may put V_i = a small regular neighborhood of F_i (resp. $\phi_i(D^2)$). Then we call the n-tuple $(V_1, V_2, \cdots, V_n)$ with the above properties (1), (2) and (3), a system of companion handlebodies of L with genera $(g_1, g_2, \cdots, g_n)$.

So we can restate Proposition 5 as follows.

PROPOSITION 5'. *If $L = (K_1, K_2, \cdots, K_n)$ is a boundary link and if there is a system of companion handlebodies $(V_1, V_2, \cdots, V_n)$ of L with genera $(1, \cdots, 1)$, then L is a h-split link.*

DEFINITION 7. *For a boundary or h-split link, respectively, $L = (K_1, K_2, \cdots, K_n)$, let $\{(V_1, V_2, \cdots, V_n)\}$ be the set of all companion handlebodies of L with the set of genera $\mathcal{G} = \{(g_1, g_2, \cdots, g_n)\}$. We introduce an order in $\mathcal{G}$ such that for $(g_1, g_2, \cdots, g_n)$, $(g'_1, g'_2, \cdots, g'_n) \in \mathcal{G}$ we define $(g_1, g_2, \cdots, g_n) < (g'_1, g'_2, \cdots, g'_n)$ if $\sum_{i=1}^n g_i < \sum_{i=1}^n g'_i$ or if $\sum_{i=1}^n g_i = \sum_{i=1}^n g'_i$ and $max\{g_1, \cdots, g_n\} < max\{g'_1, \cdots, g'_n\}$.*

For example $(1,2) < (2,2)$ and $(2,2) < (1,3)$. So if $\sum_{i=1}^n g_i = ng$, then the minimum element of $\{(g_1, g_2, \cdots, g_n)\}$ with $\sum_{i=1}^n g_i = ng$ is equal to $(g, \cdots, g)$ if such system of companion handlebodies exists.

EXAMPLE 8. *(Figure 10.) There is an example which is h-split but does not seem to be the genera equal to (1,1,1).*

Figure 10.

5. n-contractible curves

Let us introduce an idea of n-contractible curve. Let M^3 be a 3-manifold and α be a simple closed curve in M^3. If α is homotopically trivial,then there exists a map ϕ of a disk D^2 into M^3, $\phi : D^2 \longrightarrow M^3$ such that

1. $\phi|_{\partial D^2}$ is an embedding,
2. $\phi(\partial D^2) = \alpha$
3. the type of singularities are only clasp types.

We call ϕ (or $\phi(D_2)$) a spanning singular disk of α. If the induced homomorphism by inclusion, $i_* : \pi_1(\phi(D^2)) \longrightarrow \pi(M^3)$, is 0-map, we call $\phi(D^2)$ an inessential singular 2-disk.

DEFINITION 8. *Let α be a simple closed curve in M^3. If α is contractible in M^3, we call α 0-contractible (i.e. usual contractible). If α is 0-contractible and if there is a spanning singular disk of α which is inessential, we call α 1-contractible. And next if there is a spanning singular disk of α on which every simple closed curve is 1-contractible, we call α 2-contractible. Inductively if there is a spanning singular disk of α on which every simple closed curve is $(n-1)$-contractible, α is n-contractible. For every positive integer n if α is n-contractible, we call α ∞-contractible.*

EXAMPLE 9. *(Figure 11.) n-contractible curves*

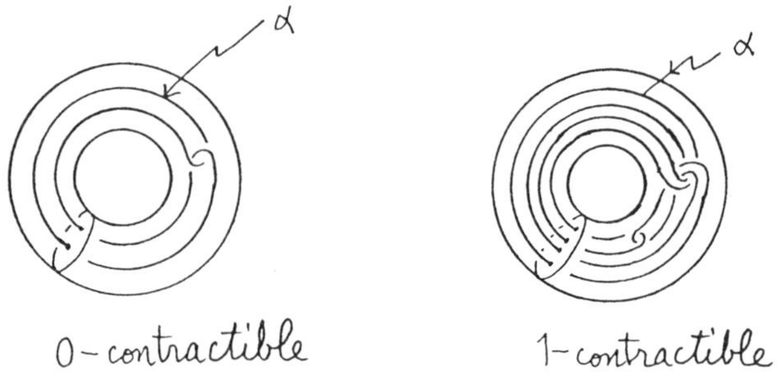

Figure 11.

We shall change the above definition formulating in the analogical way as Smythe [S2].It is as follows;

DEFINITION 9. *Let A and B be disjoint polyhedra in $\mathbf{S}^3$ such that A and $\mathbf{S}^3 - B$ are connected .*

A is 0-contractible in $\mathbf{S}^3 - B$ if the induced homomorphism $\pi_1(A) \longrightarrow \pi_1(\mathbf{S}^3 - B)$ is trivial.

A is 1-contractible in $\mathbf{S}^3 - B$ if there is a connected polyhedron A_1 in $\mathbf{S}^3 - B$ such that

1. *$A \subset A_1$,*
2. *the induced homomorphism $\pi_1(A) \longrightarrow \pi_1(A_1)$ is trivial, and*
3. *A_1 is 0-contractible in $\mathbf{S}^3 - B$.*

Inductively A is n-contractible in $\mathbf{S}^3 - B$ if there is a connected polyhedron A_1 in $\mathbf{S}^3 - B$ such that

1. *$A \subset A_1$,*
2. *the induced homomorphism $\pi_1(A) \longrightarrow \pi_1(A_1)$ is trivial, and*
3. *A_1 is $(n-1)$-contractible in $\mathbf{S}^3 - B$.*

THEOREM 6. *Let $L = (K_1, K_2)$ be a 2-component link with K_i $(i = 1, 2)$ trivial knots. If L is h-split, then K_1 is 1-contractible in $\mathbf{S}^3 - IntN(K_2)$ and K_2 is 1-contractible in $\mathbf{S}^3 - IntN(K_1)$.*

Proof. Suppose that K_1 is not 1-contractible in $\mathbf{S}^3 - IntN(K_2)$. Then for any spanning singular disk $\phi_1(D^2)$ of K_1 there is a singular closed curve c_1 on $\phi_1(D^2)$ which is not contractible in $\mathbf{S}^3 - IntN(K_2)$, that is, $[c_1] \neq 1 \in \pi_1(\mathbf{S}^3 - IntN(K_2)) \cong \mathbf{Z} \cong H_1(\mathbf{S}^3 - IntN(K_2))$. So $lk(c_1, K_2) \neq 0$. Hence $[K_2] \neq 1 \in \pi_1(\mathbf{S}^3 - c_1)$. It means that there are no disjoint spanning singular disks $\phi_1(D^2)$, $\phi_2(D^2)$ of K_1 and K_2 respectively. □

PROPOSITION 7. *Let $L = (K_1, K_2)$ be a 2 components link with K_i $(i = 1, 2)$ trivial knots. If K_i $(i = 1, 2)$ are both 1-contractible in the complement of one of components, and if there is a spanning singular disk ϕ_1 of K_1 such that*

(1) $\#\phi_1^{-1}(Sing(\phi_1))$ is one pair of preimage of a clasp singularity and

(2) any simple closed curve c on $\phi_1(D^2)$ is contractible in $\mathbf{S}^3 - K_2$, then L is h-split.

Proof. Since K_1 is a trivial knot, a simple closed curve c_1 on $\phi_1(D^2)$ which intersects $Sing(\phi_1(D^2))$ one point transversally is also a trivial knot. So $N(\phi_1(D^2)) = N(c_1)$ is a trivial solid torus. And $c_1 \simeq *$ in $\mathbf{S}^3 - IntN(K_2)$ by the assumption. So

$lk(c_1, K_2) = 0$ and $[K_2] = 1$ in $\pi_1(\mathbf{S}^3 - IntN(c_1)) \cong \mathbf{Z} \cong H_1(\mathbf{S}^3 - IntN(c_1))$. Hence there is a spanning singular disk $\phi_2(D^2)$ in $\mathbf{S}^3 - IntN(c_1) = \mathbf{S}^3 - IntN(\phi_1(D^2))$, that is, $\phi_1(D^2) \cap \phi_2(D^2) = \phi$. Therefore L is h-split. □

EXAMPLE 10. *(Figure 12.) L is h-split and K_i $(i = 1, 2)$ are both 1-contractible in the complement of another component.*

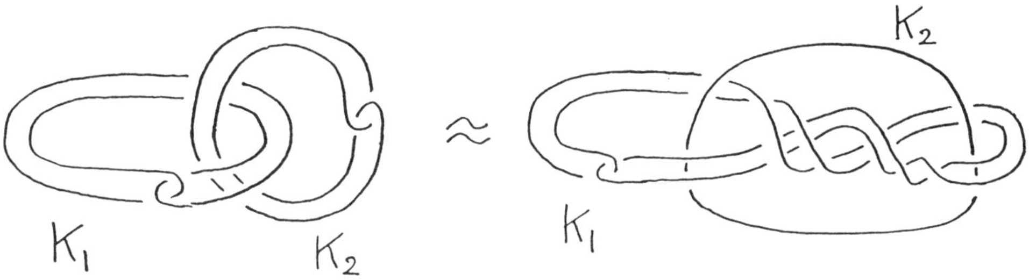

Figure 12.

6. Relation between n-contractible, n-linking and n-splitting

In 1937 [E],Eilenberg introduced the notion of n-linking for a link $L = (K_1, K_2)$ of two components in the 3-sphere. The definition is inductive;

(0) k_1 is 0-linked with k_2 if $lk(k_1, k_2) \neq 0$

(n) k_1 is n-linked with k_2 if every polyhedron in $\mathbf{S}^3 - k_2$ in which k_1 is null homologous contains a simple closed curve which is $(n-1)$-linked with k_2.

In 1970 [S2], Smythe introduced the notion of n-splitting for a link $L = (K_1, K_2)$ of two components in the 3-sphere. The definition is also inductive;
Let $A, B \subset \mathbf{S}^3$. We say that A is (-1)-split from B if $A \cap B = \phi$. We say that A is n-split from B if there is a compact polyhedron A_1 containing A such that the induced map $H_1(A) \longrightarrow H_1(A_1)$ is the trivial map, and A_1 is $(n-1)$-split from B. Thus a link $L = (K_1, K_2)$ has K_1 0-split from K_2 if and only if $lk(K_1, K_2) = 0$ (and therefore if and only if K_1 is not 0-linked with K_2). K_1 is 1-split from K_2 if and only if the link L is a boundary link (and therefore if and only if K_1 is not 1-linked with K_2).

First n-linking and n-splitting are based at homology by definition and n-contractibility is based at homotopy (especially based on the fundamental group). So n-linking and n-splitting are useful tools to study boundary links but they do not work to distinguish h-split link from the boundary link. But under the assumption that the link has Brunnian property, these concepts, i.e. n-linking, n-splitting and n-contractible, are closely related. In fact we can say the following. Let X be a subspace of Y. Since the induced homomorphism $\pi_1(X) \longrightarrow \pi_1(Y)$ 0-map implies $H_1(X) \longrightarrow H_1(Y)$, for a 2 components link $L = (K_1, K_2)$ K_2 is n-contractible in $\mathbf{S}^3 - K_1$ implies that K_2 is n-splitting from K_1. And they coincide for sufficiently large n because of Haken's finiteness theorem. Smythe proved the following, using Haken's finite theorem;

THEOREM 8. *(Smythe [S2]) Let A and B be polyhedra in $\mathbf{S}^3$, with both B and $\mathbf{S}^3 - B$ connected. Then A is geometrically splittable from B if and only if A is n-split from B for every integer B.*

For the n-splitting, the duality holds, that is:

THEOREM 9. *(Smythe [S2]) A is n-split from B if and only if B is n-split from A, for any A, B in $\mathbf{S}^3$.*

On the other hand for the n-contractibility, there is no duality in general. Example 1 says that K_1 is 0-contractible in $\mathbf{S}^3 - K_2$ but K_2 is not 0-contractible in $\mathbf{S}^3 - K_1$. Under the assumption that $L = (K_1, K_2)$ is Brunnian, there is the duality for the 0-contractibility using the linking number (i.e. homological property). However, I do not know whether or not the duality exists for the n-contractibility with $n \geq 1$ even if L has the Brunnian property.
In the following Example 11 and 12 we do not show that there is a duality. However, in fact K_1 and K_2 are both 0-contractible in Example 11. In Example 12 K_1 and K_2 are both 2-contractible in the complement of another component.

EXAMPLE 11. *(Figure 13) $L = (K_1, K_2)$ is a boundary link and h-trivial link in the strong sense. So there are disjoint spanning surfaces F_1, F_2 of K_1, K_2 respectively. Furthermore K_2 bounds a singular disk $\phi_2(D^2)$ (shaded region) in $\mathbf{S}^3 - F_1$. Although it cannot be seen easily, that K_1 bounds a singular disk in the complement of K_2 since K_1 is contractible in $\mathbf{S}^3 - K_2$. And K_1 is not 1-contractible (with respect to $\phi_2(D^2)$) in $\mathbf{S}^3 - K_2$ because $lk(K_1, c_2) \neq 0$ for a simple closed curve c_2 on $\phi_2(D^2)$. And K_2 is not 1-contractible in $\mathbf{S}^3 - K_1$ for the same reason.*

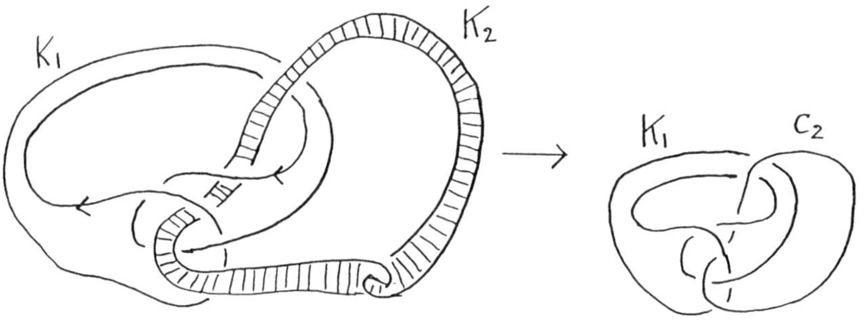

Figure 13.

EXAMPLE 12. *(Figure 14.) $L = (K_1, K_2)$ is h-split link. So both K_i $(i = 1, 2)$ are 1-contractible in the complement of another component by Theorem 6. K_2 is 2-contractible in $\mathbf{S}^3 - K_1$,since we take a simple closed curve c_2 on $\phi_2(D^2)$. However, the closed curve $c_2 \simeq *$ in $\mathbf{S}^3 - K_1$, and it bounds a singular disk $\phi_{22}(D^2)$. Then, the simple closed curve c_{22} on $\phi_{22}(D^2)$ is also contractible in $\mathbf{S}^3 - K_1$. So K_2 is 2-contractible in $\mathbf{S}^3 - K_1$. Although K_1 does not seem to be 2-contractible in $\mathbf{S}^3 - K_1$, it is also 2-contractible. The reason is that we take a spanning singular disk $\phi_1(D^2)$ of K_1. Then the simple closed curve c_1 is contractible in $\mathbf{S}^3 - K_2$ and c_1 bounds a singular disk $\phi_{11}(D^2)$. But the only one simple closed curve c_{11} passing through $Sing(\phi_{11})$ is contractible in $\mathbf{S}^3 - K_2$. Hence K_1 is also 2-contractible in $\mathbf{S}^3 - K_2$.*

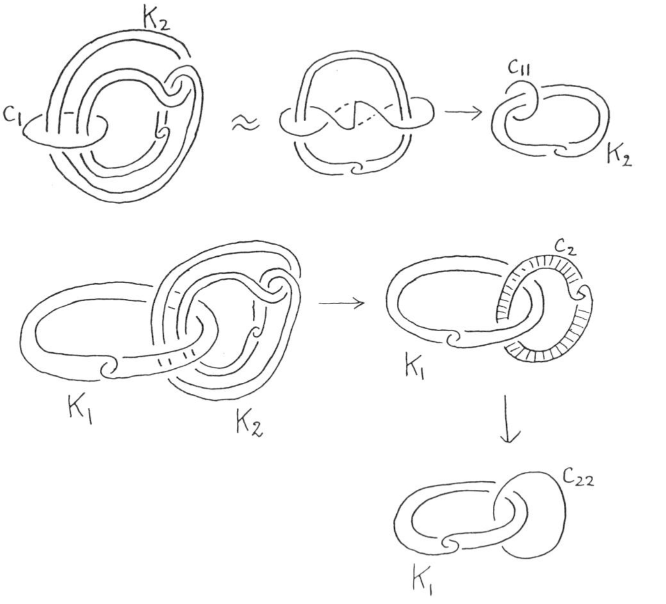

Figure 14.

Furthermore, for the Seifert surfaces of a link, there is a useful equivalence, i.e. this equivalence, called

S-equivalence, induces the among Seifert matrices of a link. There is no such effective equivalence among the spanning singular disks of a link. So, we need to introduce the following definition .

DEFINITION 10. *Let $L = (K_1,\ K_2)$ be a two component link which is h-trivial in the strong sense. So there are spanning singular disks $\phi_i(D^2)$ $(i = 1,\ 2)$ of K_i in the complement of another component of the link L. We call $L = (K_1,\ K_2)$ $h-split$ in the strong sense if for a fixed spanning singular disk $\phi_1(D^2)$ of K_1 (resp. $\phi_2(D^2)$ of K_2) there is a spanning singular disk $\phi_2(D^2)$ of K_2 in $\mathbf{S}^3 - \phi_1(D^2)$ (resp. $\phi_1(D^2)$ of K_1 in $\mathbf{S}^3 - \phi_2(D^2)$. We also say that $L = (K_1,\ K_2)$ is h-split with respect to ϕ_1 (or ϕ_2) when ϕ_1 (or ϕ_2) is fixed.*

In Example 11 $L = (K_1,\ K_2)$ is a boundary link and h-trivial in the strong sense. But L is not h-split with respect to ϕ_2 (a shaded region). Because for a simple closed curve c_2 on $\phi_2(D^2)$, $lk(K_1,\ c_2) \neq 0$. So K_1 cannot bound a singular disk in the complement $\mathbf{S}^3 - c_2 \simeq \mathbf{S}^3 - \phi_2(D^2)$.

It is easy to show that the following proposition holds.

PROPOSITION 10. *Let $L = (K_1,\ K_2)$ be a 2 components link which is h-trivial in the strong sense. If L is h-split with respect to a spanning singular disk ϕ_1 of K_1 in $\mathbf{S}^3 - K_2$. Then K_2 is 0-contractible in $\mathbf{S}^3 - c_1$ for any simple closed curve c_1 on $\phi_1(D^2)$.*

7. Dual category for the fundamental group

In general, in the homotopy theory there is no duality as in the homology theory . However, under some restrictions, there are some circumstances such that the duality holds. So, we introduce the concept of duality between fundamental groups.

DEFINITION 11. *Let* $\mathcal{A} = \{A_i\}$, $\mathcal{B} = \{B_j\}$ *be two families of sets in* $\mathbf{S}^3$ *such that for every* $A_i \in \mathcal{A}$ *and* $B_j \in \mathcal{B}$, $\mathbf{S}^3 - A_i$ *and* $\mathbf{S}^3 - B_j$ *are both connected. We say that* $\mathcal{A}$ *and* $\mathcal{B}$ *are in dual category for the fundamental group if for the pair of* $A_i \in \mathcal{A}$, $B_j \in \mathcal{B}$ *with* $A_i \cap B_j = \phi$, *the induced homomorphism* $\pi_1(A_i) \longrightarrow \pi(\mathbf{S}^3 - B_j)$ *trivial implies the induced homomorphism* $\pi_1(B_j) \longrightarrow \pi(\mathbf{S}^3 - A_j)$ *trivial where* $\pi_1(A_i)$ *(or* $\pi_1(B_j)$ *) means the fundamental group for each component of* A_i *(or* B_j*) if* A_i *(or* B_j*) is disconnected.*

THEOREM 11. *Let* $L = (K_1,\ K_2)$ *be a h-trivial link in the strong sense such that* K_1 *is 1-contractible in* $\mathbf{S}^3 - K_2$. *And let* $\phi_1(D^2)$ *be a spanning singular disk of* K_1 *in* $\mathbf{S}^3 - K_2$ *such that every simple closed on* $\phi_1(D^2)$ *is contractible in* $\mathbf{S}^3 - K_2$. *If* $\mathcal{A} = \{K_1,\ \phi_1(D^2)\}$, $\mathcal{B} = \{K_2\}$ *are in dual category for the fundamental group, then* L *is h-split.*

Proof. Since $\pi_1(\phi_1(D^2)) \longrightarrow \pi(\mathbf{S}^3 - K_2)$ is trivial and $\mathcal{A}$ and $\mathcal{B}$ are in dual category, $\pi_1(K_2) \longrightarrow \pi_1(\mathbf{S}^3 - \phi_1(D^2))$ is trivial. So there is a singular disk $\phi_2(D^2)$ of K_2 in $\mathbf{S}^3 - \phi_1(D^2)$. This means L h-split. □

8. Characterization of disjoint spanning surfaces and Seifert matrices of them for the h-split links

If $L = (K_1, K_2, \cdots, K_n)$ is a h-split link, every component K_i of L bounds disjoint spanning singular disks $\phi_i(D^2)$. Surgering these singular disks (see Figure 3) we get disjoint spanning surfaces F_i $(i = 1, 2, \cdots, n)$ of K_i with special properties and Seifert matrices A of L from these surfaces as follows. Let $g_i = g(F_i)$ be the genus of F_i and let $\{a_{i1}, b_{i1}, \cdots, a_{ig_i}, b_{ig_i}\}$ be a symplectic basis of $H_1(F_i)$. We also use the same symbols a_{ij}, b_{ij} simple closed curves on F_i which represent a_{ij}, b_{ij} respectively.

PROPOSITION 12. *Under the above situation for any* i, j $(i = 1, 2, \cdots, n; j = 1, 2, \cdots, g_i)$, a_{ij} *or* b_{ij} *bound a non-singular disk in* $\mathbf{S}^3 - (L - \{K_i\})$ *and* $lk(a_{ij}, K_i) = 0$ *or* $lk(b_{ij}, K_i) = 0$.

Proof. We may assume that every disjoint spanning singular disk $\phi_i(D^2)$ $(i = 1, 2, \cdots, n)$ has only clasp type of singularities. So by Figure 15, it is obvious. □

We may assume that $a_{i,j}$ bounds a non-singular disk in $\mathbf{S}^3 - (L - \{K_i\})$ and $lk(a_{ij}, K_i) = 0$ in the above Proposition 12 by changing symbol if necessary.

The following definition of Seifert matrix of a boundary link $L = (K_1, K_2, \cdots, K_n)$ for disjoint spanning surfaces F_i $(i = 1, \cdots, n)$ is due to ([K1]).

DEFINITION 12. *([K1]) A matrix* $A = (A_{ij})$ $i, j = 1, 2, \cdots, n$ *of* n^2 *matrices is called a Seifert matix of type n if*

(i) for $i = 1, 2, \cdots, n$ A_{ii} *is a* $(2g_i \times 2g_i)$*-matrix with the property that* $A_{ii} - A_{ii}^T$ *is unimodular and*

(ii) for $i, j = 1, 2, \cdots, n$ *and* $i \neq j$ $A_{i,j}$ *is a* $(2g_i \times 2g_j)$*-matrix with the property that* $A_{ij} = A_{ji}^T$

In fact the entry of the matrix A of the link L is defined by the Seifert pairing $\sigma : H_1(F) \times H_1(F) \longrightarrow \mathbf{Z}$ defined by $\sigma(\alpha, \beta) = lk(\alpha, i_+\beta)$ where F is the disjoint union of F_i, $F = \coprod_{i=1}^n F_i$.
We will arrange rows of A from the 1st to the last as corresponding to a_{11}, $b_{11}, \cdots$, a_{1g_1}, b_{1g_1}, a_{21}, $b_{21}, \cdots, a_{ng_n}$, b_{ng_n} and arrange columns from the 1st to the last same as rows.

If $L = (K_1, K_2, \cdots, K_n)$ is a h-split link, we get the following.

PROPOSITION 13. *If L is a h-split link, for the Seifert matrix A of L defined above, the row corresponding to a_{ij} is $(0\,0 \cdots 0\;1\;1\;0 \cdots 0)$ (resp. $(0\,0 \cdots 0\;1\;0 \cdots 0)$) where 1 1 (resp. 1) correspond to the columns corresponding to a_{ij}, b_{ij} (resp. b_{ij}) and the column corresponding to a_{ij} is $(0 \cdots 0\;1\;0 \cdots 0)^T$(resp. $(0\,0 \cdots 0\;1\;1\;0 \cdots 0)^T$) where 1 (resp. 1 1) corresponds to "a_{ij}"-row (resp. "a_{ij} and b_{ij}"-row) and $(\;)^T$ means the transpose of the matrix ( ). Furthermore if a_{ij} corresponds a ribbon singularity of a singular disk, the row corresponding to a_{ij} is $(0 \cdots 0\;1\;0 \cdots 0)$ (resp. $(0 \cdots 0)$) where 1 is corresponds to "b_{ij}"-column and the column corresponding to a_{ij} is null, $(0 \cdots 0)^T$ (resp. $(0 \cdots 0\;1\;0 \cdots 0)$ where 1 corresponds to "b_{ij}"-row).*

For example if $n = 2$ and $g_1 = g_2 = 1$, A is a form

$$\begin{pmatrix} 1 & 1 & 0 & 0 \\ 0 & a_{22} & 0 & a_{24} \\ 0 & 0 & 1 & 1 \\ 0 & a_{42} & 0 & a_{44} \end{pmatrix}$$

or

$$\begin{pmatrix} 1 & 0 & 0 & 0 \\ 1 & a_{22} & 0 & a_{24} \\ 0 & 0 & 1 & 0 \\ 0 & a_{42} & 1 & a_{44} \end{pmatrix}$$

where $a_{42} = a_{24}$, etc.
And if $n = 2$ and $g_1 = g_2 = 1$ and if the singular disks ϕ_i $(i = 1, 2)$ have the type of singularities only ribbon type, A is a form

$$\begin{pmatrix} 0 & 1 & 0 & 0 \\ 0 & a_{22} & 0 & a_{24} \\ 0 & 0 & 0 & 1 \\ 0 & a_{42} & 0 & a_{44} \end{pmatrix}$$

$$\begin{pmatrix} 0 & 0 & 0 & 0 \\ 1 & a_{22} & 0 & a_{24} \\ 0 & 0 & 0 & 0 \\ 0 & a_{42} & 1 & a_{44} \end{pmatrix}$$

where $a_{42} = a_{24}$, etc.

Proof. It is obvious by Figure 15. □

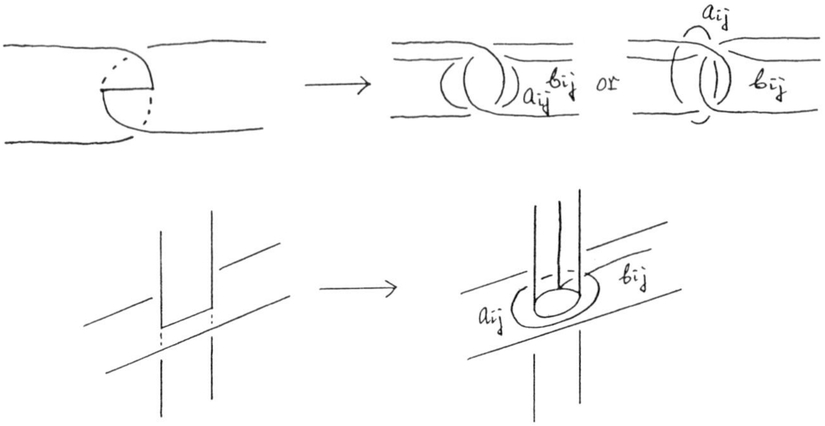

Figure 15

References

[C1] Cochran,T.D. : Derivatives of links : Milnor's concordance invariants and Massey's products, Memoirs A.M.S. vol.84 no.427 Providance (1990)

[C2] —— : Homology boundary links and Branchfield forms : Concordance classification and new tangle-theoretic construction, Topology 33(3) (1994) 397-427

[C-O] Cochran,T.D. and Orr,K. : Not all links are concordant to boundary links, Ann. of Math. 138 (1993) 519-554

[E] Eilenberg,S. : Sur les espaces multicohérent II, Fundamenta Mathematica 29 (1937) 101-122.

[Ha] Hain,R.M. : Iterated integrals, intersection theory and link groups, Topology 24(1) (1985) 45-66

[Hi] Hillman,J.A. : Alexander ideals of links, Lect. Note Math. 895 Springer-Ver. (1981)

[K1] Ko,K. : Seifert matrices and boundary link cobordism, Trans.A.M.S. 299 (1987) 657-681

[K2] Kobayashi,K. : ω-contractible curves and $C(\omega)$-condition, Kobe J. Math. 1 (2) (1984) 133-146

[M1] Milnor,J. : Link groups, Ann. of Math. 59 (1954) 177-195

[M2] ——— : Isotopy of links, Algebraic Geometry and Topology. A Symposium in Honor of S.Lefshetz, Princeton Univ. Press. Princeton, N.J. (1957) 280-306

[O1] Orr, K : New invariants and applications, Comm. Math. Helv. 62, (1987) 542-560

[O2] —— : Homotopy invariants of links, Invent. Math. 95(2) (1989) 379-394

[O3] —— : Link concordance invariants and Massey products, Topology 30(4) (1991) 699-710

[S1] Smythe,N.E. : Boundary links, In Topology Seminar, Wisconsin 1965 (ed. by R.H.Bing) AS60 (1966) 69-72

[S2] Smythe,N.E. : n-linking and n-splitting, Amer.J.Math. 92 (1970) 272-282

Department of Mathematics, Tokyo Woman's Christian University, 167-8585 Suginamiku Zempukuzi 2-6-1, Tokyo, JAPAN

E-mail address: `kazuaki@twcu.ac.jp`

Contemporary Mathematics
Volume **233**, 1999

Finite Type Invariants of Ribbon 2-Knots

Kazuo Habiro, Taizo Kanenobu, and Akiko Shima

ABSTRACT. We construct a 'Vassiliev-like' filtration on the free abelian group generated by the set of ribbon 2-links in 4-space in two equivalent ways. This determines a notion of finite type invariants of ribbon 2-links. We prove that the higher derivatives of the normalized Alexander polynomial of ribbon 2-knots at $t = 1$ are of finite type.

Vassiliev [**6**] has discovered a new family of invariants for 1-knots, which are called *Vassiliev invariants* or *finite type invariants.* They are characterized by their extension to singular knots. In this paper, we generalize these invariants to a higher dimension. A 2-knot in a 4-space is a ribbon 2-knot if it bounds an immersed 3-disk with only ribbon singularities. Making use of them, we give a notion of finite type invariants of type n, $n = 0, 1, 2, \ldots$, for a class of ribbon 2-knots. Then we show the existence of a finite type invariant for a ribbon 2-knot. We denote the normalized Alexander polynomial of a ribbon 2-knot K by $\widetilde{\Delta}_K(t)$, which satisfies $\widetilde{\Delta}_K(1) = 0$ and $d/dt\widetilde{\Delta}_K(1) = 0$. Then we have the following:

MAIN THEOREM (Theorem 2.2). *The nth derivative of the normalized Alexander polynomial of a ribbon 2-knot at $t = 1$, $d^n/dt^n\widetilde{\Delta}_K(1)$, is a finite type invariant of type n.*

One may expect the existence of finite type invariants for a ribbon 2-knot which do not arise from the Alexander polynomial such as the Jones polynomial of a 1-knot; cf. [**2**]. However, Habiro and Shima [**4**] have shown that the Alexander polynomial determines all the finite type invariants of a ribbon 2-knot.

In Sect. 1, we construct a 'Vassiliev-like' filtration on the free abelian group generated by the set of ribbon 2-knots, and define a finite type invariant for a ribbon 2-knot. In Sect. 2, we prove the main theorem above.

A ribbon 2-knot has a projection onto a generic $\mathbb{R}^3 \subset \mathbb{R}^4$ such that the singular set consists of only double points. In Sect. 3, we construct another filtration of

1991 *Mathematics Subject Classification.* Primary 57Q45; Secondary 57Q35.

Key words and phrases. Finite type invariant, ribbon 2-knot, Alexander polynomial.

The second author was partially supported by Grant-in-Aid for Scientific Research (No. 09640125), Ministry of Education, Science and Culture, Japan.

The third author was partially supported by Research Fellowships of the Japan Society for the Promotion of Science for Young Scientists.

ribbon 2-knots making use of this singular set, which also defines finite type invariants. This definition might seem a natural extension of a Vassiliev invariant to a higher dimension. In fact, we may prove that this filtration coincides with the first one. The proof will be given in a forthcoming paper.

Finally, we note that the notion of the finite type invariants of ribbon 2-knots may be extended to ribbon 2-links, and that the results of this paper may be generalized to higher dimensions.

1. Ribbon 2-knots

A *ribbon* $(n+1)$*-disk*, $n \geq 1$, is an immersed $(n+1)$-disk of D^{n+1} into $\mathbb{R}^{n+2}$ with only transverse double points such that the singular set consists of ribbon singularities, that is, the preimage of each ribbon singularity consists of a properly embedded n-disk in D^{n+1} and an embedded n-disk interior to D^{n+1}. An n-knot is a *ribbon n-knot* if it bounds a ribbon $(n+1)$-disk in $\mathbb{R}^{n+2}$.

DEFINITION 1.1. A *ribbon handlebody* H is a ribbon 2-disk, which is a 2-dimensional handlebody in $\mathbb{R}^3$ consisting of $p+1$ 0-handles $D_1, D_2, \dots, D_{p+1}$, p 1-handles $B_1, B_2, \dots, B_p$ and no 2-handle such that the preimage of each ribbon singularity consists of an arc in the interior of a 0-handle and a cocore of a 1-handle; see Figure 1. We put $\mathcal{D} = D_1 \cup D_2 \cup \cdots \cup D_{p+1}$ and $\mathcal{B} = B_1 \cup B_2 \cup \cdots \cup B_p$. Then the set of ribbon singularities in H is the connected components of the intersection $\mathcal{D} \cap \mathcal{B}$.

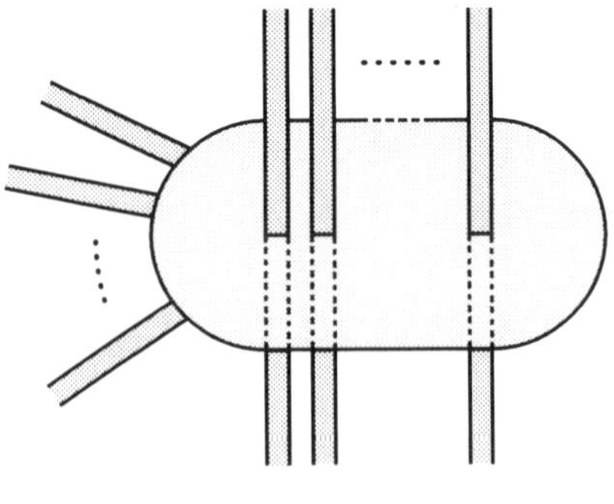

FIGURE 1

We associate to H an immersed 3-disk V_H in $\mathbb{R}^4$ defined by

$$V_H = \mathcal{D} \times [-2, 2] \cup \mathcal{B} \times [-1, 1].$$

Then V_H is a ribbon 3-disk for the ribbon 2-knot $K_H = \partial V_H$ in $\mathbb{R}^4$.

By using [**8**] and [**10**], for any ribbon 2-knot K in $\mathbb{R}^4$, there exists a ribbon handlebody H such that K is ambient isotopic to the associated 2-knot K_H. Therefore, the set of ambient isotopy classes of associated 2-knots of ribbon handlebodies is the same as set of ambient isotopy classes of ribbon 2-knots.

DEFINITION 1.2. A ribbon handlebody is *simple* if each 0-handle D_i contains at most one ribbon singularity and, moreover, if D_i contains one, then D_i is incident to just one 1-handle.

The following is easy; see Figure 2.

LEMMA 1.3. *For any ribbon handlebody H, there exists a simple ribbon handlebody H_0 such that the associated ribbon 2-knots K_H and K_{H_0} are ambient isotopic.*

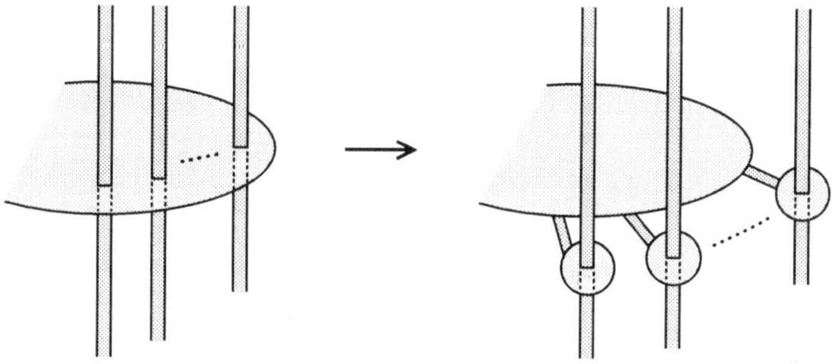

FIGURE 2

DEFINITION 1.4. A *marked ribbon handlebody of order k*, (H, σ^k), is a pair of a simple ribbon handlebody H and k ribbon singularities σ^k of H.

Let $\mathcal{R}_0$ be the $\mathbb{Z}$-module generated by the set of ambient isotopy classes of ribbon 2-knots. For a marked ribbon handlebody (H, σ^k) with $\sigma^k = \{s_1, s_2, \dots, s_k\}$, let $H_{\varepsilon_1,\varepsilon_2,\dots,\varepsilon_k}$ be a ribbon handlebody obtained from H by modifying near each marked singularity $s_i \in \sigma^k$, $1 \leq i \leq k$, according to the integer $\varepsilon_i \in \{0,1\}$ with the rule illustrated in Figure 3, and $K_{\varepsilon_1,\varepsilon_2,\dots,\varepsilon_k}$ be the associated ribbon 2-knot determined from $H_{\varepsilon_1,\varepsilon_2,\dots,\varepsilon_k}$. We define an element $e(H, \sigma^k)$ of $\mathcal{R}_0$ by

$$e(H, \sigma^k) = \sum_{\varepsilon_1,\varepsilon_2,\dots,\varepsilon_k \in \{0,1\}} (-1)^{\varepsilon_1+\varepsilon_2+\cdots+\varepsilon_k} [K_{\varepsilon_1,\varepsilon_2,\dots,\varepsilon_k}],$$

where $[K_{\varepsilon_1,\varepsilon_2,\dots,\varepsilon_k}]$ is the ambient isotopy class of the ribbon 2-knot $K_{\varepsilon_1,\varepsilon_2,\dots,\varepsilon_k}$.

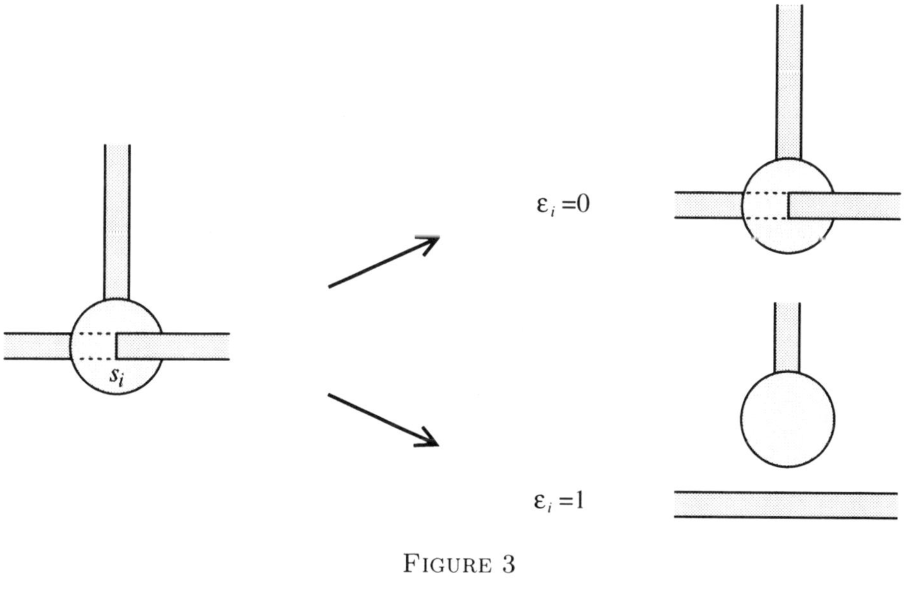

FIGURE 3

Let $\mathcal{R}_k$ denote the $\mathbb{Z}$-submodule of $\mathcal{R}_0$ generated by the elements $e(H, \sigma^k)$. Then we have a filtration of ribbon 2-knots:

$$\mathcal{R}_0 \supset \mathcal{R}_1 \supset \mathcal{R}_2 \supset \cdots \supset \mathcal{R}_k \supset \cdots .$$

DEFINITION 1.5. Let v be an invariant of ribbon 2-knots taking values in an abelian group A. Then we may extend v to a $\mathbb{Z}$-linear map $\mathcal{R}_0 \to A$. We call v a *finite type invariant of type* k, if the restriction $v|\mathcal{R}_{k+1}$ is a zero map.

2. Finite type invariants of ribbon 2-knots extracted from the Alexander polynomials

Let K be a ribbon 2-knot. The Alexander polynomial $\Delta_K(t) \in \Lambda = \mathbb{Z}[t, t^{-1}]$ of a ribbon 2-knot K is well-defined up to multiplication by a unit of the ring Λ, i.e., $\pm t^m$, where m is an integer. It is known that $\Delta_K(t)$ satisfies $\Delta_K(1) = \pm 1$, and conversely, Kinoshita [5] has shown that every polynomial $f(t) \in \Lambda$ with $f(1) = \pm 1$ is realized as the Alexander polynomial of a ribbon 2-knot. Therefore we may assume $\Delta_K(1) = 1$ by multiplying -1 if necessary. The ambiguity of multiplication by t^m, $m \in \mathbb{Z}$, is killed by requiring that $d/dt\Delta_K(1) = 0$. That is, if an Alexander polynomial $\Delta_K(t) \in \Lambda$ with $\Delta_K(1) = 1$ is given, we set

$$\widetilde{\Delta}_K(t) = t^{-d/dt\Delta_K(1)}\Delta_K(t)$$

and call it the *normalized Alexander polynomial* of the ribbon 2-knot K. Then Kinoshita's theorem is restated as follows.

PROPOSITION 2.1. *For any $f(t) \in \Lambda$ with $f(1) = 1$ and $d/dtf(1) = 0$, there is a ribbon 2-knot K with $\widetilde{\Delta}_K(t) = f(t)$.*

We define an infinite family of integer invariants $\alpha_2(K), \alpha_3(K), \ldots$ of a ribbon 2-knot K by expanding the normalized Alexander polynomial of K as follows:

$$\widetilde{\Delta}_K(t) = 1 + \alpha_2(K)(t-1)^2 + \alpha_3(K)(t-1)^3 + \cdots .$$

The coefficient $\alpha_n(K)$ is also defined by

$$\alpha_n(K) = \frac{1}{n!}\frac{d^n}{dt^n}\widetilde{\Delta}_K(1).$$

By Proposition 2.1, the invariants $\alpha_2, \alpha_3, \ldots$ are independent in the sense that for any finite sequence of integers $p_2, p_3, \ldots, p_n$, $n \geq 2$, there is a ribbon 2-knot K with $\alpha_2(K) = p_2$, $\alpha_3(K) = p_3, \ldots, \alpha_n(K) = p_n$.

The rest of this section is devoted to the proof of the following theorem.

THEOREM 2.2. *The invariant α_n, $n = 2, 3, \ldots$, is a finite type invariant of type n.*

REMARK 2.3. For a classical knot K in S^3, we can also define the normalized Alexander polynomial $\widetilde{\Delta}_K(t) \in \Lambda$ in a similar way. It is known that $\widetilde{\Delta}_K(t)$ satisfies $\widetilde{\Delta}_K(t^{-1}) = \widetilde{\Delta}_K(t)$. So it is natural to expand $\widetilde{\Delta}_K(t)$ in $t - 2 + t^{-1}$. Setting $z^2 = t - 2 + t^{-1}$, we have

$$\widetilde{\Delta}_K(t) = 1 + a_2(K)z^2 + \cdots + a_{2n}(K)z^{2n} \in \mathbb{Z}[z]$$

for some n. This polynomial is called the Alexander-Conway polynomial of the knot K. The coefficient a_{2k} is known to be a Vassiliev invariant of type $2k$ [1], which is proved by using a skein relation for the Alexander-Conway polynomial of classical links. The proof in the ribbon 2-knot case (Theorem 2.2) is harder than that in the classical case mainly because of lack of such a simple skein relation.

2.1. Presentations of ribbon 2-knot groups. Let $H = \mathcal{D} \cup \mathcal{B}$, where $\mathcal{D} = D_1 \cup D_2 \cup \cdots \cup D_{p+1}$ and $\mathcal{B} = B_1 \cup B_2 \cup \cdots \cup B_p$, be a simple ribbon handlebody.

2.1.1. *Generators.* By an *edge*, we mean a component of $\mathcal{B} \setminus \mathcal{D}$, which is homeomorphic to $[0,1] \times (0,1)$. Then each 1-handle B_i is partitioned into some edges. Suppose that there are $n(\geq p)$ edges $E_1, E_2, \ldots, E_n$ in H. For each edge E_i we choose a meridian curve μ_i of a boundary component of E_i in such a way that the intersection number of μ_i and E_i is $+1$; see Figure 4. We choose a base point b_0 in $\mathbb{R}^3 \setminus (\mathcal{D} \cup \mathcal{B})$, and choose a simple curve γ_i from b_0 to a point in μ_i for each i. Then the path $\gamma_i \cdot \mu_i \cdot \bar{\gamma}_i$ determines an element x'_i in $\pi_1(\mathbb{R}^3 \setminus \partial(\mathcal{D} \cup \mathcal{B}), b_0)$, and hence $x_i = \iota_\#(x'_i)$ in the ribbon knot group $G(K) = \pi_1(\mathbb{R}^4 \setminus K_H, b_0)$. Here we regard $\mathbb{R}^3 = \mathbb{R}^3 \times \{0\} \subset \mathbb{R}^4$ and $\iota : \mathbb{R}^3 \hookrightarrow \mathbb{R}^4$ is the inclusion map. It is known that the elements $x_1, x_2, \ldots, x_n$ generate the group $G(K)$.

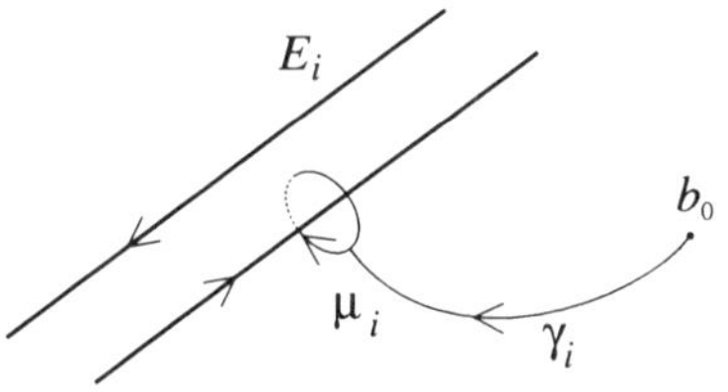

FIGURE 4

2.1.2. *Relators.* There are two types of 0-handles in H: a 0-handle of type I contains only one ribbon singularity, and that of type II contains no ribbon singularity. Let D_i be a 0-handle of type I. Then there are two edges E_{i_1} and E_{i_2} separated by D_i as depicted in Figure 5. Let E_{i_3} be the edge incident to D_i. (Here we exclude the possibilities that $i_3 = i_1$ and that $i_3 = i_2$ by inserting a 0-handle into the edge E_{i_3} if necessary.) Then it is easy to see that there is a relator

$$r^{\mathrm{I}}_{D_i,0} = x_{i_1}^{-1} x_{i_3} x_{i_2} x_{i_3}^{-1}$$

in $G(K)$, which we call a relator of type I.

Let D_j be a 0-handle of type II in H. If there are m edges $E_{j_1}, E_{j_2} \ldots E_{j_m}$ incident to D_j, $m \geq 1$, then there are $m-1$ relators

$$r^{\mathrm{II}}_{D_j,1} = x_{j_1}^{-1} x_{j_2},\ r^{\mathrm{II}}_{D_j,2} = x_{j_2}^{-1} x_{j_3},\ \ldots,\ r^{\mathrm{II}}_{D_j,m-1} = x_{j_{m-1}}^{-1} x_{j_m}$$

in $G(K)$, which we call relators of type II.

Thus we obtain a presentation of $G(K)$ with $n-1$ relators; see [**7**], [**9**]:

$$\langle\ x_1, x_2, \ldots, x_n \mid r^{\mathrm{I}}_{D_i,0}\ (D_i \text{ is of type I}),\ \ r^{\mathrm{II}}_{D_j,l}\ (D_j \text{ is of type II})\ \rangle,$$

which we call a presentation associated with the simple ribbon handlebody H.

Now we illustrate how the presentation change if we deform a 0-handle of type I to that of type II by taking off a 0-handle of type I from a 1-handle. Let D_i, E_{i_1},

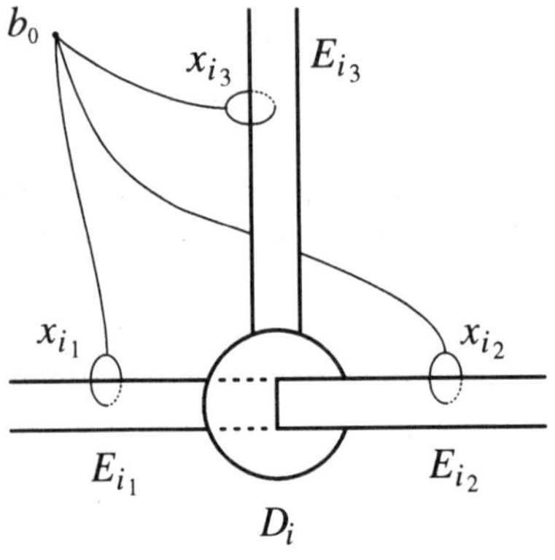

FIGURE 5

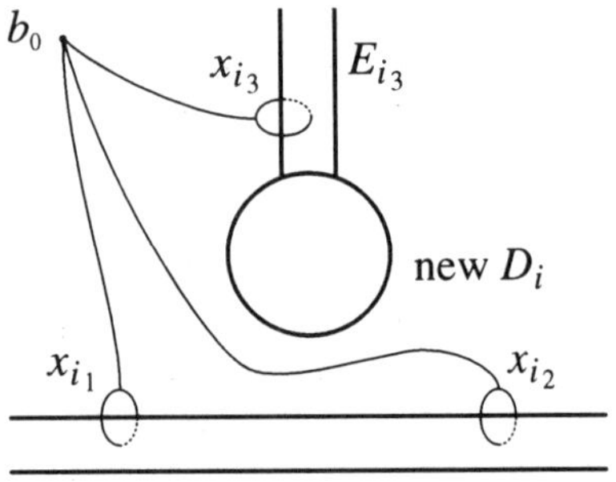

FIGURE 6.

E_{i_2} and E_{i_3} be as given in the explanation of relators of type I above. If we take off D_i from the 1-handle containing E_{i_1} and E_{i_2} as depicted in Figure 6, then the relator $r^{\mathrm{I}}_{D_i,0} = x_{i_1}^{-1}x_{i_3}x_{i_2}x_{i_3}^{-1}$ is replaced with $r^{\mathrm{I}}_{D_i,1} = x_{i_1}^{-1}x_{i_2}$.

2.2. Alexander polynomials of ribbon 2-knots. Let (H, σ^k), where $H = \mathcal{D} \cup \mathcal{B}$ with $\mathcal{D} = D_1 \cup D_2 \cup \cdots \cup D_{p+1}$, $\mathcal{B} = B_1 \cup B_2 \cup \cdots \cup B_p$, and $\sigma^k = \{s_1, s_2, \ldots, s_k\}$, be a marked ribbon handlebody. We assume that the first k 0-handles $D_1, D_2, \ldots, D_k$ are of type I which contain the ribbon singularities $s_1, s_2, \ldots, s_k$, respectively. Given $\varepsilon_1, \varepsilon_2, \ldots, \varepsilon_k \in \{0, 1\}$, let $H_{\varepsilon_1,\varepsilon_2,\ldots,\varepsilon_k}$ denote the simple ribbon handlebody obtained from H by the rule depicted in Figure 3. Then the group of the associated ribbon 2-knot $K_{\varepsilon_1,\varepsilon_2,\ldots,\varepsilon_k}$ of $H_{\varepsilon_1,\varepsilon_2,\ldots,\varepsilon_k}$ is presented by

$$\langle\, x_1, x_2, \ldots, x_n \mid r_{1,\varepsilon_1} = r^{\mathrm{I}}_{D_1,\varepsilon_1},\ r_{2,\varepsilon_2} = r^{\mathrm{I}}_{D_2,\varepsilon_2},\ \ldots,\ r_{k,\varepsilon_k} = r^{\mathrm{I}}_{D_k,\varepsilon_k},\ r_{k+1}, \ldots, r_{n-1} \,\rangle.$$

From this presentation, we compute the Alexander polynomial $\Delta_{\varepsilon_1,\varepsilon_2,\ldots,\varepsilon_k}(t)$ of the ribbon 2-knot $K_{\varepsilon_1,\varepsilon_2,\ldots,\varepsilon_k}$ by using the Fox differential [**3**]. For a word r in the free group $F_n = \langle x_1, x_2, \ldots, x_n \rangle$, let $\tilde{r}$ denote the row matrix of length $n-1$

$$\tilde{r} = \left(\phi\left(\frac{\partial r}{\partial x_1}\right), \phi\left(\frac{\partial r}{\partial x_2}\right), \ldots, \phi\left(\frac{\partial r}{\partial x_{n-1}}\right) \right),$$

where $\partial/\partial x_i \colon \mathbb{Z}F_n \to \mathbb{Z}F_n$ is the Fox differential operator in the variable x_i, and $\phi \colon \mathbb{Z}F_n \to \Lambda$ is the homomorphism of rings defined by $\phi(x_1) = \cdots = \phi(x_n) = t$. Thus if $1 \le i \le k$, then

$$\tilde{r}_{i,0} = \begin{pmatrix} \overset{i_1}{-t^{-1}} & \overset{i_2}{1} & \overset{i_3}{-1+t^{-1}} & \overset{\text{others}}{0} \end{pmatrix},$$

and

$$\tilde{r}_{i,1} = \begin{pmatrix} \overset{i_1}{-1} & \overset{i_2}{1} & \overset{i_3}{0} & \overset{\text{others}}{0} \end{pmatrix}$$

for some i_1, i_2, $i_3 \in \{1, 2, \dots, n\}$. Therefore we obtain

$$\Delta_{\varepsilon_1,\varepsilon_2,\dots,\varepsilon_k}(t) = \det \begin{pmatrix} \tilde{r}_{1,\varepsilon_1} \\ \tilde{r}_{2,\varepsilon_2} \\ \vdots \\ \tilde{r}_{k,\varepsilon_k} \\ \tilde{r}_{k+1} \\ \vdots \\ \tilde{r}_{n-1} \end{pmatrix},$$

where $\tilde{r}_{i,\varepsilon_i}$, $1 \le i \le k$, and $\tilde{r}_j$, $k+1 \le j \le n-1$, are row matrices with entries in Λ. Given δ_1, $\delta_2, \dots, \delta_k \in \{0, 1, *\}$, we define $\Delta_{\delta_1,\delta_2,\dots,\delta_k}(t)$ by requiring

$$\Delta_{\delta_1,\dots,\delta_{i-1},*,\delta_{i+1},\dots,\delta_k}(t) = \Delta_{\delta_1,\dots,\delta_{i-1},0,\delta_{i+1},\dots,\delta_k}(t) - \Delta_{\delta_1,\dots,\delta_{i-1},1,\delta_{i+1},\dots,\delta_k}(t).$$

Set $\tilde{r}_{i,*} = \tilde{r}_{i,0} - \tilde{r}_{i,1}$, which is $(t-1)w_i$ for some row matrix w_i with entries in Λ. Then we have

$$\Delta_{\delta_1,\delta_2,\dots,\delta_k}(t) = \det \begin{pmatrix} \tilde{r}_{1,\delta_1} \\ \tilde{r}_{2,\delta_2} \\ \vdots \\ \tilde{r}_{k,\delta_k} \\ \tilde{r}_{k+1} \\ \vdots \\ \tilde{r}_{n-1} \end{pmatrix}.$$

Up to now we have proved the following lemma.

LEMMA 2.4. *Let δ_1, $\delta_2, \dots, \delta_k \in \{0, 1, *\}$ and q be the number of $*$'s in δ_1, $\delta_2, \dots, \delta_k$. Then we have*

$$\Delta_{\delta_1,\delta_2,\dots,\delta_k} \in (t-1)^q \Lambda.$$

Note that we may assume that $\Delta_{0,0,\dots,0}(1) = +1$ by changing the order of generators if necessary. We define a $\mathbb{Z}$-linear map $d : \Lambda \to \mathbb{Z}$ by $d(f(t)) = d/dt f(1)$. We set $a_0 = d(\Delta_{0,0,\dots,0}(t))$, and $a_i = d(\Delta_{0,\dots,\underset{i}{1},\dots,0}(t))$ for $i = 1, 2, \dots, k$. Then we have the following:

LEMMA 2.5. *If $\varepsilon_1, \varepsilon_2, \dots, \varepsilon_k \in \{0, 1\}$, then*

$$d(\Delta_{\varepsilon_1,\varepsilon_2,\dots,\varepsilon_k}(t)) = a_0 + \sum_{i=1}^{k} \varepsilon_i (a_i - a_0).$$

PROOF. We prove by induction on the number of 1 in ε_1, $\varepsilon_2, \dots, \varepsilon_k$, l, say. If $l = 0, 1$, the lemma is trivial. Suppose that the lemma is true for l (≥ 1). We prove for the case $\varepsilon_{k-1} = \varepsilon_k = 1$. From the linearity of d, we have

$$\begin{aligned} d(\Delta_{\varepsilon_1,\dots,\varepsilon_{k-2},*,*}(t)) =& d(\Delta_{\varepsilon_1,\dots,\varepsilon_{k-2},0,0}(t)) - d(\Delta_{\varepsilon_1,\dots,\varepsilon_{k-2},0,1}(t)) \\ & - d(\Delta_{\varepsilon_1,\dots,\varepsilon_{k-2},1,0}(t)) + d(\Delta_{\varepsilon_1,\dots,\varepsilon_{k-2},1,1}(t)), \end{aligned}$$

which is zero, since $\Delta_{\varepsilon_1,\dots,\varepsilon_{k-2},*,*} \in (t-1)^2\Lambda$ by Lemma 2.4. Also, from the inductive hypothesis, we have

$$d(\Delta_{\varepsilon_1,\dots,\varepsilon_{k-2},0,0}(t)) = a_0 + b,$$
$$d(\Delta_{\varepsilon_1,\dots,\varepsilon_{k-2},0,1}(t)) = a_k + b,$$
$$d(\Delta_{\varepsilon_1,\dots,\varepsilon_{k-2},1,0}(t)) = a_{k-1} + b,$$

where $b = \sum_{i=1}^{k-2} \varepsilon_i(a_i - a_0)$. This completes the proof. □

LEMMA 2.6. *Let $f_{\varepsilon_1,\varepsilon_2,\dots,\varepsilon_k}(t) \in \Lambda$ for $\varepsilon_1,\varepsilon_2,\dots,\varepsilon_k \in \{0,1\}$. Suppose that for each subset H of $\{1,2,\dots,n\}$, the alternating sum $F_H(t)$ defined by*

$$F_H(t) = \sum_{\substack{\varepsilon_1,\varepsilon_2,\dots,\varepsilon_k \in \{0,1\} \\ \varepsilon_i = 1 \text{ if } i \notin H}} (-1)^{\varepsilon_1+\varepsilon_2+\dots+\varepsilon_k} f_{\varepsilon_1,\varepsilon_2,\dots,\varepsilon_k}(t)$$

is contained in $(t-1)^{|H|}\Lambda$, where $|H|$ is the number of elements in H. Let a_1, $a_2,\dots,a_k$ be integers. Then the polynomial $G(t)$ defined by

$$G(t) = \sum_{\varepsilon_1,\varepsilon_2,\dots,\varepsilon_k \in \{0,1\}} (-1)^{\varepsilon_1+\varepsilon_2+\dots+\varepsilon_k} t^{\varepsilon_1 a_1+\varepsilon_2 a_2+\dots+\varepsilon_k a_k} f_{\varepsilon_1,\varepsilon_2,\dots,\varepsilon_k}(t)$$

is contained in $(t-1)^k\Lambda$.

PROOF. Set $u_i = t^{a_i} - 1 (\in (t-1)\Lambda)$. Then we have

$$\begin{aligned} G(t) &= \sum_{\varepsilon_1,\varepsilon_2,\dots,\varepsilon_k \in \{0,1\}} (-1)^{\varepsilon_1+\varepsilon_2+\dots+\varepsilon_k}(1+u_1)^{\varepsilon_1}(1+u_2)^{\varepsilon_2}\dots(1+u_k)^{\varepsilon_k} f_{\varepsilon_1,\varepsilon_2,\dots,\varepsilon_k}(t) \\ &= \sum_{H \subset \{1,2,\dots,k\}} u^H F_H(t), \end{aligned}$$

where $u^H = \prod_{1 \le i \le k, i \notin H} u_i \in (t-1)^{k-|H|}\Lambda$. By assumption, $F_H(t) \in (t-1)^{|H|}\Lambda$, and thus we have $G(t) \in (t-1)^k\Lambda$. □

PROOF OF THEOREM 2.2. The normalized Alexander polynomial $\widetilde{\Delta}_{K_{\varepsilon_1,\varepsilon_2,\dots,\varepsilon_k}}(t)$ of $K_{\varepsilon_1,\varepsilon_2,\dots,\varepsilon_k}$ is calculated from

$$\widetilde{\Delta}_{K_{\varepsilon_1,\varepsilon_2,\dots,\varepsilon_k}}(t) = t^{-d(\Delta_{\varepsilon_1,\varepsilon_2,\dots,\varepsilon_k}(t))}\Delta_{\varepsilon_1,\varepsilon_2,\dots,\varepsilon_k}(t).$$

We set

$$\widetilde{\Delta}_{*,*,\dots,*}(t) = \sum_{\varepsilon_1,\varepsilon_2,\dots,\varepsilon_k \in \{0,1\}} (-1)^{\varepsilon_1+\varepsilon_2+\dots+\varepsilon_k} \widetilde{\Delta}_{K_{\varepsilon_1,\varepsilon_2,\dots,\varepsilon_k}}(t).$$

Then we have only to prove $\widetilde{\Delta}_{*,*,\dots,*}(t) \in (t-1)^k\Lambda$. We may assume $d(\Delta_{0,0,\dots,0}) = 0$ by multiplying $t^{-d(\Delta_{0,0,\dots,0}(t))}$ to all the $\Delta_{\varepsilon_1,\varepsilon_2,\dots,\varepsilon_k}(t)$ if necessary. Setting $a_i = d(\Delta_{0,\dots,\underset{i}{1},\dots,0})$, we have $d(\Delta_{\varepsilon_1,\varepsilon_2,\dots,\varepsilon_k}) = \varepsilon_1 a_1 + \varepsilon_2 a_2 + \dots + \varepsilon_k a_k$ from Lemma 2.5. Therefore,

$$\widetilde{\Delta}_{*,*,\dots,*}(t) = \sum_{\varepsilon_1,\varepsilon_2,\dots,\varepsilon_k \in \{0,1\}} (-1)^{\varepsilon_1+\varepsilon_2+\dots+\varepsilon_k} t^{\varepsilon_1 a_1+\varepsilon_2 a_2+\dots+\varepsilon_k a_k}\Delta_{\varepsilon_1,\varepsilon_2,\dots,\varepsilon_k}(t).$$

Then Lemmas 2.4 and 2.6 yield $\widetilde{\Delta}_{*,*,\dots,*}(t) \in (t-1)^k\Lambda$, completing the proof. □

3. Another filtration of ribbon 2-knots

A 2-knot K in $\mathbb{R}^4$ is *simply knotted* if, after an ambient isotopy of K, the image $p(K)$ under an orthogonal projection $p : \mathbb{R}^4 \to \mathbb{R}^3$ is an immersed sphere whose singular set consists of only double points. It is known ([**8**], [**10**]) that a 2-knot is a ribbon 2-knot if and only If it is simply knotted. We will define another filtration for ribbon 2-knots using immersions of 2-spheres into $\mathbb{R}^3$ whose singular set consists of double points.

DEFINITION 3.1. A *diagram* D is an immersion from a 2-sphere into $\mathbb{R}^3$ satisfying the following two conditions:

(i) The singular set $S(D)$ of D in $\mathbb{R}^3$ has only transverse double points; each component of $S(D)$ is a circle.

(ii) For any component C of the singular set $S(D)$, let A_1, A_2 be two annuli such that $A_1 \cup A_2$ is the intersection of a regular neighborhood of C and the image of D. Then one of the two annuli A_1 and A_2 is marked either by 'a' (for 'above') or by 'b' (for 'below'); see Figure 7.

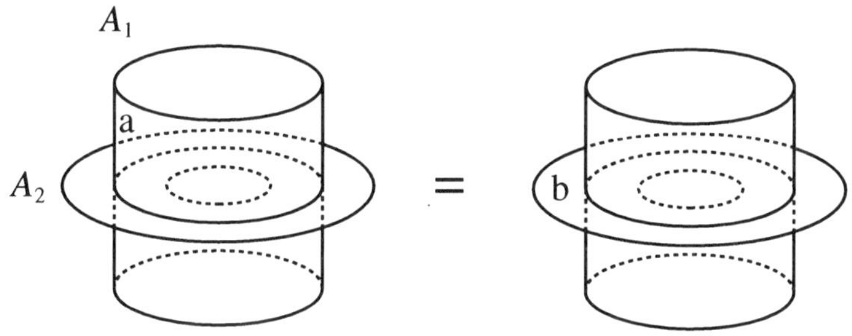

FIGURE 7

We define the associated 2-knot K'_D of a diagram D by the following properties:

(i) $p(K'_D) = D$.

(ii) $K'_D \cap (\mathbb{R}^3 \times \{0\}) = (D \setminus \text{Int(a-annuli in } D)) \times \{0\}$, and $K'_D \subset \mathbb{R}^3 \times [0, \infty)$.

These conditions determine a 2-knot K'_D up to ambient isotopy. Let $\mathcal{R}'_0$ be the $\mathbb{Z}$-module generated by the set of ambient isotopy classes of the 2-knots obtained from diagrams as above. Then we have $\mathcal{R}'_0 = \mathcal{R}_0$.

DEFINITION 3.2. A *marked diagram of order* k, (D, Γ^k), is a pair of a diagram D and a subset Γ^k of the components of the singular set $S(D)$ consisting of k elements. In practice, we describe by writing the mark '$*$' near the components of Γ^k; see Figure 8(a).

Let (D, Γ^k), $\Gamma^k = \{C_1, C_2, \dots, C_k\}$, be a marked diagram. For $\varepsilon_1, \varepsilon_2, \dots, \varepsilon_k \in \{0, 1\}$, we define the diagram $D_{\varepsilon_1, \varepsilon_2, \dots, \varepsilon_k}$ from (D, Γ^k); see Figure 8(b). If $\varepsilon_i = 0$, then delete the mark '$*$' of C_i. If $\varepsilon_i = 1$ and if there is a mark 'a' (resp. 'b') on the annulus in a regular neighborhood of C_i, then delete the mark '$*$' of C_i and replace 'a' (resp. 'b') with 'b' (resp. 'a'); see Figure 8(c). Then the diagrams $D_{\varepsilon_1, \varepsilon_2, \dots, \varepsilon_k}$ and D are identical outside of the regular neighborhood N_{C_i} of C_i, and inside of N_{C_i} they look as in Figure 8(b) or 8(c). We define an element $e'(D, \Gamma^k)$ of $\mathcal{R}'_0$ by

$$e'(D, \Gamma^k) = \sum_{\varepsilon_1, \varepsilon_2, \dots, \varepsilon_k \in \{0,1\}} (-1)^{\varepsilon_1 + \varepsilon_2 + \cdots + \varepsilon_k} [K'_{\varepsilon_1, \varepsilon_2, \dots, \varepsilon_k}],$$

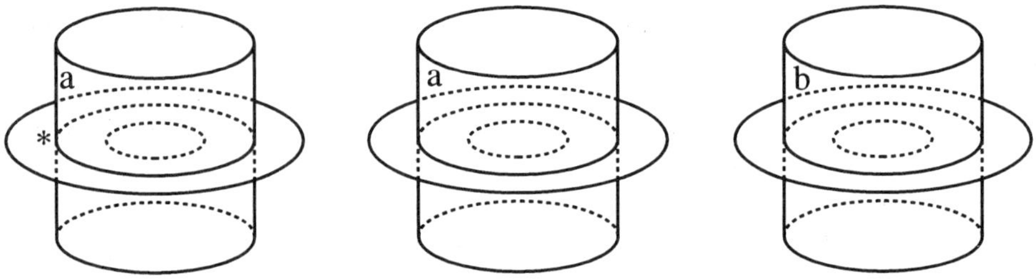

(a) (b) $\varepsilon_i = 0$ (c) $\varepsilon_i = 1$

FIGURE 8

where $[K'_{\varepsilon_1,\varepsilon_2,\dots,\varepsilon_k}]$ is the ambient isotopy class of the associated ribbon 2-knot $K'_{\varepsilon_1,\varepsilon_2,\dots,\varepsilon_k}$ of the diagram $D_{\varepsilon_1,\varepsilon_2,\dots,\varepsilon_k}$.

We define the submodule $\mathcal{R}'_k$ of $\mathcal{R}'_0$ generated by the set of all $e'(D,\Gamma^k)$. Then we obtain another filtration of a ribbon 2-knots:

$$\mathcal{R}'_0 \supset \mathcal{R}'_1 \supset \mathcal{R}'_2 \supset \cdots \supset \mathcal{R}'_k \supset \cdots.$$

We can prove the following; the proof will be given in a forthcoming paper.

THEOREM 3.3. *The $\mathbb{Z}$-modules $\mathcal{R}_k$ and $\mathcal{R}'_k$ coincide for each $k(\geq 0)$.*

References

[1] D. Bar-Natan, *On the Vassiliev knot invariants*, Topology **34** (1995), 423–472.
[2] J. S. Birman and X-S. Lin, *Knot polynomials and Vassiliev's invariants*, Invent. Math. **111** (1993), 225–270.
[3] R. H. Fox, *Free differential calculus. II. The isomorphism problem of groups*, Ann. of Math. **59** (1954), 196–210.
[4] K. Habiro and A. Shima, in preparation.
[5] S. Kinoshita, *On the Alexander polynomials of 2-spheres in a 4-sphere*, Ann. of Math. **74** (1954), 518–531.
[6] V. A. Vassiliev, *Cohomology of knot spaces*, Advances in Soviet Mathematics (V. I. Arnold, ed.), vol. 1, Amer. Math. Soc., Providence, RI, 1990, pp. 23–69.
[7] T. Yajima, *On the fundamental groups of knotted 2-manifolds in the 4-space*, Osaka Math. J. **13** (1962), 63–71.
[8] ———, *On simply knotted spheres in* $\mathbf{R}^4$, Osaka J. Math. **1** (1964), 133–152.
[9] ———, *On a characterization of knot groups of some spheres in* $\mathbf{R}^4$, Osaka J. Math. **6** (1969), 435–446.
[10] T. Yanagawa, *On ribbon 2-knots. The 3-manifold bounded by the 2-knots*, Osaka J. Math. **6** (1969), 447–464.

GRADUATE SCHOOL OF MATHEMATICAL SCIENCES, UNIVERSITY OF TOKYO, TOKYO 153-8902 JAPAN
E-mail address: `habiro@ms.u-tokyo.ac.jp`

DEPARTMENT OF MATHEMATICS, OSAKA CITY UNIVERSITY, OSAKA 558-8585 JAPAN
E-mail address: `kanenobu@scisv.osaka-cu.ac.jp`

GRADUATE SCHOOL OF MATHEMATICAL SCIENCES, UNIVERSITY OF TOKYO, TOKYO 153-8902 JAPAN
E-mail address: `shima@ms.u-tokyo.ac.jp`

Contemporary Mathematics
Volume **233**, 1999

Arrangement of Markov moves for 2-dimensional braids

SEIICHI KAMADA

ABSTRACT. Every surface link type can be represented by a closed 2-dimensional braid and such a representative is unique up to (2-dimensional) Markov move equivalence. We investigate quasi-commutativity of such Markov moves. As an application, we prove that if two simple 2-dimensional braids represent the same surface link type then they are related by a Markov move equivalence through simple 2-dimensional braids; and give a relation between unknotting numbers of surface links and those of 2-dimensional braids.

The Alexander and Markov theorems state that every link type is represented by a closed braid and such a braid is unique up to Markov move equivalence, [**Al; B1; Ma; Mor2; Sk; T; Vo; Y**]. For a surface link, that is a closed oriented surface embedded in R^4, an analogue holds. L. Rudolph defined the notion of a braided surface and proved that every ribbon surface in a 4-ball is deformed into a braided surface [**R2; R4**]. It is related to (1-dimensional) knot theory, cf. [**R1; R2; R3; R4; R5; R6; R7; R8**]. On the other hand, O.Ya. Viro [**Vi**] defined a similar notion called a 2-dimensional braid. His definition is more restrictive than Rudolph's, however it is related to 2-dimensional knot theory. In [**Vi**], he announced that the following theorem is proved by an analogous argument to Alexander's.

THEOREM 0.1 ([**Vi**]). *Every surface link type is represented by the closure of a 2-dimensional braid.*

By an argument on normal forms of surface links [**KSS**] and by the Markov theorem on classical dimensional braids, the following theorem is proved in [**K2**].

1991 *Mathematics Subject Classification.* Primary 57Q45, 57M25.
Key words and phrases. 2-dimensional braids and links, surface braids, Markov move equivalence.
The author is partially supported by Grant-in-Aid for Scientific Research (No. 09740066), Ministry of Education, Science and Culture, Japan.

THEOREM 0.2 ([**K2**]). *Every surface link type is represented by the closure of a simple 2-dimensional braid.*

Since there exist infinitely many non-simple (closed) 2-dimensional braids which are not deformed into simple ones by local perturbations [**K4**], Theorem 0.2 is not a direct consequecnce of Theorem 0.1. (The existence of such strange non-simple 2-dimensional braids is proved by use of Morton's curious example [**Mor1**] and its extension by Fiedler [**F**] of 4-braid representatives of the unknot which cannot be simplified to the trivial 1-braid representative without stabilizations.)

An analogue of the Markov theorem holds for 2-dimensional braids:

THEOREM 0.3 ([**K3; K7**]). *If 2-dimensional braids S and S' represent the same surface link type, then there exists a sequence of 2-dimensional braids $S = S_0, S_1, \ldots, S_s = S'$ such that for each i ($i = 1, \ldots, s$), S_i is obtained from S_{i-1} by a braid ambient isotopy, a conjugation, a stabilization or the inverse operation of a stabilization.*

In § 2 we give a lemma on rearrangement of (generalized) 2-dimensional Markov moves; braid ambient isotopies, conjugations, (generalized) stabilizations and the inverses of (generalized) stabilizations. That is our key lemma. Then we strengthen Theorem 0.3 into our main theorem as follows:

THEOREM 0.4. *If 2-dimensional braids S and S' represent the same surface link type, then there exists a sequence of 2-dimensional braids $S = S_0, S_1, \ldots, S_s = S'$ and an integer k such that S_i ($i = 1, \ldots, s$) is obtained from S_{i-1} by*

(1) *a conjugation or a stabilization if $i < k$,*
(2) *a braid ambient isotopy if $i = k$, and*
(3) *a conjugation or the inverse of a stabilization if $i > k$.*

Using Theorem 0.4, we have the following results; Theorems 0.6—.0.9.

THEOREM 0.5. *Let S and S' be simple 2-dimensional braids whose closures represent the same surface link type. There exists a sequence $S = S_0, S_1, \ldots, S_s = S'$ as in Theorem 0.4 such that every 2-dimensional braid S_i ($i = 0, 1, \ldots, s$) is simple.*

Simple 2-dimensional braids have been studied well and are easy to treat. For instance, they are represented by graphics called charts (cf. [**K1**]) and it is known when two charts represent the same simple 2-dimensional braid (cf. [**K5**]). So it is convenient to consider a sequence as in Theorem 0.5 instead of a a sequence as in Theorem 0.3.

THEOREM 0.6. *Every non-simple 2-dimensional braid is braid ambient isotopic to a simple 2-dimensional braid after applying some conjugations and stabilizations.*

THEOREM 0.7. *Let S be a 2-dimensional braid. If the closure of S is a ribbon surface link, then S is braid ambient isotopic to a ribbon 2-dimensional braid after applying some conjugations and stabilizations.*

THEOREM 0.8. *Let S be a 2-dimensional braid. If the closure of S is an unknotted surface knot, then S is braid ambient isotopic to an unknotted 2-dimensional braid after applying some conjugations and stabilizations.*

A *surface knot* is a surface link which is connected. The notions of a *ribbon* 2-dimensional braid and an *unknotted* 2-dimensional braid are defined in [**K1**]. In § 4, we give alternate definitions to them in terms of braid systems.

Every surface link F is transformed into an unknotted one by surgery along some 1-handles attached to F (cf. [**Bo; HK**]). The *unknotting number of F*, denoted by $u(F)$, is defined by the minimum number of such 1-handles (cf. [**Bo; HK; HMS; K8; KM; Mi; Si**]). For a simple 2-dimensional braid S, the *unknotting number of S*, denoted by $u(S)$, is defined by the minimum number of "simple" 1-handles attached to S along which the surgery transforms S into an unknotted 2-dimensional braid. If the closure of a simple 2-dimensional braid S represents the surface link type of F (we say that S is a *simple braid presentation* of F), then we have

$$u(F) \leq u(S)$$

and hence

$$u(F) \leq \min\{u(S) \mid S \text{ is a simple braid presentation of } F\}.$$

Using Theorem 0.4, we have the following:

THEOREM 0.9. *For any surface knot F, we have*

$$u(F) = \min\{u(S) \mid S \text{ is a simple braid presentation of } F\}.$$

In general, for a surface knot/link F, it is difficult to find 1-handles unknotting F (cf. [**HK**]). However, for a simple 2-dimensional braid S, there exists an easy method to find 1-handles unknotting S (cf. Lemma 3.4 of [**K5**]). Thus Theorem 0.9 and the inequality above the theorem are useful to calculate the unknotting number $u(F)$.

Our approach to Theorem 0.4 is natural even for classical Markov moves.

THEOREM 0.10. *If braids b and b' represent the same link type, then there exists a sequence of braids $b = b_0, b_1, \ldots, b_s = b'$ and an integer k such that b_i ($i = 1, \ldots, s$) is obtained from b_{i-1} by*

(1) *a conjugation or a stabilization if $i \leq k$, and*
(2) *a conjugation or the inverse of a stabilization if $i > k$.*

This theorem is mentioned in [**BM4**] without proof and it is proved in [**B2**] for a special case that the link type is an unlink. In § 6, we give an elementary proof to this theorem.

1. Markov moves for 2-dimensional braids

Let $D_1^2 \times D_2^2$ be the product of two disks and $pr_i : D_1^2 \times D_2^2 \to D_i^2$ $(i = 1, 2)$ the ith factor projection. Let X_m be a set of m distinct interior points of D_1^2. A surface S embedded properly and locally flatly in $D_1^2 \times D_2^2$ is called a *2-dimensional braid* of degree m if (i) the map $pr_2|_S : S \to D_2^2$ is a degree-m branched covering and (ii) $\partial S = X_m \times \partial D_2^2$.

Two 2-dimensional m-braids S and S' are said to be *equivalent* if there exists an isotopy $\{h_t\}_{t\in[0,1]}$ of $D_1^2 \times D_2^2$ such that $h_0 = \mathrm{id}$, $h_1(S) = S'$, for each t $(t \in [0, 1])$ h_t is fiber preserving (when we regard $D_1^2 \times D_2^2$ as a D_1^2 bundle over D_2^2) and $h_t|_{D_1^2 \times \partial D_2^2} = \mathrm{id}$. In this definition, replace the condition that h_t $(t \in [0, 1])$ is fiber preserving by a condition that $h_t(S)$ is a 2-dimensional m-braid. Then we say that S and S' are *braid ambient isotopic*. It is obvious that if S and S' are equivalent then they are braid ambient isotopic. However the converse is not true in general. For instance, a branch point may split some branch points by braid ambient isotopies, although it may not split by equivalence (cf. [**K4**]). This significant feature comes from the existence of branch points. (Notice that a 2-dimensional braid without branch point is always trivial, [**K1**]. We also notice that two classical braids b and b' are equivalent if and only if they are braid ambient isotopic [**Ar**].) We usually regard equivalent 2-dimensional braids as the same.

Let S^2 be a 2-sphere obtained from D_2^2 and $\overline{D_2^2}$, a copy of D_2^2 with the opposite orientation, by identifying their boundaries. A 2-dimensional braid S of degree m is extended to a closed oriented surface $\widehat{S}$ in $D_1^2 \times S^2$ with $pr_1(\widehat{S} \cap D_1^2 \times \overline{D_2^2}) = X_m$. Fix an identification of $D_1^2 \times S^2$ with a tubular neighborhood of a standard 2-sphere in R^4. We call the image of $\widehat{S}$ in R^4 the *closure* of S.

In [**K3; K7**], conjugations and stabilizations are defined in a geometrical way. In the rest of this section we describe them in terms of braid systems.

Let S be a 2-dimensional braid of degree m, $\Sigma(S) \subset D_2^2$ the branch point set of the branched covering $pr_2|_S$ and y_0 a fixed point of ∂D_2^2. For a loop $c : ([0, 1], \{0, 1\}) \to (D_2^2 \backslash \Sigma(S), y_0)$, we define a loop $\rho_S(c) : ([0, 1], \{0, 1\}) \to (C_m, X_m)$ by

$$\rho_S(c)(t) = pr_1(S \cap D_1^2 \times \{c(t)\}),$$

where C_m is the configuration space of m interior points of D_1^2. Identify the m-braid group B_m with $\pi_1(C_m, X_m)$. The *braid monodromy*

$$\rho_S : \pi_1(D_2^2 \backslash \Sigma(S), y_0) \to B_m$$

of the 2-dimensional braid S is defined by sending $[c]$ to the element of the m-braid group $B_m = \pi_1(C_m, X_m)$ represented by the loop $\rho_S(c)$.

Let $\Sigma(S)$ consist of n points. A *Hurwitz arc system* $\mathcal{A} = (a_1, \ldots, a_n)$ for $\Sigma(S)$ with base point y_0 is an ordered set of n simple arcs in $D_2^2 \backslash \Sigma(S)$ such that (i) $a_i \cap a_j = \{y_0\}$ for $i \neq j$, (ii) for each i $(i = 1, \ldots, n)$, $\partial a_i = \{y_0, y_i\}$ for some point y_i of $\Sigma(S)$, and (iii) $a_1, \ldots, a_n$ are ordered in the positive direction around y_0. For each i $(i = 1, \ldots, n)$, let c_i be a loop in $D_2^2 \backslash \Sigma(S)$ with base point y_0

such that it runs along a_i, goes around y_i in the positive direction and comes back along a_i. Let η_i be the element of $\pi_1(D_2^2\backslash\Sigma(S), y_0)$ represented by the loop c_i. We call the ordered set $(\eta_1, \dots, \eta_n)$ the *generator system of* $\pi_1(D_2^2\backslash\Sigma(S), y_0)$ *associated with the Hurwitz arc system* $\mathcal{A}$. The *braid system of* S *associated with* $\mathcal{A}$ is $(\rho_S(\eta_1), \dots, \rho_S(\eta_n)) \in P^n(B_m)$, where $P^n(B_m)$ is the Cartesian product of n copies of B_m.

Not every element of $P^n(B_m)$ is a braid system of a 2-dimensional braid.

LEMMA 1.1 ([**K4**]). *Let A_m be the subset of B_m defined in* [**K4**]. *An element $\overrightarrow{b} = (b_1, \dots, b_n)$ of $P^n(B_m)$ is a braid system of a 2-dimensional braid if and only if the product $b_1 \cdots b_n$ is the identity $1 \in B_m$ and each b_i belongs to A_m.*

We recall some properties of the subset A_m, which are assumed in this paper. The subset A_m satisfies the following:

(1) Standard generators σ_j $(j = 1, \dots, m-1)$ of B_m and their inverses σ_j^{-1} are elements of A_m.
(2) If $b \in A_m$ and $\beta \in B_m$, then $\beta^{-1}b\beta \in A_m$.
(3) If $b \in A_m$, then $b \cdot \sigma_m^{\epsilon} \in B_{m+1}$ is an element of A_{m+1}, where $\epsilon \in \{\pm 1\}$.

(We usually use the same symbols for m-braids and their images by the natural injection $B_m \to B_{m+1}$, unless it makes confusion.)

For a group G, we denote by $P^n(G)$ the Cartesian product of n copies of G. The *slide action*, from the right side, of B_n on $P^n(G)$ is defined by

$$\mathrm{slide}(\sigma_j)(g_1, \dots, g_n) = (g_1, \dots, g_{j-1}, g_j g_{j+1} g_j^{-1}, g_j, g_{j+1}, \dots, g_n).$$

Two elements of $P^n(G)$ are called *slide equivalent* if they are in the same orbit of the slide action.

LEMMA 1.2 ([**R2; R4; K4**]). *Two 2-dimensional braids of degree m are equivalent if and only if their braid systems are slide equivalent.*

For an element $\overrightarrow{b} = (b_1, \dots, b_n) \in P^n(B_m)$ and $\gamma \in B_m$, we put

$$\mathrm{conj}_\gamma(\overrightarrow{b}) = (\gamma^{-1} b_1 \gamma, \dots, \gamma^{-1} b_n \gamma) \quad \in P^n(B_m),$$

$$\mathrm{stab}(\overrightarrow{b}) = (b_1, \dots, b_n, \sigma_m^{-1}, \sigma_m) \quad \in P^{n+2}(B_{m+1}),$$

and

$$\mathrm{stab}_\gamma(\overrightarrow{b}) = (b_1, \dots, b_n, \gamma^{-1}\sigma_m^{-1}\gamma, \gamma^{-1}\sigma_m\gamma) \quad \in P^{n+2}(B_{m+1}).$$

By Lemma 1.1 and the properties of A_m stated above, we see that if $\overrightarrow{b}$ is a braid system of a 2-dimensional braid (of degree m), then $\mathrm{conj}_\gamma(\overrightarrow{b})$, $\mathrm{stab}(\overrightarrow{b})$ and $\mathrm{stab}_\gamma(\overrightarrow{b})$ are braid systems of 2-dimensional braids (of degree m and of degree $m+1$). We say that $\mathrm{conj}_\gamma(\overrightarrow{b})$, $\mathrm{stab}(\overrightarrow{b})$ and $\mathrm{stab}_\gamma(\overrightarrow{b})$ are obtained from $\overrightarrow{b}$ by a *conjugation associated with* γ, by a *stabilization* and by a *generalized stabilization associated with* γ, respectively.

A 2-dimensional braid S' is said to be obtained from S by a *conjugation*, a *stabilization* or a *generalized stabilization* if S' has a braid system obtained from a braid system of S by a corresponding move. By Lemma 1.2, this condition is equivalent to that any braid system of S' is obtained from any braid system of S by a corresponding move modulo slide equivalence. The notions of a conjugation and a stabilization defined this way are equivalent to those in [**K4; K7**] defined geometrical way.

LEMMA 1.3. *Let a 2-dimensional braid S' be obtained from a 2-dimensional braid S by a conjugation, a stabilization or the inverse of a stabilization. The closures of S and S' belong to the same surface link type.*

PROOF. It is easy to see that if S' is a conjugate of S, then their closures $\widehat{S}$ and $\widehat{S'}$ are ambient isotopic in $D_1^2 \times S^2$ and hence in R^4. Suppose that S' is obtained from S by a stabilization. Let $\overrightarrow{b} = (b_1, \ldots, b_n) \in P^n(B_m)$ and $\mathrm{stab}(\overrightarrow{b}) = (b_1, \ldots, b_n, \sigma_m^{-1}, \sigma_m) \in P^{n+2}(B_{m+1})$ be braid systems of S and S'. Consider a 2-dimensional braid S^+ of degree $m+1$ with braid system

$$(b_1, \ldots, b_n) \in P^n(B_{m+1}).$$

This is the union of S and a trivial disk in $D_1^2 \times D_2^2$, and the closure $\widehat{S^+}$ of S^+ is the union of $\widehat{S}$ and a trivial 2-sphere in R^4. Then S' is obtained from S^+ by surgery along a simple 1-handle corresponding to $(\sigma_m^{-1}, \sigma_m)$, which is a trivial 1-handle connecting $\widehat{S}$ and the trivial 2-sphere in R^4 (cf. [**K1**]). (If the reader is familiar with chart description, the 1-handle corresponds to a free edge of label m in the chart.) Thus $\widehat{S}$ and $\widehat{S'}$ are ambient isotopic in R^4. □

We shall prove in the next section that generalized stabilizations are decomposed into conjugations and stabilizations. Hence if S' is obtained from S by a generalized stabilization, then the closures belong to the same surface link type.

By a *2-dimensional Markov move* (or a *generalized 2-dimensional Markov move*, resp.), we mean a braid ambient isotopy, a conjugation, a stabilization (resp. a generalized stabilization) or the inverse of a stabilization (resp. the inverse of a generalized stabilization). A finite sequence of 2-dimensional braids, $S_0, S_1, \ldots, S_s$, is a *2-dimensional Markov sequence* (or a *generalized 2-dimensional Markov sequence*, resp.) if for each i ($i = 1, \ldots, s$), S_i is obtained from S_{i-1} by a 2-dimensional Markov move (resp. a generalized 2-dimensional Markov move). If two 2-dimensional braids are related by such a sequence, then their closures in R^4 belong to the same surface link type. Theorems 0.3–0.5 states that the converse is true.

2. Rearrangement of 2-dimensional Markov moves

In this section we investigate quasi-commutativity of generalized 2-dimensional Markov moves and prove the main theorem.

LEMMA 2.1. *If a 2-dimensional braid S' is obtained from S by a generalized stabilization, then there is a 2-dimensional Markov sequence $S = S_0, S_1, S_2, S_3 =$*

S' such that $S_0 \mapsto S_1$ is a conjugation, $S_1 \mapsto S_2$ is a stabilization and $S_2 \mapsto S_3$ is a conjugation.

PROOF. Let $\overrightarrow{b} \in P^n(B_m)$ and $\overrightarrow{b'} \in P^{n+2}(B_{m+1})$ be braid systems of S and S' such that $\overrightarrow{b'} = \mathrm{stab}_\gamma(\overrightarrow{b}) \in P^{n+2}(B_{m+1})$ for some $\gamma \in B_m$. Consider 2-dimensional braids S_1 and S_2 determined by $\overrightarrow{b_1} = \mathrm{conj}_{\gamma^{-1}}(\overrightarrow{b}) \in P^n(B_m)$ and $\overrightarrow{b_2} = \mathrm{stab}(\overrightarrow{b_1}) \in P^{n+2}(B_{m+1})$, respectively. Since $\overrightarrow{b'} = \mathrm{conj}_\gamma(\overrightarrow{b_2})$, we have the result. □

LEMMA 2.2. *Let S_0, S_1, S_2 be a generalized 2-dimensional Markov sequence.*

(1) *If $S_0 \mapsto S_1$ is a braid ambient isotopy and $S_1 \mapsto S_2$ is a conjugation, then there is a generalized 2-dimensional Markov sequence S_0, S_1', S_2 such that $S_0 \mapsto S_1'$ is a conjugation and $S_1' \mapsto S_2$ is a braid ambient isotopy.*
(2) *If $S_0 \mapsto S_1$ is a conjugation and $S_1 \mapsto S_2$ is a braid ambient isotopy, then there is a generalized 2-dimensional Markov sequence S_0, S_1', S_2 such that $S_0 \mapsto S_1'$ is a braid ambient isotopy and $S_1' \mapsto S_2$ is a conjugation.*

PROOF. (1) If necessary modifying S_0, S_1 and S_2 up to equivalence, we may assume that S_1 is obtained from S_0 by a braid ambient isotopy of $D_1^2 \times D_2^2$ keeping $D_1^2 \times N(\partial D_2^2)$ fixed and that the restrictions of S_1 and S_2 to $D_1^2 \times (D_2^2 \backslash N(\partial D_2^2))$ are the same, where $N(\partial D_2^2)$ is a collar neighborhood of ∂D_2^2 in D_2^2. These moves commute each other. The assertion (2) is proved similarly. □

LEMMA 2.3. *Let S_0, S_1, S_2 be a generalized 2-dimensional Markov sequence.*

(1) *If $S_0 \mapsto S_1$ is a braid ambient isotopy and $S_1 \mapsto S_2$ is a generalized stabilization, then there is a generalized 2-dimensional Markov sequence S_0, S_1', S_2 such that $S_0 \mapsto S_1'$ is a generalized stabilization and $S_1' \mapsto S_2$ is a braid ambient isotopy.*
(2) *If $S_0 \mapsto S_1$ is a conjugation and $S_1 \mapsto S_2$ is a generalized stabilization, then there is a generalized 2-dimensional Markov sequence S_0, S_1', S_2 such that $S_0 \mapsto S_1'$ is a generalized stabilization and $S_1' \mapsto S_2$ is a conjugation.*

PROOF. Let $\overrightarrow{b_0} \in P^n(B_m)$, $\overrightarrow{b_1} \in P^{n'}(B_m)$ and $\overrightarrow{b_2} \in P^{n'+2}(B_{m+1})$ be braid systems of S_0, S_1 and S_2 such that $\overrightarrow{b_2} = \mathrm{stab}_\gamma(\overrightarrow{b_1})$ for some $\gamma \in B_m$. (1) Let S_1' be a 2-dimensional braid whose braid system is $\mathrm{stab}_\gamma(\overrightarrow{b_0})$. Using a braid ambient isotopy carrying S_0 to S_1, we obtain a braid ambient isotopy carrying S_1' to S_2. (2) Suppose that $\overrightarrow{b_1} = \mathrm{conj}_\beta(\overrightarrow{b_0})$, where $\beta \in B_m$. Let $\overrightarrow{b_1'} = \mathrm{stab}_{\gamma\beta^{-1}}(\overrightarrow{b_0})$ and S_1' a 2-dimensional braid whose braid system is $\overrightarrow{b_1'}$. Since $\overrightarrow{b_2} = \mathrm{conj}_\beta(\overrightarrow{b_1'})$, we have the result. □

LEMMA 2.4. *Let S_0, S_1, S_2 be a generalized 2-dimensional Markov sequence such that $S_0 \mapsto S_1$ is the inverse of a generalized stabilization and $S_1 \mapsto S_2$ is a generalized stabilization. Then there is a generalized 2-dimensional Markov sequence S_0, S_1', S_1'', S_2 such that $S_0 \mapsto S_1'$ is a generalized stabilization, $S_1' \mapsto S_1''$ is a conjugation, and $S_1'' \mapsto S_2$ is the inverse of a generalized stabilization.*

PROOF. Let $\overrightarrow{b_0} \in P^{n+2}(B_{m+1})$, $\overrightarrow{b_1} \in P^n(B_m)$ and $\overrightarrow{b_2} \in P^{n+2}(B_{m+1})$ be braid systems of S_0, S_1 and S_2 respectively such that $\overrightarrow{b_0} = \mathrm{stab}_\beta(\overrightarrow{b_1})$ and $\overrightarrow{b_2} = \mathrm{stab}_\gamma(\overrightarrow{b_1})$ for some $\beta, \gamma \in B_m$. Let $\overrightarrow{b'_1} \in P^{n+4}(B_{m+2})$ be $\mathrm{stab}_{\sigma_m\gamma}(\overrightarrow{b_0})$, i.e.,

$$\overrightarrow{b'_1} = (b_1, \ldots, b_n, \beta^{-1}\sigma_m^{-1}\beta, \beta^{-1}\sigma_m\beta, (\sigma_m\gamma)^{-1}\sigma_{m+1}^{-1}(\sigma_m\gamma), (\sigma_m\gamma)^{-1}\sigma_{m+1}(\sigma_m\gamma)),$$

where $\overrightarrow{b_1} = (b_1, \ldots, b_n)$. Let $\overrightarrow{b''_1} \in P^{n+4}(B_{m+2})$ be $\mathrm{conj}_{\sigma_{m+1}}(\overrightarrow{b'_1})$. Since elements in the image of $B_m \to B_{m+2}$ commute with σ_{m+1} in B_{m+2}, we have

$$\overrightarrow{b''_1} = (b_1, \ldots, b_n, (\sigma_m^{-1}\beta)^{-1}\sigma_{m+1}^{-1}(\sigma_m^{-1}\beta), (\sigma_m^{-1}\beta)^{-1}\sigma_{m+1}(\sigma_m^{-1}\beta), \gamma^{-1}\sigma_m^{-1}\gamma, \gamma^{-1}\sigma_m\gamma).$$

This is slide equivalent to

$$(b_1, \ldots, b_n, \gamma^{-1}\sigma_m^{-1}\gamma, \gamma^{-1}\sigma_m\gamma, (\sigma_m^{-1}\beta)^{-1}\sigma_{m+1}^{-1}(\sigma_m^{-1}\beta), (\sigma_m^{-1}\beta)^{-1}\sigma_{m+1}(\sigma_m^{-1}\beta)),$$

which is $\mathrm{stab}_{\sigma_m^{-1}\beta}(\overrightarrow{b_2})$. Let S'_1 and S''_1 be 2-dimensional braids determined by $\overrightarrow{b'_1}$ and $\overrightarrow{b''_1}$. □

For a 2-dimensional braid S, we denote by $\deg(S)$ the degree of S. Let $S_0, S_1, \ldots, S_s$ be a generalized 2-dimensional Markov sequence. Then $\deg(S_i) = \deg(S_{i+1})$, $\deg(S_i) < \deg(S_{i+1})$ or $\deg(S_i) > \deg(S_{i+1})$ implies that S_{i+1} is obtained from S_i by a braid ambient isotopy/conjugation, a generalized stabilization or the inverse of a generalized stabilization, respectively.

LEMMA 2.5 (Rearrangement of generalized Markov moves).

(1) *For a generalized 2-dimensional Markov sequence* $S_0, S_1, \ldots, S_s$ *with* $\deg(S_0) = \deg(S_1) = \cdots = \deg(S_s)$, *there is a generalized 2-dimensional Markov sequence* S_0, S'_1, S_s *with* $\deg(S_0) = \deg(S'_1) = \deg(S_s)$.

(2) *For a generalized 2-dimensional Markov sequence* S_0, S_1, S_2 *with* $\deg(S_0) = \deg(S_1) < \deg(S_2)$, *there is a generalized 2-dimensional Markov sequence* S_0, S'_1, S_2 *with* $\deg(S_0) < \deg(S'_1) = \deg(S_2)$.

(3) *For a generalized 2-dimensional Markov sequence* S_0, S_1, S_2 *with* $\deg(S_0) > \deg(S_1) = \deg(S_2)$, *there is a generalized 2-dimensional Markov sequence* S_0, S'_1, S_2 *with* $\deg(S_0) = \deg(S'_1) > \deg(S_2)$.

(4) *For a generalized 2-dimensional Markov sequence* S_0, S_1, S_2 *with* $\deg(S_0) > \deg(S_1) < \deg(S_2)$, *there is a generalized 2-dimensional Markov sequence* S_0, S'_1, S''_1, S_2 *with* $\deg(S_0) < \deg(S'_1) = \deg(S''_1) > \deg(S_2)$.

PROOF. It follows from Lemmas 2.2–2.4. □

Now we prove the main theorem.

PROOF OF THEOREM 0.4. By Theorem 0.3, there exists a 2-dimensional Markov sequence (hence a generalized 2-dimensional Markov sequence) between S and S', say $S = S_0^{(0)}, S_1^{(0)}, \ldots, S_t^{(0)} = S'$. Using Lemma 2.5, we have a new generalized 2-dimensional Markov sequence $S = S_0^{(1)}, S_1^{(1)}, \ldots, S_u^{(1)} = S'$ such that $\deg(S_i^{(1)}) < \deg(S_{i+1}^{(1)})$ $(0 \le i < j)$, $\deg(S_j^{(1)}) = \deg(S_{j+1}^{(1)}) = \deg(S_{j+2}^{(1)})$ and $\deg(S_i^{(1)}) > \deg(S_{i+1}^{(1)})$ $(j+2 \le i \le u-1)$ for some j. For each i with

$0 \leq i < j$, $S_i^{(1)} \mapsto S_{i+1}^{(1)}$ is a generalized stabilization. By Lemma 2.1, it is replaced by a 2-dimensional Markov sequence consisting of two conjugations and a stabilization. For each i with $j+2 \leq i \leq u-1$, $S_i^{(1)} \mapsto S_{i+1}^{(1)}$ is the inverse of a generalized stabilization, which is replaced by a Markov sequence consisting of two conjugations and the inverse of a stabilization. Then we have a 2-dimensional Markov sequence $S = S_0^{(2)}, S_1^{(2)}, \ldots, S_v^{(2)} = S'$ such that for i with $0 \leq i < k$, $S_i^{(2)} \mapsto S_{i+1}^{(2)}$ is a conjugation or a stabilization, $S_k^{(2)} \mapsto S_{k+1}^{(2)}$ is a braid ambient isotopy, and for i with $k+1 \leq i \leq v-1$, $S_i^{(2)} \mapsto S_{i+1}^{(2)}$ is a conjugation or the inverse of a stabilization. □

3. Simple 2-dimensional braids

A *simple* 2-dimensional braid is a 2-dimensional braid S such that the branched covering map $pr_2|_S : S \to D_2^2$ is a simple branched covering.

An element of the braid group B_m is called a *band* if it is a conjugate of a standard generator σ_j ($j \in \{1, \ldots, m-1\}$) or its inverse. Recall that every band is an element of the subset A_m of B_m.

LEMMA 3.1 ([**R2; R4; K4**]). *Let S be a 2-dimensional braid of degree m and $\overrightarrow{b} = (b_1, \ldots, b_n) \in P^n(B_m)$ a braid system of S. The 2-dimensional braid S is simple if and only if each b_i $(i = 1, \ldots, n)$ is a band.*

LEMMA 3.2. *Let S be a simple 2-dimensional braid. Any 2-dimensional braid obtained from S by a conjugation or a stabilization is also simple.*

PROOF. It follows from Lemma 3.1 and the definition of conjugations and stabilizations in terms of braid systems. □

PROOF OF THEOREM 0.5. Consider a 2-dimensional Markov sequence $S = S_0, S_1, \ldots, S_s = S'$ between S and S' as in Theorem 0.4. Since S is simple, $S = S_0, S_1, \ldots, S_{k-1}$ are simple (Lemma 3.2). Since S' is simple, $S' = S_s, S_{s-1}, \ldots, S_k$ are simple (Lemma 3.2). □

PROOF OF THEOREM 0.6. Let S be a non-simple 2-dimensional braid. By Theorem 0.2, there exists a simple 2-dimensional braid S' whose closure represents the same surface link type as the closure of S. Consider a 2-dimensional Markov sequence $S = S_0, S_1, \ldots, S_s = S'$ between S and S' as in Theorem 0.4. Since S' is simple, $S' = S_s, S_{s-1}, \ldots, S_k$ are simple (Lemma 3.2). The 2-dimensional braid S_{k-1} is obtained from S by applying some conjugations and stabilizations. Since $S_{k-1} \mapsto S_k$ is a braid ambient isotopy, we have the result. □

4. Ribbon 2-dimensional braids and unknotted 2-dimensional braids

The notions of a *ribbon* 2-dimensional braid and an *unknotted* 2-dimensional braid are defined in [**K1**]. Here we give alternative definitions to them in terms of braid systems.

A 2-dimensional braid S of degree m is *ribbon* (or *unknotted*, resp.) if it has a braid system $\overrightarrow{b} = (b_1, \dots, b_n) \in P^n(B_m)$ such that

(1) each b_i $(i = 1, \dots, n)$ is a band (resp. a standard generator of B_m or its inverse),
(2) $b_{2j} = b_{2j-1}^{-1}$ for every $j = 1, \dots, n/2$.

LEMMA 4.1. *Let S be a ribbon 2-dimensional braid. Any 2-dimensional braid obtained from S by a conjugation or a stabilization is also ribbon.*

PROOF. It follows from the definition of ribbon 2-dimensional braids and the definition of conjugations and stabilizations in terms of braid systems. □

LEMMA 4.2. *Let S be an unkotted 2-dimensional braid.*

(1) *If S is connected, then any 2-dimensional braid obtained from S by a conjugation is equivalent to S itself.*
(2) *The 2-dimensional braid obtained from S by a stabilization is unknotted.*

PROOF. Let $\overrightarrow{b} = (b_1, \dots, b_n) \in P^n(B_m)$ be a braid system of S such that each b_i $(i = 1, \dots, n)$ is a standard generator of B_m or its inverse and that $b_{2j} = b_{2j-1}^{-1}$ for every $j = 1, \dots, n/2$. If S is connected, then all standard generators $\sigma_1, \dots, \sigma_{m-1}$ of B_m must appear in the braid system. Then $\overrightarrow{b}$ is slide equivalent to $(\sigma_1, \sigma_1^{-1}, \sigma_2, \sigma_2^{-1}, \dots, \sigma_{m-1}, \sigma_{m-1}^{-1}, \sigma_1, \sigma_1^{-1}, \dots, \sigma_1, \sigma_1^{-1})$. Thus we may assume $\overrightarrow{b}$ to be such a braid system. If S' is obtained from S by a conjugation associated with $\gamma \in B_m$, then by definition S' has a braid system $\overrightarrow{b'} = \mathrm{conj}_\gamma(\overrightarrow{b}) = (\gamma^{-1} b_1 \gamma, \dots, \gamma^{-1} b_n \gamma)$. In a special case that $\gamma = \sigma_k$ $(k \in \{1, \dots, m-1\})$ or σ_k^{-1}, it is verified that $\overrightarrow{b'}$ is slide equivalent $\overrightarrow{b}$. Hence for a general case of $\gamma \in B_m$, we see that $\overrightarrow{b'}$ is slide equivalent $\overrightarrow{b}$. This implies that S' is equivalent to S. The second assertion is obvious. □

REMARK. The first assertion of Lemma 4.2 is easily proved if we use chart description of a simple 2-dimensional braid (refer to [**K1; K5**] or [**CS1; CS2; CS3; K6; K8**]). Since S is unknotted, it is presented by an *unknotted chart*; that is a chart consisting of free edges, say Γ. (In fact, this is the original definition of an unknotted 2-dimensional braid in [**K1**].) If S is connected, every integer of $1, 2, \dots, m-1$ appears as the label of a free edge of Γ, where m is the degree of S. A conjugation associated with σ_k (or σ_k^{-1}) makes an effect to add a loop with label k surrouding Γ. Using a C-move of type I (cf. [**K1; K5**]), we can eliminate the loop. This implies that the 2-dimensional braid obtained from S by a conjugation associated with σ_k (or σ_k^{-1}) is equivalent to S. The second assertion is also obvious, since a stabilization makes an effect to add a free edge with label m.

THEOREM 4.3 ([**K1**]). *A surface link is ribbon (or unknotted, resp.) if and only if it has a simple braid presentation which is ribbon (resp. unknotted).*

PROOF OF THEOREMS 0.7 AND 0.8. Let S be a 2-dimensional braid whose closure is a ribbon surface link (resp. an unknotted surface knot). By Theorem 4.3, there exists a ribbon (resp. unknotted and connected) 2-dimensional braid S' whose closure represents the same surface link type as the closure of S. Consider a 2-dimensional Markov sequence $S = S_0, S_1, \ldots, S_s = S'$ between S and S' as in Theorem 0.4. Since S' is ribbon (resp. unknotted and connected), by Lemma 4.1 (Lemma 4.2), we see that $S' = S_s, S_{s-1}, \ldots, S_k$ are ribbon (resp. unknotted and connected). The 2-dimensional braid S_{k-1} is obtained from S by applying some conjugations and stabilizations. Since $S_{k-1} \mapsto S_k$ is a braid ambient isotopy, we have the result. □

5. Unknotting number of a 2-dimensional braid

Let F be a surface link. A simple arc α in R^4 is a *chord* attached to F if $F \cap \alpha = \partial\alpha$. Two chords α and α' are *equivalent* (or *equivalent in the strict sense*, resp.) if α is ambient isotopic to α' in R^4 keeping F (resp. both F and $F \cap \alpha$) setwise fixed. We define a chord attached to a 2-dimensional braid in $D_1^2 \times D_2^2$ similarly except to suppose an additional condition that the arc α is contained in the interior of $D_1^2 \times D_2^2$.

A 3-ball h in R^4 is a *1-handle* attached to F if $F \cap h$ is a pair of disks in ∂h and the surface $F' = \mathrm{cl}(F \cup \partial h - F \cap h)$ is orientable. We call F' the surface link obtained from F by *surgery* along h. Two 1-handles h and h' are *equivalent* (or *equivalent in the strict sense*, resp.) if h is ambient isotopic to h' in R^4 keeping F (resp. both F and $F \cap h$) setwise fixed. A chord α is called a *core* of h if it is a deformation retract of h.

LEMMA 5.1 ([**Bo; HK**]).

(1) *Two cores α and α' of the same 1-handle are equivalent. Moreover if $\partial\alpha = \partial\alpha'$, then they are equivalent in the strict sense.*
(2) *Two 1-handles h and h' with the same core are equivalent. Moreover if $\partial h = \partial h'$, then they are equivalent in the strict sense.*
(3) *Let α and α' be cores of 1-handles h and h', respectively. The chords α and α' are equivalent if and only if h and h' equivalent.*

Let S be a 2-dimensional braid in $D_1^2 \times D_2^2$. A chord α attached to S is said to be *simple* if α is contained in a fiber $D_1^2 \times \{y\}$ for some $y \in D_2^2$. A 1-handle h attached to S is said to be *simple* if a core of h is simple and the surgery along h changes S into a 2-dimensional braid.

REMARK. If h is a 1-handle attached to S whose core α is simple, then h is equivalent (in the strict sense) to a simple one. Moreover, if S is simple then the surgery result is also simple. [By a fiber-preserving ambient isotopy of $D_1^2 \times D_2^2$, we can deform α into a straight segment α'. Let S' be the 2-dimensional braid obtained from S by this ambient isotopy. There exists a simple 1-handle whose core is α' attached to S' as in [**K1**], say h'. The image of h' by the inverse of the ambient isotopy is a simple 1-handle whose core is α attached to S. By Lemma 5.1, we see the assertion.]

LEMMA 5.2 ([**K1**]). *For any simple 2-dimensional braid S, there exists a set of simple 1-handles attached to S along which the surgery transforms S into an unknotted 2-dimensional braid.*

For a simple 2-dimensional braid S, the *unknotting number* $u(S)$ of S is defined by the minimum number of simple 1-handles attached to S along which the surgery transforms S into an unknotted 2-dimensional braid.

By Theorem 4.3, we have

$$u(F) \leq u(S)$$

for any simple braid presentation S of a surface link F.

THEOREM 5.3 ([**K6**]). *Let F be a surface link and $\{\alpha_1, \ldots, \alpha_n\}$ a set of mutually disjoint chords attached to F. There exists an ambient isotopy of R^4 which carries F into the closure of a simple 2-dimensional braid S and $\{\alpha_1, \ldots, \alpha_n\}$ into a set of simple chords attached to S.*

PROOF OF THEOREM 0.9. It is sufficient to show that there exists a simple braid presentation S of F with $u(S) \leq u(F)$. Put $n = u(F)$. Let $\{h_1, \ldots, h_n\}$ be a set of 1-handles attached to F such that an unknotted surface link, say $\widetilde{F}$, is obtained from F by surgery along $\{h_1, \ldots, h_n\}$. Let $\alpha_1, \ldots, \alpha_n$ be cores of $h_1, \ldots, h_n$. By Theorem 5.3, without loss of generality, we may assume that F is the closure of a simple 2-dimensional braid S and $\alpha_1, \ldots, \alpha_n$ are simple chords attached to S. Moreover, by the remark below Lemma 5.1, we may assume that $h_1, \ldots, h_n$ are simple 1-handles, namely, the surgery result of S along $h_1, \ldots, h_n$ is also a simple 2-dimensional braid, say $\widetilde{S}$. By Theorem 4.3, there exists an unknotted 2-dimensional braid $\widetilde{S}'$ which is a simple braid presentation of $\widetilde{F}$. Consider a 2-dimensional Markov sequence $\widetilde{S} = \widetilde{S}_0, \widetilde{S}_1, \ldots, \widetilde{S}_s = \widetilde{S}'$ between $\widetilde{S}$ and $\widetilde{S}'$ as in Theorem 0.4. Since $\widetilde{S}'$ is unknotted and connected, by Lemma 4.2, we see that $\widetilde{S}' = \widetilde{S}_s, \widetilde{S}_{s-1}, \ldots, \widetilde{S}_k$ are unknotted. Let $S = S_0, S_1, \ldots, S_k$ be a 2-dimensional Markov sequence such that S_i is obtained from S_{i-1} by the same operation as $\widetilde{S}_i$ is obtained from $\widetilde{S}_{i-1}$. Then the simple 1-handles $h_1, \ldots, h_n$ attached to $S = S_0$ is transformed into simple 1-handles attached to S_k, and the surgery result is $\widetilde{S}_k$. Thus $u(S_k) \leq n$. Since S_k is a simple braid presentation of F, we completes the proof. □

6. Rearrangement of 1-dimensional Markov moves

For braids $b, \beta \in B_m$ and $\epsilon \in \{\pm 1\}$, we say that a braid $b' \in B_{m+1}$ is obtained from b by a *generalized stabilization* of type ϵ associated with β if

$$b' = b \cdot \beta^{-1} \sigma_m^{\epsilon} \beta,$$

where we denote by the same symbol b the m-braid and its image by the natural injection $B_m \to B_{m+1}$. If $\beta \in B_m$ is the identity, the generalized stabilization is a stabilization in the ordinary sense. A *(generalized) Markov sequence* is a finite sequence of braids such that each braid is obtained from the previous one by a conjugation, a (generalized) stabilization or the inverse of a (generalized) stabilization.

LEMMA 6.1. *Let b' be obtained from b by a generalized stabilization. Then there is a Markov sequence $b = b_0, b_1, b_2, b_3 = b'$ such that $b_0 \mapsto b_1$ is a conjugation, $b_1 \mapsto b_2$ is a stabilization and $b_2 \mapsto b_3$ is a conjugation.*

PROOF. Let $b' = b \cdot \beta^{-1}\sigma_m^{\epsilon}\beta$, where $b, \beta \in B_m$ and $\epsilon \in \{\pm 1\}$. Put $b_1 = \beta b \beta^{-1}$ and $b_2 = \beta b \beta^{-1}\sigma_m^{\epsilon}$. □

LEMMA 6.2. *Let b_0, b_1, b_2 be a generalized Markov sequence. If $b_0 \mapsto b_1$ is a conjugation and $b_1 \mapsto b_2$ is a generalized stabilization, then there is a generalized Markov sequence b_0, b_1', b_2 such that $b_0 \mapsto b_1'$ is a generalized stabilization and $b_1' \mapsto b_2$ is a conjugation.*

PROOF. Let $b_1 = \beta^{-1}b_0\beta$ and $b_2 = b_1\gamma^{-1}\sigma_m\epsilon\gamma = \beta^{-1}b_0\beta\gamma^{-1}\sigma_m\epsilon\gamma$, where $b_0, \beta, \gamma \in B_m$ and $\epsilon \in \{\pm 1\}$. Put $b_1' = b_0\beta\gamma^{-1}\sigma_m^{\epsilon}\gamma\beta^{-1}$. □

LEMMA 6.3. *Let b_0, b_1, b_2 be a generalized Markov sequence such that $b_0 \mapsto b_1$ is the inverse of a generalized stabilization and $b_1 \mapsto b_2$ is a generalized stabilization. Then there is a generalized Markov sequence b_0, b_1', b_1'', b_2 such that $b_0 \mapsto b_1'$ is a generalized stabilization, $b_1' \mapsto b_1''$ is a conjugation, and $b_1'' \mapsto b_2$ is the inverse of a generalized stabilization.*

PROOF. Let $b_0 = b_1\beta^{-1}\sigma_m^{\epsilon}\beta \in B_{m+1}$ and $b_2 = b_1\gamma^{-1}\sigma_m^{\delta}\gamma \in B_{m+1}$, where $b_1, \beta, \gamma \in B_m$ and $\epsilon, \delta \in \{\pm 1\}$. Put $b_1' = b_0(\sigma_m\gamma)^{-1}\sigma_{m+1}^{\delta}(\sigma_m\gamma) \in B_{m+2}$ and $b_1'' = \sigma_{m+1}^{-1}b_1'\sigma_{m+1} \in B_{m+2}$. Then

$$\begin{aligned} b_1'' &= \sigma_{m+1}^{-1}b_0(\sigma_m\gamma)^{-1}\sigma_{m+1}^{\delta}(\sigma_m\gamma)\sigma_{m+1} \\ &= \sigma_{m+1}^{-1}b_1\beta^{-1}\sigma_m^{\epsilon}\beta\gamma^{-1}\sigma_m^{-1}\sigma_{m+1}^{\delta}\sigma_m\gamma\sigma_{m+1} \\ &= \sigma_{m+1}^{-1}b_1\beta^{-1}\sigma_m^{\epsilon}\beta\gamma^{-1}\sigma_{m+1}\sigma_m^{\delta}\sigma_{m+1}^{-1}\gamma\sigma_{m+1} \\ &= b_1\beta^{-1}\sigma_{m+1}^{-1}\sigma_m^{\epsilon}\sigma_{m+1}\beta\gamma^{-1}\sigma_m^{\delta}\gamma \end{aligned}$$

(note that σ_{m+1} commutes with elements of B_{m+2} which are in the image of B_m)

$$\begin{aligned} &= b_1\beta^{-1}\sigma_m\sigma_{m+1}^{\epsilon}\sigma_m^{-1}\beta\gamma^{-1}\sigma_m^{\delta}\gamma \\ &= b_1\gamma^{-1}\sigma_m^{\delta}\gamma(\sigma_m^{-1}\beta\gamma^{-1}\sigma_m^{\delta}\gamma)^{-1}\sigma_{m+1}^{\epsilon}(\sigma_m^{-1}\beta\gamma^{-1}\sigma_m^{\delta}\gamma) \\ &= b_2(\sigma_m^{-1}\beta\gamma^{-1}\sigma_m^{\delta}\gamma)^{-1}\sigma_{m+1}^{\epsilon}(\sigma_m^{-1}\beta\gamma^{-1}\sigma_m^{\delta}\gamma). \end{aligned}$$

□

For a braid b, we denote by $\deg(b)$ the degree of b. Let $b_0, b_1, \ldots, b_s$ be a generalized Markov sequence. Then $\deg(b_i) = \deg(b_{i+1})$, $\deg(b_i) < \deg(b_{i+1})$ or $\deg(b_i) > \deg(b_{i+1})$ implies that b_{i+1} is obtained from b_i by a conjugation, a generalized stabilization or the inverse of a generalized stabilization, respectively.

LEMMA 6.4 (Rearrangement of generalized Markov moves).

(1) *For a generalized Markov sequence* $b = b_0, b_1, \ldots, b_s = b'$ *with* $\deg(b_0) = \deg(b_1) = \cdots = \deg(b_s)$, *there is a generalized Markov sequence* b, b' *with* $\deg(b) = \deg(b')$.

(2) *For a generalized Markov sequence* b_0, b_1, b_2 *with* $\deg(b_0) = \deg(b_1) < \deg(b_2)$, *there is a generalized Markov sequence* b_0, b_1', b_2 *with* $\deg(b_0) < \deg(b_1') = \deg(b_2)$.

(3) *For a generalized Markov sequence* b_0, b_1, b_2 *with* $\deg(b_0) > \deg(b_1) = \deg(b_2)$, *there is a generalized Markov sequence* b_0, b_1', b_2 *with* $\deg(b_0) = \deg(b_1') > \deg(b_2)$.

(4) *For a generalized Markov sequence* b_0, b_1, b_2 *with* $\deg(b_0) > \deg(b_1) < \deg(b_2)$, *there is a generalized Markov sequence* b_0, b_1', b_1'', b_2 *with* $\deg(b_0) < \deg(b_1') = \deg(b_1'') > \deg(b_2)$.

PROOF. The first assertion is obvious, since the composition of two conjugations is a conjugation. The other assertions follow from the previous lemmas. □

PROOF OF THEOREM 0.10. By the Markov theorem, there exists a (generalized) Markov sequence between b and b', say $b = b_0^{(0)}, b_1^{(0)}, \ldots, b_t^{(0)} = b'$. Using Lemma 6.4, we have a generalized Markov sequence $b = b_0^{(1)}, b_1^{(1)}, \ldots, b_u^{(1)} = b'$ such that $\deg(b_i^{(1)}) < \deg(b_{i+1}^{(1)})$ $(0 \le i < j)$, $\deg(b_j^{(1)}) = \deg(b_{j+1}^{(1)})$ and $\deg(b_i^{(1)}) > \deg(b_{i+1}^{(1)})$ $(j+1 \le i \le u-1)$ for some j. For each i with $0 \le i < j$, $b_i^{(1)} \mapsto b_{i+1}^{(1)}$ is a generalized stabilization. By Lemma 6.1, it is replaced by a Markov sequence consisting of two conjugations and a stabilization. For each i with $j+1 \le i \le u-1$, $b_i^{(1)} \mapsto b_{i+1}^{(1)}$ is the inverse of a generalized stabilization, which is replaced by a Markov sequence consisting of two conjugations and the inverse of a stabilization. Then we have the desired Markov sequence between b and b'. □

EXAMPLE 6.5. Let b_0, b_1, b_2, b_3 be a Markov sequence such that $b_0 \mapsto b_1$ is the inverse of a stabilization, $b_1 \mapsto b_2$ is a conjugation, and $b_2 \mapsto b_3$ is a stabilization. Then there is a Markov sequence $b_0, b_1', b_1'', b_2'', b_2', b_3$ such that $b_0 \mapsto b_1'$ is a conjugation, $b_1' \mapsto b_1''$ is a stabilization, $b_1'' \mapsto b_2''$ is a conjugation, $b_2'' \mapsto b_2'$ is the inverse of a stabilization, and $b_2' \mapsto b_3$ is a conjugation. This fact is obtained from Lemma 6.4 (or Lemmas 6.1–6.3). We describe the Markov sequence precisely.

Suppose that $b_0 = b_1\sigma_m^{\epsilon} \in B_{m+1}$, $b_2 = \beta^{-1}b_1\beta \in B_m$, $b_3 = b_2\sigma_m^{\delta} \in B_{m+1}$, where $b_1, \beta \in B_m$ and $\epsilon, \delta \in \{\pm 1\}$. Put

$$
\begin{aligned}
b_1' &= (\beta\sigma_m^{-1})^{-1} b_0 (\beta\sigma_m^{-1}) \in B_{m+1},\\
b_1'' &= b_1'\sigma_{m+1}^{\delta} \in B_{m+2},\\
b_2'' &= (\sigma_m\beta^{-1}\sigma_{m+1}\beta\sigma_m^{-\delta}\beta^{-1}\sigma_m)^{-1} b_1'' (\sigma_m\beta^{-1}\sigma_{m+1}\beta\sigma_m^{-\delta}\beta^{-1}\sigma_m) \in B_{m+2},\\
b_2' &= (\sigma_m^{-\delta}\beta^{-1}\sigma_m)^{-1} b_3 (\sigma_m^{-\delta}\beta^{-1}\sigma_m) \in B_{m+1}.
\end{aligned}
$$

Then $b_2'' = b_2' \sigma_{m+1}^{\epsilon} \in B_{m+2}$. We obtain the desired Markov sequence this way. Of course, using this result, one may prove Theorem 0.10.

7. Some open problems

QUESTION 7.1. *Let S and S' be simple 2-dimensional braids. Suppose that they are braid ambient isotopic. Are they equivalent?*

A partial answer is given in [**K4**]: If S is braid ambient isotopic to S' by an isotopy $\{h_t\}_{t\in[0,1]}$ of $D_1^2 \times D_2^2$ such that for each $t \in [0,1]$, $h_t(S)$ has the same number of branch points as S, then S and S' are equivalent. The following is a weaker version of this question, which is still open.

QUESTION 7.2. *Let S and S' be simple 2-dimensional braids. Suppose that they are braid ambient isotopic. Do there exist simple 2-dimensional braids $\widetilde{S}$ and $\widetilde{S}'$ such that they are obtained from S and S' by some conjugations and stabilizations respectively and that $\widetilde{S}$ is equivalent to $\widetilde{S}'$?*

CONJECTURE 7.3. *If simple 2-dimensional braids S and S' represent the same surface link type, then there exists a sequence of simple 2-dimensional braids $S = S_0, S_1, \ldots, S_s = S'$ such that for each i $(i = 1, \ldots, s)$, S_i is equivalent to S_{i-1} or obtained from S_{i-1} by a conjugation, a stabilization or the inverse operation of a stabilization.*

This conjecture is a strengthened version of Theorem 0.5. Of course, if Question 7.1, or 7.2, is solved affirmatively, then the conjecture is true. At present no candidate for its counterexample is known to the author.

PROBLEM 7.4. *Construct a new invariant of a 2-dimensional braid that does not change under braid ambient isotopies, conjugations and stabilizations.*

REFERENCES

[Al] J. W. Alexander, *A lemma on systems of knotted curves*, Proc. Nat. Acad. Sci. USA **9** (1923), 93–95.

[Ar] E. Artin, *Theory of braids*, Ann. of Math. **48** (1947), 101–126.

[B1] J. S. Birman, *Braids, links, and mapping class groups*, Ann. Math. Studies 82, Princeton Univ. Press, Princeton, N.J., 1974.

[B2] ———, *Studying links via closed braids*, Lecture Notes on the Ninth KAIST Mathematical Workshop **1** (1994), 1–67.

[BM1] J. S. Birman and W. Menasco, *Studying links via closed braids IV: Composite links and split links*, Invent. Math. **102** (1990), 115–139.

[BM2] ———, *Studying links via closed braids II: On a theorem of Bennequin*, Topology Appl. **40** (1991), 71–82.

[BM3] ———, *Studying links via closed braids I: A finiteness theorem*, Pacific J. Math. **154** (1992), 17–36.

[BM4] ———, *Studying links via closed braids V: The unlink*, Trans. Amer. Math. Soc. **329** (1992), 585–606.

[BM5] ———, *Studying links via closed braids VI: A non-finiteness theorem*, Pacific J. Math. **156** (1992), 265–285.

[BM6] ———, *Studying links via closed braids III: Classifying links which are closed 3-braids*, Pacific J. Math. **161** (1993), 25–113.

[Bo] J. Boyle, *Classifying 1-handles attached to knotted surfaces*, Trans. Amer. Math. Soc. **306** (1988), 475–487.

[Br] E. Brieskorn, *Automorphic sets and braids and singularities*, Contemp. Math. **78** (1988), 45–115.

[CS1] J. S. Carter and M. Saito, *Knotted surfaces, braid moves and beyond*, Knots and Quantum Gravity, ed. J. Baez, Oxford U. Press, to appear.

[CS2] ———, *Knot diagrams and braid theories in dimension 4*, Real and complex singularities, Papers from the 3rd international workshop, (São Carlos, Brazil, 1994), Pitman Res. Notes Math. Ser. 333, 1995, pp. 112–147.

[CS3] ———, *Braids and movies*, J. Knot Theory Ramifications **5** (1996), 589–608.

[F] T. Fiedler, *A small state sum for knots*, Topology **32** (1993), 281–294.

[G] F. González-Acuña, *A characterization of 2-knot groups*, preprint.

[HK] F. Hosokawa and A. Kawauchi, *Proposals for unknotted surfaces in four-spaces*, Osaka J. Math. **16** (1979), 233–248.

[HMS] F. Hosokawa, T. Maeda, S. Suzuki, *Numerical invariants of surfaces in 4-space*, Math. Sem. Notes Kobe Univ. **7** (1979), 409–420.

[K1] S. Kamada, *Surfaces in R^4 of braid index three are ribbon*, J. Knot Theory Ramifications **1** (1992), 137–160.

[K2] ———, *A characterization of groups of closed orientable surfaces in 4-space*, Topology **33** (1994), 113–122.

[K3] ———, *Alexander's and Markov's theorems in dimension four*, Bull. Amer. Math. Soc. **31** (1994), 64–67.

[K4] ———, *On braid monodromies of non-simple braided surfaces*, Math. Proc. Camb. Phil. Soc. **120** (1996), 237–245.

[K5] ———, *An observation of surface braids via chart description*, J. Knot Theory Ramifications **4** (1996), 517–529.

[K6] ———, *Surfaces in 4-space: A view of normal forms and braidings*, Lectures at Knots 96 (ed. S. Suzuki), World Scientific Publishing Co., 1997, pp. 39–71.

[K7] ———, *Generalized Alexander's and Markov's theorems in dimension four*, preprint.

[K8] ———, *Unknotting immersed surface-links and singular 2-dimensional braids by 1-handle surgeries*, Osaka J. Math., to appear.

[KM] T. Kanenobu and Y. Marumoto, *Unknotting and fusion numbers of ribbon 2-knots*, Osaka J. Math., to appear.

[KSS] A. Kawauchi, T. Shibuya, S. Suzuki, *Descriptions on surfaces in four-space, I. Normal forms*, Math. Sem. Notes Kobe Univ. **10** (1982), 75–125.

[Ma] A. A. Markov, *Über die freie Äquivalenz der geschlossner Zopfe*, Rec. Soc. Math. Moscou **1** (1935), 73–78.

[Mi] K. Miyazaki, *On the relationship among unknotting numbers, knotting genus and Alexander invariant for 2-knots*, Kobe J. Math. **3** (1986), 77-85.

[Moi] B. G. Moishezon, *Stable branch curves and braid monodromies*, Algebraic Geometry, Lect. Notes in Math. 862, 1981, pp. 107–192.

[Mor1] H. R. Morton, *An irreducible 4-string braid with unknotted closure*, Math. Proc. Cambridge Philos. Soc. **93** (1983), 259–261.

[Mor2] ———, *Threading knot diagrams*, Math. Proc. Cambridge Philos. Soc. **99** (1986), 247–260.

[R1] L. Rudolph, *Algebraic functions and closed braids*, Topology **22** (1983), 191–202.

[R2] ———, *Braided surfaces and Seifert ribbons for closed braids*, Comment. Math. Helv. **58** (1983), 1–37.

[R3] ———, *Some topologically locally-flat surfaces in the complex projective plane*, Comment. Math. Helv. **59** (1984), 592–599.

[R4] ———, *Special positions for surfaces bounded by closed braids*, Rev. Mat. Iberoamericana **1** (1985), 93–133.

[R5] ———, *Quasipositivity and new knot invariants*, Rev. Mat. Univ. Complut. Madrid **2** (1989), 85–109.

[R6] ———, *A characterization of quasipositive Seifert surfaces (Constructions of quasipositive knots and links, III)*, Topology **31** (1992), 231–237.

[R7] ———, *Quasipositive annuli (Constructions of quasipositive knots and links, IV)*, J. Knot Theory Ramifications **1** (1992), 451–466.

[R8] ———, *Quasipositivity as an obstruction to sliceness*, Bull. Amer. Math. Soc (N.S.) **29** (1993), 51–59.

[Si] D. Silver, *Free group automorphisms and knotted tori in S^4*, J. Knot Theory Ramifications **6** (1997), 95–103.

[Sk] R. K. Skora, *Closed braids in 3-manifolds*, Math. Z. **211** (1992), 173-187.

[T] P. Traczyk, *A new proof of Markov's braid theorem*, preprint.

[Vi] O. Ya. Viro, Lecture given at Osaka City University, September, 1990.

[Vo] P. Vogel, *Representation of links by braids: A new algorithm*, Comment. Math. Helv. **65** (1990), 104-113.

[Y] S. Yamada, *The minimal number of Seifert circles equals the braid index of a link*, Invent. Math. **89** (1987), 347-356.

Department of Mathematics, Osaka City University, Sumiyoshi-ku, Osaka 558-8585, Japan

E-mail address: kamada@sci.osaka-cu.ac.jp

Contemporary Mathematics
Volume **233**, 1999

A CRITERION FOR SIGNED UNKNOTTING NUMBER

Paweł Traczyk

Abstract. We consider various versions of signed unknotting number. We give tables for prime knots up to nine crossings.

1. Definitions

In this note we investigate some notions of signed unknotting number. First of all let us recall that the classical unknotting number $u = u(K)$ of a knot K is the minimum number of crossings that must be reversed to transform K into the unknot. The minimum is taken over all possible projections of K. Now, each crossing change is of $+ \longrightarrow -$ or $- \rightarrow +$ type — the sign of a crossing is independent of the choice of orientation of the knot (of course this is not true for links of more that one component.) For shortness we will later denote such changes as $+-$ or $-+$ changes. Now we can formulate the following definitions.

Definition 1.1. *Suppose K is a knot of unknotting number $u_K = u$. Then u_{+-}, is the minimum number of $+-$ changes of crossings in a sequence of u changes transforming the given knot into the unknot. Here, $-+$ changes are also allowed but it is assumed that the total number of changes be equal to the unknotting number u.*

Definition 1.2. *Suppose K is a knot. Then u_{++}, is the minimum number of $+-$ changes necessary to transform K into the unknot. Here we assume that only such changes are allowed. As a convention we say that $u_{++} = \infty$ if it is not possible to make such transformation at all.*

Definition 1.3.. *Suppose K is a knot. Then u_+ is the minimum number of $+-$ changes necessary to transform K into the unknot (in a sequence of both $+-$ and $-+$ changes). Here, there is no restriction on the number of $-+$ changes).*

The notation is chosen in the following way: u_+ means that we are counting $+-$ operations only, and we allow any number of $-+$ operations in the process. The double $+$ sign in the u_{++} symbol means that we change only positive crossings, that is we insist on using the $+-$ operation only. And the u_{+-} symbol indicates

1991 *Mathematics Subject Classification.* Primary 57M25.

supported by Polish KBN grant No 2P03A1113.

that we use and count both $+-$ and $-+$ operations and we prefer to minimize the use of $+-$. The symbol u_+ is used in the same sense as in [**CL**].

We will also use symbols u_-, u_{-+} and u_{--}, which are defined similarly.

REMARK.. To make the definitions more clear let us explain that for the figure eight knot $u_+ = u_- = 0$, while $u = 1$ — the figure eight knot may be transformed into the unknot by reversing a single positive crossing or a single negative crossing — neither one *has* to be used. At the same time $u_{++} = u_{--} = 0$.

2. u_{++}, u_+ and u_{+-} are all different

We start with the following observation.

PROPOSITION 2.1.. $u_{++} \geq u_{+-} \geq u_+$.

PROOF. . This is immediately obvious from the definitions.

We now go on to showing that the three notions are different.

First of all, it may happen that $u_{++} = \infty$. A sufficient condition for this is that $\sigma_K > 0$, where σ is the Murasugi signature of a knot — it is well known that the $+-$ operation does not decrease the signature. On the other hand, u_+ and u_{+-} are by definition always finite.

It is less obvious that u_+ may be different from u_{+-}. The smallest example is the prime knot $K = 6_1$. We will show that $u_+(K) = 0$, while $u_{+-}(K) = 1$. It is obvious that one can transform K into the trivial knot by a single $+-$ change (so $u_{+-} \leq 1$) and also by a sequence of two $-+$ changes (in which case $+-$ change is not used at all, which means that $u_+ = 0$). What remains to prove is that it is not possible to transform K into the unknot by a single $-+$ change (which would mean $u_{+-}(K) = 0$, of course $u(K) = 1$). Suppose that it is possible to transform K into the unknot by a single $-+$ operation. Then there exists a skein triple $(K_+, K = K_-, K_0)$ with K_+ the trivial knot. Now, the Jones polynomial of K is $V_K(t) = t^{-4} - t^{-3} + t^{-2} - 2t^{-1} + 2 - t + t^2$, and its evaluation at $\omega = exp(\frac{2\Pi i}{6})$ equals $\sqrt{3}i$. As $V_{K_+}(t) = 1$ we can calculate the hypothetical value of $V_{K_0}(\omega)$ and this turns out to be $\sqrt{3} - 2i$. This is a contradiction because $V(\omega)$ has to be of the form $\pm i(\sqrt{3})^k$ (see [**LM**], Theorem 3).

3. A criterion for signed unknotting number

The argument used in Section 1 to show that 6_1 cannot be transformed into the trivial knot by one $-+$ operation may be also used in a more general context. We have the following Theorem.

THEOREM 3.1. *Let K be a knot and $V_K = V$ its Jones polynomial. Let $V(\omega) = (-1)^s \cdot (i\sqrt{3})^d$. If $u(K) = d$, then $u_{+-} \equiv (-1)^s \pmod 2$.*

PROOF. We know that $V_{L_+}(\omega)$, $V_{L_-}(\omega)$ and $V_{L_0}(\omega)$ all have to be of the form $\pm i(\sqrt{3})^k$. It is obvious that the exponents (in the expression $\pm i(\sqrt{3})^k$) cannot differ by more than 1. Assume that $K = K_d, \ldots, K_1 =$ *unknot* is an unkotting sequence for K. Then the exponents of K_i and K_{i-1} must differ by 1 or -1. It may be checked by direct calculation that if (K_i, K_{i-1}) is a pair of (L_-, L_+) type, then the sign in the expression $\pm i(\sqrt{3})^k$ is not changed, and in the opposite case $((L_+, L_-)$

type) the sign *is* changed. Therefore it is the *mod* 2 number of $+-$ operations that determines the sign in $V(\frac{2\Pi i}{6}) = (-1)^s \cdot (i\sqrt{3})^d$.

4. Examples

The information contained in Theorem 3.1 is often helpful in determining u_{+-} and u_{-+}, especially when combined with the signature. We have already considered one special case of this — the determination of u_{+-} and u_{-+} for 6_1.

Now we consider several examples illustrating how the criterion can be used.

EXAMPLE 4.1. Let $K = 8_{18}$. The Jones polynomial of K is $t^{-4} - 4t^{-3} + 6t^{-2} - 7t^{-1} + 9 - 7t + 6t^2 - 4t^3 + t^4$ and $V(\omega) = 3$, which means that the d parameter of Theorem 2.1 equals 2. It is known that $u(K) = 2$, therefore by the Theorem $u_{+-} = u_{-+} = 1$.

EXAMPLE 4.2. Let $K = 9_{48}$. The Jones polynomial is $V_K(t) = -2t^{-6} + 3t^{-5} - 4t^{-4} + 6t^{-3} - 4t^{-2} + 4t^{-1} - 3 + t$ and $V(\omega) = -3$. It is known that $u(K) = 2$, so $d = u$. It follows that in every sequence of length two of operations unknotting K both $+-$ and $-+$ change must appear an even number of times. However, $\sigma_K = 2$. This means that a $-+$ change *must* appear at least once in every sequence unknotting K. It follows that $u_{-+} = 2$ and $u_{+-} = 0$.

It is obvious from the definition that

$$u \geq u_{+-} + u_{-+}.$$

In some cases this provides sufficient information to determine the unknotting number u.

EXAMPLE 4.3. Let $K = 7_4$. The unknotting number of this knot is equal to 2. This was established in [**L**] in 1985. It turns out that $V_K(\omega) = i\sqrt{3}$. We show how Theorem 3.1, the above inequality and the signature can be used together to show $u = 2$. Suppose $u = 1$. Then Theorem 3.1 implies that $u_{+-} = 1$. But $\sigma_K = 2$, which implies that $u_{-+} \geq 1$. It follows that $u \geq u_{+-} + u_{-+} \geq 2$ — a contradiction.

REMARK. In the above example the assumption that $u_K = 1$ lead us to the conclusion that to unknot K we have to perform both $+-$ and $-+$ change which is a contradiction. This proves that in fact $u_K = 2$. It should be stressed that the conclusion that changes of both types have to be made is no longer valid. In fact one can unknot 7_4 with two $-+$ changes.

EXAMPLE 4.4. Let $K = 7_7$. It is known that $u = 1$ and it may be checked directly that $u_{--} = u_{-+} = 1$. On the other hand $V(\omega) = -i\sqrt{3}$, which proves that the number of $-+$ operations in any unknotting sequence of length one is odd, and therefore equal to one. This implies that $u_{++} > 1$. But one *can* unknot K using a sequence of two $+-$ operations, hence $u_{++} = 2$. Of course the signature of K equals 0.

EXAMPLE 4.5. Let $K = 9_{35}$. As of this writing u_K is not known (at least to the author). It is known that $u \leq 3$ but $u = 2$ is still not excluded. It is possible to unknot K using a sequence of three $-+$ operations, so that $u_{--} \leq 3$. On the other hand $V(\omega) = 3$ which proves that if u is equal to 2, then in any unknotting sequence of length 2 both $-+$ and $+-$ operation *must* be used. It follows that in fact $u_{--} = 3$.

EXAMPLE 4.6.. Let $K = 9_{46}$. This time we know that $u_K = 2$. Like in the previous example $V(\omega) = 3$ and we can use the same arguments to prove that $u_{-+} = 3$.

EXAMPLE 4.7.. Let $K = 9_{37}$. Again, we know that $u_K = 2$. In this case $V(\omega) = -3$, which means that in any sequence of unknotting operations of length 2 an even number of both $-+$ and $+-$ operations must be used. It turns out that it is possible to unknot K using two $-+$ operations and it is also possible with two $+-$ operations. At the same time it is *not* possible to unknot K with a sequence of one $-+$ operation followed by one $+-$ operation.

EXAMPLE 4.8.. Let $K = 10_{67}$. In this case $\sigma = 2$ and $V(\omega) = i\sqrt{3}$. If the unknotting number were 1, then $\sigma = 2$ would imply $u_{-+} = 1$ and $V(\omega) = i\sqrt{3}$ would imply $u_{+-} = 1$. This is a contradiction, so in fact $u_K = 2$ (it is easily checked that $u \leq 2$. This is one of undecided cases in the table in [**K**].

5. Tables

In this Section we give a table containing some very general information about signed unknotting numbers. We use notation as in [**CL**]. We consider a function $f_K(n)$ (K being a knot) defined in the following manner: Suppose we are allowed to perform n $+-$ operations on the considered knot K. Then $f_K(n)$ is the minimum number of $-+$ operations still necessary to transform K into the trivial knot. In the table below quatation marks indicate cases that are undecided and the degree of uncertainty. For example 3? means "probably 3, but 2 has not been excluded" and 3?? means "probably 3, but 1 has not been excluded." The function is non–increasing, so identical values for 2 and ∞ give complete information for all arguments ≥ 2. The last two columns show the signature and the parameter $v = V(\omega)$. $V(\omega)$ is indicated by showing the sign and the exponent. So an entry like " $-$ 1" means $V(\omega) = -\sqrt{3}$.

Table

knot	$f(0)$	$f(1)$	$f(2)$	$f(\infty)$	σ	v	knot	$f(0)$	$f(1)$	$f(2)$	$f(\infty)$	σ	v
3_1	1	1	1	1	2	$-$ 1	7_1	3	3	3	3	6	$-$ 0
4_1	1	0	0	0	0	$-$ 0	7_2	1	1	1	1	2	+ 0
5_1	2	2	2	2	4	$-$ 0	7_3	2	2	2	2	4	$-$ 0
5_2	1	1	1	1	2	$-$ 0	7_4	2	2?	2?	2?	2	+ 1
6_1	2	0	0	0	0	+ 1	7_5	2	2	2	2	4	$-$ 0
6_2	1	1	1	1	2	+ 0	7_6	1	1	1	1	2	$-$ 0
6_3	1	0	0	0	0	+ 0	7_7	2	0	0	0	0	+ 1

knot	$f(0)$	$f(1)$	$f(2)$	$f(\infty)$	σ	v
8_1	3?	0	0	0	0	+ 0
8_2	2	2	2	2	4	− 0
8_3	2	2?	0	0	0	− 0
8_4	2	1	1	1	2	− 0
8_5	2	2	2	2	4	− 1
8_6	2	1	1	1	2	+ 0
8_7	1	1	1	1	2	+ 0
8_8	2	1	1?	1?	0	+ 0
8_9	1	0	0	0	0	+ 0
8_{10}	2?	1	1	1	2	− 1
8_{11}	1	1	1	1	2	− 1
8_{12}	2	1	0	0	0	− 0
8_{13}	1	1	1	1	2	− 0
8_{14}	1	1	1	1	2	− 0
8_{15}	2	2	2	2	4	+ 1
8_{16}	2?	1	1	1	2	+ 0
8_{17}	1	0	0	0	0	+ 0
8_{18}	∞?	1	1?	1?	0	+ 2
8_{19}	3	3	3	3	6	+ 1
8_{20}	1	1	1?	1?	0	− 1
8_{21}	1	1	1	1	2	− 1
9_1	4	4	4	4	8	+ 1
9_2	1	1	1	1	2	− 1
9_3	3	3	3	3	6	− 0
9_4	2	2	2	2	4	− 1
9_5	2	2?	2?	2?	2	+ 0
9_6	3	3	3	3	6	+ 1
9_7	2	2	2	2	4	− 0
9_8	2	1	1	1	2	− 0
9_9	3	3	3	3	6	− 0
9_{10}	3?	3?	3?	3?	4	− 1
9_{11}	2	2	2	2	4	+ 1
9_{12}	1	1	1	1	2	+ 0
9_{13}	3?	3?	3?	3?	4	+ 0

knot	$f(0)$	$f(1)$	$f(2)$	$f(\infty)$	σ	v
9_{14}	3?	0	0	0	0	+ 0
9_{15}	2	1	1	1	2	+ 1
9_{16}	3	3	3	3	6	+ 1
9_{17}	2	1	1	1	2	+ 1
9_{18}	2	2	2	2	4	− 0
9_{19}	1	1?	0	0	0	− 0
9_{20}	2	2	2	2	4	− 0
9_{21}	1	1	1	1	2	− 0
9_{22}	1	1	1	1	2	− 0
9_{23}	2	2	2	2	4	+ 1
9_{24}	1	1	1?	1?	0	− 1
9_{25}	2	1	1	1	2	+ 0
9_{26}	1	1	1	1	2	+ 0
9_{27}	2?	0	0	0	0	+ 0
9_{28}	1	1	1	1	2	− 1
9_{29}	2?	1	1	1	2	− 1
9_{30}	2?	0	0	0	0	− 0
9_{31}	2	1	1	1	2	− 0
9_{32}	2?	1	1	1	2	+ 0
9_{33}	2?	1	1	1	0	+ 0
9_{34}	2	0	0	0	0	+ 1
9_{35}	3	3?	3?	3?	2	+ 2
9_{20}	2	2	2	2	4	− 0
9_{21}	1	1	1	1	2	− 0
9_{22}	1	1	1	1	2	− 0
9_{23}	2	2	2	2	4	+ 1
9_{24}	1	1	1?	1?	0	− 1
9_{25}	2	1	1	1	2	+ 0
9_{26}	1	1	1	1	2	+ 0
9_{27}	2?	0	0	0	0	+ 0
9_{28}	1	1	1	1	2	− 1
9_{29}	2?	1	1	1	2	− 1
9_{30}	2?	0	0	0	0	− 0
9_{31}	2	1	1	1	2	− 0

knot	$f(0)$	$f(1)$	$f(2)$	$f(\infty)$	σ	v
9_{32}	2?	1	1	1	2	$+0$
9_{33}	2?	1	1	1	0	$+0$
9_{34}	2	0	0	0	0	$+1$
9_{35}	3	3?	3?	3?	2	$+2$
9_{36}	2	2	2	2	4	$+0$
9_{37}	2	2	0	0	0	-2
9_{38}	3?	3?	3?	3?	4	$+1$
9_{39}	1	1	1	1	2	-0
9_{40}	2	1	1	1	2	$+1$

knot	$f(0)$	$f(1)$	$f(2)$	$f(\infty)$	σ	v
9_{41}	3?	1	1?	1?	0	$+0$
9_{42}	1	1	1	1	2	-0
9_{43}	2	2	2	2	4	$+0$
9_{44}	1	0	0	0	0	-0
9_{45}	1	1	1	1	2	$+0$
9_{46}	3	1	1?	1?	0	$+2$
9_{47}	3	1	1	1	2	$+2$
9_{48}	2	2	2?	2?	2	-2
9_{49}	3?	3?	3?	3?	4	$+0$

References

[K] A. Kawauchi, *A survey of knot theory*, Birkhauser Verlag, Basel, 1981.

[L] W.B.R. Lickorish, *The unknotting number of a classical knot*, Contemp. Math. **44** (1985), 117–121.

[LM] W.B.R. Lickorish and K.C. Millett, *Some evaluations of link polynomials*, Comment. Math. Helv. **61** (1986), 349–359.

[CL] T.D. Cochran and W.B.R. Lickorish, *Unknotting information from 4-manifolds polynomials*, Trans. Amer. Math. Soc. **297** (1986), 25–142.

Department of Mathematics, Warsaw University ul. Banacha 2, 02-097 Warszawa, POLAND

E-mail address: traczyk@mimuw.edu.pl

Contemporary Mathematics
Volume **233**, 1999

On Some Extensions of Artin's Braid Relations

H. Nencka

ABSTRACT. A class of relations among the generators of the braid group B_n is proposed. This class, which generalize the Artin relations and solve partially the Xu's problem, provides new relations for quantum groups.

1. Introduction - Definitions and Notions.

This note is devoted to some extensions of the relations introduced by Artin [**Ar1 - 4**] for braid groups. The group B_n which is the group on n strings $\gamma_1, \dots, \gamma_n$ under the product operation, can be described as follows.

DEFINITION 1.1. *A geometric n-braid β is a pattern on n interwoven oriented strings γ in a slab of 3-space* $^2 \times [0,1] \in {}^3$, *connecting n distinct initial points $(x_1, x_2, \dots, x_n)$ in the plane* $P_0 = {}^2 \times \{0\}$ *to the corresponding end points $(x'_1, x'_2, \dots, x'_n)$ in the plane* $P_1 = {}^2 \times \{1\}$, *intersecting each intermediate plane* $P_t = {}^2 \times \{t\}, t \in [0,1]$ *in exactly n distinct points.*

We may visualize an n-braid β via 2D-pattern distinguishing overpass ("left" interwoven) from underpass ("right" interwoven) of string γ.

Reidemeister [**Re1,2**] has shown that two braids are equivalent if one can be continously transformed into the other without breaking any strings. More precisely by the following isotopic transformations (Fig. 1).

The "left" interwoven of the string $x_i x'_{i+1}$ over the string $x_{i+1} x'_i$ is denoted by σ_i, and the "right" interwoven is denoted by σ_i^{-1}. Then a braid is completely determined as a power product of the σ's.

The Reidemeister equivalence class of n-braids can be equipped with a product operation consisting in juxtaposition, erasure of the middle plane, and rescaling.

LEMMA 1.1. *Under the product operation B_n is a group, calls the braid group.*

Artin [**Ar1**] has shown that the $\sigma_i \sigma_j$ braid is equivalent to $\sigma_j \sigma_i$ braid if $|i-j| > 1$, and $\sigma_i \sigma_j \sigma_i$ braid to $\sigma_j \sigma_i \sigma_j$ one provided that $|i-j| = 1$. The braid group B_n formed by n strings with σ as a generator is widely considered in the literature (Artin [**Ar1**], Bohnenblust [**Bo**], Chow [**Ch**], Coxeter [**Co**]).

1991 *Mathematics Subject Classification.* 57M25,20F36,20F38,16A24,.
Key words and phrases. Artin relations, braid groups, links.
This work is partially supported by the pluriannual project PRAXIS N°XXI, Portugal.

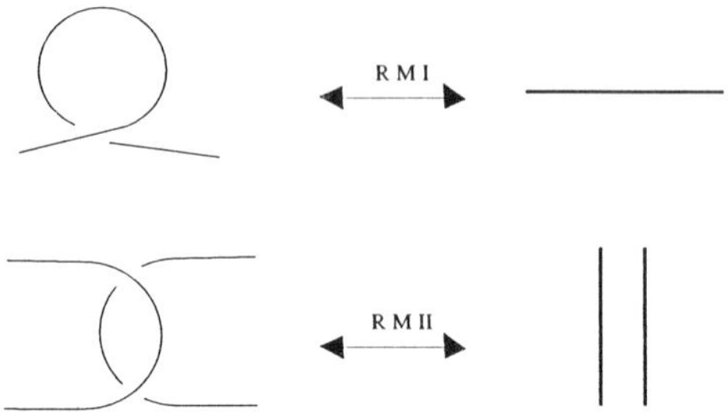

FIGURE 1. I and II Reidemeister Moves

LEMMA 1.2. *The following relations:*

$$\sigma_i\sigma_j = \sigma_j\sigma_i\,, \qquad \sigma_i\sigma_{i+1}\sigma_i = \sigma_{i+1}\sigma_i\sigma_{i+1}$$

define a braid group B_n, where σ_i, $i = 1,\ldots,n$ are the generators of an infinite group.

Let us notice that we have used the same symbol σ for an overcrossing and the associated generator.

Closed n-braids are obtained from (open) n-braids by identification of the end points with the initial points in such a manner that $x_i' = x_i$. The study of the closed braid is equivalent to this one of knots or links. Indeed Alexander's theorem [**Al**] shows that every knot or link may be represented as a closed braid, for some n. The first proof of the theorem was given by Alexander [**Al**]. However, there exist another three constructive proofs of it given by Yamada [**Ya**], Vogel [**Vo**] and Morton [**Mo**].

A basic tool to extract topological properties of knots and links [**Ma**] is given by Markov [**Ma**]. The proof is given by Birman [**Bi1**] and Traczyk [**Tr**].

DEFINITION 1.2. *Let $\sim$ be the equivalence relation on B generated by the following transformation:*

$$\begin{array}{lrcll} (I) & (bb',n) & \sim & (b'b,n), & \forall\, b,b' \in B_n, \\ (II) & (b,n) & \sim & (\sigma b^{\pm}, n+1), & \forall\, b \in B_n. \end{array}$$

The conjugacy MMI *and the stabilization* MMII *(see figure 2) are called I and II Markov moves, respectively. The equivalence classes by the relation $\sim$ are called Markov classes.*

DEFINITION 1.3. *Two elements of B_n are Markov equivalent if they can be deformed into each other by a finite sequence of Markov moves* MMI *and* MMII.

DEFINITION 1.4. *The equivalent braids corresponding to the same links are mutually transformable by applications of Markov moves. The* MMI *and* MMII *transformations respectively, are determined as follows:*

MMI - $C_\beta : B_n \to B_n\,,\ \alpha \mapsto C_\beta(\alpha) = \beta\alpha\beta^{-1}\,,\ \beta \in B_n$

MMII - $i_n : B_n \to B_{n+1}\,,\ \alpha \mapsto i_n(\alpha) = \alpha\sigma_n^{\pm}\,.$

The equivalence relation on B_∞ where $B_\infty = \bigsqcup_{n=1}^{\infty} B_n$ generated by MMI and MMII determines equivalence classes for all $\beta \in B_\infty$ for a given link. The

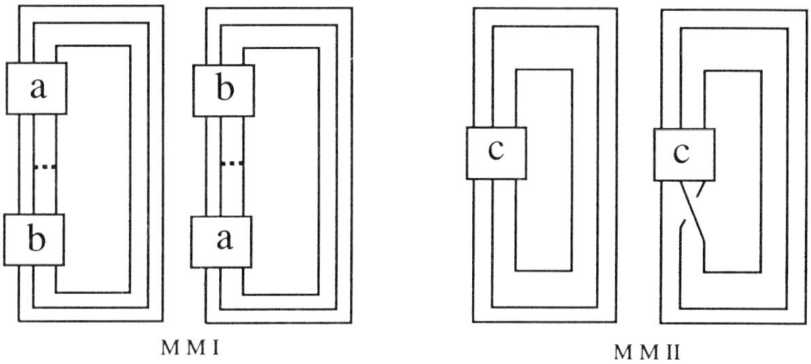

FIGURE 2. Markov Moves

classification of a link types is equivalent to an algebraic problem of classification of Markov classes $[\beta]$ in B_∞ .

The Markov theorem can be stated in terms of Markov equivalence in the following way:

THEOREM 1.1 (Markov Theorem). *Let B_∞ be the disjoint union of the braid groups: $B_\infty = \bigsqcup_{i=1}^{\infty} B_i$ and let $\beta, \beta^* \in B_\infty$ are Markov equivalent in the sense that the corresponding closed braids represent the same link type.*

See for example Birman [**Bi2**].

A Markov Theorem without stabilization for the unlink was given by Birman [**Bi3**]

THEOREM 1.2 (Birman). *Let L be a closed braid representative of the r-component unlink K_{0r} and U the closure of trivial braid of r-strings. Then there is a finite sequence of closed braids*

$$L = L_1 \to L_2 \to \dots \to L_q = U ,$$

with the same axis A such that each L_{i+1} is obtained from L_i by one of
a) deletion of a trivial loop,
b) a complexity reducing exchange move
c) a complexity reducing isotopy in the complement of A.

The proof is given in Birman [**Bi3**].

Artin, as one of the first to have realized that the braid theory could be used as a tool to study knots and links. After the Alexander's and Markov's discovery, Garside [**Ga**] has presented a solution to the conjugacy problem in braid groups. Let S_n be the monoid of positive words in B_n. Than the inclusion $S_n \hookrightarrow B_n$ becomes injective. If, the word $\Delta = (\sigma_1\sigma_2 \dots \sigma_{n-1})(\sigma_1\sigma_2 \dots \sigma_{n-2}) \dots (\sigma_1\sigma_2)\sigma_1 \in B_n$, and taking account its properties, namely that

$$\begin{aligned} i) \quad & \sigma_i^{-1} = \Delta^{-1}\Pi , \\ ii) \quad & \sigma_i \Delta = \Delta\, \sigma_{n-i} , \\ iii) \quad & \Delta \text{ can begin and end with any } \sigma_i . \end{aligned}$$

We see that every word $\beta \in B_n$ can be written as $\beta = \Delta^k \Pi$, where Π is some positive word. It has been shown [**Ga**] as well that any conjugate of a positive word can be obtained by a sequence of conjugations by permutations braid.

Elrifai and Morton [**El-Mo**] have developped an algorithm to obtain a unique normal form from a given word . Xu [**Xu**] has shown a new solution to the conjugacy problem in a case of the group B_3. The solution bases upon new set of generators of B_3 and shortest words in those generators

Using generators $a_1 = \sigma_1, a_2 = \sigma_2, a_3 = \sigma_1^{-1}\sigma_2\sigma_1$, the braid group B_3 has a new presentation

$$B_3 = \langle a_1, a_2, a_3 | \alpha = a_3 a_2 = a_2 a_3 = a_1 a_3 \rangle .$$

The word α satisfies the following properties:

$$a_i^{\pm}\alpha = \alpha a_{i+1}^{\pm} ,$$

$$a_i^{\pm}\alpha^{-1} = \alpha^{-1} a_{i-1}^{\pm} .$$

If β is:

i) a positive word , there is a unique normal form $\beta = \alpha^k P$, where $k \geq 0$ and P is a nondecreasing positive word,

ii) a negative word, there is a unique normal form $\beta = N\alpha^k$, where $k \leq 0$ and N^{-1} is a nondecreasing positive word,

iii) a mixed word , a normal form is $\beta = NP$ where P and N^{-1} are nondecreasing positive words.

In this way the conjugacy problem and the shortest word problem in conjugacy classes for positive and negative words can be solved. In eighties, Thurston [**Th**] has given a solution to conjugacy problem in mapping class groups. Nextly, Bestvina and Handel [**Bes-Ha**] and Los [**Lo**] have presented solutions of the conjugacy problem for B_n. In 1983 Bennequin [**Be**] proved that if a knot was everywhere transverse to the kernel of the 1-form $dz + r^2 d\theta$,(where (r, θ, z) are cylindrical coordinates in ^{3}),that is to a standard contact structure , then it could be deformed through transverse knots to a closed braid relative to the z-axis. Certain incompressible spanning surfaces bounded by these closed braids and using the foliations on those surfaces found invariants of knots transverse to the contact structure under transverse isotopy

Birman and Menasco [**Bi-Me1-6**] using the combinatorics of the same foliations on spanning surfaces as these which were studied by Bennequin as a tool, initiated a study of the closed braid representatives of knot or link. They have extended the applications to the study of surfaces other than spanning surfaces for links. It has been shown by Birman and Finkelstein [**Bi-Fi**] that incompressible spanning surfaces for links and closed incompressible surfaces in link complements admit a singular foliation.

2. Basic Lemmas.

LEMMA 2.1. *Given B_n, the word $\alpha_1 = \sigma_1^{-1}\sigma_2 ... \sigma_{n-2}\sigma_{n-1}\sigma_{n-2} ... \sigma_2\sigma_1$, satisfies the following relations:*

$$\alpha_1 = \sigma_{n-1}\sigma_{n-2} ... \sigma_2 \sigma_1 \sigma_2^{-1} \sigma_3 \sigma_4 ... \sigma_{n-3} \sigma_{n-2} \sigma_{n-1} .$$

REMARK 2.1. Let $n = 3$,then $\alpha_1 = \sigma_1^{-1}\sigma_2\sigma_1$ which is identical to Xu's a_3 . The following identity holds

$$\alpha_1 = \sigma_2\sigma_1\sigma_2^{-1}.$$

PROOF.

$$\alpha_1\sigma_2 = \sigma_1^{-1}\sigma_2\sigma_1\sigma_2,$$

applying the Artin's relation, we have

$$\alpha_1\sigma_2 = \sigma_1^{-1}\sigma_1\sigma_2\sigma_1.$$

Since

$$\alpha_1 = \alpha_1\sigma_2\sigma_2^{-1},$$

then

$$\alpha_1 = \sigma_1^{-1}\sigma_1\sigma_2\sigma_1\sigma_2^{-1}.$$

LEMMA 2.2. *Given B_n, the word α_1 verifies as well the following auxiliary relations:*

$$\begin{aligned}
\alpha_1 &= \sigma_1^{-1}\sigma_2\ldots\sigma_{n-3}\sigma_{n-1}\sigma_{n-2}\ldots\sigma_2\sigma_1\sigma_{n-1}\\
&= \sigma_1^{-1}\sigma_2\ldots\sigma_{n-4}\sigma_{n-1}\sigma_{n-2}\ldots\sigma_2\sigma_1\sigma_{n-2}\sigma_{n-1}\\
&= \sigma_1^{-1}\sigma_2\ldots\sigma_{n-5}\sigma_{n-1}\sigma_{n-2}\ldots\sigma_2\sigma_1\sigma_{n-3}\sigma_{n-2}\sigma_{n-1}\\
&= \sigma_1^{-1}\sigma_2\ldots\sigma_{n-6}\sigma_{n-1}\sigma_{n-2}\ldots\sigma_2\sigma_1\sigma_{n-4}\sigma_{n-3}\sigma_{n-2}\sigma_{n-1}\\
&= \ldots\\
&= \sigma_1^{-1}\sigma_2\ldots\sigma_{n-i}\sigma_{n-1}\sigma_{n-2}\ldots\sigma_2\sigma_1\sigma_{n-i-2}\sigma_{n-i-1}\sigma_{n-i}\sigma_{n+i}\ldots\sigma_{n-1}\\
&= \ldots\\
&= \sigma_1^{-1}\sigma_{n-2}\sigma_{n-1}\sigma_{n-2}\ldots\sigma_2\sigma_1\sigma_4\ldots\sigma_{n-3}\sigma_{n-2}\sigma_{n-1}\\
&= \sigma_1^{-1}\sigma_{n-1}\sigma_{n-2}\ldots\sigma_2\sigma_1\sigma_3\ldots\sigma_{n-3}\sigma_{n-2}\sigma_{n-1}\\
&= \sigma_{n-1}\sigma_{n-2}\ldots\sigma_2\sigma_1\sigma_2^{-1}\sigma_3\sigma_4\ldots\sigma_{n-3}\sigma_{n-2}\sigma_{n-1}.
\end{aligned}$$

LEMMA 2.3. *Given B_n let $\alpha_2 = \sigma_1\sigma_2\ldots\sigma_{n-2}\sigma_{n-1}\sigma_{n-2}\ldots\sigma_2\sigma_1$, then the following relations are satisfied :*

$$\begin{aligned}
\alpha_2 &= \sigma_1\sigma_2\ldots\sigma_{n-3}\sigma_{n-1}\sigma_{n-2}\ldots\sigma_2\sigma_1\sigma_{n-1}\\
&= \sigma_1\sigma_2\ldots\sigma_{n-4}\sigma_{n-1}\sigma_{n-2}\ldots\sigma_2\sigma_1\sigma_{n-2}\sigma_{n-1}\\
&= \sigma_1\sigma_2\ldots\sigma_{n-5}\sigma_{n-1}\sigma_{n-2}\ldots\sigma_2\sigma_1\sigma_{n-3}\sigma_{n-2}\sigma_{n-1}\\
&= \sigma_1\sigma_2\ldots\sigma_{n-6}\sigma_{n-1}\sigma_{n-2}\ldots\sigma_2\sigma_1\sigma_{n-4}\sigma_{n-3}\sigma_{n-2}\sigma_{n-1}\\
&= \ldots\\
&= \sigma_1\sigma_2\ldots\sigma_{n-i}\sigma_{n-1}\sigma_{n-2}\ldots\sigma_2\sigma_1\sigma_{n-i-2}\sigma_{n-i-1}\sigma_{n-i}\sigma_{n+i}\ldots\sigma_{n-1}\\
&= \ldots\\
&= \sigma_1\sigma_{n-2}\sigma_{n-1}\sigma_{n-2}\ldots\sigma_2\sigma_1\sigma_4\ldots\sigma_{n-3}\sigma_{n-2}\sigma_{n-1}\\
&= \sigma_1\sigma_{n-1}\sigma_{n-2}\ldots\sigma_2\sigma_1\sigma_3\ldots\sigma_{n-3}\sigma_{n-2}\sigma_{n-1}\\
&= \sigma_{n-1}\sigma_{n-2}\ldots\sigma_2\sigma_1\sigma_2\sigma_3\sigma_4\ldots\sigma_{n-3}\sigma_{n-2}\sigma_{n-1}.
\end{aligned}$$

REMARK 2.2. Let $n = 3$ the Artin's relations are deduced from the above given equations, namely:

$$\begin{aligned}\alpha_2 &= \sigma_1 \sigma_2 ... \sigma_{n-2} \sigma_{n-1} \sigma_{n-2} ... \sigma_2 \sigma_1\\ &= \sigma_1 \sigma_2 \sigma_1\\ &= \sigma_{n-1} \sigma_{n-2} ... \sigma_2 \sigma_1 \sigma_2 \sigma_3 \sigma_4 ... \sigma_{n-3} \sigma_{n-2} \sigma_{n-1}\\ &= \sigma_2 \sigma_1 \sigma_2 .\end{aligned}$$

LEMMA 2.4. *Given B_n, the word $\alpha_3 = \sigma_1^{\delta_1} \sigma_2^{\delta_2} ... \sigma_{n-2}^{\delta_{n-2}} \Theta$, where $\delta_i = \pm 1$ ($i = 1, 2, ... n-2$) and $\Theta_{n-1} = \sigma_{n-1} \sigma_{n-2} ... \sigma_2 \sigma_1$ satisfies the following relations:*

$$\begin{aligned}\alpha_3 &= \sigma_1^{\delta_1} \sigma_2^{\delta_2} ... \sigma_{n-3}^{\delta_{n-3}} \Theta_{n-1} \sigma_{n-1}^{\delta n-2}\\ &= \sigma_1^{\delta_1} \sigma_2^{\delta_2} ... \sigma_{n-4}^{\delta_{n-4}} \Theta_{n-1} \sigma_{n-2}^{\delta_{n-3}} \sigma_{n-1}^{\delta_{n-2}}\\ &= \sigma_1^{\delta_1} \sigma_2^{\delta_2} ... \sigma_{n-5}^{\delta_{n-5}} \Theta_{n-1} \sigma_{n-3}^{\delta_{n-4}} \sigma_{n-2}^{\delta_{n-3}} \sigma_{n-1}^{\delta_{n2}}\\ &= \sigma_1^{\delta_1} \sigma_2^{\delta_2} ... \sigma_{n-6}^{\delta_{n-6}} \Theta_{n-1} \sigma_{n-4}^{\delta_{n-5}} \sigma_{n-3}^{\delta_{n-4}} \sigma_{n-2}^{\delta_{n-3}} \sigma_{n-1}^{\delta_{n-2}}\\ &= ...\\ &= \sigma_1^{\delta_1} \sigma_2^{\delta_2} ... \sigma_{n-i}^{\delta_{n-i}} \Theta_{n-1} \sigma_{n-i-2}^{\delta_{n-i-3}} \sigma_{n-i-1}^{\delta_{n-i-2}} \sigma_{n-i}^{\delta_{n-i-1}} \sigma_{n+i}^{\delta_{n+i-1}} ... \sigma_{n-1}^{\delta n-2}\\ &= ...\\ &= \sigma_1^{\delta_1} \sigma_2^{\delta_{n-2}} \Theta \sigma_4^{\delta_3} ... \sigma_{n-3}^{\delta_{n-4}} \sigma_{n-2}^{\delta_{n-3}} \sigma_{n-1}^{\delta_{n-2}}\\ &= \sigma_1^{\delta_1} \Theta_{n-1} \sigma_3^{\delta_2} ... \sigma_{n-3}^{\delta_{n-4}} \sigma_{n-2}^{\delta_{n-3}} \sigma_{n-1}^{\delta_{n-2}}\\ &= \Theta_{n-1} \sigma_2^{\delta_1} \sigma_3^{\delta_2} \sigma_4^{\delta_3} ... \sigma_{n-3}^{\delta_{n-4}} \sigma_{n-2}^{\delta_{n-3}} \sigma_{n-1}^{\delta_{n-2}} .\end{aligned}$$

LEMMA 2.5. *Given B_n, for Θ_{n-1} defined in Lemma 2.4 we have the shift relations:*

$$\begin{aligned}\sigma_i^{\pm} \Theta_{n-1} &= \Theta_{n-1} \sigma_{i+1}^{\pm} ,\\ \sigma_i^{\pm} \sigma_{i+1}^{\pm} \Theta_{n-1} &= \Theta_{n-1} \sigma_{i+1}^{\pm} \sigma_{i+2}^{\pm} ,\end{aligned}$$

where $i = 1, ..., n-2$, and

$$\sigma_i^{\pm} \sigma_{i+1}^{\pm} \sigma_{i+p}^{\pm} \Theta_{n-1} = \Theta_{n-1} \sigma_{i+1}^{\pm} \sigma_{i+2}^{\pm} \sigma_{i+p+1}^{\pm} ,$$

where $i + p \leq n - 2$.

THEOREM 2.1. *Given B_n , $\Phi_j = \sigma_{n-1} \sigma_{n-2} ... \sigma_j^{-1} ... \sigma_1$ and $j = 1, ... n-2$ we have the following :*

$$\begin{aligned}\sigma_i^{\pm} \sigma_{i+1}^{\pm} \sigma_{i+p}^{\pm} \Phi_j &= \Phi_{j+1} \sigma_{i+1}^{\pm} \sigma_{i+p}^{\pm} \sigma_{i+p+1}^{\pm} ,\\ &= \sigma_i^{\pm} \sigma_{i+1}^{\pm} \sigma_{i+p}^{\pm} \Phi_{n-1}\\ &= \Phi_1 \sigma_{i+1}^{\pm} \sigma_{i+p}^{\pm} \sigma_{i+p+1}^{\pm} ,\end{aligned}$$

PROOF. Let $n = 5$, and $\Theta = \sigma_4\sigma_3\sigma_2\sigma_1$ considering $j = n-2$, $\Phi_{n-2} = \sigma_4\sigma_3^{-1}\sigma_2\sigma_1$ and we have:

$$\begin{aligned}\sigma_1\sigma_2\sigma_3\Phi_2 &= \sigma_1\sigma_2\sigma_3\sigma_4\sigma_3^{-1}\sigma_2\sigma_1\\ &= \sigma_1\sigma_2(\sigma_3\sigma_4\sigma_3^{-1})\sigma_2\sigma_1\\ &= \sigma_1\sigma_2(\sigma_4^{-1}\sigma_3\sigma_4)\sigma_2\sigma_1\\ &= \sigma_4^{-1}\sigma_1(\sigma_2\sigma_3\sigma_2)\sigma_1\sigma_4\\ &= \sigma_4^{-1}\sigma_1(\sigma_3\sigma_2\sigma_3)\sigma_1\sigma_4\\ &= \sigma_4^{-1}\sigma_3(\sigma_1\sigma_2\sigma_1)\sigma_3\sigma_4\\ &= \sigma_4^{-1}\sigma_3\sigma_2\sigma_1\sigma_2\sigma_3\sigma_4\\ &= \Phi_4\sigma_2\sigma_3\sigma_4\,,\end{aligned}$$

so for an arbitrary n and $j = n-2$ we have

$$\begin{aligned}&\sigma_1\sigma_2\ldots\sigma_(n-2)\Phi_{n-2}\\ &\quad= \sigma_1\sigma_2\ldots\sigma_(n-2)\sigma_(n-1)\sigma_(n-2)^{-1}\sigma_(n-3)\ldots\sigma_3\sigma_2\sigma_1\\ &\quad= \sigma_1\sigma_2\ldots[\sigma_(n-2)\sigma_(n-1)\sigma_(n-2)^{-1}]\sigma_(n-3)\ldots\sigma_3\sigma_2\sigma_1\\ &\quad= \sigma_1\sigma_2\ldots\sigma_(n-1)^{-1}\sigma_(n-2)\sigma_(n-1)\sigma_(n-3)\ldots\sigma_3\sigma_2\sigma_1\\ &\quad= \sigma_(n-1)^{-1}\sigma_1\sigma_2\ldots\sigma_(n-3)\sigma_(n-2)\sigma_(n-3)\ldots\sigma_3\sigma_2\sigma_1\sigma_(n-1)\\ &\quad= \ldots\\ &\quad= \sigma_(n-1)^{-1}\sigma_(n-2)\sigma_(n-3)\ldots\sigma_3\sigma_1\sigma_2\sigma_1\sigma_3\ldots\sigma_(n-3)\sigma_(n-2)\sigma_(n-1)\\ &\quad= \sigma_(n-1)^{-1}\sigma_(n-2)\sigma_(n-3)\ldots\sigma_3\sigma_2\sigma_1\sigma_2\sigma_3\ldots\sigma_(n-3)\sigma_(n-2)\sigma_(n-1)\\ &\quad= \Phi_{n-1}\,.\end{aligned}$$

Let $j = p$ where $p \leq n-2$, then

$$\begin{aligned}&\sigma_1\sigma_2\ldots\sigma_{n-2}\Phi_p\\ &\quad= \sigma_1\sigma_2\ldots\sigma_{n-2}\sigma_{n-1}\sigma_{n-2}\ldots\sigma_{p+1}\sigma_p^{-1}\sigma_(p-1)\ldots\sigma_3\sigma_2\sigma_1\\ &\quad= \sigma_1\sigma_2\ldots\sigma_{n-1}\sigma_{n-2}\sigma_(n-1)\ldots\sigma_(p+1)\sigma_p^{-1}\sigma_(p-1)\ldots\sigma_3\sigma_2\sigma_1\\ &\quad= \sigma_(n-1)\sigma_1\sigma_2\ldots\sigma_(n-2)\ldots\sigma_p\sigma_(p+1)\sigma_p^{-1}\sigma_{p-1}\ldots\sigma_3\sigma_2\sigma_1\sigma_{n-1}\\ &\quad= \ldots\\ &\quad= \sigma_{n-1}\sigma_{n-2}\ldots\sigma_{p+2}\ldots\sigma_1\sigma_2\sigma_3\ldots\sigma_{p-1}\sigma_p\sigma_{p+1}\sigma_p^{-1}\sigma_{p-1}\ldots\sigma_3\sigma_2\sigma_1\ldots\sigma_{n-2}\sigma_{n-1}\\ &\quad= \sigma_{n-1}\sigma_{n-2}\ldots\sigma_{p+2}\ldots\sigma_1\sigma_2\sigma_3\ldots\sigma_{p-1}\sigma_{p+1}^{-1}\sigma_p\sigma_{p+1}\sigma_{p-1}\ldots\sigma_3\sigma_2\sigma_1\ldots\sigma_{n-2}\sigma n-1\\ &\quad= \sigma_{n-1}\sigma_{n-2}\ldots\sigma_{p+1}^{-1}\sigma_p\ldots\sigma_3\sigma_2\sigma_1\sigma_2\sigma_3\ldots\sigma_p\sigma_{p+1}\ldots\sigma_{n-2}\sigma_{n-1}\\ &\quad= \Phi_{p+1}\sigma_2\sigma_3\ldots\sigma_{n-1}\,.\end{aligned}$$

NOTE 2.1. Let $n = 3$, the following identities hold:

$$\sigma_1^{\pm}\sigma_2^{\pm}\sigma_1^{\pm} = \sigma_2^{\pm}\sigma_1^{\pm}\sigma_2^{\pm}\,,$$

(the Artin relations),

$$\begin{array}{llcl} i) & \sigma_1^{-1}\sigma_2\sigma_1 & = & \sigma_2\sigma_1\sigma_2^{-1}, \\ ii) & \sigma_1\sigma_2^{-1}\sigma_1^{-1} & = & \sigma_2^{-1}\sigma_1^{-1}\sigma_2, \\ iii) & \sigma_1\sigma_2\sigma_1^{-1} & = & \sigma_2^{-1}\sigma_1\sigma_2, \end{array}$$

PROOF.
i)

$$\begin{aligned} \sigma_1^{-1}\sigma_2\sigma_1 &= \sigma_1^{-1}\sigma_2\sigma_1\sigma_2\sigma_2^{-1} \\ &= \sigma_1^{-1}\sigma_1\sigma_2\sigma_1\sigma_2^{-1} \\ &= \sigma_2\sigma_1\sigma_2^{-1}, \end{aligned}$$

ii)

$$\begin{aligned} \sigma_1\sigma_2^{-1}\sigma_1^{-1} &= \sigma_1\sigma_2^{-1}\sigma_1^{-1}\sigma_2^{-1}\sigma_2 \\ &= \sigma_1\sigma_1^{-1}\sigma_2^{-1}\sigma_1^{-1}\sigma_2 \\ &= \sigma_2^{-1}\sigma_1^{-1}\sigma_2, \end{aligned}$$

iii)

$$\begin{aligned} \sigma_1\sigma_2\sigma_1^{-1} &= \sigma_2^{-1}\sigma_2\sigma_1\sigma_2\sigma_1^{-1} = \sigma_2^{-1}\sigma_1\sigma_2\sigma_1\sigma_1^{-1} \\ &= \sigma_2^{-1}\sigma_1\sigma_2. \end{aligned}$$

NOTE 2.2. Let $n = 4$, the following identities hold:

$$\begin{array}{llcl} i) & \sigma_1^{\pm}\sigma_2^{\pm}\sigma_3^{\pm}\sigma_2^{\pm}\sigma_1^{\pm} & = & \sigma_3^{\pm}\sigma_2^{\pm}\sigma_1^{\pm}\sigma_2^{\pm}\sigma_3^{\pm}, \\ ii) & \sigma_1^{-1}\sigma_2\sigma_3\sigma_2\sigma_1 & = & \sigma_3\sigma_2\sigma_1\sigma_2^{-1}\sigma_3, \\ iii) & \sigma_1\sigma_2^{-1}\sigma_3\sigma_2\sigma_1 & = & \sigma_3\sigma_2\sigma_1\sigma_2\sigma_3^{-1}, \\ iv) & \sigma_1\sigma_2\sigma_3^{-1}\sigma_2^{-1}\sigma_1^{-1} & = & \sigma_3^{-1}\sigma_2^{-1}\sigma_1^{-1}\sigma_2\sigma_3, \\ v) & \sigma_1\sigma_2\sigma_3^{-1}\sigma_2^{-1}\sigma_1 & = & \sigma_3^{-1}\sigma_1\sigma_2^{-1}\sigma_1\sigma_3, \\ vi) & \sigma_1\sigma_2\sigma_3\sigma_2^{-1}\sigma_1^{-1} & = & \sigma_3^{-1}\sigma_2^{-1}\sigma_1\sigma_2\sigma_3, \\ vii) & \sigma_1\sigma_2\sigma_3\sigma_2^{-1}\sigma_1 & = & \sigma_3^{-1}\sigma_2\sigma_1\sigma_2\sigma_3, \\ viii) & \sigma_1\sigma_2\sigma_3\sigma_2\sigma_1^{-1} & = & \sigma_3\sigma_2^{-1}\sigma_1\sigma_2\sigma_3. \end{array}$$

PROOF.
i) Using the Artin's relations we have,

$$\begin{aligned} \sigma_1^{\pm}\sigma_2^{\pm}\sigma_3^{\pm}\sigma_2^{\pm}\sigma_1^{\pm} &= \sigma_1^{\pm}\sigma_3^{\pm}\sigma_2^{\pm}\sigma_3^{\pm}\sigma_1^{\pm} \\ &= \sigma_3^{\pm}\sigma_1^{\pm}\sigma_2^{\pm}\sigma_1^{\pm}\sigma_3^{\pm}. \end{aligned}$$

ii) Using the Artin's relations and Note 2.1 we write

$$\begin{aligned} \sigma_1^{-1}\sigma_2\sigma_3\sigma_2\sigma_1 &= \sigma_1^{-1}\sigma_3\sigma_2\sigma_3\sigma_1 \\ &= \sigma_3\sigma_1^{-1}\sigma_2\sigma_1\sigma_3 = \sigma_3\sigma_2\sigma_1\sigma_2^{-1}\sigma_3, \end{aligned}$$

iii)

$$\begin{aligned} \sigma_1\sigma_2^{-1}\sigma_3\sigma_2\sigma_1 &= \sigma_3\sigma_2\sigma_1\sigma_2\sigma_3^{-1} = \sigma_1\sigma_3\sigma_2\sigma_3^{-1}\sigma_1 \\ &= \sigma_3\sigma_1\sigma_2\sigma_1\sigma_3^{-1} \\ &= \sigma_3\sigma_2\sigma_1\sigma_2\sigma_3^{-1}, \end{aligned}$$

iv)

$$\begin{aligned}\sigma_1\sigma_2\sigma_3^{-1}\sigma_2^{-1}\sigma_1^{-1} &= \sigma_1\sigma_3^{-1}\sigma_2^{-1}\sigma_3\sigma_1^{-1}\\ &= \sigma_3^{-1}\sigma_1\sigma_2^{-1}\sigma_1^{-1}\sigma_3\\ &= \sigma_3^{-1}\sigma_2^{-1}\sigma_1^{-1}\sigma_2\sigma_3\,,\end{aligned}$$

v)

$$\begin{aligned}\sigma_1\sigma_2\sigma_3^{-1}\sigma_2^{-1}\sigma_1 &= \sigma_1\sigma_3^{-1}\sigma_2^{-1}\sigma_3\sigma_1\\ &= \sigma_3^{-1}\sigma_1\sigma_2^{-1}\sigma_1\sigma_3\,,\end{aligned}$$

vi)

$$\begin{aligned}\sigma_1\sigma_2\sigma_3\sigma_2^{-1}\sigma_1^{-1} &= \sigma_1\sigma_3^{-1}\sigma_2\sigma_3\sigma_1^{-1}\\ &= \sigma_3^{-1}\sigma_1\sigma_2\sigma_1^{-1}\sigma_3\\ &= \sigma_3^{-1}\sigma_2^{-1}\sigma_1\sigma_2\sigma_3\,.\end{aligned}$$

vii)

$$\begin{aligned}\sigma_1\sigma_2\sigma_3\sigma_2^{-1}\sigma_1 &= \sigma_1\sigma_3^{-1}\sigma_2\sigma_3\sigma_1\\ &= \sigma_3^{-1}\sigma_2\sigma_1\sigma_2\sigma_3\,,\end{aligned}$$

viii)

$$\begin{aligned}\sigma_1\sigma_2\sigma_3\sigma_2\sigma_1^{-1} &= \sigma_1\sigma_3\sigma_2\sigma_3\sigma_1^{-1}\\ &= \sigma_3\sigma_1\sigma_2\sigma_1^{-1}\sigma_3\\ &= \sigma_3\sigma_2^{-1}\sigma_1\sigma_2\sigma_3\,.\end{aligned}$$

THEOREM 2.2. (Generalization of Artin Relations) *Consider a braid group B_n for any $i = 2, ..., n-2$, we have the following tower relations:*

$$\sigma_1\sigma_2\ldots\sigma_{i-1}\sigma_i\sigma_{i-1}\ldots\sigma_2\sigma_1 = \sigma_i\sigma_{i-1}\ldots\sigma_2\sigma_1\sigma_2\ldots\sigma_{i-1}\sigma_i\,.$$

PROOF. This follows from above Lemmas and Notes 2.1 and 2.2

THEOREM 2.3. *In the B_n the following relations holds for an arbitrary i :*

i) $$\begin{aligned}&[\sigma_{i+1}\sigma_i\sigma_{i+2}\sigma_{i+1}]^{\pm}[\sigma_{i+3}\sigma_{i+4}\sigma_{i+2}\sigma_{i+3}]^{-1}[\sigma_{i+1}\sigma_i\sigma_{i+2}\sigma_{i+1}]^{\pm}\\ &= [\sigma_{i+3}\sigma_{i+4}\sigma_{i+2}\sigma_{i+3}]^{\pm}[\sigma_{i+1}\sigma_i\sigma_{i+2}\sigma_{i+1}]^{\pm}[\sigma_{i+3}\sigma_{i+4}\sigma_{i+2}\sigma_{i+3}]^{\pm}\,,\end{aligned}$$

ii) $$\begin{aligned}&[\sigma_{i+1}\sigma_i\sigma_{i+2}\sigma_{i+1}]^{-1}[\sigma_{i+3}\sigma_{i+4}\sigma_{i+2}\sigma_{i+3}][\sigma_{i+1}\sigma_i\sigma_{i+2}\sigma_{i+1}]\\ &= [\sigma_{i+3}\sigma_{i+4}\sigma_{i+2}\sigma_{i+3}][\sigma_{i+1}\sigma_i\sigma_{i+2}\sigma_{i+1}][\sigma_{i+3}\sigma_{i+4}\sigma_{i+2}\sigma_{i+3}]^{-1}\,,\end{aligned}$$

iii) $$\begin{aligned}&[\sigma_{i+1}\sigma_i\sigma_{i+2}\sigma_{i+1}][\sigma_{i+3}\sigma_{i+4}\sigma_{i+2}\sigma_{i+3}][\sigma_{i+1}\sigma_i\sigma_{i+2}\sigma_{i+1}]^{-1}\\ &= [\sigma_{i+3}\sigma_{i+4}\sigma_{i+2}\sigma_{i+3}]^{-1}[\sigma_{i+1}\sigma_i\sigma_{i+2}\sigma_{i+1}][\sigma_{i+3}\sigma_{i+4}\sigma_{i+2}\sigma_{i+3}]\,,\end{aligned}$$

iv) $$\begin{aligned}&[\sigma_{i+1}\sigma_i\sigma_{i+2}\sigma_{i+1}][\sigma_{i+3}\sigma_{i+4}\sigma_{i+2}\sigma_{i+3}]^{-1}[\sigma_{i+1}\sigma_i\sigma_{i+2}\sigma_{i+1}]^{-1}\\ &= [\sigma_{i+3}\sigma_{i+4}\sigma_{i+2}\sigma_{i+3}]^{-1}[\sigma_{i+1}\sigma_i\sigma_{i+2}\sigma_{i+1}]^{-1}[\sigma_{i+3}\sigma_{i+4}\sigma_{i+2}\sigma_{i+3}]\,.\end{aligned}$$

PROOF. Using sucessivly the Artin relation and the these one given in Note 2.1, we show by induction that for $i = 1$, all the given relation hold. Since i is an arbitrary number than the general relation hold are fulfilled as well.

3. Calculus on closed braids.

Let us introduce shortly a qualitative method, called here string calculus, to recognize a closed braid. The method is presented in Nencka [**Ne**]. Consider a closed braid β' in ^{3}. Let us put:

$$\gamma_x = \gamma((x,y)|S_{(x,y)}^{(x+1,y+1)\epsilon}|(x+1,y+1))$$

if the string γ_x goes over the other one, and where $\epsilon = +1$,

$$\gamma_x = \gamma((x,y)|S_{(x,y)}^{(x+1,y+1)\epsilon'}|(x+1,y+1))$$

if the string γ_x goes below the other one, and where $\epsilon' = -1$,

$$\gamma_x = \gamma((x,y)||(x+1,y+1))$$

if there is no crossings. Consider a string family $(\gamma_i)_{i=1,\dots,k}$, where k is a word lenght. For any i, $\gamma_i = \gamma_{i_1}\triangle\gamma_{i_2}\triangle\dots\triangle\gamma_{i_m}$, where "$\triangle$" defines an operation of gluing a terminal (initial) point of γ_{i_m} with initial (terminal) point of $\gamma_{i_{(m+1)}}$ $(\gamma_{i_{(m-1)}})$.

DEFINITION 3.1. *The string family can be equipped with a multiplication law* "$\circ$" *defined as follows:*

$$\gamma((x,y)|S_{(x,y)}^{(x\pm1,y+1)\epsilon_1}|(x\pm1,y+1))\circ\gamma((x\pm1,y+1)|S_{(x\pm1,y+1)}^{(x\pm2,y+2)\,\epsilon_2}|(x\pm2,y+2))$$

In [**Ne**] we prove that

THEOREM 3.1. *The family $(\gamma,\circ)$ equipped with multiplication law defined above is a grupoid.*

DEFINITION 3.2. *To have a closed braid, we introduce an additional string system defined as:* $\gamma' = (\gamma((0,0)||(0,k),\gamma((1,0)||(1,k))\dots\gamma((n,0)||(n,k)))$.
The system

$$(\hat\gamma) = (\gamma,\gamma')$$

is associated to a given closed braid.

The I and II Reidemeister moves are described in the following way.

LEMMA 3.2. *The I Reidemeister move can be represented in the string calculus as :*

$$\gamma((x,y)|S_{(x,y)}^{(x+1,y+1)\epsilon_1}|(x+1,y+1))\circ\gamma((x+1,y+1)||(x+1,y))$$
$$\circ\gamma((x+1,y)|S_{(x+1,y)}^{(x,y+1)\epsilon_1'}|(x,y+1)) = \gamma((x,y)||(x,y+1))$$

where $\epsilon_i' = -\epsilon$

LEMMA 3.3. *The II Reidemeister move is*

$$\gamma((x,y)|S_{(x,y)}^{(x+1,y+1)\epsilon_1}|(x+1,y+1))\circ\gamma((x+1,y+1)||(x+1,y+2))$$
$$\circ\gamma((x+1,y+2)|S_{(x+1,y+2)}^{(x,y+3)\epsilon_1'}|(x,y+3))\circ\gamma((x,y+3)||(x,y))$$
$$=\gamma((x,y)||(x,y+3)),$$

and for the second string

$$\gamma((x+1,y)|S_{(x+1,y)}^{(x,y+1)\epsilon_1'}|(x,y+1))\circ\gamma((x,y+1)||(x,y+2))$$
$$\circ\gamma((x,y+2)|S_{(x,y+2)}^{(x+1,y+3)\epsilon_1'}|(x+1,y+3))\circ\gamma((x+1,y+3)||(x+1,y))$$
$$=\gamma((x+1,y)||(x,y+3)).$$

We note that :

THEOREM 3.2. *For a given link the following properties hold:*

i) $$\gamma((x,y)|X_{(x,y)}^{(x\pm 1,y+1)}|(x\pm 1,y+1))\circ\gamma((x\pm 1,y+1)||(x\pm 1,y)) = \gamma((x,y)||(x,y+1))$$

ii) $$\gamma((x,y)|X_{(x,y)}^{(x\pm 1,y+1)}X_{(x\pm 1,y+1)}^{(x,y+2)}|(x\pm 1,y+2)) = \gamma((x,y)||(x,y+1))$$

where $X = S^{+} or S^{-1}$.

PROOF. It is easily seen that after applying the IRM in the string calculus representation we obtain the right side.

4. Properties of considered closed braids.

THEOREM 4.1. *Let us consider a word:*

i) $\sigma_1 ...\sigma_{n-2}\Theta_{n-1}$ *with* $\Theta_{n-1} = \sigma_{n-1}...\sigma_1$. *The associated closed braid is* $n-2$*-component link, with a link number* $lk = n-2$.

ii) $\alpha_1^{n-1} = \sigma_1^{-1}\sigma_2 ...\Theta_{n-1}$ *with* $\Theta_{n-1} = \sigma_{n-1}...\sigma_1$. *The closed braid associated to* α_1^{n-1} *is composed from one unknot and* $n-2$*-component link*

iii) $\sigma_1\sigma_2 ...\Phi_j^{n-1}$ *with* $\Phi_j^{n-1} = \sigma_{n-1}...\sigma_1$. *The closed braid associated to* α_1^{n-1} *is composed from one unknot and* $n-2$*-component link.*

iv) $\sigma_1\sigma_2 ...\Phi_{n-1}^{n-1}$. *The closed braid associated to it is composed from* $n-1$*-component link.*

v) $\sigma_1^{-1}\sigma_2^{-1}...\sigma_{n-2}^{-1}\Theta_{n-1}$ *with* Θ_{n-1} *defned previously, than the closed braid associated to it is* $n-1$*- component unknot.*

PROOF. The proof as not very much constructive will be omitted. It is sufficient to apply the string calculus to any of the case. We need simply note that in any of the case, the word

$$\sigma_1^{\epsilon_k}\sigma_2^{\epsilon_{k-1}}...\sigma_{n-1}^{\epsilon_{n-1}}...\sigma_2^{\epsilon_2}\sigma_1^{\epsilon_1}$$

can be simplified by the II Markov move. Applying the I Reidemeister move (section 3) in the string representation

$$\gamma_0(...((n-2,n-2)|S_{(n-2,n-2)}^{(n-1,n-1)\epsilon_{n-1}}|(n-1,n-1))$$
$$\circ\gamma'((n-1,n-1||(n-1,0))\circ\gamma_{n-1}((n-1,0)||(n-1,n-2))$$
$$\circ\gamma_{n-1}((n-1,n-2)|S_{(n-1,n-2)}^{(n-2,n-1)\epsilon_{n-1}'}|(n-2,n-1))\circ ...$$
$$=\gamma_0(...((n-2,n-2)||(n-2,n-1))...).$$

In the case i) the word $\sigma_1 ... \sigma_{n-2}\Theta_{n-1}$ with $\Theta_{n-1} = \sigma_{n-1} ... \sigma_1$ has the following part in the string representations, namely:

$$\gamma_0((0,0)|S_{(0,0)}^{(1,1)\epsilon_1'} ... S_{(n-3,n-3)}^{(n-2,n-2)\epsilon_{n-3}'} S_{(n-2,n-2)}^{(n-1,n-1)\epsilon_{n-2}'} S_{(n-2,n-1)^{(n-3,n)\epsilon_n}} \cdots$$
$$\dots S_{(1,k-1)}^{(0,k)\epsilon_k}|(0,k)) \circ \gamma'((0,k)||(0,0)) .$$

which is just a $n-2$-component link. Applying the same rules to the next cases we obtain the theorem.

5. References

[Al] J.M. Alexander, *Topological invariants of knots and Links*, Trans. Math. Soc. **30** (1923), 275–306.

[Ar1] E. Artin, *Theorie der Zoepfe*, Abh. Math. Sem. Univ. Hamburg **4** (1926), 47–72.

[Ar2] E. Artin, *Theory of Braids*, Ann. Math. **48** (1947), 101-126.

[Ar3] E. Artin, *Braids and Permutations*, Ann. Math. **48** (1947), 643–649.

[Ar4] E. Artin, *The Orders of the Calssical Simple Groups*, Comm. Pure Appl. Math. **8** (1955), 295–472.

[Be] D. Bennequin, *Entrelacements et equations de Pfaff*, Asterisque, **107** (1983), 87–161.

[Bes-Ha] M. Bestvina and M. Handel, *Train tracks and automorphism of free groups*, Ann. Math. **135** (1992), 1–51.

[Bi1] J. Birman, *Braids, links and mapping class groups*, Ann. Math. stud. **82** (1974).

[Bi2] J. Birman, *New Points of View in Knot Theory*, Bull. Amer. Math.Soc. **28** (1993), 253–287.

[Bi3] J. Birman, *Studying Links via Closed Braids*, Lect. Notes of the Ninth Kaist Math. Workshop **1** (1994), 1–64.

[Bi-Me1] J. Birman and W. Menasco, *Studying Links via Closed Braids IV: Split Links and Composite Links*, Inv. Math. **102** (1990), 115–139.

[Bi-Me2] J. Birman and W. Menasco, *Studying Links via Closed Braids II: On a theorem of Bennequin*, Topology and its Applications **40** (1991), 71–82.

[Bi-Me3] J. Birman and W. Menasco, *Pacific J. Math.* **154** (1992), 17–36.

[Bi-Me4] J. Birman and W. Menasco, *Studying Links via Closed Braids V: The Unlink*, Trans. AMS **329** (1992), 585–606.

[Bi-Me5]J. Birman and W. Menasco, *Studying Links via Closed Braids VI: A non-finiteness theorem*, Pacific J. Math. **156 .** (1992), 265–285.

[Bi-Me6] J. Birman and W. Menasco, *Studying Links via Closed Braids III: Classifying links which are closed 3-braids*, Pacific J. Math. **161 .** (1993), 25–113.

[Bi-Me7] J. Birman and W. Menasco, *Special positions for essentisal tori in link complements*, Topology **33 .** (1994), 525–556.

[Bi-Fi] J. Birman and E. Finkelstein, *Studying surfaces via closed braids*, J. Knot Theory and its Ramifications (1997).

[Bo] F. Bohnenblust, *The Algebraical Braid Group.*, Ann. Math. **48** (1947), 127–136.

[Ch] W. Chow, *On the Algebraical Braid Group.*, Ann. Math. **49** (1948), 654–658.

[Co] H.S.M. Coxeter, *Factor Groups of the Braid Groups .*, Proc. Fourth Canad. Math. Congress. (1959), 95–122.

[El-Mo] E.A. Elrifai and H. R. Morton, *Algorithm for positive braids*, Quart. J. Math. Oxford . **45** (1994), 479–497.

[Ga] F. Garside, *The braid groups and other groups .*, Quart. J. Math. Oxford, **20** (1968), 235–254.

[Lo] J.E. Los, *Knots, braid index and dynamical type.*, Topology **33** (1994), 257–270.

[Ma] A.A.Markov, *Uber die freie Aequivalenz Geschlossener Zoepfe.*, Recueil Math. Moscow **1** (1935), 73 – 78.

[Mo] H.R.Morton, *Threading Knot Diagrams.*, Math. Proc. Cambridge Philos. Soc. **99** (1986), 247–260.

[Ne] H. Nencka, *Some methods to identify knots.*, Meth. Funct. Anal. Appl. (1998).

[Re1] K. Reidemeister, *Einfuhrung in die Kombinatorische Topologie.*, Braunschweig (1932).

[Re2] K. Reidemeister, *Knotentheorie.*, Erg. d. Math. **1** (1932), 1.

[Th] W.P.Thurston, *On the geometry and dynamics of diffeomorphisms of surfaces*, Bull. Amer. Math. Soc. **19** (1988), 417–431.

[Tr] P. Traczyk, *A new proof of Markov braid theorem.*, preprint.

[Vo] P. Vogel, *Representation of links by braids: A new algorithm.*, Comm. Math. Helv. **65** (1990), 104–113.

[Xu] P. Xu, *The Genus of Closed 3-Braids.*, J. Knot Theory and its Ramifications **1** (1992), 303–326.

[Ya] S. Yamada, *The minimal number of Seifert circles equals the braid index of a link.*, Invent. Math. **89** (1987), 347–356.

H. NENCKA, UNIVERSIDADE DA MADEIRA, 9000-FUNCHAL, MADEIRA, PORTUGAL
E-mail address: nencka@dragoeiro.uma.pt,nencka@cptsu2.univ-mrs.fr

Contemporary Mathematics
Volume **233**, 1999

Asymptotic behavior of energy of polygonal knots

JUN O'HARA

ABSTRACT. We study *regularization of "modified electrostatic energy"* of charged knots, which was introduced to produce a good-looking representative for each knot type. We define new energy functionals for smooth knots, and give formulae of asymptotic behavior of energy of polygonal knots which we can apply to numerical experiments.

1. Introduction

A notion of *"energy of knots"* was proposed by Fukuhara [F] and Sakuma [S]. Their motivation is to define a beautiful representative embedding for each knot type. Their approach is to define a suitable functional on the space of knots and define the *canonical configuration* for each knot type as the embedding that attains the minimum value of the functional within its ambient isotopy class. Suppose we can deform a knot along the gradient flow so as to decrease the value of the functional until it comes to a critical point. Assume that the knot type is kept unchanged during this deformation process. Assume also that the number of critical points is finite for each knot type. Then we can detect the knot type of a given knot. In order to guarantee the invariance of the knot type during the deformation process, crossing changes should not be allowed. Thus we require that our functional should satisfy the following property:

($*$) If a knot degenerates to an immersion with double points then the functional blows up.

We call our functional a *knot energy functional* if it satisfies the above property. We remark that this property is not sufficient to guarantee the existence of the canonical configuration for each knot type because a knot might degenerate to a map that does not belong to the original knot type.

1991 *Mathematics Subject Classification.* 57M25, 49Q10, 53A04.
Key words and phrases. knot, energy.

If a knot f_0 attains the minimum value of a knot energy functional $\mathcal{E}$ among its ambient isotopy class, we call f_0 an *$\mathcal{E}$-minimizer.*

One of the most natural candidates for a knot energy functional would be the electrostatic energy of charged knots, but it turns out to be infinite for any knot. We can *regularize* it to obtain a finite valued functional $E^{(1)}$ as in (2.9), but $E^{(1)}$ does not satisfy the above property $(*)$. Thus we are obliged to consider the "*$r^{-(\alpha+1)}$-modified electrostatic energy*" defined under the non-physical assumption that Coulumb's repelling force between a pair of point charges of distance r is proportional to $r^{-(\alpha+1)}$ where $\alpha > 1$. A finite valued functional $E^{(\alpha)}$ can be defined after regularization, which satisfy the property $(*)$ if and only if $\alpha \geq 2$. Freedman, He, and Wang showed that $E^{(2)}$ is invariant under a Möbius transformation, i.e. a composition of reflections with respect to spheres in $\mathbb{R}^3$, and showed that there is an $E^{(2)}$-minimizer for any knot type of a *prime* knot ([FHW]). We remark that there are infinitely many $E^{(2)}$-minimizers for each non-trivial prime knot type since an image of an $E^{(2)}$-minimizer by a Möbius transformation, or its mirror image, is again an $E^{(2)}$-minimizer for the same knot type. On the contrary Kusner and J. Sullivan conjectured that there are no $E^{(2)}$-minimizers for any knot type of a *composite* knot ([KS]). Their numerical experiments imply that both component tangles corresponding to K_1 and K_2 of a composite knot $K_1 \sharp K_2$ shrink to points as the knot evolves itself to decrease its value of $E^{(2)}$, so that the knot degenerates to the trivial knot. When $\alpha > 2$ we showed that there is an $E^{(\alpha)}$-minimizer for any knot type if $2 < \alpha < 3$ ([O3]). We conjecture that the number of $E^{(\alpha)}$-minimizer $(2 < \alpha < 3)$ is finite for any knot type.

On the other hand, numerical experiments to find a good-looking representative embedding for each knot type have been carried out ([A], [BO1], [BS], [F1,2], [Gu], [KS], [LS], [S1,2], [Su], [Hu], [Sc], [Wu], and [GA]). Since $E^{(2)}$ is unable to evolve a small tangle because of the Möbius invariance property, a higher power index $\alpha > 2$ is used or the total squared curvature functional is added in such computer programs as [F2], [Gu], and [Sc]. In [S2] Simon observed that the value of $E^{(2)}$ of a regular polygon with n vertices is asymptotic to $\frac{\pi^2}{6}n$.

This paper is arranged as follows. In §2 we define the regularization of "$r^{-(\alpha+1)}$-modified electrostatic energy", $E^{(\alpha)}$, which was given in [O1,3] when $2 \leq \alpha < 3$ and was announced in [O4,5] when $\alpha \geq 3$.* We show that $E^{(\alpha)}$ $(\alpha \geq 3)$ is a knot energy functional. In §3 we give formulae of the asymptotic behavior of the "$r^{-(\alpha+1)}$-modified electrostatic energy" $\check{E}_n^{(\alpha)}$ of a polygonal knot with n vertices which are equally dotted on a smooth knot f as n goes to infinity. (These formulae were also announced in [O4,5].) One of these formulae implies the above observation by Simon. In §4, using these formulae, we consider the asymptotic behavior of polygonal knots with n vertices in a given knot type that attain the (local) minimum value of $\check{E}_n^{(\alpha)}$ as n goes to infinity. We give conjec-

---

*Part of the results of this paper were reported with errors in the author's talk at the AMS Meeting, Iowa, March 1996 and KNOTS'96, Waseda University, Tokyo, July 1996.

tures on the limit conformations, which are naturally derived from the formulae in §3.

The reader is referred to [O4,5] for the survey of this subject.

In §2 and §3, for the sake of simplicity, we assume that a knot $f : S^1 = \mathbb{R}/\mathbb{Z} \to \mathbb{R}^3$ is smooth and is parametrized by arc length, namely, $|f'(t)| = 1$ $(\forall t \in S^1)$.

2. Definition of regularization of "$r^{-(\alpha+1)}$-modified electrostatic energy", $E^{(\alpha)}$

2.1 Subtracting terms for regularization.

Let $f : S^1 = \mathbb{R}/\mathbb{Z} \to \mathbb{R}^3$ be a smooth embedding that is parametrized by arc-length, namely, $|f'(t)| = 1$ $(\forall t \in S^1)$. Suppose that the knot is uniformly electrically charged. (We need this assumption of uniformity for the following reason. Suppose electrons can move freely along a knot. Then the energy of a knot which is almost same as the standard circle except for a small tangle can be close to that of the standard circle itself if the tangle part carries a small charge. Thus every knot will degenerate to the circle in order to decrease its energy.) Assume that Coulumb's repelling force between a pair of point charges of distance r is proportional to $r^{-(\alpha+1)}$ where $\alpha \geq 1$. Then the $r^{-(\alpha+1)}$-modified electrostatic energy of a knot f is given by

$$\iint_{S^1 \times S^1} \frac{dxdy}{|f(x)-f(y)|^\alpha}, \tag{2.1}$$

which turns out to be infinite for any knot because the integral blows up around the diagonal set $\triangle = \{(x,y) \in S^1 \times S^1 | x = y\}$. We regularize (2.1) by subtracting a term which blows up to the same order as $|f(x)-f(y)|^{-\alpha}$ at the diagonal set $\triangle$ to obtain a finite valued functional $E^{(\alpha)}$. For that purpose let us expand $|f(x)-f(y)|^{-\alpha}$ by $|x-y|$.

CLAIM 2.1. *Let $f : S^1 = \mathbb{R}/\mathbb{Z} \to \mathbb{R}^3$ be a smooth embedding with $|f'(t)| = 1$ $(\forall t \in S^1)$. For $i \geq 4$ define $a_i = a_i(f; x)$ by*

$$a_{2m} = \frac{1}{2m-1}\left\{2\sum_{k=2}^{m-1}\frac{k(2m-k)-2m+1}{k!(2m-k)!}(f^{(k)}, f^{(2m-k)}) + \frac{(m-1)^2}{(m!)^2}|f^{(m)}|^2\right\}$$

$$a_{2m+1} = \frac{1}{m}\left\{\sum_{k=2}^{m}\frac{k(2m+1-k)-2m}{k!(2m+1-k)!}(f^{(k)}, f^{(2m+1-k)})\right\},$$

where $f^{(k)} = f^{(k)}(x)$. Then for $4 \leq N \in \mathbb{N}$

$$|f(x+t)-f(x)|^2 = t^2 - \sum_{n=4}^{N} a_n t^n + O(t^{N+1})$$

for any t $(|t| \leq \frac{1}{2})$.

PROOF. Suppose $x = 0$ and $f(0) = 0$. Let $f^{(k)}$ denote $f^{(k)}(0)$. Since $(f'(y), f'(y)) = 1$ for all $y \in S^1$,

$$\frac{d^{n-2}}{dy^{n-2}}(f'(y), f'(y))\Big|_{y=0} = 0 \qquad \forall n \geq 3.$$

(1) If $n = 2m+1$ for some $m \geq 1$ then

$$\sum_{k=1}^{m} {}_{2m-1}C_{k-1}(f^{(k)}, f^{(2m+1-k)}) = 0,$$

hence

$$(f', f^{(2m)}) = -\sum_{k=2}^{m} {}_{2m-1}C_{k-1}(f^{(k)}, f^{(2m+1-k)}). \tag{2.2}$$

(2) If $n = 2m$ for some $m \geq 2$ then

$$2\sum_{k=1}^{m-1} {}_{2m-2}C_{k-1}(f^{(k)}, f^{(2m-k)}) + {}_{2m-2}C_{m-1}(f^{(m)}, f^{(m)}) = 0,$$

hence

$$(f', f^{(2m-1)}) = -\sum_{k=2}^{m-1} {}_{2m-2}C_{k-1}(f^{(k)}, f^{(2m-k)}) - \frac{{}_{2m-2}C_{m-1}}{2}|f^{(m)}|^2. \tag{2.3}$$

On the other hand, by Taylor's theorem

$$f(t) = \sum_{k=1}^{2l} \frac{f^{(k)}}{k!}t^k + O(t^{2l+1}),$$

which implies

(2.4)
$$\begin{aligned}|f(t)|^2 &= (f(t), f(t)) \\ &= t^2 + 2\sum_{m=1}^{l}\left(\sum_{k=1}^{m}\frac{(f^{(k)}, f^{(2m+1-k)})}{k!(2m+1-k)!}\right)t^{2m+1} \\ &\quad + \sum_{m=2}^{l}\left\{2\sum_{k=1}^{m-1}\frac{(f^{(k)}, f^{(2m-k)})}{k!(2m-k)!} + \frac{|f^{(m)}|^2}{(m!)^2}\right\}t^{2m} + O(t^{2l+2}).\end{aligned}$$

Remark that the coefficient (f', f'') of t^3 is 0. Using (2.2) and (2.3) to delete the terms of the form $(f', f^{(n-1)})t^n$ in (2.4), we have

$$\begin{aligned}|f(t)|^2 &= t^2 \\ &- \sum_{m=2}^{l}\frac{1}{2m-1}\left\{2\sum_{k=2}^{m-1}\frac{k(2m-k)-2m+1}{k!(2m-k)!}(f^{(k)}, f^{(2m-k)}) + \frac{(m-1)^2}{(m!)^2}|f^{(m)}|^2\right\}t^{2m} \\ &- \sum_{m-2}^{l}\frac{1}{m}\left\{\sum_{k=2}^{m}\frac{k(2m+1-k)-2m}{k!(2m+1-k)!}(f^{(k)}, f^{(2m+1-k)})\right\}t^{2m+1} + O(t^{2l+2}),\end{aligned}$$

which completes the proof. □

Let us calculate some of the a_n's.

$$
\begin{aligned}
a_4 &= \frac{|f''|^2}{12}, \\
a_5 &= \frac{(f'', f^{(3)})}{12}, \\
a_6 &= \frac{(f'', f^{(4)})}{40} + \frac{|f^{(3)}|^2}{45}, \\
a_7 &= \frac{(f'', f^{(5)})}{180} + \frac{(f^{(3)}, f^{(4)})}{36}, \\
a_8 &= \frac{(f'', f^{(6)})}{1008} + \frac{2(f^{(3)}, f^{(5)})}{315} + \frac{|f^{(4)}|^2}{448},
\end{aligned} \tag{2.5}
$$

and so on. By Claim 2.1 for $m \geq 3$

(2.6)

$$
\begin{aligned}
|f(x+t) - f(x)|^2 &= t^2 - \sum_{n=4}^{m+1} a_n t^n + O(t^{m+2}) \\
&= t^2\{1 - (a_4 t^2 + a_5 t^3 + \cdots + a_{m+1} t^{m-1} + O(t^m))\}.
\end{aligned}
$$

On the other hand Taylor's formula implies

$$
(1+t)^a = 1 + at + \frac{a(a-1)}{2} t^2 + \frac{a(a-1)(a-2)}{3!} t^3 + \cdots
$$

for $|t| < 1$. Therefore $|f(x+t) - f(x)|^{-\alpha}$ is expanded by t as

$$
\begin{aligned}
|f(x+t) - f(x)|^{-\alpha} &= t^{-\alpha}\{1 - (a_4t^2 + a_5t^3 + \cdots + a_{m+1}t^{m-1} + O(t^m))\}^{-\frac{\alpha}{2}} \\
&= t^{-\alpha}\Big\{1 + \frac{\alpha}{2}\left(a_4t^2 + a_5t^3 + \cdots + a_{m+1}t^{m-1} + O(t^m)\right) \\
&\quad + \frac{\alpha(\alpha+2)}{8}\left(a_4^2t^4 + 2a_4a_5t^5 + (a_5^2 + 2a_4a_6)t^6 + \cdots + O(t^{m+2})\right) \\
&\quad + \frac{\alpha(\alpha+2)(\alpha+4)}{48}\left(a_4^3t^6 + \cdots + O(t^{m+4})\right) + \cdots\Big\} \\
&= t^{-\alpha}\Big\{1 + \frac{\alpha}{2}a_4t^2 + \frac{\alpha}{2}a_5t^3 + \left(\frac{\alpha}{2}a_6 + \frac{\alpha(\alpha+2)}{8}a_4^2\right)t^4 \\
&\quad + \left(\frac{\alpha}{2}a_7 + \frac{\alpha(\alpha+2)}{4}a_4a_5\right)t^5 \\
&\quad + \left(\frac{\alpha}{2}a_8 + \frac{\alpha(\alpha+2)}{8}(a_5^2 + a_4a_6) + \frac{\alpha(\alpha+2)(\alpha+4)}{48}a_4^3\right)t^6 \\
&\quad + \cdots + O(t^m)\Big\}
\end{aligned}
$$

(2.7)

$$
\begin{aligned}
&= t^{-\alpha} + \frac{\alpha|f''(x)|^2}{24}t^{2-\alpha} + \frac{\alpha(f''(x), f^{(3)}(x))}{24}t^{3-\alpha} \\
&\quad + \frac{\alpha}{2}\left\{\frac{(f''(x), f^{(4)}(x))}{40} + \frac{|f^{(3)}(x)|^2}{45} + \frac{(\alpha+2)|f''(x)|^4}{576}\right\}t^{4-\alpha} \\
&\quad + \frac{\alpha}{72}\left\{\frac{(f^{(2)}, f^{(5)})}{5} + (f^{(3)}, f^{(4)}) + \frac{(\alpha+2)|f''|^2(f'', f^{(3)})}{8}\right\}t^{5-\alpha} \\
&\quad + \cdots + O(t^{m-\alpha})
\end{aligned}
$$

for $0 < t \ll 1$, which implies a sufficient subtraction term for regularizing (2.1).

Let $\delta(x,y)$ be the shortest arc-length between $f(x)$ and $f(y)$ along a knot $f(S^1)$, which is same as $\min\{|x-y|, 1-|x-y|\}$ since f is parametrized by arc length.

Put $n_\alpha = \max\{2, [\alpha]\}$. Let $\Phi_f^{(\alpha)}(x,y)$ denote the sum of the first n_α terms of

(2.8)

$$
\begin{aligned}
&|f(x) - f(y)|^{-\alpha} - \delta(x,y)^{-\alpha} - \frac{\alpha|f''(x)|^2}{24}\delta(x,y)^{2-\alpha} - \frac{\alpha(f''(x), f^{(3)}(x))}{24}\delta(x,y)^{3-\alpha} \\
&\quad - \left\{\frac{\alpha}{2}\left(\frac{(f''(x), f^{(4)}(x))}{40} + \frac{|f^{(3)}(x)|^2}{45}\right) + \frac{\alpha(\alpha+2)|f''(x)|^4}{1152}\right\}\delta(x,y)^{4-\alpha} \\
&\quad - \frac{\alpha}{72}\Big\{\frac{(f^{(2)}(x), f^{(5)}(x))}{5} + (f^{(3)}(x), f^{(4)}(x)) \\
&\qquad\qquad + \frac{(\alpha+2)|f''(x)|^2(f''(x), f^{(3)}(x))}{8}\Big\}\delta(x,y)^{5-\alpha} - \cdots .
\end{aligned}
$$

Since

$$\int_{-\frac{1}{2}}^{\frac{1}{2}} t^p dt < \infty \quad \text{if} \quad p > -1,$$

$\Phi_f^{(\alpha)}(x,y)$ is integrable on $S^1 \times S^1$.

DEFINITION 2.2. Define the regularization of "$r^{-(\alpha+1)}$-modified electrostatic energy" $E^{(\alpha)}(f)$ of a knot f by

$$E^{(\alpha)}(f) = \iint_{S^1\times S^1} \Phi_f^{(\alpha)}(x,y)dxdy.$$

We call α the power index of $E^{(\alpha)}$.

Therefore $E^{(\alpha)}(f)$ is given according to the value of α as follows.

When $1 \le \alpha < 3$ $E^{(\alpha)}$ is given by

(2.9)

$$E^{(\alpha)}(f) = \iint_{S^1\times S^1} \left(\frac{1}{|f(x)-f(y)|^\alpha} - \frac{1}{\delta(x,y)^\alpha} \right) dxdy,$$

when $3 \le \alpha < 4$ $E^{(\alpha)}$ is given by

$$E^{(\alpha)}(f) = \iint_{S^1\times S^1} \left(\frac{1}{|f(x)-f(y)|^\alpha} - \frac{1}{\delta(x,y)^\alpha} - \frac{\alpha|f''(x)|^2}{24\delta(x,y)^{\alpha-2}} \right) dxdy,$$

when $4 \le \alpha < 5$ $E^{(\alpha)}$ is given by

$$E^{(\alpha)}(f) = \iint_{S^1\times S^1} \left(\frac{1}{|f(x)-f(y)|^\alpha} - \frac{1}{\delta(x,y)^\alpha} - \frac{\alpha|f''(x)|^2}{24\delta(x,y)^{\alpha-2}} - \frac{\alpha(f''(x), f^{(3)}(x))}{24\delta(x,y)^{\alpha-3}} \right) dxdy,$$

and so on.

2.2 When is $E^{(\alpha)}$ a knot energy functional?

THEOREM 2.3. *$E^{(\alpha)}$ is a knot energy functional if and only if $\alpha \ge 2$.*

PROOF. *Necessity.* Suppose $\bar{f}$ is a "singular knot" which intersects itself perpendicularly at a point P. Suppose that $\bar{f}(S^1)$ is a union of crossed straight line segments around P. Then the contribution of the neighborhood of P to the integral (2.9) is given by

$$\int_{-c}^{c}\int_{-c}^{c} \frac{dsdt}{(s^2+t^2)^{\frac{\alpha}{2}}} \sim 4\int_0^{c'}\int_0^{\frac{\pi}{2}} \frac{d\theta dr}{r^{\alpha-1}} \tag{2.10}$$

which is infinite if and only if $\alpha \ge 2$. Thus $E^{(\alpha)}$ takes a finite value for a "singular knot" with a self-intersection when $\alpha < 2$ and hence $E^{(\alpha)}$ is not a knot energy functional if $\alpha < 2$.

Sufficiency. We showed that $E^{(\alpha)}$ is a knot energy functional if $2 \le \alpha < 3$ in [O3]. (We remark that $E^{(\alpha)}$ is equal to $\alpha e_\alpha{}^1$ in [O2,3] when $\alpha < 3$.) Let us prove that $E^{(\alpha)}$ is a knot energy functional when $3 \le \alpha < 4$ here. The argument is parallel when $\alpha \ge 4$.

Gromov's distortion ([Gr]) of a knot f is given by

$$\mathrm{Distor}(f) = \sup_{x \neq y} \frac{\delta(x,y)}{|f(x) - f(y)|}.$$

Then the theorem is reduced to the following proposition. □

PROPOSITION 2.4. *Let $3 \le \alpha < 4$. For any $C > 0$ and $b > 0$ there is $A_\alpha > 0$ such that if a knot f with*[1]

$$|f^{(n)}(x)| \le C \quad \forall x \in S^1, n = 1,2,3,4 \tag{2.11}$$

satisfies

$$\mathrm{Distor}(f) \ge A_\alpha$$

then

$$E^{(\alpha)}(f) \ge b.$$

PROOF. By (2.7) there are positive constants ϵ_1 with $\epsilon_1 \le \frac{1}{2}$ and C' such that for any f that satisfies (2.11), for any t with $0 < t \le \epsilon_1$, and for any $x \in S^1$

$$\left| |f(x+t) - f(x)|^{-\alpha} - t^{-\alpha} - \frac{\alpha |f''(x)|^2}{24} t^{2-\alpha} \right| \le C' t^{3-\alpha}. \tag{2.12}$$

By (2.5) and (2.6) there is a positive constant $\epsilon_2 >$ with $\epsilon_2 \le \frac{1}{2}$ such that for any f that satisfies (2.11) and for any x, y with $0 < \delta(x,y) \le \epsilon_2$

$$\frac{\delta(x,y)}{|f(x) - f(y)|} \le 2. \tag{2.13}$$

Let $\epsilon = \min\{\epsilon_1, \epsilon_2\}$. Then there is a positive constant C'' such that for any f that satisfies (2.11)

$$\iint_{\delta(x,y) \ge \epsilon} \left| \frac{1}{\delta(x,y)^\alpha} + \frac{\alpha |f''(x)|^2}{24 \delta(x,y)^{\alpha-2}} \right| dx dy \le C''. \tag{2.14}$$

On the other hand (2.12) implies that for any f that satisfies (2.11)

$$\iint_{\delta(x,y) \le \epsilon} \left| \frac{1}{|f(x) - f(y)|^\alpha} - \frac{1}{\delta(x,y)^\alpha} - \frac{\alpha |f''(x)|^2}{24 \delta(x,y)^{\alpha-2}} \right| dx dy \le \frac{2\epsilon^{4-\alpha}}{4-\alpha} C'. \tag{2.15}$$

Suppose a knot f satisfies (2.11) and that $\mathrm{Distor}(f) = a > 2$. Then (2.13) implies that there is a pair of points (x_0, y_0) with $\delta(x_0, y_0) \ge \epsilon$ and

$$\frac{\delta(x_0, y_0)}{|f(x_0) - f(y_0)|} = a.$$

[1]We remark that the condition on $|f^{(n)}|$ is not necessary when $\alpha < 3$.

Put $d_0 = |f(x_0) - f(y_0)|$. If $|t|, |s| \leq \frac{\epsilon}{4}$ then

$$|f(x_0 + t) - f(y_0 + s)| \leq d_0 + |t| + |s|.$$

Therefore

$$\begin{aligned}\iint_{\delta(x,y)\geq\epsilon} \frac{dxdy}{|f(x) - f(y)|^\alpha} &\geq \int_{x_0-\frac{\epsilon}{4}}^{x_0+\frac{\epsilon}{4}} \int_{y_0-\frac{\epsilon}{4}}^{y_0+\frac{\epsilon}{4}} \frac{dxdy}{|f(x) - f(y)|^\alpha} \\ &\geq 4 \int_0^{\frac{\epsilon}{4}} \int_0^{\frac{\epsilon}{4}} \frac{dtds}{(d_0 + t + s)^\alpha} \\ &= \frac{4}{(\alpha-1)(\alpha-2)} \left\{ \frac{1}{{d_0}^{\alpha-2}} - \frac{2}{(d_0 + \frac{\epsilon}{4})^{\alpha-2}} + \frac{1}{(d_0 + \frac{\epsilon}{2})^{\alpha-2}} \right\}\end{aligned}$$

Since $d_0 = a^{-1}\delta(x_0, y_0) \leq \frac{1}{2a}$

(2.16)
$$\geq \frac{4}{(\alpha-1)(\alpha-2)} \left\{ (2a)^{\alpha-2} - \frac{2}{(\frac{\epsilon}{4})^{\alpha-2}} \right\}.$$

The estimates (2.14), (2.15), and (2.16) imply that if f satisfies (2.11) and that $\mathrm{Distor}(f) = a$ then

$$E^{(\alpha)}(f) \geq \frac{2^\alpha a^{\alpha-2}}{(\alpha-1)(\alpha-2)} - \frac{2^{2\alpha-1}}{(\alpha-1)(\alpha-2)\epsilon^{\alpha-2}} - \frac{2\epsilon^{4-\alpha}}{4-\alpha}C' - C'',$$

which completes the proof of Proposition. □

2.3 Remarks and examples

The first example $E^{(2)}$ was defined in [O1] (in fact $E_{[O1]}(f) = \frac{1}{2}(E^{(2)}(f) - 4)$) and is called the **Möbius energy** in [FHW] *et al.*

EXAMPLE 2.5. Let $\iota_0(t) = (\frac{1}{2\pi}\cos 2\pi t, \frac{1}{2\pi}\sin 2\pi t, 0)$ be the standard planar circle. Then its energy is given by

$$\begin{aligned}E^{(2)}(\iota_0) &= \int_0^1 dx\, 2 \int_0^{\frac{1}{2}} \left(\frac{1}{(\frac{1}{\pi}\sin \pi y)^2} - \frac{1}{y^2} \right) dy \\ &= 2 \lim_{\epsilon \downarrow 0} \Big[-\pi \cot \pi y + \frac{1}{y} \Big]_\epsilon^{\frac{1}{2}} \\ &= 4.\end{aligned}$$

$$
\begin{aligned}
E^{(3)}(\iota_0) &= \int_0^1 dx\, 2\int_0^{\frac{1}{2}} \left(\frac{1}{(\frac{1}{\pi}\sin \pi y)^3} - \frac{1}{y^3} - \frac{3\cdot(2\pi)^2}{24y} \right) dy \\
&= 2\pi^2 \int_0^{\frac{\pi}{2}} \left(\frac{1}{\sin^3 t} - \frac{1}{t^3} - \frac{1}{2t} \right) dt \\
&= 2\pi^2 \lim_{\epsilon \downarrow 0} \left[\frac{1}{4}\left(\log(1-\cos t) - \log(1+\cos t) - \frac{1}{1-\cos t} + \frac{1}{1+\cos t} \right) \right. \\
&\qquad \left. + \frac{1}{2t^2} - \frac{1}{2}\log t \right]_{\epsilon}^{\frac{\pi}{2}} \\
&= 4 + \pi^2 \log \frac{4}{\pi} - \frac{\pi^2}{6} \approx 4.7392\cdots .
\end{aligned}
$$

3. Asymptotic formulae of "$r^{-(\alpha+1)}$-modified electrostatic energy" of polygonal knots

In order to make connection to numerical experiments, let us consider corresponding finite dimensional counterpart of $E^{(\alpha)}$.

Let $\mathbb{P}(n)$ be a set of polygonal knots with n vertices:

$$\mathbb{P}(n) = \{\mathbf{f} = (\mathbf{x}_1, \mathbf{x}_2, \cdots, \mathbf{x}_n) \in (\mathbb{R}^3)^n \ : \ \mathbf{x}_i \neq \mathbf{x}_j \quad (i \neq j)\}$$

We identify $\mathbf{f} \in \mathbb{P}(n)$ with a piecewise linear map $\mathbf{f} : S^1 \to \mathbb{R}^3$ with $\mathbf{f}(\frac{i}{n}) = \mathbf{x}_i$. Put $\mathbb{L}(n)$ to be its subset consisting of equilateral polygonal knots:

$$\mathbb{L}(n) = \{\mathbf{f} = (\mathbf{x}_1, \mathbf{x}_2, \cdots, \mathbf{x}_n) \in \mathbb{P}(n) : |\mathbf{x}_i - \mathbf{x}_{i-1}| = \frac{1}{n} \quad \forall i\}$$

where suffixes are considered mod n, namely, $\mathbf{x}_n = \mathbf{x}_0$. Let $\check{E}_n^{(\alpha)} : \mathbb{P}(n) \to \mathbb{R}$ ($\alpha > 2$) be the "$r^{-(\alpha+1)}$-modified electrostatic energy" of a polygonal knot each of whose vertices has equal charge $\frac{1}{n}$:

$$\check{E}_n^{(\alpha)}(\mathbf{f}) = \frac{1}{n^2} \sum_{i \neq j} \frac{1}{|\mathbf{x}_i - \mathbf{x}_j|^\alpha}.$$

Put $\widehat{E}_n^{(\alpha)} = \check{E}_n^{(\alpha)}|_{\mathbb{L}(n)} : \mathbb{L}(n) \to \mathbb{R}$ to be the restriction of $\check{E}_n^{(\alpha)}$ to $\mathbb{L}(n)$.

Let $f : S^1 \to \mathbb{R}^3$ be a smooth knot. Consider a polygonal knot with n vertices

$$\check{\mathbf{f}}_n = \left(f(0), f(\frac{1}{n}), \cdots, f(\frac{n-1}{n})\right) \in \mathbb{P}(n).$$

Put

$$
\begin{aligned}
\widetilde{E}_n^{(\alpha)}(f) &= \check{E}_n^{(\alpha)}(\check{\mathbf{f}}_n) \\
&= \frac{1}{n^2} \sum_{i \neq j} \frac{1}{|f(\frac{i}{n}) - f(\frac{j}{n})|^\alpha}. \qquad (3.1)
\end{aligned}
$$

Then $\widetilde{E}_n^{(\alpha)}(f)$ blows up as n goes to infinity. We give formulae of asymptotic behavior of $\widetilde{E}_n^{(\alpha)}(f)$. For the sake of simplicity we use notation $A(n) \sim B(n)$ by which we mean $\lim_{n\to\infty}(A(n) - B(n)) = 0$.

PROPOSITION 3.1. *Asymptotic behavior of* $\widetilde{E}_n^{(\alpha)}(f)$ *as* n *goes to infinity is given according to the value of* α *as follows.*

(1) *When* $\alpha = 1$

$$\widetilde{E}_n^{(1)}(f) \sim E^{(1)}(f) + 2(\log \frac{n}{2} + C),$$

where C *is Euler's constant*

$$C = \lim_{n\to\infty} \left(1 + \frac{1}{2} + \cdots + \frac{1}{n} - \log n\right) \approx 0.5772\cdots,$$

which appears in the formulae only when α *is an integer except for 2.*

(2) *When* $1 < \alpha < 3$

$$\widetilde{E}_n^{(\alpha)}(f) \sim E^{(\alpha)}(f) + 2\left(\zeta(\alpha)n^{\alpha-1} - \frac{2^{\alpha-1}}{\alpha-1}\right), \tag{3.2}$$

where $\zeta(\alpha)$ *is the Riemann's zeta function*

$$\zeta(\alpha) = 1 + \frac{1}{2^\alpha} + \frac{1}{3^\alpha} + \cdots.$$

(3) *When* $\alpha = 3$

$$\widetilde{E}_n^{(3)}(f) \sim E^{(3)}(f) + 2(\zeta(3)n^2 - 2) + \frac{1}{4}\left(\int_0^1 |f''(x)|^2 dx\right)(\log \frac{n}{2} + C). \tag{3.3}$$

(4) *When* $3 < \alpha < 4$

$$\widetilde{E}_n^{(\alpha)}(f) \sim E^{(\alpha)}(f) + 2\left(\zeta(\alpha)n^{\alpha-1} - \frac{2^{\alpha-1}}{\alpha-1}\right) + \frac{\alpha}{12}\left(\int_{S^1} |f''(x)|^2 dx\right)\left(\zeta(\alpha-2)n^{\alpha-3} - \frac{2^{\alpha-3}}{\alpha-3}\right).$$

(5) *When* $\alpha = 4$

$$\widetilde{E}_n^{(4)}(f) \sim E^{(4)}(f) + 2(\zeta(4)n^3 - \frac{8}{3}) + \frac{1}{3}\left(\int_{S^1} |f''(x)|^2 dx\right)(\zeta(2)n - 2) + \frac{1}{3}\left(\int_{S^1} (f''(x), f^{(3)}(x)) dx\right)(\log \frac{n}{2} + C),$$

and so on.

In particular when $\alpha = 2$

$$\widetilde{E}_n^{(2)}(f) \sim E^{(2)}(f) + \frac{\pi^2}{3}n - 4.$$

Let $\hat{\mathbf{f}}_n$ be an equilateral polygonal knot with n vertices which lie on a knot which is similar to $f(S^1)$. Then the ratio $\widehat{E}_n^{(2)}(\hat{\mathbf{f}}_n)/\widetilde{E}_n^{(2)}(f)$ approaches 1 as n goes to infinity. Thus we have $\widehat{E}_n^{(2)}(\hat{\mathbf{f}}_n) \sim \frac{\pi^2}{6}n + E_{[O1]}(f)$, which implies the observation by Simon.

Since

$$E^{(\alpha)}(f) = \iint_{S^1\times S^1} \Phi_f^{(\alpha)}(x,y)dxdy = \lim_{n\to\infty}\frac{1}{n^2}\sum_{i\neq j}\Phi_f^{(\alpha)}(\frac{i}{n},\frac{j}{n}).$$

the proposition is reduced to the following lemma.

LEMMA 3.2.

(1) *When* $\alpha = 1$,

$$\frac{1}{n}\sum_{j=1}^{n-1}\frac{1}{\delta(0,\frac{j}{n})} \sim 2\Big(\log\frac{n}{2} + C\Big).$$

(2) *When* $\alpha > 1$

$$\frac{1}{n}\sum_{j=1}^{n-1}\frac{1}{\delta(0,\frac{j}{n})^\alpha} \sim 2\left(\zeta(\alpha)n^{\alpha-1} - \frac{2^{\alpha-1}}{\alpha-1}\right).$$

PROOF. We prove in case when n is an odd integer $n = 2m+1$.
(1) The left-hand-side is equal to

$$\begin{aligned}\frac{2}{n}\sum_{j=1}^{m}\frac{1}{\frac{j}{n}} &= 2\sum_{j=1}^{m}\frac{1}{j}\\ &= 2\Big\{\Big(\sum_{j=1}^{m}\frac{1}{j} - \log m\Big) + \log m\Big\}\end{aligned}$$

which is asymptotic to

$$2\Big(C + \log\frac{n}{2}\Big).$$

(2) The left-hand-side is equal to

$$\frac{2}{n}\sum_{j=1}^{m}\frac{1}{(\frac{j}{n})^\alpha} = \frac{2}{n}\Big(\sum_{j=1}^{\infty}\frac{1}{(\frac{j}{n})^\alpha} - \sum_{j=m}^{\infty}\frac{1}{(\frac{j}{n})^\alpha}\Big)$$

which is asymptotic to

$$2\left(\zeta(\alpha)n^{\alpha-1} - \int_{\frac{1}{2}}^{\infty}\frac{dt}{t^\alpha}\right). \qquad \square$$

PROOF OF PROPOSITION 3.1. We prove the case when $\alpha = 3$. The argument is parallel when $\alpha \neq 3$.

$$\begin{aligned}
E^{(3)}(f) &= \iint_{S^1\times S^1}\left(\frac{1}{|f(x)-f(y)|^3}-\frac{1}{\delta(x,y)^3}-\frac{|f''(x)|^2}{8\delta(x,y)}\right)dxdy \\
&= \lim_{n\to\infty}\frac{1}{n^2}\sum_{i\neq j}\left(\frac{1}{|f(\frac{i}{n})-f(\frac{j}{n})|^3}-\frac{1}{\delta(\frac{i}{n},\frac{j}{n})^3}-\frac{|f''(\frac{i}{n})|^2}{8\delta(\frac{i}{n},\frac{j}{n})}\right) \\
&= \lim_{n\to\infty}\left\{\widetilde{E}_n^{(3)}(f)-\frac{1}{n}\sum_{j=1}^{n-1}\frac{1}{\delta(0,\frac{j}{n})^3}\right. \\
&\qquad\left.-\frac{1}{8}\left(\frac{1}{n}\sum_{i=1}^{i=n}\left|f''(\frac{i}{n})\right|^2\right)\left(\frac{1}{n}\sum_{j=i+1}^{i+n-1}\frac{1}{\delta(\frac{i}{n},\frac{j}{n})}\right)\right\} \\
&= \lim_{n\to\infty}\left\{\widetilde{E}_n^{(\alpha)}(f)-2(\zeta(3)n^2-2)-\frac{1}{8}\left(\int_{S^1}|f''(x)|^2dx\right)\cdot 2\left(\log\frac{n}{2}+C\right)\right\}. \square
\end{aligned}$$

4. Asymptotic behavior of energy minimizing polygonal knots

Numerical experiments have been carried out to evolve a given polygonal knot so as to decrease its value of $r^{-(\alpha+1)}$-modified electrostatic energy until it comes to a critical conformation. Most of computer programs allow us to change the number of vertices of a polygonal knot at each step of knot-evolving process. Usually we start with small number of vertices because it is faster, and after a knot comes (close) to a critical point we increase the number of vertices to produce a smoother configuration. One might ask what the limit configuration is. We give conjectures that are naturally derived from the formulae in the previous section.

Let $[K]$ be a non-trivial knot type. Suppose $\mathbf{f}_n^{(\alpha)} = \mathbf{f}_n^{(\alpha)}([K]) \in \mathbb{L}(2^n)\cap[K]$ attains the minimum value of $\widehat{E}_{2^n}^{(\alpha)}$ among $\mathbb{L}(2^n)\cap[K]$. Assume $\mathbf{f}_n(0)=0$ for any n. We can choose a subsequence $\{\mathbf{f}_{n_1(1)}^{(\alpha)},\mathbf{f}_{n_2(1)}^{(\alpha)},\cdots\}$ of $\{\mathbf{f}_1^{(\alpha)},\mathbf{f}_2^{(\alpha)},\cdots\}$ so that $\mathbf{f}_{n_i(1)}^{(\alpha)}(\frac{1}{2})$ converges as i goes to infinity. Then we can choose a subsequence $\{\mathbf{f}_{n_1(2)}^{(\alpha)},\mathbf{f}_{n_2(2)}^{(\alpha)},\cdots\}$ of $\{\mathbf{f}_{n_1(1)}^{(\alpha)},\mathbf{f}_{n_2(1)},\cdots\}$ so that $\mathbf{f}_{n_i(2)}^{(\alpha)}(\frac{j}{4})$ converges as i goes to infinity for any j. Inductively for any k we can choose a subsequence $\{\mathbf{f}_{n_1(k)}^{(\alpha)},\mathbf{f}_{n_2(k)}^{(\alpha)},\cdots\}$ out of $\{\mathbf{f}_{n_1(k-1)}^{(\alpha)},\mathbf{f}_{n_2(k-2)}^{(\alpha)},\cdots\}$ such that $\mathbf{f}_{n_i(k)}^{(\alpha)}(\frac{j}{2^k})$ converges as i goes to infinity for any j $(1\le j\le 2^k)$. Let $f_\infty^{(\alpha)} = f_\infty^{(\alpha)}([K]) : S^1 \to \mathbb{R}^3$ be a continuous map with

$$f_\infty^{(\alpha)}(\frac{j}{2^k}) = \lim_{i\to\infty}\mathbf{f}_{n_i(k)}(\frac{j}{2^k})$$

for any k and j. One might ask the meaning of $f_\infty^{(\alpha)}$. Assume that $f_\infty^{(\alpha)}$ is sufficiently smooth so that $E^{(\alpha)}(f_\infty^{(\alpha)})$ can be defined.

When $2\le\alpha<3$ the formula (3.2)

$$\widetilde{E}_n^{(\alpha)}(f) \sim E^{(\alpha)}(f) + 2\left(\zeta(\alpha)n^{\alpha-1}-\frac{2^{\alpha-1}}{\alpha-1}\right),$$

naturally leads us to the following conjectures:

(4.1) If $\alpha = 2$ and K is a prime knot then $f_\infty^{(2)}$ is one of the $E^{(2)}$-minimizers which are guaranteed to exist by [FHW].

(4.2) If $\alpha = 2$ and K is a composite knot then $f_\infty^{(2)}$ is the standard circle as was conjectured by [KS].

(4.3) If $2 < \alpha < 3$ then $f_\infty^{(\alpha)}$ is an $E^{(\alpha)}$-minimizer which is guaranteed to exist by [O3].

On the other hand, if $\alpha \geq 3$ we conjecture that $f_\infty^{(\alpha)}$ is a "singular knot" with double points for the following reason. When $\alpha = 3$ we have

$$(3.3)\quad \widetilde{E}_n^{(3)}(f) \sim E^{(3)}(f) + 2(\zeta(3)n^2 - 2) + \frac{1}{4}\left(\int_0^1 |f''(x)|^2 dx\right)(\log\frac{n}{2} + C).$$

As n increases the contribution of the total squared curvature

$$\kappa^{(2)}(f) = \int_0^1 |f''(x)|^2 dx$$

dominates the right hand side. Suppose $f_\infty^{(3)}$ is an embedding that does not attain the local minimum value of the total squared curvature functional. Then there is an embedding $\bar{f} \in [K]$ in the neighbourhood of $f_\infty^{(3)}$ with respect to the C^2-topology with $\kappa^{(2)}(\bar{f}) < \kappa^{(2)}(f_\infty^{(3)})$. Since $E^{(3)}(\bar{f})$ is finite, (3.3) implies that $\widetilde{E}_n^{(3)}(\bar{f}) < \widetilde{E}_n^{(3)}(f_\infty^{(3)})$ for $n \gg 1$. Suppose that $\bar{\mathbf{f}}_n \in \mathbb{L}(2^n) \cap [K]$ is a polygonal knot whose vertices lie on a smooth knot that is similar to $\bar{f}(S^1)$. As $\widehat{E}_{2^n}^{(3)}(\mathbf{f}_n)$ and $\widehat{E}_{2^n}^{(3)}(\bar{\mathbf{f}}_n)$ converge to $\widetilde{E}_{2^n}^{(3)}(f_\infty^{(3)})$ and $\widetilde{E}_{2^n}^{(3)}(\bar{f})$ respectively, $\widehat{E}_{2^n}^{(3)}(\mathbf{f}_n^{(3)}) > \widehat{E}_{2^n}^{(3)}(\bar{\mathbf{f}}_n)$ for $n \gg 1$, which is a contradiction. Thus if $f_\infty^{(3)}$ is an embedding then it must attain the local minimum value of the total squared curvature functional. On the other hand, Langer and Singer showed that there are no stable critical point of the total squared curvature functional except for the standard circle. Therefore $f_\infty^{(3)}$ can not be an embedding.

This conjecture is also supported by the following observation that the gradient of $\widehat{E}_n^{(\alpha)}$ in the tangent space of $\mathbb{L}(n)$ is dominated by local data as n goes to infinity if $\alpha \geq 3$, which can be shown by the same argument as in [O6].

Since the leading terms of the right hand sides of the formulae of Theorem 3.1 which are dependent on f are the total squared curvature functional for all $\alpha \geq 3$, we conjecture that

(4.4) If $\alpha \geq 3$ then $f_\infty^{(\alpha)} = f_\infty^{(\alpha)}([K])$ satisfies the following conditions:

(1) $f_\infty^{(\alpha)}$ is independent of α.

(2) $f_\infty^{(\alpha)}$ attains the minimum value of the total squared curvature functional among $\overline{[K]}$, where $\overline{[K]}$ is a closure of $[K]$ in the set of immersions with respect to the C^2-topology.

(3) $f_\infty^{(\alpha)} \in \overline{[K]} \setminus [K]$.

REFERENCES

[A] K. Ahara, Computer program and video : *Energy of a knot* (1990).

[BO1] G. Buck and J. Orloff, *Computing canonical conformations for knots*, Topology Appl. **51** (1993), 247–253.

[BS] G. Buck and J. Simon, *Knots as dynamical systems*, Topology Appl. **51** (1993), 229–246.

[FHW] M. H. Freedman, Z.-X. He, and Z. Wang, *Möbius energy of knots and unknots*, Ann. of Math. **139** (1994), 1–50.

[F1] S. Fukuhara, *Energy of a knot*, A Fête of Topology, Y. Matsumoto, T. Mizutani, and S. Morita, eds., Academic Press, 1987, pp. 443–452.

[F2] ———, computer program : *Energy of a knot.*

[GA] "http://www.gang.umass.edu/docs/archive/knots.html".

[Gr] M. Gromov, *Filling Riemannian manifolds*, J. Differential Geom. **18** (1983), 1–147.

[Gu] C. Gunn, Computer program *"Linkmover"* at Geometry Supercomputer Project (1988-1989).

[Hu] M. Huang, "http://www.eecs.uic.edu/ mhuang/".

[KNS] Kusner, R., Norman. P. and Sullivan, J., *Möbius Energy Minimizing Knot Catalog*, preprint III.18, GANG/UMass Amherst (1993).

[KS] R. B. Kusner and J. M. Sullivan, *Möbius energies for knots and links, surfaces and submanifolds*, Geometric Topology (Proceedings of the 1993 Georgia International Topology Conference), AMS/IP Studies in Adv. Math., W. H. Kazez ed., Amer. Math. Soc. and International Press, Cambridge, MA., 1996, pp. 570–604.

[LS1] J. Langer and D. A. Singer, *Knotted elastic curves in* $\mathbb{R}^3$, J. London Math. Soc. **30** (1984), 512–520.

[LS2] ———, *The total squared curvature of closed curves*, J. Differential Geom. **20** (1984), 1–22.

[LS3] ———, *Curve straightening and a minimax argument for closed elastic curves*, Topology **24** (1985), 75–88.

[LS] Ligocki, T.J. and Sethian, J.A., *Recognizing Knots Using Simulated Annealing*, J. Knot Theory Ramifications **3** (1994), 477–495.

[O1] J. O'Hara, *Energy of a knot*, Topology **30** (1991), 241–247.

[O2] ———, *Family of energy functionals of knots*, Topology Appl. **48** (1992), 147–161.

[O3] ———, *Energy functionals of knots II*, Topology Appl. **56** (1994), 45–61.

[O4] *Energy of knots*, in Japanese, Suugaku **49-4** (1997), 131–144, English translation is to appear in Suugaku Exposition.

[O5] ———, *Energy of knots*, Ideal knots, L. H. Kauffman, V. Katritch, and A. Stasiak eds., World Scientific, Singapore, to appear around May 1998. The updated version is available at "http://www.comp.metro-u.ac.jp/õhara/texfiles/esurvey.dvi".

[O6] ———, *Criticality for the knot energy*, preprint (1994).

[Sc] R. Scharein, Computer program. Information is available at "http://www.cs.ubc.ca /spider/scharein/".

[S1] J. K. Simon, *Energy functions for polygonal knots*, J. Knot Theory Ramifications **3** (1994), 299-320.

[S2] ———, *Energy functions for knots; beginning to predict physical behavior*, to appear in Mathematical approaches to biomolecular structure and dynamics (J.P. Mesirov, K. Schulten, and D.W. Sumners Eds.), IMA Vol. Math. Appl. Vol. 82, Springer-Verlag, NY, 1996.

[Su] J. M. Sullivan, *Knots minimizing a Möbius-invariant energy*, VHS video tape (11/93) available from the Geometry Center, Minneapolis.

[W] Y.-Q. Wu, Computer program *"Ming"*, Information is available at "http://www.math. uiowa.edu/ ~wu/".

DEPARTMENT OF MATHEMATICS, TOKYO METROPOLITAN UNIVERSITY, 1-1 MINAMI-OHSAWA, HACHIOUJI-SHI, TOKYO 192-03, JAPAN.

E-mail address: ohara@comp.metro-u.ac.jp

Selected Titles in This Series

(Continued from the front of this publication)

207 **Yujiro Kawamata and Vyacheslav V. Shokurov, Editors,** Birational algebraic geometry: A conference on algebraic geometry in memory of Wei-Liang Chow (1911–1995), 1997

206 **Adam Korányi, Editor,** Harmonic functions on trees and buildings, 1997

205 **Paulo D. Cordaro and Howard Jacobowitz, Editors,** Multidimensional complex analysis and partial differential equations: A collection of papers in honor of François Treves, 1997

204 **Yair Censor and Simeon Reich, Editors,** Recent developments in optimization theory and nonlinear analysis, 1997

203 **Hanna Nencka and Jean-Pierre Bourguignon, Editors,** Geometry and nature: In memory of W. K. Clifford, 1997

202 **Jean-Louis Loday, James D. Stasheff, and Alexander A. Voronov, Editors,** Operads: Proceedings of Renaissance Conferences, 1997

201 **J. R. Quine and Peter Sarnak, Editors,** Extremal Riemann surfaces, 1997

200 **F. Dias, J.-M. Ghidaglia, and J.-C. Saut, Editors,** Mathematical problems in the theory of water waves, 1996

199 **G. Banaszak, W. Gajda, and P. Krasoń, Editors,** Algebraic K-theory, 1996

198 **Donald G. Saari and Zhihong Xia, Editors,** Hamiltonian dynamics and celestial mechanics, 1996

197 **J. E. Bonin, J. G. Oxley, and B. Servatius, Editors,** Matroid theory, 1996

196 **David Bao, Shiing-shen Chern, and Zhongmin Shen, Editors,** Finsler geometry, 1996

195 **Warren Dicks and Enric Ventura,** The group fixed by a family of injective endomorphisms of a free group, 1996

194 **Seok-Jin Kang, Myung-Hwan Kim, and Insok Lee, Editors,** Lie algebras and their representations, 1996

193 **Chongying Dong and Geoffrey Mason, Editors,** Moonshine, the Monster, and related topics, 1996

192 **Tomek Bartoszyński and Marion Scheepers, Editors,** Set theory, 1995

191 **Tuong Ton-That, Kenneth I. Gross, Donald St. P. Richards, and Paul J. Sally, Jr., Editors,** Representation theory and harmonic analysis, 1995

190 **Mourad E. H. Ismail, M. Zuhair Nashed, Ahmed I. Zayed, and Ahmed F. Ghaleb, Editors,** Mathematical analysis, wavelets, and signal processing, 1995

189 **S. A. M. Marcantognini, G. A. Mendoza, M. D. Morán, A. Octavio, and W. O. Urbina, Editors,** Harmonic analysis and operator theory, 1995

188 **Alejandro Adem, R. James Milgram, and Douglas C. Ravenel, Editors,** Homotopy theory and its applications, 1995

187 **G. W. Brumfiel and H. M. Hilden,** $SL(2)$ representations of finitely presented groups, 1995

186 **Shreeram S. Abhyankar, Walter Feit, Michael D. Fried, Yasutaka Ihara, and Helmut Voelklein, Editors,** Recent developments in the inverse Galois problem, 1995

185 **Raúl E. Curto, Ronald G. Douglas, Joel D. Pincus, and Norberto Salinas, Editors,** Multivariable operator theory, 1995

184 **L. A. Bokut′, A. I. Kostrikin, and S. S. Kutateladze, Editors,** Second International Conference on Algebra, 1995

183 **William C. Connett, Marc-Olivier Gebuhrer, and Alan L. Schwartz, Editors,** Applications of hypergroups and related measure algebras, 1995

182 **Selman Akbulut, Editor,** Real algebraic geometry and topology, 1995

181 **Mila Cenkl and Haynes Miller, Editors,** The Čech Centennial, 1995

(See the AMS catalog for earlier titles)

The Development and Use of Standards by Non-state Actors

A Study of the Dynamics of Regulating Sustainability Assurance

av

Amanda Tan-Sonnerfeldt

AKADEMISK AVHANDLING

som för avläggande av filosofie doktorsexamen
vid Ekonomihögskolan, Lunds universitet,
kommer att offentligen försvaras i EC3:210, Holger Craafoords Ekonomicentrum,
torsdagen den 8 december 2011, kl. 10.00.

Department of Business Law
School of Economics and Management
Lund University

ISBN 978-91-7473-192-7

Printed by Media-Tryck 2011
Lund, Sweden

Acknowledgements

This thesis was inspired by many interdisciplinary projects taken at a time of change. Changes in global financial regulation, the rise of standards, the "come back" of the sustainability agenda... gave many, from different academic disciplines around the world, a common ground to discuss regulatory issues in contemporary society. The journey to complete this thesis had been a challenging, meaningful and at times an exhilarating one.

I owe my deepest gratitude to my supervisors, Professor Krister Moberg and Professor Anne Loft, who have shared their knowledge and expertise, and read the many initial drafts with patience and understanding. I would also like to extend my sincere thanks and appreciation to Professor Claes Norberg who started this project and for encourage me to pursue this topic from an interdisciplinary perspective. I would also like to thank Mr Reidar Peters and Far for helping fund this project; Professor Håkan Hydén, Professor Karin Jonnergård, Professor Robert Scapens, Dr Fredrik Ljungdahl, Dr Adam Diamant, Dr Thomas Riise Johansen and Pernilla Broberg for commenting on my previous drafts and for sharing theoretical and practical insights that shaped the project. I am also grateful to the participants of the CSEAR conference in 2007 at St. Andrews and Öresund Auditing Research Network conference in 2008, who have shared their ideas and views. Last but not least, I sincerely thank all my colleagues for their support and my family for their patience and understanding over these years.

To the Sonnerfeldts, Tans and Songs . . . especially for Sam

Abbreviations

AICPA	American Institute of Certified Public Accountants
AS	Assurance Standard
ASB	Accounting Standards Board
APB	Auditing Practices Board
APS	AccountAbility Principles Standards
BCBS	Basel Committee on Banking Supervision
BFN	The Swedish Accounting Standards Board
CAG	Consultative Advisory Group
CAP	Compliance Advisory Panel
CEO	Chief Executive Officer
CSR	Corporate Social Responsibility
DJSI	Dow Jones Sustainability Index
DnR	Norwegian Institute of Public Accountants
ED	Exposure Draft
EITI	Extractive Industries Transparency Initiative
EMAS	Eco-Management and Audit Scheme
EU	European Union
ERP	Environmental Performance Review
EWP	Environmental Working Party
FEE	European Federation of Accountants
GAAP	Generally Accepted Accounting Principles
GRI	Global Reporting Initiative
G250	Fortune Global 250
HSE	Health, Safety and Environment

IAASB	International Auditing and Assurance Standards Board
IIRC	International Integrated Reporting Committee
IIRF	International Integrated Reporting Framework
IAIS	International Association of Insurance Supervisors
IAPC	International Auditing Practices Committee
IAS	International Accounting Standards
IASC	International Accounting Standards Committee
ICAEW	The Institute of Chartered Accountants in England and Wales
IDW	Institute of Public Auditors in Germany
IES	International Education Standards
IESBA	International Ethics Standards Board for Accountants
IFAC	International Federation of Accountants
IFAE	International Framework for Assurance Engagements
IFRS	International Financial Reporting Standards
IOSCO	International Organization of Securities Commissions
IRCA	International Register of Certificated Auditors
ISA	International Standards on Auditing
ISAE	International Standards on Assurance Engagements
SEAP	Sustainability Expert Advisory Panel
SES	Stakeholder Engagement Standard
ISRE	International Standards on Review Engagements
ISO	International Organization for Standardization
ISQC	International Standard on Quality Control
MNEs	Multinational Enterprises
NIVRA	Royal Dutch Institute of Registered Accountants
NGO	Non-Governmental Organization
OECD	Organization for Economic Co-operation and Development
PIOB	Public Interest Oversight Board
RMS	Resource Management System
RN	The Supervisory Board of Public Accountants

SCAS	Special Committee on Assurance Services
SEAAR	Social and Ethical Accounting, Auditing and Reporting
TBL	Triple Bottom Line
TFEU	Treaty of the Functioning of the European Union
TEU	Treaty of the European Union
UK	United Kingdom
UN	United Nations
US	United States of America
WB	World Bank

Table of Content

Table of Figures

1. Introduction

1.1 Introduction

The globalization, liberalization and privatization waves in the last two decades have significantly reshaped the regulatory landscape and environment in which businesses operate. The increase in the size and scope of cross border business activities and growing interdependence of economies have resulted in problems and challenges that cannot be tackled solely by states at a national level.[1] Policy initiatives are increasingly being taken at regional, transnational and global levels to address these problems. Although states continue to play an active role in the setting of international standards[2]; non-state actors such as international and civil society organizations have emerged to set "alternatives to legal rules"[3] to fill perceived regulatory gaps. These initiatives have been taken particularly in areas where their concerns are inadequately addressed.[4] These "alternatives to legal rules", such as "soft law", "best practice codes" and "standards"[5] have been adopted by members or groups in society due to various reasons. These include, the authority given to it by hard law; requirements to gain access to memberships, resources or certifications; or simply socialization, acculturation or normative pressures.[6] Limited by resources and regulatory capacity to set rules on a global scale, regulatory initiatives no longer operate within the exclusive

---

[1] Peters, A., Koechlin, L., Förster, T. & Zinkernagel, G. (2009) Non-state Actors as Standard Setters, Cambridge University Press, p. 1-4; see also, Picciotto, S. (2002) "Introduction: Reconceptualization regulation in the era of globalization", Journal of Law and Society, 29(1), p. 1.

[2] For example, states play an active role in the standard setting process of the Extractive Industries Transparency Initiative (EITI). The EITI is a public-private-partnership (PPP), a partnership between states, businesses and civil society organizations. It sets a global standard for transparency in oil, gas and mining. For more information, refer to http://eiti.org/eiti.

[3] Kerwer, D. (2005) "Rules that Many Use: Standards and Global regulation" Governance: An international Journal of Policy, Administration, and Institutions, 18(4), p. 611.

[4] Peters, A. et al. (2009), p. 2.

[5] Kerwer, D. (2005), p. 611.

[6] Djelic, M.; Sahlin-Andersson, K. (2007) Transnational Governance, Cambridge University Press, p. 6. See also Henning, R. (2000) "Selling Standards" in Brunsson, N.; Jacobsson, B. (2000) "A World of Standards", Oxford University Press, p. 115-124 for examples.

competence of nation states. They are also undertaken increasingly by different polymorphous non-state actors, i.e. non-state actors organised in different forms and constitutions.[7]

Sustainability assurance is an area where non-state standard setters have rapidly developed standards. Sustainability assurance standards carry the notion of improving the quality of assurance engagements, which result in more credible sustainability reports.[8] Until 2003, there were no internationally accepted standards providing guidance to assurance providers in this field. Today, sustainability assurance engagements are performed according to standards set at the international, national levels or a combination of them. Important international standards include the International Standard on Assurance Engagements (ISAE3000) by the International Auditing and Assurance Standards Board (IAASB) of the International Federation of Accountants (IFAC) and AA1000 Assurance standard (AA1000AS) by AccountAbility. National standards include RevR6 by Far in Sweden, 3410N by Royal NIVRA in the Netherlands (NIVRA)[9] among others.[10] These standard setting initiatives have been keenly supported by various parties including the European Federation of Accountants (FEE)[11] and the Global Reporting Initiative (GRI)[12]. As the accountancy profession is one of the major actors performing sustainability assurance engagements[13], the setting of sustainability assurance standards has recently found its way to the 2009-2011 strategy and work program IFAC sets. IFAC has

---

[7] Peters, A. et al. (2009) Preface and p. 1-4.

[8] Refer to section 1.3.5 of this thesis.

[9] Royal NIVRA in the Netherlands has recently merged with the Nederlandse Orde van Accountants-Administratieconsulenten (NOvAA) and has adopted the abbreviation NBA. In the thesis, due to the timeframe of this study, NIVRA is used.

[10] Country specific standards have been published in the Netherlands, Sweden, Germany, France, Japan and Australia.

[11] FEE Position Paper: Call for Action: Assurance for Sustainability, June 2004, p. 32. FEE Press Release: "European Accountants Call for a Specific International Standard on Sustainability Assurance" 13th June 2006.

[12] Global Reporting Initiative "Sustainability Reporting Guidelines" 2002, p. 17-18. GRI is a private standard setting organization that set sustainability reporting standards through a large multi-stakeholder network of experts worldwide to guide triple bottom line reporting. It encourages independent assurance and development of standards and guidelines for the assurance process.

[13] The term sustainability assurance engagement has been chosen over assurance on sustainability reports commonly referred to in professional literature as it captures a wider range of engagements.

expressed its intention to develop international standards through the IAASB to provide guidance to its members in this area.[14]

Who and why are the non-state actors taking regulatory initiatives? What are the characteristics of standards set by non-state actors? How do they emerge and function? This thesis concerns regulation by non-state standard setters. Research will be centred on the development and use of sustainability assurance standards by the accountancy profession, with the inclusion of a case study on their development and use in Sweden.

1.2 Regulation in the era of globalisation

1.2.1 Contemporary use of the term 'regulation'

Law manifests itself in different forms.[15] This assertion can be extended to regulation. The term "regulation", has become "sufficiently ubiquitous"[16], yet "notoriously difficult" to generally agree on its meaning and scope.[17] As the term is not "owned" by a particular academic discipline or institution, it is not surprising that it has captured the attention of legal, political and social sciences scholars, many of whom suggest multidisciplinary or interdisciplinary studies to understand regulation in this modern era.[18]

Black illustrated the multiplicity of meanings by capturing the contemporary usage of the term regulation by academia, governments, civil society organizations and

---

[14] IFAC (2008) "Basis for Conclusions: International Auditing and Assurance Standards Board Strategy and Work Program, 2009-2011", p. 10-11.

[15] Banakar, R. (2009) "Law through sociology's looking glass. Conflict and competition in sociological studies of law" in Denis, A., Kalekin-Fishman, D. (2009) "The New ISA handbook in Contemporary International Sociology: Conflict, competition, cooperation", London: Sage. http://papers.ssrn.com/sol3/papers.cfm?abstract_id=1574821 p. 59.

[16] Picciotto, S. (2002), p . 1.

[17] Morgan, B.; Yeung, K. (2007) "An Introduction to Law and Regulation" Cambridge University Press, p. 3; Black, J. (2001) "Decentring Regulation: Understanding the Role of Regulation and Self-Regulation in a 'Post-Regulatory' World", Current Legal Problems, 54(1) p. 128-144.

[18] Peters, A. (2009), p. 11. See also Mörth, U. (2004) "Soft law in Governance and Regulation" Cheltenham: Edward Elgar. The author approached her analysis from an interdisciplinary perspective encompassing sociology, political science, law and organisational studies, among others.

multinational enterprises (MNEs). [19] The term has been used to depict a legal instrument; an intentional or goal-oriented process to control, direct, alter, enable, coordinate or render constant; or an outcome or property.[20] The term regulation, described by Black as an "ever-expanding concept" has been used to depict both initiatives originating from both state as well as non-state actors via a whole array of instruments which may be legally or non-legally binding.[21] This scenario reaffirms the shifts and emergence of regulatory initiatives and control systems in other, multiple non-state locations. Regulation in the era of globalisation challenges the traditional legal conceptual understanding of the term 'regulation' and points to the importance of finding an appropriate theoretical lens to analyse the dynamics of regulating sustainability assurance.

1.2.2 Conceptions and definition of regulation

Baldwin and Cave identified three conceptions of regulation.[22] From the narrowest perspective, regulation is regarded as the promulgation of an authoritative set of rules, often accompanied by some mechanism for monitoring and enforcing compliance[23]. This conception centres on the deliberate attempt by the state to influence socially valued behaviour through setting, monitoring and enforcing rules.[24] A broader conception regards regulation as deliberate state initiative to influence industrial or social behaviour to steer the economy, for example through taxation and subsidies.[25] These two conceptions place the state at the regulatory locus and preclude from its definition other forms of social control by non-state actors. The third conception however "breaks the connection with the state"[26] and conceives regulation as all forms of social control.[27] It adopts a broad perspective as it includes within the scope of

[19] Black, J. (2001), p. 133-144. Black "artificially" grouped the use of the term regulation into: What is assumed "regulation"? Who or what is performing it? What organizational or institutional form it takes? What actors or area of social life it applies? Via what mechanisms, instruments or techniques are regulation conducted?

[20] Ibid.

[21] Ibid.

[22] Baldwin, R.; Cave, M. (1998) "Understanding Regulation" Oxford University Press, p. 2.

[23] Ibid.

[24] Ibid., p. 35-39.

[25] Ibid., p.2.

[26] Black, J. (2002) "Critical Reflections on Regulation", CARR; London School of Economics p. 8.

[27] Baldwin, R.; Cave, M. (1998), p. 2.

regulation, any form of social control or influence independent of the hierarchical forms of state law.

Heeding caution that a too broad a definition risks the danger of conceptual incoherence and too narrow a definition fails to capture the emergence of non-state actors, multiple levels and sites of regulation, this thesis adopts Black's definition of regulation.[28] Black combined the task aspects of a cybernetic definition of regulation with the temporal, intentional and problem solving aspects to develop an essentialist definition.[29] According to Black, "regulation is the sustained focused attempt to alter the behaviour of others according to defined standards or purposes with the intention of producing a broadly identified outcome or outcomes, which may involve mechanisms of standard setting, information-gathering or monitoring and behaviour-modification".[30]

This definition is suitable to be used to analyse the initiatives to regulate sustainability assurance for the following reasons. First, the definition does not situate the state at the centre of regulatory activity; the involvement of the state in regulation can lie on the continuum of being the sole regulator, one of the participating actors, or playing no role at all. It allows a more pluralistic, less hierarchical account of regulation.[31] Applied to this thesis, it captures within its scope the regulatory initiatives by the two prominent non-state standard setters at the international level: the IAASB and AccountAbility. The former is a standard setting board of IFAC a non-state civil society organization that develops international standards for auditing, quality control, review, other assurance and related services.[32] AccountAbility, the latter, positions itself as a leading global non-state organization that sets standards through a multi-stakeholder process to help organizations become more accountable, responsible and sustainable.[33]

---

[28] Black, J. (2002), p. 19-20.

[29] Ibid., p. 20.

[30] Ibid. Note for Chapter 2 Hancher, L.; Moran, M.(1989) "Organizing Regulatory Space" in Hancher, L.; Moran, M.(1989) "Capitalism, Culture, and Economic Regulation", Clarendon Press, p. 271. The authors identified common feature of regulation to includes: design of rules, creation of responsible institutions responsible for implementation, clarify meaning and rules in part circumstances; enforce rules.

[31] Ibid., p. 21.

32 http://www.ifac.org/auditing-assurance.

33 http://www.accountability.org/.

It is important to note that there are other prominent standard setters in the field of sustainability e.g., International organisation for Standardization (ISO), Social Accountability International which have been excluded from the scope of this study as they are certification schemes which the companies

Second, the crux of regulation is placed on attempts to alter behaviour. It recognises the relational nature of regulation, the existence of interactions and interdependencies between social actors and between these social actors and the government.[34] Third, the intention to regulate and mechanisms of standard setting, information gathering and behavioural modification have been built into the definition, suggesting the presence of some form of institutionalised mechanism that allows a distinction between regulation and "all forms of social control" such as market, social and cultural forces.[35]

1.2.3 Cognitive frame of regulation

The background presented in section 1.1, depicts regulation in contemporary society as consisting of multiple levels and sites of regulation originating from state and non-state actors, emerging and operating concurrently and in overlapping ways. Heterogeneity of regulation challenges traditional conceptions of legal legitimacy and traditional formulations of the rule of law.[36] This suggests a need for a broader and more contextual theoretical framework to analyse the dynamics of regulating sustainability assurance.

Decentred regulation involves "a shift (and recognition of such a shift) in the locus of the activity of 'regulating' from the state to other, multiple, locations, and the adoption on the part of the state of particular strategies of regulation".[37] The notion of decentring "opens up the cognitive frame of regulation" forcing the examination of "what is it we are looking at, and looking for, when we seek to analyse regulation."[38] Building on the work of Black in socio-legal studies, and Deljic and Sahlin-Andersson in transnational governance, this thesis revisits the usually taken for granted fundamental questions of regulation: Who is regulating? What is the character of

---

utilises to serve a different function from the reporting and assurance. This will be further explained in chapter 5.

[34] Smith, K. (2002) "What is Regulation: A Reply to Julia Black" Australian Journal of Legal Philosophy, 27, p. 42-43; Black, J. (2002), p. 6.

[35] Black, J. (2002), p. 22-26.

[36] Ibid.

[37] Black, J. (2001), p. 112.

[38] Black, J. (2002), p. 27.

standards? And how are standards interpreted and enforced?[39] This is presented in table 1.1.

When regulation is state-centred, existing legal institutions in a country determine how rules are established, interpreted and enforced. For example, in many Westphalia nations, the state adopts legally binding rules that are constitutionally valid, i.e. originating from a legal basis and subject to a legitimate due process. The courts are the final interpreters of these rules, the sources of law and material that aid interpretation are clearly distinguished, and these rules are enforced by sanctions.[40] In comparison, regulation by non-state organisations operates external to the existing legal institutions. As such, the regulatory design is more responsive to the context in which it is operating.[41]

Fundamental questions	State-centred regulation	Non-state-centred regulation
Who is regulating?	State	Non-state actors
What is the character of rules[42]	Legally binding	Legally non-binding standards but could be binding through other means
How are they interpreted?	Formal (interpreted by authorities)	Less formal (greater room for user interpretation)

[39] Black, J. (2001), Black, J. (2002) referred to above. Djelic, M.; Sahlin-Andersson, K. (2006), p. 5-6, the authors built on Baldwin and Cave's conceptions on regulation.

[40] De Cruz, P. (1999) "Comparative Law in a Changing World", Second edition, London: Cavendish Publishing Limited, p. 46, 66-68, 99-102, 243-244, 267-269. See also, Nergelius, J. (2000) "Constitutional Law" p. 73-75; Vogel, H-H. (2000) "Sources of Swedish Law", p. 61-62; 49-54 in Bogdan, M. "Swedish Law in the new millennium", Norstedt Juridik for a description of the legal system in Sweden.

[41] Black, J. (2002), p. 27.

[42] Djelic and Sahlin-Andersson used the term "dimension of rules" when revisiting the four areas in transnational governance. The authors used the term "Mode of regulation" to describe if a rule is legally or non-legally binding and "nature of rules" to depict the formality of interpretation. This thesis adopts terminology in line with Black, J. (1997) Rules and Regulators, Oxford University Press.

How are they enforced?	Legal sanctions	Social mechanisms (memberships, normative pressures)

Figure 1.1: The fundamental questions of regulation

Although decentred regulation captures regulation in the modern era, much remains to be understood as to: why non-state actors get involved, what gives them or their standards legitimacy, etc. These questions have to be analysed and understood in the context within which non-state regulation emerges in order to understand the activities, the growing interdependence and plurality of actors involved in the regulatory process.[43] It has become an empirical question of interest to other disciplines than law, however the boundary as to which academic discipline it falls, is unclear and it does not simply fall into one academic discipline, but several.

1.2.4 Research approach

This thesis adopts an interdisciplinary approach drawing from the socio-legal, social science and legal disciplines. The approach adopted draws heavily from applied socio-legal studies[44] reviewing the four fundamental questions as a point of departure to understand the emergence and functioning of standards by non-state actors. Focus is placed on regulation in context as well as the nature and character of the regulatory instrument.

As regulating sustainability assurance is new and developing, little is known about this phenomenon and its features are likely to undergo constant redefinitions. This justifies an interdisciplinary explorative approach to research. The time frame of study is set from 1997 to 2010 to capture the emergence and development of sustainability assurance standards.

---

[43] Hutter, B. (2006) "The Role of Non-state Actors in Regulation", in: Schuppert, Gunnar Folke, (ed.) Global governance and the role of non-state actors, Nomos Publishers, Berlin, pp. 63.; Keohane, R., Nye, J. (1989) Power and Interdependence: World Politics in Transition, Second edition, Boston, Little, Brown, chapter 1.

[44] Banakar, R. (2009), p .68.

1.3 Sustainability assurance

This section provides background information on the subject matter of regulation through a descriptive account of the developments in sustainability reporting and assurance. The purpose is to draw attention to the key issues and context within which sustainability assurance has developed to stand in better stead to structure the purpose, research questions and make choices with regard to the scope of this study.

1.3.1 The sustainability agenda

Negative environmental and social impact that results from the activities pursued by MNEs has intensified global concerns on climate change, efficient use of resources and protection of society and culture. Public distrust of corporations has been further fuelled by negative media publicity, high profile public protest and lawsuits against several MNEs.[45] Sustainability is one key policy area where there are a growing number of regulatory initiatives by state and non-state actors at international, transnational, regional and national levels.[46] The shifts in regulatory focus towards self-assessment and control, which utilizes the resources of the regulated parties, have created the need for new instruments and institutions of monitoring[47]. Corporations are increasingly utilizing these tools to legitimize their existence and operations in society[48].

A popular tool used to aid legitimacy is external reporting by corporations.[49] Corporations measure and communicate the environmental and social impact of their operations to external parties through publication of reports. Sustainability issues are

---

[45] Among some of the high profile events were, the Battle of Seattle in 1999, Law suits against Nike, Shell, Monsanto, Exxon, British Petroleum, etc.

[46] Zadek, S.; Raynard, R. (2004) "The Future of Sustainability Assurance" ACCA research report 86, p. 39. For example, the UN Global Compact and multi-stakeholder initiatives such as GRI, ETI the international level; EMAS at the regional level; Chapter 6 of this thesis for examples in Sweden. See also, Djelic, M. and Sahlin-Andersson, K. (2006), p. 4 for the concept of transnational governance.

[47] Power, M. (2007) The Audit Society: Rituals of Verification, Oxford University Press. The book covers in-depth discussion of three new areas that supported the audit explosion thesis.

[48] Tomkins, C. (2001) "Interdependencies, Trust and Information in Relationships, Alliances and Networks", Accounting, Organizations and Society, 26(2), p. 161.

[49] This form of reporting is distinguished from internal reports, reports by external parties to the corporation and social audits. See also, Gray, R. (2000) "Current Developments and Trends in Social and Environmental Auditing, Reporting and Attestation: A Review and Comment, International Journal of Auditing, 4(3), p. 251 which makes a distinction between these forms of reporting.

commonly reported in stand-alone sustainability reports; part of the annual report; part of the management or board of directors' report or web-based report[50]. This form of accountability is based on the ideals of increased transparency, accountability, participative democracy and the reduction of informational asymmetries between the management and interested parties.[51] Sustainability reporting is on an upward trend and has recently become more important to businesses and actors in the capital market, in addition to the more traditional "other stakeholders" such as environmental and human rights non-governmental organisations (NGOs). In practice, the number of Global Fortune 250 (G250) companies reporting on environmental and social performance increased from 45% in 2002, to 52% in 2005, to 80% in 2008.[52]

1.3.2 The business conception of 'sustainability'

In the last two decades, the term "sustainability" has been used increasingly by policy makers, academics, businesses and NGOs among others. It has however been interpreted in different ways. A widely accepted definition of sustainability originates from the Brundtland report. Sustainable development is defined as the "development that meets the needs of the present without compromising the ability of future generations to meet their own needs"[53]. This development-oriented conception regards sustainability as a global concept concerning eco-efficiency and fair distribution of resources between the current and future generations; suggesting a scale of economic activity relative to its ecological life support systems.

Corporations that publish sustainability reports have borrowed the term sustainability to refer to their economic, social and environmental performance.[54] The business

---

50 Refer to chapter 6. See also Scott, P., Jackson, R. (2002) "Environmental, Social and Sustainability Reporting on the Web: Best Practices" Corporate Environmental Strategy, 9(2), p. 193-202. The authors covered three approaches to web-based report include: piggy-backed, standalone and integrated approach.

51 Refer to studies conducted in this light such as, Adams, C. (2004) "The Ethical, Social and Environmental Reporting-performance portrayal Gap" Accounting, Auditing & Accountability 17(5) p. 731-757; Adams, C.; Evans R. (2004) 'Accountability, Completeness, Credibility and the Audit Expectations Gap', Journal of Corporate Citizenship,14, p. 97-115.

52 KPMG International Survey of Corporate Environmental Reporting, June 1999; KPMG "International Survey of Corporate Sustainability Reporting" June 2002, 2005, 2008.

53 United Nations (UN) Department of Economic and Social Affairs, Division for Sustainable Development: Agenda 21.

54 The term economic used in this context does not refer to financial accounting information but more to firms economic efforts to better society e.g. philanthropy.

conception of sustainability in the reporting context commonly refers to and has been used interchangeably with 'triple bottom line'[55]. The incongruence in definitions brought about by the importation of a broad-based sustainability concept carrying both a moral and normative imperative into an area of practice characterized by measurement of impact and performance sets foundation for expectation gaps between the reporting organization and its various stakeholders.[56]

1.3.3 The notion of 'third party assurance'

Sustainability reports have been critiqued as a form of corporate 'green washing', adding value to the corporation instead of accountability and transparency towards the reporting organisation stakeholders[57]. Researchers have increasingly scrutinised these reports finding reporting-performance portrayal gaps and indicating its potential to reduce rather than increase the visibility of corporate social environmental impact.[58] Studies have also pointed out corporate managements' dominant influence over the contents and scope of reporting, the immaturity of reporting standards, the lack of stakeholder engagement, the exclusion of and favouritism towards particular stakeholder groups. This all contribute to undermining the credibility of reports.[59]

---

[55] O'Dwyer, B.; Owen, D. (2005) "Assurance Statement Practice in Environmental, Social and Sustainability Reporting: a Critical Evaluation" The British Accounting Review, 37(2), p. 207.

[56] Ibid, p. 207 footnote 4, see also, Gray, R.; Milne, M. (2002), Sustainability reporting: Who's Kidding Whom? Chartered Accountants Journal of New Zealand, 81(6), p. 66–70.

[57] 'Green washing' is a term coined by Jay Westerveld in 1986 that refers to companies reporting their practices as "green," when, in all reality; they used common practices shared throughout the industry. The term was originated by combining the words "green" and "whitewashing" Cowan, D.; Dopart, P. et al. (2010) "A Cross-sectional Analysis of Reported Corporate Environmental" Regulatory Toxicology and Pharmacology; Owen, D., Swift, T., Humphrey, C., Bowerman, M. (2000) "The new social audits: accountability managerial capture or the agenda of social champions?. European Accounting Review 9 (1), p. 81–98.

[58] Gray, R. (2010) "Is accounting for sustainability actually accounting for sustainability and how would we know" An exploration of narratives of organisations and the planet, Accounting, Organisations and Society, 35(1), p. 4-61; Adams, C. (2004), p. 731; Adams, C.; Evans R. (2004), p. 97.

[59] Studies include O'Dwyer, B. (2003) "Conceptions of corporate social responsibility: The nature of management capture" Accounting, Auditing and Accountability Journal, 16(4) p. 523-557. Also O'Dwyer and Owen (2005) Thomson, I. and Bebbington, J. (2005), "Corporate Social Reporting: A Pedagogic Evaluation", Critical Perspectives on Accounting, 16(5), p. 507-533, Unerman, J.; Bennett, M. (2004) "Increased stakeholder dialogue and the internet: towards greater corporate accountability or reinforcing capitalist hegemony?" Accounting, Organizations and Society *29*(7), p. 685-707, Lee Moerman, Sandra Van Der Laan, (2005) "Social reporting in the tobacco industry: all smoke and mirrors?, Accounting, Auditing & Accountability; Gray, R. & Milne, M. (2002), p. 66-70. Swift, T.; Dando, N., "From Methods to Ideologies: Closing the Assurance Expectations Gap in

The apparent credibility gap motivated NGOs' to publish reports on companies from the stakeholders' point of view. For example, Friends of the Earth, an NGO published on their own accord, a report on Shell. The report, entitled "The Other Shell Report" highlighted the activities of Shell from the perspective of particular stakeholders who felt that their concerns had been unfairly reported by the company.[60]

Michael Power, in his book 'the audit explosion', observed that the notion and definitional fuzziness of the term "audit" has allowed the percolation of the body of audit knowledge into a variety of contexts.[61] It has allowed "audits" to extend beyond its traditional role to institutionalize different forms of accountability in an era of deregulation. How the word "audit" is used is often "an attempt to claim a particular mission, social status or policy objective for a practice of accountability and control."[62] The term audit has been used to label other evaluation services on activities, conduct or performance of others where acknowledged and legitimate interest can be claimed.[63]

The essence of the well-institutionalised idea of an audit is that information is more credible if it has been subjected to independent examination by a third party. From another standpoint, audits bring 'comfort' to parties who would otherwise be vulnerable to accounts prepared in the self-interest of management. With reference to sustainability reporting, third party assurance has been demanded by reporting organisations (pull), put forward by assurance providers (push) and supported by a series of other groups, as a vital part of the credibility and trust building process. These groups include the European Commission, FEE, GRI, AccountAbility and academics.[64] The increase in sustainability assurance engagements can therefore be explained by a "pull-push-endorsement" mechanism involving multiple actors, for different purposes, based on the notion that third party assurance increases the

---

Social and Ethical Accounting Auditing and Reporting" Journal of Corporate Citizenship, 8, p. 81–90.

60 A copy of the report can be obtained from www.foe.co.uk/resource/reports/lessons_not_learned.pdf.

61 Power, M. (1997b), p. 5.

62 Courville, S. et al. (2003) "Introduction: Auditing in Regulatory Perspective" Law and Policy, Vol. 25, No. 3 July 2003 p. 179.

63 Refer to footnote .

64 The European Commission recognised the benefits of assurance in the Green Paper, "Promoting a European Framework for Corporate Social Responsibility" July 2001, p. 18. See also, footnote 11 and 12 of this thesis.

credibility of sustainability reports. This resulted in an "explosion" of services undertaken by not only accounting firms but other assurance providers. The percentage of G250 firms commissioning assurance according to the KPMG survey has increased steadily from 29% in 2002, 30% in 2005 and 40% in 2008.[65]

1.3.4 Assurance providers and sustainability assurance

The assurance market today is characterised by the presence of a variety of assurance providers including professional accounting firms, quality and corporate social responsibility assurance consultancy firms, civil society assurors and NGOs providing a variety of assurance services.[66] Although commonly referred to as "assurance engagements", these engagements can take the form of data verification, consulting, or endorsement of a report aimed at various different users.[67] Reporting organisations, on a case-by-case basis, exercise choices on the forms of assurance services and assurance provider based on their corporate philosophy, stage of development, management values, etc.[68] As such, supported by research, the potential for management capture is prevalent and the understanding of sustainability assurance is complicated by the nature and purpose of the engagement; and how, by whom and who for it is performed.

1.3.5 The credibility of sustainability assurance

Academic studies have critically scrutinised the practice of sustainability assurance making broad inferences about practice through content analysis of assurance reports. The following section highlights a few key studies in this area. Studies on environmental (sustainability) assurance engagements performed prior to the development of sustainability assurance standards in the mid to late 90s reflected inconsistencies in practice. Based on sixty-eight companies across 11 countries in the mid to late 1990s, Kamp-Roelands detected variations in the scope, objective, criteria, subject matter of the assurance engagement; work carried out by assurance providers; level of assurance and wording of the assurance providers' opinions.[69] With a focus on

---

65 Refer to footnote .

66 Zadek, S. & Raynard, R. (2004), p. 9.

67 Ibid.

68 Refer to footnote 65.

69 Kamp-Roelands, N. (2002) Towards a Framework for Auditing Environmental Reports, Tillburg University. The 11 countries were Australia, France, Germany, Denmark, Finland, Italy, Netherlands, Norway, Sweden, Switzerland and the UK. Refer to p.325 for the sample companies.

the Top 50 reporters identified by the SustainAbility, Ball et al. expressed concerns on the independence of assurance providers and the extent of corporate commitment to "external transparency and accountability". Findings indicated evidence of management controlling the process, and assurance providers emphasizing environmental management systems rather than environmental performance of the firm. For these reasons, soft audits could be performed on selective disclosures by management.[70]

The development of guidance in assurance provided by the GRI, FEE and the AA1000 series were taken into account in a study by O'Dwyer and Owen which provided an update of Ball's findings.[71] With a focus on transparency and accountability, the authors found improvements in the extent of work undertaken, independence and a greater focus on the performance dimensions attributed to the AA1000 methodology.[72] The issues of independence (despite improvements) and management capture however, prevailed. The study showed evidence of assurance providers' reluctance to address a specific constituency and the lack of stakeholder input and involvement in the assurance process. The authors also observed the more conservative approach adopted by assurance providers belonging to the accountancy profession as compared to consultancy providers. Consultancy providers who were more evaluative and provided positively phrased conclusions implying a higher level of assurance.[73] To extend the analysis of the practice of assurance beyond the "best reports", Deegan et al., based their findings on a broad sample of European and UK third-party assurance statements that accompanied the release of sustainability reports.[74] This study illustrated that despite the developments in sustainability reporting and guidance for assurance providers, there is still a high level of variability in practice. The study reaffirmed the issues of independence, variations in practice and importantly, established variations in the contents of assurance statements both within and across countries.

70 Ball, A. et al. (2000), p. 17-19.

71 Owen, D.; O'Dwyer, B. (2005) The sample studied included 41 reports with assurance short listed for the ACCA ESRA scheme in 2002 consisting of 28 environmental, 13 social sustainability reports, 46% assured by auditors, 54% by consultancies.

72 Ibid., p. 224.

73 Ibid., p. 210-212, 223-227.

74 Deegan, C. Cooper, B & Shelly, M. "An investigation of TBL Report Assurance Statements: International Evidence" European Accounting Association Conference Paper, May 2005. Deegan et al 2005 cautioned against the focus on best reporters to reflect best assurance practice and performed analysis of assurance practice from 170 reports available internationally at the time the research was undertaken, 8 are from 2000, 86 from 2001, 65 from 2002, and 11 from 2003.

Several other studies have supported the above key findings and in addition attributed the following factors to undermining the credibility of sustainability assurance. These factors include the lack of a common denominator of the subject matter, the lack of developed criteria, diversity in standards and methodology of assurance providers, and the ambiguity of assurance reports and the lack of assurance standards.[75]

The issues pertaining to the elements of assurance, such as the nature of the subject matter and criteria were not only highlighted in academic publications but also publications by the accountancy profession and other organisations.[76] The presence of a perceived regulatory gap led academics, assurance providers and other organisations to call for assurance standards which carry the notion of being able to transverse space and potentially increase comparability, uniformity, enhancing quality, cooperation and coordination to mitigate these "failures".[77]

In 2008, 62% and 33% of the assurance engagements performed on sustainability reports published by G250 have been conducted according to the ISAE3000 and AA1000AS respectively compared to 24% and 18% in 2005.[78]

1.4 Purpose of the thesis

Non-state actors have been increasingly engaged in regulation, operating and setting rules outside of and parallel to existing legal institutions. These "alternative to legal rules" have been increasingly influential in domestic law and policy, with growing evidence of them penetrating legal, professional as well as social orders.[79] De Bellis

[75] Owen, D.; O'Dwyer, B. (2007) "Corporate social reporting and stakeholder accountability: The missing link" Accounting, Organizations and Society 32(7-8) p. 649-667; Park, J. and Broson, T. (2004), "Experiences of and views on third-party assurance of corporate environmental and sustainability reports" Journal of Cleaner Production, 13(10-11), p, 1095-1106; Wallage, P. (2000) "Assurance on sustainability reporting: An auditor's view" Auditing, p, 53-65; Dando, N.; Swift, T.(2003) "Transparency and assurance: Minding the credibility gap" Journal of Business Ethics, 44(2-3), p. 195-200.

[76] Reports published by AccountAbility, KPMG, FEE. These documents will be analysed in chapter 5.

[77] Braithwaite and Drahos (2000) Global Business Regulation Cambridge University press. Brunsson, N., and B. Jacobsson "A World of Standards" year: 2000 Oxford: Oxford University Press. Tamm-Hallström, K. "Organizing international standardization" year: 2004 Cheltenham: Edward Elgar.

[78] KPMG(2008), refer to footnote .

[79] De Bellis,M (2006) "Global Standards for Domestic Financial regulators" Global Jurist Advances, Vol. 6 (3) p. 1.

puts forth: "...domestic regulatory policies are increasingly becoming the result (and sometimes, the pure application) of rules established at the global level."[80] In research on global financial regulation, Humphrey and Loft have noted the rise of standards in financial reporting and auditing and, governance through standards is reaching new heights of complexity.[81] Despite the growing importance of standards regulating the accountancy profession, little is understood about the emergence, essential features and how standards work in areas that are not the core service of the accountancy profession.

This thesis has a two-fold purpose.

> *First, it aims to contribute to our understanding of regulation by non-state standard setters through analysing the dynamics of regulating sustainability assurance through standards made by the accountancy profession.*

With reference to section 1.3.5, the nature of sustainability assurance engagements and the standards adopted to perform such an engagement are largely dependent of the type of provider. In this light, a conscious choice has been made to narrow down this study to focus on one type of provider or interpretive community defined as, a group which members share technical knowledge, norms and interests to analyse how sustainability assurance is regulated.

A conscious choice has been made to narrow the scope of this study to the accountancy profession for the following reasons. First, large professional accounting firms hold about 70% of the global sustainability assurance market share. Assurance providers from the accountancy profession have been under obligation through membership of IFAC since 2005 to adopt ISAE3000. Consultancy and other providers, on the other hand, adopt standards voluntarily. Second, the financial audit is the core service performed by the accountancy profession which has defined its role as a "profession" in society. Compared to the financial audit regulation that is institutionalized and well documented at the international, regional and national levels, little is known about the regulation of non-financial assurance services, where the accountancy profession does not possess "profession jurisdiction" in the market. Third, sustainability assurance is "somewhat similar" to the financial audit but has a broader and less defined subject matter and operates in a different context, one which

---

80 Ibid.

81 Humphrey, C; Loft, A (2011 forthcoming), "Standardizing global financial reporting: Global governance without global government", p. 26.

is undergoing rapid development. This makes the interplay between the initiatives regulating sustainability assurance, and those regulating the financial audit interesting both at the international and national levels.

To preclude differences in country-specific institutional features that could have a bearing on the different standards and how standards are used in different jurisdictions; the analysis of how sustainability standards work has been restricted to the case of Sweden. The choice of Sweden has been motivated by three observations. First, the Swedish state is the first in the world to impose on state-owned companies to adopt the GRI reporting guidelines and commission assurance on their sustainability report to enhance credibility.[82] This initiative has put sustainability assurance termed by the Swedish audit profession[83] themselves, as a "minority sport"[84] high on the policy agenda. Second, Far[85] has replaced the contents of its original sustainability assurance standard with the Dutch standard drafted by NIVRA in December 2008, making only minor adjustments. Third, the scope of the statutory audit and legal concept of audit are defined more broadly in Sweden than most jurisdictions.[86] This raises questions as to whether sustainability assurance already falls within the scope of hard law. If it does, how will it affect the role of standards set by Far?

The second purpose has been motivated by the Swedish case.

> *It aims to contribute to our understanding of the potential interplay between standards set by the accountancy profession and hard law in the regulation of sustainability assurance in Sweden.*

To summarise, this thesis carries the ambition of providing an understanding of how standards by the accountancy profession, in an area outside its core competence, are developed and how these standards work when they are applied by users that operate with national jurisdiction. It makes no attempt to solve legal problems or the problems of decentred regulation. The findings however may have useful policy

---

[82] Ministry of Enterprise, Energy and Communications (2007) Guidelines for external reporting by state owned companies, 27/11/2007.

[83] In Sweden, due to specific legislation regulating auditors and the historical context, the accountancy profession involved in auditing is commonly referred to as the "audit profession". The term accountancy profession will be adopted in this thesis in the context of the Swedish case.

[84] Sweden's report to the European Sustainability Reporting Association, 2006.

[85] Far is the professional institute for authorized public accountants (auktoriserade revisorer), approved public accountants (godkända revisorer), and other highly qualified professionals in the accountancy sector in Sweden. It is not on the abbreviation list as Far is the name of the organisation.

[86] This will be analysed in section 8.4 of the thesis.

implications. Findings may be of interest to standard setters, supervisory bodies, policy makers and those who have a keen interest in regulating the accountancy profession and sustainability assurance. It may also serve as a platform to discuss how regulation can be improved and how new emerging forms of regulation can be used alongside or with "state-centred" regulation to achieve policy objectives.[87]

1.5 Research questions and scope of the thesis

1.5.1 Research questions

The four fundamental questions presented in section 1.2.3 serve as a building block to formulate the research questions to realise the two-fold purpose. To analyse the dynamics of regulating sustainability assurance through standards, the questions highlight the importance of analysing the context which standards are developed and used as well as the contents and character of standards as a regulatory tool.

Sustainability assurance is a well-supported, rapidly developing concept and practice that is high on policy agendas. Both concepts 'sustainability' and "assurance" have been construed differently by different communities which leave the concept of 'sustainability assurance' open to different interpretations. With little known about sustainability assurance, precaution is taken to avoid being hostage to the accountancy profession's concepts and claims to expertise. Preceding an attempt to unfold and analyse the dynamics of regulating sustainability assurance by the accountancy profession, there is a need, first to grasp how 'assurance' has been conceived by the profession and coded by the IAASB in ISAE3000. This background information does not intend to contribute to theoretical discussions but provide readers with a basis for understanding the issues on assurance and serve as an explanatory source for findings of the other research questions that follow.

[87] Black, J. (2001), p. 105-112,145-146.

Research question 1:

How has the "assurance" concept been framed and codified by the IAASB?

The literature reviewed in section 1.3.5 has shown the presence of a regulatory gap pertaining to the quality of sustainability assurance engagements performed. With the absence of state initiatives in this area, the loci of regulating sustainability assurance has been shifted to non-state actors. The second research question is built on "who is regulating?" Non-state actors (unless provided for in hard law) are not under legal obligation nor are they delegated legal authority to take regulatory initiatives or participate to fill the perceived regulatory gaps. Their actions are motivated by the environment and conditions which they operate; their acceptance is based on their legitimizing strategies.[88] Strict organizational boundaries therefore cannot be drawn between these actors (standard setters or participants) and their social and political environment even if professional bodies set standards to regulate their own members.[89]

Although the study focuses on the accountancy profession, the IAASB is only one important actor that has an interest in setting standards to guide sustainability assurance engagements at the international level. It is therefore important to look beyond the internal procedures and organisational borders of the IAASB to understand the context which standards are developing. To capture the standard setting dynamics, the presence of other actors, (including non-standard setters) with an interest to set standards, determine and influence regulatory issues have to be recognised. The second research question has been structured not only to map out the actors involved in filling the perceived regulatory gap but to gain an understanding of the interest they represent, the objective of their involvement and their attempts to gain legitimacy to analyse the interactions and patterns of influence on regulatory issues.

---

[88] Young, J. (1994) "Outlining regulatory space: Agenda issues and the FASB" Accounting, Organizations and Society, 19(1), p. 86-89.

[89] ibid.

Research question 2

How has the perceived regulatory gap on sustainability assurance been filled by standards?

The third research question has been motivated by the fundamental questions pertaining to the character and interpretation of rules. Sustainability assurance standards are a codification of the standard setter's guidance; they function as regulatory tools. Standards cannot apply themselves but rely on users to apply them.[90] Assurance providers that adopt standards however operate in a wider social environment where external forces are constantly shaping practice. Acknowledging that standards set by non-state actors are not legally binding, and enforcement structures are not as established or well-defined, users are given greater flexibility in the application and use these standards compared to hard law.[91]

Research question 3

How do standards constructed by the accountancy profession work to guide sustainability assurance in Sweden?

Ascribing to the view that regulation by state and non-state actors co-exist and the presence of non-state regulation does not dislodge the state nor displace law (understood in its traditional sense).[92] Regulated parties may have to comply with a web of state and non-state standards that conflict or overlap with each other. There is a lack of clarity as to the "status" sustainability assurance standards set by the profession have when applied in a national context or the reciprocity between these

[90] Hart, H. (1961) "The Concept of Law" Oxford, p. 123. According the Hart, rules cannot apply themselves. In this thesis, an extension was made to standards.

[91] Mörth, U (2004) "Soft law in Governance and Regulation: An Interdisciplinary Analysis, Cheltenham: Edward Elgar.

[92] Debates exist particularly in the political science field with regard to the role of the state in light of emerging plurality of new actors. Refer to Rosneau, J 2006 "The Study of World Politics: Globalisation and Governance, London: Routledge for further information.

standards and hard law. The second purpose of the thesis can be realised through an analysis of how these standards interplay to regulate sustainability assurance.

Research question 4

How do standards set by the accountancy profession interplay with hard law to regulate sustainability assurance in Sweden?

1.5.2 Scope of the thesis

The heart of this thesis is centred on regulation of sustainability assurance by the accountancy profession through standards and how such standards interplay with hard law. The purpose excludes from the thesis's scope other state and non-state initiatives taken that contribute to the sustainability policy objective.[93]It also excludes legal control mechanisms such as contract law, tort law and the supervision of auditors by the state that have a significant influence on the behaviour of assurance providers.

The conscious choice of placing the focal point on the how standards as a regulatory tool are set and how they work makes clear that this thesis is not a study about the profession and also excludes from its scope "professional judgment" by assurance providers. The researcher acknowledges the importance of studies on the accountancy profession, jurisdictional competition, the backstage of assurance practice and the profession as a site of regulation. Relevant studies have been reviewed in this thesis to contribute to the understanding of the context within which the accountancy profession sets standards and operates. As this thesis has been built on reviewing the four fundamental questions of regulation presented in section 1.2.3 of this chapter, it recognises but excludes from its scope social science literature on the translation of standards.

This study only focuses on regulation of one type of sustainability assurance provider therefore, standards on financial and sustainability reporting, sustainability assurance standards by other standard setters (e.g., AA1000AS) and assurance standards pertaining to other subject matter by the accountancy profession (e.g., ISAE 3410 Assurance Engagements on Greenhouse Gas Statements") are taken into account only to the extent that is relevant to this case.

---

[93] The former include the many initiatives taken in corporate law, tax law, environmental law, labour law and financial market regulation to achieve sustainability policy objectives.

1.6 The role of theory, positioning and potential contributions

1.6.1 The role of theory

Regulation as explained in section 1.2 is an area of study that is not confined within a particular academic discipline. This study does not attempt to review the vast and dispersed literature on regulation but has chosen to adopt a decentred view of regulation as a point of departure to analyse the dynamics of regulating sustainability assurance through standards. The interdisciplinary character of the topic has required the researcher to draw on theory and literature grounded on different philosophies. Literature on regulation in legal, socio-legal, political science and organisational literature was reviewed to sensitise the researcher to the issues and challenges of regulation in contemporary society. The review has served as a foundation to formulate the problems and research questions of this thesis.

Regulation theory has been defined by Morgan and Yeung as a "set of propositions or hypotheses about why regulation emerges, which actors contribute to that emergence, and typical patterns of interaction between regulatory actors."[94] This thesis has drawn on the explanatory and prescriptive elements of decentred regulation to provide a framework to analyse forms of control by non-state actors and the structuring and facilitation of particular patterns of ordering with respect to the regulation of sustainability assurance.[95] Decentred regulation and the dimension of rules have been adopted to construct a theoretical framework in a systematic manner to perform the analysis and make sense of the findings to identify patterns that potentially contribute to theory.[96]

Relevant analytical frameworks have been reviewed, as have studies specific to sustainability assurance. Both these contributed to improving understanding of the context of the regulatory environment and specific issues pertaining to sustainability assurance. This helps to avoid the potential pitfalls of "reinventing the wheel" especially when the subject matter of study is of interest to several disciplines.

---

[94] Morgan, B.; Yeung, K. (2007), p. 16.

[95] Ibid, p.17.

[96] Ibid.

1.6.2 Positioning and potential contributions of the thesis

Regulation by non-state standard setters is not on main agenda of traditional legal research.[97] In light of the globalization / deregulation debates, legal researchers have begun to focus on the drafting and interpretation of international standards, the regulatory gaps resulting from the increase of cross border activities and development in technology[98], the role of non-state actors in the process of "law making".[99] The concept of soft law appears to be conceived as a grey area between of law and political science. There is an increase in academic journals along the lines of "Modern Law Review"; "Journal of Law and Society", "Law and Policy" etc. reflect that groups of legal researchers have begun to embrace the challenges of this modern era in a more interdisciplinary fashion. The study of the profession and technical standards developed in a specialised context however, this continues to be an area that is much neglected by legal academics despite the active role of professions in the construction of regulatory arrangements in a decentred regulatory world[100]. This study attempts to provide empirical evidence on the development and functioning of standards by the accountancy profession vis-a-vis revisiting the fundamental questions of regulation.[101]

Current research on the regulation of the "global audit profession" in the social science field depicts the global regulatory arena as a complex, ever changing domain "poorly depicted in regulation papers"[102]. Papers published, particularly in the journal Accounting, Organizations and Society, bring to the attention of researchers the wide range of institutions with whom the accountancy profession interacts in the global sphere which "increasingly set boundaries for audit practice and ideas and thought processes that shape practice".[103] There are also shifts in the focus of audit research from national to transnational regulation, involving new actors other than state and

---

[97] Peters, A et al (2009) p. 1-12.

[98] For example, a symposium "Legal perspectives in a globalised business environment" was held at Lund University in 2004 exploring the effects of globalisation and the legal challenges in this era in the areas of tax, CSR, labour law etc.

[99] For example, research on business actors' involvement in *Lex mercatoria*, Public Private Partnerships (PPP) and corporate self-regulation (particularly in the areas of corporate social responsibility (CSR)). See also, Hutter, B. (2006) "The role of non-state actors in Regulation" in Schuppert Global governance and the role of non-state actors, Baden Baden, p. 63-79.

[100] Williams highlights that with the exception of Dezalay and Garth 1996 regulation through professional standards are largely neglected

[101] Refer to section 1.2.3

[102] Humphrey, C.; Loft, A.; and Woods, M. (2009)

[103] Humphrey, C.; Loft, A. (forthcoming)

professional organisations[104]. A rich literature on the context which financial audit is regulated focuses on the new institutional arrangements. They concern how standards are set, their passage towards international status and translations into national standards. There are few empirical studies of non-financial assurance services.[105] Those which are done, centre on the types of assurance; determinants of assurance, effectiveness of communication of assurance and their effect on auditor independence. Academic research in this area appears to be fragmented and narrow, focusing on particular issues with the exception of the study conducted by a consortium of academics commissioned by IFAC which resulted in the publication of a paper "Determination and communication of levels of assurance other than high" in June 2002.

Studies specific to sustainability assurance have mainly been conducted adopting normative and / or critical perspectives in light of stakeholder and legitimacy theories.[106] Questions of for whom and why assurance is commissioned dominate the research agenda.[107] Other studies carried out using quantitative methods found associations between the demand for assurance and country specific institutional features namely: stakeholder orientation and strength of legal and enforcement systems[108]; the level of awareness of sustainability in a country as well as the visibility of the industry in terms of its environmental and social impact. These studies however do not go into greater depth on how these institutional features work to influence the demand of sustainability assurance. The study of how sustainability assurance is regulated and the role of standard setters have been neglected by researchers in this field.

Standards in social science have been regarded much as *one of the many* constraints (others include e.g., negotiation with management, professional ideology) that affect the behaviour of assurance providers. Academic studies on standards guiding sustainability assurance have focused on its impact on the assurance report published by assurance providers, and the differences between it and those for financial audits.[109]

---

[104] Reviewed in greater detail in chapter 2

[105] Refer to Hasan, M; Maijoor, S; Theodore, J; Roebuck, P. et al. "Different Types of Assurance Services and Levels of Assurance Provided" International Journal of Auditing, 9(2) p. 91-102 for a literature review.

[106] Owen , D. (2005); Owen, D. and O'Dwyer, B. (2007), Deegan, C. et al. (2005), Adams, C. et al. (2004).

[107] Kamp-Roeland's uses audit theory and postulates as a main tool for analysis.

[108] Simnett, R.;Vanstralen, A.; Chua.W. (2009) 'Assurance of Sustainability Reports' The Accounting Review, 84 (3) p. 937-967.

[109] Refer to section 1.3.5 of this thesis.

The analysis of the contents of assurance standards however has been left to organisations, which are themselves involved directly in the standard setting process.[110]

This thesis has two potential contributions. First, it attempts to contribute to the above studies by focusing on the regulation of sustainability assurance through standards. It attempts to open the black box of standards by analysing the context where standards are set, the context within which it functions, and the contents of standards. Second, it aims to bridge the understanding between legal and social science scholars on how sustainability assurance is regulated by analysing the potential interplay of standards set by the accountancy profession with hard law to regulate sustainability assurance.

1.7 Structure of the thesis

Purpose	Area of focus	Research Question	Level
To analyse the dynamics of how sustainability assurance has been regulated by the accountancy profession	Context of standard setting	How has the "assurance" concept been framed and codified by the IAASB?	International
	Context of standard setting	How has the perceived regulatory gap on sustainability assurance been filled by standards?	International
	Context of standard setting Content of Standard Context which standards are used	How do standards constructed by the accountancy profession work to guide sustainability assurance in Sweden?	Sweden

[110] For example, in Iansen-Rogers and Oelschlagel (2005) "Assurance standards briefing: AA1000 assurance standard and ISAE3000" Accountability and KPMG Sustainability Netherlands. ISAE3000 was compared to AA1000AS and NIVRA's ED3410.

To analyse the potential interplay between standards set by the accountancy profession and hard law in the regulation of sustainability assurance in Sweden	Character of standard	How do standards set by the accountancy profession interplay with hard law to regulate sustainability assurance in Sweden?	Sweden

Figure 1.2: Snapshot of this thesis

This thesis is divided into four main parts depicted in the figure below.

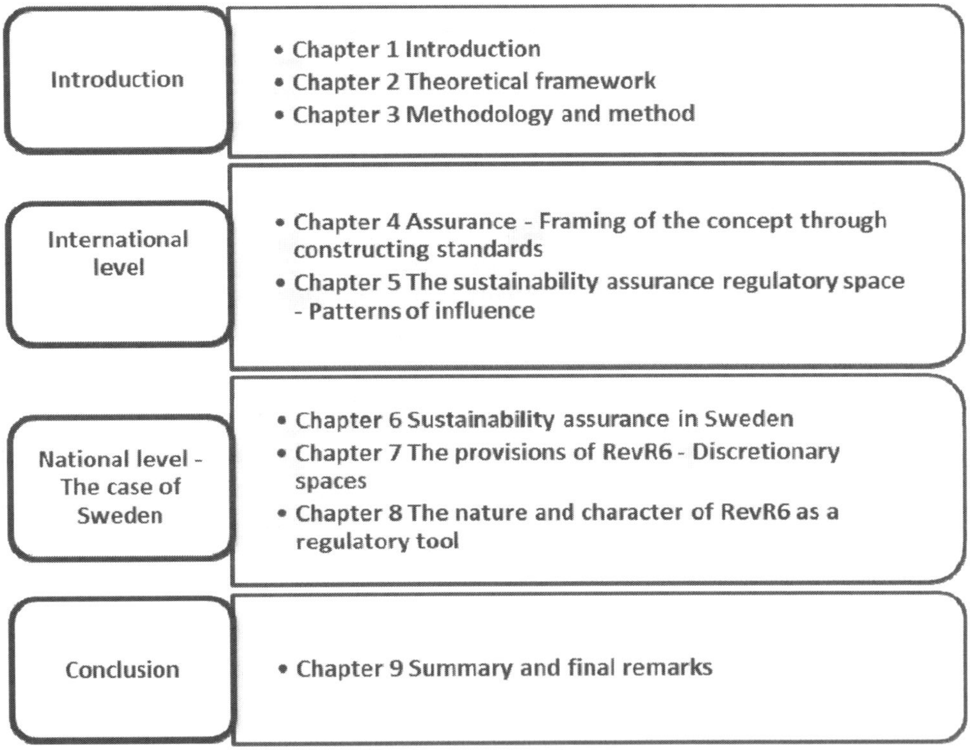

Figure 1.3 Structure of the thesis

Following the introduction above, chapter 2 presents the theoretical framework adopted to analyse the research questions. Chapter 3 presents the methodology, research design and a discussion of the challenges of interdisciplinary studies.

Empirical findings are presented in the second and third parts of the thesis. The second part presents the analysis of the context of standard setting at the international level. Chapter 4 provides essential background knowledge on how the generic assurance concept has been framed and codified by the accountancy profession in ISAE3000. Findings of this chapter provides information on how the accountancy profession understands and defines assurance, aspects to be regulated and the function of assurance standards in the context they operate. Chapter 5 looks into non-state actors at work in developing sustainability assurance standards to fill the regulatory gap.

The third part of the thesis presents the Swedish case study. Chapter 6 describes relevant institutional characteristics of Sweden that determine the context in which sustainability assurance standards are set, and function It highlights in particular, the role and standard setting activities of Far. Chapter 7 and 8 focuses on how standards work. Chapter 7 presents the analysis on how ISAE3000 and events at the international level provide guidance to Far in the setting of sustainability assurance standards in Sweden. Chapter 8 focuses the how RevR6 has worked to guide sustainability assurance engagements in Sweden and its interplay with hard law to regulate sustainability assurance engagements carried out by the audit profession.

The conclusion presented in chapter 9 integrates the findings of this study and revisits the purpose that had motivated the thesis. Finally, contributions and potential areas of research are discussed in light of current developments.

2. Theoretical Framework

2.1 Chapter objectives

The objective of this chapter is to explain the theoretical foundations and framework constructed to analyse how sustainability assurance is regulated. The decentred view of regulation has provided a theoretical framework to understand, to develop research questions, and make choices of the tools of inquiry to analyse how sustainability assurance is regulated. This chapter starts with a recapitulating and highlighting the conceptual core of decentred regulation and its appropriateness to analyse the regulation of sustainability assurance. It is followed by an explanation of how decentred regulation serving as a starting point allows a departure from the traditional legal lens, opening the possibility of integrating other non-legal theories to develop suitable tools to perform the analysis of the four questions. The tools sculpted to analyse the context of standard setting and the provisions of standards are then presented. The final section of this chapter discusses the challenges and issues pertaining to the adoption of this theoretical framework.

2.2 Decentred regulation

The decentred view of regulation is depicted by the shift of regulatory loci from the state.[111] It recognizes the changing capacity of state and departs from the assumption that the state has monopoly over the exercise of power and control over social actors.[112] This suggests the existence of other, and possibly multiple, sites issuing regulatory initiatives not originating from the state, but with the impetus to fill any perceived regulatory gaps. The cartography of decentred regulation thereby extends the map of regulation to forms of control which are not formally part of the state

---

[111] Black, J. (2001), p. 112.

[112] Ibid.

centric legal system.[113] It places importance on the understanding of the function and essential characteristics of these non-state forms of control that operate outside the existing legal institutional structures. At the same time, this perspective does not dislodge the state.[114]

2.2.1 Conceptual core of decentred regulation

The following section highlights Julia Black's conception of decentred regulation. According to Black, the conceptual core of a decentred view of regulation can be depicted by five notions: complexity, fragmentation, un-governability, interdependence and the rejection of a clear public and private divide.[115]

Decentred regulation recognizes the regulatory initiatives of different actors and / systems. These actors, due to the different interest they represent and different philosophical groundings, have different ideas of what they intend to achieve with regulation, and what powers or resources they have to achieve them.[116] Without a hierarchical structure or central authority body to coordinate these efforts, or pre-established constitutionally agreed on procedures of rule-making, regulation occurs in 'many rooms'.[117] Regulation can be grounded on different objectives and different regulators, and can prescribe different courses of action to a regulated community. Decentred regulation functions in a dynamic sphere characterized by complexity arising from interactions, patterns of influences and power relationships between, the actors and / systems.[118]In decentred regulation, Black argues that governments do not have monopoly on the exercise of power and control.[119] In fact, the complexity and dynamic nature of problems, in light of how contemporary society operates makes it difficult for one single actor to be equipped with the knowledge, information and instruments required to regulate effectively. With the rise of corporations and challenges of modern society, regulatory power is dispersed between social actors and

---

[113] Smith, D. (2002) "What is Regulation? A Reply to Julia Black" Australian Journal of Legal Philosophy 27, p. 39.

[114] Morgan, B.; Yeung, K. (2007) p. 4.; Black, J. (2001), p. 145-146.

[115] Black, J. (2001), p. 104-112; Black, J. (2002), p. 2-7.

[116] Black, J. (2002), p. 3.

[117] Ibid., p. 6.

[118]Ibid., p. 4.

[119] ibid.

between social actors and the state leading to fragmentation of knowledge, power and control, the second notion of decentred regulation.[120]

The third, un-governability refers to actors being self-regulating and autonomous. The idea originates from Foucauldian notion of governance, the "conduct of conduct"[121]. This carries the idea that no one single actor or system can dominate or act directly upon another. Regulated parties are not "static"; they are dynamic actors that respond in different ways to regulation in different contexts. Regulation can therefore result in an unintended regulatory outcome.[122] Black puts forth the prescription that regulatory design should be contextual, taking into account the regulated parties and alternative forms of control that are not direct interventions. The fourth notion of decentred regulation refers to interaction and interdependencies between actors and systems. Contemporary society does not ascribe to the more state-centric assumptions of regulation that society has a need and the state has the solution and capacity to regulate, rather, each party has problems and solutions.[123] Related to the above notions is the collapse of the public and private dichotomies built on regulation being viewed as a product of interaction between social actors and not of formal constitutional authority of the state. From a decentred view, "regulation is not so much an activity as a product of activity" [124] this points to the idea that regulation is co-produced and the importance of the understanding and defining the context which regulation takes place.

2.2.2 Challenges of decentred regulation

The five notions of decentred regulation provide a diagnostic tool to observe the presence of regulation in areas not within the scope of the traditional legal discipline. It also serves a source to explain regulatory failures.

Decentred regulation cannot avoid challenges resulting from the fragmentation and interdependence of actors involved. Decentred regulation is challenged functionally and systematically by coordination problems, regulatory gaps and overlap, resulting

---

[120] Black, J. (2002) drawing on the work of Foucault (1991) "Governmentality" in Burchell,G, Gordon, C and Miller "The Foucault effect: studies in governmentality, Rose and Miller(1992) Political power beyond the state: problematics of government British journal of sociology 43 (23) p. 173.

[121] Foucault,M. (1991) "Governmentality" in Burchell,G. Gordan, C. and Miller, P. "The Foucault effect: Studies in Governmentality", London: Harvester Wheatsheaf, p. 48.

[122] Black, J. (2002), p. 5.

[123] Ibid.

[124] Ibid., p. 6.

from the absence of an appointed central co-ordinator. [125] For example, there is no accepted hierarchy of regulation nor is there a final interpreter of standards to deal with any conflicts between or within standards. Although Black draws on the systems theory by Luhmann; weak theoretical links had been made between decentred regulation and autopoiesis. Black instead focuses on policy contributions suggesting prescriptive elements to meet systemic challenges. These elements include coordinating, steering, influencing and balancing; and to create new patterns of interaction to enable social actors to organize themselves as alternative solutions, to work towards achieving policy objectives. Decentred regulation also faces democratic and normative challenges due to the lack of agreement as to who should participate in standard setting process? Who should decide on the goals of regulation? What goals and whose goals should be pursued.[126]

The decentred regulatory framework challenges the traditional legal conception of regulation and the role of the state and the law.[127] The conceptual core of decentred regulation, point to the need to draw from non-legal theories to sculpted tools to analyse regulation by non-state actors to enable the understanding of both the context and content of standards.

2.3 The choice of a decentred view of regulation

2.3.1 Self-regulation from the socio-legal perspective

Self-regulation is a "normatively loaded term"[128] and can be defined in many ways in regulatory literature. It has been used to denote soft law, non-government collective arrangements, bilateral arrangements such as contracting between parties; industry involvement of rule formation, unilateral adoption of standards; intra-firm regulation; neo-corporatist arrangements.[129] Its various forms make a single satisfactory definition impossible. Gunningham and Rees identified three dimensions through which self-

[125] Black, J. (2009) "Legitimacy, accountability and polycentric regulation: dilemmas, tri-lemmas and organisational response" in Peters, A. (2009) p. 242.

[126] Ibid, p. 241-244.

[127] Black, J. (2001), p. 145-146; Black, J. (2002),p. 3-7.

[128] Ibid., p. 115.

[129] Ibid., p. 121.

regulation can be better understood. First, self-regulation can take place at two levels: individual and industry. The former refers to normative orders of private initiatives such as individual corporations' voluntary compliance to a standard or code of conduct. The latter refers to collective efforts by business networks or professional communities to regulate its members. These efforts could include adoption of standards, monitoring as well as enforcement actions to govern the conduct of firms as well as individuals. Second, self-regulation could have an economic objective where concerns are centred on the effective functioning of markets or other facets of economic life; or social where the normative orders are formed to protect a broader constituency. Thirdly, self-regulation can be distinguished by the degree of government regulation. It has been observed that states have been involved to a different degree of formality and extent of standard setting and / or enforcement in supporting this form of non-state regulation.[130] In its purest state, self-regulation can be understood as non-state actors regulating the behaviour of their own organisation members or network of firms without state intervention. In a 'less pure' state, the state can be involved in regulating different aspects and to different extents. For example, the state can have no role in standard setting and enforcement processes but officially sanctions, monitors and takes steps to ensure the effectiveness of the program. The state can also take a larger role with active involvement in standard setting or take part in enforcing of privately established standards.[131] The former is also referred to mandated partial self-regulation and the latter two which are subject to some form of state restructuring or oversight, are termed enforced self-regulation.[132]

The term self-regulation has been a common term adopted by researchers in both the legal and the social science fields to define regulation where the locus of regulation is not centred on the state. This raises the question as to why decentred regulation has been adopted in this thesis and not self-regulation as a theoretical lens, and how self-regulation concurs with decentred regulation.

From the above, one agreed characteristic of 'pure' self-regulation is that it is non-state, the steering does not come from the state but others: others regulate themselves. As such, it does not conflict with the theoretical construct of the decentred thesis and shares the common emphasis of the importance of the context within which

---

[130] Gunningham, N. and Rees, J. (1997) "Industry self-regulation: An institutional perspective", Law and Policy. 19(4), p. 364-366.

[131] Sinclair, D. (1997) "Self-regulation Versus Command and Control? Beyond False Dichotomies" Law and Policy. 19(4), p. 531-532, 534.

[132] Baldwin, R. and Cave, M. (1999), p.133-136.

regulation emerges and operates. The choice of decentred regulation is preferred to self-regulation as it takes the focus away from definitional debates of what is self-regulation, and offers a broader lens to analyse issues pertinent to regulation in contemporary society.[133] Self-regulation as Gunningham and Rees observe, often operates in the shadow of rules provided by general law or other co-regulatory strategies e.g., government acts as a facilitator or prescribes particular outcomes desirable to achieve its policy objectives.[134] This hints that self-regulation is often built on the assumption that the state has a monopoly in the exercise of control as such self-regulation takes place in areas where the state has chosen not to exercise jurisdiction.[135]

Building on the ideas of Peters and Black, it is recognised here that regulation can emerge from multiple sites; a decentred perspective of regulation would enable the capture of the dynamics of regulation and the interplay between the regulatory instruments from the state and non-state actors so as to better deal with the complexity of today's problems.

2.3.2 Self-regulation by the accountancy profession

A profession is a community or group of experts that share a particular set of interests, skills and role in society, which provide a value-based foundation for the actions of members.[136] When professional status is obtained, the specialized occupational group may be granted legal exclusive protection of their work, the right to engage in self-regulation among other privileges.[137] The accountancy profession thereby has its own corpus of knowledge, expertise and vocabulary and regulates its members though relevant standards and disciplinary procedures. As such, we can conceive that the profession has its own discursive strategies with regards to self-regulation.

IFAC uses the term self-regulation to describe one of the two complementary ways which could be used to regulate the audit profession. It refers to the regulation of the members of the profession by the professional body in each country, which is recognized and delegated responsibility to serve such that function. The other being

[133] Black, J. (2001), p. 112-114.

[134] Gunningham, N; Rees, J. (1997), p. 544.

[135] Black, J. (2001), p. 118.

[136] Abbott, A "The system of professions: an essay on the division of expert labour" Chicago University Press, p. 7

[137] Arena, M.; Jeppesen, K. (2010) "The Jurisdiction of Internal Auditing and the Quest for Professionalization: The Danish Case" International Journal of Auditing, 14(2), p. 111

"external regulation" which is regulation by the government either by one of its bodies or through an agency delegated regulatory powers by the government e.g., oversight boards.[138]

In the era of globalization, structural shifts in the audit regulatory field were observed. The accounting firms have increased in size and cross border audit services and expanded their services beyond the traditional audit domain.[139] With less restriction on capital movements across national boundaries and growing interdependencies of economies, harmonization in auditing standards is called upon. These developments have challenged the capacity and resources of the state to regulate the large accounting firms that have outgrown the existing regulatory structures.[140] It created the need for a shift in the locus of regulation from one that is dominated by professional associations and the state at a national level, to standard setting at an international level.

In 1962, institutional initiatives emerged to standardize auditing practice at an international level[141] which set foundation for the establishment of IFAC. IFAC has since risen as a global organisation setting international standards for the accountancy profession.[142] Its role as an international standard setter was strengthened in the light of the Asian financial crisis through its role in the "New financial architecture"[143]. The regulation of the global accountancy profession is now represented by interlocking relationships between IFAC; international regulators comprising the World Bank (WB), the International organisation of securities commission (IOSCO), the International Association of Insurance Supervisors (IAIS), the Basel Committee on Banking Supervision (BCBS) and the European Commission; and the large audit firms namely: KPMG, PWC, Ernst and Young and Deloitte.[144]

---

138 IFAC policy position 1: IFAC's position on professional regulation, December 2007, p. 3.

139 Suddaby, R.; Cooper, D.; and Greenwood, R. (2007) "Transnational regulation of professional services: Governance dynamics of field level organizational change" Accounting, Organizations and Society, 32(4-5), p. 335.

140 Ibid., p. 337-343.

141 http://www.ifac.org/about-ifac/organization-overview/history IFAC history.

142 Loft, A.; Humphrey, C.; Turley, S.(2006), "In pursuit of global regulation: Changing governance and accountability structures at the International Federation of Accountants (IFAC)" Accounting, Auditing & Accountability, 19(3), p. 430-432.

143 Humphrey, C.; Loft, A. and Woods, M. (2009) "The global accountancy professionand the international financial architecture: Understanding regulatory relationships at a time of financial crisis" Accounting, Organizations and Society, 34(6-7), p. 812-813.

144 Ibid.

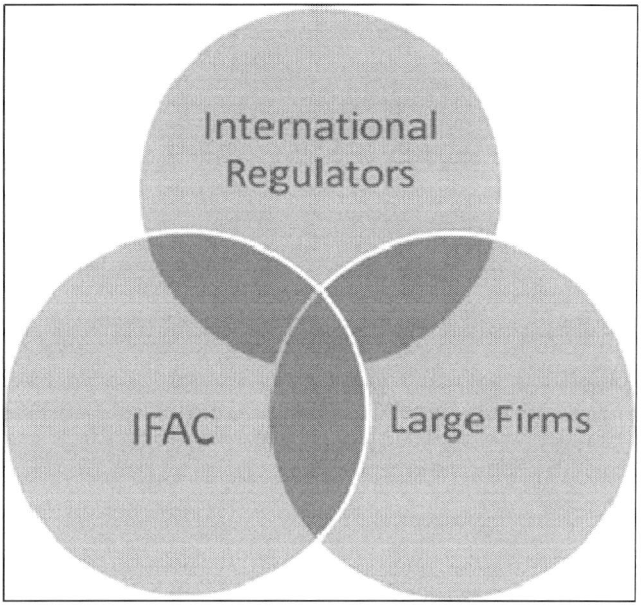

Figure 2.1 Interlocking relationships in global audit regulation[145]

Its global standard setting status was further anchored when the European Union (EU) made a provision on international standards on auditing in the Statutory Audits of Annual Accounts and Consolidated Accounts Directive[146]. The directive did not provide for the adoption of the International Standards on Auditing (ISA) by the IAASB, but gave the commission the option to adopt such standards within the EU provided certain criteria are fulfilled.[147]

The changes in regulatory structures and involvement of new actors brought changes in the dominant institutional logic of audit regulation from one based on professional trusteeship to one based on commercial exchange of expertise.[148] In addition, the strong representation of the financial community is noted. The shifts in the

[145] Ibid., figure 1, p. 813.

[146] Directive 2006/43/EC of the European Parliament and of the Council 17 May 2006 on statutory audits of annual accounts and consolidated accounts, amending Council Directives 78/660/EEC and 83/349/EEC and repealing Council Directive 84/253/EEC, Article 26(1).

[147] Ibid, Article 26(2).

[148] Suddaby, R. (2007), p. 335. See also, Caramanis, C. (2002) "The interplay between professional groups, the state and supranational agents: Pax-American in the age of globalisation" Accounting, Organisations and Society, 27, p. 401-404.

regulatory structure and regulatory logic of the accountancy profession provide a background to enable understanding of the complex global regulatory arena within which sustainability assurance standards are being developed and used. These developments also justify the need for a theoretical framework that looks beyond state-centric notions of regulating the accountancy profession. Crucial to the discussion in this case, the accountancy profession has not been granted sole jurisdiction to carry out sustainability assurance engagements, which pinpoints and gives priority to the need to take into account efforts and influences of other institutions interested to fill this perceived regulatory gap.

2.4 Sculpting tools to analyse the context of standard setting

The decentred view of regulation provides a conceptual lens to analyse the dynamics of regulating sustainability assurance and further builds the research questions and focal points. The notions of complexity, fragmentation and interdependence of decentred regulation points to the importance of analysing who the actors occupying the sites of regulation are, their regulatory purpose, their patterns of interaction and the influences present between the actors when regulating sustainability assurance.

2.4.1 Decentred regulation and regulatory space

The above brings to our attention the complexity of standard setting in a decentred regulatory regime, indicating a need for a "device" to analyse the inner workings and the environment from which standards emerge, develop and are legitimised or de-legitimised. The "regulatory space"[149] metaphor is utilised in this thesis to analyse the standard setting context.

The analytical construct "Regulatory space" was introduced by Hancher and Moran in the study of economic regulation under advanced capitalism[150]. Regulatory space is a device depicting the 'place' where regulation occurs, defined by the range of regulatory issues subject to public decision.[151] In a decentred regulatory context, this device is thus particularly useful to enhance our understanding of the nature of the

---

[149] Morgan, B.; Yeung, K. (2007), p. 59-68.

[150] Hancher, L.; Moran, M. (1989) "Regulatory Space" in Hancher, L.; Moran, M. Capitalism, Culture and Economic Regulation. Clarendon Press, p. 271.

[151] ibid, p.277.

shared space, power and influences.[152] This includes the terms on which actors can enter into the space (rules of admission or exclusion); the relationships between actors and their influence on the regulatory issues negotiated.

The image of space connotes the idea that there is a gap available for occupation and it may be unevenly divided between different participants: state or non-state; major or minor. Actors have the ability to construct not just themselves, but also regulatory issues in and out of the space, defining and redefining the boundaries. Actors interpret, construct and reconstruct their interest while constructing and linking problems, actions and solutions within the space.[153] Through these strategies territory is carved out of the existing space.[154]

Boundaries that demarcate the regulatory space can be defined by a range of issues.[155] The range of application is therefore not constrained. In this case, the regulatory space is defined as the standard setting space which sustainability assurance standards are being constructed at both the national and international levels.

Hancher and Moran determined five forces that shape the regulatory space and allocate power within it: the national political and legal setting, historical timing, organizational structure, character of markets, and nature of issues.[156] Particularly relevant is the nature of regulatory issues as it affects the composition of participants in the space. For example, if regulatory issues are constructed to be highly technical, a different group of actors is likely to control the regulatory process than if the issues are political. Changes in regulatory issues are indicative of the shifting balance of power in regulatory space. As issues are negotiated, actors gain, maintain, or lose space.[157]

Although some of the five forces refer to its application at a national level, this does not deter the regulatory space metaphor to be applied at an international level.[158] In addition, similar to decentred regulation, it does not make distinct the private and public divide. The state is viewed as one of the actors in the regulatory space.

---

152 Morgan, B.; Yeung, K. (2007), p. 310-311.

153 Young, J. (1994) "Outlining regulatory space: Agenda issues and the FASB" Accounting, Organizations and Society, 19(1), p. 86-89.

154 MacDonald, L.; Richardson, A. (2004) "Identity, appropriateness and the construction of regulatory space: the formation of the Public Accountant's Council of Ontario "Accounting, Organizations and Society 29(5-6) Issue: 5-6, p. 489-494.

155 Hancher, L.; Moran, M. (1989), p.278-279.

156 Ibid. 279-297.

157 Morgan, B.; Yeung, K. (2007), p. 63.

158 Ibid.

2.4.2 The use of regulatory space

The regulatory space metaphor has been applied to study the complexity and interdependence in regulation.[159] In the field of accounting, it has been applied by Young to shed light on existing institutional arrangements (formal or informal) and the complexity of agenda and standard setting.[160] Young identified a series of constructions which took place within the regulatory space: first, the condition of practice was constructed as a problem by the participants of the space. Second, rationality for standard setting action was established which "gets the issue onto the standard setter's agenda". Third, solutions were constructed to be able to resolve problems. This thesis draws on Young's constructionist approach which is particularly suitable due to the fact sustainability assurance is regulated by non-state actors and operates in a market shared by many institutions with diverse roots, pursuing different objectives and grounded in different philosophical and functional pursuits. Furthermore, the sustainability assurance concept and standards are still "in the making"; definitions and issues remain to be negotiated.

The regulatory space device supports looking beyond the organizational boundaries of the IAASB as "the standard setter" and takes into account the shifting and competing influences within the decentred regulatory environment. Along the lines of Young, this thesis adopts the regulatory space metaphor as a conceptual space where international standards develop to regulate sustainability assurance. It serves to map out what activities actors engage in to legitimise their entry, non-entry or exit from the regulatory space. Going beyond this, it enables the examination of how problems and issues have been constructed, which has a role in determining the legitimate solutions negotiated in the space.

2.4.3 Legitimacy in decentred regulation

Non-state standard setters are often not delegated the authority to set standards. Legitimacy has to be established for both the *standard setting organisation* and the *standards*.[161] The former calls for analysis of how actors construct legitimacy of their role in the regulatory space. The latter calls for analysis of the ideological persuasive

159 MacDonald, L.; Richardson, A. (2004), p. 489-498.

160 Young, J. (1994), p. 85.

161 Tamm-Hallström, K. (2004) "Organizing international standardization" Cheltenham: Edward Elgar, p. 28- 31.

arguments put forth by actors of usefulness and benefits of standards and the standard setting process.[162]

The term legitimacy has been used in several distinct literatures to address normative and cognitive forces that empower, construct and constraints organizational actors.[163] This thesis adopts Suchman's definition looking at legitimacy as a "*generalized perception or* assumption *that the actions of an entity are desirable, proper,* or *appropriate within some socially constructed system of norms, values, beliefs, and definitions".*[164] In decentred regulation, standard setters are not granted legal authority, their acceptance is based on *"freely exercised reason and collective decisions reached through rational reasoning or deliberations"*[165]. *The* adoption and compliance to the standard are dependent on the regulated parties' willingness to limit the exercise of free will. It is thus important to analyse how legitimacy is constructed as institutions act to establish, maintain, and defend their position in the regulatory space. Just as important is how legitimate arrangements can be "de-legitimized" through complex interactions in the regulatory space.

In the case of sustainability assurance, standards have emerged from more than one site. The regulatory space device provides an additional dimension to enable the analysis of how standard setters respond to competing legitimacy demands by stakeholders and competition for legitimacy between standard setters.[166] Within the regulatory space, the legitimacy of non-state standards can be analysed in terms of input, throughput and output legitimacy.[167] Input legitimacy takes into account the participatory quality of parties involved in rule making. It points to the consideration of how legitimacy can be enhanced through the composition of participants in the standard setting process; the extent decisions are based on expert knowledge, technical agreement and consensus, and the extent of resource symmetry between different participants.[168] Throughout legitimacy, focuses on the standard setting process. It

---

[162] Ibid.

[163] Suchman, M (1995) "Managing Legitimacy: Strategic and Institutional Approaches" Academy of Management Review Volume: 20 Issue: 3 Pages: 571-610. Refer to Suchman synthesis of different concepts of organisation legitimacy. See also, Baldwin, R.; Cave, M. (1998), p. 82-88, five sources of legitimacy from a socio-legal view point; Peters, A. et al, (2009), p. 18-21.

[164] Ibid., p. 574.

[165] Boli, J. (1999), "conclusion: world authority structures and legitimations" in Constructing world culture internal non-governmental organisations. Stanford University Press, p. 273.

[166] Black, J. (2009) "Legitimacy, accountability and polycentric regulation: dilemmas, tri-lemmas and organisational response" in Peters, A. (2009) p. 255-267.

[167] Quack, S. (2010) "Law, expertise and legitimacy in transnational economic governance: an introduction" Socio-Economic Review, 8, p. 6-9.

[168] Ibid.

provides a basis for analyzing if the standard setting procedures have been legitimately drawn up and implemented. Evaluation can be based on whether the agenda setting and voting procedures are transparent, whether there are opportunities for different actors in the regulatory space to debate and engage in dialogues, etc. Output legitimacy refers to the perceived efficiency and results achieved through the standard.

With standards emerging from multiple sites, each standard setter engages in the process of legitimizing its standard setting role and standards to their relevant stakeholders with potentially conflicting demands point to the importance of analyzing how actors at site level construct and maintain legitimacy and also how these standard setters and standards interact and influence each other in building legitimacy with their relevant stakeholders.

To summarise, decentred regulation provides a theoretical lens to analyse the context of standard setting. The regulatory space metaphor is used as a device to map out how the perceived regulatory gaps concerning sustainability assurance are filled by standards from the mid-1990s to 2010. The theoretical framework and literature further specifies four sub-questions crucial to the analysis of how these gaps are filled. They include: Which actors are included and excluded from the space? What activities have actors engaged in, in order to legitimise their entry, non-entry or exit from the regulatory space? How have problems been constructed, issues raised and legitimate solutions negotiated to create a need for sustainability assurance standards and to establish regulatory issues? What are the patterns of influence and the outcome of the construction process?

2.5 Sculpting tools to analyse how standards work

2.5.1 Standards as a regulatory tool

Standards have been defined in a number of different ways using different theoretical lenses. Legal theorists, in common law countries, tend to define standards by distinguishing them from rules. Standards allow the users wider discretion whereas rules are more specific.[169] Political scientists tend to look upon standards as mostly addressing technical products and processes, playing a marginal role compared to international law in global regulation.[170] The neo-institutional organization theorists

---

[169] Schlag, "Rules and Standards" UCLA Law Review, 33, p. 379-430.

[170] Kerwer, D. (2005), p. 162.

adopt the view that "standards constitute rules about what those who adopt them should do". Standards are distinguished from binding directives.[171] Although these are over generalizations and oversimplification of the different disciplines, the common element is that standards provide guidance.

This thesis defines standards in two ways. On a more general level, it defines standards as 'guides for behaviour'. This broad definition removes the debate away from the provenance and administration of the standard. This definition allows voluntary best practice rules, professional standards, codes of conduct etc. to be included within its scope. In the process of standard setting, standard setters codify rules, the "general norms guiding conduct or action in a given type of situation"[172], into the provisions of standards. From a narrower perspective, standards refer to regulatory instruments by state or non-state standard setters to regulate behaviour. Standards referred to in this thesis include AA1000AS by AccountAbility, Global reporting standards by GRI, ISAE3000 by IAASB etc. It is interesting to note that the term "standards" was only adopted by IFAC for auditing guidelines, in 1991. According to IFAC, standards better reflected the position and authority of its technical pronouncements then its previous term 'guidelines'.

The way contemporary society has been organised has given rise to the prominence of standards as a regulatory tool particularly where initiatives have been taken by quasi-state or non-state actors to fill perceived regulatory gaps. Standards by non-state actors have a different character from hard law. These standards are, in themselves not legally binding and more flexible in terms of how it is set, published, applied and interpreted. The increasing prominence of standards in contemporary society can be partly attributed to its greater flexibility, timeliness and ability to penetrate national borders.[173] These standards rely on adopter's willingness to comply and are often accompanied by soft enforcement mechanisms.

---

171 Brunsson, N.; Jacobsson, B.(2000), p. 1, 5.

172 Twining, E. and Miers, D.(1985), How to Do Things With Rules, second edition p.127.

173 Kerwer, D (2005), p. 614.

2.5.2 Understanding the Principles versus Rules debate

2.5.2.1 Principles versus Rules debate

The term 'Rules' is another term that has several conceptions. It is to be noted there that the conception of "rules" in the principles versus rules debate in legal literature and accounting and audit literature differ from the definition of rules presented in section 2.5.1. The "principles" versus "rules" debate has been significant in both academic and policy discussions. Within academia, there are also differences on how principles and rules are conceived within and between different disciplines.

From a legal perspective, the distinction between principles and rules is primarily based on the degree of specificity.[174] However, differences in conceptions do prevail. According to Dworkin, "If the facts a rule stipulates are given, then either the rule is valid, in which case the answer it supplies must be accepted, or it is not, in which case it contributes nothing to the decision."[175] Principles, on the other hand "state a reason that argues into one direction, but does not necessitate a particular decision"[176]. Following this line of reasoning, conflicts can occur in principles, which are less specific compared to rules. Other scholars argue against this *strict* approach to the application of rules and the idea of conflict. Focus is placed more on the degree of specificity of the provisions in question. Principles have been regarded as to merely limit and not dictating consideration to be taken.[177] This perspective informs that the degree of specificity of a provision has an effect on the conceptual and functional dimensions of provisions in standards.

Consistent with legal literature, the linguistic structure has been used here to distinguish principles from rules. The context which these debates take place is centred on the issues of the "rules-based" US Generally Accepted Accounting Principles and the arguments for a more principles-based approach to regulation. This literature refers to principles as broad guidelines derived from a complete and internally consistent framework. The framework requires users of the standard to exercise professional judgment in applying the principles to specific transactions and

---

[174] Dworkin, R. (1967) "The Model of Rules", University of Chicago Law Review p. 26; Raz, J. (1972), "Legal principles and the limits of law" Yale Law Journal, p. 838; Hart, H. (1994) The Concept of Law, Claredon Press, p. 260; Schauer (1997) "Prescriptions in three dimensions", Iowa Law Review, p. 914.

[175] Dworkin, R. (1967), p. 25.

[176] Ibid. p. 26.

[177] Raz, J. (1972), p. 846.

events.[178] Rules based regimes (typically referring to US GAAP) on the other hand are perceived to entail detailed guidance. Guidance comes in the form of implementation guidance quantitative thresholds, qualifications and exception clauses.[179] The concept of rules derived from this context differs from the legal perspective, as rules from a legal point of view may be derived from "a complete and internally consistent conceptual framework or set of high level principles".[180]

The arguments against rules-based regulation are that it fosters creative compliance or that it might encourage users of the standards regard the detailed rules as a final port of call and as a result lose sight of the principles or purpose of regulation. Principles-based regulation on the other hand, allows scope for professional judgment, flexibility and innovation but is critiqued for offering inadequate protection. Research has shown that the distinction between principles-based and rules-based regulation is more theoretical than practical.[181] For example, US GAAP is based on principles but regarded as rules-based due to its scope exceptions, treatment exceptions, and implementation guidance.[182] The IAASB standards are regarded as principles-based, standards are however often accompanied by guiding provisions, which making them more rules-based. This discussion leads to the conclusion that it is not meaningful to engage in principles versus rules debate viewing the two as dichotomies[183], the two competing models tend to be combined in practice to achieve the "right balance of discretion" standard setting bodies decide to grant users.

Black puts forth the notion that principles-based regulation sits well within the dynamism of contemporary society, with decentred regulation.[184] Whilst rules

---

[178]Tweedie, D. (2007) 'Take it from the Top', 'Can Global Standards be Principle Based?', The Journal of Applied Research in Accounting and Finance, July 2007. p. 8.

[179] Refer to Dickey, J.; Scanlon, M. "Principles-Based" Accounting Standards—an Accident Waiting to Happen?', Insights, February 2006. Tweedie (2007), p.8, Wustemann, J; Wustermann, S.(2010) Abacus, 46(1). McBarnet, D. and Whelan, A."The Elusive Spirit of the Law: Formalism and the struggle for legal control (1991) Modern Law Review, 54, p. 848, 850-851.

[180] Wustemann, J.; Wustemann, S (2010) and Benett, B; Bradbury, M Prangnell, H. (2006) "Rules, principles and judgement in accounting standards, Abacus, 42(2) p. 189-203 showed through analysis of FASB standards with two principles based standard pertaining to Research & Development expenditure that all standards contained rules are based on principles and require judgement.

[181] Schipper,K. (2003) "Principles based Accounting Standards", Accounting Horizons, 16(3), p. 219-232.

[182] ibid.

[183] Benett, B; et al. (2006) p. 189-203.

[184] Black, J. (2007) " Making a success of principles-based regulation" Law and Financial Market Review, p. 193-196.

emphasise the rhetoric of "directing and controlling" to achieve the regulatory purpose, principles-based regulation moves the operations of the provisions to the sites of application where discretion is exercised by the users of standards to a greater extent compared to a rules-based regime.

Principles based regulation has been the "preferred" approach adopted by policy makers in contemporary society as it is broadly applicable and can be applied purposively to achieve regulatory objectives.[185] However, it is not without its limitations and paradoxes. Principles based regulation has been criticised as 'laxed', vague and lacking the prescriptive elements for regulation to work effectively.[186] It can also lead to more rules, as uncertainty can result in implementing parties, internal management, supervisory or enforcement bodies setting specific detailed guidance that serve as a final port of call. In such cases, the key characteristics of flexibility and "purposiveness" upon which principles-based regulation is founded on, are lost.[187] Principles based regulation also creates potential for a communication paradox. Instead of enhancing communication between parties, regulation can be hindered due to incoherent interpretations of the principles by the standard setter.[188] The ultimate paradox according to Black, is the trust paradox. The mutual trust and responsibility which principles-based regulation facilitates is as mentioned above, a precondition to its effective functioning.[189]

2.5.2.2 IFAC and the concept of principles-based standard

IFAC is conceived to be a proponent of principles-based standards. In light of the public interest objective and IAASB's role to set high-quality international auditing and assurance standards, the Clarity project was adopted in 2004. This involved reviewing the drafting conventions used in making auditing standards, in recognition that standards need to be understandable, clear, and capable of consistent application.[190] One objective of the project is to ensure auditing standards are

---

[185] Black, Julia (2008) *Forms and paradoxes of principles-based regulation.* LSE law, society and economics working papers, 13-2008. Department of Law, London School of Economics and Political Science, London, UK, 3(4), p. 425.

[186] Black, J. (2007), 196-199.

[187] Ibid., p. 446.

[188] Ibid., p. 447.

[189] Ibid., p. 456.

[190] http://www.ifac.org/IAASB/ProjectHistory.php?ProjID=0024

principles-based.[191] Dennis, based on his study of the Clarity project argues that adequate consideration had not been given to the conceptualisation of "principles-based" standards and the relevant precepts such as objectives, requirements and principles of auditing and explanatory guidance.[192] His work established that the term principles-based was two different conceptions. First, standards have been derived from principles in a conceptual framework that was not completely realised. Second the objective to promulgate principles-based standards was challenged by the lack of clarity between objectives and objective-based standards.

2.5.2.3 To engage or not to engage in the principles versus rules debate?

The above draw our attention to three key points. First, from a theoretical viewpoint, legal, accounting and auditing literature regard principles and rules as competing modes of rationality. In reality they represent a false dichotomy; standards consist of provisions which can be principles or rules-based. Standards tend to be balancing the tensions between certainty and flexibility, uniformity and idiosyncrasies of each case. Second, the work of Dennis has highlighted that the conceptual foundation of "principles-based" as used by IFAC is not concrete, as many precepts remain to be conceptualised in greater depth. The framework, within which standards are drafted does not appear to be clear and is constantly being reconstructed. Third, the paradoxes of principles-based regulation highlight the potential that more rules can emerge and serve as the final port of call, detrimental to the functioning of the spirit or purpose of principles, which principles-based regulation promulgates.

This thesis will not engage in this debate, as classifying standards into principles or rules is not regarded as helpful to the understanding of how standards as regulatory tool work. Instead, the focus will be placed on analysing the provisions of standards themselves.

2.5.3 "Rules" and their limits

In this section, the "principles versus rules debate" is put aside, and the definition of rules as "general norms guiding conduct or action in a given type of situation"[193] is re-adopted. The broader definition includes its applicability to hard law but also voluntary best practice rules. The use of the term "rules" according to the above

---

[191] Ibid.

[192] Dennis, I. (2010) "Clarity Begins at Home: An Examination of the Conceptual Underpinnings of the IAASB's Clarity Project" International Journal of Auditing, 14, p.298-316.

[193] Twining, W.; Miers, D., (1985) How to Do Things With Rules, p. 127.

definition is also broader than "rules" as referred to in the "principles versus rules" debate, which focuses on only the linguistic structure.[194]

Scholars that write about regulation have noted that in the construction of rules, that the operative basis is built by the process of grouping together particular attributes of an object or occurrence, then making generalisations about them to build up a category or definition.[195] This process involves simplification, suppressions and assessment of probable causations contributing to the limits of rules.[196]

The limits of rules are well documented in legal literature on enforcement. The following section draws heavily on the work of Black. Black points out three unanimously agreed problems pertaining to the use of rules in a general context. These are: rules tend to be over or under inclusive, indeterminate and require interpretation.[197] These problems are rooted in the nature of rules and the nature of language. The former are a consequence of the rule making process and the latter and based on the fact that rules are interpreted by users, and thereby contingent upon their understanding and interpretation of the language used in the construction of the provision.[198]

In the course of construction, due to how rules are made, certain subsequently relevant properties may be neglected (rule is under-inclusive); similarly, irrelevant properties may be drafted in the rules (rule is over-inclusive), resulting in a mismatch between the scope of the rule and its purpose. Black identifies two other reasons of this mismatch: first, the causal relationship between the event and goal is only an approximate one and the dynamism of events make certain the imperfect causal match between the generalisation and the aim of the rule.[199]

Rules are indeterminate. This arises out of the anticipatory nature of rules, the linguistic structure and its reliance on users to apply them. Rules can never be sufficiently precise or explicit enough to cover every potential situation. Rule makers can set rules that are: incomplete in that they omit certain facets; imprecise or evaluative which call for subjective judgment. On the other hand, rules can be over precise, which tend to encourage creative compliance which can undermine the objective of the regulator.[200] Social jurisprudence scholars have proposed the increase

---

[194] It will be introduced in section 2.5.6 in this thesis that rules have other dimensions.

[195] Baldwin, R. and Cave, M. (1999), Black, J. (1997) among others.

[196] Ibid.

[197] Black, J. (1997), p. 6-19.

[198] Ibid.

[199] Ibid.

[200] Morgan, B.; Yeung, K. (2007), p.153-160; McBarnet, D. and Whelan, A. (1991), p. 848, 850-851.

in rule precision to overcome problems of over or under-inclusiveness and legal uncertainty.[201] However, contrary to what one might expect, rules and discretion are not opposing entities[202]. Whilst the structure and complexity of rules have implications on discretion, rules due to its limits could fail to control discretion[203] regardless of the level of precision.

2.5.4 Discretion – the notion of choice

Discretion starts with the notion of choice. Users of standards exercise discretion in the process of translating rules to rules, or rules to action; as such abstract rules become reality.[204] Hawkins describes discretion to be the "scope for the play of individual judgment afforded within a structure of rules."[205] It exists in circumstances when more than one decision is considered proper, when no known rule covers the situation, or when rules do not yield clear answers as to the course of action to be taken. Discretion can be viewed as attractive as it allows users to: determine which rule, or parts of the rule to apply. It can also fill gaps; resolve conflicts within and between rules; and allow a purposive approach to the application of rules.[206] On the contrary, discretion can also be viewed as a type of power that potentially corrupts, leading to questions on legitimacy and the risk of users abusing the discretion. This could lead to the lack of consistency and uniform application of the rule.[207]

Discretion and interpretation are two related but distinct concepts. Regulation studies have in addition pointed to the importance of distinguishing between legal and regulatory interpretation.[208] The theoretical distinction between these terms is beyond the scope of this thesis. This thesis refers to discretion as the scope in which users of

---

201 Black, J. (1997) refers to this as the first generation rule talk. Scholars in the past argue for an increase in rule precision as a means of ensuring socially undesirable behaviour is discouraged.

202 Hawkins "using legal discretion" p .35.

203 Hawkins, K (1992b) "The uses of discretion", Oxford U Press; Galligan, D. (1990) Discretionary powers Oxford.

204 Hawkins, K. (1992a), p. 11.

205 Ibid., Preface.

206 Schneider, C (1992) "Discretion and Rules: A Lawyer's View" in Hawkins, K. The Uses of Discretion, Oxford University Press, p. 47-88.

207 Ibid.

208 Georgosouli, A. "Regulatory Interpretation: Conversational or Constructive? Oxford Journal of Legal Studies, Vol. 30, No. 2 (2010), pp. 361–384. The author stresses the importance of distinguishing between legal and regulatory interpretation as due to the arguably broader scope of regulation as defined above and that regulatory regimes tend to operate separately from the courts.

assurance standards exercise choices within the provisions of standards. Interpretation on the other hand, is used to describe the outcome as to how legal rules are read through legal sources.

The indeterminacy of provisions allows for the exercise of discretion and opportunities of individuals in society to exercise power.[209] The space in which discretion is exercised, whether formally conferred or assumed within an institutional context, is however constrained by social and organizational forces.[210]

2.5.5 The context of applying rules

Socio-legal studies have taken up the notion of the reader and the text. From the perspective of Fiss, interpretation can be viewed as "neither a wholly discretionary nor a wholly mechanical activity" but a "dynamic interaction between the reader and text. This definition takes into account both the "text" and the "reader" placing between the two a set of disciplinary rules derived from the specific institutional setting where interpretive activity takes place.[211] Rules, therefore can be conceived as constraints. Breaking from the legal sphere, contextualists such as Fish however argue that rules cannot serve as constraints as they need to be interpreted.[212] Text and readers are "structures of constraints", "components of and structures in the larger structure of a field of practices". This viewpoint suggests one to look beyond the text of rules to assess what is tacit in the rule that cannot be textually expressed.[213]

Rules rely on users to interpret and apply them; how a rule is interpreted is dependent on interpretive communities, i.e., communities who share culture, norms and goals, constituted by institutional practices.[214]. This perspective suggests that a rule is only as good as its interpretation and the presence of a level of shared understanding, diminishes the need for explicit rules. It would also allow the standard setter to rely more on tacit understanding in regulation.[215]

---

[209] Hawkins, K. (1992a), p. 12.

[210] Ibid.

[211] Middle path. Positivist view meaning as property off and embedded in text and subjectivist view meaning is discretionary.

[212] Fish, S. (1984) "Fish v. Fiss" Standford Law Review, 36(6) p. 1339.

[213] Ibid.

[214] Black, J. (1997) p. 31.

[215] Ibid., p. 30-32.

2.5.6 Dimension of rules – the discretionary space

To sculpt an analytical framework to understand how standards function, the above literature channels our focus on analysing the linguistic construction of standards, the discretionary space granted to users applying the standard, and how the discretion is exercised within the context in which users operate.

This thesis draws on socio-legal studies' concept of dimension of rules[216] to develop a toolbox to analyse the status and nature of sustainability assurance standards by the accountancy profession.

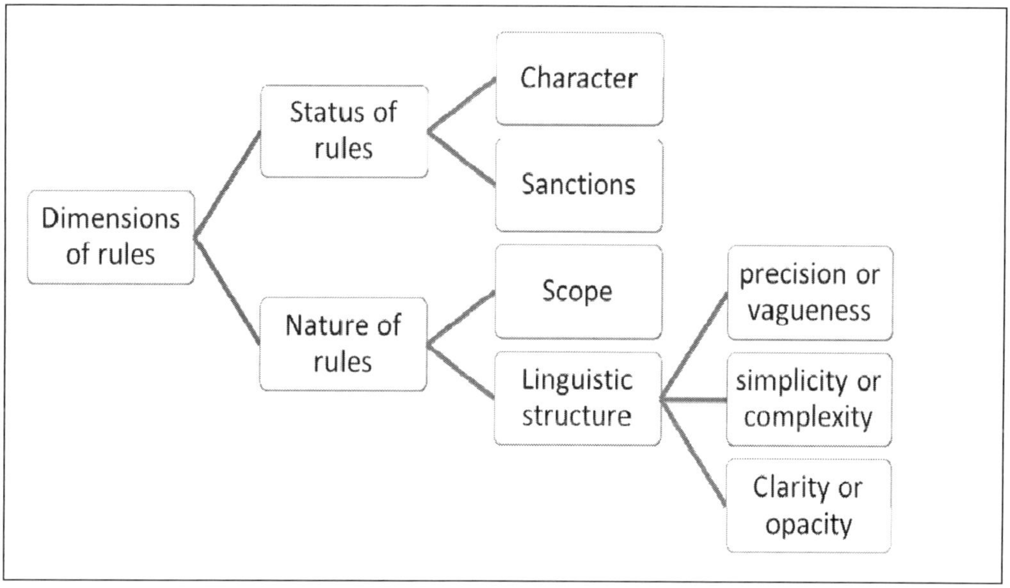

Figure 2.2 Dimensions of rules[217]

With reference to the table below, the provisions of standards can be analysed through the linguistic structure along three sub-dimensions: level of precision; level of clarity and level of complexity; and the scope of the standard to determine the discretion granted to users.

---

216 Rules referring to section 2.5.3 are "general norms guiding conduct or action in a given type of situation" which defines within its scope voluntary standards

217 Source: Black, J. (1997), p. 6-19.

The sub-dimensions and characteristics to be used in the evaluation process are summarised in the table below.

Dimension	Characteristics
Linguistic structure • precision or vagueness	Degree to which action to be performed, time or place, applicability of the provision is precise? Degree to which the provisions contains evaluative and generic terms?
• simplicity or complexity	Number of factual situations or assessments involved in determining the rules applicability?
• clarity or opacity	Degree to which in contains word with well- defined and universally accepted meanings
Substance and scope	Degree which provision match the objective of the standard (over or under-inclusive)

Figure 2.3 The toolbox to analyse provisions of standards[218]

From the above discussion, the dimensions of rules provide a theoretical basis and tools to analyse how standards set by the accountancy profession to guide sustainability assurance engagements. The above literature emphasizes the importance of the contents of standards (comprising the scope and linguistic structure), the character of standards (whether it is legally binding) and the context within which they operate.

[218] The toolbox was built on the dimensions of rules in Black, J. (1997) p. 6-12.

2.6 Conclusion

2.6.1 A snapshot of the theoretical framework and focal points

Research questions	Theoretical framework	Focal points	Chapters
How has the "assurance" concept been framed and codified by the IAASB?	Global audit regulation	Context of standard setting What are the effects of developments of the regulation of the global accountancy profession on the framing of the assurance concept?	4
How has the perceived regulatory gap on sustainability assurance been filled by standards?	Decentred regulation utilising the regulatory space construct and legitimacy theory	Which actors are included and which are excluded from the space? What activities have actors engage in to legitimise their entry, non-entry or exit from the regulatory space? How have problems been constructed, issues raised and legitimate solutions negotiated to: create a need for sustainability assurance standards and to establish regulatory issues. What are the patterns of influence and the outcome of the construction process?	5

How do standards constructed by the accountancy profession work to guide sustainability assurance in Sweden?	Dimension of rules and the use of discretion	Content of standards How have assurance standards guiding sustainability assurance engagements been constructed? How have they been used by the different groups of users of the accountancy profession? More specifically: How much discretion has been granted by ISAE3000 to Far? Are provisions precise or vague? Clear or opaque? Simple or vague? How has Far exercised discretion granted by ISAE3000 in revising RevR6? How have Swedish practitioners exercised discretion granted by RevR6 when carrying out sustainability assurance engagements	7 8
How do standards set by the accountancy profession interplay with hard law to regulate sustainability assurance in Sweden?	Decentred regulation legal framework	What is the status of sustainability assurance standards? Is there law directly governing sustainability assurance? If not, does sustainability assurance fall under the current legal framework regulating auditors? What are the implications on the status and functioning of standards by the accountancy profession in guiding sustainability assurance?	8

Figure 2.4 The theoretical framework and focal points of the thesis

2.6.2 Challenges and choices

Definitions are based on the theoretical lens adopted. Drawing theory and literature from different disciplines brought challenges when terms are ubiquitous, terms that have strong ties with particular disciplines and terms that do note travel well in different legal jurisdictions. In addition, the accountancy profession has its own language and discursive strategies adopting these terms in a different way, from that of the academic literature. This chapter has attempted to explain how these terms, such as regulation, standards, rules and law are used in the context of this study.

The way regulation is conceptualised in academia is often dependent on the problem or issue the researcher is writing on. In developing the decentred view of regulation, Black has adopted mainly an essentialist view focusing on the characteristics of regulation. In the course of applying this theoretical framework, it was discovered that a more functionalist and conventionalist definition of regulation, seeking to inquire what function regulation plays in society, and how regulation is understood by the accountancy profession, is important to conduct this analysis. The first research question analyses how the "assurance" concept has been framed and codified by the IAASB, added understanding of not only the subject matter of regulation but the accountancy profession perspective of regulation in the context it operates.

Many debates surround the scope of what is regulation. This raises the question of whether patterns of interaction of rational actors, social norms and other factors such as technology that influence regulation should fall within the scope of regulation. With reference to Black's essentialist definition of regulation presented in chapter 1, which places focus on the intentionality to alter behaviour and mechanisms present in regulation, these factors are not likely to fall within the definition of regulation but rather regulation in terms of Baldwin's third conception, under "all forms of social control". However, it does not dismiss the importance of these factors on the development and functioning of regulation. This thesis analyses these factors with the context of standard setting and context which standards are applied.

Decentred regulation is depicted by the shift of regulatory loci from the state recognizing multiple sites which regulation can take place. This also invokes the questions as to what is the difference of state centred regulation and law? And how is it distinguished from legal pluralism? A more relevant question is perhaps if these interesting debates are essential or relevant to the purpose of the thesis?

The distinction between regulation and law is unclear as much depends on the theoretical lens adopted in such debates. Black suggests that "there is as little and as much agreement as to what law is as there is as to what regulation is"[219] and in the broadest sense, the two terms could be "re-descriptions of each other". In line with legal pluralists, the former describes "patterns of social ordering" and the latter refer to, "forms of social control". Academics writing in different contexts have also subordinated one to the other. Some writers claim law as one of the many instruments used to regulate a particular phenomenon whilst others tend to look at regulation as one particular source of law e.g., in EU law[220]. To avoid disputes on

---

[219] Black, J. (2002), p. 22.

[220] Treaty of the Functioning of the European Union (TFEU).

definitions, some researchers analyse these forms to influence behaviour along a spectrum of legality.[221] The purpose of the thesis is to contribute to our understanding of regulation by non-state standard setters and examine how this form of decentred regulation interplays with hard law. The focus is therefore placed on how regulation emerges and functions outside existing legal institutions and the reciprocity between the decentred regulation and state-centred regulation. The difference between law and state-centred regulation thereby lies outside the scope of the thesis.

Legal pluralism rejects the 'the ideology of legal centralism' challenging the conceptions of the states' monopolistic claim to power over the recognition, legitimacy and validity of law.[222]It therefore has in common with decentred regulation that rules may emanate not only from the state. Legal pluralism has not been explored further in this thesis as there is a lack of agreement as to what is a recognized source of law. Decentred regulation captures within its scope regulation of non-state standard setters and avoids centring the thesis on whether assurance standards can be regarded a source of law.

---

[221] For example, refer to Abbott, K; Keohane, R; Moravcsik, A; Slaughter, AM and Snidal, D (2000) "The concept of legalization", International Organization, 54(3), p. 401-419.

[222] Griffiths, John (1986) "What is Legal Pluralism" in Journal of Legal Pluralism 24: 1-55.

3. Methodology and Method

3.1 Chapter objectives

The objective of this chapter is present the methodology and method of this thesis. It aims to provide readers with a descriptive account of how the research is conducted. Chapter 1 presented the challenges placed on legal scholarship to understand regulation in contemporary society. As regulation shifts from the state to polymorphous non-state actors, decentred regimes emerged to fill the regulatory gaps. How standards are set and how they work differ from hard law. The authority and sanction structures of these decentred regimes are not as clearly defined, and dependent on a wider socio-political context. This stresses the importance of looking beyond the traditional legal methodology to explore how sustainability assurance is regulated.

3.2 Methodology

Methodology characterizes the intellectual and philosophical underpinnings of research.[223] It refers to the ontological and epistemological assumptions held when carrying out the research. Methodology has an effect on how research questions are formulated and the choice of methods applied.[224] Ontology refers to the assumptions about the nature of the reality that is being researched.[225] Epistemology is the study of knowledge and justified belief.[226] Epistemological concerns lie with the grounds for

[223] Scapens, R. (2007) "Using Case Studies in Accounting Research: The Role of Theory", Lund University Workshop, p. 1.

[224] Bryman, A.; Bell, E. (2007) "Business Research Methods" Second Edition, Oxford University Press, p. 25.

[225] Ibid., p. 4.

[226] Ibid. See also, Scapens, R. (2007), p. 4.

making knowledge claims. They determine the necessary and sufficient conditions of knowledge and its sources.[227]

In western societies, legal theory is founded ontologically on the rule model.[228] The study of law and state-centred regulation are often methodologically restricted to this authority paradigm.[229] Paradigms represent a general perspective or way of thinking that reflects the fundamental beliefs or assumptions of a discipline or an area of a discipline,[230] Doctrinal research dominating the discipline tends to operate within a closed system of rules, structures and concepts.[231] Although academic commentators' critique the drafting, lack of coherence and content of law or court judgements, they do not dispute the status of the text of the law and court decisions.[232]

Traditional legal research methods are characterised by rigorous textual analysis of the sources of law. They include legislatures, preparatory works, judgments of the courts and other legal authorities, etc. Legal reasoning stems from this self-referential system.[233] The legal context is structured as a body of rules that can be studied in isolation, away from the broader social context in which legal systems exist.[234] As such, traditional legal research conforming to the authority paradigm consists of an inbuilt enclosure where law can be studied introspectively, independent of social forces.[235] Simply put, it is how law sees itself, explains itself to itself and continually reaffirming itself, without furthering societies' understanding of law.[236] The authority model used in traditional legal research therefore provides too narrow a perspective to capture the dynamics of decentred regulation. To understand the context in which standards by non-state actors are developed and used, a different means to penetrate and comprehend social reality is required.

---

227 Ibid.

228 Samuel, G. (2009) "Interdisciplinarity and the Authority Paradigm: Should Law Be Taken Seriously by Scientists and Social Scientists?", Journal of Law and Society, 36(4), p. 431-432.

229 Ibid., Tomlins, C. (2007). "How Autonomous is Law?" Annual Review of Law and Social Science, 3, p. 46-49.

230 Gioia, D.; Pitre, E. (1990) "Multiparadigm Perspectives on Theory Building" The Academy of Management review; 15(4), p. 585.

231 Samuel, G. (2009), p. 455.

232 Samuel, G. (2009), p. 437.

233 Vick, D. (2004), "Interdisciplinarity and the Discipline of Law", Journal of Law and Society, 31(2), p. 164.

234 Ibid., p.181.

235 Samuel, G. (2009), p. 432.

236 Samuel, G. (2009), p. 456.

3.2.1 Socio-legal studies

This study turns to a socio-legal perspective to extend the assumptions of reality beyond the legal text, court judgments and legal modes of reasoning. The "socio-legal way" however is relatively new and in itself problematic. Socio-legal studies draw from the traditional disciplines of law and sociology. Its interdisciplinary nature requires juxtaposing legal and social science episteme.[237] Banakar identifies two broad categories of socio-legal research.[238] The first is socio-legal studies, reflecting the concerns of sociology. Research conducted in the tradition of the sociology of law, for example, studies law and legal institutions from an external perspective drawing on sociological theories and uses social science methods.[239] A related "school" is the 'law and society' movement in the United States[240]. The two schools of thought adopt a similar approach; the latter however, is less restricted to theory and methods of sociology and draws to a greater extent from other social sciences.

The second category refers to applied socio-legal studies. Applied socio-legal studies from, which this thesis draws on, reflect the concerns of law and policy makers. It adopts an interdisciplinary approach to study "law in context" and policy research reflecting the concerns of law and policy makers[241]. Applied socio-legal studies originated in the United Kingdom, a movement by law schools' attempt to become interdisciplinary. This explains the often positioning of its research in light of its relationship to, and oppositional role within, law.[242]

This thesis draws extensively from the applied socio-legal approach to study "regulation in context". As socio-legal research is not confined within the authority paradigm, a wider definition of rules intended to influence behaviour can be embraced within its scope.[243] The study of "law in context" has been extended to

---

[237] Banakar, R. (2009) "Law Through Sociology's Looking Glass: Conflict and Competition in Sociological Studies of Law" in Denis, A.; Kalekin-Fishman, D. (2009)The new ISA Handbook in contemporary international sociology: conflict competition and cooperation, Sage, p. 58.

[238] Ibid., p. 64-65.

[239] Ibid., p. 64-67.

[240] Ibid. p. 68.

[241] Ibid.

[242] Thomas, P. (1997) 'Socio-Legal Studies: The Case of Disappearing Fleas and Bustards', in Thomas, P., *Socio-Legal Studies*. Aldershot: Dartmouth, p. 3.

[243] Picciotto, S. (2002), p. 1.

"regulation in context" for the purpose of this thesis. It allows the context of regulating sustainability assurance to be analysed from a broader perspective, to take into account the socio-political, economic, market, legal and other considerations affecting how standards are developed and used.[244] Policy research adopts a pragmatic view of regulation.[245] It has as a focal point, the gap between the intention of regulators and reality. This brings to attention the importance of determining the goals of the non-state standard setters, issues pertaining to implementation and enforcement of standards.

There is no general agreement as to whether "socio-legal is a discipline, sub-discipline or methodology". Due to the fact that methods of investigation have not been developed specifically to perform this type of study, socio-legal researchers employ a wide variety of social scientific methods.[246] Methodological debates have been a neglected field in socio-legal studies.[247] Socio-legal studies are often been critiqued for having a weak method, lacking rigour and detailed structure in comparison with research by 'pure' social scientists.[248] At the same time, those doing socio-legal research are criticised for being empiricists, serving primarily policy needs rather than making theoretical contributions.[249] This thesis attempts to draw from social science methodology to explain the basis of how the research questions have been formulated, and the research designed to perform the analysis.

3.2.2 Ontological and epistemological assumptions

With reference to chapter 2, the understanding of how sustainability assurance standards are set and function involves the analysis of both the context and content of standards using an interdisciplinary approach. As disciplines are grounded on different assumptions, this thesis requires more than one paradigm to be adopted. Multi-paradigm research involves applying divergent paradigm lenses empirically. This thesis does not try to integrate social science and legal paradigms, but use the different insights and distinct frame of references of each paradigm in sequence, or in

---

[244] Black, J. (2007), p. 214-215.

[245] Travers, M. (2001) "Sociology of Law in Britain", American Sociologist, 32, p.26.

[246] Banakar, R. (2009), p. 70.

[247] Banakar, R.; Travers, M. (2005) Theory and Method in Socio-legal Research, Hart Publishing, p.ix.

[248] Ibid. p.23-25.

[249] Travers, M. (2001), p.27.

parallel, to understand the regulation of sustainability assurance from competing versions of reality.[250]

The context of standard setting is analysed drawing on a constructionist perspective inspired by the work of Young, in "Outlining regulatory space: Agenda issues and the FASB", published in the journal "Accounting, Organisations and Society".[251] In this thesis, standards are regarded as a product of the construction process. In the analysis of questions 1 and 2, a social science, interpretive paradigm is adopted. Interpretive research focuses on understanding social actors meaning and definition of a situation.[252] This paradigm accepts as the status quo understanding "things as they are". It does not attempt to promote radical change which critical researchers or the radical humanists attempt,[253] or take a normative stance in legal research. This research offers decision makers an objective information source, which can be utilised to facilitate change. The logic motivating the use of this paradigm stems from the fact that much remains to be understood about the use of standards to regulate the accountancy profession, in relatively new area of service, in an open market where other non-accountancy standard setters and providers are present.

The assumption that standards are a product of construction originates from the view that reality is a negotiated order.[254] In this light, meanings are established, reinforced and changed on an on-going basis through interactions, negotiations and relationships between the parties involved. Knowledge comes through subjective interpretation, rather than objective observation. This assumption ascribes to the existence of multiple realities, socially constructed by different actors. Different actors or groups of actors can interpret a particular term or concept differently.[255]

This perspective has several implications on the design of the method of inquiry. First, this perspective sensitizes the researcher to the fact that true meanings are seldom represented on the surface of an event or "texts" which are products of construction.[256] This establishes the importance of tracing the origins and development of certain often 'taken for granted terms' or discourses used. To gain in-

---

[250] Lewis, M.; Grimes, A. (1999) "Metatriangulation: Building Theory from Multiple Paradigms" Academy of Management Review, 24(4), p.673-675.

[251] Young, J (1994) "Outlining regulatory space: Agenda issues and the FASB" Accounting, Organizations and Society, 1994, vol. 19, issue 1, pages 83-109.

[252] Saunders, M.; Lewis, P.; Thornhill, A. (2009) Research Methods for Business Students, Fifth Edition, Prentice Hall, p.118-121.

[253] Ibid., p.121.

[254] Bryman, A.; Bell, E. (2007), p.23.

[255] Ibid., p.22-25.

[256] Saunders, M.; Lewis, P.; et al. (2009), p.119.

depth insights on the development of the regulation of sustainability assurance through standards, the shared realities that were formed, sustained and had changed over time has to be deconstructed. Second, this assumption notes the importance of not neglecting any unanticipated phenomenon or influences. Sustainability assurance is at the initial stages of its development, from the literature reviewed in chapter 1, it appears that many actors are involved in the development and application of sustainability assurance standards. To capture their interaction and interdependences on, how they shape these standards, there is a need to maintain an open, flexible attitude in the data collection process. There is a need for constant reflection and modification of the research design to take into account the "un-looked for" or unforeseen interactions or negotiations that could shape meaning or context.[257] Third, epistemology is placed on meanings and 'processes leading to the output' rather than the 'output itself' (e.g. the coded standard).[258] This favours qualitative inductive methods of research and case studies.[259] An inductive method of inquiry to analyse the context of standard setting is preferred to normative oriented theories of regulation, as it allows the researcher to maintain neutrality towards to values and goals of standard setters.

In the research process, it was found that although this paradigm was well-suited to research into the context which standards is set (research questions 1, 2), it does not provide the means to analyse the product of the construction process to capture the multifaceted nature of how standards, as a regulatory tool work to influence or guide behaviour.

3.2.3 Bringing in multi-paradigms

The literature reviewed in chapter 2, allows our understanding that interpretation is partly a product of the structure of rules.[260] For example, evaluative terms constructed within rules would grant users a wider discretionary compared to clear and precise terms. In operating external to existing legal institutions however, the instrumental rationality of decentred regulation does not parallel that of traditional legal reasoning in curial or civil law. Standards operate in a wider context, which affects its interpretation.[261]

---

[257] Ibid., p.121.

[258] Ibid., p.119.

[259] Bryman, A.; Bell, E. (2007), p. 11-15.

[260] Morgan, B. ; Young, K. (2007) p. 158 with reference to the work of Black, J. (2007)

[261] Refer to the literature reviewed in chapter 2.

Black drew on the work of, among others, Wittgenstein to look into the interpretation of rules.[262] Wittgenstein argued in his work on language and mathematics that the automatic, unreflective following of rules arose from shared judgments with regards to the meaning and application of that rule. If language is to be a means of communication there must be agreement, not only in definitions, but also in judgments.[263] Clarity of interpretation, of a rule is contingent upon the agreement as to the meaning of the rule originating from shared norms and forms of life. Black pointed out three aspects applicable to understanding the nature and functioning of legal rules which this thesis observes. First, a rule or word with a literal meaning simply refers to the meanings members of a community would assign to it unreflectively. This assigned meaning may not be similar to that of other communities. Second, the application of rule is not an objective fact but depends on which interpretive community reads the rule. Third, interpretation constructed by the communities can serve as a framework to explain empirical observations.[264]

In pursuit of the third research question on how standards by the accountancy profession work to guide sustainability assurance, a multi-paradigm sequential technique is therefore used.[265] This technique applies paradigm lenses in succession where the output of one paradigm is used as input for "the next part" of the study. Each paradigm contributes a layer of meaning to develop a more holistic view to understand the phenomenon.[266] The understanding of the standard setting context established from research questions 1 and 2 adopting the interpretive paradigm, serves as input to analyse the content of the standard. The provisions of standards are analysed with reference to their linguistic structure from a socio-legal perspective. The objective is to determine the discretionary space granted to users of these standards. The discretionary space then serves as a basis to analyse the extent of discretion exercised by regulated parties in the specific context they operate using the social science interpretive paradigm. This approach grasps both disparate and complementary focal points of the two paradigms illustrated in the diagram below. The research design is therefore epistemologically plural concerning the evidence needed to understand how standards work.

---

[262] Black, J. (2007), p.17.

[263] Ibid.

[264] Ibid., p. 17-18

[265] Lewis, M.; Grimes, A. (1999), p. 675.

[266] Ibid.

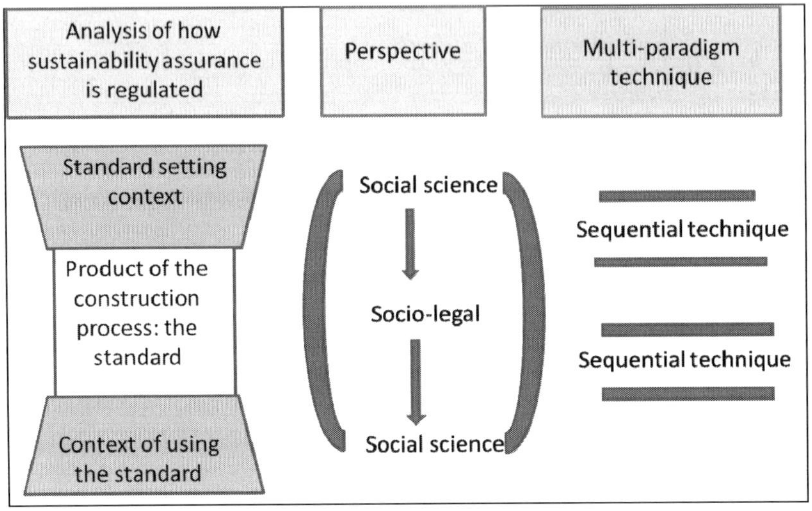

Figure 3.1 Multi-paradigm used in sequence

The second purpose of the thesis is to examine how standards set by the accountancy profession interplay with hard law in the regulation of sustainability assurance engagements in Sweden. As users of standards by non-state actors operate under national jurisdictions, a legal analysis using the legal authority paradigm has to be performed to determine if sustainability assurance falls already within the scope of hard law. Adoption of the authority paradigm temporary shuts out the broader context of standard setting efforts by the profession that are not driven by state impetus. The analysis focuses on the relevant legal sources in Sweden. Findings from research questions 1 to 3 in section 1.4, can then be put alongside findings from the legal analysis to analyse the reciprocity of the standards by the accountancy profession and hard law that determine how sustainability assurance is regulated. This approach preserves the theoretical conflict of the two paradigms.[267] The parallel study is reflected in the diagram below.

[267] Ibid.

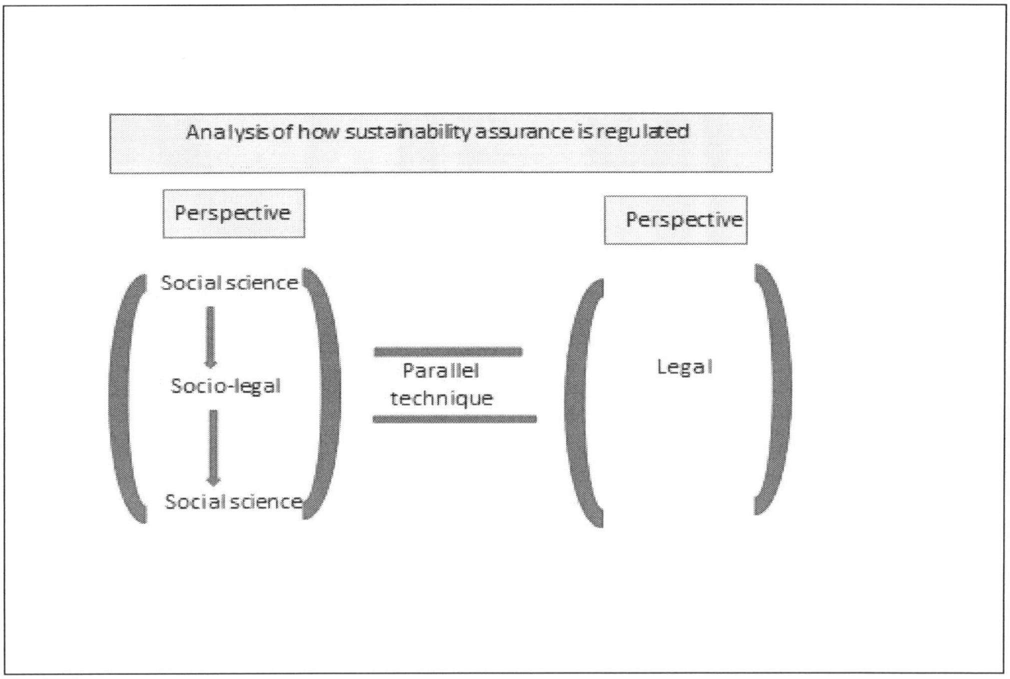

Figure 3.2 Multi-paradigms used in parallel

3.3 Method

Research method refers to specific techniques applied to collect data under the philosophical assumptions of the researcher. [268] As the method is embedded in the research process, it is presented adopting a process-oriented approach.

To capture the dynamics of the development and functioning of standards in this new domain, a longitudinal approach is required to map changes and detect developmental influences. This thesis is an explorative study that seeks predominantly to explain "what has been going on" [269] from 1997-2010. Longitudinal studies also allow the capture of "contemporary knowledge and perspectives" of different actors at different points in time.[270]

---

[268] Bryman, A.; Bell, E. (2007), p. 40.

[269] Saunders, M.; Lewis, P.; et al. (2009), p. 139-140.

[270] Ibid., p. 155. See also, Bryman, A.; Bell, E. (2007), p. 62.

A reflective approach has been adopted to take into account the rapid development and unanticipated influences. The design for research questions 1 to 3 had not been performed in sequence. A circular and iterative research process had been adopted which involved continuous analysis and reflection through stages of exploration. It is important to point out that each research question serves a different purpose contributing to the understanding of the development and use of standards. The time period within which data collection efforts are centred is not completely consistent for the three questions. The purpose of the first question is to provide an understanding of how the assurance concept has been framed by the IAASB through standards. The period of focus is set between the mid-1990s and 2005 when the profession started to intentionally construct the assurance concept, to the time the standard was made obligatory for its members. Although efforts to revise the ISAE3000 started in 2008, the 2005 version of ISAE3000 is the guidance for national standard setters to develop subject specific standards.

The second question analyses the context in which sustainability assurance standards are set, the analysis places focus on the entire period of study from 1997 to 2010.

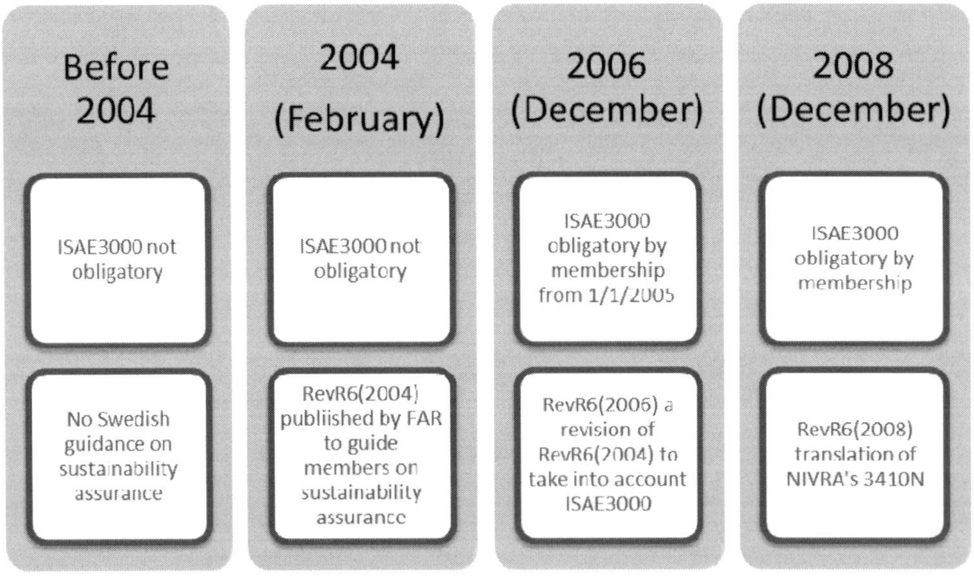

Figure 3.3 The revisions of RevR6

Referring to the diagram above, Far had drafted its first sustainability assurance standard in 2004. At that point in time, national standard setters were not obliged to use the ISAE3000. In 2006, Far's standard was revised in order to align its revisions with ISAE3000. A second revision was made in 2008, the second revision, adopted the provisions of the NIVRA standard guiding assurance engagements on sustainability reports. This illustrates that the Swedish interpretation of the ISAE3000 standard only occurred at the first revision. In this light, the RevR6(2006) will be adopted as a source to analyse how the discretion granted by ISAE3000 was exercised by Far in 2006. To inquire into how the discretion granted by RevR6(2006) has been exercised by the assurance providers, data collected are focused on the period before RevR6(2008) come into effect.

The sources of data and methods of analysis for each research question are presented below.

RQ	Sources of information	Method of analysis
1	Exposure drafts and standards of IAASB Published and open archival electronic and physical documents of IAASB and IFAC	Context of standard setting through interpretive paradigm Analysis of documents and longitudinal analysis of the provisions of standards <u>Corroborating evidence:</u> Documents from other sources
2	Published and open archival electronic and physical documents of relevant institutions and correspondence between them in the regulatory space Exposure drafts and standards of relevant standard setters	Context of standard setting through interpretive paradigm Analysis of documents and longitudinal analysis of relevant provisions of standards Audio recordings of IAASB meetings and documents from other sources <u>Corroborating evidence:</u> Informant interviews with experts
3	IFAE, ISAE3000, RevR6 (2004, 2006, 2008) Findings from research questions 1 and 2 Assurance reports by assurance providers of selected Swedish companies Interviews with representatives of sample companies	Interaction between content and context of the standard using a multi-paradigm sequential approach Case study of Sweden Analysis of the contents of relevant standards focusing on the dimensions of rules Analysis of the contents of assurance reports Informant interviews with a selected sample of companies *<u>Corroborating evidence:</u>* Informant interviews with experts Information from other sources, e.g. publications by PWC, Deloitte and KPMG, ESRA etc.
4	Companies Act (2005:551) Auditors Act (2001:883) Other legal sources Findings from research questions 1 to 3	Legal analysis Legal analysis of the relevant sources of law

Figure 3.4 Research design

3.3.1 Inquiry into the context which rules are made

The first two research questions have been structured taking on a constructionist perspective. The first question is concerned with how the generic assurance concept has been constructed and coded by IAASB from 1997, when the first draft of the standard was exposed for comment, to 2005 when the standard was made effective. The second question enquires into the how sustainability assurance standards have emerged and developed to fill the regulatory gap. First, the activities in the regulatory space marking the context within which assurance and sustainability assurance standards develop are mapped out. An inductive approach has been adopted focusing on identifying new issues and new patterns of relationships to capture the dynamic context of standard setting. Explanations are sought through evidence collected rather than through theory.[271] The regulatory space has been adopted to provide a construct to analyse the activities and events that occur from 1997 to 2010. The documents pertaining to the revision of ISAE3000 have not been included in this thesis, as the revision process has just begun.[272]

Documents serve as an important source of data to perform the analysis pertaining to research questions 1 and 2. They include both unsolicited electronic and physical data that are contained in open archives or are publicly available. The corpus of documents includes those published in the course of running the institutions and those with a specific communicative intent.[273] Examples of the former include minutes of meetings, working papers, task force or committee reports, annual reports and websites. Obtainable documents from the following institutions: IFAC, IAASB, FEE, GRI, AccountAbility, and NIVRA are analysed. These documents are artefacts that give insights to institutional routines; they serve as a record of the decisions taken and a log describing the goals, processes and procedures of each of these actors. Documents serve as an unobtrusive data collection venue of snap shots or records of official memories defined by cultural and organisational dictates of the past.[274] Documents with a communicative intent include: invitations to comment on exposure drafts, standards and explanatory memorandums, press releases, transcripts of speeches, discussion and position papers among others. [275] These documents in particular, provide a dynamic view of the interactions that take place in the regulatory

---

[271] Bryman, A.; Bell, E. (2007), p. 11-15.

[272] The IAASB published the revised ISAE3000 for exposure on the 28/4/2011.

[273] Saunders, M.; Lewis, P.; et al. (2009), p. 258-280

[274]Ibid., p.268. Amdou, T (2008) For the record? Understanding regulatory processes through archival materials: the example of planning agreements the author, issue compilation, Cardiff U law school, p. 183-185.

[275] Saunders, M.; Lewis, P.; et al. (2009), p. 258-280.

space. The analysis of the contents of correspondence between different actors, allows the capture of the intentions of actors issuing the documents and the reception of these documents by other parties. Crucial to understanding the context in which standards are set, the contents of documents provides insights on each actor's strength of influence in constructing, negotiating and shaping regulatory issues as well as both the assurance and sustainability assurance standards.

Documentary analysis is particularly suited for longitudinal studies; document serves as written representations of positions and discussions at different points in time[276]. Documentation reflects human cognition coded into text, which informs the researcher of the cognitive schemas of groups at study.[277] This enables the researcher to seek out a paper / and electronic trail that informs about the narratives, the tensions and negotiations, critique and learning that occur within and between different actors.[278] A longitudinal analysis of documents allows the patterns of interactions and influences to be identified and traced.[279] In this case, documents are used to identify the regulatory objectives of the institutions and patterns of influence in filling the perceived regulatory gaps, to provide an understanding of the context within which standards develop at the international level. The use of documents primarily as a source of data however has limitations. The researcher has no control over the quality in the preparation process of the documents. The evidential trail establishing reality, in this case how sustainability assurance standards emerge and develop, is tied with the availability, quality and appropriateness of documents used.[280]

During the course of collecting data, a number of difficulties were encountered. Difficulties include, obtaining web pages and on-line documents that had been removed and were not retrievable from web archives; access to private archives and documents published in other languages. The transparency of actors has not been consistent between organisations or within organisations over time. For example, when the IAPC became the IAASB in 2002, minutes of meetings and working documents from 2002 were made available publicly. From 2005, audio recordings of the meetings became accessible through the website which further enhanced the availability and quality of data used collected to perform the analysis. Audio recordings grant interested parties' access to the verbal interaction and negotiations by various parties that the minutes of the meetings fail to capture. The quality of

---

[276] Miller & Dingall,"Context and Method in Qualitative Research" 1997, Sage Publications, p. 37-56.

[277]Amdou, T (2008), p. 185.

[278]Ibid.

[279] Saunders, M.; Lewis, P.; et al. (2009), p. 269.

[280] Ibid. p. 268-272.

evidence collected through documents in this thesis therefore, vary in quality. The reliability of the data is mitigated through the collection of corroborative evidence. Such evidence is collected primarily using two ways. First, documents published by different organisations on a particular event are collected for the purpose of reliability checks. Second, corroborative evidence is collected through informant interviews with expert customised to fill the information gaps of documents. These interviews were conducted with three experts to enhance the understanding of cross sectional and temporal dimensions on the development of standards.[281] The interview questions were broad, semi-structured, open-ended questions organised along three broad themes: technical issues, standard setting process and standard setting context. During the interview process, the interviewer attempted to be primarily the receiver of information, allowing the expert to speak with minimum interruption. The purpose was to create a relaxed environment of exchange to avoid the interviewer being viewed as a potential critic and to guard against interpretational bias the interviewer might possess when steering the interview.[282] The experts are promised anonymity as therefore their views will be presented collectively in relevant parts of the thesis. They also add reliability to information from other sources.

3.3.2 Inquiry into how standards work

To look into how standards work, empirical data was collected from both socio-legal and social science perspectives, to inquire into the provisions of standards and the context within which they are used. Inquiring into the discretionary space granted by standards requires knowledge on the nature of the regulatory tool (standards) and the context in which they are applied. To more narrowly define the context and interpretive communities adopting sustainability assurance standards, Sweden has been chosen as a case country. Far as member of IFAC, is required by virtue of its membership to align its sustainability assurance standards to ISAE3000, when it was made obligatory to its members in 2005.

The analysis of how sustainability assurance standards work has been conducted in the following sequence. First, an analysis was conducted on how ISAE3000 has been used by Far in the revision of RevR6(2006). The socio-legal lens has been adopted to analyse the scope and linguistic structure of relevant provisions of the International Framework for Assurance Engagements (IFAE) and ISAE3000, to determine the

---

281 Meuser, M.; Nagel, U. (2010) "The Expert Interview and Changes in Knowledge Production", in Bogner,A.; Liittg, B.; Menz, C. Interviewing Experts, p. 25-28.

282 Bogner, A and Menz, W(2010) "The theory generating expert interview: Epistemological Interest, forms of knowledge and interaction. From Chapter 2 of Bogner, Littig and Menz, "Interviewing Experts", PalgraveConnect p. 67-69.

discretion granted to Far. A two-part analysis is then conducted to determine how the provisions in RevR6(2006) "work" on a national level.

The first part takes a similar approach as the afore-mentioned, to establish the discretion granted by RevR6(2006) to assurance providers using the standard. The second part, examines the exercise of that discretion by assurance providers in the context of practice, within which organisational and market related forces come into play. Content analysis has been carried out on assurance reports of selected companies making inferences to practice to determine the effects RevR6(2006) on the elements of assurance. This analysis has been supplemented by interviews with a representative / representatives from each the reporting organisations in the sample. Interviews were conducted with the aim of gaining deeper insights with regard to the scope of the sustainability assurance engagements and the emphasis placed on the work performed by assurance providers. The details of the method used to select the sample of companies, and the structure of the interview questions will be presented in when discussing the Swedish case in Chapter 8.

3.3.2.1 Content analysis of assurance reports

Content analysis has been broadly defined as "any methodological measurement applied to text or other symbolic materials for social science purposes"[283] It is a method that is widely used to analyse the practice trends of sustainability assurance internationally.[284] The analysis of the content of assurance reports over a period of time serves two purposes. First, it acts as an unobtrusive way to obtain an overview of the common trends and practices in this relatively new area.[285] Second, it serves as a source of data to enable both longitudinal as well as cross-sectional analysis to trace the development of sustainability assurance practices and evaluate how standards work. Although software programs are available to facilitate the processing of data, the size of the sample and purpose of this analysis allowed the content analysis to be carried out manually.

The contents of assurance reports were analysed according to coded categories developed using the elements of assurance. They included, the parties involved, the subject matter, the criteria, the scope of the engagement, work performed and the

---

[283] Shapiro, G.; Markoff, F. (1997)"Text analysis for the social sciences: methods for drawing statistical inferences form text and transcripts p.14 in Duriau, V. et al. 2007 "A content analysis of the content analysis literature in organisation studies: Research themes, data sources and methodological refinements," Organisational research methods 10(5) p.6.

[284] Refer to section 1.3.5 of the thesis.

[285] Duriau, V. et al. 2007, p.23.

providers' conclusions.[286] In the coding process, categories were revised to capture the developments in standards and to provide for "unexpected" categories interpreted by the researcher to be significant to the purpose of carrying out this analysis. Following, this the results were interpreted to enable a comparative analysis between companies and also detect trends on how the introduction and revision of standards have affected the form of the engagement. This allows conclusions on how discretion has been exercised by assurance providers. Although content analysis provides a non-intrusive, replicable way to perform this inquiry, this method only reflects the communication from the assurance provider to the reader of reports.

3.3.2.2 Interviews with companies

Content analysis of assurance reports is supplemented by interviews with representatives of the companies. Findings from interviews provides insights on the use of standards by assurance providers, it also sheds light on the level of transparency, i.e., what is communicated to the reader and what takes place behind closed doors. For example, details of the work carried out, the duration, focus of the engagement, etc. Details of the interview questions and process will be presented with the Swedish case in section 8.1.2.2 of the thesis.

On a more general note and for the timeframe 2008-2010, the researcher relied on documentation issued by Far, Ministry of Enterprises, Energy and Communications, the Balans, a journal by Far (but does represent Fars position) on debates, news and comments on accounting and auditing, business law and other issues of interest to the profession. It also relied on GRI and the Corporate register to obtain statistics and retrieve information concerning sustainability reporting and assurance trends in Sweden. Informant interviews (distinguished from expert interviews) were arranged to clarify certain actions or policies of actors. The reliability and completeness of the databases was contingent on reporting organisations' onus to submit their reports and information. Information contained was cross checked with reports issued by ESRA, PWC and Deloitte.

3.4 Methodological concerns

The methodological concerns of this thesis centre on its interdisciplinary and multidisciplinary nature. A discipline can be defined as a "comparatively self-

[286] Refer to chapter 4.

contained and isolated domain of human experience which possesses its own community of experts who share a distinctive set of goals, concepts, facts, tacit skills and methodologies"[287] Each discipline is made up of members forming a social community with a common distinctive knowledge base sharing their own mode of discourse and jargon. This shared discourse not only facilitates communication but forms a power structure through which distinguishes accepted members and those that are excluded from the community. [288]

There is no single definition of interdisciplinary. In performing interdisciplinary research academics borrow to different degree from each discipline as such can be placed on a continuum ranging from a lesser to a greater degree of synthesis.[289] Banakar and Travis, stressed that interdisciplinary *"indicates an ambition to understand and integrate aspects of two or several disciplinary perspectives into one single approach".*[290] Interdisciplinary stands in contrast with multidisciplinary approaches, which refer to research that, make no integration of knowledge and perspectives between the disciplines and studies that transgress discipline boundaries respectively.[291] Balkin, even labelled interdisciplinary studies as a "rebellious, even romantic" pursuit as traditional boundaries are challenged.[292] Interdisciplinary studies encourage researchers to look beyond the methodological restrictions and standards of each discipline for the quest of new knowledge or explain a new phenomenon.[293] Associated with this type of research are connotations of originality and innovation which is often challenged by traditionalist for lacking theoretical foundation, rigour and reliability. It is strongly stressed that this paper does not serve such ideals, nor does the researcher have any intention of engaging in methodological debates. The motivation has been merely pragmatic, built on the regulatory challenges of contemporary society to understand how sustainability assurance is regulated. By drawing on the socio-legal approach and socio-legal theories of regulation, the interdisciplinary element is introduced.

---

[287] Nassani, M. (1995) "Fruits, Salads and Smoothies: A working definition of interdisciplinarity" Journal of Educational thought, p. 125.

[288] Vick, D. (2004), p. 166.

[289] Roslender, R.; Dillard, J. (2003) Reflections on the Interdisciplinary perspectives on accounting project" Critical Perspectives on Accounting, 14, p. 327-328.

[290] Banakar, R.; Travers, M. (2005), p. 5.

[291] Ibid.

[292] Balkin, J. (1996) Interdisciplinary as Colonization,

http://www.yale.edu/lawweb/jbalkin/writings.htm p. 957.

[293] Banakar, R.; Travers, M. (2005), p. 5.

This thesis builds its analysis using multi-paradigms sequentially and one after the other. Scholars do not have a common position as to whether paradigms can be combined.[294] A review by Lewis and Grimes show that multi-paradigm research in the social science field through parallel and sequential techniques is not new and had been used for decades.[295] The multi-paradigm approach across disciplines in this thesis has been approached "consciously and intentionally" with the awareness of both the disparities and the complementary nature of the various perspectives.[296] These paradigms are regarded as a means to understand a phenomenon, each contributing a layer of meaning on how sustainability assurance is regulated. Knowledge of the theories, methods and techniques of both disciplines were considered and discourses familiarised to appreciate, draw from and make contributions in return to both disciplines.[297]

---

[294] Bryman, A.; Bell, E. (2007), p. 26-27.

[295] Lewis, M and Grimes, A (1999), p. 674.

[296] Ibid. p. 673-675.

[297] Banakar, R.; Travis, M. (2005) p. 5.

4. Assurance- Framing the Concept through Constructing Standards

"Do you agree with the use of the term 'assurance engagement' in connection with environmental work? If not, what term do you prefer?"298

4.1 Introduction

4.1.1 Background

The word 'assurance' originates from the Latin term 'ad securus' i.e. to make safe. In laymen's terms assurance is taken to mean 'a positive declaration intended to give confidence', 'a freedom from doubt', 'a binding commitment to give or refrain from something' or 'a type of insurance'.[299] With no profession granted jurisdictional monopoly of the assurance market, the term "assurance" has been interpreted differently by the various actors involved in and receiving assurance services. [300]

The financial audit is the core service performed by the accountancy profession, who incidentally engages in 'audit-like' engagements on other subject matter. These engagements are commonly referred to today as 'assurance engagements'. Although the term assurance has been, and continues to be used today, in a general sense to mean the degree of confidence attached to an engagement, its use to depict a particular type of engagement was only intentionally framed and defined by the accountancy profession in the mid-90s. The term 'assurance' used to describe

---

298 FEE (1999) Discussion Paper: "Providing Assurance on Environmental Reports" p. 8.

299 These explanations were obtained from the Cambridge and Oxford on-line dictionaries.

300 The term jurisdiction refer to professional jurisdiction as referred to in Abbott, A. (1988) "The system of professions: an essay on the division of expert labour" Chicago University Press.

engagements is therefore an artificially created construct by accountants to describe the nature of the engagements they perform.[301]

4.1.2 Chapter objectives

The first purpose of the thesis is to analyse the dynamics of regulating sustainability assurance by the accountancy profession. The focus on the accountancy profession warrants a descriptive chapter to explain how the profession, as an interpretive community, frame the assurance concept and codify it into standards to guide the performance of such engagements by its members.

Prior to analysing the interaction and patterns of influence to set sustainability assurance standards in the regulatory space, this chapter provides a description of how the generic assurance concept has been framed by the IAASB (formerly the International Auditing Practices Committee[302] (IAPC)) over time, through the construction of standards in the context which it operates. The objective is to provide an explanatory source for readers to understand the subject matter of regulation (assurance) and the common stand, activities and constraints the accountancy profession adopts in the interaction with other non-accountancy actors or standard setters in the regulatory space for setting sustainability assurance standards.

This is performed by analysing the initial drafts of the assurance standards drawn up by IAASB, comments on the exposure drafts (ED), relevant archival documentation and publications by IFAC and other organisations. The choice of using standards as a main source of information is based on the assumption that standards are a codification of what the assurance process should entail. Following the drafting of standards not only provides insights on the explicit conceptualisation of assurance at particular points in time, it provides insights into the problems and issues, and how they have been 'resolved'.

This chapter is structured in two parts. The first part presents the accountancy profession's attempts to redefine the assurance concept in the mid-1990s. This is followed by a longitudinal account of the development of assurance standards from the period between 1997, when the IAASB embarked on the project to set standards to guide engagements to increase credibility of information, to 2005 when the

---

301 FEE (2003b) Issues Paper: Principles of assurance: Fundamental theoretical issues with respect to assurance in assurance Engagements - Full-Length Version, P y, 89.

302 The IAASB was set up in 2002 to replace the IAPC. This thesis adopts the current name IAASB in the analysis.

ISAE3000 was made effective. Developments after 2005 are outlined in the conclusion of the chapter.

4.2 Redefining assurance in the mid-1990s

Two standard setting bodies, the American Institute of Certified Public Accountants (AICPA) and the IAASB (known as the IAPC then)[303], on separate initiatives made attempts to frame the 'assurance' concept. This thesis primarily focuses on the efforts taken by the latter body. It is however, important to note how AICPA framed the assurance concept during that same time period in the mid-nineties. Although the concepts of both standard setting bodies overlap, there are significant differences pertaining to the elements of assurance. In addition, the extensive promotion of the 'Elliot Report' in North America and the international accountancy profession, have popularised the term assurance.[304]

4.2.1The Elliott Report redefines assurance

In the mid-1990s, audit firms were faced the challenge of intense competition in a saturated market driving the need to lower prices and engage in cost cutting measures.[305] The AICPA, based on recommendations from the 1993 Santa Fe conference[306], set up the Special Committee on Assurance Services (SCAS), also referred to as the Elliott Committee, to "develop a strategic plan for an expanded assurance function and to act as a catalyst for change to achieve its implementation."[307] The efforts resulted in the publication of the 'Elliott report', within which formally introduced a new concept of assurance services built by reconceiving the attest model.[308] The report identified and defined specific services for

---

303 The IAPC has since 2002 been renamed the IAASB.

304 Nugent, M. (1999) "The Australian accounting role in developing assurance services", Australian Accounting Review, 9(2), p. 4.

305 Elliot, R. (1994) "Confronting the future: Choices for the attest function" Accounting Horizons, 8(3), p. 107.

306 The Santa Fe conference on audit / assurance was held in May 1993 organised by the AICPA

307 Elliott, R. (1994) "Confronting the future: Choices for the attest function" Accounting Horizons, 8(3) p. 106.

308 AICPA (1997) "Report of the Special Committee on Assurance Services"; See also, Elliott (1994), p. 117-119.

practitioners to deliver and create on-going mechanisms to develop opportunities for the profession in the future.[309] With a market seeking objective, they expounded arguments for the profession to break out of their traditional financial auditing along three dimensions of services performed. The dimensions include: Who are the customers? How are technologies employed? What services? (Refer to figure 4.1)

This allowed potential extension of the financial to non-financial engagements; attestations to systems and the audit objectivity of improving reliability to relevance of an engagement. The objective was to develop new markets for auditor expertise and a framework for a broader range of assurance services[310].

---

[309] AICPA (1997); Elliott, R. (1997) "Assurance Services Opportunities: Implications for Academia", Accounting Horizons 11(4), p. 61. Elliot, R., Pallais. D (1997) "First know your market" Journal of Accountancy, online issues, July. www.aicpa.org/PUBS/JOFA/july977elliot.htm. 15th October 2007.

[310] AICPA (1997), under Concept of assurance services; Elliott, R.; Pallais. D. (1997) "Are You Ready for New Assurance Services?", Journal of Accountancy, June, p. 47–51. Pallais, D. (1996) "Assurance services: where are we going: where we're going", Journal of Accountancy, 182(3), p. 16-17.

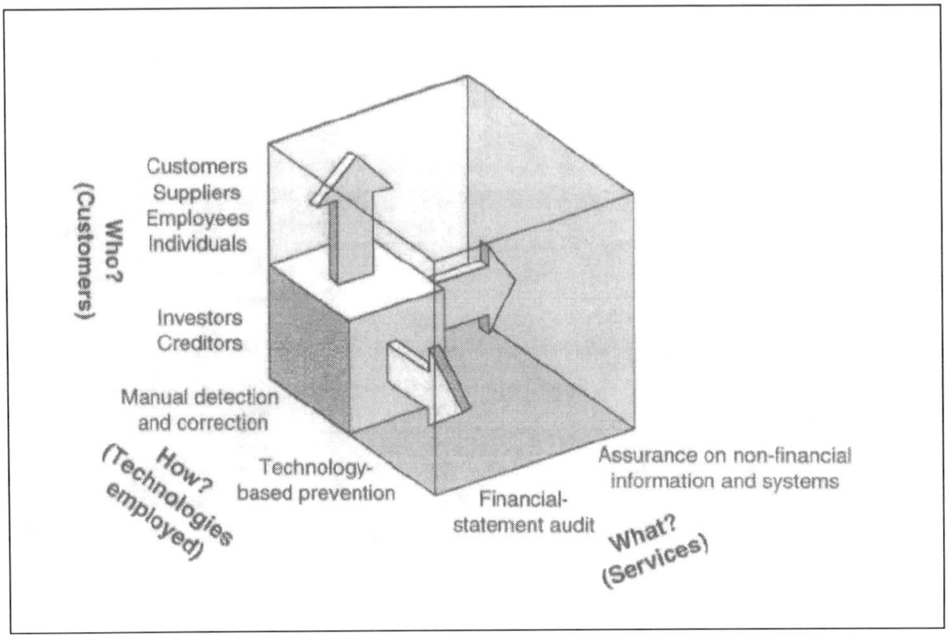

Figure 4.1 Strategic expansion of audit space[311]

The Elliott committee defined assurance services to be *"any independent professional services that improve the quality of information or its context, for decision makers".*[312] Context referring to other dimensions of information e.g., way it is produced, reported, aggregated or used. It positioned assurance services against attestation by stressing the goal of information improvement rather than the issuance of a report on it.[313] Where attestation services required the attester to issue a report on the reliability of a written representation or assertion that is the responsibility of another party; the committee regarded the written report as *"unnecessarily constricting".*[314] This perspective also suggests that the service itself rather than the report provided value.

311 Elliot, R. (1997a), p. 62.

312 AICPA (1997) under: Concept of assurance services.

313 Nugent, M.(1999) "The Australian accounting role in developing assurance services" Australian Accounting Review 9(18) p. 3-4.

314 AICPA (1997) under: Concept of assurance services.

Hence, a report was put synonymously with other symbolic communication methods e.g., a logo or stamp.[315] The report also made a distinction between assurance and consultancy. It was stressed that assurance serves to improve decision making information; the engagement is therefore focused on the decision maker, the information they use, and require assurance providers to be independent.[316] Consultancy engagements on the other hand, focus on outcomes; consultants can thereby be involved in the decision making process of the firm.[317]

From the diagram below, the attestation services, *which include* audit, review and agreed- upon procedures are a subset of assurance services.[318] Within the assurance domain (excluding attestation), four categories of assurance services were developed. They included relevance services, systems reliability assurance, navigation services and context services. From these categories, the committee developed services in the areas of risk measurement, business performance measurement, information systems reliability, electronic commerce (e.g. WebTrust), healthcare performance measurement and Eldercare (now PrimePlus).[319]

---

[315] Ibid.

[316] Ibid.; See also Alford, S. (1999) “A critical appraisal of assurance services development in Australia” Australian Accounting Review 9(18) p. 11-16.

[317] AICPA (1997), under: Concept of assurance services

[318] Ibid.

[319] Ibid., under: New Assurance services.

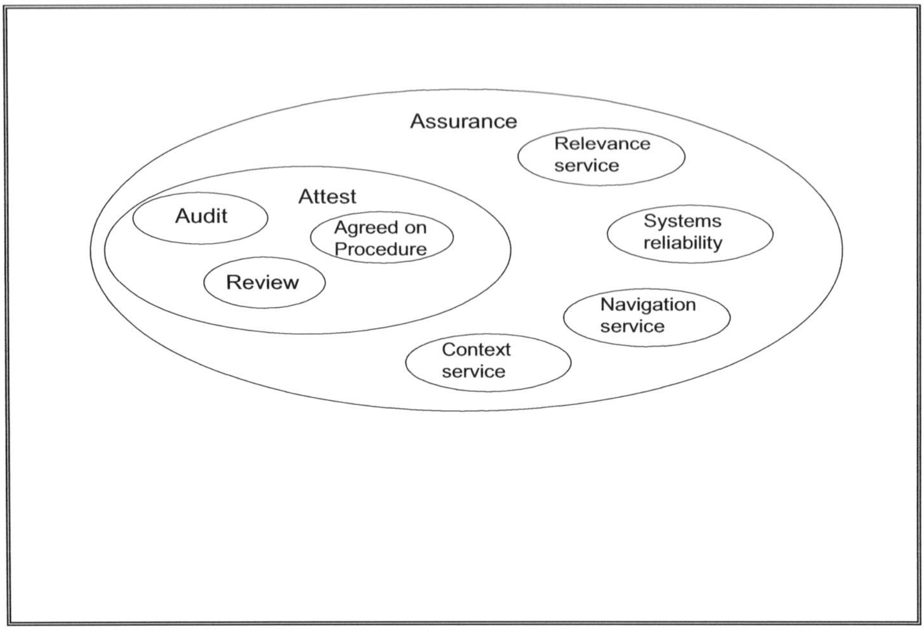

Figure 4.2 Concept of assurance services[320]

The Elliott committee's concern was the strategic expansion of the audit space. Due to its initial standpoint that frameworks and written reports can be restricting, the general consensus on the work effort model[321] and that standards were to be set for each product; the factors determining the assurance was less of an issue for the AICPA, as it was for the IAASB.[322] Hence, although all six service areas were defined by SCAS, the provisions with regard to elements of engagement were not consistent across these services.[323]

320 AICPA (1997), under: Concept of assurance services.

321 The work effort model will be explained in sec. 4.3.2 of the thesis

322 IFAC (2002) "The Determination and Communication of Levels of Assurance Other than High", p. 39.

323 Ibid, p. 38-40.

4.2.2 IAASB's approach to assurance

The IAASB is one of the standard setting boards of IFAC. IFAC was founded in 1977, at the 11th World Congress of Accountants in Munich, with sixty-three members. Its initial efforts were guided by a twelve point work program of which the first was to develop statements which can serve as international auditing guidelines. [324]

The IAASB did not share AICPA's definition of assurance nor did it engage in similar market creation strategies. Rather, it presented its view that the accounting firms in practice, driven by market demand and / legislation were already providing services to enhance credibility of information (E.g., audit of internal controls or environmental information). This created the need for the IAASB to embark on a project to set standards to guide "professional accountants'[325] on the provision of such services.

Engagements commissioned to provide "assurance" on reports were referred to as "reporting service engagements" prior to 1999. It was expressed by IAASB that the term assurance had been intentionally avoided in recognition that the term was ambiguous and had no commonly accepted definition.[326] The objective of a reporting service engagement was:

"... to enable the professional accountant to report on the credibility of information, referred to as subject matter that is the responsibility of another party, by evaluating the information against identified suitable criteria and expressing a conclusion about that subject matter."[327]

To distinguish these services from consulting and other services, the ED highlighted seven elements of assurance which had to be present for an engagement to fall under the scope of assurance engagement. Elements included: the professional accountant; the subject matter (data, systems and processes, behaviour); the responsible party; the intended user; the criteria; the engagement process and the report.[328]

Whilst AICPA proactively identified specific assurance services the profession could undertake, IAASB took the position of serving an existing needs of the market through proven expertise in financial audits. Although both approaches have in

---

324 http://www.ifac.org/History/work_program.php

325 The choice to use "professional accountant" was a conscious one adopted by IFAC so as not to confuse providers with the financial auditor.

326 ED 1: Reporting on the credibility of information, explanatory memorandum, sec. 7.

327 Ibid., general principles, sec. 2. See also sec. 5.

328 Ibid., framework for reporting by professional accountants on the credibility of information, sec. 7.

common, an extension of the audit (or attest) function, IAASB's narrower and more 'financial audit or audit-related' way of defining assurance distinguished, its approach from the AICPA. As such, the IAASB's conception of assurance precludes some that of the new assurance services that were developed by the AICPA.

The following sections present the structure and standard setting process of IAASB and describe how the assurance concept had been framed over time in the course of IAASBs standard setting activities.

4.3 The Development of assurance standards

The framing of the assurance concept and the development standards on assurance occurred most intensively from 1997 to 2003. The drafting process of IFAE and ISAE3000 spanned over 6 years and had been exposed for comments three times before the standard was approved, and made obligatory to the members of IFAC. The table below presents a summary of the developments from 1997 to 2008.

Year	Key Documentation	Activities by IAASB
1997	ED 1: Reporting on the credibility of information	Issued: August 97 Due date for comments: February 98
1999	ED 2: Assurance Engagement	Issued: April 1999 Due date for comments: June 1999
2000	Proposed International Framework for Assurance Engagements and ISAE100	Issued: June 2000
2003	Proposed IFAE and ISAE2000 Assurance engagement on subject matters other than historical financial information	Issued: March 2003 Due date for comments: June 2003
2003	IFAE and ISAE3000	Approved: December 2003
2005	IFAE and ISAE3000	Effective from January 2005
2008	Revision of ISAE 3000	On going

Developments of the IFAE and ISAE3000 from 1997

4.3.1 Development towards ISA100: 1997-2000

4.3.1.1 IFAC and IAASB:1997-2000

In 1997, IFAC had gained substantial influence through an increase in its membership. It had 140 member bodies, representing 3 million accountants in 101 countries. In 1998, IFAC council approved significant allocation of extra funds for a five year period to increase the output of IAASB by about 60% over five years in recognition of demands for tougher auditing rules that are purported to enhance the reliability of financial statements and consistency of approach across countries.[329]

[329] http://web.archive.org/web/19990427125226/www.ifac.org/LastestDevelopments/PressReleases/1998-10-22.html 25th October 2011. Press release: Global Accountancy Profession Gives Boost to

Although this initiative was mainly to revise core financial auditing standards, the development of standards for "assurance type" engagements was identified as one of the priority areas, reflecting the move from "attest" to a broader "assurance" framework.[330]

In 1997, IFAC referred to itself as a world-wide organisation for the accountancy profession with the mission to develop and enhance the profession and enable it to provide high quality services in the public interest.[331] Its organisation consisted of the IFAC assembly which made constitutional decisions and elected the Council. The council appointed the executive committee, sets policies and oversees the operations, implementation of programs, and the work of technical committees and task forces. The secretariat provided overall direction and administration. The IAASB was one of the seven committees of IFAC whose role was to "improve the degree of uniformity of auditing practices and related services throughout the world by issuing pronouncements on a variety of audit and attest functions and promoting their acceptance worldwide"[332] . It operated under the Terms of reference and Preface. Standards were drafted with input from IFAC member bodies, their members and a consultative group that represented regulators, preparers, and users of financial statements.[333]

4.3.1.2 Logic of IFAC's project

The deregulation phase during 80s and 90s led to development of new monitoring and control mechanisms where Michael Power, Professor at the London School of Economics described as explosion of demand for audit like services.[334] Studies conducted by the profession showed that its members were increasingly engaged in providing these "audit like" services on broad range of subject matters. There was, however, no common terminology adopted.[335] These engagements were termed

Development of International Standards on Auditing including which included quotations from Frank Harding, President of IFAC in 1998.

330 ibid.

331 Mission statement from http://web.archive.org/web/19970414073307/http://www.ifac.org/

332 http://web.archive.org/web/19980530032048/www.ifac.org/Committees/index.html

333http://web.archive.org/web/19980530032400/http://www.ifac.org/Committees/IAPC/index.html.

334 Power, M. (1997b), p. 1.

335 IFAC(1995) Discussion Paper. "The accountancy profession and the environment" p. 2-5.

"audit", "assurance", "review", "limited review" etc. The IAASB believed that a standard at institutional level would serve a two-fold purpose. First, it enabled consistency in performance of assurance engagements by the profession globally; and second, it would establish the accountancy profession as primary providers of such services.[336]Read in conjunction with IFAC's mission statement, the two-fold objective was clear. The "enhancement of the profession" could be achieved through "establishing accountants as the primary providers of these assurance services" and public interest could be served by standards that improve the quality of assurance services. [337]

To establish a need for assurance standards, the IAASB drew on the logic that decision quality was dependent relevance and reliability of information. It questioned the quality of decisions using unaudited data, emphasizing problems that supported the need for the well-accepted institutionalized financial audit and review.[338] This construction enabled the IAASB to stress the 'programmatic elements'[339] i.e. the normative broad claims of an audit, and offer as a solution assurance that could enhance confidence and reliability of a subject matter. By stressing similarities with the need for financial audits, and the importance of reliability and relevance; it built on the existing public acceptance that independent verification of information and system operations enhances the credibility of information[340]. The above illustrates the process lifting the institutionalised trust in the financial audit system established through the profession's claim to expertise in independent validation, to allow similar expectations to be re-embedded into the assurance domain. [341]

IAASB further legitimized the accountancy profession as "the choice provider" of these services in the explanatory memorandum stressing that attributes the profession possessed was recognized as "equally valuable in enhancing credibility of information (providing assurance) about a on broader range of subject matter"[342]. It also claimed

---

[336] ED 1: Reporting on the credibility of information, explanatory memorandum, sec. 6.

[337] In 1996, the mission statement of IFAC was constructed as: *"The International Federation of Accountants (IFAC) is the world-wide organization for the accountancy profession. The mission of IFAC is the development and enhancement of the profession to enable it to provide services of consistently high quality in the public interest."*

[338] ED 1: Reporting on the credibility of information, preface.

[339] Power, M (2007b) Programmatic elements refer ideas and concepts shapes the mission of practice. P. 6

[340] AICPA Thomas Ray: international assurance standards exposed for comment. October 1997

[341] This section draw ideas from the work of Power, M. (1997a) "Expertise and the construction of relevance: accountants and environmental audit" Accounting, Organizations and Society, 22(2) p. 125.

[342] ED 1 Explanatory Memorandum, sec. 3.

that the profession was equipped with the strengths and competencies due to its audit methodology, standards, evidence gathering techniques and evaluation skills that could be extended to other information.[343] Further, the explanatory memorandum differentiated the profession from other assurance providers recognising the attributes of the profession's objectivity, independence and skills[344].

4.3.1.3 Development towards ISAE100

Two drafts had been exposed to the public which worked towards the International Standard on Assurance Engagements 100 (ISAE 100[345]). ED 1: Reporting on the Credibility of information (thereafter, ED1), approved by IFAC in June 1997 was exposed for comments from August to February 1998. IAASB published that respondents of ED 1 indicated clear support of its development of the framework and standards, despite the issues of concern raised.[346] A second ED was warranted in light of the significant changes drafted to take into account issues raised by commentators.[347]

ED2: Assurance Engagements (ED2, thereafter) was issued and exposed for comment from March 1999 to July 1999.

The structure and scope of initial EDs

ED 1 was structured in four parts: explanatory memorandum; framework; general principles and examples. A broad approach had been adopted by the IAASB to include within its scope a spectrum of engagements.

The framework put forth by IAASB suggested that the international standards on reporting service engagements had been built drawing from the traditional financial audit. The development entailed lifting the elements and general principles of the financial auditing to a higher more abstract or generic level, which could be re-embedded to apply to an array of different subject matters. The framework and

---

343 Ibid.

344 Ibid sec. 4.

345 The International standard on assurance engagement 100 was abbreviated as ISA100 in many documents. This thesis adopts the abbreviation ISAE100 to avoid confusion with the current ISA100.

346 ED2 Assurance Engagements sec. 9: "an overwhelming majority of commentators commended the IAPC for being forward looking and the initiative shown in dealing with this issues given its importance to the future of the profession".

347 ED2, Preface.

general principles in ED1 therefore covered within their scope the financial audit and review. In addition, potential areas where subject specific standards could be drafted were identified. Specifically mentioned in the framework were information (e.g., prospective financial information, statistical information, performance indicators); systems and processes (e.g., internal controls, environmental management systems) and behaviour (e.g., corporate governance, compliance with regulation, and human resource practices).

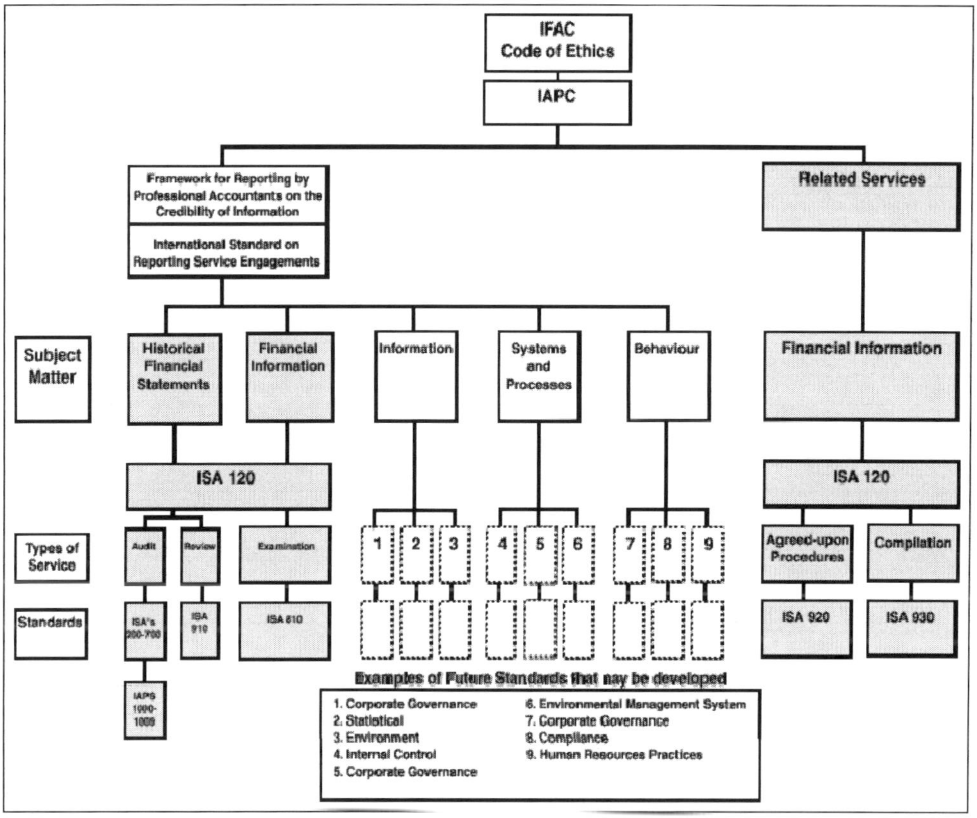

Figure 4.3: The IAPC's pronouncements presented in ED1 [348]

The first ED provided a definition of reporting service engagements built on the seven elements of assurance presented in a diagram illustrating the process and relationship between the elements. As guidance had been built on the financial audit

[348] ED 1, Framework, s.3.

model (and included within its scope the financial audit) it adopted the structure and vocabulary used in the financial audit.

> Guidance was provided in four key areas: conditions of acceptance; behaviour of the practitioner; work to be performed such that a conclusion can be reached, and reporting of the conclusion obtained.[349]

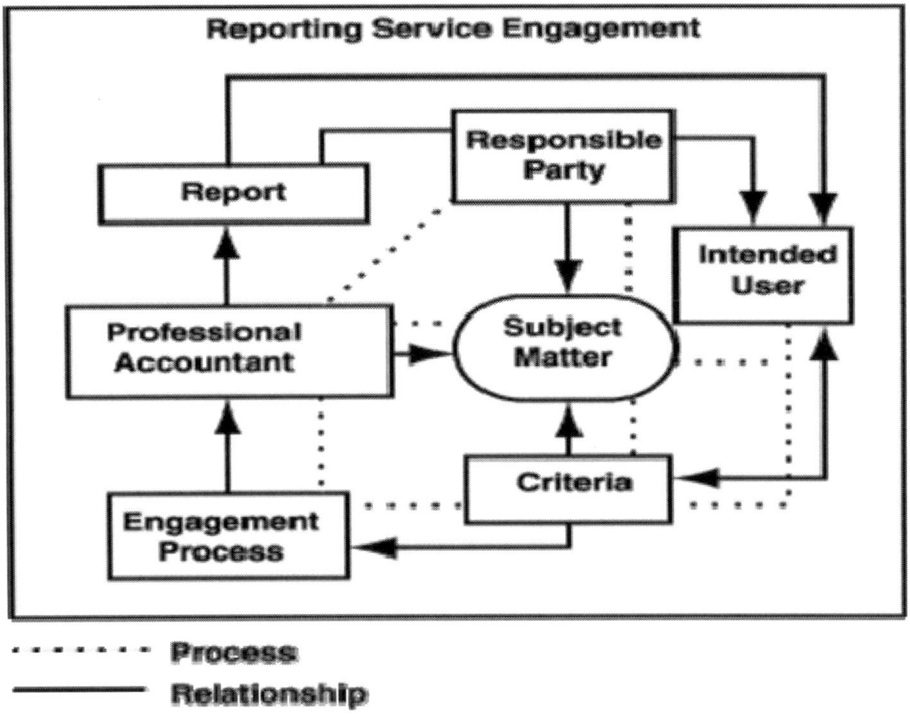

Figure 4.4: Process and relationship between the seven elements of assurance

Strong emphasis was placed on the exclusion list, which identified what assurance was not. In ED 1 efforts were made to distinguish between the type of engagements covered by the standard with agreed-upon procedures and compilations guided by ISA 920 and 930 respectively. Agreed upon procedures required the practitioner to report on factual findings but not express a conclusion on the subject matter or report, and compilations did not require the gathering and evaluating of evidence and therefore there was no need to express conclusions.[350] The ED excluded from the

---

[349] ED 1 General principles of a reporting service engagement, sec. 3-39.

[350] ED2, sec. 8 and 21.

scope the preparation of tax returns, management consulting and other advisory services.[351] It further clarified in ED2 that an agreed upon procedure which resulted in a report of factual findings was precluded from the standard whereas an "engagement that is similar to an agreed upon procedures engagement" which resulted in "the expression of a conclusion that provides a level of assurance" was not precluded.[352]

Debated issues in the initial drafts

Four issues dominated the debates in the standard setting process. The IAPC then did not have the same transparent structures as IAASB has today, the standard setting space was closely confined to member organisations (national and regional), and its members and those directly involved in regulating the profession.

The first issue pertains to use of the term assurance. ED 2 was entitled "Assurance Engagements". The IAASB decided on this course of action in recognition of the wide usage of the term in practice (IAASB and in many countries) and that the key function of the engagements was the provision of assurance. Its decision took also into consideration respondents view that "reporting on the credibility of information" was "too restrictive" and understated the breadth of services that could fall under the scope the standard. [353]

The second and most debated on issue was the determinants of assurance. There have been two competing models adopted by different standard setters in how assurance should be determined. They are the work effort model predominantly held by the North American members of the profession and the interaction of variables (Multiple determinants) view mainly held by members from the UK. Jurisdictions that adopt the work effort model have as a prerequisite for accepting an assurance engagement, that the subject matter and criteria be suitable or appropriate. The level of assurance of an accepted engagement is thus depends on users' needs and the nature, timing and extent of procedures performed i.e. the work effort. Jurisdictions that adopt the interaction of variables approach requires the profession use their judgment on the level of assurance based on the inter-relationship between the subject matter, criteria, procedures adopted to gather evidence and the quantity and quality of appropriate evidence.[354]

---

[351] ED2, sec. 8.

[352] ED 2 sec. 6 and 7.

[353] ED 2, sec. 11.

[354] IFAC(2002), p.8 and 9

The differences between the two methodologies (in its pure state) have implications for work to be carried out to determine level of assurance (explained above) and acceptance of an engagement. With regards to the work effort model, due to the fact that suitability of subject matter and criteria are prerequisites, an engagement that is rejected for a "high level" of assurance to be performed, cannot be accepted and performed at a "moderate level" of assurance. For the latter, as subject matter and criteria form part of the variables that determine level of assurance, these engagements may be acceptable offering a lower level of assurance. For example, the criteria could be less objective (qualitative or evolving criteria) or subject matter measurement may be less reliable.

Although ED1 stated in its guidance, the minimum prerequisites for subject matter and criteria, which require the subject matter had to be "identifiable", "be subject to evidence gathering procedures" and "capable of evaluation"[355], it is inclined towards the interaction of variables model. This was evident from the seven elements and their relationship expressed in the first ED and provisions in paragraph 33 which stated that the level of assurance (agreed by the provider and appointee) is based on the: "nature and form of subject matter; nature and form of the criteria applied to the subject matter; nature and extent of the process used to collect and evaluate evidence; and sufficiency and appropriateness of evidence likely to be available."

The third issue of concern was the level of assurance. Level of assurance can be expressed as a continuum (low to absolute), specified single level assurance (high or moderate) or dual level (high and moderate). ED 1 introduced the concept of level of assurance that could range from a low level to an absolute level depending on the interaction of the four variables.[356] The standard gave flexibility, as it took into account the idiosyncrasies of engagements. It gave assurance providers and commissioners freedom to identify the elements upon which assurance is to be provided. Although respondents indicated support of the concept of continuum of assurance, respondents voiced difficulties in its application. In particular, it was difficult in practice to convey the level of assurance with clarity and precision required by paragraph 34 of ED1.[357] Concerned were raised that a spectrum reduced comparability, increased litigation risk and motivated conservative reporting which, is contrary to both the profession and in public interest. In ED2, IAASB retained the continuum of assurance concept which was generally agreed that to be conceptually

[355] ED 1 General principles sec. 7

[356] Ibid.

[357] ED2 Introduction, sec. 16-20.

sound, but introduced two distinct levels of assurance: high (audit) and moderate (review) determined by the interaction of variables.[358]

The fourth issue concerned output of the engagement. There had been variations as to what forms of representation were allowed, how 'less than high' assurance findings should be communicated or expressed to enhance the user understanding. There was also discussion if the long or short form of the report should be used. The framework allowed written, oral, electronic or symbolic representations though it specified that written communication was most effective as it avoided misunderstandings. In addition, details of work performed and evidence leading to the conclusion could be provided.[359] In ED1, assurance providers could specify in their conclusion, a level of assurance within a range offering additional explanation on how his / her conclusions were derived based on interrelationship between the variables in the report.[360] In ED2, to differentiate reporting under the two levels of assurance, assurance providers were guided to convey a high level of assurance through a positive form of expression. For example, in the assurance provider's opinion *"whether subject matter conforms, in all material respects, to suitable criteria."*[361] In the case of moderate assurance, conclusion is expressed in a negative statement. For example, *"whether anything has come to the professional accountant's attention that causes the professional accountant to believe that the subject matter does not conform, in all material respects with the identified suitable criteria."*[362] The draft maintained the importance of flexibility and did not impose a standardized form of reporting.

ISAE 100 – a temporary solution?

Despite the redrafting of ED1, three issues remained on the agenda in the framing of the assurance concept. First, the "moderate assurance" concept was ambiguous compared to a "high level" of assurance. The term moderate assurance could represent a level "just below high" to "just above low" and it was difficult to determine the amount of work to be carried out.[363] This was evident in the survey carried out by the consortium of academics commissioned by IFAC in 2002. From 52 respondents, the study found that the percentage of confidence assurance providers believed they provided for engagements performed at a moderate level, ranged from 10% to 88% with a mean average of 60%. In comparison, for high level assurance engagements, it

---

358 ED2 sec. 26-36.

359 ED2 Introduction, sec. 24-26.

360 ED1 General principles sec. 34

361 ED 2 sec. 33

362 Ibid sec. 36.

363 IFAC(2002), p. 33.

ranged from 55% to 98% with a mean of 88%. The differences in the standard deviation indicated the greater variation in the understanding of moderate level assurance by assurance providers.[364] Second, the standard neither provided sufficient guidance nor explained adequately how the interaction of variables could result in a moderate level of assurance hence, assurance providers given similar circumstances could make different conclusions. Respondents voiced that although greater degree of uniformity and standardization in reporting brought about consistency, the flexibility that was provided for in ED1 was lost. [365]The third issue was derived from respondents supporting the "work effort model". These respondents expressed concerns about having a criteria such as 'best practices' as they are evolving and could lead to inconsistencies in practice.

The IAASB introduced ISAE 100 in June 2000, providing a framework for all assurance engagement (high and moderate) but provided the basic principles and essential procedures only to engagements intended to provide a high level of assurance. Although the concept of moderate assurance was introduced with financial review in 1988 and therefore not new, the IAASB in the light of the problems and issues defining moderate assurance, commissioned research into the unresolved issues and deferred providing guidance in this area.[366]

---

[364] Ibid, p. 97, 145.

[365] Ibid 146-148.

[366] IFAC commissioned research from international group of academics (International symposium for audit research group) resulted in monograph "Determination and communication of levels of assurance other than high" in June 2002

4.3.2 Development towards ISAE 3000 (2001-2003)

4.3.2.1 IFAC and IAASB 2001-2003

"In developing and implementing our vision for the future, we not only looked inward at the profession's needs, we also looked outward. ..IFAC is one that not only effectively represents the international accountancy profession, but also anticipates the responds to changes in the global environment in which accountants work."

Tsuguoki (Aki)Fujinuma President of IFAC 2001

In 2001, IFAC's membership increased to 156 representing, 2.4 million accountants in 114 countries. Its leadership set forth to build a new foundation, strengthening IFAC's infrastructure to support its mission of taking into account the needs of IFAC stakeholders including governments, regulators, investors and other users and addressing those concerns from an international perspective, making public interest central to the IFAC mission.[367] The architects of IFAC's reform structured two essential elements. First, they reformed governance structures and standard setting activities to allow greater transparency, and public and regulatory input into those processes. Second, it embarked on creating a global regulatory structure for the profession.

The IAPC was reconstituted as the IAASB[368] in April 2002 "to better reflect its role mandate and structure"[369]. Although IAASB's functional role is primarily the same as that of its predecessor, the setting of "audit and attest" pronouncements (stated in the goal of IAASB) has been defined more broadly as the setting of "auditing, assurance, quality control and related services standards". IFAC attributed this to the changing needs of the market increasing demands for the audit "to extend beyond the traditional attest function into assurance services".[370] Drawing from IFAC's public interest mandate, public interest has been built into the objectives of IAASB which defines its standard setting role to contribute to public interest through promoting global acceptance and convergence of standards.

---

[367] IFAC Annual report (2001) p. 3.

[368] The IAASB's goal is to serve the public interest by setting high quality auditing, assurance, quality control and related services standards and by facilitating the convergence of international and national standards, thereby enhancing the quality and uniformity of practice throughout the world and strengthening public confidence in the global auditing and assurance profession.

[369] IAASB Annual report (2002), p. 8.

[370] http://web.archive.org/web/20010107211700/www.ifac.org/News/Initiatives.tmpl

IAASB's role was expressed as: "to serve the public interest by setting high quality auditing, assurance, quality control and related services standards and by facilitating the convergence of international and national standards, thereby enhancing the quality and uniformity of practice throughout the world and strengthening public confidence in the global auditing and assurance profession"[371] The governing document of IAASB the Interim Term of Reference and Preface to International Standards was approved by IFAC board in July 2003 and states the standard setting due process.

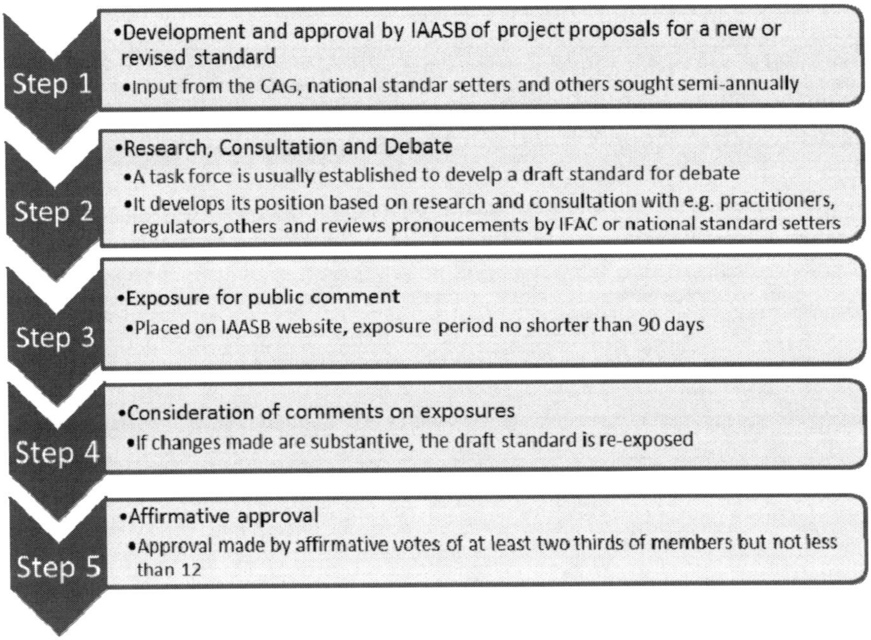

Figure 4.5 IAASB standard setting due process[372]

Transparency of this process is enhanced through making meetings open to the public where agenda, background papers, minutes and audio recordings of meetings are available on the website. Similarly, EDs and comments on them can be accessed on-line.

---

[372] IAASB "Interim Term of Reference and Preface to International Standards"

In 2003, provision was made to establish the Public Interest Oversight Board (PIOB) independent of the accountancy profession[373]. The board would comprise representatives from international organisations, national regulators and academics responsible to oversee standard setting processes of the IAASB, IAESB, IESBA, their respective Consultative Advisory Groups (CAGs) as well as the Compliance Advisory Panel (CAP) of IFAC ensure processes were conducted in public interest. This "two-prong approach"[374] capitalised on the profession's expertise at the same time ensured public interest was presented. This mitigated the pitfalls of the lack of independence of the profession setting its own rules. For the IAASB, the CAG acts as a source of external stakeholder view and advice on IAASB's agenda, project priorities and time table; project content, technical and other relevant matters, the composition of the board has however been since 1993 dominated by organisations who have an interest in the functioning of the financial markets.[375]

4.3.2.2 International Framework of Assurance and ISAE 2000

In the IAASB meeting in June 2002, the board took a stand that audit and reviews were not engagements unique to themselves but considered types of assurance engagements with special characteristics. This warranted a separate series of standards dealing with audits and review of financial statements and with other assurance engagements.[376] IAASB proceeded to revise ISAE 100 to provide guidance to both high and moderate assurance engagements, following the publication of the commissioned research.

---

[373] The PIOB was officially established in 2005

[374] PIOB Chairman speech 2010 p. 1

[375] List of 2002 IAASB CAG members: Association for Investment Management and Research (AIMR); Basel Committee on Banking Supervision; Council of the Bars and Law Societies of the EU ; European Commission; European Federation of Accountants and Auditors for SMEs ; European Federation of Financial Executives' Institutes; European Federation of Investment Funds & Companies ;European Savings Banks Group ;European Financial Reporting Group; Federation des Experts Comptables Europeens (FEE);Federation of European Stock Exchanges; Information Systems Audit and Control Association (ISACA); Institute of Internal Auditors (IIA); International Accounting Standards Board (IASB) ; International Actuarial Association; International Association of Insurance Supervisors ;International Bar Association ;International Monetary Fund ;International Organization of Securities Commissions (IOSCO) ;International Organization of Supreme Audit Institutions (INTOSAI) ;United Nations Conference on Trade & Development (UNTAD) ;World Bank ;World Federation of Stock Exchanges from http://web.archive.org/web/20021215082436/http://www.ifac.org/IAASB/CAG.php

[376] IAAAB Minutes of the Meeting, June 2002 par. 8

The IAASB CAG suggested that the IAASB developed a document articulating the theoretical underpinnings of the assurance function.[377] The task force appointed by IAASB outlined 11 principles articulating the foundation of a framework for assurance engagements (assurance building blocks) to IAASB board members and technical advisors.[378] In December 2002, the "Assurance building blocks" was published as an agenda item outlining how the task force dealt with the principles in the revised framework and ISAE2000.

At the same time, it embarked on a project to restructure the auditing and assurance handbook which resulted in the division of ISAE 100 into two separate pronouncements. The IFAE (derived from paragraphs 1 to 30) and ISAE 2000 (derived from paragraphs 31-74 of ISAE100). The framework would apply to all assurance engagements, including the financial audit. ISAE2000 would provide guidance on subject matter other than historical financial information where no subject matter specific ISAE exists (in line with the first principle)[379]. This framework allowed ISAEs to develop to provide guidance on issues pertaining to all subject matters e.g., materiality; and also subject specific standards.

---

[377] IAAAB Minutes of the Meeting June 2002 and October 2002.

[378] Feedback was received and discussed by the Task force 3 conference calls and a physical meeting.

[379] Proposed Terms of Reference, Preface to the International Standards on Quality Control, Auditing, Assurance and Related Services and Operations Policy No 1 – Bold Type Lettering,

Assurance building blocks

Principle 1
The current ISAE 100 will be split into 2 documents: Framework and ISAE 2000

Principle 2
The Framework will apply to all professional accountants, when performing assurance engagements. ISAE 2000 will apply only to public practitioners.

Principle 3
Practitioners obtain assurance and provide a conclusion that conveys that assurance.

Principle 4
Criteria must always be suitable for the engagement

Principle 5
If there is any inconsistency between the Framework or ISAE 2000 and the requirements of a specific standard, the requirements of the specific Standard will take precedence.

Principle 6
Neither the "work effort view" nor the "interaction of variables view" will be specifically referred to.

Principle 7
Detailed (audit-type) procedures will always lead to a positive expression of assurance. Limited procedures (ordinarily inquiry and analytical procedures) will always lead to a negative expression of assurances

Principle 8
The labels "high assurance" and "moderate assurance" will not be used.

Principle 9
Relevant characteristics of the subject matter (whether it is, for example: qualitative, prospective, subjective) are to be described in the practitioner's report as appropriate to the circumstances.

Principle 10
There will be some circumstances in which an assurance report cannot meaningfully add to the credibility of the subject matter because of the combined effect of:• limited procedures,• the particular characteristics of the subject matter, and/or• limitation(s) in the available evidence.These engagements should not be accepted.

Principle 11
I n all cases, any limitation in the quantity and quality of evidence will be either: particular to the circumstances of the engagement, in which case a qualified report is to be issued (scope limitation), or • inherent in the nature of the subject matter

Figure 4.6: Highlighting the key points of the assurance building blocks[380]

4.3.2.3 Dealing with old "debated issues"

Three issues were left unresolved in ISAE100 which had to be addressed in the new draft standard: the work effort versus Interaction of variables view; the level of assurance and its reporting. These issues had been "resolved" in the revised draft the following way.

The research commissioned by IAASB "discounted" the pure work effort model.[381] It was found that in countries ascribing to that model, distinctions between different subject matters were drawn to determine different engagement rules and communication reports. Users also drew distinctions associating different subject matter with levels of assurance. The research supported the interaction of variables view and identified three variables: subject matter, specific criteria and work effort as determinants of and interact to determine a given level of assurance.[382] In contrast to the pure interaction of variables view, quantity and quality of evidence was regarded as a function of the three aforementioned variables rather than an independent variable in this regard. The survey also found other factors that affected the level of assurance including: user needs, litigation risk, availability of standards, distribution of reports, regulatory requirements, adequacy of fees and industry practice.[383] In light of the above, the consortium of academics argued that each of the three determinants had to be properly addressed in the assurance report to make enable effective communication between the provider and users.[384]

The study also highlighted that there is a general agreement suitable criteria affects the level of assurance. The origins in the debate lie with the fact that certain jurisdictions regard the factor as a dichotomous variable; it therefore becomes a determining factor as to whether an engagement should be accepted. Other jurisdictions regard suitability of criteria to lie on a continuum and as a factor determining level of assurance that can be provided on any engagement.

---

380 IAAAB Minutes of the Meeting December 2002 Agenda item 7A.

381 IFAC(2002) p. 112.

382 IFAC(2002) p. 121.

383 Ibid., p. 99, 110.

384 Ibid., p. 121.

Work effort versus Interaction of variables view

Although the study was highlighted in the IAASB Technical Director's report, to the author's knowledge, no direct references had been made to this study in publicly available standard setting documents. The task force introduced Principle 6 which accepted the work effort, the interaction of variables as well as other existing models that had their own conceptual justification. It made a conscious attempt to avoid specific references to the names of the models to prevent "*blind support*". The task force incorporated elements from both models in a model which it thought was not necessary to name.

Principles 8 and 9 illustrated that assurance incorporated elements of both the "work effort" and "interaction of variables" models. Instead of equating assurance with "work effort" or the "interaction of the four variables", the "new model" equated work effort i.e. the level of evidence gathering procedures, with a specific form of expression used in the providers' conclusion (Principle 8). Relevant characteristics of the subject matter, as appropriate to the circumstances had to be described in the assurance report which readers could take into account in their decision making based on "assured information" (Principle 9).

The principles specified circumstances which engagements should not be accepted. Mentioned specifically were when subject matter[385] and criteria[386] were not suitable, when there was a lack of available evidence and limited procedures (Principles 4 and 10). [387]

Level of Assurance

The ED avoided the phrase "level of assurance" to guard against over emphasis of the importance of quantitative aspects of assurance and oversimplification of the complex relationship between the different variables. Instead, the terminology "two distinct levels of evidence gathering procedures" was utilized, namely, "audit-level" and "review-level" (Principle 7).[388] For example, an "audit-type" engagement requires detailed procedures to be carried out and the assurance provider provides a positive

---

[385] IFAE sec. 25 ISAE2000 sec. 90(c).

[386] Suitability of criteria has been a prerequisite for engagement acceptance as suitability was essential to allow consistent evaluation or measurement of the subject matter within the context of professional judgment. Variations or evolving criteria thus will not affect the assurance obtained by the assurance provider. IFAE, sec. 27, 29 & 31; ISAE2000 sec 65, 90(h).

[387] ISAE2000 sec. 57.

[388] "Audit-level" and "review-level" are used instead of audit and review to indicate the range of engagements that could take other names than audit or review. Sec. 4, 39, 45-47 of the Framework.

expression of assurance conveying reasonable assurance. Comparatively, a "review-type" engagement requires ordinary inquiry and analytical procedures which leads to a negatively phrased conclusion conveying limited assurance. This addressed the dilemma faced by the board of whether to allow both positive and negative assurance for moderate assurance engagements.[389]

Furthermore, high and moderate assurance will be replaced by reasonable and limited assurance (Principle 8).[390]

Engagement type	Level of assurance risk	Work	Conclusion	Assurance conveyed
Review-level engagement	Moderate	Ordinary inquiry and analytical procedures, based on sound understanding of entity, subject matter and objectives of the engagement	Negative	Limited
Audit-level engagement	Low	Include obtaining corroborating evidence through e.g. assessment of internal control systems, test of records and responses to inquiries using techniques e.g., inspection, observation, confirmation and computation.	Positive	Reasonable

Figure 4.7 Types of assurance engagements[391]

389 IAASB Meeting June 2002 sec. 5. The board subsequently decided on negative assurance except "to the extent that standards dealing with specific topics allowed or required the use of a positive statement".

390 IFAE sec. 45-47.

391 IFAE sec. 4.

Figure 4.8 Assurance reporting

There had been a lack of agreement from standard setters if the assurance provider should provide assurance to a constituency or obtain assurance for him or herself and convey it through an assurance report which has implications standard setting, planning and performing the assurance engagement. Principle 3 stated that *"Practitioners **obtain** assurance and **provide** a conclusion that conveys that assurance".* This principle breaks down the assurance process into steps:

The practitioner first obtains assurance and forms a conclusion; he then reports to convey that assurance. The reader then reads the report to obtain assurance which could differ from that of the assurance provider. This principle draws a line on where the assurance provider's responsibility in that the perception of assurance conveyed by readers are removed from the scope of the standards. These efforts coincide with efforts to position the audit or assurance function at the end of the reporting chain.[392]

Principle 11 provided that limitation in sufficiency of evidence due to circumstances in engagement should result in the issue qualified report. However, if it is due to inherent nature of subject matter it should disclosed in the assurance report.[393] The framework required all assurance reports issued in writing, withdrawing possibility of oral or symbolic representations.[394] Other principles covered: parties obliged to follow the framework and standards (Principle 2); precedence of standards and conditions which the framework can be overridden (Principle 5).

4.3.2.4 New issues of debate in the framework and ISAE2000

In March 2003, the IFAE and ISAE2000 were exposed for a 3 month comment period. Twenty-nine comment letters were received and underwent detailed review by the task force in four meetings.[395] The following new issues of debate surfaced.

The relationship between the framework, ISAs and ISAEs

The respondents expressed concerns about the lack of clarity of the relationship between the framework, ISAs and ISAEs. Uncertainty presided on the authority of imperatives that were in the framework but not drafted in ISAE2000. Questions were raised as to whether ISA's should or could be used as a guide in areas not covered by the ISAE, for example, use of work of experts and joint responsibility that were mentioned in paragraph 24 of ISAE2000 without further guidance. It was also mentioned that there was duplication in the provisions between the framework and standards[396]. This led the task force to clarify the role of the framework and standards in the revised standard ISAE3000. It was clarified that the former did not in its own right establish standards or provided procedural guidance; the role was that of the ISAs and ISAEs. The relationship between the ISAs and ISAEs was made clear with the provision that although ISAs did not apply to engagements covered by ISAE they could "nevertheless provide guidance to practitioners"[397]

---

392 IFAC (2001) Rebuilding Confidence in Financial Reporting.

393 IFAE sec 40-42, ISAE2000 sec. 93(b).

394 IFAE sec 43 ISAE2000 sec. 87-88.

395 IAAAB Minutes of the Meeting Agenda item 2, two physical meetings and two conference calls were held.

396 Comment letters by Deloitte and the Charterer Accountants of Canada.

397 ISAE3000, sec. 3.

Lack of specific guidance

The national standards setting bodies such as the APB, expressed concerns regarding the broad scope of ISAE2000 as it could result in an insufficiently precise "framework" standard that creates confusion and contribute to the expectation gap[398]. The respondent called for subject specific standards instead, to provide more extensive guidance areas such as e.g. materiality, engagement risk, etc.; to enable quality and consistency to be achieved. Other respondents commented on the lack if guidance in specific issues such as engagement risks and for specific subject matter. For example, it was pointed out that the standard refer to key concepts such as materiality and engagement risk to be considered when planning and performing the engagement however, there had been no further guidance on these areas.[399] Other respondents also call for subject matter specific ISAEs particularly in sustainability reporting.[400]

IAASB maintained that a generic standard was essential to fill the void that existed in absence of subject specific evolving assurance engagements. It was regarded as *"appropriate to protect public interest and helpful to practitioners"*.[401] The development of subject specific ISAEs providing more specific guidance were to follow in the near future.[402] The IAASB also felt that subject matter specific ISAEs would better address topics such as joint responsibility and materiality etc.

What is assurance?

The respondents voiced concerns on clarity and precision of the concept "assurance engagement". Several issues were raised. It was expressed that the distinction between assurance engagements in general and those that can be accepted is unclear.[403] Respondents also raised the question whether assurance engagements that do not fall with the scope of assurance can be taken on as consulting engagements which do not require the same stringent independence requirements?[404] These commentators proposed a narrower definition of assurance engagements to include only those that were within the scope addressed by the document and discuss the acceptance criteria

---

398 Comments letter by APB.

399 Comments letter by Far.

400 IAAAB Minutes of the Meeting, Oct 2003

401 IAAAB Minutes of the Meeting Oct 2003

402 ibid

403 Comment letters by KPMG, APB, Southern society of chartered accountants, South African institute of chartered accountants.

404 Comment letters by South African institute of chartered accountants and IDW.

separately. The task force noted that the need for professional judgment under the above mentioned circumstances and identified the clarification of the assurance concept as a potential issue to be discussed at the revision of ISAE3000. Respondents noted that despite changes in terminology which had implications on the "philosophical" underpinnings of the understanding of assurance. E.g. the use of phrases the practitioner "obtains assurance" "provide a conclusion that conveys that assurance" the terms were not reflected in the definition.[405]

The issue on whether assurance applied to evaluation of a subject or subject matter information was raised. IAASB was in agreement that the framework should reflect the intention to focus on information that is the output of the process and it was the subject matter information which practitioner gathers sufficient appropriate evidence.[406] The incoherence of the framework and ISAE which regards to the objective of an assurance was also highlight. It was brought up that whilst objective stated in the framework focuses on the reduction of assurance risk to appropriate level, the drafting of ISAE2000 had greater focus on catering to the needs of intended users.[407]

Levels of assurance

Respondents expressed concern that aligning "reasonable and limited assurance" with "low and moderate" risk which was inversely related to assurance could be lead to users' perception that they were pure terminology substitutes for "high" and "moderate" levels respectively.[408] Responding to commentators, the task force articulated the objectives of both types of engagements, made significant changes in the discussion of risk and description of the nature, timing and extent of procedures in the framework. The task force kept the links between the levels of assurance and risk but noted the phrase "acceptably low" which required engagement circumstances including subject matter to be taken into account in determining acceptable level of assurance of a particular level of engagement.[409]

Respondents pointed to the lack of guidance on how much evidence was required for limited assurance. Concerns were expressed in particular as these engagement could have a level of risk from *"just below that is necessary for reasonable assurance to a floor level defined as at least sufficient for assurance obtained by the practitioner to be*

---

405 Comment letter by Far.

406 IAASB meeting December Further guidance pertaining to direct reporting engagements and assertion based engagements were needed.

407 Comment letters by AICPA, Far.

408 Comment letters by IDW, FEE.

409 IAAAB Minutes of the Meeting October 2002

meaningful." For example, corroborating evidence was required if the subject matter in a review engagement did not conform in all material aspects with identified criteria, no further guidance however was provided on corroborating evidence[410]

The task force recognized that in drafting standards for generic assurance, the nature, timing and extent of procedures could vary. Hence, it attempted to significantly enhance requirements for the description of work performed to ensure users awareness of procedures taken by the assurance provider, retain negative expression. It also change review level engagements to limited assurance engagements to avoid confusion with review engagements which pertains to historical financial information.[411] It was decided that a statement be added that third tier standards were expected to provide more detailed guidance.

IAASB approved the issuance of the IFAE and ISAE 3000 by ballot vote on December 31, 2003. It was noted that dissenting views were expressed by two members[412] pertaining to the introduction of terms "subject matter", "subject matter information" introduced after the ED and the "confusing and illogical" provisions and practitioners responsibilities associated with "assertion-based engagements" and "direct reporting engagements". Concerns were also expressed on the lack of guidance on limited assurance engagements.

4.4 The IFAE and ISAE3000

The IAASB issued IFAE and ISAE3000 in January 2004 accompanied by a media release stating: "As the provision of broader assurance services is an evolving field, the IAASB intends to keep the practical application of the framework and ISAE3000 under review, and invites feedback from practitioners and others to share their experience with applying them.[413]

Basic principles and essential procedures were coded and published to guide professional accountants in public practice guidance to perform assurance

---

[410] Sec. 34 ISAE2000, sec. 40 of framework. APB

[411] IAAAB Minutes of the Meeting October 2002

[412] Dec 2003 IAASB meeting minutes: Mrs Esdon and Dr Yang Agenda item 1A p. 2

[413] IFAC News Jan2004 sec. 56. Initial intentions were to make a request for feedback in the framework however, the task force suggested the request be through a press release instead or on the IAASB website as the original intention could affect perception of the permanence and certainty of the framework (Dec2003, Agenda 3 IAASB meeting)

engagements in general (not specific to sustainability assurance). The scope of the standard included engagements other than audits or reviews of historical financial information covered by International standards of auditing (ISA) and review engagements (ISRE) and precluded those covered by International Standards for Related Services such as Agreed on Procedures, compilation of financial and other information, preparation of tax returns and consulting[414]. It is read in conjunction with the IFAE and had to be applied observing the Code of ethics for professional accountants and International standards on quality control (ISQC1). The figure below summarise diagrammatically the structure of the standards. IFAC members (accounting networks) were obligated to apply this standard effective January, 2005.

ISAE3001-3399 had been set aside for potential development of additional standards applicable to all assurance engagements i.e. topic specific ISAEs e.g., planning, documentation etc. ISAE3400-3999 set aside for subject specific ISAE e.g., ISAE 3402: Assurance reports on controls at a third party service organisation.

In accordance with IAASB pronouncements in 2003, term assurance is not defined in the glossary of its pronouncements. The term assurance engagements include within its scope audits of financial statements as well as other subject matter both guided by the International framework of assurance. Assurance Engagements are also used to distinguish the latter "audit like" engagements on subject matters other than the financial audit from the financial audit.

The provisions contained in ISAE3000 can classified into four areas of guidance. The first area covers behaviour of the assurance provider and quality control of the engagement which are cross referenced to the code of ethics and ISQC1 respectively.[415] The second area of guidance centre on the types, conditions of engagement acceptance and terms of engagement.[416] This is followed by guidance on processes related to planning and performing the engagement and the resulting output.[417]

The drafting of ISAE3000 was not drafted and has not been revised in line with the "clarity principles". It nevertheless is principles based "framework standard". Its provisions will be analysed in chapter 7.

---

[414] Assurance though may be part of a consulting engagement. IFAE sec. 13.

[415] ISAE 3000 sec. 4-6.

[416] Ibid sec. 7-11.

[417] Ibid sec. 12-44, 45-56.

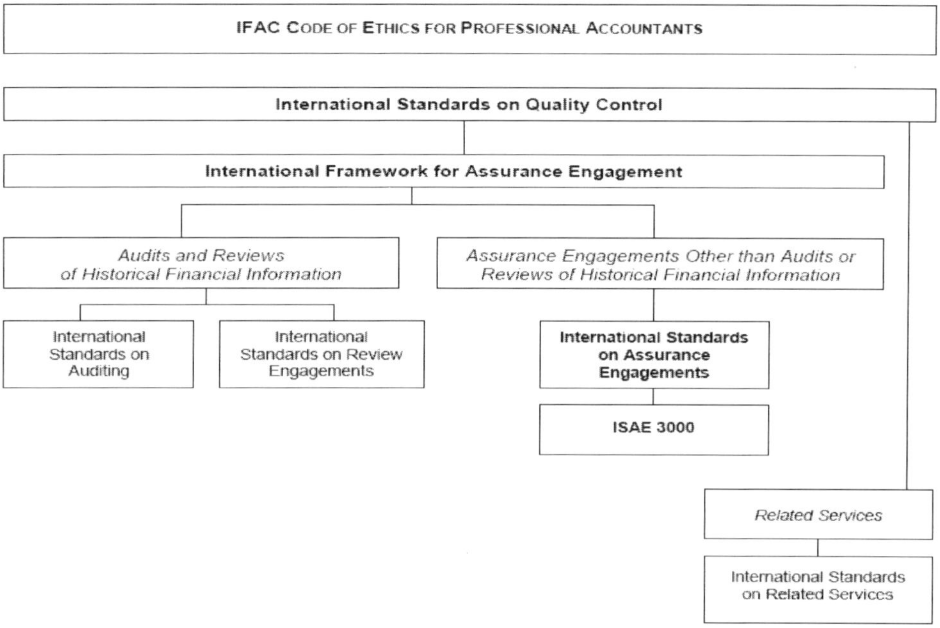

Figure 4.7 IAASB pronouncements 2003

4.5 Discussion and conclusion

This chapter has provided a descriptive account of how the assurance concept has been framed by IAASB through the construction of standards, in the context within which IFAC operates from 1997-2005. This account provides an explanatory source as to the position taken, as well as the opportunities and constraints faced by the accountancy profession in the interaction with other non-accountancy actors or standard setters in the regulatory space for setting sustainability assurance standards.

The IAASB is a standard setting body of IFAC, which represents and sets standards providing guidance to its members. The core service performed by the profession is the financial audit, IFAC strategy to position itself as a global standard setter within the new financial architecture have an impact on the actors involved in its standard setting process, the issues and the interest it represents. Its activities are constrained by what its key stakeholders, who are predominantly interested in the functioning of the capital market, regard as appropriate. Although IFAC in its recent ED on a public interest framework for the accountancy profession in November 2010 define "public interest" as the "common benefit that all citizens share from the services provided by the accountancy profession…" adhering to three criteria: assessing costs and benefits for society as a whole; democratic principles and processes and respect for cultural and ethical diversity.[418] This did not apply during the drafting of ISAE3000.

The framing of the assurance concept took place prior to and during the time period when IFAC made crucial developments in the governance and organisation structures introducing a standard setting due process involving greater transparency and participation of stakeholders. The assurance concept had been constructed through lifting the broad idealistic notions of, and decontextualizing knowledge from the financial audit domain to a global abstract level which can be re-embedded into other subject matter retaining terminology used in financial audits. This is evident from its initial draft of having an umbrella framework and general principles in order to guide assurance on reporting service engagements (including the financial audit) and IAASB's constant struggle to position the financial audit within the more general assurance framework.

The glossary of IAASB's handbook of International Quality Control, Auditing, Review, Other Assurance, and Related Services Pronouncements in 2010 does not define the term assurance but defines assurance engagement, reasonable and limited assurance engagements and assurance engagement risks. In the IFAE, assurance engagement is defined by a list of general requirements needed to carry out the engagement. Findings show that there had not been a strong conceptual underpinning of what is an assurance engagement from the start. In the process of tracing how the assurance concept is framed by IAASB, the chapter identified problems and issues that emerged during the course of developing the framework and ISAE3000 and how IAASB had attempted to find a common ground or "resolve" them. The accountancy profession had been divided as to the prerequisites of accepting an engagement, the determinants of assurance and the related level and reporting of assurance engagements. These areas had been debated on and negotiated as the standard evolved over time. It was not till 2002 after a suggestion was made by the IAASB CAG that a task force was formed to explicitly lay down the "building

[418] *Exposure draft IFAC Policy Position Paper #4:* A Public Interest Framework for the Accountancy Profession November 2010

blocks" of assurance where conscious choices of certain terminology and phrases that imparted particular meaning were made.

The lack of agreement of "moderate assurance" and the loss of confidence in the financial audit resulting in the developments in the regulatory environment created a need for a clearer distinction between the financial audit and assurance services on other subject matter. IAASB structured the ISAs and ISAEs as two separate sets of standards containing the general principles of assurance which are to be read in conjunction with the IFAE. There was however a lack of drafting conventions as well as clarity as to the guidance provided in the framework and general principles (specific ISAEs). This highlights the potential problem of duplication between the two pronouncements and the danger that the ISAE is isolated from revisions made to the ISAs. For example, ISA 300 planning, ISA700 reporting, ISA320 and 450 on materiality as well as ISA315 and 330 on the audit risk model have undergone revisions.[419] Hence, even though assurance services (financial and non-financial) adopt common terminology, the developments in financial auditing are not reflected in engagements performed according to ISAE3000 leading to inconsistencies within the profession. On the other hand, this separation allows the IAASB a freer rein to make decisions on the general principles pertaining to non-financial assurance. The evolving assurance concept, the positioning away from the financial audit creates an opportunity for the accountancy profession to interpret ISAE3000 more flexibly allowing a more dynamic response to other actors in the regulatory space for sustainability assurance.

The findings of this chapter also provide insights to the limits of ISAE3000. Awareness of the problems and issues pertaining to the assurance concept negotiated within the profession and the decisions made by the IAASB to build a stronger foundation and "resolve" some of the standing issues provide an explanatory source to some of the provisions of the IFAE and ISAE. Its generic nature and intended wide scope of application and the lack of clarity between framework, principles, requirements and guidance provide insights to the analysis the provisions ISAE3000 and how it functions.

The following table presents a summary of key developments at various stages illustrating how assurance is framed at particular points in time.

In March 2009, the IAASB published a project proposal to revise ISAE3000 to adopt clarity drafting conventions[420] , to incorporate enhanced requirements and guidance and to consider if separate topical ISAEs were needed in areas such as planning,

---

[419] Other examples include: ISA260 communication with those charged with governance; ISA580 written representations; ISA620 Experts.

[420] Not drafted according to clarity conventions

documentation, use of work of experts.[421] The amendment had no intention of revising settled conceptual issues from 2003.

Project stage	Timing
Approval of project proposal	March 2009
High level directions paper	September 2009
Issues paper	March 2010
First reading of ED	June 2010
Approval of ED	September 2010
Exposure period	October 2010 - March 2011
Review and reading of post ED	September 2011
Approval and revised ISAE3000	March 2012

Figure 4.8: IAASB ISAE3000 project proposal March 2009

The project was supported by calls from stakeholders who required more guidance applying the standard including national auditing standard setters applying ISAE3000 to set more specific standards. A survey conducted by IAASB in 2008 concluded that 9 of 17 respondents supported the revision project. According to IAASB, the planned review was in line with the goal of serving public interest in light of the increasing demand of such services.[422] The revision of ISAE3000 was to provide an up-to-date guidance and ensure consistency. The scope of the project include: dealing with revised and redrafted ISAs; considering the extent which topical ISAEs were "practical and relevant "; considering the developments and results of studies within IAASB and external to but related to the work of IAASB; considering if drafting should take into account non accountants (issuing assurance reports) understanding to the standard[423] and; the extent to which additional and more clearly articulated guidance is required. For example, the call for additional guidance regarding materiality, engagement risk, use of experts, documentation and joint assurance engagements.

---

421 IAASB Main Agenda (March 2009) Item 4D "Project Proposal: Revision of ISAE3000" sec. A1

422 UK, Japan, USA, Australia commented on continuing increase in these services

423 The IAASB would also consider non-professional accountants understanding of ISAE3000 (written with the presumption that it would be applied by accountants) when performing such engagements

5. The Sustainability Assurance Regulatory Space - Patterns of Influence

5.1 Chapter objectives

This chapter presents an analysis of how the perceived gap regulatory gap pertaining to the quality of sustainability assurance has been filled by standards. It focuses on the actors involved in standard setting, how, why are they involved and their influences in the development of sustainability assurance standards. This is carried out using the regulatory space construct, which in this thesis depicts the 'place' where standard setting occurs. The literature reviewed in chapter 2, focuses the attention of the research on which actors are included and excluded from the regulatory space; what activities the actors engage in to legitimise their entry, non-entry or exit from the regulatory space; how problems have been constructed; how issues have been raised and how legitimate solutions were negotiated in the standard setting process.

The objective of this analysis is to obtain a richer understanding of the context in which standards are set, and identify the patterns of influence in the regulatory space that shape the regulatory issues, which determine the way sustainability assurance standards develop. The findings from this chapter serve to enrich our understanding of the terms and discourses used by various actors and provide an explanatory source to analyse choices made by national standard setters and assurance providers applying these standards.

This chapter starts with providing an account of the significant events in the mid-nineties that led to the accountancy profession's involvement in sustainability assurance. This is followed by the mapping of developments of sustainability assurance standards in three phases, each discussed as a separate sub-section:

Phase 1 (1999-2002) - the period prior to the existence of internationally accepted sustainability assurance;

Phase 2 (2003-2005) - the period after the first international sustainability assurance standard AA1000AS was published by AccountAbility and ISAE 3000 was made obligatory to IFAC members;

Phase 3 (2006-2010) – the period when standard setters engage in the first revisions of their sustainability assurance standards

This chapter concludes with a discussion of the standard setting context in the international regulatory space and highlights the on-going efforts pursued by actors involved in developing sustainability assurance standards.

5.2 Background

The Environmental theme became increasingly important in the global policy agenda in the early 90s and was brought to the accountants′ agenda through the work of the Intergovernmental working group of experts on International Standards of Accounting and Reporting (ISAR) and the Organisation for Economic Co-operation and Development (OECD).[424] In this context, the FEE was commissioned in 1992 by the European Commission, to carry out a study on the status of environmental accounting in the member states of the European Union(EU). This was presented to the EC Accounting Advisory Forum.[425] The Environmental Working Party (EWP) of the FEE was formed in December 1993 due to its growing profile and involvement in environmental issues. It served to encourage and facilitate the exchange of information among member bodies and other interested parties; promote the role of the accountancy profession; stimulate developments in environmental auditing, accounting and reporting and encourage pan-European studies and research programs in this field.[426] The EWP, today the named the Sustainability Group of the FEE, subsequently became one of the most prominent groups in influencing the development of sustainability assurance standards.

The FEE's objective that this time (prior to a revision made in 1998) was to serve as a representative body for Europe's leading accountancy institutions, to promote and advance the interest of the European accountancy profession which stands in contrast to today's emphasis on public interest. To develop its own position, it orchestrated

---

[424] IFAC Discussion Paper: The Accountancy professionand the Environment, IAPC, May 1995. ISAR published a report on Environmental disclosures: International survey of corporate reporting practices in 1992. The OECD organised a round table on Environmental accounting and auditing for its work group on accounting standards in May 1993.

[425] Collison, D.; Slomp, S. (2000) "Environmental accounting, auditing and reporting in Europe: the role of FEE" European Accounting Review, 9(1) p. 114. The Accounting Advisory Forum is a consultative body which the European Commission set up in 1990. FEE was a member, other members include representatives of the accounting standard-setting bodies in Member States of the EU.

[426] Ibid. For more information, refer to the original source document is Terms of reference of the Environmental Working Party, 1998.

the workshop "Environmental stewardship and management: positioning the accountancy profession" organized by the FEE's Danish member body, the Institute of State Authorized Public Accountants. The workshop was targeted at senior partners of leading accounting firms and senior technical experts working in environmental field within accountancy profession[427]

The conclusion of the workshop was published in 1995 with theme "wake up, do, communicate, protect" laying grounds for professionalization and quests for professional jurisdiction in this field.[428] The workshop created awareness and prepared accounting firms to extend their activities, applying their unique expertise to environment aspects e.g., designing internal control systems, environmental reporting and third party verification of information. The accountancy profession was also encouraged to develop appropriate review and reporting standards on environmental issues, cooperate with experts and advisers from other disciplines, as well as participate actively in development of environmental legislation and regulations. These include the development of Eco-Management and Audit Scheme (EMAS) and ISO.[429] The profession sought involvement through communication of its position in environmental field to governments and other policymakers, industry and other stakeholders and strongly emphasized that the public should be protected against "experts" who do not have necessary education in environmental accounting and auditing and exclude them from practicing in this area.[430]

In specific countries large audit firms were already involved in environmental auditing. The scope of these engagements varied widely as there was no standardized verification, audit or reporting practices.[431] A survey by FEE in 1995 reflected that there was growing tendency for national professional institutes and other standard setting bodies to be involved in environmental auditing.[432] Accountancy bodies and research institutions in various countries undertook studies on environmental reporting, accounting and auditing[433]. There were initiatives in sponsorship of

---

[427] Ibid. p. 115.

[428] "Final Conclusions of the FEE Workshop: Environmental Stewardship and Management: Positioning the Accountancy Profession" in FEE (1995) *Environmental Accounting, Reporting and Auditing: Survey of Current Activities and Developments within the Accountancy Profession*, p. 85-87.

[429] Ibid, p. 16.

[430] Ibid, p. 86-87.

[431] Kamp-Roelands, N.(2002) "Towards a Framework for Auditing Environmental Reports", Tilburg University, 2002.

[432] FEE (1995), p.17.

[433] Ibid. Appendix 2 of FEE(1995) report gives examples in the UK, including ICAEW and ACCA. Other institutions include Canadian Institute of Chartered Accountants (CICA), Foreningen af Statsautoriserede Revisorer of Denmark, Limperg Institute of the Netherlands.

academic research centres[434] and conferences[435]; introduction of courses in tertiary institutions[436]; formation of networks and associations for financial auditors trained in environmental issues to exchange practical experiences, and participation in discussions on environmental accounting, reporting and the related auditing activities[437].

Studies carried out under the aforementioned initiatives concluded that the financial auditors could make important contributions to the training of environmental auditors and implementation of environmental audits in light of the adequacy of the accounting techniques and internal control procedures to audit EMS. However, the financial audit expertise only formed a part of the required expertise.[438]

The events highlighted in this section laid foundation to the activities undertaken and the way key issues were defined by the FEE, professional bodies and the large accounting firms in the sustainability assurance regulatory space.

5.3 Development of sustainability assurance standards

Several actors have issued guidance and standards to guide assurance engagements on sustainability reporting that had a significant impact on the development of standards at an international level. These institutions can be grouped in three levels.

---

434 For example, the Centre for Environmental and Social Accounting Research in University of Dundee was at that point in time sponsored by KPMG and ACCA.

435 ACCA, ICAEW and NIVRA cosponsored the Environmental management, accounting and auditing (EMAA) network workshop focusing on environmental accounting and sustainable development for researchers and practitioners: discussing research and practice in environmental management, accounting and auditing.

436 For example, the University of Amsterdam 1990, introduced environmental accountancy in its curriculum to enable accountancy profession to specialize in environmental services.

437 For example, Vereniging voor Miliue Accountancy (VMA) Association for Environmental Accountancy, a professional organization was founded in 1991, EMAA was formed in 94 by the Limperg institute to encourage research within and across Europe on environmental management, accounting and auditing.

438 Blokdijk, J.; Drieenhuizen, F. (1992) "The Environment and the Audit Profession" European Accounting Review, 1(2), p. 437-438.

Levels	Actors	Standards	Users
International	IAASB of IFAC	IFAE and ISAE3000	Members of IFAC are obligated to apply ISAE3000 effective 1/1/2005
	AccountAbility	AA1000 AS (2003 and 2008)	Open source standard[439]
	GRI**	Overarching Principles of Assurance (2001)[440]	Not targeted at particular users GRI has since withdrawn the document
Regional	FEE*	Framework for providing assurance on sustainability reports (2002)	Not intended to be used as a standard
National***	NIVRA	Exposure drafts 3410, 3010 (2005) 3410N Sustainability assurance standard (2006)	Members of NIVRA

Figure 5.1 Standard setting institutions in the regulatory space[441]

*FEE is not a standard setting body but issued initial sustainability assurance guidance that had a significant influence in the regulatory space. **GRI sets reporting standards. One of its work groups, the verification work group published the Overarching guidance on sustainability assurance in 2003. ***Note that sustainability assurance standards at the national level are present in Sweden, Germany, France, Australia and Japan to provide guidance to relevant assurance providers at a national level but the standards do not have the same impact as NIVRA standards in the regulatory space.

---

439 The AA1000AS(2008) imposed new licensing restrictions for its use. For more information, refer to http://www.accountability.org/standards/aa1000as/licensing-fee.html.

440 GRI (2001), *Overarching Principles for Providing Independent Assurance on Sustainability Reports*, Global Reporting Initiative Verification Working Group, Boston, MA.

441 Source: own.

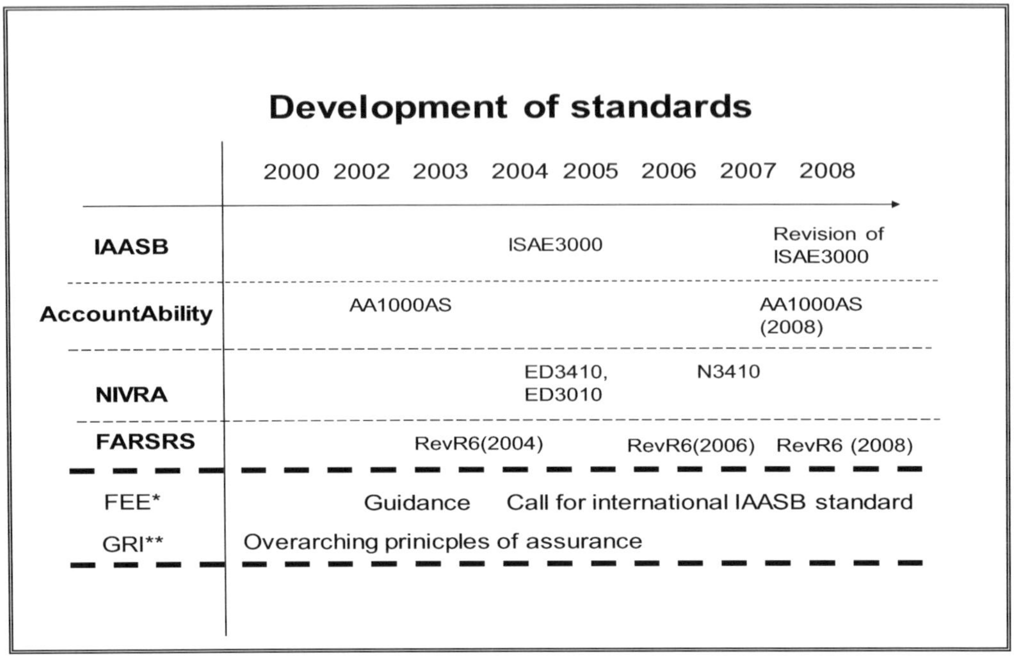

Figure 5.2 Development of standards in the regulatory space

The diagram above presents a snapshot of the developments of assurance and sustainability assurance standards over time.

At this juncture, it is perhaps important to explain the absence of the European Union in the above list. The European Commission published a Fifth Action program on Environment "Towards Sustainability" in 1992. The EMAS is a voluntary initiative designed to improve companies' environmental performance. The aim is to recognize and reward those organizations that go beyond minimum legal compliance, continuously improve their environmental performance, and external reporting of results of verification by an external party.[442] 'Verification' used in EMAS covers a different subject matter of assurance. According to the Act, the scope of verification includes the organisation's environmental review, environmental policy, environmental management system and internal environmental audit.[443] Apart from

---

442 REGULATION (EC) No 1221/2009 OF THE EUROPEAN PARLIAMENT AND OF THE COUNCIL of 25 November 2009 on the voluntary participation by organisations in a Community eco-management and audit scheme (EMAS), repealing Regulation (EC) No 761/2001 and Commission Decisions 2001/681/EC and 2006/193/EC.

443 Ibid., Act 2(24).

the 1996 Green paper on CSR in support of stakeholder involvement contributing to the quality of assurance, the EU has remained till today dormant in its support for assurance on sustainability reports.[444]

5.3.1 Phase 1(1999-2002): Establishing the regulatory issues

5.3.1.1 Establishing a regulatory space

FEE published a discussion paper in October 1999 for two purposes: To stimulate international debate on assurance engagements on environmental reports, and to provide input to the IAASB and the GRI to develop an internationally accepted assurance framework in the environmental area.[445] At that point in time, the IAASB had received comment letters for ED2 Assurance standards, and was in the process of developing a framework and generic assurance standard which could be used to develop subject matter specific assurance standards. Assurance on environmental reports had been listed as an example. The GRI, a standard setting body for sustainability reporting had created a Verification Working Group, a *"small but informed and representative international forum"*[446] with the objective of reaching a consensus on a series of questions. Questions crucial to this discussion include the role of verification and its use to enhance reports prepared according to GRI standards, and the role GRI can play in the process.[447]

FEE established the discussion by stressing the relationships between these concepts. Reporting is part of communication; communication needs credibility; credibility needs guidelines on contents and assurance procedures. By establishing that relevance and accuracy are foundations of credibility, it put forward that *"room for differences in the scope of engagements and in subsequent reporting is undesirable to both entities and the users of environmental reports and increases the expectations gap."*[448] It supported the "room for differences" with findings from its research paper from 1996.[449] The paper highlighted the lack of harmonization in environmental assurance as variations were

444 European Commission "Promoting a European Framework for Corporate Social Responsibility" Green Paper, July 2001, p. 18.

445 FEE (1999) Discussion paper: "Providing assurance on Environmental Reports", p. 6.

446 http://web.archive.org/web/20010710133538/globalreporting.org/News/June1999Update.htm downloaded 2/2011.

447 Ibid.

448 FEE(1999), p. 11-12.

449 FEE(1996) Research paper "Expert statements in Environmental Reports"

found on subject matter, objectives of the engagements, criteria, procedures required and reporting practices which potentially led to differences between what societies expected assurance providers to achieve and what they could and should reasonably expect to accomplish. To further magnify the problem, the report mentions the lack of mandatory requirement, generally accepted reporting guidelines as well as the presence of other providers than accountants issuing the assurance report in this area. By constructing diversity as a problem, FEE called for the development of internationally accepted guidance on reporting and assurance as "solutions" to mitigate the diversity. The discussion paper was intended to act as a catalyst towards an internationally accepted assurance framework in the environmental reporting field.[450]

The discussion paper provided a space for those interested to raise and discuss issues pertinent to assurance on environmental reporting. The definition of environmental report adopted by the FEE was not restricted to separate reports but included environmental information published in site, health and safety, sustainability or annual reports in order to stimulate interest, thoughts and ideas from the larger pool of potential respondents.[451] The core group invited to comment were professional accountants, FEE also invited other interested groups such as stakeholders, companies, environmental consultants and other experts into the regulatory space.[452] By identifying the subject matter as providing assurance on environmental reports, the FEE excludes from the space parties interested in assurance of systems and processes that do not result in reporting (e.g., certification schemes). Through the discussion paper, the FEE played a significant role "inviting and excluding participants" into the regulatory space, putting forth issues and stimulating debates on how to guide assurance engagements for environmental (which was later extended to sustainability) information.

The discussion received 9 responses: 6 from National standard setting bodies or bodies representing the accountancy profession within Europe, 2 from Danish organizations and 1 from a practitioner.[453]

---

450 FEE issued two other relevant papers: FEE(1999) Position Paper "Review of International Accounting Standards for Environmental Issues" 1999 and FEE(1999)Position Paper "Towards a Generally Accepted Framework for Environmental Reporting", which contributed significantly to the development of the GRI guidelines.

451 FEE(1999), p. 3.

452 Ibid.

453 Responses from ICAEW, IDW, APB, Royal NIVRA, CECCAR, DnR, Danish Commerce and Companies agency, Danish Confederation of Trade Unions and Carter, a CPA. The comments were considered and published by FEE in its Position paper (2000) Analysis to responses to FEE discussion

5.3.1.2 Introducing regulatory issues to the regulatory space

The FEE raised seven issues which set the agenda for debates, discussions and negotiations in the regulatory space.

The term assurance and "impediments" to a high level of assurance

The FEE raised the issue that different terms had been used to describe assurance services. Engagements to assure environmental reports were also referred to as verification, review, limited review, audit or attestation. The term "environmental audit" was also defined and used differently by various actors that had practical and conceptual differences. For example, environmental audits could refer to the audit of environmental activities, events, conditions, management systems or information about these matters[454]; verification procedures by EMAS; audit of environmental management systems under the ISO14000 series; energy consumption studies; waste impact assessments; etc.

FEE proposed as a solution the term "assurance" to describe the audit of environmental information. The FEE strengthened the appropriateness of using assurance by differentiating it from "audits" and "verifications" that had more specialized institutionalized meanings.[455] The term assurance would be a solution to avoid confusion and expectations gap with the aforementioned established services. The term was supported by IDW, NIVRA and DnR based on the fact that it was in line with the terminology adopted by IAASB in its second ED on assurance engagements. Two UK respondents voiced concerns on the use of the term assurance. Due to its wide application in practice, ICAEW voiced concerns on its clarity and the degree of comfort and responsibility of accountants. APB asserted the importance that the term applied should reflect the nature and scope of the engagement e.g. agreed upon procedures or review in light of the immaturity of environmental reporting and assurance. Despite some concerns raised, FEE concluded that there was general agreement to the use of the term assurance.[456]

The FEE brought the debate on the determinants of the levels of assurance into the environmental domain by questioning the impediments to high level assurance. Two key issues it highlighted were the lack of reporting standards which serve as assurance

---

paper "Providing assurance in Environmental Reports" published in October 1999 and Discussion Paper (2002) "Providing assurance on sustainability reports".

[454] FEE(1999), p. 12.

[455] FEE(2000), p. 8,9.

[456] FEE(2000) p. 7, 8.

criteria and the qualitative nature of the subject matter which made obtaining sufficient and appropriate evidence difficult. Respondents raise various concerns pertaining to the lack of and immaturity of criteria resulting in inconsistencies and completeness in reporting. Whilst the issues raised by the national accountancy bodies much resembled the debate IAASB was involved in, in the development of ISAE300, the non-accountancy respondents raised other concerns. One example is the Danish confederation of trade unions who cautioned against too much stress on quantitative environmental information which could result in the neglect of relevant qualitative data, and the development of competences. It expressed concerns of an audit led rather than stakeholder led development in environmental reporting.[457]

FEE attached primary importance on developing the reporting criteria and related assessment standards. The final position adopted by the FEE was that the level of assurance was a matter between the "client" and assurance provider and the impediments remained in its initial position to be resolved. NIVRA pointed out several matters to address including engagement risk, reporting relationship with clients and knowledge competence which at that point did not appear to be taken up by the FEE.[458]

Responsibility for assurance

FEE raised issues pertaining to the approach required to perform environmental assurance and responsibility of the assurance providers. It gained consensus that multidisciplinary teams and a flexible approach was warranted. The term multidisciplinary carries the connotation of the need for an orchestrator to coordinate activities of different experts (defined in the paper as independent individuals, employees of the accounting firm or reporting firm); it also provided a basis for subordinating other forms of expertise to the audit process.[459]

The questions constructed by FEE were: "who should be the signatory? How many reports should be issued? Who and to what extent should the accountant and expert bear responsibility?"[460] These questions can be conceived as a way to seek legitimate consensus as to who the orchestrator should be. There was no consensus as to who the legitimate "coordinator" to the multidisciplinary teams should be. Most of the respondents felt that the ultimate responsibility depended on the nature and scope of the engagement though some respondents preferred the professional accountant bear

---

[457] Ibid, p.4-6.

458Ibid.

459 Power, M. (1997) "Expertise and the construction of relevance: accountants and environmental audit" Accounting, Organizations and Society, year: 1997, volume: 22, issue: 2, p. 123-146;

460 FEE(2000) p. 15-18.

the ultimate responsibility. Respondents were of the view that a single report issued and sighed by either the auditor or environmental expert clearly stating the responsibilities of each party in the report; a jointly signed report within the framework of the legal system and professional standards or one that is signed under the firms overall name (both auditor and experts are employees of the firm providing assurance) were acceptable and preferred to the alternative of having more than one report. The APB and IDW made reference to ISA 620 "Using the work of an expert". The FEE gain consensus and held the view that a single report is preferred; responsibilities of parties should be clearly spelt out and the report should be signed by either or both parties. The legitimate coordinator and the guidance provided in ISA620 were issues to be further debated.

Acceptance of Engagement

On engagement acceptance, the FEE raised the issues as to whether scope limitations to the boundaries of the reporting organization, the subject matter and obtaining proper evidence were acceptable. Respondents generally agreed that scope limitations on subject matter should not deter engagement acceptances due to the non-mandatory developing status of environmental reporting and assurance.[461] However, precautions had to be taken on clarity to avoid misleading readers of the report and issue a qualified report if needed.

The APB however felt scope limitations on obtaining evidence could not be assessed due to lack of standards in determining material and "reasonable evidence"[462] required. Concerns were also raised by IDW pertaining to an assurance provider's responsibility on the information contained but not subject to assurance and the applicability of ISA720 "Other information in documents containing audited financial statements". The FEE adopted the position that the responsibility lie with the provider to transparently declare the unavailability of evidence to support disclosures and the providers' role to acquire adequate evidence to conclude that the report was not misleading. Although there was general consent that assurance could be provided on qualitative information, FEE cautions that the level of assurance depended on the type of information and there could be instances that assertions were not verifiable and should not be included in the report.

The FEE brought forth the question on how suitable of criteria could be enhanced to provide an effective determinant in providing assurance on sustainability reports. In this context, a few respondents point to the need for environmental reporting framework or standards. NIVRA put forth that development of "suitable criteria"

---

[461] DnR specifically mentioned, scope limitations are acceptable in limited assurance engagements.

[462] The term was raised in APB's comments.

involve discussion and agreement between the reporting organization, stakeholders of the reporting organization and assurance provider. The APB and the ICAEW also recommended needs of stakeholders to be established to determine relevant information to disclose, the latter made reference to the audit committees role. This led FEE to conclude that the GRI and FEE generally accepted framework for environmental reporting could be supplemented by stakeholder opinions to determine suitable criteria.

Risk assessment and internal control

The FEE sought the opinion of respondents on actions available to assurance providers in cases where the reporting organizations environmental reporting system were inadequate to ensure reasonable reliability of disclosed information (a significant problem in practice) and whether providers should perform work to determine if the systems in operation were in line with the reporting organizations reporting policy.

Both the respondents and FEE felt that the agreed objectives and scope of the engagement determined this matter, providers should only evaluate those procedures within the EIS or EMS relevant to the engagement objectives be stated it clearly in the assurance report (or environmental report). The role of provider was to plan and perform the engagement to reduce to an acceptable level risk of expressing inappropriate conclusion. The FEE was of the opinion that withdrawal from an engagement should be an action of last resort. The APB however questioned if the traditional audit risk model was suitable and pointed to alternative business risk approaches adopted by the large accountancies could be of greater value to environmental assurance.

Compliance with law and regulation

The FEE raised the issue on an assurance provider's responsibility with regards to a breach of environmental law or regulation the reporting organization could have incurred which is unreported. The respondents had a common view that compliance of laws was the responsibility of management and management representations had to be obtained but were cautious as to the role of assurance providers. FEE structured its final position in line with NIVRA's respond that the assurance provider in case of non-disclosure, had to assess the impact on the report and information materiality to users, ascertain management's reasons, consider termination and need to disclose matter to third parties and seek legal advice if required.

Materiality

Both FEE and respondents agreed on the importance of the materiality concept which warrants further discussion. The respondents comments centred on the definition; what constituted materiality; from whose perspective it should be defined and the significance of a legal threshold. NIVRA pointed out importance of stakeholder consultation as a "major process" to determine materiality. Taking into consideration respondents′ views, FEE stressed importance of provider understanding factors that influenced decisions of users. It provided a list of issues relevant in assessing materiality including the relevance of stakeholder views, legal thresholds, basis to decide materiality of an item, relation between size and impact of an entity and changing expectations etc. It noted the difference between financial and environmental assurance and indicated a possible for explanations in the engagement letter.

Reporting

FEE suggested that the lack of a clear structured approach to reporting contributed to an expectations gap between what assurance providers intended to convey and users' perception of assurance. There was consensus that clarity in reporting was fundamental to close the expectations gap and the report should be standardized to contain certain elements.[463] In line with APB's suggestion, FEE concluded that rather than premature standardization, "current practices" could be analysed to better understand assurance providers and users perspectives with regard to style and content of the assurance reports on different engagements and disseminate best practices. It however maintained that a structured approach explaining the scope of engagement, responsibilities of parties and level of assurance provided would facilitate better understanding of assurance engagements.

The table in appendix 1 presents a summary of the key issues where general consensus had been reached, those that required further evaluation and those that had not been taken up in FEE's final position. It is noted that many unresolved issues such as materiality, the assurance concept, criteria, etc. from setting the generic assurance standard ISAE3000 re-emerge in these discussions. It also explains IAPC's choice not to stake a claim in the regulatory space on sustainability assurance in 1999 due to the lack of consensus among members with regard to moderate assurance.

[463] The FEE identified 11 elements that should be included: Title, addressee, description of engagement and identification of subject matter, objective and level of assurance, responsibility of respective parties, criteria, description of work performed, description of inherent and specific limitations on engagement, conclusion based on work performed, name and address of provider, date which report is signed.

5.3.1.3 Extending from environment assurance to the sustainability domain

The GRI published its first sustainability reporting guidelines in 1999 which drew on FEE's position paper 1999 "Towards a Generally Accepted Framework for Environmental Reporting ". Adopters increased from 20 in 1999 to 150 in 2002. The term sustainability was increasingly used by reporting organisations to reflect the expansion of their environmental reports to include economic and social elements. The developments in reporting create a need for assurance services filled by assurance providers extending their services to cover the three domains of triple bottom line reports.

In 2001, the Environmental Working Party of the FEE changed its name to Sustainability Working Party to illustrate its extension of activities from the environmental to the sustainability reporting and assurance domain. In 2002, it engaged in two initiatives that contributed to developments in assurance on sustainability reports. First it held a sustainability conference[464], which it perceived led to support and consensus amongst standard setters, NGOs, accountants, investors and consultants present that third party assurance raised quality and credibility of sustainability reports.[465] Second, it published a Discussion Paper in 2002: "Providing assurance on sustainability reports" with a focus on independent third party assurance prior to the World Summit on Sustainable Development held in Johannesburg. FEE legitimised this initiative as being in line with its goal as raising the quality and credibility of sustainability reporting and claiming the belief that its members had resources and proven expertise in the provision of reporting and assurance services fundamental to strengthening sustainability reports. [466]

The FEE redefined participants invited to participate in the regulatory space in comparison to its 1999 initiative. In light of the increasing number of stand-alone sustainability reports being issued, FEE limited the discussion paper to providing assurance on stand-alone sustainability reports.[467] On one hand, it broadened its field of potential respondents to include the reporting organization, users or stakeholders, standard setters, other assurance providers and experts. These parties were invited into the space in recognition that the accountancy profession had no monopoly status in

---

[464] The Sustainability conference was held In April 2002, organised by FEE and its Danish member body.

[465] FEE Annual report 2002, p. 9

[466] President of FEE in 2002 Göran Tidström in Discussion paper "Providing assurance on sustainability reports".

[467] FEE(2002) p. 6.

this voluntary field and other parties report bounded by different codes of practice provided assurance services and issued assurance reports published with sustainability reports. Further, two significant shifts in discourse took place between 1999 and 2001 which impelled FEE to encouraged participation of reporting organisations and users of reports respectively. First, as part of the process to rebuild confidence of the audit, IFAC instituted the view that "audit is part of the reporting process" which extended the responsibility of the reporting organisation.[468] Second, IAASB's construction that assurance providers merely obtain assurance and conveys assurance in the report put responsibility on the users to understand and interpret the report issued by the provider. On the other hand, FEE excluded from the space actors with an interest in assurance of sustainability information reported in other areas than the stand alone report. The FEE put forward 30 questions identifying the following key issues in assurance of sustainability reports: the objectives, approach to and key elements of assurance, expectations gap and standard setting.

Objectives of assurance

The new actors invited brought the "business case" discussion into the regulatory space. To establish the business case, FEE raised questions on the benefits and drawbacks for sustainability reporting and assurance to enhance the discussion of the objectives of assurance from "commissioners" as well as stakeholders' points of view.[469] To stimulate discussions in this field, the FEE raised questions as to who should appoint the assurance provider, for who is the assurance commissioned and objectives of a "full scope" assurance engagement [470] e.g., whether the information accuracy or the completeness and balance of the report was the objective of engagements.

Issues from 1999 in the context of 2002

FEE raised the issues from the 1999 paper bringing into discussion their relevance in the sustainability domain including a discussion of the appropriateness of the term assurance.[471]

In recognition of other providers of assurance services, FEE brought into the debate the different non-provider specific approaches to assurance including the consultancy, accountancy, social audit, internal audit and other judgment approaches such as

---

[468] IFAC (2001) Rebuilding Confidence in Financial Reporting.

[469] FEE (2002), questions 4 to 6.

[470] Ibid., questions 17-19, 21-22.

[471] Ibid. questions 1 to 3.

rating agencies and use of independent experts or "famous faces". The paper introduced the "comprehensive approach" which principally was the "accountancy approach" incorporating the elements of appeal of the social and consultancy approaches taking into account stakeholder dialogue and understanding of management systems and processes. FEE brought to attention the use of terms such as "external" and "third party" by both accountancy and non-accountancy assurance providers depicting the notion of independence. It then stressed that no established ethical, procedural or reporting standard for providing assurance on sustainability reports were in place and raised the issue of whether and how special emphasis should be given to the quality of independence, a well-established and institutionalized prerequisite of the financial audit.

The FEE brought into discussion again the acceptability of scope limitations imposed by the reporting organisation and the assurance provider's responsibility to deal with information in the report that is not part of the engagement with reference to ISA720. It raised the need for reporting standard setters to consider "auditability" when drawing up reporting guidelines questioning whether subject matter that cannot be subject to any or any "acceptable" level of assurance should be included in the report or segregated from disclosures that assurance could be performed?

The 2002 discussion paper defined materiality as a threshold quality of the significance of a disclosure such that, above it, omission or misstatement of a disclosure would affect the users of the report, and should be considered on an overall basis. FEE raises the question if a conceptual framework for sustainability reporting was required to address the concept of materiality and what it should include.[472] It also brought up the role and issue of stakeholders in determining materiality. For example, what role stakeholders could play in appointing assurance providers? What is meant by balancing need of stakeholders; should and if so, how stakeholders can be involved in an objective way in assurance? Should stakeholder dialogue be provided for in assurance standards and whether it is a process which should be examined to address concerns of management capture? The paper also invited comments on key issues to address in assurance on environmental, social and economic subject matter within sustainability report.[473]

The FEE re-raised the issues of assurance reporting. E.g., how assurance could be communicated in the sustainability report? If there was more than one assurance provider how their responsibilities should be conveyed and whether reporting

---

[472] Ibid. question 5.

[473] Ibid., questions 21-23 , 26-28

according to the minimum contents specified in ISA100 was sufficient or if a longer narrative form was required.[474]

Standard setting

The FEE raised a core question if the time was appropriate to set standard on sustainability assurance. It identified, thereby invited into the regulatory space two potential standard setters the IAPC and AccountAbility. Given that assurance providers were drawing from ISA100 in practice, which depicted the "accountancy approach", FEE questioned it this approach which did not take into account dialogue with stakeholders explicitly was sufficient or flexible enough to accommodate assurance engagements on sustainability reports or if more specific pronouncements were needed. With increasing number of firms and assurance providers who were familiarity with AA1000 series, FEE brought into discussion if the new standard AA1000AS would meet the need for an assurance standard and the potential role AA1000 series would have on provision of assurance services.[475]

5.3.1.3 Other related activities

Efforts from the GRI Verification Committee

GRI published a revised version of its reporting standards in 2002. Two accountancy bodies the ACCA (member of FEE) and CICA were amongst bodies represented on GRI steering committee. The GRI advocated the use of assurance and encouraged the development of assurance guidelines.[476] A Verification Working Group which FEE has representation, was formed as a forum to discuss how verification was to be addressed in the GRI as well as needs, expectations and best practices in assurance on sustainability reports. Its role was to encourage standardization of verification methodologies; development of tools and it created a forum for reporters, users and verifiers to reach mutual understanding. A working paper "Overarching Principles for providing independent assurance on sustainability reports" was published as a discussion or feedback mechanism.[477]

---

474 Ibid., questions 5, 29-31

475 Ibid., questions 14-16

476 http://web.archive.org/web/20010710133538/globalreporting.org/News/June1999Update.htm Feb 2011.

477 GRI(2002) Glossary and Annexes Annex 1 "Overview of the Global Reporting Initiative".

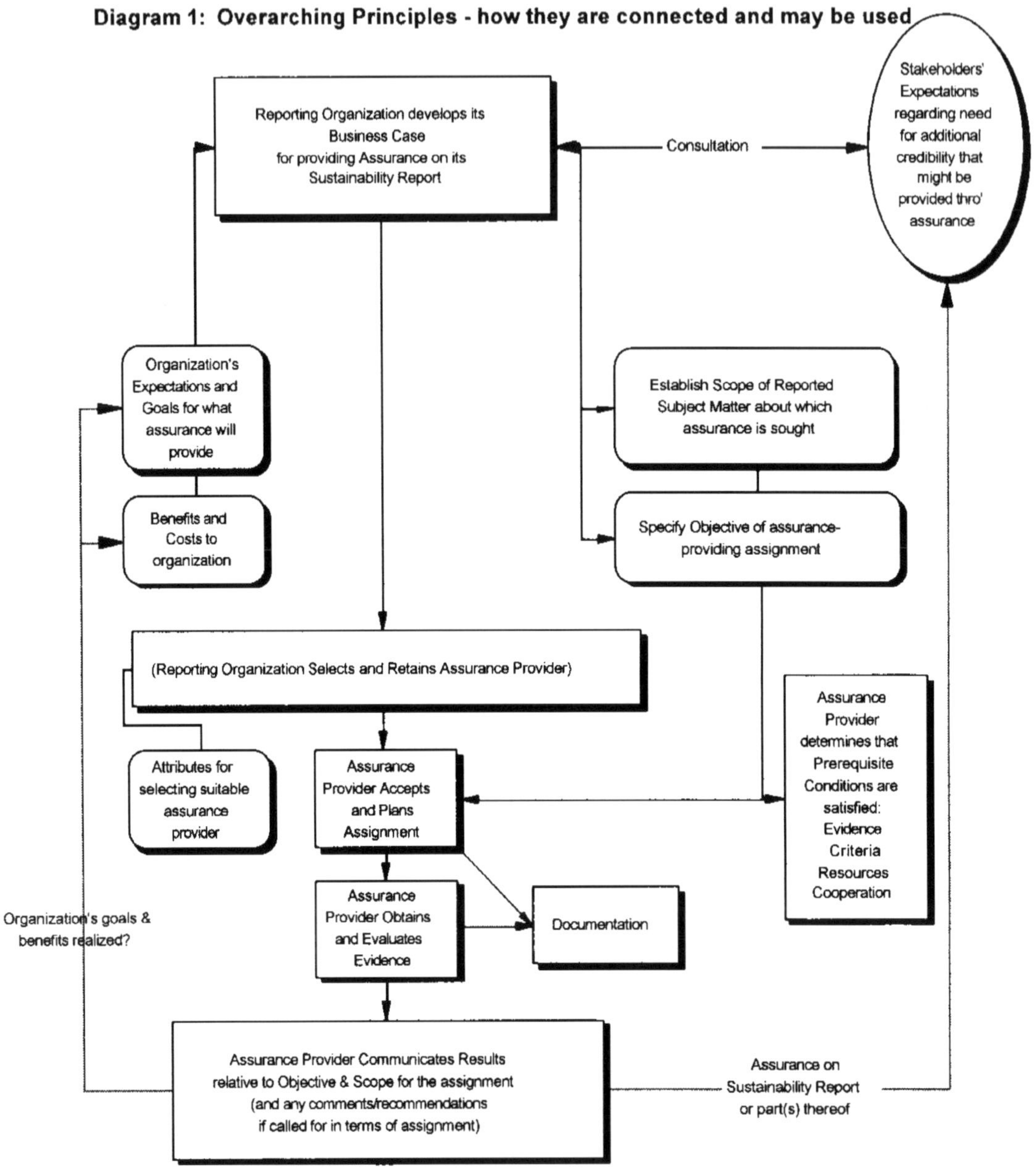

Figure 5.3 Overarching principles of assurance

One could draw several parallels of the issues raised by GRI with FEE's publications. Both FEE and GRI stressed that the assurance process should be considered in terms of the value they might bring to the reporting organization i.e. the business case. Both organisations also brought forth conditions for assurance acceptance, approaches and procedures to assurance, communication of results and attributes of assurance providers. The GRI however adopted a more stakeholder oriented approach and looked upon assurance as "one approach" that reporting organization could use to enhance credibility of sustainability report. GRI encouraged the reporting organizations to clarify with assurance providers if all categories of stakeholders were recognized, if any significant stakeholders had been excluded and if the organization had ascertained expectations of stakeholders with regard to reporting and improving credibility. In assessing evidence and determining materiality, GRI suggested that consideration should be given on whether corroborative statements and other evidence from external stakeholders were collected (where necessary) and if stakeholders were consulted about the usefulness and credibility of report content and usefulness of assurance. Organizations were also encouraged to continually assess the results of the assurance process where possible in consultation with stakeholders, to further improve value and its credibility.[478]

In communicating their conclusions, GRI provided minimum requirements in assurance reporting. In comparison with the accountancy approach ISAE 100, the key differences lie in GRI's support for assurance providers' opinion regarding the accuracy, completeness, reliability and balance of the sustainability report relative to the scope and the objective to the engagement and the use of constructive reporting. The GRI also encourage assurance providers to include in describing the work carried out the extent different categories of stakeholders that participated in the planning and execution of the assurance process and indicate any constraints to this process. In addition, a statement assuring assurance provider's independence and freedom from bias and conflicts of interest was recommended.

The GRI asserts the importance of the assurance providers' ability to balance the consideration of the interest of different stakeholders and independence when making a choice of an assurance provider. It defines as independent those who have not been involved in the design and development or implementation of the reporting organization's monitoring and reporting systems or assisted in compiling of the sustainability report.

[478] Annex to the GRI G2 guidelines.

AccountAbility's introduction of the AA1000 series

AccountAbility was incorporated in UK in 1996 as a non-profit organization with the mission of promoting accountability for sustainable development, to rebuild trust between corporation and stakeholders and address trust deficit. Its work is based on "collaborative governance" by international membership of business, service providers, civil society organization and research institutes and is focused on addressing the need for organizations to embed their stakeholder engagement process into daily activities.[479] In 1999, it introduced the AA1000 series, a framework to guide organizational learning, performance and progress with the aim of improving the quality of social and ethical accounting, reporting and auditing. Although it was not particularly active in the regulatory space for sustainability assurance between 1999 and 2002, AccountAbility was the first organisation to publish an international standard to guide sustainability assurance engagement in 2003. AccountAbility is not an organization representing the accountancy profession; however, several members of the accountancy profession have been actively involved in the technical committee developing AA1000 standards.

The European Union

The European Union provided guidance on the reporting of environmental issues by building on annual report requirements. On 30 May 2001, the European Commission issued a recommendation 2001/453/EC on the recognition, measurement and disclosure of environmental issues in the annual accounts and annual reports of companies.[480] This recommendation which is not legally binding derives its legal basis from The Amsterdam Treaty[481] and the 1992 Fifth action program on the environment "Towards sustainability". The aim was set to improve decision usefulness of the annual report by providing meaningful comparable information to allow users to assess more completely environmental risk and liability's impact on the financial health of the company. The recommendation requires the disclosure of environmental issues to the extent that they are material to the financial performance or the financial position of the reporting entity in the annual and

---

[479] http://www.accountability.org/

[480] Commission Recommendation 2001/453/EC of 30 May 2001 on the recognition, measurement and disclosure of environmental issues in the annual accounts and annual reports of companies (notified under document number C(2001) 1495)

[481] In line with Article 6, a key element for promoting sustainable development is the principle of integrating environmental requirements into other policies. The Internal Market Council was invited to consider how integration of environmental objectives into all Community policies and actions could be achieved in this particular domain through strategies adopted by the Cardiff and Vienna European Councils where a Community initiative in the area of accounting was proposed.

consolidated annual report, notes to the annual or consolidated accounts or balance sheet depending on the item providing more specific guidance on recognition, measurement, disclosure of environmental issues in annual accounts and reports of companies, taking into account IAS 36, 37 and 38.

The path pursued by the European Commission focuses on reporting in the annual accounts and therefore excluded from the regulatory space established by FEE that limited the discussion paper to providing assurance on stand-alone sustainability reports. The recommendation however has an effect on separate environmental reports (the subject matter of assurance) as it recommends the alignment of the scope, boundaries, the reporting dates and period; cross referencing and consistency checks of the reporting entity's environmental report with the annual report to facilitate readers. It also carries implications on the financial audit and auditors extending the financial auditors domain to include auditing or checking consistency of environmental information within the annual report depending on national law.

5.3.1.5 Summary of Phase 1 1999-2002

1999-2002 marked a period where the number and scope of sustainability reports were on a rapid increase and GRI reporting guidelines were gaining international acceptance. The development for sustainability assurance standards lagged behind sustainability reporting. Up till 2002, no standard had been published internationally to guide sustainability assurance engagements.[482]

FEE through the publication of discussion and position papers created a space for identifying and discussing core regulatory issues serve to support and catalyse standard setters' efforts towards setting sustainability assurance standards. The logic behind their active role in this field can be summed up by the following quote:

"Sustainability reporting is in its infancy and accountants, in particular, have an opportunity to influence the way that it develops in parallel with financial reporting. Accounting profession should also recognize urgency of engaging in those quickly developing areas to ensure expertise is usefully put to work for the public interest and to maintain a position of influence and leadership for the benefit of its members."[483]

---

[482] Some countries had standards guiding the environmental audit e.g., IDW. Firm specific standards were developed, usually customized to clients' needs

[483] FEE Update on Sustainability Issues Nov 2001: Sustainability and the accounting profession (with part ref to GRI)

Acknowledging the need to establish the "business case" for sustainability assurance, it invited not only the accountancy profession but also other actors into the regulatory space whilst excluding others. It played a significant role identifying issues to be debated which consequently sought out issues where a consensus arose and those that remained to be resolved. The FEE also gained influential representation in the work of GRI and had been active in responding to the invitation for comments in IAPC's (IAASB) work towards a generic assurance engagement standard. Whilst GRI's verification committee elaborated on the role of stakeholders in sustainability assurance and developed broad principles published in the annex of its reporting standards, the IAPC refrained from involvement.

From the activities in the regulatory space, the following table builds on table 2 to identify the key regulatory issues have been established. Although assurance is applied the sustainability assurance context, the issues debated when setting a generic assurance standard by IAPC continue to resurface particularly in the planning and performing assurance engagements, levels and communication of assurance.

Issues pertaining to the assurance provider	Competence, ethics and independence of the assurance providers? Use of experts and division of responsibilities. Should ISA620 provide guidance? Who should be the legitimate coordinator?
Objectives of assurance	Is there a business case for assurance? Who should appoint assurance provider? What should be the objective of an assurance engagement? Accuracy or "balance and completeness"? Acceptable scope limitations? Applicability of a piecemeal approach?
Planning and performing the assurance engagement	Subject matter and criteria prerequisites? Role of stakeholders? How to determine materiality? Applicability of traditional financial audit risk model? Relationship between elements and level of assurance?
Levels of assurance	Gaps towards achieving a "high level" of assurance? How it can be addressed through reporting or scope limitations?
Communication of assurance	Narrative or standard form of reporting? Should minimum contents to specified, is it sufficient? Is constructive reporting more favourable?

Figure 5.4 Phase 1 key regulatory issues

5.3.2 Phase 2 (2003-2005): Competition and cooperation within the regulatory space

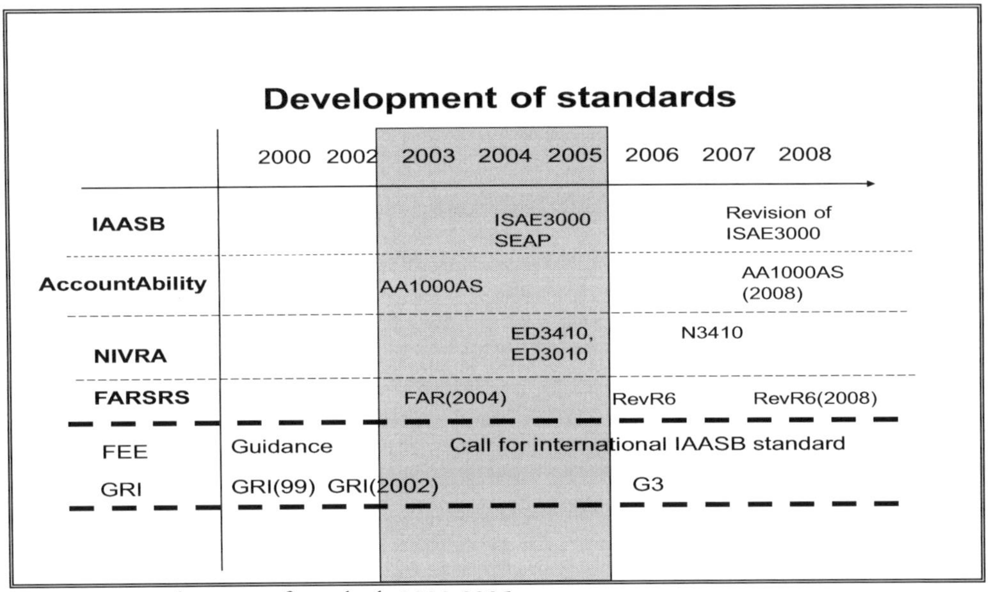

Figure 5.5 Development of standards 2003-2005

Phase 2 depicted a time period where standard setting institutions both at the global and national levels significantly increased their involvement in the regulatory space.

Three significant paths of development occurred in parallel that influenced and drew from one another through interaction and representations between actors where technical expertise was shared, shaping the development of sustainability assurance. First, in 2003, the IAASB approved the ISAE3000, which members of IFAC were obligated to apply from 1st January 2005 (Chapter 4). Sustainability assurance engagements carried out by the accountancy profession fell within the scope of ISAE3000. Second, AccountAbility staked a claim in the regulatory space publishing the first "internationally accepted" sustainability assurance standard AA1000 Assurance Standard (AA1000AS). Third, four European national assurance standard setters took a proactive stance to draft and published national guidance for assurance on sustainability reports to guide their members to performed such engagements.

Country	Standard setter	Standard	Status
Sweden	Far	Independent review of voluntary separate sustainability report	Issued: February 2004
Netherlands	NiVRA	Assurance engagements relating to sustainability reports (N 3410)	ED Issued: January 2005[484]
France	Compagnie Nationale des Commissaires aux Comptes (CNCC)	France Pratique professionnelle relative au rapport du Commissaire aux Comptes sur certaines données ou informations environnementales et sociales ou sur les procédures d'établissement de ces données ou informations, contenus dans le rapport rendant compte en matière de développement durable (not translated)	Finalised: November 2004
Germany	Institut der Wirtschaftsprüfer in Deutschland e.V. (IDW)	Generally Accepted Assurance Principles for Audit or Review of Sustainability Reports (ED ASS 821)	Issued: May 2005 Exposure draft

Figure 5.6 National sustainability assurance standards[485]

5.3.2.1 Claiming a stake in the regulatory space with AA1000AS

In March 2003, AccountAbility claimed a stake in the regulatory space. It attempted to gain legitimacy as a standard setter through persuasive assertions that it is a *"leading international professional institute, offers professional development and certification, and undertakes leading-edge research and related public policy advocacy"* and that it adopted an *"innovative, multi-stakeholder governance model"*. Efforts were taken to heighten its profile through influencing the on-going development of the architecture of sustainability standards through collaboration with the GRI, Transparency International and the ISO Social Responsibility Working Group, to ensure that

[484] Practitioners working with subject matter experts from other disciplines on non-financial assurance engagements (ED 3010) was issued together with ED 3410 but has been withdrawn.

[485] Source: FEE (2006)

standards took a stakeholder focused approach. In addition, it collaborated with CSR network rating yearly to rate the extent which the Fortune Global 100 companies integrated responsible business practices into their core activities.

AccountAbility entered the space with the launch of a non-proprietary, open source standard AA1000AS, the first international sustainability assurance. The standard was positioned as part of the AA1000 series to reach out to existing users of the AA1000 series and a stand-alone standard for assurance. Two supporting documents, guidance notes and practitioner-2-practitioner notes were published. The former offered technical guidance to provide operational interpretation and clarifications of the standard; the latter provided practitioners' accounts and analysis of using AA1000AS.

AccountAbility supported the notion that trust building efforts through the development of reporting standards would be undermined without a generally accepted profession standard for assurance. It constructed as "problem" that few assurance providers adopted a forward looking approach that gave adequate perspective on the broader context and materiality to stakeholders.[486] It carved out its position in the regulatory space pointing out existing standards were either too narrowly focused on specific aspects such as carbon emissions, too broad or focused primarily on the accuracy of quantitative data and robustness of accounting systems. It also pointed out that these standards were commercially proprietary which made it difficult to scrutinise, compare or replicate.[487]

AccountAbility emphasized the participatory quality of the standard setting process to legitimize the AA1000AS. AA1000AS was based on a, multi-governance forward looking model which integrated key learning with emerging practice of sustainability management and accountability, reporting and assurance practices. AccountAbility constantly emphasised the use of *"extensive international consultation process (direct participation)"* and *"drawing on the practical experience and perspectives of the business, public and civil society sector"*[488]. It further stressed input legitimacy of the standard based on the representation and composition of its technical committee which included the accountancy profession e.g., ACCA and members belong to Big 4), other assurance providers, companies, civil organizations dealing with sustainability issues.[489] In the process, it demonstrated expertise and competence drawing from and building on mainstream financial, environmental and quality-related assurance. It

---

486 AA1000AS (2003) sec. 2.4.

487 AA1000AS (2003)

488 http://www.accountability.org/

489 AA1000 technical committee composition: 14 members. 5 practitioners, 2 representing big 5, 2 representing accounting organization, 1 independent consultant, 4 other organizations and research center FEE participated but stopped after 2005

stressed output legitimacy i.e., efficiency and effectiveness of standards drawing attention to that it was complementary to the GRI Reporting Guidelines and also other company-specific approaches to disclosure. In addition, it claimed that the standard could be applied by all providers in conjunction with their current frameworks including the ISAE3000. It was not till 2007, that this claim was openly challenged.

Further strategies to legitimise AA1000AS were adopted after publication of the standard through referral networks or collaboration with businesses (e.g., Novo Nordisk, British Petroleum), civil society organisations (e.g., IRCA, CSR network), the accountancy profession (e.g., KPMG Sustainability), and academia (e.g. Harvard). Six months after the publication of the standard, AccountAbility collaborated with Novo Nordisk co-hosting an event to bring together *"business leaders from leading adopting companies"* and issued Practitioner Notes to assist in the practical development of the standard both for and by companies and assurance providers. It also collaborated with International Register of Certificated Auditors (IRCA) to launch the Certified Sustainability Assurance Practitioners Program in November 2004 combining its "sustainability assurance expertise" with IRCA's expertise in personnel certification programs.[490]

5.3.2.2 Shifts in the focus of regulatory issues from phase 1

In AA1000AS, AccountAbility defined the subject matter of assurance to include not only the sustainability report but sustainability performance i.e. the reporting organisations policy, decisions, actions that affect sustainability outcomes[491] thereby redefining the boundaries of the regulatory space. AA1000AS was described as *"generally applicable standard for assessing, attesting to, and strengthening the credibility and quality of organizations' sustainability reporting, and their underlying processes, systems and competencies."*[492] The AA1000 series placed emphasis on the issues of materiality and the role of stakeholders. Organisations that applied AA1000 series had to be committed to the practice of inclusivity i.e., to understand sustainability

---

[490] Certification as a Sustainability Assurance Practitioner Including the Accountability Sustainability Assurance Competency Framework Built on AA1000 Assurance Standard platform (p7). There were three grades of certification(Associate sustainability assurance practitioner, sustainability and lead sustainability assurance practitioner) targets not only internal and external sustainability assurance providers but also consultants, novices and members of CSR departments working with this issue with the aim to improve stakeholders confidence in expertise and provide better understand of standards as such.

[491] AA1000AS Annex A3.

[492] AA1000 AS sec.. 2.2.

performance, impact and views of stakeholders; to consider and coherently respond to aspirations and needs of stakeholders in policies and practices and account to stakeholders for its decision, action and impacts. Though materiality and the role of stakeholders were raised in FEE's discussion papers at phase one, AccountAbility explicitly constructed three "reporting" principles: materiality, completeness and responsiveness to support inclusivity which assurance providers had to take into account when performing the engagement.

Following the publication of AA1000AS, it issued a paper "Redefining materiality" in July 2003 that reassessed materiality in three aspects: Intention (to whom?), subject (about what?) and calibration (significance to fall within the bounds of materiality) comparing the traditional perspective of materiality and proposed the need for a more structured approach to deal with materiality in sustainability reporting.

Dimensions	Classical definition	Materiality reassessed
Intention	Owners of financial capital or their representatives	Wider range of shareholders interested in environmental and social impact of company's performance
Subject	Financial performance	Financial and Non-financial performance
Calibration	Judgment of a reasonable person would be changed or influenced by inclusion or correction of an item	Extension of intention and subject changes the basis a reasonable person would decide on materiality

Figure 5.7 Reassessing materiality

Being in a regulatory space where no profession had been granted monopoly status allowed AccountAbility to reassess and challenge existing terminology introduced in the regulatory space. It assessed and re-defined materiality as whether the reporting organisation has included in the public report adequate information about its sustainable performance for its stakeholders to be able to make informed judgments, decisions and actions. It introduced a five part materiality test to provide organisations and assurance providers a guide to assess materiality giving consideration not only to direct short term financial impact but policy based performance, business Peer based norms, stakeholder behaviour and concerns and societal norms.

	Part	Test
1	Direct short term financial impact	Determine if social and environmental performance affects financial performance in the short term e.g. expenses or taxes
2	Policy based performance	Determine if company has agreed policy statements of a strategic nature in this area
3	Business Peer based norms	Determine similar companies' view of what issues or aspects of performance are material
4	Stakeholder behaviour and concerns	Determine stakeholder view of what issues or aspects of performance are material
5	Societal norms (regulatory and non-regulatory)	Determine if the company considered societal norm which could be embedded in regulation, are to be regulated or emerging within the investment community

Figure 5.8 Five part materiality test

Completeness and the role and importance of stakeholders were raised in discussions in the regulatory space in 1999-2002. Difficulties with the completeness concept were attributed to the lack of a general reporting framework and the immaturity of reporting standards at that point in time. AccountAbility defined completeness as "the extent to which reporting organization can understand material aspects of sustainability performance". The role of the provider was to identify material shortfalls in that understanding and encourage the reporting organisation to address those issues or admit their shortcomings in the report which otherwise would be reflected in the assurance report. AccountAbility also introduced the concept of responsiveness defined as the "evaluation of whether reporting organization has responded to stakeholder concerns, policies and relevant standards and communicated these responses in its report."[493] Assurance providers were guided to

[493] AA1000AS Sections 4.2-4.4 respectively.

evaluate if processes and resources for developing, communicating and follow up of a response were available. The AA1000AS also required assurance providers to provide additional commentary in the assurance report to inform stakeholders on the progress made and areas for improvement of the reporting organisation. In September 2005, it introduced an exposure draft AA1000 Stakeholder Engagement Standard (AA1000SES) a principles-based standard providing a robust basis for designing, implementing, evaluating and assuring the quality of stakeholder engagement was published

The above illustrates AccountAbility's attempt to construct materiality and the role of stakeholders at the forefront of the agenda in determining the objective, the planning, performance and communication of sustainability assurance engagements. Through its definition of the subject matter of assurance and role of stakeholders, it re-invited into the regulatory space assurance providers that provide assurance on information other than those contained in a separate sustainability report. It also down played strongly debated topics by accountancy profession on the competence and independence of assurance providers, the rigour of assurance process, subject matter and criteria prerequisites, determinants and the levels of assurance. Through staking a claim and redefining terminology in the space, AccountAbility had significant influence on the development of sustainability assurance standards and practice changing the balance of power of actors and regulatory issues at focus.

The AA1000AS however is not without its limitations. Issues relating to the behaviour of the assurance provider and work performed were provided for in AA1000AS however no detailed guidance had been given[494]. It provided that the level of assurance should increase over time and was a function of information available, sufficient evidence, underlying systems and processes, internal assurance systems, existing assurance, resources allocated, legal or commercial constraints and competencies of provider. No further guidance was given.

[494] AA1000AS section 5.

5.3.2.3 Response by the accountancy profession

The FEE

In 2003, FEE published a short paper "The benefits of assurance" developed by the Sustainability Assurance Subgroup which was presented at the launch of the AA1000AS by AccountAbility. The paper defined assurance as a "provision of confidence" and assurance engagement according to IFAE and ISAE3000. It stressed the both internal and external benefits of independent assurance and pointed out:

"The intended users of the report are the immediate eternal beneficiaries, but the organisation frequently benefits indirectly" The indirect benefits attributed "improved public perception of its activities" and "positive changes, for example in control procedures and reporting systems."[495]

In 2004, it published the Call for Action: Assurance for sustainability. The paper built on and updated the issues raised in the 2002 discussion paper and called for action by standard setters, reporters and other stakeholders (government, investors, NGOs) to develop reporting and assurance. In this paper, FEE focused not on the regulatory issues previously defined but on the action and behaviour desired of participants in the regulatory space. A shift in focus on technical issues and the elements of assurance is replaced by the capital market and stakeholder discourses. The purpose was stated as to promote sustainability assurance to businesses and the capital markets and the sharing of knowledge across national and corporate boundaries, breaking technical barriers. To highlight, FEE stated "it is hoped that the paper will provoke debate that will enhance the ability of national and international standard setters, notably the IAASB, to meet the needs of the market place for suitable standards of assurance that will increase the credibility of sustainability reports."[496] In recognition that AA1000AS was a sustainability assurance standard but had not issued supporting ethical codes and guidance on quality standards[497], it called on IFAC and AccountAbility "to co-operate and not to compete so as to move speedily to provide high quality standards of assurance on sustainability"[498].

FEE also called for standard setters to cater to piecemeal approach to reporting and improve disclosure requirements on internal and external assurance; inform and educate by making available simple language explanations of pronouncements; adopt

---

[495] FEE(2003) p. 4.

[496] FEE Annual report 2005 p. 27.

[497] FEE 2004 p. 20.

[498] FEE 2004, p. 109

a more proactive approach to stakeholder engagement and reduce the democratic deficit through higher participation of stakeholders in development of standards.

Efforts of NIVRA

Royal NIVRA is the professional body representing accountants in the Netherlands. Its responsibility and duties are laid down in the Dutch *Registeraccountants* Act. Its statutory objective has been interpreted as to advance proper practice of the profession, foster their joint interest and protect their reputation.[499]

NIVRA has been involved actively in the work of IAPC and FEE pertaining to environmental reporting and auditing and had been an active participant in the regulatory space. In January 2005, NIVRA published in English, two drafts: ED 3410 "Assurance engagements relating to sustainability reports" and ED 3010 "Practitioners working with Subject Matter Experts from other disciplines on Non-Financial Assurance Engagements"[500] exposed for a six month comment period.

NIVRA published IAASB's intention of using the exposure drafts as a source standard to build for wider international use, if the exposure drafts were well received. The NIVRA draft was the first standard on sustainability assurance issued by a standard setting body representing the accountancy profession and based on ISAE3000 in Europe.[501] The efforts of NIVRA brought to discussion the accountancy approach in the context of AA1000AS a non-accountancy standard used by a number of assurance providers belonging to the accountancy profession. For example, how materiality could be defined within the scope of ISAE3000, role of assurance provider, what constituted suitable criteria and subject matter, level of work required to make conclusions and reporting requirements.

ED3410 provides guidance to engagement on a separate sustainability report. It defines Sustainability report as: "a report in which a reporting organization provides information to relevant targeted groups of stakeholders and accounts for the

---

[499] NIVRA (2006) Role and Task of NIVRA: NIVRA's position in society and profession. "NIVRA's membership comprises approximately 14,000 registeraccountants, working in positions in which the quality of information and reporting is paramount. NIVRA's task is to advance the quality of the professional services provided by registeraccountants, both for the public good and to foster registeraccountants' own mutual interests."

[500] ED3010 had as its objective to distinguish three possible models for cooperation with experts: model 1 where the assurance provider bears undivided responsibility (based on ISA620 and RAC 621N "Cooperation between auditors and actuaries); model 2 based on joint responsibility and model 3 covered two separate engagements were new.

[501] The draft recommendation published by Far in 2004 was not based on ISAE3000 but drew from ISA100 and FEE discussion paper .

economic, environmental and social aspects of the organization and the effects of its business activities on society. The sustainability report consequently offers a platform for dialogue between the organization and its stakeholders." It puts emphasis on the intended user identified as forming the "core aspect" both for the reporting organization and the assurance provider. It also provides guidance on the practitioner's responsibility with regards to reporting materiality and completeness of a report two issues that were emphasized by the GRI and AccountAbility.

Accordingly, the auditor should: "assess the decision making process of the organizations management concerning the selection and depth of information in the sustainability report (reporting materiality) and, at a minimum, verify that the report is not misleading, or could be, owing to under-emphasis or over-emphasis on certain groups of intended users, topics and/or information on these topics, and that it satisfies the legitimate information needs of the report's intended users"[502] defined as "information need of an average representation of the intended users or user groups can demand based on legislation, general accepted reporting standards, case law or agreements between the reporting organization and these intended users or user group."[503]

The draft thus distinguished reporting materiality (for the targeted group of stakeholders and the reporting organisation) and tolerance (materiality for planning and performing the assurance engagement which according to the standard was a matter of professional judgment). It also introduced guidance for testing for completeness to assess if the report addresses the legitimate information needs of 6 user groups: suppliers, employees, customers; environmental, financial and social stakeholders.

The IAASB

The IAASB had refrained from involvement in the regulatory space of assurance standards for sustainability reporting. In April 2004, IAASB steering committee formed the Sustainability Experts Advisory Panel (SEAP) in recognition of its increased popularity and lack of harmonisation and quality control in activities among assurance providers. It identified as "significant" the fact that prior to ISAE3000 being made obligatory to members: 50% of reports issued by accounting firms (who had 50% market share at that point in time) failed to cite reporting

---

[502] ED3410, Sec. 13. defined *as "information need of an average representation of the intended users or user groups can demand based on legislation, general accepted reporting standards, case law or agreements between the reporting organization and these intended users or user group."*

[503] Provisions of the Dutch standard will be analysed in the context of RevR 6 (2008) in chapter 8.

criteria and 55% of assurance reports issued by made no reference to standards in accordance with which the engagement has been performed. The IAASB attributed the problems to the absence of suitable reporting criteria (and the potential of the G3) and standard on sustainability assurance issued by a recognized standard setting body of the accountancy profession. "Anything we can do the raise the quality of assurance services"[504] was viewed to improve public interest.

The SEAP had two crucial functions: first, to channel efforts and expertise to contribute to the development of sustainability reporting standards by the GRI[505] and second, contribute to development of sustainability assurance standards by commenting on NIVRA's exposure drafts. The Panel consists of 16 members: 2 from IFAC; 5 from Big 4; 3 consultants, 3 from organisations representing the accountancy profession and 1 retired individual. It is also interesting to note that 8 members were affiliated to GRI, 5 with AccountAbility and 4 were active in the FEE. Analysis of the geographical composition, half the panel were form Europe (United Kingdom, Netherlands and Scandinavia), 3 from Asia, 2 from Canada and Oceania and only one member was from the United States.

SEAP's analysis and comments of the NIVRA exposure draft served a two-fold purpose. First, it was in line with IAASB's strategy to build on the efforts of national standard setters. In the March 2005 meeting, NIVRA's ED3410 was viewed as "*one upon which IAASB would be able to build for wider international use*" as sustainability reporting and assurance were not a legal requirement in the Netherlands and therefore the standards would not be tailored the Dutch regulatory environment.[506] Furthermore, it was "well received" and NIVRA had been involved in IAASB's work in "that area".[507] Second, the IAASB in line with its commitment during the publication of ISAE3000 to keep the IFAE and ISAE3000 under review indicated

---

[504] Audio meeting June 2005

[505] *(a) Engaging directly with the Chair of the GRI Board and the Chair of the GRI Stakeholder Council, and GRI staff;(b) Nominating members for the GRI Technical Advisory Committee (TAC), which provides the GRI Board with high-level technical guidance on the Guidelines revision process, and working groups dealing with specific issues for the revision;(c) Providing technical background material for the TAC about what is meant by "suitable criteria," and how the accounting framework could relate to sustainability reporting; (d) Responding to a GRI exposure draft about determining a reporting organization's sustainability reporting boundaries; and (e) advise on the development of guidance for assurance on sustainability reports.*

[506] IAASB March 2005 meeting, audio recording, discussion held on Thursday morning, March 10, 2005 Agenda item 7.

[507] ibid. It was raised in the meeting if comments would be given to IDW which would release its standard in May the same year. The board however clarified that also the board would be interested in the document, comments should not be an expectation. It would however encourage people in the expert panel to submit feedback to efforts of national standard setters in that respect.

that an analysis of NiVRA's ED3410 based on IAF and ISAE3000 would highlight the practical implementation issues arising from guidance provided by IFAE and ISAE3000.[508]

A task force[509] set up by the IAASB steering committee for this purpose, with the advice from the Sustainability Experts Advisory Panel (SEAP)[510] were asked to comment on the draft. The SEAP was asked to provide initial comments in less than one week, which were collated and discussed at a Task Force conference call on 1 February where detailed comments were present in June 2005.[511] The task force raised two matters to be discussed at the board meeting: the form of the feedback to be given to NIVRA and the major issues of concern identified by the task force and SEAP members.

The task force proposed feedback in the form of a comment letter from the board using precautionary language to avoid expectations that IAASB is doing more than offering comments. With deliberation, a member of the board commented:

"What struck me particularly was the number of issues and the very significant issues which the task force identified in a short time...I certainly have big concerns about the stakeholder group ...the identification of stakeholders... response needs to be fairly firm...probably the response should not be batched as a response from the board as such but perhaps from the panel ... learning as we go along...not necessarily have full authority of the board but input from the board and you will have the support of the board when sending the response"

The task force proposed in the next meeting that feedback to NIVRA in the form of a comment letter was to be signed by the Chairman of the task force / SEAP Roger Simnett and Technical manager of the IAASB Michael Nugent[512]. A board member expressed that it was "entirely satisfactory" as the signatories gave recognition that the

---

[508] IAASB March 2005 meeting, Agenda Item 7, par 7.

[509] Sustainability Task force has the task of developing formal response to ED 3410 3010 liaising with SEAP. Members of the Task Force are: • Roger Simnett, chair • Denise Esdon • Rogério Gollo • Cedric Gelard.

[510] Sustainability Experts Advisory Panel (SEAP) as set up in April 2004. It's tasks include: to give advice on the development of guidance for assurance on sustainability reports and provide expertise to the sustainability task force in providing useful feedback to enhance NIVRA's developmental process pertaining to the 2 exposure drafts, 3410 and 3010, to provide IAASB with insights as to application of ISAE 3000, to liaise with GRI to have G3 as generally accepted reporting criteria by accounting community.

[511] The ED was available in English 25/1/2005, initial comments were submitted 31/1/2005 for the meeting 1/2/2005 documented in par 8. IAASB March 2005 meeting, agenda item7.

[512] And staff instead of Michael Nugent (minutes) inconsistency with audio recordings

draft had been gone through thoroughly by "Roger and his group" and the board had not gone through the drafts in as great detail as one of their own drafts.[513] It was worded in the comment letter to NIVRA that "while members of IAASB have reviewed these comments they did not necessarily represent the views of IAASB".[514] IAASB also encourage other national auditing and assurance standard setters and other interested parties to respond to the draft. Among the respondents was the FEE which stressed the importance of a more principles based approach to standard setting.[515] The structure of the feedback to NIVRA clearly indicated IAASB's interest in the area but also intentions to exclude itself from the standard setting role in the regulatory space.

The second matter raised by the task force within the week were the major issues pertaining to NIVRA's ED: the scope of guidance, suitability of criteria, scope of engagement, stakeholder engagement, materiality and completeness, joint responsibility, reasonable and limited assurance, reporting and other matter such as quality, drafting style and fraud. The details were not discussed at the meeting (refer to the left two columns of the table 7). The IAASB board and technical members were encouraged to submit comments by 26/3[516].

The task force met in April, took into account issues identified by SEAP and feedback letters they received and prepared the comment letter paying attention to the EDs consistency with IFAE and ISAE3000.[517] The task force noted that the issues of criteria, engagement acceptance and responsibilities of practitioners were among the issues in the feedback received and were incorporated in the response to NIVRA.

In both meetings, details of the major issues raised by the task force were not discussed. However, meeting attendees were given the opportunity to raise specific issues. In the March meeting, issues surrounding transparency of the criteria; supplementary criteria and identification of stakeholders were brought up as areas of concern. In the response to NIVRA, the task force supported the fact that suitable criteria are context sensitive and did not support having one sole criterion. It emphasized that the evaluation of the suitability of criteria should take into consideration the purpose of the engagement rather than for the purpose of the report and to perform assurance engagement. It opined that criteria which reports are prepared should be complete and made available to users of the report. A

---

513 Prior to this comment, Roger Simnett commented "I am getting mini nods" suggesting the general consensus of this approach by members attending the meeting

514 IAASB main agenda 2005, Agenda item 7 and Item 3A, par. 3, p. 1

515 Far contributed to the comment letter through its participation in FEE.

516 Draft minutes of 14th Meeting of the IAASB March 7-11 Agenda Item 1-A, par. 9 p. 14-15.

517 Draft minutes of 15th Meeting of the IAASB June 13-17 Agenda Item 1-A par. 9 p. 17.

supplementary criterion is therefore not consistent with IFAE and ISAE w.r.t role of criteria, assurance standards and assurance practitioners.

The role of stakeholder engagement was discussed in relation to the completeness assertion and implications on level of assurance. There was consensus that completeness was viewed to be broader than a cost benefits considerations. Concerns were raised as to whether completeness made it impractical to conclude reasonable assurance. It was noted in the meeting that completeness consists two components: identifying relevant information which stakeholder engagement play a role and full disclosure in those areas. The latter was identified as a significant area of concern.[518]

The task force prepared extensive comments to NIVRA in the area of engagement Acceptance, practitioners' responsibility, completeness of the standard, editorial and other matters, drafting style structure and terminology in addition to the points raised in the March meeting which influenced the revision of ED3410. Among key issues are inconsistencies detected with regard to evidence gathering procedures, E.g. review engagements according to ED3410 does not have to test effectiveness of internal controls; the drafting style; the terms audit, review, materiality and relevance and the possibility to issue a qualified conclusion accompanied by negative assurance were not consistent with IFAE. The points are summarised the table presented in appendix 4.

The IAASB through this initiative constructed NIVRA into the regulatory space, giving legitimacy NIVRA's sustainability standards, this initiative also essentially side-line the efforts of other national standard setters such as IDW which was issuing an exposure draft according to ISAE3000 a few months later. The board confirmed the marginal role IAASB intended to play[519] with claims that this type of engagement fell under the scope of ISAE3000 and given the IAASB workload, there was no guarantee that it would initiate assurance standards specific to sustainability reporting.[520] NIVRA's ED3410 and the comments made by the task force raises discussions on how the approach adopted by AccountAbility and GRI could be used within the framework of IAASB pronouncements.

Its intention to contribute to the development of GRI and comment on NIVRA's exposure drafts set boundaries and the direction which the accountancy profession was to take. An interesting observation was that the task force recommended the sustainability report to be defined along the lines of *"comprehensive information prepared by an entity about its economic, environmental and social performance. This information may be presented in a stand-alone report, or included as a separately identifiable part of a broader collection of information, e.g., in the entity's annual report,*

---

[518] Ibid.

[519] Ibid.

[520] Ibid., p. 18.

or in a suitable form on its website".[521] The task force drew on the work of GRI in at least three separate points. First, it proposed NIVRA adopt the term "performance" in its definition of sustainability report instead of "aspects of the organization" and the" effects of its business activities on society" thereby aligning its terminology with GRI. Second, the definition made clear its applicability with regard to the source (where sustainability information is reported) and form of reporting and favored the "incremental approach" which GRI promotes. Third, in defining the scope of the report the task force made reference to the "boundary protocols" for guidance on supply chains issues.

KPMG Sustainability (Netherlands)

From 2003, two international standards provide guidance to assurance providers from the accountancy profession to perform assurance on sustainability reports: the ISAE3000 and AA1000AS. The joint application of standards gained popularity which led to concerns if the standards were technically complementary or conflicting due to the different philosophical underpinnings, ethical standards providers were subject to, the objective and approach to assurance.

This led AccountAbility and KPMG Sustainability Netherlands to issue a report in April 2005 endorsing that: "based on comparison of the values, approach, methodology, conclusion and communication of the two standards, combined use of the two standards would likely deliver enhanced results". This conclusion was supported by the claimed facts that "the two standards are technically complementary, there is no methodological conflict, both approaches draw from the financial audit model and the combination will deliver a win-win situation as different objectives lead to different emphasis."[522] . Accordingly, each standard was settled to have a function that complemented each other. ISAE3000 provided guidance on the assurance process whilst AA1000AS aligned the process to take into account stakeholders' material interest. The authors gave an example of how the 2 standards could be applied presented in the diagram in the below:

---

[521] IAASB meeting June 2005 Agenda item 3B.

[522] Iansen-Rogers & Oelschlagel, 2005.

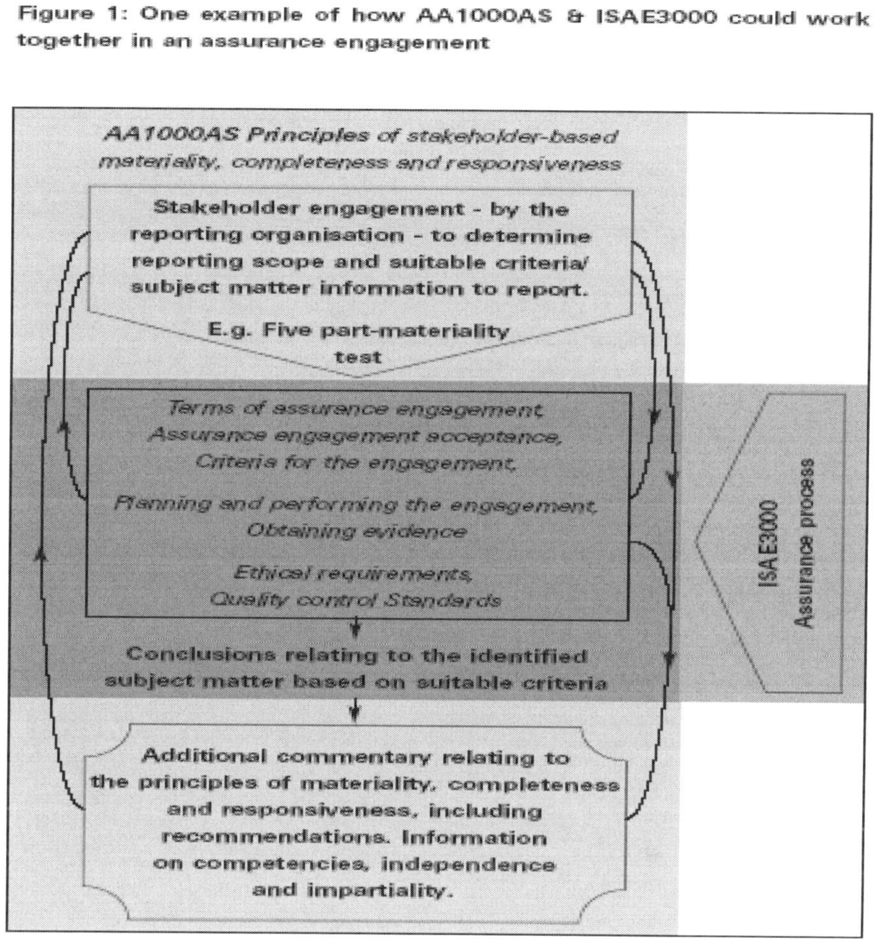

Figure 5.9 Sustainability Assurance Model[523]

The accountancy profession did not take a common stand with regard to how AccountAbility or its standard should be viewed. As assurance providers increasingly applied the standards jointly, FEE voiced concerns on the risk of inconsistent application of AA1000AS in the absence of a code of ethics, quality control standards and Guidance notes on AA1000AS[524] and stated that a common position has not been agreed.[525] National standard setting organisations revealed they had "no

[523] Ibid.

[524] FEE (2004)

[525] Informant interview with FEE.

comments" or stated that "we only recognise our global assurance standard i.e. ISAE 3000 but qualifies that having said that, we have no problems with our members using our standard with AA1000AS so long as it is not in conflict with our standard (ISAE 3000)."[526]

5.3.2.4 Other related activities

The European Union continued to building on the reporting requirements of sustainability matters in the annual reports of companies. The Fourth[527] and Seventh[528] Company Law Directives based on Article 50 par 2(g) of the Treaty of the functioning of the European Union (TFEU) form the arsenal of Community legal acts governing company accounts, coordinating Member States' provisions on the presentation and content of annual and consolidated accounts, annual reports, the valuation methods used and their publication applying to all companies with limited liability. The provision specifies the minimum equivalent legal requirements pertaining to the extent of publicly available financial information giving a "true and fair view" of the company's position and auditing requirements to providing safeguards to investors and third parties. These directives were amended by Directive 2003/51EC[529] to be implemented by member states by 1/1/2005[530]. Provisions were made stating the requirement for companies to disclose *"to the extent necessary for an understanding of the company's development, performance or position, the analysis shall include both financial and, where appropriate, non-financial key performance indicators relevant to the particular business, including information relating to environmental and employee matter*[531] to provide a fair review of the development of the company's business and of its position in the company's annual report.[532] Member states of the

526 Interview with Experts.

527 Fourth Council Directive 78/660/EEC of 25 July 1978 on the annual accounts of certain types of companies.

528 Seventh Council Directive 83/349/EEC of 13 June 1983 on consolidated accounts.

529 Directive 2003/51EC of the European Parliament and of the Council of 18 June 2003 amending Directives 78/660/EEC, 83/349/EEC, 86/635/EEC and 91/674/EEC on the annual and consolidated accounts of certain types of companies, banks and other financial institutions and insurance undertakings.

530 Ibid. par 1 The amendment is the result of efforts by the Lisbon European Council (March 2000) to accelerate the completion of the internal market for financial services urging steps to be taken to enhance comparability of financial statements which called for further initiatives both in financial reporting and auditing.

531 Ibid, par 14b.

European Union through transposition of the directive could influence sustainability reporting practices which had a potential effect on the development of assurance and assurance standards.

5.3.2.5 Summary of phase 2 (2003-2005)

From 2003-2005, sustainability reporting and adoption of the GRI sustainability reporting guidelines continue to increase rapidly.[533] For both G250 and N100[534] firms sustainability reporting increase by 15% and 43% respectively.[535] The extension of the EU reporting requirements to include non-financial issues, led to more companies integrating sustainability information into their annual reports.

In this phase, stakes in the regulatory space were claimed. Several standards were published to provide guidance on sustainability assurance. At the international level, the ISAE3000 provided the accountancy profession generic guidance of assurance services and the AA1000AS provided an open source standard applicable to all providers on sustainability assurance. At a national level in Europe, NIVRA, IDW, Far and CNCC published pronouncements to guide their members on providing assurance on sustainability reports. Both drafts from NIVRA and IDW were translated to English and exposed to the public for comments. The IAASB who refrained from involvement, took its first step by setting up a task force and appointing the SEAP to comment on NIVRA's exposure draft. The logic of its involvement could be tied to its objective to mitigate diversity of practice to enhance the quality of audit and assurance services performed by its members. It was also in

---

532 It is also noted that Council Directive 2004/101/EC of 27 October 2004, amending the Directive 2003/87/EC establishing a scheme for greenhouse gas emission allowance trading within the Community, in respect of the Kyoto Protocol's project mechanism though carries no direct provisions on financial reporting on emissions trading, it nevertheless impacts the financial reporting and audit framework. Issues arise from legal requirements and need to determine fair value, account for emissions allowances, government grants, penalties and the liability to surrender that arises as emissions are made. (FEE 2005 FEE Alert: Emissions trading). Also related to this issue, the prospectus directive requires environmental risk to be disclosed.

533 The trend has been reflected in both academic studies e.g., Kolk 2003, 2010 and surveys carried out by consulting companies, KPMG and ESRA reports.

534 N100 refers to the 100 largest companies by revenue in 22 countries.

535 KPMG survey 2008 Reporting trends however tend to be country (with Japan and United Kingdom taking the lead in reporting) and sector specific (with forestry, electronic, utilities oil and gas, chemicals industries being more likely to publish reports).

line with its strategy to build on the work of national standard setters, in which case NIVRA's ED3410 was then identified as being a potential starting point when it engages in setting standards in this specific subject matter. The FEE as a non-standard setter continued its influence through the sustainability subgroup's comment letters on the drafts issued by AccountAbility, Royal NiVRA (ED3410 and ED3010) and GRI (G3 guidelines) as well as its representations e.g., in IFAC's SEAP.

In the regulatory space, there is evidence of both competition and cooperation by actors. As no party own professional jurisdiction of the market for assurance services, debates centre on the definition of terms and the weight of regulatory issues. At the same time, there was evidence of learning as members of actors within the regulatory space are cross represented in committees. Actors gain legitimacy stressing the "complementary" features of standards. For example, AccountAbility claimed that AA1000AS complemented the GRI, both standards placed heavy weightage on stakeholder interest. Key regulatory issues pertaining to the competence and ethics of assurance providers, objectives, planning, performance, level and communication of assurance continue to be debated on. The weight of the issues however shifted heavily towards debates on stakeholder materiality in the assurance process. The cooperation of AccountAbility and KPMG Sustainability further lifted materiality on to a "meta-level" suggesting that stakeholder materiality had influence on the assurance process i.e. AA1000AS and the five part materiality test could be used to determine subject matter, objective, criteria, planning, performance and communication of assurance engagements. The introduction of AA1000AS thus shifted the focus of the debate (gave greater prominence to) towards the issues of materiality and responsiveness to stakeholders and completeness or balance of the reports.

The accountancy profession did not take a common stance on AA1000AS. The approach taken by NIVRA in ED3410 tested the interpretation boundaries of ISAE3000 with regard to reporting materiality, test for completeness and responsibilities of assurance providers which lead to comments by the SEAP and FEE. The above also showed evidence of competition and collaborative efforts by standard setters in particular the joint adoption of accountancy based standards with AA1000AS.

5.3.3 Phase 3 (2006-2010): Reflections and defending the regulatory space

Activities in the regulatory space were intensified in period of 2006-2010 that set a stage for significantly developments in sustainability assurance.

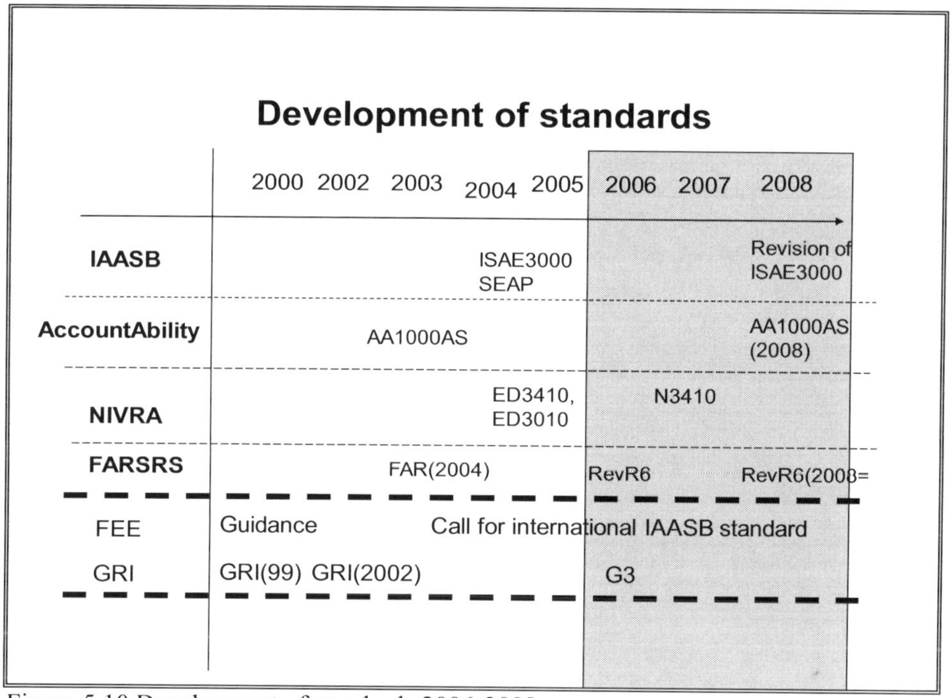

Figure 5.10 Development of standards 2006-2008

5.3.3.1 Intensified efforts by AccountAbility

AccountAbility intensified efforts in the regulatory space through the publication of further guidelines[536] on application of the principles of materiality, completeness and responsiveness in 2006 and on how the different approaches to assurance can be combined in 2007. The latter report focused on the application of AA1000AS within the context of GRI's reporting standard G3 and ISAE 3000. It also provided a sample of an assurance statement that met both AA1000AS and ISAE3000 standards.

[536] E.g., AA1000 Guidance Note on the Principles of Materiality, Completeness and Responsiveness as they Relate to the AA1000 Assurance Standard 2006.

AccountAbility continued building legitimacy as an "international standard setter" through publications on key issues within the regulatory space,[537] collaboration with reputable academic institutions in researching issues pertaining to accountability and publication of the Accountability rating in the well-regarded business magazine Fortune.

AccountAbility embarked on a project to revise AA1000AS in 2007. It launched Wiki standard setting where interested parties could participate, raise issues and voice concerns on current assurance practices through the web. In addition, it held worldwide consultations in about 20 countries to obtain feedback on concerns which were then published on the website. Although the Technical Committee[538] was ultimately responsible for the text construction of the revised standard[539], AccountAbility invited interested parties to comment during the three periods for public participation[540]. Concurrently, efforts were made to align AA1000AS to ISAE3000.

The revision of AA1000AS resulted in the drafting of the AccountAbility Principles Standard (AA1000APS) and the Assurance Standard (AA1000AS(2008)). The former defined and established the criteria for the AccountAbility Principles (previously the principles of materiality, completeness and responsiveness) that were essentially reporting principles[541]. The latter provided the requirements for conducting the assurance engagement. Guidance on the use of AA1000AS(2008) was drafted to provide guidance to assurance providers, reporting organisations and users of assurance statements.

The revision introduced three key changes. First, completeness was replaced by inclusiveness as one of the AccountAbility principles. Second, AA1000AS (2008) distinguished between two types of engagements, Type 1 required the evaluation of publicly disclosed information, systems and processes of the organisation to ensure adherence to the three AccountAbility principles and Type 2, required the additional evaluation of the reliability of specified performance information against a reporting

---

537 For example, the stakeholder manual and reports such as "what assures consumers", "What assures: listening to the words of assurance" etc.

538 The 2008 technical committee made up of the following representatives: 2 from Big 4, 1 from accounting organization, I from GRI, 2 from other civil society organisations, 2 consultants, 1 academic, 6 form companies

539 The Consultation issues register include: What is assurance, sustainability assurance? For who? To whom is the duty of care, role in relation to financial audits and other certifications, relation to other standards and guidelines, barriers to commissioning an sustainability assurance

540 Jan24 – April 4, 2008; April 30 – June 13th 2008, July 4th – Sept 19th 2009

541 Interview with experts

framework. Third, the introduction of two levels of assurance: high and moderate which could be used in combination. High level of assurance referred to where "sufficient evidence has been obtained to support their statement such that the risk of their conclusion being in error was very low but not zero" whereas moderate referred to "not reduced to very low but not zero". The evidence required to make conclusions were specified in the standard. For example, high level of assurance required extensive depth of evidence gathered including corroborative evidence and sufficient sampling at lower levels gathered from internal and external sources from all levels of the organisation. In addition, the AA1000AS specified in greater detail the perquisites of accepting an engagement (e.g., appropriate subject matter and criteria); requirements imposed on the assurance provider (e.g., clearer ethical and competence requirements, due care); assurance process (e.g., planning requirements, evaluation processes adherence for Type 1 & 2 engagements); output requirements (e.g. content of assurance report) and other requirements (e.g., , terms of engagement, engagement contract, use of standard).

AccountAbility introduced a licencing fee to assurance providers for each commercial use of AA1000AS (2008). A fee of £500 (otherwise £200) is charged if the turnover of the reporting organisation is greater than £2 million.

5.3.3.2 FEE's Response in the regulatory space

The FEE focused its efforts in two areas: call to IAASB to develop an international standard on sustainability assurance and respond to AccountAbility's publications.

FEE's call for an international assurance standard

FEE persisted on its efforts to call IAASB for a specific standard on sustainability assurance, responding to IAASB's intention of issuing merely short practice statement to guide sustainability assurance. FEE reiterated that ISAE 3000 was "too general and high level for practical use"[542] and highlighted that it did not provide guidance on suitable criteria, particular to engagements of this nature as criteria could be determined through stakeholder dialogue observations which could only be assessed in the midst of the engagement. [543]It then made persuasive arguments in line with IAASB's objectives of protecting public interest and the interest of the accountancy profession. It asserted the benefits of a sustainability assurance standard and made

[542] FEE comment letter to IAASB and SEAP on "Consultation paper on assurance aspects of G3 –The GRI 2006 Draft Sustainability Reporting Guidelines" May 2006, par. 4, p. 2.

[543] ibid., p. 3-4.

links between standards, quality of assurance services and the credibility and quality of sustainability reporting in this voluntary field.[544] It also stressed that "competing standards are finding their way to the market"[545], AccountAbility was regarded as a leader in this field and users needed to distinguish between the type of provider and assurance provided. The FEE called for "IAASB to allocate resources as soon as they become available to fulfil its professional leadership role in this area by promulgating quality practices and sharing knowledge in developing a specific assurance standard in relation to sustainability".[546]

The call for standard was supported with a discussion paper published in June 2006 on "Key Issues in Sustainability Assurance - An overview" highlighting similarities and differences of country specific standards of four European countries (Sweden, Netherlands, Germany and France). The discussion paper also provided constructive comments the national standards from FEE's viewpoint which could potentially contribute to the work of SEAP and IAASB standard setting activities in the future.[547]

Following the discussion paper, a round table was held in November the same year consisting members from FEE, IAASB, GRI, AccountAbility, a representative from European Parliament and leading corporations. The following conclusions were reached. First, there is demand for assurance from the investors' community and companies to raise the credibility of the sustainability reporting which completeness of the report and concentration on the relevant information were needed. Second, the assurance report was identified as a key element to an effective approach, the problem was that reports issued by the accounting profession at that time were not always clear to the readers. There was a need to work on the content of the report, what is meant, explain the work the accountants do and what the report can bring to the stakeholders. It was agreed that different levels of assurance for different elements of the sustainability report could be considered and assurance work should be carried out with multidisciplinary teams comprising experts on sustainability, specific sectors and accountants. Third, it was agreed that clear and recognized criteria of sustainability reporting need to be in place on which to provide assurance and the GRI G3 guidelines could serve as a reference. The accounting profession was to assess G3 as suitable criteria. Fourth, it was agreed that a global framework on sustainability assurance should be at principle level and different standards may be developed with the aim of converging over time.

---

544 Ibid., par 12, p. 3

545 Ibid. par. 13, p .3.

546 Ibid.

547 FEE (2006)

The Round Table gave a particular stakeholder, the investor community special mention though no further elaboration was made, having a "dominant" stakeholder had implications on how relevant issues would be defined. It also reiterated the importance of the regulatory issues it introduced into the regulatory space including criteria, levels and communication of assurance and reaffirmed the accountancy approach through the use of multidisciplinary teams. This initiative reaffirmed regulatory issues (in phase one) in a new context taking into consideration developments and involving new actors in the regulatory space.

FEE expressed the professions disappointment in a comment letter to the member of European Parliament who participated at the Round Table on the EC communication "Implementing the partnership for growth and jobs: making Europe a Pole of Excellence on CSR" 2006. The comment letter stated that the document "*unfortunately did not call for global standards in relation to CSR reporting and assurance*" but only stressed that reporting transparency of CSR information would better serve stakeholders' understanding on organizations non-financial performance.[548]

To facilitate activities and undertake a more prominent role in the sustainability field in 2007, the work group was restructured into 3 groups: Sustainability Advisory Panel consisting of external high level individuals e.g., CEOs to advise the FEE's work in the sustainability area at a strategic level; the Sustainability Policy Group, to address political, policy and strategic issues and set direction for the group and the Sustainability Working Party to undertake technical work such as preparation of discussion papers, comment letters etc.[549]

Responding to AccountAbility

The FEE had taken a collaborative position in 2004 recommending IAASB not to compete but work with AccountAbility in standard setting, it also participated in technical committee drafting the AA1000 standards. In July 2007, the FEE took a different stand. Following the AccountAbility's publication of guidance notes on combining ISAE3000 and AA1000AS in 2007, FEE responded by questioning the nature of assurance engagements performed according to AA1000AS, and the compatibility of using both standards. It commented that the guidance notes included many characteristics which resembled 'agreed upon procedures' performed

[548] FEE Comment Letter to Mr. Richard Howitt, MEP Assurance on Sustainability Reports 6/12/2006.

[549] Informant interview with FEE.

by the accountancy profession under IAASB's International Standards on Related Services 4400 and therefore did not view the two standards as complementary.[550]

FEE also commented on revision draft of AA1000AS published on 24 January and 17 July 2008 but made clear that comments did not constitute FEE's endorsement or any form of support for AccountAbility standard.[551] FEE's comments centred on: the different meanings of common terminology between AA1000AS and ISAE3000; the lack of distinction between reporting and assurance principles in standards and areas that were neglected or lacked clarity.

FEE commented on term "materiality" which had been commonly used in both IAASB standards and AA1000AS but had different meanings. It commented that AccountAbility did not distinguish between reporting and assurance materiality. FEE pointed out that AccountAbilty's definition of materiality involved identifying issues relevant to performance and evaluating how important the issues are; relevance was therefore a subset of materiality.[552] In comparison, materiality in financial audits could be viewed as threshold, tolerance or "tolerable rate of deviation or misstatement"[553]which put materiality as a subset of relevance. This point of view was not shared by Accountability. Based on its definition (reporting materiality) it regarded definitions as compatible. The difference AccountAbility claimed rested on the fact that ISAE3000 defined materiality in terms of the intended user whereas AA1000As referred to stakeholders which required materiality to be defined from broader perspective.[554]

FEE commented on the lack of distinction between reporting and assurance principles in standards AA1000AS. Assurance principles had been equated with reporting principles and the differentiation was needed to determine the suitability of criteria. It also commented on the lack of clarity as to whether subject matter referred to the content of sustainability report or process to compile the report; whether the standard catered to full report or parts that were subject to assurance. It expressed concerns with regard to leaving the reporting organisation to determine their own performance criteria in Type 2 assurance engagements which did not meet criteria stated in ISAE3000. It held the view that the requirement imposed on oversight by

---

550 FEE Comment Letter to AccountAbility on Consultation on Guidance Note on Assurance Levels & Statements as they relate to AA1000 Assurance Standard 26/7/2007.

551 FEE Consultation on AA1000AS revision draft 24/1/2007 par. 2.

552 FEE Comment letter to AccountAbility, par. 9-14, 22/9/2008.

553 ED ISA 530 Audit Sampling. Tolerance in NIVRA ED3410, FEE Comment Letter to Mr. Alan Knight, Head of Standards and Related Services, AccountAbility Comment Letter on Consultation on AA1000 Assurance Standard Revision Draft issued on 24/1/2008, par 11-13.

554 AccountAbility (2010) Frequently asked questions www.accountability.org.

the assurance provider was insufficient in the revision draft, and there was a need for a system of quality control and minimum requirements to be set out.

The FEE further requested that the FEE Discussion paper:" Key issues in sustainability assurance – an overview" not to be referred to in AccountAbility's publication as the discussion paper did not constitute guidance and the proposed guidance had been "taken out of context further the text had been paraphrased". [555]

To improve the compatibility with ISAE3000 FEE suggested AccountAbility require the level of assurance to be decided upfront rather than allow the scope of an engagement to define an anticipated level of assurance adjusted during the engagement which tend to result in clean reports. It also suggested AccountAbility introduce common standard headings in assurance reports and recommend different forms of conclusion leaving the reporting organisation to report on implementation of previous recommendations. The revision draft introduced two levels of assurance, high and moderate. FEE opined that the chosen terminology were "outdated" and brought back the debate that the term "high" denoted achieving a level higher than the financial audit which was misleading. It recommended the terms reasonable and limited in line with IFAE and ISAE3000.

5.3.3.3 Actions by national standard setters

Standards published by NIVRA and Far underwent revisions in this phase. In the Netherlands, NiVRA from the comment letters received during the exposure period redrafted the ED3410 and adopted a more "principles based" "clarity style" of drafting. The standard N3410 "Assurance Engagements Relating to Sustainability Reports" was made effective July 2007. Far revised RevR6(2004) and aligned the new RevR6 "Assurance on Sustainability Reporting" standard with ISAE3000 in December 2006 due to its membership obligations. In December 2008, it made a second revision to RevR6 and replaced its contents with the Dutch standard. (RevR6 will be analysed in greater detail in chapters 7)

---

[555] FEE Comment Letter to Mr. Alan Knight, Head of Standards and Related Services, AccountAbility Comment Letter on Consultation on AA1000 Assurance Standard Revision Draft issued on 24/1/2008, par. 12.

5.3.3.4 Other related developments in the regulatory space

GRI published the G3 reporting guidelines

In 2006, the third version of GRI reporting guidelines was published following a 3 year consultation process[556]. It introduced a reporting framework containing Sustainability Reporting Guidelines, Protocols (definitions for key terms in the indicator, compilation methodologies scope and other technical references) and Sector Supplements (sector specific guidelines) that guide reporting organisations on what and how to report. Adopters of the standards increased to over 750 organisations.

The G3 provided stakeholder oriented principles and guidance to reporting organisations. Guidance determining the content of the report was built on materiality, stakeholder inclusiveness, the sustainability context and completeness. The GRI defined quality of the sustainability report as a function of balance, comparability, accuracy, timeliness, reliability, and clarity. Provisions were also made on how to set reporting boundaries and standard disclosures including strategy and profile, management approach and performance indicators. The GRI developed a grid which reporters could declare to what extent the GRI framework and been applied (A, B or C) and if external assurance had been commissioned (+). For example, reporting organisation fulfilling requirements of level B could "self-declare" its compliance to the GRI framework, commission a GRI application level check (the extent which GRI reporting framework has been used) and / or third party assurance.

The G3 took into consideration the "auditability of the report" and encouraged both internal and external assurance to enhance report credibility. It defined external assurance as *"activities designed to result in published conclusions on the quality of the report and the information contained within"*[557] as distinguished from certifications of sustainability performance. It however did not express preference of a particular assurance provider recognizing professional assurance providers, stakeholder panels, other external groups or individuals on equal footing. It withdrew earlier working paper on overarching principles of assurance and instead published key qualities of

---

[556] The G3 draft was released for public comment in the beginning of 2006, 60 Stakeholder Council members were appointed and the Organizational Stakeholder membership program was launched in 2003. The structured Feedback Process concludes having engaged 450 experts worldwide in 2004 and a Technical Advisory Committee (TAC) appointed in 2005 guiding major technical revisions and development of a third version of the guidelines. 270 responses were received. It was also previewed in 28 cities worldwide to over 3000 participants. The comments were considered by the GRI board, technical advisory committee and stakeholder council and integrated in the final version of the G3.

[557] GRI *(2006), G3 p. 38*

external assurance. The list reflected a combined strength of both the accountancy approach and that of AccountAbility.

key qualities of external assurance	competence of assurance providers in subject matter and assurance practices
	systematic, documented, evidence-based, characterized by defined procedures;
	assessment of whether the report provides a reasonable and balanced presentation of performance, taking into consideration the veracity of data in a report as well as the overall selection of content;
	independence and impartiality of assurance providers to the reporting organisation;
	extent GRI reporting framework had been used as a criteria in reaching conc usion;
	conc usion are published and a statement from the assurance provider on their relationship to the report preparer are made available

Figure 5.11 Key qualities of external assurance

IAASB's consultation paper on G3

The development in GRI reporting guidelines had a significant impact on assurance standards. The key issue was whether G3 constituted a "suitable criteria" reflected in IAASBs consultation paper: Assurance Aspects of G3 published February 2006 The purpose of the consultation paper was to stimulate interest in G3, encourage members to respond to G3, assess if G3 would be acceptable as a suitable criteria. The publication of G3 lifted the discussion on whether the long standing problem of "lack of suitable criteria" could be addressed with the G3. IFAC addressed in the consultation paper a discussion of assurance, stakeholder engagement, relevance and materiality. In particular, it evaluated G3 against "suitable criteria" according to IFAE to assess if G3 fulfilled the requirements of relevance, completeness, reliability, neutrality and understandability.[558] The table below provides a summary of IAASB's evaluation.

Characteristics of suitable criteria	International Framework for Assurance	Evaluation of G3
Relevance	Criteria contribute to conclusions that assist decision-making by the intended users	Multi-stakeholder process: Is it sufficient to ensure relevance? Are there circumstances where guidelines are irrelevant?
Completeness	When relevant factors that could affect the conclusions in the context of the engagement circumstances are not omitted	Are disclosure requirements complete? How will stakeholder engagement affect completeness Immaterial core indicators may be omitted Do such omissions undermine completeness?
Reliability	Allow reasonably consistent evaluation or measurement of the subject matter	G3 allow a number of choices on e.g. reporting boundary, restating current disclosures, deciding on level of aggregation in the report Will such choices undermine reliability?
Neutrality	Contribute to conclusions that are free from bias	Do components of G3 limit extent which preparers can introduce undue bias into report preparation that the professional accountant could still conclude in accordance with G3?
Understandability	Clear, comprehensive, and not subject to significantly different Interpretations.	Is each component understandable and drafted clearly?

Figure 5.12 IAASB consultation paper on the G3

[558] IFAE sec. 36

With reference to the consultation paper, the FEE reaffirmed its position that criteria is dependent on specific needs of reporting company and intended users and therefore different sets of criteria would be appropriate for different types of engagements rather than a "one size fits all reporting criteria". It also stressed the importance of consistent evaluation in assurance engagements.[559] Although a consultation paper "Assurance aspects of G3" was published in February 2006 for comments by March, according to FEE, SEAP informed that IFAC would not pursue an IFAC/SEAP response to GRI.

IFAC's projects

IFAC's priorities in fulfilling its role in the "new financial architecture" channelled significant resources to the Clarity project quoted as: "our highest standard setting priority"[560]. The clarity project involved the redrafting of existing IAASB standards to improve auditors' understanding of the pronouncements, through greater clarity enabling more consistent application to enhance the quality and uniformity of practice worldwide.[561] IFAC carried out extensive consultation with its stakeholders on whether resources were to be allocated to new areas of assurance or consolidating and improving current existing standards.[562]

In September 2007, IAASB stated its intention in 2009-2011 draft strategic plan to contribute to the development in sustainability reporting through the world capital market perspective. The IAASB justified its involvement stating *that* "international investors expect greater focus on sustainability from participants in the world capital markets, annual reports include sustainability information and often auditors are involved in assurance engagements on corporate sustainability reporting. IAASB was called on to consider how to contribute to the development of such reporting and plans to do so." [563]

---

[559] Par 14-17 FEE correspondence with IFAC ref IAASB consultation paper on assurance aspects of G3 23*5/2006

[560] IAASB Annual report 2006, p. 2

[561] Improvements made through: •Identifying the auditor's overall objectives when conducting an audit in accordance with ISAs; •Setting an objective in each ISA and establishing the auditor's obligation in relation to that objective; •Clarifying the obligations imposed on auditors by the requirements of the ISAs and the language used to communicate such requirements; •Eliminating any possible ambiguity about the requirements an auditor needs to fulfil; and •Improving the overall readability and understandability of the ISAs through structural and drafting improvements. (IAASB)

[562] Reflected in IAASB Annual report 2006, p. 6.

[563] Draft consultation paper IAASB strategy 2009-2011 p. 11 Section A.3.2 p. 7 (Agenda item 1.1 IAASB (March 2008).

It proposed revision of ISAE3000 as well as reviewing NiVRA's 3410N to determine whether it provided an appropriate basis for an international pronouncement in the third quarter of 2009 among the projects to contribute to developments in sustainability reporting. The project timetable was presented as follows:

Action	ISAE3000	Review N3410
Project proposal	Q1 2009	Q3 2009
Review Issues	Q3 2009	Q2 2010
ED	Q1 2010	Q4 2010
ED Comments	Q4 2010	Q3 2011
Approve final ED	Q2 2011	-

Figure 5.13 IAASB's strategic plan

52% of respondents commented on IAASB's involvement in sustainability assurance standards and 18% commented on using NIVRA standards as a point of departure.[564] Respondents that supported IAASB's involvement[565] felt the need to consolidate the leading role of the accountancy profession and gain market share[566], improve quality[567], increase harmonisation[568] and supported the public interest argument[569].

564

	SA	NIVRA
Member bodies	10/21*	3/21
Firms	4/6	1/6
Others	5/6	4/6
Regulators	1/4	0/4
Government organisation	3/7	0/7
	23	8

*10 of 21 member bodies commented on the issue

565 CNR, KPMG, ACCA, Far, Australian Treasury

566 Institut des Reviseurs d'Entreprise, ICA Scotland, EY, Canadian public accountability board, European association of listed companies, FEE, PWC, Florida institute of CPA

567 Comment letter from PWC.

568 Comment letters from European Association of listed companies, Deloitte, EY. ICAEW

569 Comment letters from ICA Australia, EY, European Association of listed companies

There were also respondents that viewed it as a second priority (first to be given to the financial audit)[570] which need a separate strategic initiative[571]. Three respondents were in favour of IAASB reviewing N3410 as an initial point of departure reasons were to avoid duplication of efforts[572] and respond the threat of competition[573]. Other members caution that it could be early to set standards in this field and that could impede development[574]; IAASB should adopt a wider view taking into account other approaches by other jurisdictions[575] as well as outside the profession[576]; and one respondent felt that there was no clear user demand of the service and thus did not meet the criteria to develop an international standard.[577]

5.3.3.5 Summary of phase 3 (2006-2010)

At phase three, GRI reporting standards continue to gain international acceptance at an increasing rate with 80% and 45% of the G250 and N100 firms adopting GRI standards in sustainability reporting.[578] At the international level, standard setting efforts continue to intensify. IAASB announced indication to revise ISAE3000 as well as look into NIVRA's N3410 as a starting point to develop standard to guide assurance on sustainability reports. AccountAbility continued to develop its standards constructing input and throughput legitimacy through wiki providing transparency and access to the revision of the AA1000AS. National standard setters engage in revision of their exposure drafts or standards in light of the increasing acceptance of GRI standards and activities in the regulatory space.

At phase 3, activities centre less on cooperation and learning. From the observations above, actors at the phase refocus on their own goals and engage in defending and protecting their own terminology through publications and comment letters. Emphasis is stressed on conflicting elements, stressing regulatory issues that are significant to the objectives of the respective institutions.

---

[570] Comment letters from IDW, International Organization of securities commissions (Spain)

[571] Comment letter from Australian government auditing and assurance standards board

[572] Comment letters from FEE and Institut des Reviseurs d'Entreprise

[573]Comment letters from FEE, Far

[574] Comment letter from PWC

[575] Comment letters from ACCA, PWC

[576] Comment letter from Canadian Public accountability board

[577] Comment letter from APB

[578] Comment letter from KPMG 2008

5.3.4 Patterns of influence in the regulatory space

Tracing activities over the three phases show patterns of actions engaged by institutions within the regulatory space. The actions by institutions and emphasis of activities in the regulatory space summarised at the end of each phase is presented in the following table.

Phases	Action by institutions	Emphasis of activities
Phase 1 1997-2002 Attempts to establish the regulatory issues and the regulatory space	Problem construction Identification of issues Establish legitimacy through building links with the established financial audit Invitation to space: who/ what are in and who/what are out	Seek out issues agreed upon and issues that remain to be debated
Phase 2 2003-2005 Competition, learning and Cooperation	Codification of standards Exploratory "what other models can offer?" Learning Establish legitimacy through cooperation	Broad based standards Identify similar terminology Cooperation, exploratory approach Make reference to strength of other standards…stress complimentary elements
Phase 3 2006-2010 Competition, reflection and consolidation	Importance of own institutions objectives (reason for their existence) Protection of terminology "Us vs. them" approach Greatly reduced cooperation More detailed codification in revised standards Learnt from and taken what's best from other standards	Highlighting weaknesses of competing standards Force other standard setters to address issues that are "your strengths" Stress conflicting elements More defined approach "take and make it your own" Harder for joint application of standards

Figure 5.14 Summary of the three phases

In the first phase, the FEE raised the key regulatory issues built on extending auditing concepts to a broader sustainability domain. It also built on the work of IAPC applying the generic assurance concept into the sustainability domain. Regulatory issues raised pertain to the assurance provider, objectives of sustainability assurance, the planning, level and commination of assurance. Unlike the financial audit that is mandated by law to be carried out by qualified auditors, sustainability assurance remains voluntary and is carried out by assurance providers, other than members of the accountancy profession. FEE brought the business case into the regulatory space and invited other actors including reporting organisations, users of reports, other assurance providers and standard setters to participate in the space. The introduction of the business case indicates a shift of regulatory logic from that of professional trusteeship one that is driven by commercial logic.

In the second phase, AccountAbility staked a claim in the regulatory space publishing the first international sustainability assurance standard. The different conceptions of "sustainability", "assurance" and "sustainability assurance" allowed AccountAbility to stake claims in the space broadening sustainability assurance to reflect the level sustainabilty is embedded in a reporting organisation's structure and activities. Without the legal mandate, terminology used by the accountancy profession in this field could also be adopted and redefined by other institutions. AccountAbility brought to prominence the importance of reporting the "right "information in addition to the accuracy of information. It redefined materiality, completeness and the role of stakeholders in the assurance process which the accountancy profession was forced to address. As sustainability assurance was at its initial stages of development, fragmentation of knowledge was recognised and there was an emphasis on cooperation and learning by institutions in the space. The dormant status of IAASB in setting standards on sustainability assurance after the publication of ISAE3000 led to different opinions surfacing from different institutions of the accountancy profession on how AA1000AS should be regarded. Decentred nature of regulation led to two international standards duplicating, overlapping and competing with each other. The obligation imposed on IFAC members to adopt the ISAE3000 in 2005 created a hierarchy of standards within standard setting bodies of the accountancy profession. National standard setting bodies setting subject specific assurance standards from then had to draft national standards based on ISAE3000.

In the third phase, cooperative efforts between different actors were downplayed. Actors in particular those from the accountancy profession focused on reiterating the issues it had established and protecting key definitions and terminology. The profession stressed conflicting definitions and elements of assurance and highlighted the weaknesses of "other" standards and made ISAE3000 obligatory to members. This made it more difficult for users to jointly apply standards and directed the discussions in the regulatory space towards addressing issues that are the strengths of the profession. The IAASB announced intention of using sustainability standards by NIVRA as a point of departure to set international standards. This initiative brought

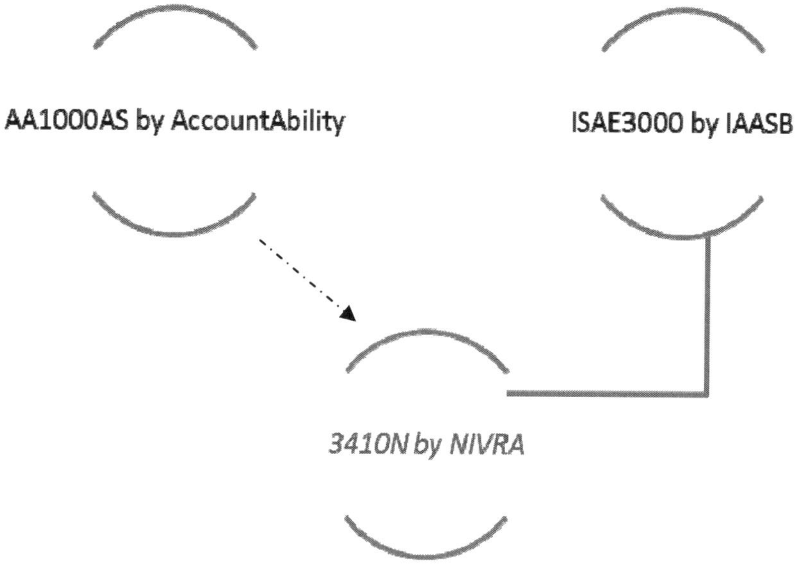

Figure 5.15 Sustainability assurance standards

NIVRA and 3410N into the international regulatory space as well as NIVRA's interpretation of ISAE 3000.

5.4 Conclusion

5.4.1 Who is regulating?

The broad conception of both the terms 'sustainability' and 'assurance' explain the presence of various types of providers operating under different ethical frameworks providing a myriad of services to cater to the needs of reporting organisations. The presence of a regulatory gap pertinent to the quality of engagements led to various initiatives from non-state actors to call for or /and set standards to improve the quality and harmonise sustainability assurance.

At the international level, standards that guide sustainability assurance can be depicted by the diagram above. The AA1000AS remain the only international standard drafted specifically to guide sustainability assurance. ISAE3000 provides generic guidance to assurance services that is being applied to sustainability assurance engagements performed by the accountancy profession. NIVRA's 3410N although drafted in line with ISAE3000 and specifically for sustainability assurance is not in its own right an international standard. It can be regarded "pseudo-international" by virtue of IAASB's intentions with the standard.

5.4.2 Filling the perceived gaps in the regulatory space

Standards on sustainability assurance are set by non-state actors. The regulatory space metaphor has been used to map out and trace patterns of influence of various actors engaged in filling the perceived regulatory gap to gain a better understanding of the context which standards and its contents develop. As the accountancy profession has not been award monopoly status over sustainability assurance engagements; the regulatory space has been occupied by multiple actors with diverse roots pursuing different objectives grounded by different philosophical and functional pursuits. The development of sustainability assurance standards have been shaped mainly by the dynamic interactions between FEE, IFAC, GRI, AccountAbility and national bodies representing the accountancy profession, creating the international context by which standard setting takes place.

Opposed to state-centred regulation where the role of legislators and due processes for law making are provided in statutes, non-state actors (unless appointed or given authority by law) are neither legally obligated nor delegated authority to engage in regulatory initiatives. Observation of the activities illustrate that their involvement or non-involvement in the space, their interest or non interest in particular regulatory issues as well as timing of involvement are determined by the "logic of appropriateness". This observation supports Young's and MacDonald's study that an institution's involvement is dependent how it perceives its legitimate role and purpose and expectations of others in response to the negotiations in the regulatory space rather than a mere self-interest or response to pressure. In positioning their response to "fulfil" their legitimate role, they signal willingness to "play by the rules" and "uphold norms". Decisions of participating institutions involvement and the regulatory issues, brought forth are not made in isolation but are embedded in a social and historical context.[579] Findings also highlight that organisations that do not set standards standard setters such as the FEE can have significant influence in inviting

[579] Hancher, L ; Moran; M.(1989), chapter 10

and excluding institutions in and out of the space through the construction of problems, raising of issues and negotiation of legitimate solutions. Actors can also evoke legitimacy claims through support of third party organisations.

The fact that these are non-state institutions means that the legitimacy of their involvement in the regulatory space and standard setting process are not mandated or provided for in hard law. Just as securing legitimacy is central to 'audit explosion' non-state actors engage in activities to create and secure legitimacy as standard setters and of their standards. Similarly institutions who are not standard setters engaged in activities to legitimise their involvement and input in discussions or debates in the regulatory space. Strategies engaged to establish legitimacy vary according to the type and objectives of the organisation, the communities which efforts are targeted at, influences in the regulatory space and time. The profession had the advantage of relying on the established institutional structures and discourses of regulating the financial audit. Initial efforts by institutions representing the accountancy profession create legitimacy through construction of overlaps between the issues and expertise required for financial audit and environmental / sustainability assurance stressing the need for flexible multidisciplinary teams and calling for a legitimate coordinator putting the accountancy profession in a favourable position. In contrast, AccountAbility, new to the assurance field in 2003 engage in input, throughout and output legitimacy strategies stressing expertise, representations, transparency, democratic participation, usability of standards and alignment with the GRI to gain legitimacy. The difference in initial legitimizing strategies coming into the regulatory space can be attributed to the "newness" the institution is in the assurance regulatory space but also its appeal to different communities which standards are targeted.

The next three chapters provide an analysis of how standards have been constructed and used at a national level using Sweden as a case study. Professions legitimize their services provided and the way tasks are carried out by stressing that the results they produce are culturally valued and produced in a culturally approved manner.[580] Standard setters' of the profession potentially contributes to the maturity and legitimacy of these practices[581] coding guidance as to what is assurance and how assurance engagements should be carried out.

---

[580] Abbott (1988)

[581] O'Dywer, B.(2011 forthcoming). The Case of Sustainability Assurance: Constructing a New Assurance Service. Contemporary Accounting Research

6. Sustainability Assurance in Sweden

6.1 Introduction

6.1.1 Chapter objectives

This chapter introduces the Swedish case. It is structured into two parts. The first part presents the development of sustainability reporting and assurance in Sweden from the mid-1990s to 2010. This is followed by an account of the role played by Far in the development of Swedish sustainability assurance standards.

6.1.2 Background

The Swedish economy is characterised by the strong presence of both private and state- owned companies. The key industries: telecommunications, transport, engineering and production, forestry products, iron & steel, pharmaceuticals and food processing operate in environmentally and socially sensitive areas. On the international front, Sweden ranks favourably among the countries that are 'doing most' to advance its business competitiveness through responsible business practices.[582] Sweden had been under a long social democratic reign in modern history, which built and assigned a central place to the state as an institution.[583] The pursuit of the social democratic ideology resulted in strong institutionalization of environmental and social issues on the public policy agenda. It is a welfare society with high labour union affiliations. Since 2006, Sweden has been governed by the Moderate Party, allied with the Centre party, Liberal People's party, and the Christian Democrats, with a common political platform.[584] The state maintains an active role in promoting the sustainability agenda. Its long history of policies oriented towards the

---

[582] AccountAbility Research Report "The State of Responsible Competitiveness 2007: making sustainable development count in global markets" July 2007, p. 22.

[583] The Social democrats had been in government with the exception of the following periods (1976-82), (1991-94) and (2006- present).

[584] http://www.regeringen.se/sb/d/9395/a/86398 25th October 2011.

environmental agenda and later corporate social responsibility(CSR), backed by strong legislative and regulatory frameworks to ensure that companies meet their responsibilities, creates the foundation for Swedish companies to develop competence to take into account sustainability issues. These issues include climate change, gender, human rights and anticorruption when conducting business. In 2010, Sweden was ranked fourth for having the best environmental policy and ranked first for ethical behaviour of companies[585].

For example, the Swedish government pursues an environmental policy with the objective of passing on to the next generation a society in which the major environmental problems have been solved.[586] The Swedish parliament, following the OECD Environmental Performance Review (EPR) Programme which provided an independent assessment of Sweden's progress in achieving domestic and international environmental policy commitments, adopted the Swedish Environmental Code (1998:808) in 1998 which entered into force 1 January 1999. The code contains rules pertaining to 15 amalgamated acts to promote sustainable development to assure a healthy and sound environment for present and future generations. It delegates the right to relevant authorities to impose information requirements on firms (or individuals) operating in environmentally sensitive areas that may require licenses or permits. For example, the code requires operators needing permits in pursuit of environmentally hazardous activities to file a report to the supervisory authority annually containing measures taken to comply with the conditions laid down in a decision granting a permit and of the results of those measures.[587] Government regulatory bodies are also empowered to impose sanctions in the case of breaches.[588] Other examples include regulation in the area of worker protection and public pension funds. The former includes a realm of Acts enacted to protect employees in the work place such as the Board Representation Act(1987:1245), Prohibition of Discrimination of Employees Working Part Time and Employees with Fixed-term Employment Act (2002:293), Work Environment Act (1977:1160), Parental Leave Act(1995:584), Employment (Co-Determination in the Work Place Act(1976:580), etc. The latter refers to the National Public Pension Funds Act (2000:192) enacted in 2000. The Act requires National Pension Funds to disclose in the development of their investment activities and placement of funds in the Board of Directors

---

585 Source: Advantages of Sweden, World Economic Forum, European Commission 2010.

586 http://www.sweden.gov.se/sb/d/574/a/25800 march 2011. In the strategy for Sustainability Development in 2004, the government identified eight core areas making up the elements of a sustainable society including the environment, climate change, public health, social cohesion, learning, sustainability economic growth, community planning and regional development.

587 The Swedish Environmental Code (1998:808), ch. 26, sec. 20.

588 Ibid., ch. 30, sec. 2.

Report.[589] The previous blacklisting of high profile MNEs for unethical behaviour by Pension fund, serves as a strong incentive for firms to engage in ethical practices.

In the area of CSR, the Swedish government extended an invitation to the corporate community in 2002 to participate in the Swedish Partnership for Global Responsibility Initiative to express support and strive to fulfilment of the OECD Guidelines and the Global Compact Principles[590]. This initiative and other private initiatives such as CSR Sweden, a business driven network of Swedish companies that offer practical solutions to CSR practices and maintain strong links to CSR Europe and other networks, play an essential role in stimulating and creating an environment for CSR. These initiatives brought sustainability further on the agenda of the private and state-owned companies.

6.2 Sustainability reporting and assurance in Sweden

Sweden gained accession into the European Union in 1994, the Act on Sweden's accession to the EU (1994:1500) binds Sweden to the treaties and other sources of law enacted prior to the accession. Its membership, specific to sustainability reporting and assurance makes relevant the Fourth, Seventh[591], the Accounts Modernisation Directives and the Directive on statutory audit of annual accounts and consolidated accounts.[592] These secondary legislations of the EU had been transposed into Swedish law in an effort to harmonise reporting and auditing in the EU.

---

589 National Public Pension Funds Act (2000:192), ch. 6, sec. 3.

590 The following 18 companies have committed to this initiative: ICA, Löfbergs Lila, The Body Shop, Folksam, H&M, OMX, ITT Flygt, Vattenfall, KPA, Sweco, Banco, V&S Group, Lernia, Apoteket, Sveaskog, SJ, Sweroad och Akademiska Hus.

591 Fourth Council Directive 78/660/EEC of 25 July 1978 based on Article 54(3)(g) of the Treaty on the annual accounts of certain types of companies. Seventh Council Directive 83/349/EEC of 13 June 1983 based on Article 54(3)(g) of the Treaty on consolidated accounts. Directive 2003/51/EC of the European Parliament and of the Council of 18 June 2003 amending Directives 78/660/EEC, 83/349/EEC, 86/635/EEC and 91/674/EEC on the annual and consolidated accounts of certain types of companies, banks and other financial institutions and insurance undertakings.

592 Directive 2006/43/EC on statutory audit of annual accounts and consolidated accounts and amending Council Directives 78/660/EEC and 83/349/EEC. Directive 2008/30/EC of the European Parliament and of the Council of 11 March 2008 amending Directive 2006/43/EC on statutory audits of annual accounts and consolidated accounts, as regards the implementing powers conferred on the Commission.

6.2.1 Reporting framework on sustainability issues

The European council adopted the Accounts Modernization Directive 2003/51EC in 2003 to be implemented by member states by 1/1/2005. Recapitulating from Chapter 5, the Directive provides the requirement for certain companies to disclose *"to the extent necessary for an understanding of the company's development, performance or position, the analysis shall include both financial and, where appropriate, non-financial key performance indicators relevant to the particular business, including information relating to environmental and employee matter"*[593] to provide a fair review of the development of the company's business and of its position in the company's annual report. This provision had been transposed in the Annual Accounts Act (1995:1554) in 2005. The Act placed relevant disclosure requirements of non-financial information in the Board of Directors' reports required for the purpose stated in the directive.[594] The Annual Accounts Act applies to both state-owned and private businesses with the exception of certain financial institutions, credit intuitions and insurance companies.[595]

Mandatory reporting requirements on specific social and environmental issues in the notes of the accounts and directors' report existed prior to the transposition of the directive. The Annual Accounts Act (1995:1554) was amended in 1999 to require environmental information in the Board of Directors' report.[596] It placed disclosure requirement on corporations on issues of important for the appraisal of its status, result and future performance. Corporations which are operating under licenses or have an obligation to report according to the Swedish environmental code were required to provide disclosures on the impact of the corporations' production activities (and not the impact on the use or reuse of its product) on the natural environment.[597]

The Swedish Accounting Standards Board (Bokföringsnämden (BFN)), a governmental body with the main objective of promoting the development of generally accepted accounting principles through issuing of general advice and information material on accounting matters and accounting practices published a

---

593 DIRECTIVE 2003/51/EC OF THE EUROPEAN PARLIAMENT AND OF THE COUNCIL of 18 June 2003 amending Directives 78/660/EEC, 83/349/EEC, 86/635/EEC and 91/674/EEC on the annual and consolidated accounts of certain types of companies, banks and other financial institutions and insurance undertakings, Art. 1(14b).

594 Årsredovisningslagen (1995:1554) ch. 6, sec. 1.

595 Årsredovisningslagen (1995:1554) ch. 1, sec. 2.

596 Ibid., ch. 6, sec. 1

597 Ibid.

statement to provide guidance on environmental information in the Board of Directors.[598] Provisions include disclosures of activities stated in the license, importance of the license, renewal of the license and prospective injunctions.[599] The transposition of the EU directive provided a more generalised provision, broadening the scope of disclosures within which the existing legislations were examples.

In Sweden, although state-owned companies are subject to the same legislation as privately-owned companies, state ownership policy imposes on state-owned companies certain important principles with regard to their administration. According to state ownership policy, the boards of the state-owned companies are responsible for "matters relating to ethical issues, the environment, human rights, gender equality and diversity. It is an important part of the Government's ownership policy for companies to have a well-considered strategy for these sustainability issues. This work is to be communicated both internally and externally."[600] State-owned companies thereby are subject to guidelines issued by the state in addition to the general framework applicable to corporations in general. One such area relate to reporting and auditing requirements.

In 2006, the Ministry of Enterprises, Energy and Communication, catalysed by NGO (among others, Amnesty International) campaigns, for greater accountability engaged Samhällsnytta AB (SN), a CSR consulting firm, and PwC to perform a study of the reporting on sustainability issues by state owned companies. The consultants highlighted the State's risks of breaching international conventions ratified due to delegating the onus of managing sustainability issues to the board of directors without a systematic monitoring system.[601] By constructing the problems as one of "control", the consultants proposed "better reporting" as a solution which in addition brought about the advantages of transparency about performance, democracy and efficient use of resources of the state owned companies.

In 2007, the Swedish state owned 55 companies either operating under market conditions (value creation) or serving special societal interest (create value through achieving social goals). There were 40 wholly owned and 15 partly owned (four of which are listed in the stock exchange). The total value of the companies were

---

[598] BFN U 98:2 Environmental information in the Board of Directors report. (Miljöinformation i förvaltningsberättelsen) For more information see, http://www.bfn.se/.

[599] Ibid.

[600] Swedish Ministry of Enterprise, Energy and Communications "Guidelines for external reporting by state-owned companies", 29th November 2007, p. 1.

[601] "Genomlysning av de statligt ägda företagens hållbarhetsredovisning 2006" Samhällsnytta AB and PwC One example quoted in the report concerned LKAB's mining operations in China in Amnesty's report.

estimated to be worth 770 billion SEK and had about 190 000 employees.[602] At the end of 2007, the Ministry of Enterprise, Energy and Communications issued "Guidelines for external reporting by state-owned companies" to update the 2002 guidelines to compliment current accounting legislation and generally accepted accounting principles to be applied to 2008 reports. The guideline imposes on state owned companies to report their non-financial information according to the GRI guidelines on a "comply or explain basis".[603] In 2010, further guidance was provided for state-owned companies. For example, state-owned companies have to include a brief analysis of the sustainability issues considered as important, publish a clear report of the stakeholder analysis and stakeholder dialogue to identify and take a position on the significant risk and opportunities.[604] An overview of the principles for reporting financial and non-financial information for state-owned companies in Sweden is presented in appendix 2.

Companies that are not state-owned that opt to voluntarily disclosure more than what is required by law, can draw on EU recommendation 2001/453/EC on the recognition, measurement and disclosure of environmental issues in the annual accounts and annual reports of companies[605]; normative frameworks by International organisations such as the UN Global Compact and the OECD MNE Principles that encourage sustainability reporting or multi-stakeholder or private frameworks such as the GRI and AA1000APS that provide more direct guidance on the framework and principles on sustainability reporting. Although the abovementioned frameworks and reporting guidelines originate from state and non-state initiatives, the challenges of fragmentation, overlap and conflict of these provisions are increasing being alleviated by cross references made to GRI as the sustainability reporting standard.[606]For example, the UN Global Compact Office has mentioned that the GRI G3 Guidelines offer a "globally recognized reporting guidance that can help produce an excellent COP"[607], and are therefore recommending it for use by the Global

---

[602] Guidelines for external reporting by state-owned companies p. 1.

[603] Swedish Ministry of Enterprise, Energy and Communications "Guidelines for external reporting by state-owned companies", 29th November 2007, p. 2.

[604] Annual report state-owned companies 2010.

[605] Commission Recommendation 2001/453/EC of 30 May 2001 on the recognition, measurement and disclosure of environmental issues in the annual accounts and annual reports of companies (notified under document number C(2001) 1495).

[606] UNPRI in paragraph 3 refer to the use of GRI reporting standards, UN Global compact's COP encourage use of GRI, OECD has invited GRI's involvement in the 2009 revision of the OECD MNE Principles. AccountAbility positions its standards as being complimentary to GRI. Also to note, GRI has taken initiative to distinguish its standards from ISO 26000 and SA 8000 that are management rather than reporting standards.

[607] http://www.unglobalcompact.org/COP/Performance_Indicators.html 3/3/2007

Compact. The OECD has invited GRI's involvement in the 2009 revision of the OECD MNE Principles.[608] Most importantly, the GRI is involved in developing the International Integrated Reporting Framework (IIRF).[609] In the light of this trend, and the guidance for external reporting by state-owned companies, it can be expected that more Swedish companies may draw guidance from GRI guidelines in the future.

The above explains that the current reporting framework on environmental and social issues of a corporation is regulated by a mix of mandatory requirements and non-state standards and guidelines or best practice recommendations.[610] Sustainability information can therefore be found in the annual report (Board of Directors report, Notes of the accounts), in a separate report or website that can be stand alone, referenced to or attached to the annual report.

6.2.2 Assurance of sustainability information

Companies are not required by law to commission assurance on their sustainability reports. State owned companies are mandated by the state as an owner, to commission assurance on their reports using a "comply or explain" policy. The guideline for external reporting by state-owned companies states that "the sustainability report shall be quality assured by independent scrutiny and assurance"[611], requiring these reports to be subject to third party assurance. Sustainability reports, like financial reporting are therefore to be reviewed by a third party. The difference, however, lies with the fact that the government has not expressed a preference on the type, nature and provider of assurance.

Far issues guidance in two areas pertaining to the review of sustainability information. It has issued RevU5, is a statement that provides guidance to auditors in carrying out the statutory audit of a company pertaining to non-financial information required by the Annual Accounts Act.[612] This provides guidance on the planning, work to be carried out, working with environmental experts and responsibilities of providers.[613] It

---

608 "OECD Guidelines for MNE's and the GRI Sustainability Reporting" www.oecd.org/dataoecd/25/26/35150230.pdf

609 http://www.theiirc.org/about/. Refer to section 9.4 of this thesis.

610 At the international level, KPMG report Carrot and Stick Chapter 4 provides a summary of mandatory and non-mandatory guidance on reporting of various countries.

611 Swedish Ministry of Enterprise, Energy and Communications "Guidelines for external reporting by state-owned companies", 29th November 2007, p. 4.

612 RevU5 "Revisorns beaktande av icke-finansiella upplysningar vid granskning av årsredovisningen"

613 Ibid. sec. 3-6.

has also issued RevR6 to guide members to provide assurance on sustainability reports. RevR6 will be analysed in the following chapters.

6.2.3 Sustainability reporting trends

The presentation of sustainability reporting and assurance trends below has been divided in two time periods: mid 1990s to 2008[614] and 2009-2010. This is due to the enactment of state external reporting guidelines by state-owned companies in November 2007, applicable to 2008 reports.

The environmental and sustainability themes featured prominently in the global policy agenda in the 1990s. Interest in environmental / sustainability reporting in Sweden started in the 1990s and the publication of such reports (separate reports and separate section in the annual report) was on an increasing trend from 1994-2003, levelling off in the period from 2004 to 2008. In the initial stages, the reporters were the large usually listed companies (e.g., Volvo, H&M), state owned companies (e.g., Green Cargo, SJ) or high profile private companies (e.g., IKEA) with operations in sustainability sensitive areas. Statistics from 2004-2008 ranked Sweden as the top 8th reporter by output (number of reports) behind UK, Germany, Spain, Italy, France, Netherlands and Switzerland.

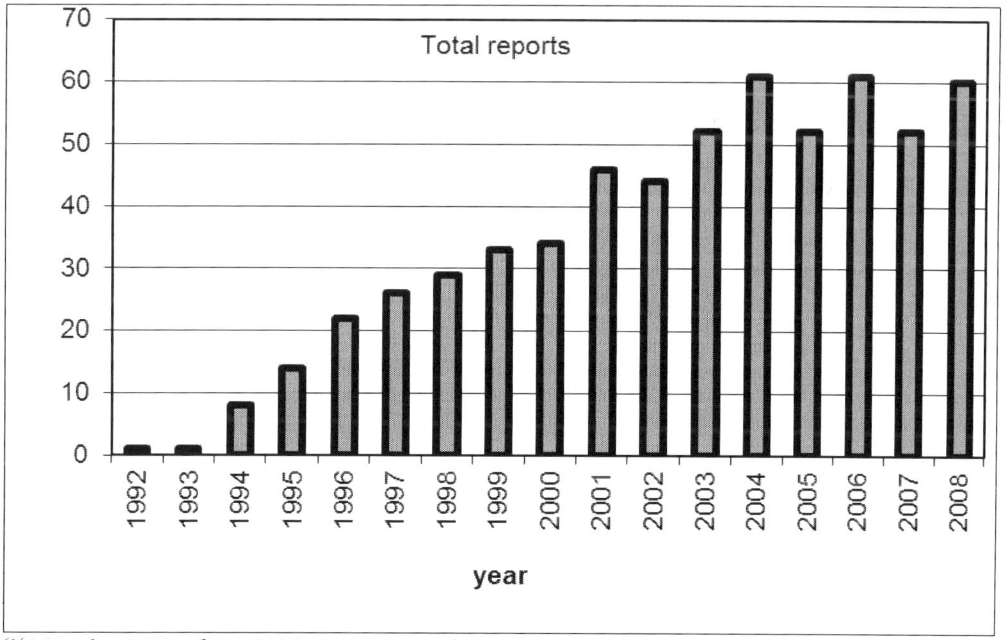

[614] Note that 2008 refer to 2007 reporting period

Figure 6.1 Sustainability reporting trends 1992-2008

Corporations had been motivated by different forces over time to engage in sustainability reporting. Three companies interviewed in the sample studied, attributed environmental /sustainability reporting in the mid-90s, to the need for a fair representation of their work in the environmental area to be given to stakeholders. The need was driven by the environmental agenda in public policy, the negative media publicity of large corporations and results of surveys, which corporations felt did not reflect the extent of their environmental work. Sustainability reporting in mid-2000s has been attributed by Far to the increasing interest of the financial community towards such reports, and state ownership policy.[615] Driving factors of sustainability reporting in Sweden were linked the pressure from the financial investment community for example, Sustainability Sweden Index by NASDAQ OMX launched the OMX GES Sustainability index series, DJSI and FTSE4Good; the Principles for Responsible Investment was launched and analyst and rating agencies were increasingly interested in the area.[616]

The increase in environmental / sustainability reporting was accompanied by a gradual increase in the scope of the contents of these reports to include more environment, social as well as economic indicators. Until the turn of the century, reports were however heavily weighted towards environmental aspects. The term sustainability among others, such as corporate social responsibility, corporate responsibility, environmental and social responsibility, was among the titles adopted to mark the expanded scope of the report. Since 2001, there has been an increasing trend of companies integrating the reporting of sustainability issues in the annual report. In many cases, the sustainability reports were attached to the end of the annual report. Far attributes this to the Swedish Financial Analyst Association's recommendation for companies to insert the sustainability report into the annual report, provide references in the annual report to a separate sustainability report or to place sustainability information on the company's web page.

---

[615] European Sustainability Reporting Association report for Sweden 2006, 2007, 2008.

[616] Ibid.

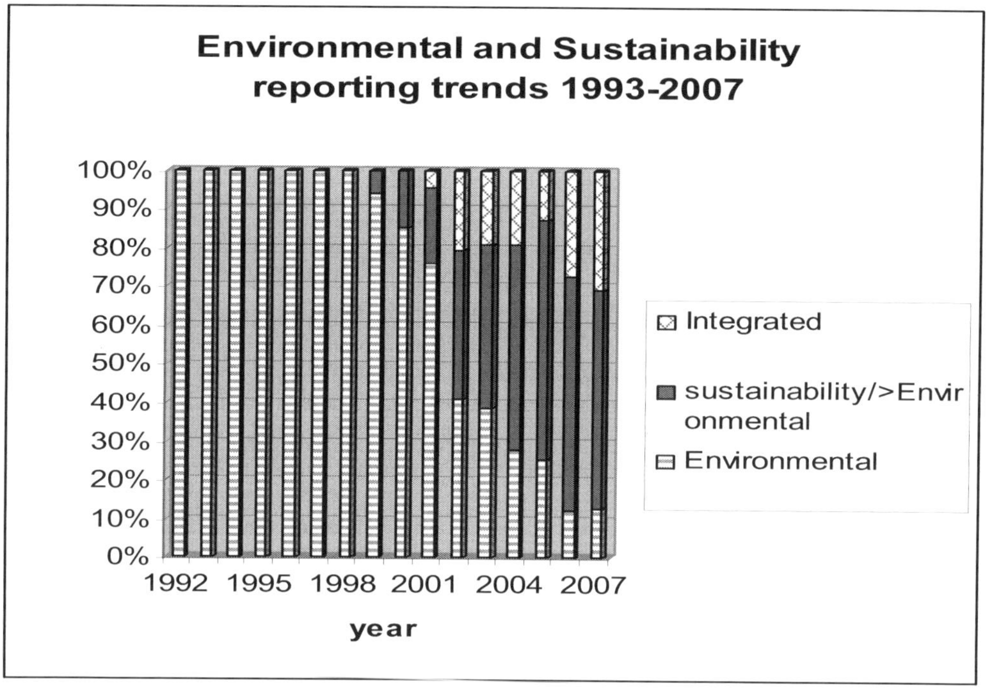

Figure 6.2 Environment and sustainability trends 1992-2007[617]

Prior to the publication of GRI in 2000, firms structured reports based on their internal guidelines or with the aid of consultants' tools. Early adopters of GRI attributed its usefulness as a tool to increase credibility of the sustainability reporting and serve as an inspiration or a template to design their sustainability report.[618] Research found that the guidelines served well as a learning experience and an internal rather than external communication tool.[619] The adoption of GRI by Swedish companies increased from 2003 as GRI gained increasing international acceptance.[620] This trend was given a further push with a translation of GRI to Swedish in 2008.

---

617 Integrated refers to the reporting of sustainability issues in the annual report

618 Hedberg and Von Malmborg (2003) The Global Reporting Initiative and corporate sustainability reporting in Swedish companies" Corporate Social Responsibility and Environmental Management, 10(3), p. 153.

619 Ibid.

620 KPMG "International Survey of Corporate Sustainability Reporting" June 2002.

The guidelines on external reporting for state-owned companies altered the reporting and statistics significantly. The report to ESRA indicated that 88% of 172 companies surveyed disclosed some form of sustainability information.[621] From 2007-2009, companies with "more extensive sustainability information" defined by the aforementioned study as more than 10 pages increased from 22% to 49%.[622] Adoption of GRI guidelines increased from 16% to 47%. The statistics show that the increase in reporters and the adoption of GRI is not only attributed to state-owned companies but also large and mid-cap companies.[623]

The above table below the impact of the guidelines on reporting practices of state-owned companies. From 2007-2009, the percentage of state-owned reporters increased from 20% to 96%, with only one state-owned company not reporting according to GRI. State-owned companies that did not issue a report attributed it to the limited extent of activities and the lack of resources. This government initiative and the state's attempt to put forth that state owned companies as role models in the sphere of environmental and social responsibility continue to play a part catalysing the increase in the adoption of GRI by not only large corporations but SMEs.

---

621 The sample consisted large cap, mid cap and 43 state owned companies.

622 Refer to Far 2010 Report ESRA report for Sweden 2009 Large cap, mid cap, state owned companies increased from 33% to 50%; 10%-14% and 26% to 93%.

623 In 2007, 9 state-owned companies adopted the GRI guidelines; these figures do not correspond to the report to ESRA as it includes Vin & Sprit and Vasakronen which were sold in 2008. The 2008 statistics based on 47 out of 54 state owned companies including those listed in the stock exchange. 7 state owned companies are not included including dormant companies (ExStattum, Zenit Shipping, Preaktio), partly owned Russian company (Dom Shvetsil) and newly established companies (Vectura, Svevia and Fouriertransform), 91% of state-owned companies issued separate sustainability report (or attached a sustainability report within the annual report) of which 98% adopted GRI guidelines. The four companies that did not comply with the reporting requirements include Arbetslivsresur, Arlandabanan infrastructure, Kasernen and Swedish ships Mortgage bank.

It is noted however, that despite improvements in 2009, 78% of state-owned companies continue to apply GRI at level C, 18% and 4% report at apply guidelines at levels B and A respectively. Aggregating both state-owned and other reporters, 9%, 14% and 64% of reporters adopt GRI at application levels A, B and C respectively.

	2007		2008		2009	
Number of state-owned companies	50		47		48	
Reporters	10	20%	43	91%	46	96%
GRI	8	80%	42	98%	45	98%
GRI application level A			2	5%	2	4%
GRI application level B			3	7%	8	18%
GRI application level C			31	74%	35	78%
GRI application not stated			6	14%	0	0%
Assured	4	40%	34	79%	40	87%

Figure 6.3 State-owned companies reporting and assurance practices from 2007-2009[624]

[624] Note statistics may differ from the report to ESRA due to the state-owned companies taken into account when calculating statistics. The above table adopts figures from Annual report for state owned companies 2007, 2008 and 2009.

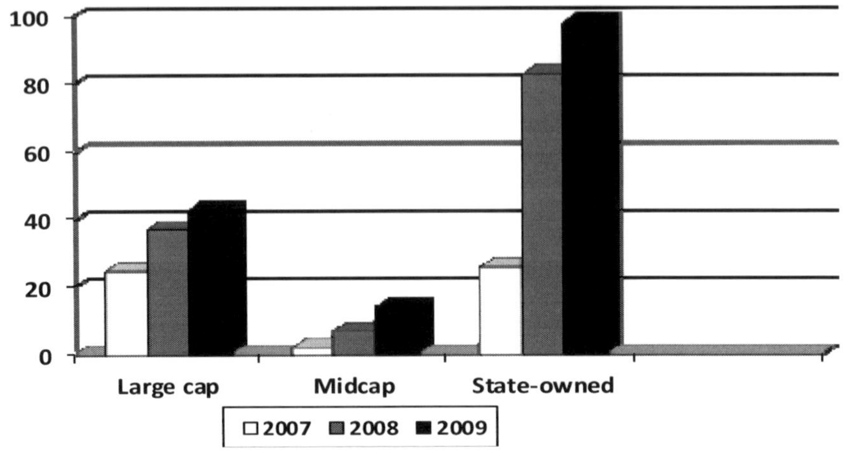

Figure 6.4 Companies adopting GRI guidelines[625]

6.2.4 Sustainability assurance trends

Third party assurance on voluntary sustainability reports has not been a common practice in Sweden.[626] From 1995-2007, only about 20 reporting organisations had commissioned third party assurance, 18 of which are corporations. Third party assurance has been more common amongst companies from the engineering, forestry and transport sectors than service sectors such as banking, due to the environmental impact of their operations. The table below shows assurance engagements commissioned by corporations in Sweden from 1995 to 2007.

---

625 European Sustainability Reporting Association report for Sweden 2009.

626 European Sustainability Reporting Association report for Sweden 2006. The percentage of assured reports in Sweden stood at 8% the lowest among the 11 European countries that participate in ESRA.

	Company	1995	1996	1997	1998	1999	2000	2001	2002	2003	2004	2005	2006	2007
1	SAS			Deloitte	Deloitte	Deloitte	Deloitte	Deloitte	Deloitte	Deloitte	Deloitte	Deloitte	Deloitte	Deloitte
2	SKF								Deloitte	Deloitte	Deloitte	KPMG	KPMG	KPMG
3	SCA	Sustainability				Institute of Environmental Management and Economics (Austria)	Deloitte	Deloitte	Deloitte	Deloitte	Deloitte	Deloitte	Deloitte & PWC	Deloitte & PWC
4	Cargo											Deloitte		
5	Cerealia					KPMG	KPMG	KPMG	KPMG	KPMG				
6	Telia					EY	EY							
7	Perstop	EY	EY	EY										
8	Meats			KPMG	KPMG	KPMG	KPMG	KPMG	KPMG					
9	g												PWC	PWC
10	Sprit												PWC	PWC
11	Ericsson												PWC	DNV
12	Kberg	KPMG	KPMG	KPMG										
13	Sandvik													KPMG
14	pel													PWC
15	SI													PWC
16	ASG	SustainAbility	SustainAbility	SustainAbility	SustainAbility									
17	Energy				Deloitte									
18	Scania					DNV	DNV							
Total engagements		4	3	5	4	6	6	4	5	4	3	4	7	10
Accountancy providers		2	2	4	3	4	5	4	5	4	3	4	7	5
Other providers		2	1	1	1	2	1	0	0	0	0	0	0	1

Figure 6.5 Assurance Engagements from 1995-2007[627]

The first assurance engagements were commissioned by Perstorp and Stora Kopparbergs Bergslags for their 1995 reports. From 1995-2007, the yearly average of assurance stood at 5 engagements per year. Prior to the introduction of guidance issued by Far to guide assurance on sustainability reports, only 12 companies had engaged assurance. 7 of the 12 companies had terminated this practice due to:

[627] Source: own

changes in reporting policy resulting from the merger and acquisition[628], substantial improvements in internal control processes diminishing the need for external assurance[629] or than sustainability information was reported in the Board of Directors report in the annual report.

From 2005, the sustainability assurance engagements have been on an increasing trend. In 2007, 9 companies had commissioned assurance on their 2007 report. 4 of the 9 companies were state owned SAS, Vin & Spirit, SJ and Svenskaspel.[630] Three interesting developments are observed. First, Green Cargo made changes to their reporting practice. Sustainability information that was reported in a separate section of the annual report in 2005 was incorporated in the Board of Directors report in the annual report in 2006. The effect on assurance is noted as the sustainability assurance report published with the annual in 2005 is not been published with 2006 and 2007 reports. Second, SCA in 2006 and 2007 commissioned two assurance engagements by different providers, PwC and Deloitte assured the Human resources section and the environmental section of the report respectively. Third, the increase in the number of assurance engagements coincided with international trends supporting the 'incremental approach' and the publication of RevR6(2004) by Far that allowed the acceptance of engagements that are limited in scope. There appears to be a link between the increase in assurance engagements with the possibility of scope limitations as companies that have newly commissioned assurance after the publication of standards are more narrow in scope.[631]

---

628 Stora Kopparberg merged with Enso Oyj 1/1/99 to become Finnish company Stora Enso Oyj. Assurance on sustainability reports are performed by Finnish providers. Birka Energy in 1999 issued reports under Fortum Corporation a Finnish company, the merger took place in 2002. Their reports had been verified by KPMG Finland merger in 2002; Telia merged to become Teliasonero, ASB taken over by German Company Danzas.

629 Swedish Meats Annual report 2003

630 Vin & Spirit is no longer state owned. It was sold to Pernod Ricard since 23/7/2008.

631 Refer to findings in section 8.2.1.1 of this thesis.

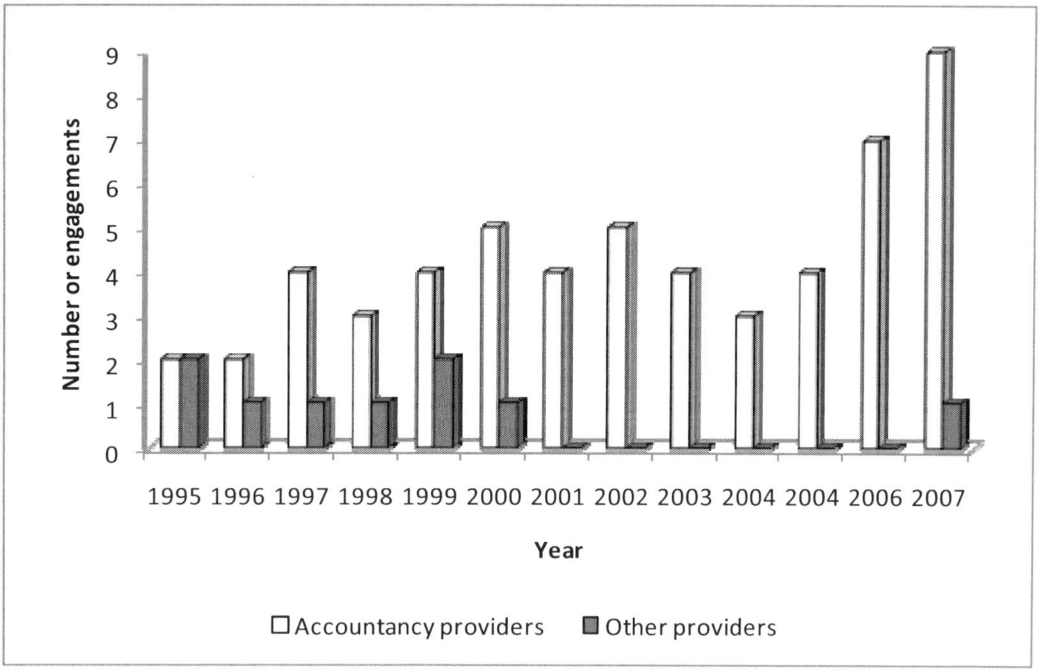

Figure 6.6 Assurance trends from 1995-2007[632]

The external reporting guidelines issued by the state had a significant impact on the number of assurance engagements commissioned in 2008. Forty-three reports (both state owned and privately owned companies) were assured by third party providers. 81% were state-owned, 19% were privately owned. The number of state owned firms that engage assurance services increased from 4 companies in 2007 to 35 companies in 2008.[633] For corporations that are not state-owned, assurance on sustainability reports remains a "minority sport" with 8 companies commissioning assurance services.[634]

For the reporting year 2009, although statistics reflect an increase (from 26% in 2007 to 30% in 2008) in companies that have their reports assured by a third party. Since 2007, the new companies engaging sustainability assuarance that are not state-owned are three large cap listed companies: ICA, Holmen and Swedish Match.

[632] Source: own

[633] Annual report State-owned Companies 2008.

[634] European Sustainability Reporting Association report for Sweden 2008.

6.2.5 Market share of assurance providers

In the period from 1995 to 2010, although Swedish companies have engaged different types of providers to assure their sustainability reports, the accounting firms had and continue to possess a dominant share of the assurance market in Sweden. In 2002, Deloitte and KPMG were the only providers that shared the market for assurance of sustainability reports (separate or separate section in the annual report) until 2006 when PwC was engaged to perform sustainability engagements for 4 companies of the 6 companies. This could be attributed to PwC's knowledge of the companies through other engagements and the movement of significant individuals to PwC in 2005. Apart from DNV that was engaged by Ericsson to provided assurance on the 2007 Corporate Responsibility Report (previously assured by PwC, the market share of assurance providers in Sweden from 1995-2007 has been dominated by three of the Big four firms.

It is important to note that both accounting firms and consulting firms do not only offer assurance services on sustainability reports in this area. Consulting firms also provide certification and quality assurance services. Likewise, the accounting firms perform agreed upon procedures and other consultancy services with regards to both reporting and risk management on this subject matter. For example, Deloitte had been SAS's consultants from (1995-1996) before assuring SAS reports in 1998.

In 2009, most state owned companies engaged their financial auditors to assure their sustainability reports with the exception of three companies Bostadguaranti, Visit Sweden and Vokenåsen that commissioned DNV and Respect sustainable business (the latter two). The Big 4 thus have 85% of the market of the state owned companies and 86% of the total market.

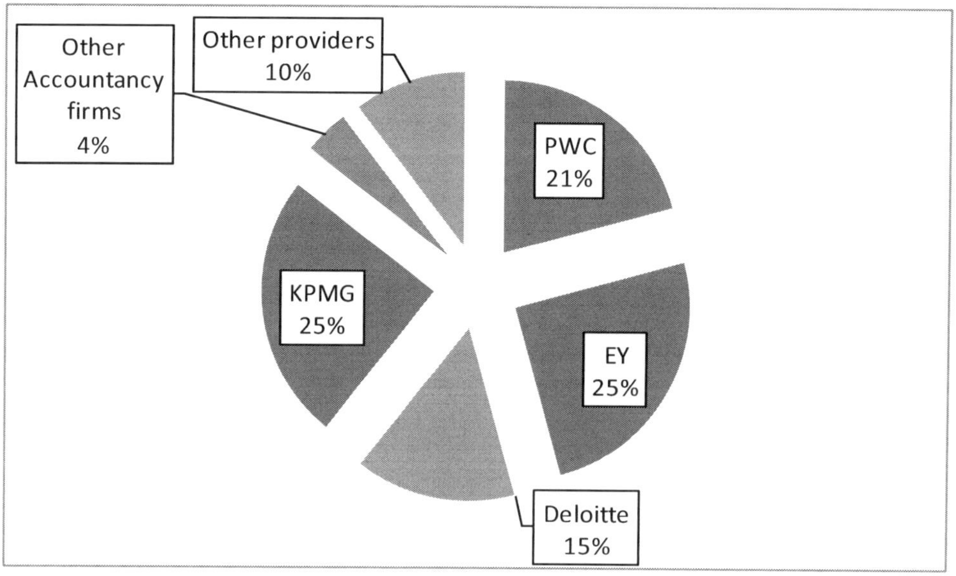

Figure 6.7 Market share of assurance providers in 2010[635]

6.3 Far and the development of sustainability assurance standards

6.3.1 Far and its representation in the regulatory space

The Föreningen Auktoriserade Revisorer was founded in 1923 by authorized public accountants. It merged with Svenska revisorsamfundet (SRS) in September 2006 to form FARSRS. It is today named Far, and is a non-profit private organization operating as the professional institute representing approximately 6500 authorized public accountants, approved public accountants and other highly qualified professionals or specialist, in the accountancy sector in Sweden.[636] The vision of Far

[635] Source: own (reporting year 2009)

[636] www.far.se (October 2011)

has evolved over time.[637] In 2011, the institute has the vision to serve its member needs through commitment and initiative, to deliver the right support at the right time to members, whether it involves global industry issues or practical tools. The three core values: leadership, confidence and involvement, guides the organisation to creating value, confidence and trust between business and community through showing leadership by pursuing and express clear opinions on issues concerning its members and developing members' professional roles and services.[638]

Far is an active member of IFAC, FEE and the Nordic Federation of Accountants. Its representatives have held key positions in the committees and work groups.[639] Far has participated in the sustainability assurance regulatory space in its own capacity, it commented on various exposure drafts including IAASB's ISAE2000. It has also participated through its membership in FEE that commented on NIVRA's ED3410 published in January 2005. The representative of Far held the chair for FEE's Sustainability Assurance Subgroup from 2002-2006 when FEE played a prominent role in standard setting activities in the regulatory space. He also serves as a member of the Sustainability Experts Advisory Panel (SEAP) and the Management Committee of the European Sustainability Reporting Association. Fars current representatives continue to participate in the Sustainability Policy Group and Working Party of the FEE.

Far's board, auditors, nominating committee, membership council, and other boards (examinations, quality and disciplinary) are appointed in the annual general meeting. It works through its board; policy groups, standing committees, permanent or ad-hoc working groups in four areas, the audit of small firms, large firms, taxes and consultancy. A general secretary and permanent secretariat is located in Stockholm.[640]

6.3.2 Far's guidance on assurance of sustainability reports

Far's involvement and the role undertaken by the Far Specialist in the sustainability assurance regulatory space has puts Far in position to be familiar with the issues and debates surrounding sustainability assurance. Far claims to be the first standard setter

---

637 Wallerstedt, E. (2001), "The emergence of the Big Five in Sweden", European Accounting Review, 10(4), p. 848. The author highlighted the driving factors behind Far's formation, one of which was to gain a common platform to debate on issues pertaining to authorisation of auditors.

638 Ibid.

639 Göran Tidström is president of the International Federation of Accountants (IFAC). Prior to this role, he served as IFAC deputy president (2008-2010) and FEE president, Lars-Olle Larson was Chairman of sustainability subgroup and member of SEAP.

640 Articles of Association of Far, sec. 1.

of the accountancy profession to publish and make available to members a standard to guide assurance engagements on sustainability reports. The standard "förslag till recommendation: Oberoende översiktlig granskning av frivillig separat hållbarhetsredovisning" translated as "Proposed recommendation: Independent review of voluntary separate sustainability report" was published February 2004. It is important to highlight that the "proposed recommendation" carries the same authority as a standard and had been drafted following a due process.[641] A clarification sought with Far revealed that the term 'proposed' suggest that it is newly introduced an adaptation period and revisions may be needed.[642]

Far drew on the international developments and its institutionalised role of setting generally accepted auditing standards in Sweden to legitimise the need for a standard. In section 1 of RevR6(2004) the increase in publication of separate sustainability reports and the demand for sustainability assurance was noted.[643] It also gave recognition to the harmonization efforts of actors in the regulatory space, specifically mentioned were FEE's discussion paper "Providing assurance on sustainability reports" and AccountAbility's AA1000AS that was providing guidance in this area.[644]

Through Balans, a journal published by Far (via their subsidiary publishing company Far Förlag, but does represent Fars position), debates, news and comments on accounting and auditing, business law and other issues of interest to the profession are published. It highlighted the developments at the international and its effect on Swedish practice. Through the journal, Far raised awareness of issues such as the growing international acceptance of GRI[645] as a sustainability reporting standard which could be used as a criteria. In one article, the authors stressed that purpose of the RevR6 is to express 'generally accepted auditing standards' in this audit-related area.[646] It also referred to the practice by leading reporters in Sweden of publishing the sustainability report in the annual report accompanied by both a financial auditors report and an assurance report by sustainability assurance providers drawing of the institutionalized ideas of the audit to legitimize its actions and RevR6.[647]

Far's pioneering efforts to set sustainability assurance standards had little impact in the regulatory space for setting international sustainability assurance standards as it

---

[641] FEE(2006); Informant interview with Far.

[642] Informant interview with Far.

[643] RevR 6 (2004), sec. 1 and 2.

[644] RevR 6(2004), sec. 1.

[645] "Far bidrar till hållbar utveckling" Balans nr 2 2004.

[646] *Ibid.*

[647] Ibid.

was drafted in Swedish. Unlike the approach by NIVRA, Far did not expose RevR6 for comment internationally. The standard was drafted reflecting "best practice" and drawing from the work of IFAC and FEE guidelines at that time. *"We had to use the experience we had that time in the market"* making references to international practices and applying it in a local context.[648]

The first revision of the proposed recommendation was published in December 2006 to align RevR6(2004) with ISAE3000. The elements of RevR6 mirrors ISAE3000 with the exception that RevR6 provided guidance to a specific subject matter i.e. the sustainability report compared to ISAE3000's generic applicability to all assurance engagements. Another difference was that RevR6 (2006) only guided limited assurance engagements in comparison with ISAE3000 that provides guidance also on reasonable assurance engagements. In 2005, expert interview reveal the lack of demand for sustainability to be performed at a reasonable level of assurance and the lack of controls in place of the reporting organisations commissioning assurance on their reports.[649] RevR6(2006) however did not preclude providers from using ISAE3000 should reasonable assurance engagements be commissioned.[650]

A detailed analysis of the contents of RevR6(2004) and RevR6(2006) reflect that amendments to the contents of the standard appear to be insignificant. The definition of sustainability report, purpose of engagement, elements of assurance, materiality, subject matter[651], criteria[652], responsibility of parties, work procedures remained predominantly the same. This is attributed to common sources which Far and IAASB have drawn from in drafting the standards. Similarly, RevR6 (2004) had been drafted drawing from ISAE100, the work of FEE as well as best practice in Sweden. Recapitulating Chapter 4 and 5, the ISAE3000 is a revised version of ISAE100, FEE had actively contributed to IAASB standard setting and IAASB had as a reference point assurance (audit) on environmental reports in developing the assurance concept in the course of standard setting.

Changes were however reflected in the terminology and positioning of RevR6(2006) revised in line with IFAE and ISAE applied by Far. For example, throughout the standard, the term assurance engagement (bestyrkandeuppdrag) constantly replaced limited review engagement (översiktligt gransknings uppdrag). This revision constructs into the recommendation the demanded by various interest groups among them the financial markets' expectation of relevant information pertaining to

---

648 Interview with Expert

649 Interview with Expert.

650 RevR 6(2006), sec. 1.3.

651 RevR 6(2006), sec. 2.2 Code of conduct has been added as an example of subject matter.

652 RevR 6(2006), sec. 2.3 Reporting criteria has been updated to include G3.

sustainability development performance[653] in addition to the demand and international harmonisation efforts mentioned in the 2004 version. Pertaining to the provisions, the revision brought in line terms of agreement[654], work of an expert and quality control with ISAE3000.[655] It also elaborated on legal standards and ethical rules which assurance providers are obliged to follow[656]; skills and knowledge of the practitioner[657]; engagement risk; after date events and the requirements of the assurance report.[658] In December 2007, the standard was named: Rekommendation RevR6 "Oberoende granskning av frivillig separat hållbarhetsredovisning". In the title of the recommendation, the term proposed was taken out. Provisions of RevR6(2006) are analysed in the next chapter.

The second revision was completed in December 2008. The revision entailed replacing the contents of RevR6 with a direct translation of the NIVRA standard N3410 "Assurance Engagements Relating to Sustainability Reports" making only minimal changes to apply to the Swedish context. Interviews with the expert involved in the revision indicated that the revision was necessary due to improvements in the reporting and internal control measures of reporting organisations in Sweden warranting guidance on reasonable assurance engagements as well as international developments.[659] The adoption of NIVRA's 3410N could be attributable to IFAC's announcement in its strategic plan to use the NIVRA's standard as an initial point of departure to develop an international standard for this subject matter and that the Dutch model was more "tuned towards the GRI in terms of the stakeholder engagements"[660]. The RevR6 (2008) was renamed "förslag till recommendation: Bestyrkande av hållbarhetsredovisning"" translated to "proposed recommendation "Assurance on Sustainability Reports". The term proposed had been re-adopted in view that feedback would be sought from affected stakeholders in 2009 and the proposed recommendation would be revised to better adapt to Swedish practice.[661]

The table below summarises the significant events in the regulatory space, analysed in chapter 5 on the development of RevR6.

---

653 RevR 6(2006) sec.1.1.

654 RevR 6(2006) sec. 4.2. ISAE3000, sec. 11.

655 RevR 6(2006) sec. 4.5. ISAE3000, sec. 6. 26-32.

656 RevR 6(2006) sec. 1.4.

657 RevR 6(2006) sec. 4.4.

658 RevR 6(2006), sec. 5.1 and 5.3.

659 Interview with expert.

660 Interview with expert.

661 RevR 6 (2008), footnote 1.

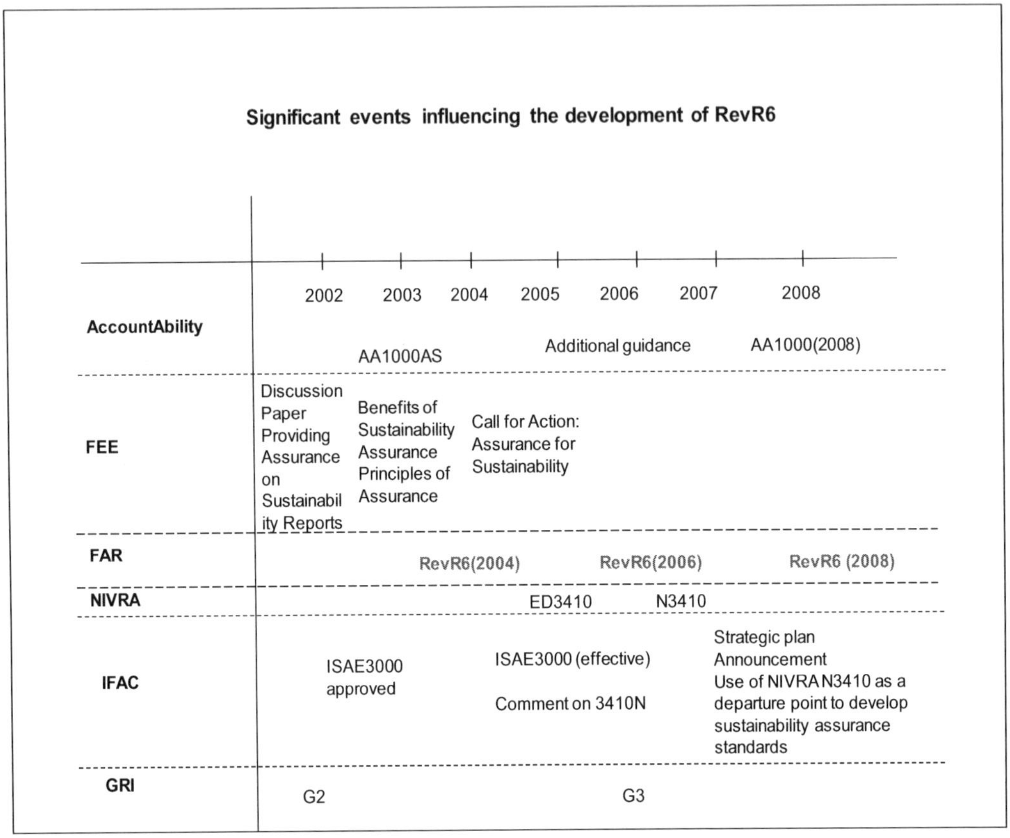

Figure 6.8 Significant events influencing the development of RevR6[662]

6.3.3 Standard setting process and legitimacy of RevR6

Standard setting on assurance on sustainability reports has been initiated and drafted by the reference group on sustainability development. The reference group comprised mainly Far Specialists in the sustainability area working in one of the Big four accounting firms.[663] To be accepted by Far as a Specialist member (in the area

[662] Source: Own

[663] In 2007, one of the specialists was not from the Big 4 in 2007.

sustainability in this case) , a person should have at least a 3 year university education, at least have worked within the field of sustainability assurance for five years, three of which are in a position with advisory capacity"[664]. Far does not qualify the specialist, the training and experience of these specialists lie mainly with the firm he or she works. In the standard setting process, a working group is assembled to draw up a draft standard which is then reviewed by a language expert. As a private organisation, Far is not obliged to expose drafts to the public for comment or to make all records public. The draft is approved by the audit committee and the board who decides if public comment is warranted. Comments on exposure drafts are by invitation though uninvited parties comments would be taken into account. Should the board deem that no comment is necessary, feedback received after the publication of the standard would be taken into account for the next revision.[665]

The legitimacy of the RevR6 is built upon experts' knowledge contribution to the input of the standard and representation by the large accounting firms who are predominantly users of the standard. The standard also draws to a large extent from the IFAE and ISAE3000 of IFAC that had been exposed several times for comment over the years. There is however little transparency on the process of standard setting to observe how different competing interests have been tackled (or avoided). The public is neither granted access to the minutes of the meetings and working papers by the standard setting group nor had there been established channels giving opportunities to interested parties such as commissioners, beneficiaries and other parties to engage in dialogue or participate in the standard setting process. Interview with one expert in 2005 attributed this to technical complex and usage of terminology by the profession and the lack of knowledge of the market in this field (at that point of development). Further, time to educate them may undermine the standard setting given that it operates under tight time constraints.[666]

In June 2009, the Far working group organised a Round Table inviting large companies such as Atlas Copco, Sandvik, SAS, Vattenfall, ICA, SKF and Teliasonera and analysts from the Robur 4th AP fund, Alecta, Meta asset management and the Swedish society for financial analyst to consult on the RevR6(2008). The round focused on the purpose and value of assuring sustainability reports. The responsibilities of the board, the value assurance statement, and what auditor can do to create value, were among the topics covered. This initiative was one of the first, by Far, to openly consult on its sustainability assurance standards.[667]

---

[664] Informant interview with Far.

[665] Ibid.

[666] Informant interview with Expert.

[667] European Sustainability Reporting Association report for Sweden 2009.

6.3.4 Other activities of Far shaping the interest in sustainability reporting and assurance

Far has over the last 10 years given recognition to "best reporters". According to Far, the organisation and categories and criteria of the awards had been tailored to the development of sustainability reporting practices and trends.[668] Prior to 2003, the best environment reports were given recognition. From 2003-2005, the awards to the best Sustainability reports were jointly organised by Far, the International Chamber of Commerce (ICC) and the Confederation of Swedish Enterprises. Reflecting the increasing interest of the financial community on sustainability reporting both in the international arena and in Sweden[669], Far has since held the "Best sustainability report" awards at FinForum, an event participated by accountants, finance directors, analysts and other finance related groups to bring sustainability reporting closer to finance related groups.

Award winners were nominated for the European Sustainability Reporting Awards held in Brussels annually with the purpose of promoting "best practice" in the sustainability reporting field. Far adopted the judging criteria used by ESRA, developed by the Association of Chartered Certified Accountants (ACCA) of the UK, incorporating GRI reporting guidelines in the judging criteria in the last few years. The ESRA initiative included a website where participating countries reported on their practices and statistics. Far has been a participant since 2005. The jury comprises sustainability experts and a secretary employed by Far. The jury members are practitioners of the Big 4 often directly involved in assurance of the nominated reports.

The prizes awarded in 2005 and 2006 were categorised according to the report which sustainability information was published, 2007 by the type of company and 2008 & 2009 by the size and type of company. The changes reflect the government external reporting guidelines to state-owned companies as well as the international trends and development of GRI's guidance to encourage SMEs to publish sustainability reports.

---

[668] http://www.far.se/portal/page?_pageid=114,388932&_dad=portal&_schema=PORTAL (September 29, 2011)

[669] Section 6.2.3 and 6.2.4

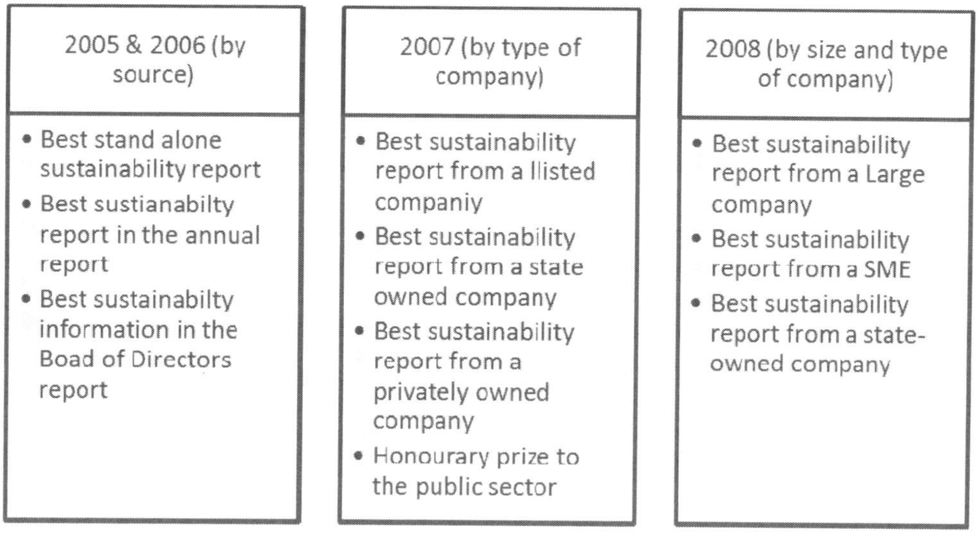

Figure 6.9 Best sustainability reporting award categories[670]

The autonomy Far has in deciding on the criteria, categories and jury provides it the opportunity to define "best practice" and drive sustainability assurance practices. For example, the 2005 Vattenfall report[671] was not assessed by the Fars judges, due to the fact that "an independent assurance statement was a requirement for inclusion in the evaluation process."[672] Its exclusion was based on the failure to fulfil part of the ACCA UK Awards criteria on credibility constituting 35% of the total score.[673] In the

---

[670] Source: own (data from: http://www.far.se/portal/page?_pageid=114,388932&_dad=portal&_schema=PORTAL

[671] European Sustainability Reporting Association report for Sweden 2006; "State-owned Vattenfall's report has been very positively evaluated by the Robert Environmental Centre. Vattenfall's vision is to be a leading European energy company. The outcome of the evaluation is: "This is an extremely well thought-out and executed report with a clear introductory section on how such a report should be read and used and an easy-to-read 50-page visionary section that is very clear on the types of impacts that electrical utilities cause and on the approaches in use at Vattenfall to diminish them, illustrated with some of the best diagrams we have seen. All this is followed by a no-nonsense 30-page section of data keyed to the GRI reporting guidelines."

[672] Ibid.

[673] Other criteria include Completeness 40% materiality, stakeholder inclusion, strategy, org context, communication 25% presentation stakeholder inclusion and structure) ACCA UK Awards criteria can be obtained from :

http://www.far.se/portal/page?_pageid=114,388952&_dad=portal&_schema=PORTAL.

process, Far excludes 92% of reporters from the awards as only 8% of the sustainability reports were accompanied by an assurance report. In 2007 however, the criterion requiring third party assurance was removed and Vattenfall was award the "Best separate sustainability report" despite having no assurance.[674]

Far keeps its stakeholders and members abreast of the developments of sustainability reporting and assurance by making the relevant reports of GRI, IFAC and FEE available on the Far website. The ESRA report published annual not only serve to report on the practices, award winners and statistics of sustainability reporting and assurance in Sweden, it also provides stakeholders a source to understand regulatory developments and the driving forces in the market. For example, the ESRA reported on the threat of 313 companies being deleted from global compact list for lack of reporting in 2006, negatively motivating companies to report.

6.4 Conclusion

Chapter 6 provides an explanatory source to understand how sustainability assurance standards are developed and work in Sweden. It provides readers with an understanding of the sustainability assurance market, the framework which sustainability information is reported and context in which standards develop. The development of Swedish sustainability assurance standards did not take place isolated from events in the sustainability assurance regulatory space. The representatives of Far's active involvement in the international scene and familiarity with the issues and debates explain Far's working group's competence in setting the first sustainability assurance standard by the accountancy profession in 2004 and keeping with the front runners such as NIVRA and IDW in this field.

The standard however catered to a "minority sport" despite efforts by Far to encourage these services. The Ministry's external reporting guidelines to state-owned companies and the development of increasing acceptance of GRI altered the practice of reporting and assurance significantly. The number of assurance engagement increased more than four fold. The assurance market is dominated by the large accounting firms, with the Big 4 having an 86% of the market share. This trend is somewhat unsurprising due to the scope of the statutory audit in Sweden. It includes

---

674 The criterion of assurance was made clear and reintroduced in 2010 and made transparent in the Far website in the above footnote. It includes three criteria pertaining to the responsibility of the board, reference or attachment to the financial report and focus and relevance, in addition to 41 criteria from ACCA.

examination of the administration of the board which puts the financial auditors in a position to be well-versed in the non-financial issues of the reporting companies. The ISAE3000, the generic standard which RevR6 had been revised in accordance to has been set under IAASB's standard setting due process.

The IAASB has provided a generic assurance standard to guide members to perform among others sustainability assurance engagements revised through a transparent due process in light of the changes in the governance structures of IFAC. The lack of an accountability system throughout the regime suggests a possibility of different levels of transparency within a decentred regime and opportunities for regulatory capture. This leads one to question if the legitimacy established at the international level can hold, and how the quality of standards set at a national level can be maintained. Far had not exposed RevR6 for comments by interested parties, as such, how ISAE3000 is applied to the subject matter sustainability, appears to be in Sweden, determined by the audit profession.

Grasping the international and national context provide a basis for understanding the context, driving forces and constraints which shape the way sustainability assurance standards are developed and how they work.

7. The Provisions of RevR6 -Discretionary Spaces

7.1 Chapter objectives

ISAE3000 is a generic standard drafted to apply to assurance engagements that do not fall under the scope of ISA, ISRE, and when there are no ISAEs that provide specific subject matter guidance. In addition to its role of providing guidance to assurance providers, it serves as a 'framework' for both the IAASB and national standard setters, to set subject specific standards.[675] Far, as a member of IFAC, is obligated by virtue of its membership, to adopt ISAE3000 from 1/1/2005. In October 2005, the IFAE and ISAE3000 were built into Far's pronouncements. In 2006, Far published RevR6(2006), the revised version of RevR6(2004), that aligned the standard with ISAE3000.

The diagram below provides a timeline of the development of RevR6 in parallel with that of ISAE3000.[676]

The purpose of this chapter is to enrich our understanding of how standards work. It presents an analysis at two levels. First, the provisions in ISAE3000 are analysed to determine the discretion Far has been granted in the revision of RevR6(2006). Second, the provisions of RevR6 are analysed to determining how Far has exercised the given discretion, and the discretion has delegated to assurance providers using the standard. Using the dimensions of rules as a toolbox, provisions of the above standards are evaluated based on its scope and linguistic structure. More specifically, dimensions to evaluate the provisions include: over or under-inclusiveness; precision or vagueness; simplicity or complexity; clarity or opacity. The findings of this chapter contribute to our understanding of the limits of rules and how discretion can be distributed and exercised in multi-levelled regulation, such as in this case.

---

[675] For example, ISAE 3402" Assurance Reports on Controls at a Service Organization"

[676] Source: own; also presented in section 3.3 of the thesis

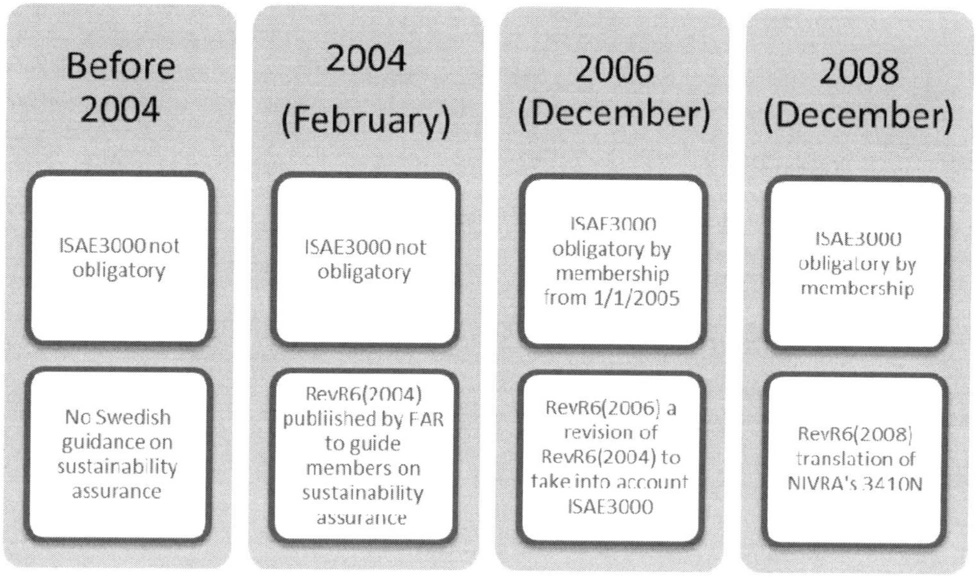

Figure 7.1: The revisions of RevR6

7.2 Classification of Far' s standards

Far issues three types of standards. The first are translations of IAASB standards, which Far translates to: Standarder för kvalitetskontroll(ISQC), Revisions standard (ISA), Standard för bestyrkande av andra sakförhållande än historisk ekonomisk information (ISAE); Standard för översiktlig granskning (SÖG); Standard för näräliggande tjänster (SNT). This is summarised in the table below.

Second, are standards issued by Far drafted specifically to the Swedish context, termed *"rekommendationer"* (RevR). The third, takes the form of an "*Uttalande*" (RevU) drafted with the purpose to clarify requirements of the law or the standards. For example, RevU5 provides guidance on assurance on environmental information in the Board of Director's report.[677]

[677] RevU 5 Revisorns beaktande av miljöfrågor vid granskning av årsredovisningen.

IAASB	Far	Abbreviation
International Standard on Quality Control (ISQC1)	Standarder för kvalitetskontroll	ISQC1
International standards on auditing (ISA)	Revisionsstandarder	ISA (previously RS)
International standards on auditing of information other than historical financial information (ISAE)	Standard för bestyrkande av andra sakförhållande än historisk ekonomisk information	ISAE (previously SBÖ)
International standards on review engagements (ISRE)	Standard för översiktlig granskning	SÖG
International Standard on Related Services (ISRS)	Standarder för näräliggande tjänster	SNT

Figure 7.2 Translation of IAASB standards[678]

As IAASB has not published a standard on assurance on sustainability reports, the standard issued by Far to provide guidance in this area has been classified as RevR, even though it has been revised according to ISAE3000. With reference to section 6.3 of this thesis, even though the term 'recommendation' translated to English by Far as 'recommendation'[679] is adopted, which can understood by some communities as non-binding, the RevRs operate and have the same effect as standards that Far translates from IAASB. In the following sections of the thesis RevR6 will be referred to as a 'standard'.

[678] Far Samlingsvolymen 2011 Del 2.

[679] Far Engelska Ordbok.

7.3 The provisions of ISAE3000 and RevR6(2006)

The ISAE3000 provides guidance in the areas of: professional ethics and quality control; prerequisites to accepting an assurance engagement; planning and performing an engagement and output of an assurance engagement.

Guidance pertaining to the code of professional ethics and quality control are cross referenced to Parts A and B of IFAC Code of Ethics for Professional Accountants (IFAC Code of Ethics) and the ISQC1 respectively. IFAC Code of Ethics adopts a minimum harmonization model, giving recognition to national differences in culture, language, legal and social systems. The IAASB therefore leaves the task of preparing, implementing and enforcing detailed ethical and quality control requirements to the member bodies in each country concerned.[680] With regard to independence rules, members are required not to set rules that are less stringent then stated in the code to ensure objectives such as integrity and objectivity, professional competence, due care and confidentiality are met.[681] The ISQC 1 requires assurance providers to set up and maintain a system that provides reasonable assurance that professional standards and legal requirements are compiled with and reports are issued in appropriate circumstances[682]. This thesis focuses only on the provisions within ISAE3000 that is read in the context of the IFAE.[683] The following sections provide an analysis the provisions of ISAE3000 and RevR6 pertaining to the other three afore-mentioned areas of guidance. A summary of the analysis is presented in table 7.5.

7.3.1 Guidance on prerequisites to accepting an engagement

The prerequisites to accepting an assurance engagement determines the scope of the standard. It identifies the elements that have to be present for the standard to apply.

7.3.1.1 Three party relationship

Section 7 of the ISAE3000 provides that an engagement can be accepted *"only if the subject matter is the responsibility of a party other than the intended users or the*

[680] Code of Ethics for Professional Accountants, sec. 1.

[681] IFAC Code of Ethics for Professional Accountants, Parts A and B.

[682] ISQC1, sec. 11. The version of ISQC1 used here, is the one prior to the revision according to clarity drafting conventions.

[683] ISAE3000, sec. 3.

practitioner". Read in conjunction with IFAE[684], an assurance engagement involves: the practitioner, responsible party and intended users, where the latter two parties can be from the same entity.[685] The term practitioner had been adopted to reflect its guidance to other parties other than to the financial auditor.[686] The responsible party is referred to as person/s delegated with the authority to manage the subject matter (direct reporting engagement), subject matter information (assertion based engagement) or both. Section 25 of the IFAE, provides that the responsible party "may or may not be the party that engages the practitioner".[687] It however does not provide further guidance on this matter. This provision adds complexity to the application of the standard, as it suggests that a fourth party could be the engaging party, which changes the balance of power and dynamics of the three-party relationship.

Intended users have been defined as consisting *"person, persons or class of persons for whom the practitioner prepares the assurance report"*.[688] The provision recognizes the difference between 'all stakeholders' and the user that assurance report is targeted at. Although it excludes 'all stakeholders', the term 'intended user' can be over-inclusive. This situation arises when the assurance provider, according to the terms of the assurance engagement does not specify the parties whom the assurance report is targeted. There is a limiting clause allowing users to be more narrowly defined according to the terms of agreement negotiated and agreed upon by the engaging party (who does not have to be the responsible party) and the assurance provider, or by law.[689] Hence, user/s of an engagement can include all intended users, targeted users, or even the responsible party as long as he/she is not the only user.[690]

The ISAE3000 allows national standard setters a wide discretion in determining the three parties required to be present in an assurance engagement. RevR6(2006) more narrowly contains the three parties in an assurance engagement on sustainability reports to: the qualified auditors and /or Far specialist; the party responsible for reported information such as the CEO, board of directors or an equivalent official that commission assurance on sustainability reports and the users of the assured

---

684 IFAE, sec. 21-30.

685 Ibid., sec. 21, 22.

686 Ibid., sec. 23.

687 Ibid., sec. 25.

688 Ibid., sec.27.

689 Ibid., sec. 28.

690 Ibid. sec. 21-28.

information.[691] Far, thereby exercise discretion to exclude from its scope the circumstances whereby the engaging party is not the responsible party.

There is a lack of clarity as to who the users are. In section 2.1 of RevR6(2006), pertaining to the conditions of acceptance, the standard refer to the user as "users of the assured information". In section 5.1 of the standard, pertaining to planning of the engagement, the standard provides that the engagement has to planned, taking into account the perspective of the relevant stakeholders. Guidance on reporting provided for in section 5.5 states that the report should be addressed to the 'intended user'. The lack of attention to the conceptual differences of 'intended user', 'relevant stakeholders' and 'users of the report' in the construction of RevR6(2006) though creates uncertainty, it grants users of the standard greater discretion to determine who the assurance report is for.

The RevR6(2006) does not guide practitioners as to who should determine the users of the report, or the process as to how this group should be identified. RevR6(2006) refers to this group of users as "usually more numerous than financial information (users)".[692] The expressed interest of the financial community stated in the preface of the standard further creates uncertainty as to whether the financial community should be the main intended user group. The vagueness in the provisions and lack of clarity in its use of terms, the discretion as to for whom the assurance is for is left to the reporting organization and the assurance provider. This has significant implications on how the purpose of engagements can be determined, how relevance and completeness of a report is defined and how an engagement is planned.

7.3.1.2 Appropriate subject matter

Subject matter is one of the key elements in an engagement as it determines to a large extent the planning and design of work to be carried out, the type of evidence available and the level of assurance. The IFAE can be applied to non-historical financial information, non-financial information, physical characteristics, systems and processes and compliance behaviour[693], the ISAE3000 applies to non-historical financial information.

The ISAE3000 refers to IFAE on guidance with regard to how assurance providers can assess the appropriateness of the subject matter as a precondition to accepting the engagement[694]. The criteria set forth in section 31 of the IFAE specify that an

---

691 RevR 6(2006), sec. 2.1.

692 Ibid., sec. 5.1.

693 IFAE, sec. 31

694 ISAE3000, sec. 8. IFAE sec. 17b(i).

appropriate subject matter is "identifiable, capable of consistent evaluation or measurement against identified criteria" and "can be subject to procedures for providers to get sufficient and appropriate evidence to support reasonable or limited assurance conclusion".[695] The provision contains evaluative terms such as "consistent evaluation and measurement" and "sufficient and appropriate evidence" that require subjective assessment by the assurance provider. Although this provision allows the exclusion certain information categories for example, prospective information, it nevertheless leaves wide discretion to the users of ISAE3000 (and the engaging party) to determine the scope and boundaries of the subject matter.

RevR6(2006) narrows the scope of ISAE3000, clearly specifying that the standard applies only to voluntary separate reports of companies or other organizations, published as a separate publication, on the company's website or be attached as a separate section of printed annual report covering "environmental, ethical and social circumstances and the financial development connected to these circumstances."[696] By defining the subject matter as such, it makes a clear distinction between information reported in the statutory financial accounts and those in the sustainability reports. RevR6(2006) provides examples of areas that should be reviewed. They include the written or numerical information about past and future developments of the company's sustainability performance, systems and processes for data collection; internal control and operations, for example, compliance with code of conduct and report on stakeholder dialogue.[697]

The RevR6(2006) does not provide specific guidance on evaluating the appropriateness of the subject matter dealing with sustainability issues. It replicates the provisions in IFAE on the criteria of appropriate subject matter. The standard allows the scope of the subject matter to be limited and requires the assurance provider to provide a clear description of the limits of the review performed in the assurance report.[698] The standard however does not specify the minimum scope that is acceptable which allows assurance providers the discretion to accept assurance engagements where the subject matter is very narrow for example, limited to one environment indicator to one that covers the full sustainability reports. Given the evaluative terms to assess appropriate subject matter and the possibility of scope limitations, assurance providers are granted very wide discretion on the evaluation of

---

[695] IFAE, sec. 33.

[696] RevR 6(2006), sec. 1.1. The RevR 6(2008) applies to sustainability reports more generally and can include other forms other than the separate sustainability report.

[697] Ibid., sec. 2.2.

[698] Ibid.

appropriate subject matter and scope of the engagement in deciding on the acceptance of sustainability assurance engagements.

7.3.1.3 Suitable criteria

ISAE3000 provides for the *"practitioner to assess the suitability of the criteria to evaluate of measure the subject matter"*[699]. Suitable criteria have been deemed important by both the framework and ISAE3000 to allow consistently evaluation of a subject matter within the context of professional judgment. The IFAE and ISAE3000 recognize that criteria are context sensitive i.e., the same subject matter can be evaluated based on different criteria. The IFAE provides guidance identifying application and inherent characteristics to guide the evaluation of suitable criteria.[700] It also contain provisions that clearly preclude the practitioners own expectations, judgments and individual experience to determination suitable criteria.[701]

The five characteristics to determine suitable criteria are as follows:

Characteristics of suitable criteria	
Application characteristics	
Relevance	criteria contribute to conclusions that assist decision-making by the intended users
Completeness	when relevant factors that could affect the conclusions in the context of the engagement circumstances are not omitted
Reliability	allow reasonably consistent evaluation or measurement of the subject matter
Inherent characteristics	
Neutrality	contribute to conclusions that are free from bias
Understandability	clear, comprehensive, and not subject to significantly differentinterpretations

Figure 7.3 Characteristics to determine suitable criteria[702]

[699] ISAE3000, sec. 19.

[700] IFAE, sec. 34-38.

[701] IFAE, sec. 36.

[702] Ibid.

The characteristics of suitable criteria have been constructed in a theoretical and vague manner, consisting evaluative and generic terms. For example, "reasonably consistent evaluation" and "contribute to conclusions that assist decision-making by the intended users" require subjective assessment by users of the standard. Furthermore, how the five characteristics are related and the weights assigned in the evaluation process are left to the discretion to users of the standard.

The IFAE and ISAE3000 classify criteria into formal and less formal benchmarks required to evaluate or measure a subject matter. Both established and specifically developed criteria with the prerequisite that it is made publicly available.[703] The former is defined as one that is legally required, set by authorized and recognized body of experts through a transparent process, whereas the latter refer to criteria that are specifically developed for the purpose of the engagement or an agreed level of performance. One example is internally developed codes of conduct. The two circumstances which specifically developed criteria can be used are when the engagement is catered to specific users with specific needs and when established criteria do not exist.[704] With regard to specifically developed criteria, the above provisions impose on the assurance provider the duty to consider if the criteria are misleading.[705] Section 20 of ISAE3000 also specifies that the intended parties or engaging party can be involved to acknowledge suitability of criteria for the intended user's purpose or if that is unattainable, consideration has to be given on work to be carried out to assess suitability of criteria and information to be provided about the criteria in assurance report.

RevR6(2006) recognizes both established as well as specifically developed criteria and its evaluation and replicates the five characteristics according to the IFAE. The standard states that in Sweden, it is the "most common situation" when the reporting organization creates the criteria which the engagement is to be based.[706] Far applies the generic standard to sustainability assurance pertaining to suitable criteria by giving to examples. First, if a reporting organization states that the report is prepared in accordance with the GRI, the GRI guidelines should serve as criteria for carrying out the assurance engagement.[707] Second, the outcome of stakeholder dialogue can "make up some of the criteria" which reporting is drawn up. If further refer to the AA1000

---

[703] Ibid., sec. 37-38. ISAE3000, sec. 20-21.

[704] Ibid.

[705] ISAE3000 sec. 20-21.

[706] RevR 6(2006), sec. 2.3.

[707] Ibid.

framework as an example of a framework an organization may apply to guide engagements with stakeholders.[708]

The guidance provided by RevR6(2006) replicates IFAE and ISAE3000. It does not provide guidance on what is required by practitioners to evaluate criteria for assurance engagements on sustainability reports. RevR6 (2006) does not provide guidance on the work the assurance provider has to perform on criteria originating from different sources (established versus specific). The standard also omits guidance on, and does not differentiate between how the assurance provider should evaluate the suitability of criteria in cases where there is no stakeholder engagement and where these engagements are present. RevR6(2006) provides that the dialogue with stakeholders can impact the reporting criteria, but does not guide providers on the evaluation of the methods and processes applied by the organization to determine the information needs of the intended users; how to determine legitimate sources of stakeholders' needs. These issues are left to the discretion of the providers. By quoting two specific under-inclusive examples reflecting the practice by leading reporters, adds to the complexity in applying the standard. For example, does the mention of GRI reporting guidelines mean that Far, giving consideration to IAASB evaluation of G3, has accepted that GRI, whether applied as a whole or in part satisfy the criteria of relevance, completeness, reliability, neutrality and understandability?[709] Furthermore, it is unclear if the two examples constitute requirements or serve as explanatory guidance. Circumstances that fall outside the examples are left to the practitioners discretion limited only by the generic categories provided for in the IFAE.

7.3.1.4 Ethical and competence requirements

ISAE3000 provides that an engagement should be accepted "only if, on the basis of a preliminary knowledge of engagement circumstances, nothing comes to the attention of the practitioner to indicate that the requirements of the Code or ISAEs will not be satisfied".[710] will not be satisfied.[711] RevR6 distinguishes between qualified auditors and other professionals who are members of Far. It states that the ethical requirements in the Auditors Act(2001:883), Auditors Ordinance(1995:665) and rules of Revisorsnämdens föreskrifter (RNFS) apply to the qualified auditor.[712] All members i.e., the qualified auditor and Far specialists are obliged to follow the

[708] Ibid.

[709] Refer to section 5.3.3.4 of this thesis on IAASB consultation paper on G3.

[710] ISAE 3000, sec. 8.

[711] Ibid.

[712] These legal sources will be explained in section 8.4 of this thesis.

professional code of ethics and quality control standards by Far. This points out that qualified auditors as compared to Far Specialist, both of who can be signatories of assurance reports, have to operate within the regulation of auditors stipulated by legislation. This analysis will be performed in section 8.4 of this thesis.

RevR6(2006) provides a 'warning provision' requiring assurance providers to take precaution not to compromise independence when engaging in stakeholder dialogue. It however leaves the discretion to determine the acceptable extent of stakeholder contact and the way stakeholders can be part of the assurance process to the assurance provider. As it stands with the guidance provided in RevR6, the assurance engagement can be carried out without contacting stakeholders of the reporting organization or directly contacting them as source of evidence.

With regards to competence, ISAE3000 places as a precondition on the acceptance of an assurance engagement the assurance provider should only accept an engagement if he or she is satisfaction that the team performing the engagement possess the necessary competencies.[713] In line with ISAE3000, RevR6(2006) leaves it up to the assurance provider to determine if such a team can be assembled.[714] The standard also replicates provision in ISAE3000 pertaining to assessing the sufficiency of evidence provided by the work of experts.[715]

7.3.1.5 Terms of engagement

The ISAE3000 require that the terms of engagement agreed between the assurance provider and the engaging party be documented in an engagement letter.[716] When the engagement is being carried out, the provider is guided not to agree to a change in terms without "reasonable justification" and not to disregard evidence obtained during the course of the engagement.[717] The standard specifies two conditions where changes would be 'reasonably justified'. They are *"a change in circumstances that affect users' requirements, or misunderstanding concerning the nature of the engagement…"*[718]. The conditions imposed are generic and potentially over-inclusive of a number of circumstances that could qualify as "reasonably justified". RevR6(2006) replicated

---

713 ISAE3000, sec. 9.

714 RevR 6(2006), sec. 4.4.

715 ISAE3000, sec. 32.

716 Ibid., sec. 10.

717 Ibid., sec.11.

718 Ibid.

this provision without further elaboration.[719] The wide discretion is therefore placed on practitioners to decide if change is justified particular when there is not a requirement for it to be evaluated by a third party or for the reasons to be transparently declared.

RevR6(2006) provides that should the above prerequisites not be met, an agreed-upon or other procedures instead of a limited assurance engagement may be performed by the assurance provider. These procedures should not result in a published assurance report expressing the assurance providers' conclusions but a report to the management of the company.[720]

7.3.2 Guidance on planning and performing an engagement

The IFAE and ISAE3000 provide that the assurance provider should consider materiality and assurance engagement risk when planning and performing assurance engagement.[721] The framework and standard puts the onus on the provider to use professional judgment, and a questioning mind in critically assessing validity of evidence that contradicts, or puts the reliability of documents presented by responsible party in question.[722] Providers are required to obtain sufficient and appropriate evidence to arrive at a conclusion.[723] Sufficient has been defined in terms of quantity of evidence whereas appropriate has been specified in terms on relevance and reliability evidence.[724]

7.3.2.1 Reporting relevance

The IFAE provides for stakeholder engagement implicitly as section 29 of IFAE states that intended users and their representatives may be involved in determining the requirements of an assurance engagement whenever it is practical but does not specify the extent, circumstances or effects and consequences of their involvement. Discretion is therefore left to users of the standard as to how intended users should be involved; to what extent and at which stage of the assurance process they should participate and whether it should be reported? Hence, stakeholder involvement can range from

---

719 RevR 6(2006), sec 4.2.

720 Ibid.

721 IFAE sec. 40-42. ISAE3000, sec. 22 to 25.

722 IFAE sec. 40. ISAE, sec. 14.

723 ISAE sec. 33. IFAE, sec. 42.

724 IFAE, sec. 42.

indirect involvement of stakeholders e.g., multi-stakeholder participation in setting GRI reporting standard to direct involvement which ranges the providers observing stakeholder dialogue as a neutral third party to having direct contact with stakeholders in assessing their needs and risk of the reporting organization.

RevR6(2006) does not define the concept of reporting materiality (relevance) in the context of sustainability assurance. The standard merely states that requires providers to gather information about material events that occur; review that material matters have been taken into account and make an overall assessment as to whether reported information reflects actual and material conditions of operations.[725] RevR6(2006) provides guidance that the engagement should be planned from the perspective of relevant stakeholders.[726] With reference to the above, there is also the lack of guidance on how 'relevant stakeholders' involvement in determining suitable criteria relates to reporting relevance. The requirements made on how to regard stakeholders in the assurance process is vague. RevR6(2006) therefore does not provide more guidance than the requirements set forth in ISAE3000 on the definition and process to define relevant stakeholders and their needs[727]. For example, RevR6 does not impose any requirement regarding the review of processes adopted by management (e.g. have the targeted groups been involved in the stakeholder dialogue and whether the reporting organization had been responsive to their needs?) and the output of the resulting engagement to determine reporting materiality.

It stands that the results of a stakeholder dialogue can form part of the criteria in determining reporting content but it is not a requirement. The providers have a wide discretion as to what extent stakeholder engagements should have in the assurance process. Furthermore, RevR6 does not establish a clear relationship between reporting materiality, audit materiality and engagement risk.

7.3.2.2 Audit materiality and engagement risk

The findings from chapter 5 show that the concept of materiality is much contested in the sustainability assurance regulatory space. Unlike materiality in financial audits which is further provided for in ISA 320 "Materiality in planning and performing an audit" and ISA 450 "Evaluation of misstatements identified during the audit", there are no established and distinct set of standards to determine materiality for assurance

---

[725] RevR 6 sec. 5.2

[726] Ibid., 5.1

[727] Interviews with the sustainability work group clarifies that the group regarded it to be the responsibility of the reporting organization.

engagements on non-historical financial information.[728] The IFAE defines engagement risk as the "risk that the practitioner expresses an inappropriate conclusion when the subject matter information is materially misstated".[729] An assurance engagement does not assure that report is free from all misstatements, the engagement assess the risk and respond to it by mitigating the engagement risk that material misstatements exist.

In the IFAE, provisions with regard to materiality are made in three areas that have been replicated.[730] First, the ISAE3000 provides the need to consider "materiality" when deciding on the "nature, timing and extent" of evidence gathering procedures and in the course of assessing whether the subject matter information is free from misstated. Second, it states that considering materiality requires for practitioners to understand and assess the factors that influence decisions of intended users. Thirdly, it provides guidance on how materiality can be assessed. Thresholds of materiality are to be "considered in the context of quantitative and qualitative factors" giving examples such as relative magnitude or the nature or extent factors impact subject matter measurement of evaluation or intended users' interests.[731] Engagement risks comprise risk of information being materially misstated prior to the assurance process and the risk that the auditor will not detect such misstatement.[732]

RevR6(2006) aligned the recommendation to ISAE3000 drafting the need for assurance providers to apply professional scepticism in conducting an engagement. It specifies the knowledge assurance providers should be equipped with to assess engagement risk. These conditions are: the assurance provider should obtain a good knowledge of the reporting organizations activities and work in the sustainability area; the reporting organization and its information technology system for gathering, processing and accounting of reported information and have knowledge of relevant

---

[728] IFAC(2010) Handbook of International Quality Control, Auditing, Review, Other Assurance, and Related Services Pronouncements. Note also, ISA 320 sec. 2. Materiality in the context of the audit is defined as "Misstatements, including omissions, are considered to be material if they, individually or in the aggregate, could reasonably be expected to influence the economic decisions of users taken on the basis of the financial statements; Judgments about materiality are made in light of surrounding circumstances, and are affected by the size or nature of a misstatement, or a combination of both; and Judgments about matters that are material to users of the financial statements are based on a consideration of the common financial information needs of users as a group. The possible effect of misstatements on specific individual users, whose needs may vary widely, is not considered."

[729] IFAE, sec. 48.

[730] ISAE3000 23, IFAE 47

[731] IFAE, sec. 48. ISAE3000, sec. 24.

[732] IFAE, sec. 49. ISAE3000, sec. 25.

law and regulations applicable to the reporting organization and / or to the industry.[733]

7.3.2.3 Level of assurance and work to be performed

The IFAE and ISAE distinguish reasonable and limited assurance by the nature, timing and extent of evidence gathering procedures. In a "reasonable assurance" engagement, the practitioner obtains sufficient and appropriate evidence to reduce risk to an acceptably low level that allows the practitioner to conclude with a positively expressed statement. Similarly, for limited assurance, the practitioner engages in evidence gathering procedures to allow risk to be reduced to a level higher than reasonable assurance yet to obtain a meaningful level of assurance as a basis for a negatively phrased conclusion.[734] The guidance to determine engagement are broadly constructed leaving wide discretion to the users of the standard based on engagement circumstances such as subject matter and level of assurance.[735]

ISAE3000 replicates the IFAE providing guidance on processes to obtain evidence for reasonable assurance engagements built on the financial audit methodology. This include: understand subject matter and other engagement circumstances; assess risk subject matter may be materially misstated; respond to risk by determining nature timing extent of further procedures linked to identified risk. These procedures include inspection, observation, recalculation, re-performance, analytical procedures and inquiry which may involve substantive procedures; corroborating inform from sources independent of responsible party, test operating effectiveness and controls and evaluate sufficiency and appropriateness of evidence.[736] The IFAE and ISAE3000 however do not provide a clear line of distinction between the amounts of work that has to be carried out to conclude with reasonable or limited assurance except for the fact that in the latter, procedures are *"deliberately limited relative to a reasonable assurance engagement"*.[737]

ISAE3000 the assurance provider has the responsibility for the reliability of documents used as evidence but disclaims responsibility with regards to authentication of documents.[738] The standard also states the circumstances when representations by the responsible party are appropriate as source of evidence. These

[733] RevR 6(2006) sec. 5.1

[734] IFAE, sec. 51-53. ISAE3000, sec. 36-37.

[735] IFAE, sec. 48-49. ISAE3000, sec. 25.

[736] IFAE, sec. 51. ISAE3000 sec. 35-36.

[737] IFAE, sec. 53. ISAE3000, sec. 37.

[738] ISAE3000, sec. 34.

circumstances are provided in section 39 of ISAE3000 which states that assurance providers need to evaluate its reasonableness and consistency with other evidence, knowledge of the source and need for corroborative evidence. The provision also clearly states that it cannot replace other evidence that is reasonable expected to be available. As the ISAE3000 only provides deliberate limitation of evidence gathering procedures relative to reasonable assurance, it does not specify the type of procedures that should be included or excluded from limited assurance engagements, giving that discretion to users of the standard.

RevR6(2006) only guide practitioners in the performance of limited assurance engagements. RevR6 excludes reasonable assurance engagements within its scope, there is no established "upper limit" which procedures can be deliberately limited. RevR6 requires procedures manifested in a *"balanced combination of review on internal control in company management systems and substantive review procedures, both analytical and detailed review procedures."*[739]

A 'non-exhaustive list' of procedures providing guidance on what should be reviewed, methods and level of detail required to reduce risk to an acceptable level to provide limited assurance has been constructed.[740] The discretion of evidence gathering procedures is determined by the risk assessment made for each engagement.

The list includes nine different procedures:

1. When reviewing both qualitative and quantitative information, the practitioner should interview representatives for the engaging party (e g representatives of the management of the company, business area management groups, managers in charge of various sustainability issues, controllers, etc.) in order to gather information about material events and activities during the time period that the sustainability report covers.

2. The practitioner should examine the reliability of the systems used to gather and compile the sustainability information (quantitative and qualitative) on which the sustainability report is based. In cases where some kind of internal audit is carried out on the systems used, the review may be based on the results from this internal audit. The practitioner also analyses the internal control system which may be built into the system and verify whether it works in practice. In cases where the systems are more complex, a specialist is consulted to assess the reliability of the systems.

---

739 RevR 6(2006) sec. 5.2.

740 Ibid.

3. The practitioner should review that the compiled sustainability data reported have been drawn up in accordance with the company's stated principles for calculation and accounting

4. The practitioner should visit one or more selected units in order to verify that material matters have been taken into account in the sustainability report and that data have been reported in a uniform manner and that they agree with established principles.

5. The practitioner should review qualitative information and statements in the report against the type of available documentation. In some cases the practitioner needs to supplement this with signed certificates from relevant officers confirming the information presented.

6. In cases where the company shows that it carries on a systematically documented dialogue with its stakeholders, this information should be reviewed by the practitioner. As one step in this review the practitioner can be present as an observer at one of these dialogue occasions. The practitioner should however be careful not to get involved in the dialogue in such a way that his impartiality or independence can be questioned.

7. The practitioner should make an overall assessment of information and data provided in the sustainability report. The purpose is to assess whether the reported information reflects actual and material conditions in the company operations.

8. The practitioner should review that all information and data provided in the voluntary separate sustainability report does not contradict the sustainability information given in the company's statutory annual report.

9. The practitioner should obtain written confirmation by the management that the facts in the sustainability report, the reported criteria, the bases for calculation and accounting principles. The management representative should also sign the original version of the sustainability report.

Figure 7.4 List of evidence gathering procedures

Practitioners have to review sufficiency and appropriateness of evidence, that serve as a basis for the practitioners' expressed conclusion.[741] The standard provides the need for additional review procedures or a qualified report in circumstances where material misstatements are detected. RevR6(2006) grants practitioner the discretion of deciding the nature and extent of procedures to be conducted to form a basis to

[741] RevR 6(2006), sec. 5.1.

provide limited assurance. By not imposing an upper limit, it could potentially be a problem when drafting procedures for reasonable assurance engagements in the future. The provisions of RevR6 leave it to the practitioners to determine stakeholder materiality and audit thresholds according to the needs and circumstances of each engagement. From the analysis above, there is also a lack of clarity on how the concept of completeness can be dealt with. Considering the scope limitations allowed and the extent of procedures, assurance providers has the discretion to accept engagements for parts of a sustainability report.

7.3.3 Guidance on assurance output

ISAE3000 recognizes the differences of each engagement and the concept of a continuum of assurance but simplifies communication of assurance to two levels: reasonable and limited.[742] It does not prescribe a standardized format accepting both the long and short forms of reporting. It prescribes a written form of the report and identifies basic elements that should make up the content of the report.[743] Section 54 of ISAE3000 also provides for the practitioner to consider appropriateness of reporting relevant issues of 'governance interest' to those 'in charge of governance'.

RevR6 requires practitioners to issues two reports: an internal review report with recommendations to the management of the reporting organization and an assurance report that the reporting organization has the option of publishing with the sustainability report.[744] It calls for a meeting between the practitioner and the reporting organization to discuss issues arising from the engagement and actions to be taken upon which the assurance process can be concluded. The requirement is in line with the definition of an "audit business" according to the Swedish Auditors.[745] The contents of the internal review report are left to the practitioners. The content of the external assurance report replicates the prescriptive clear and precise requirements of ISAE3000 as illustrated by the table below.

Content of report	ISAE3000	RevR6(2006)
Title	Title that clearly indicates that it is an independent assurance report	Title that clearly indicates that it is an independent assurance report

[742] IFAE, sec. 50.

[743] ISAE3000, sec. 45-49.

[744] RevR 6(2006) sec. 5.4.

[745] Auditors Act, sec. 2(8a).

Addressee	State the party or parties to whom the assurance report is directed. Whenever practical, the assurance report is addressed to all the intended users, but in some cases there may be other intended users	State the parties which assurance report is addressed
Name and location of assuror	Name of the firm or the practitioner, and a specific location, which ordinarily is the city where the practitioner maintains the office that has responsibility for the engagement	Signature, name, title of the practitioner
Scope and Objective of the engagement	An identification and description of the subject matter information and, when appropriate, the subject matter	Describe the subject matter and objective
Respective reporting responsibilities of reporter and assuror	A statement to identify the responsible party and to describe the responsible party and the practitioner's responsibilities:	Give account of respective reporting responsibilities of reporter and assuror
Criteria used to assess evidence and reach conclusion	Identify and make available criteria and relevant disclosures pertaining to e.g., the source, measurement method	Identify criteria which subject matter has been evaluated
Assurance standards used	A statement that the engagement was performed in accordance with ISAEs	Identify standard applied to assurance engagement
Conclusion / opinion	The practitioners' conclusion: Where appropriate, the conclusion should inform the intended users of the context in which the practitioner's conclusion is to be read. Reasonable and limited assurance should have positively and negatively phrased conclusion	Formulate conclusion based on the subject matter and purpose of engagement. Where appropriate, the conclusion should inform the intended users of the context in which the practitioner's conclusion is to be read

Performance	A summary of the work performed	A summary in bullet points of the work performed
Reporting on reservations and qualifications	Where appropriate, a description of any significant, inherent limitation associated with the evaluation or measurement of the subject matter against the criteria. Where the practitioner expresses a conclusion that is other than unqualified, the assurance report should contain a clear description of all the reasons.	Where appropriate, a description of any significant, inherent limitation associated with the evaluation or measurement of the subject matter against the criteria Where the practitioner qualifies the report, the assurance report should contain a clear description of all the reasons
Statement of limitations	When the criteria used to evaluate or measure the subject matter are available only to specific intended users, or are relevant only to a specific purpose, a statement restricting the use of the assurance report to those intended users or that purpose	When the criteria used to evaluate or measure the subject matter are available only to specific intended users a statement restricting the use of the assurance report to those intended users
Report date	Date	Date

Figure 7.5 Comparison of the reporting requirements between ISAE3000 and RevR6(2006)[746]

ISAE 3000 provides three circumstances where the practitioner should not issue an unqualified report. First, a qualified report or disclaimer should be issued when the assurance provider is restricted from obtaining evidence, to reduce engagement risk to an appropriate level which conclusions can be based. Second, the practitioner is guided to issue a qualified or adverse conclusion in the events where the conclusion of the engagement is to be worded in terms of reporting party's assertion, or if the subject matter and criterion are not fairly or misstated. Third, a qualified or disclaimer of conclusion is warranted if the subject matter or criteria were found to be unsuitable in the course of conducting the engagement.

RevR6 does not make a distinction between qualified, disclaimer and adverse conclusions. It adopts the term "bestykanderapport med avvikelse" which had been

[746] ISAE3000 sec. 45-53 and RevR6(2006) sec. 5.4

translated to "modified assurance report" to apply to the above three examples of circumstances which a qualified report may be issued. The standard requires the reporting of issues that do not affect the conclusion to be presented at the end of the report.[747]

7.3.4 Summary

The following table summarizes the above analysis on the discretion exercised by Far to revise RevR6(2006) in line with ISAE3000. Analysis of the construction of RevR6 also points out discretion it passes on to the audit firms when accepting and conducting assurance engagements in the Swedish market.

7.4 Discussion of findings and conclusion

7.4.1The exercise of discretion in a "principles based" regime

The IFAE defines and describes the elements and objectives of an assurance engagement, whereas the ISAE3000 has the role of providing the basic principles, essential procedures and related guidance consistent with the framework.[748] The ISAE 3000 was constructed in a manner to accommodate a vastly diverse range of assurance services on non-historical financial information in one pronouncement, applicable to both assertion-based and direct reporting engagements at reasonable and /or limited assurance. The broad objective has led to replication of the IFAE and the broad guidance provided due to the different requirements for engagements on different subject matter. The above analysis show empirically replications in provisions on subject matter, suitable criteria, planning and performance as well as in reporting requirements of assurance engagements.

The ISAE3000 has been criticized to be over-inclusive, difficult to apply and too general.[749] The IAASB however maintained its usefulness in providing guidance to the

---

747 RevR 6(2006) sec. 5.7.

748 IFAE, sec. 1.

749 APB Feedback statement on the 2006/7 work program consultation par 6.5.Note APB used ISAE3000 in development SIRs3000 for investment reporting. and 4000, IAASB: ISAE3000 project proposal 2009 p. 2

accountancy profession and opines that problems will be alleviated when subject specific standards are set. As such, although ISAE3000 serves to provide guidance to assurance providers, the discretion allowed to can be limited by IAASB itself or standard setters at a national level that set standards for specific subject matter according the standard. The extent which ISAE3000 is replicated reflects the respective standard setters' choice to exercise or delegate the discretion to users performing the engagement.

RevR6 (2006) is a subject specific standard drafted to guide practitioners engaged in assurance of sustainability reports in Sweden. The analysis of the contents of RevR6(2006) compared to ISAE3000 highlighted the discretion Far has chosen to exercise in the revision of standard in accordance with ISAE3000. Despite it being a subject specific standard, the above analysis detected that the IFAE and ISAE had been replicated in many areas e.g., in the suitability of criteria, work of experts, and determination of necessary competence. The findings suggest that the extent to which subject specific standards can solve the problems highlighted in chapter 4 depends on how standard setters drafting standards at the next levels on specific subject matter exercise discretion pertaining to the different issues.

Findings thus provide an empirical illustration of how discretion can decision making is distributed in multi-level regulation.

7.4.2 The importance of clarity

The background provided in chapter 6 informs that both reporting and assurance on sustainability reports are rapidly developing in Sweden. The first version of RevR6 (2004) was drafted using a "bottom up" approach which members of the standard setting committee describe as putting together the knowledge of the market and reflecting "best practice" in Sweden. Influences from the sustainability assurance regulatory space are evident as the first version of RevR6(2004) drew heavily on ISAE100 and FEE guidance. As a result the alignment of RevR6(2006) with ISAE3000 did not cause drastic changes in provisions of the standard. In RevR6(2006) references were also made to the GRI reporting guidelines and AA1000 series in the standard which leading reports in Sweden (and internationally) are adopting increasingly.

In line with Far and the Swedish Ministry's efforts to encourage more firms to engage in assurance services, given a market where reporting organizations are still developing internal control, systems and processes, flexibility had to be constructed in the guidance. It is clear from the above analysis the use of broad principles in the construction of the standard. The standard is characterized by a high level of generality, extensive use of evaluative terms and scope limitations to allow broad application and flexibility. Far has constructed a standard stressing minimum harmonization allowing wide discretion by practitioners to customize each engagement. It can be conceived that Far has taken into account the fact that GRI ascribes to an 'incremental approach' in the application of its reporting standard. The incremental approach appears to be supported by the task force of the IAASB that commented on NIVRA's exposure draft ED3410.

The provision stating the objective of an assurance engagement in IFAE, ISAE3000 and RevR6 were drafted specifying certain requirements with emphasis on the process and elements of assurance. Tracing the events in the regulatory space informs that IFAE and ISAE were drafted to promote greater consistency on how assurance engagements by the profession globally and establish the accountancy profession as primary providers of such services. The purpose of RevR6 seems to be less clear and evolving. Although RevR6(2006) state that assurance seeks to increase the users' perception of the credibility of the report, through the assurance provider's reported conclusion that the report has been drafted based on some stated criteria and there had been no material deviations[750], the purpose of the standard remain unclear as to whose interest it has been drafted. In 2004, Dan Brännström, general secretary of Far has attributed its importance to ensure better quality and transparency[751] in performance of these engagements. Legitimacy of the standard was constructed as keeping abreast with international harmonisation efforts.[752] The interest of the financial community was added in RevR6(2006) but drafted out in RevR6(2008) when adopting the provisions of NIVRA's standard 3410N.[753]

Literature reviewed in section 2.5.2 of this thesis, stressed the importance of the clarity of purpose or objectives in a principles based regulation for it to function effectively. The unclear purpose of RevR6 leaves the discretion of practitioners to be exercised without a common goal and contribute to the difficulties to drafting provisions that match the regulatory purpose of the standard.

The purpose of each engagement is left to the practitioners and the engaging party which has implications on engagement acceptance, planning, performing and reporting of an engagement. For example, practitioners vary in their decisions as to which users' needs should be considered, whether the accuracy or relevance of information should be prioritized, how suitable criteria is should be determined according to each engagement. One can preliminarily conclude that a wide variation is expected on how the guidance is used.

---

[750] RevR6 sec. 2.1

[751] Far press release, 11th March 2004.

[752] The introductory paragraph of RevR 6(2004), sec. 1. RevR 6(2006), sec. 1. RevR 6(2008).

[753] RevR 6 (2006), sec. 1.

7.4.3 The limits of provisions - the irony

Drawing from the findings presented in chapter 5, one might expect that standards are set to ensure harmonization and quality of engagements; standards serve to guide or control discretion of its users. The limits of rules reviewed in section 2.5.3 inform that rule making, (in this case, the provisions of standards) involves generalizations, simplifications and suppressions. There are therefore limits as to what standards can do. Standards cannot be perceived to cover each situation.

The analysis show empirically how the limits of rules creates discretionary margin but ironically, it is also this indeterminacy of rules that provides scope their necessary adaptation in the process of implementation to suit practical circumstances and local contexts. Evident from this analysis is that the limits of rules offer a space which weak and / evolving concepts such as materiality can develop through interactions between the standard setters, the users of standards, the market as well as other activities in the regulatory space.

In certain areas, essential procedures had been drafted by Far to limit the discretion of practitioner's in performing of engagements on sustainability assurance. The provisions however tend to be imprecise, complex and lack clarity often including evaluative terms and subjective judgment in its construction. Without guidance notes or further standards or an authority to interpret the standard, the discretion of how these vague issues should be dealt with fall on the user that is performing the engagement. The limits of rules offer the user to standards the flexibility to exercise discretion within the institutional and market context it operates.

7.4.4 Harmonisation and quality control

Standards play a role in harmonisation.[754] The level of harmonization is influenced by the discretionary space awarded to the users of the standard. The above analysis shows that the technical construction of the contents of assurance standards by the accountancy profession only reflects the explicit codification of infrastructure and processes. The standard leaves elements such as materiality and determination of engagement risk to "professional judgment". This suggests that to understand how standards work, it is important to take into consideration the tacit knowledge of particular communities within the accountancy profession.

The construction of RevR6 offers a wide discretionary space, providing users the discretion of filling in the details to suit each particular engagement. One can

---

[754] Peters, A. et al. (2009), p.12

preliminarily conclude that aside from a standardised report, it is unlikely to contribute significantly to uniformed or homogeneous practice. The disadvantage of a more precise well targeted standard in this developing field could potentially leads to creative compliance and serve as a constraint to respond effectively to evolving conditions without constant need for amendment. Constant revisions can also undermine confidence of relevant stakeholders.

The focus is therefore not how much this standard is likely to harmonize or standardize practice. Building on the work of Black presented in section 2.5.5 of this thesis, this study put forth the notion that that the extent which harmonization can be achieved with broad-based principles, lacking a clear purpose is dependent on shared understanding by an interpretive community, this case referring to Swedish practitioners. The closer the community and the more shared understanding between members would result in a higher level of harmonization.

7.4.5 Method limitations

Several difficulties were encountered in the course of the analysis. First, in the absence of detailed drafting conventions by Far, It had been difficult to identifying elements of the standards that are obligatory and parts that were guidance notes, a problem the clarity project of IAASB seeks to alleviate. Difficult was also encountered in obtaining official translations of RevR6 (2006). Second, Far pronouncements are published in Swedish and are not translated to English by a recognised appointed authority. Although there is a Swedish-English dictionary, the translation was difficult due to the evolving terminology and discourses used in this developing area. The translations in the above analysis has been based on an English version of RevR6(2004) obtained from a practitioner who is a Far expert member.

7.4.6 Conclusion

This chapter is the first of the two chapters analysing how standards work. The analysis above mapped out the discretionary space which the provisions of the standard granted to its users. RevR6 are a not self- executing, it is an explicit codification of "best practice" and are dependent on practitioners to use and apply the standard when accepting and performing assurance on separate sustainability reports. Findings of this chapter serve as input to analyse how discretion granted by assurance providers is exercised in light of the activities in the regulatory space, the Swedish market and institutional context which these engagements are performed.

8. The Nature and Character of RevR6 as a Regulatory Tool

8.1 Introduction

8.1.1 Objectives

This chapter presents the second part of the analysis of how standards work in Sweden. The analysis has been conducted using two different paradigms separately. The first section adopts a social science perspective to analyse the discretion exercised by assurance providers within the space granted to them by RevR6(2006). The second part adopts a legal lens to analyse how RevR6(2006) works within existing state-centred regulation on services provided by the accountancy profession. The findings will be put together in the conclusion, to discuss the nature and the character of RevR6 as a regulatory tool and the potential interplay of RevR6(2006) with hard law in the regulation of sustainability assurance.

8.1.2 Method

Given a better understanding of the Swedish context within which sustainability assurance operates. This section provides a more detailed account of how information has been collected to perform the analysis. To recapitulate, in the method section presented in section 3.3, two sources of data was to be collected to perform an analysis on the discretion exercised by assurance providers within the space granted to them by RevR6(2006). The contents of assurance reports and findings from the interviews with representatives from the reporting companies would be used to make inferences on the choices the assurance providers made pertaining to the elements of assurance. The latter was to serve as a supplementary information source to clarify and provide deeper insights on the assurance engagement.

8.1.2.1 Selection of sample

The assurance reports of a sample of companies have been selected to provide information to analyse how RevR6(2006) works. The selection of companies to study took into consideration the time frame, type of provider, type of company and their status as going concerns. With reference to section 6.2.4, only 18 Swedish companies had commissioned third party assurance from 1995 to 2007, of which 16 had engaged the accounting firms. Of the 16 companies, only 9 had commissioned assurance engagements from the time RevR6(2006) was published. To preclude the effects of RevR6(2008), only assurance reports published with companies' separate sustainability reports up to 2007, issued by the accountancy profession were selected. Companies refer to private companies as well as state-owned companies operating under market conditions, and who were still trading in 2007. This narrowed the sample to nine companies with a total of 40 reports to be analysed over the set timeframe. The profiles of these companies are presented in the table below.

From the above information, these companies have been further divided into two groups according to their experience with assurance engagements. Companies in Group 1 have commissioned assurance on their environmental or sustainability reports prior to the publication of assurance standards by Far. Companies in Group 2, engaged assurance services only after the introduction of RevR6(2004).

Company	Industry	Ownership	Turnover (MSEK)	Employ -ees	Years with assurance
SAS	Transport-Airlines	Public listed, state-owned 21.4%	45885	28000	11
SKF	Industrial Machinery	Public listed	58,559	42888	10
SCA	Paper and Forestry	Public listed	105913	50433	10
Vin &Sprit	Beverage	Stated owned 100%	58,599	42888	2
Ericsson	Communicatio n	Public listed	6,100	3115	2
Trelleborg	Industrial Machinery	Public listed	20,281	1588	2
Sandvik	Industrial Machinery	Public listed	86,338	47123	1
Green Cargo	Transport-rail	Stated owned 100%	6111	3050	1
SJ	Transport-rail	Stated owned 100%	8267	4053	1

Figure 8.1: Profile of companies in 2007

Group	Characteristics	Companies
1	Companies that have commissioned assurance prior to the publication of RevR6(2004)	Companies: SAS, SKF, SCA
2	Companies that have commissioned assurance after the publication of RevR6(2004)	Companies: Trelleborg, Green Cargo*, Sandvik, SJ*, Ericsson*, Vin & Sprit,

Figure 8.2: Grouping of Companies by experience with assurance engagements

*Companies whose reports have been analysed in the sample but interviews were not carried out

8.1.2.2 Data collection

An analysis was performed on the contents of the assurance reports of the nine aforementioned companies. The contents of assurance reports were analysed according to the coded categories developed initially from the elements of assurance. They include subject matter, criteria, the three parties, work performed and the assurance providers' conclusions.[755] In the coding process, categories were revised to capture the developments in standards and to provide for "unexpected" categories, these being ones that are interpreted by the researcher to be significant to the purpose of carrying out the analysis. For example, the category "standard" adopted by assurance providers was added.

The analysis has been divided into two phases: prior to and after the availability of Far's guidance. No distinction had been made between the period when RevR6(2004) was published and after its first revision, as the analysis of the contents of the standards reflect insignificant changes in the first revision. The grouping of the companies allows analysis of assurance providers' exercise of discretion in cases where the engagement is conducted on companies that have had a long standing commitment to assurance and those that are new commissions where the assurance provider is not bounded by previous engagements.

Content analysis of assurance reports allows the researcher to make inferences to practice, but does not provide the information required to understand the substance of these assurance engagements. Despite a listing of work performed by assurors, the reports do not reflect extent of work carried out: the scale of engagement, weight and emphasis placed on work performed. Hence, the above is supplemented by semi-structured interviews conducted with key representatives from six of the nine companies contacted in 2007 (Refer to appendix 3 for the list of interview questions). Interview findings provide insights on the discretion exercised by practitioners when applying the standard in the course of interactions with the reporting organizations. It also highlights the level of transparency i.e., what is communicated to the reader and what takes place behind closed doors. Three companies did not respond to the interview request. Interviews were focused on how work is performed by assurance providers for just the year 2007. This was due to practical considerations, as interviewees were better able to discuss current engagements in depth compared to previous engagements.

Telephone interviews were conducted with five of the six companies; one interview was conducted via email with two follow up interviews via telephone. Interviewees were persons responsible, decision makers in the commissioning of sustainability assurance or an appointed representative of company on issues related to sustainability

---

[755] Refer to section 4.3 of this thesis.

assurance. (Refer to appendix 3). Corroborating evidence to increase the reliability of interviews was carried out through telephone interviews with experts in 2007. This enables the possibility for one to make inferences on how discretion has been exercised by assurance providers, and how standards work and affect the elements of the sustainability assurance engagement.

8.1.2.3 Limitations of research design

Research on audit regulation and the accountancy profession highlight the importance of audit firms as a site of regulation. In the field of sustainability assurance, the risk department of the large accounting firms has been highlighted as an important party, and one which assurance providers are focused on in relation to legitimacy.[756] A relevant source of data to determine the discretion exercised by assurance providers would be the internal manuals of the large accounting firms, their methodology and working documentation (files) of the assurance providers. This was however not possible as access was denied. Alternatively, field studies where the researcher acts as a silent observer when assurance engagements are conducted could have been used, but again, access was refused. Therefore, data is collected from assurance reports and interviews with the reporting organisation are used to perform this analysis.

First, it is recognized that interviewer and interviewee biasness presides in this method of collecting information. The findings from interview with companies with regards to the work performed by assurance providers is thus contingent upon the management's perception of what the assurance provider has done in carrying out the engagement. The positions held by representatives interviewed vary between firms depending on how the reporting and assurance process has been embedded in their organisational structures. During the course of the interview, the researcher detected differences in perspectives, discourse and terminology used by interviewees.

---

[756] O'Dwyer, B., Owen, D.,Unerman, J. (2011) "Seeking legitimacy for new assurance forms: The case of sustainability assurance", Accounting, Organizations and Society , 36, (1), 31-52.

8.2 Exercise of discretion by assurance providers – findings from assurance reports

8.2.1 Discretion on engagement acceptance

8.2.1.1 Appropriate subject matter

Prior to the publication of RevR6(2004), the subject matter of assurance engagements ranged from environmental information generated by management information systems to a full environmental or sustainability reports. RevR6(2006), likewise RevR6(2004) allowed limited scope engagements to be accepted with no specification of the minimum content or a match between the scope of reporting and assurance. The scope limitations allowed by RevR6 made it clear that practitioners were granted the discretion of accepting a wider range of engagements.

The table below compares the subject matter of 9 reporting organisations in the 2007 report.[757]

Company	Assurance provider	Subject matter which assurance is provided	Scope limitation
Group 1 Companies			
SAS	Deloitte	The sustainability reporting includes the SAS Group Annual Report & Sustainability Report 2007 as well as accounting principles, GRI cross-reference list and stakeholder dialogue, found on the SAS Group website.	The review did not include the data in the table The SAS Group sick leave and accident statistics on p. 106.
SKF	KPMG	The sustainability report is presented on p. 107-123 of the SKF Annual Report 2007 including Sustainability Report, and on SKF's website in "Topics related to Annual Report 2007, including Sustainability Report", in the form of environmental performance data, Zero Accidents Award winners and compliance with GRI Guidelines.	No explicitly expressed limitation in scope

757 The assurance report by practitioners on 2007 environmental or sustainability reports is issued in 2008 performed according to RevR6(2006)

SCA	Deloitte	Information from SCA's Resource Management System (RMS) provided under the heading "Goal", on p. 8 regarding "Goal 1" and "Goal 2" and on p. 47-57 in the SCA Sustainability.	No explicitly expressed limitation in scope on environmental reporting
	PWC	All information presented in the tables, diagrams and graphs provided under the heading "Employees" and "Health and Safety" in the SCA Sustainability Report 2007	Limited to sections of the social report mentioned
Group 2 Companies			
V&S	PWC	The information presented in all tables, diagrams and graphs in the section "Responsibility" of the Annual Report for 2007 p. 22-33	Limited by type of information
Ericsson	PWC	Two performance indicators: supplier reviews "results of risk based projects, 2006" p14 & results of improvements in energy efficiency of radio base stations (Average radio base station energy consumption" p. 25	Number of indicators (specified by page numbers)
Trelle-borg	PWC	Performance indicators CO2 emissions p. 18, VOC emissions p. 19, Total waste pg. 21, work related accidents p. 22, training hours per employee p. 25, sustainability related expenditures p. 29 in the Corporate responsibility report 2007	Number of indicators (specified by page numbers)
Sandvik	KPMG	pages 19–20 and 82–88 of Sandvik's 2007 Annual Report and on Sandvik's website Under GRI index in the form of complete references G3 which have been applied in the preparation of the sustainability report	No explicitly expressed limitation in scope
Green Cargo**	Deloitte	Sustainability information related to the Scorecard areas Safety, Employees, Society, Environment and Customers presented on the inside of the cover, pages 54–65 and pages 68–69 in the Green Cargo's Annual and Sustainable Development Report 2005.	Limited to scorecard areas
SJ	PWC	Total emissions of carbon dioxide and customer satisfaction index	Limited by the number of indicators

Figure 8.3: Scope limitations of sustainability assurance engagements

*Ericsson changed the services of DNV (non-accountancy firm) in 2007. The 2006 information is presented here for the reader's information.

**Green Cargo integrated its sustainability report with its annual report in 2007. The subject matter and scope limitations are presented here for the reader's information.

The above table shows that scope and subject matter of assurance engagements on sustainability reports performed using the RevR6(2006) vary widely. Engagements can range from verification of a few performance indicators, or a section of the report, to assuring the whole published sustainability report, including accounting principles and relevant published information on the reporting companies website such as the GRI index and stakeholder dialogue. Only four of the nine companies commissioned engagements on the full sustainability report. Other engagements covered 15 to 20% of information reported. For example, SJ's engagement covered two out of ten strategic indicators, of which one was an environmental indicators and the other was a social indicator. It is important to be aware that this is not an issue specific to Sweden but one occurring internationally. Only about 22% of organisations that commission assurance worldwide engages providers to assess the entire report. Interestingly, practitioners have also defined within the scope of RevR6(2006) information pertaining to non-financial goals and performance in the report of the Board of Directors in the Annual report in Sandviks's engagement. This, from analysis in chapter 7, arguably lies outside the scope of RevR6.[758] The same pages, 19-20, were included in the financial auditors report audited at a reasonable level of assurance.

8.2.1.2 Three party relationship: the user

RevR6 clearly defines the practitioner and the reporting organisation as two of the three parties required to be present in an assurance engagement. A broad definition of user with no further guidance on how to and who should identify the user, has allowed assurance providers to keep this third party vaguely defined as the "user of the assured information" addressed as "readers" of the report[759]. The imprecise definition allows assurance providers flexibility in deciding on conditions of acceptance, scope limitations and work performed in an engagement.

The expert interviews suggested that the reporting company has the task of defining reporting materiality and identifying key stakeholders. Taking into consideration the subject matter, differences in a company's operations and the voluntary status of

---

758 It is interesting to note that these 2 pages had been audited at a reasonable level of assurance in the financial auditors report.

759 RevR6(2006), sec. 3, 5.5.

reporting, the reporting boundaries among firms differ even if assurance is performed on the full report. For example, in Sandvik's 2007 report, three key stakeholder groups: investors, customers and employees were identified and focused on, compared to SAS that include as key stakeholders suppliers, governments, regulators and aviation organisations, partnerships and networks, NGOs, the travelling public and the media in addition to the above three parties. RevR6(2006) grants assurance providers the discretion to match the map of "relevant stakeholders" with that of management or limit or extend the definition of stakeholders according to the agreed scope of the engagements. In cases where scope is limited, stakeholders that assurance is provided for may be a narrower group than that defined by management. For example, in Trelleborg's engagement, the stakeholders to whom the report is targeted by the reporting organisation are likely to be concerned with issues aside from the few specific indicators verified by assurance providers. Often it is unclear if the users defined by assurance providers match that of the reporting firm. For example, in Sandvik's engagement in 2007, the assurance report had been addressed to readers and did not make clear in the objective, scope and work performed if the assurance had been conducted the matching the key stakeholders defined by the company or had assurance providers adopted a broader definition to users.

8.2.1.3 Suitable criteria

Suitability of subject matter and reporting criteria cannot be assessed in isolation due to their inextricable relationship. The subject matter is based on reporting criteria which can serve as a basis of evaluation in an assurance engagement. Prior to the publication of GRI reporting guidelines, the criteria used in engagements that resulted in an assurance or audit report for readers included the calculation and disclosure principles drawn up by the reporting company or the assurance provider. For example, Deloitte and Touche's checklist for preparation and evaluation of environmental reports was adopted by SAS prior to SAS' reporting according to GRI guidelines.[760]

RevR6, does not provide guidance on criteria specific to assurance on sustainability reports. The recognition that criteria is context dependent, the acceptance of both established and specifically developed criteria grants the practitioner a wide discretion of deciding on its suitability based on five characteristics: relevant, complete, reliable, neutral and understandable defined in a vague manner consisting highly evaluative

---

[760] The checklist had been updated over the years that incorporated a number of principles from GRI, Social Accountability 8000 etc. reflecting the professional accounting firms' knowledge of "best practices" and developments in the regulatory space.

terms. The examples in the standard suggest that GRI reporting guidelines and stakeholder engagement fulfil the requirements of suitable criteria.

In 2007, seven reporting companies adopted the GRI reporting guidelines to different extents as the reporting criteria and the practitioners mention "applicable parts" of the GRI to be part of the assurance criteria in addition to specific accounting, calculation, measurement and disclosure principles specified and adopted by the reporting organisations. There is a lack of transparency on how suitable criteria have been assessed. For example, how partial application of GRI satisfies the application characteristics (relevance, completeness and reliability) of suitable criteria and how criteria developed by reporting organisations are neutral or "free from bias" and management capture. Stakeholder dialogue was mentioned in SAS' engagement as part of the assurance subject matter, the extent to which its output contributes to the criterion in the assurance engagement was not explicitly addressed in the assurance report.

Findings suggest assurance providers consider that criteria developed by the reporting organisation, assurance providers and partial use of GRI reporting guidelines to be suitable.

8.2.1.4 Competence of assurance providers and signatories

RevR6(2006) has left the discretion of skills warranted to carry out an engagement and assembling of the assurance team to the discretion of the assurance provider. Assurance reports do not reveal the number, skills, composition or members of the assurance team. The only available information is the signatories of the report.

Prior to the introduction of sustainability assurance standards, assurance reports were signed by one or two authorised public accountants under the name of the accounting firm. In 2005, a new group of signatory the "Expert member, Far" emerged. This group of new signatories also use other titles such as "Sustainability assurance specialist". From 2005, signatories of the assurance report could include only authorised public accountants (e.g. SAS assurance reports 2005-2007), only the expert member (Vin & Spirit 2005 assurance report) or in the other reports in this sample, signed by an authorised public accountant accompanied by a Far expert member.

The standard can be applied by both authorised public accountants and Far expert members. The qualification requirements of the two parties vary significantly. The qualification of the former is regulated at the European level by the Directive on statutory audits of annual accounts and consolidated accounts transposed into the Auditors' Act. With reference of section 6.3.3 of this thesis, there are no statutes regulating the qualification of Far Specialist (Experts). Accounting firms have the discretion of 'qualifying' experts through in-house training programs. Other members of the assurance team are assembled based on the requirements of the engagement.[761]

The discretion of qualifying experts and assembling teams with required competence to conduct a sustainability assurance engagement lie with the accounting firms.

8.2.1.5 Ethical requirements

RevR6 provides that assurance engagements should be accepted only when ethical requirements can be met. Findings reflect the trend that similar to practices in financial auditing, the independence of assurance providers are seldom made explicit in the assurance report prior to or after the introduction of RevR6. Exceptional cases in sustainability assurance where assurance providers included a statement on independence occurred when AA1000AS (2003) was jointly adopted with RevR6. For example, KPMG and Deloitte in the assurance report on SKF's engagement 2004 and 2005-2007 respectively had made explicit their independence from the reporting organization and impartiality in relation to the stakeholders of the reporting organisation in accordance with the provisions in AA1000AS (2003).

AA1000AS(2003) guides assurance providers to perform engagements taking into account materiality, completeness and responsiveness. The presence of engagements conducted using RevR6 and AA1000AS jointly confirms that assurance providers adopt the view that the standards can be used together in a way that does not compromise the ethical requirements imposed on the authorised public accountants and Far Expert members. RevR6's warning provision on the compromise of independence when engaging in stakeholder dialogue without specifying the permissible extent of stakeholder contact, enable assurance providers to interview external and internal stakeholders as of the firm as part of the evidence gathering procedures to verify the reporting organisation's responsiveness to their key stakeholders. Findings suggest that the assurance provider has the discretion of determining the extent to which stakeholder contact will compromise independence or impartiality.

---

[761] Interview with Expert

8.2.2 Planning and performance of the engagement

8.2.2.1 The choice of standards

Both RevR6(2004) and RevR6(2006) did not make provisions on whether the standard can be used with other standards and circumstances which ISAE3000 can be used except in cases involving a reasonable level of assurance. Accounting firms in Sweden by virtue of their membership in Far have applied RevR6 in performing their engagements.

The table below show the discretion of assurance providers on other standards adopted.

Company	2002	2003	2004	2005	2006	2007
SKF	Guidance by FEE	RevR6 (2004)	RevR6(2004) & AA1000AS	RevR6(2004) & AA1000AS	RevR6 & AA1000AS	RevR6 & AA1000AS
Ericsson					ISAE 3000	
Sandvik						RevR6 & AA1000AS

Figure 8.4: The choice of standards by assurance providers

Developments in the regulatory space in chapter 5 show the growing acceptance of AA1000AS an assurance standard which the accountancy providers have adopted in practice.

The above table show the joint application of RevR6 and AA1000AS to guide engagements of SKF from 2004-2007 and Sandvik in 2007. The adoption of AA1000AS by SKF was attributed to "part of the group's aim of continuous improvement"[762] It was therefore at the request of the reporting organisation that AA1000AS be applied. This is evident as despite the change in assurance providers in 2005, both standards continue to be applied jointly. The circumstance was different for Sandvik as according to its group assurance department, it had been the assurance providers who brought on board the AA1000AS. Interesting to highlight is also PWC's application of ISAE3000 applied to conducting a limit assurance engagement on Ericsson's 2006 report instead of RevR6(2006) and the change of providers to non-accountancy assurance provider the subsequent year.

[762] SKF Sustainability report 2004 p. 105

RevR6(2006) omission to provide for joint adoption of standards provides assurance providers the opportunity to accommodate reporting organisations' needs. They can thus request other standards such as AA1000AS to be employed as well as to incorporate it as part of the firm's methodology without having to explain or perform reconciliations on the different requirements by both standards.

8.2.2.2 Work performed

RevR6(2006) guides assurance providers on how to use a balanced combination of review on internal control in company management systems, and substantive review procedures, both analytical and detailed, when planning and performing an engagement from the viewpoint of the user. A non-exhaustive list was drafted that provides no ceiling on the extent of work performed in a limited assurance engagement. Assurance providers are granted wide discretion to determine what constituted "sufficient and appropriate evidence" which conclusions can be based on.

The follow presents findings of how assurance providers exercised discretion granted in three types of assurance engagements. First, in engagements where scope was limited to a few indicators or type of information; second when assurance was performed on a large part or full sustainability report and third, when assurance was performed combining RevR6 with AA1000AS. Information is based on the assurance report supplemented by information obtained from interviews with reporting companies.

Narrow scope assurance engagement

Trelleborg and Vin & Sprit were two companies in Group 2 that commissioned assurance engagements that were limited in scope. The assurance reports published with the 2006 sustainability report show that in both engagements; assurance providers carried out five of the nine listed procedures stated in RevR6. They included interviewing the management of the respective companies to understand significant events of the year; review of reliability of systems and routines of data registration, accounting and reporting of performance indicators; review of reporting, calculation principles; site visits to determine their compliance with group principles and test that information presented is based on underlying documentation. Review of qualitative statements, stakeholder dialogue, and overall assessment that reported information reflects actual and material assessment and consistency with annual report were omitted.

Interviews with companies reveal that both assurance engagements were performed by 2 to 3 active assurance providers on site. The Trelleborg engagement was limited to three key performance indicators identified by management. The engagement was performed in about a month. The assurance provider picked 5-10 production sites out of 130 which it felt were representative of the population.[763] Persons responsible were issued questionnaires (via email or phone) on how data had been collected and kept. The Vin & Sprit engagement which limited its scope to information presented in tables, diagrams and graphs of the sustainability report. The engagement was performed between 2 to 3 weeks. The assurance provider chose 10 personnel to be interviewed: 5 at site and 5 at group level. At site level, interviews were held with Environmental consultant, Site manager, Purchasing Manager, Environmental Managers from different sites. At the group level the Director Corporate Responsibility, Manager Corporate Responsibility, Consultant Corporate Responsibility, Human Resources and a Purchasing Manager HQ were interviewed.

The interviewees from both companies perceived that the assurance provider placed heavy weightage of their work performed on the verification of data and the engaging party's involvement in the assurance process to be minimal, aside from providing required information. The engagement was focused on the accuracy of facts, calculations as efforts were concentrated on tracing data back to original data sources, assessing weak links in processes and checking calculations and definitions of terms.

Wider scope assurance engagement

SAS has engaged assurance services on their environmental/ sustainability report since 1997. In 2006, assurance was commissioned on the full sustainability report as well as accounting principles, GRI cross-reference list and stakeholder dialog, found on the its Group website. The SAS engagement was the only engagement provided on a full sustainability report applying solely RevR6(2006).

The 2006 assurance report showed that assurance providers from Deloitte carried out seven of the nine listed procedures in RevR6. In addition to the five procedures performed on the two aforementioned companies' reports; assurance providers reviewed scope and limitations on the content of the report and examined if the sustainability report contain any contradictions with the annual report. The non-exhaustive list provided in the standard enabled assurance providers, in addition, to review external environmental cost and investments and SAS' compliance with law, permits and terms and conditions. Although stakeholder dialogue formed part of the subject matter, the report did not state explicitly the procedures performed to review stakeholder dialogue. Interview with the engaging party informs that stakeholders had been inactive in the assurance process as assurors draw conclusions on needs of stakeholders based on stakeholder related documents kept by the company and interviews with company personnel with regard to stakeholder issues.

---

[763] In 2007, the assurance report stated that 2 site visits were carried out, one in Houston USA and the other in Skellefteå Sweden

The 2006 assurance engagement lasted approximately 9 months involving actively four to five assurance providers on-site. The engaging party describes that the assurance cycle starts with the closing meeting (usually in May) of the last cycle where "management and practitioners discuss their experience of the year, jointly identify issues (for example, uncertainty or figures that are not to expectations that cannot be explained) and may pick sites or subsidiaries to focus on in the next assurance cycle." At the formal start of the new cycle (September), management meet with assurance providers to discuss the major changes during the year after which assurance providers start to review the underlying systems (November). Upon completion of the first draft of figures (January) by the reporting organisation, deviations are identified and sites are identified for further testing. Reporting issues are then discussed with management; agreed corrections are made before the assurance providers sign the assurance report. The process ends with the issue of a report to management with feedback on the assurance engagement, results and suggestions for improvement.

Another group 1 company, SCA, engaged two assurance providers to assure different sections of their 2006 sustainability report. Deloitte was engaged to provide assurance on the environmental section of SCA's reports and PWC was engaged by SCA to conduct assurance on parts of the social section of the same sustainability report. The analysis in this section focuses on the work performed by Deloitte on the environmental section of SCA's sustainability report to review that information from SCA's Resource Management System (RMS) in the SCA's Sustainability Report 2006, and in the PDF-file "RMS mill data" on SCA website.

With reference to the 2006 assurance report issued by Deloitte, six of the nine procedures in the recommendation were carried out. Compared to engagements limited to several indicators, worked is performed reviewing the scope and limitations of the contents of relevant parts in the sustainability report.

The engagement involved a team of four practitioners on site. The management of SCA regard assurance on the environmental section of the report to be a year-round process. Most of the assurance work was conducted three months prior to the signing of the report. The Interview reveals that, assurance providers also perform "specialised studies" on specific sites or events marked for improvement as the focus of these engagements is on the improvement of reporting accuracy and processes.[764]

[764] In SCA's case where there are two providers, the long term commitment, CEO report, Environmental Goals 3 & 4 on responsible use of wood raw material and COMPANY 3 Code of conduct, supply chain and stakeholder dialogue as well as social information that are not presented in the tables, diagrams and graphs are among strategic issues that are not covered by third party assurance.

Assurance provided on a full report combining RevR6 with AA1000AS

SKF and Sandvik had assurance engagements performed on the full sustainability report combining RevR6 and AA1000AS.

With reference to KPMG's 2006 assurance report on SKF's sustainability report, eight of the nine procedures in the recommendation were performed. In addition, three procedures were incorporated. First, interviewing certain external and internal stakeholders to "verify that the SFK report had responded to important stakeholder's concerns in public sustainability information"; second, an overall assessment was conducted to determine if the reported information, "in all material aspects, reflected stakeholder requirements regarding sustainability information" and third a review of report on the "report compliance with legislation, permits and conditions based on sustainability".

The duration of the assurance cycle lasted about nine months. An initial meeting was conducted where the assurance provider met the management (Sustainability team) to discuss key issues from the previous years and issues that SKF would like to focus on. At the meeting, the sustainability team informs the assurance provider of the significant changes that have taken place in the year. As reflected by SKF's representative: " KPMG would ask if we have any units or issues that we would like them to pay extra focus on, KPMG on the other hand will inform us how they intend to carry out the audit.

The assurance engagement was conducted by two on-site practitioners who could be local (depending on the location) or Swedish. Five sites were visited in 2006. The drafting of the sustainability report concurred with the performance of the assurance engagement. "As we write the sustainability report, KPMG would recommend or suggest where we need to add more clarification or the choice of wording to avoid misunderstanding by the readers." The assurance provider carried out procedures on accuracy and reliability of systems and reports; reviewed principles of reporting and visiting five sites in 2006. In conducting the engagement, the assurance provider determined the internal and external stakeholders to be interviewed to obtain their opinion on the report. Different internal and external stakeholders were selected annually. Stakeholder feedback and other work performed constituted evidence to be used as a basis for forming conclusions. When the work performed was completed, KPMG issued an assurance report and held a closing meeting with the Investor relations, and Head of Sustainability to discuss issues that have occurred in course of the engagement and made recommendations to the management on areas to be improved.

Sandvik commissioned sustainability assurance engagement on their sustainability reports for the first time in 2007, the subject matter of the engagement was the full sustainability report. Work performed included eight of the nine procedures in the recommendation. The engagement spanned over ten months involving three members of the assurance team from KPMG, eight sites were visited[765]. The reported procedures carried out were similar to SKF in particular with regard to stakeholders. The focus of the report by management to three key stakeholder groups was not reflected in the assurance report. Management perceived their influence to be primarily at the planning stages where there had been a "joint evaluation" of the units to visit in accordance with the level of risk.

The assurance engagement can range from two weeks to a year indicating the discretion assurance providers have on the extent of work performed and scope limitations dependent on the resources reporting organisation allocates. The engagements of SAS, SCA (environmental section), SKF and Sandvik support the typical assurance cycle drawn up by Park & Brorson[766] in 2005 depicted in the diagram below. The planning meeting between assurance providers and management of the reporting organisation marks the start of the assurance engagement for the year where information about the company's reporting principles, systems and process, internal controls are gathered and/or updated to perform risk assessment. Background work such as site visits and meeting with persons responsible for data follows. Providers then review the database performing test on integrity and accuracy of the system and consistent of definitions as data is recorded.

---

[765] Sandvik facilities in Czech republic, Germany, US, South Africa, Sweden, Italy and India

[766] Park and Brorson (2005)

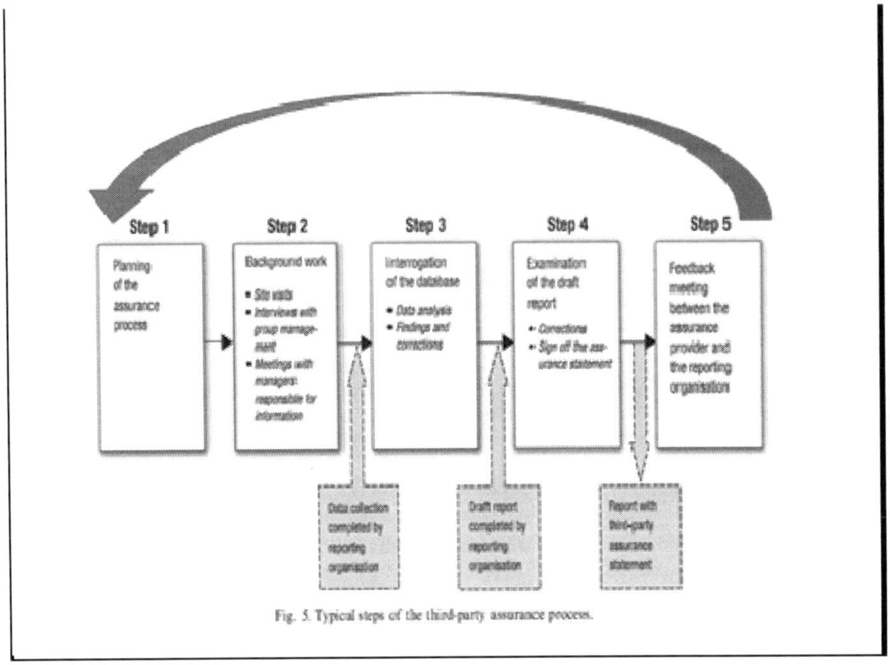

Figure 8.5 Typical steps of a third party assurance process[767]

Upon completion of the first draft of the report by the organisation, providers examine the contents against predefined criteria; suggest corrections to be made and sections to improve. Providers then put assurance on the report when amendments have been made and boundaries of the report are clearly presented. The engagement ends with a feedback meeting between providers and management. The assurance process takes place simultaneously as the reports are prepared and completed.

From interviews with SAS, SCA and SKF, it has been established that the issues discussed in the closing meeting of an assurance cycle serve as important input to the opening meeting of the following year as assurance forms part of the reporting organisations process of continuous learning and improvement. Likewise, the assurance provider continuously accumulates knowledge with regard to the relevant operations of the reporting organisation. A loop can therefore be added from step 5 to step 1 as issues and feedback discussed in the closing meeting serves as important input to the next cycle. In the case of Trelleborg and Vin & Sprit however, where the

[767] Source: Park, J.; Brorson, T. (2005), p. 1102.

scope of the engagement is significantly limited, the assurance process could commenced after the first draft of the report had been released. The location of the first two grey boxes depicting when data collection and draft report is completed are therefore not fixed but dependent on the scope of the engagement.

There is no evidence of correlation between the duration of engagement and the "newness" of assurance to an organisation. Differences in duration and stage of involvement are more attributable to the scope of engagement which is dependent on resource considerations, maturity of firms reporting and internal control systems determining the terms negotiated between the two contracting parties. The flexibility granted by RevR6 has allowed assurance providers to accept and accommodate a wide range of engagements.

8.2.3 Output of the engagement

According to RevR6, the output of the engagement is presented in two separate reports. An assurance statement is presented, usually attached to the end of the sustainability report and an internal report serving as feedback and recommendations to management which is confidential to the company. With regards to the former, RevR6 imposes a minimum content and the condition of clarity and comprehensibility. It also provides an example to guide assurance providers although how to report is at the discretion of assurance providers. The latter report is left solely to the discretion of assurance providers.

8.2.3.1 Title of report

Prior to the publication of RevR6(2004), assurance engagements were titled auditors statement, auditors report, environmental auditors report, independent auditors report, limited review reflecting the lack of harmonisation. After the publication of RevR6(2004) all reports were entitled "Statement of limited review" according to the example presented in §5.5 of RevR6(2004) The company interviewed confirmed that the deviation of the title did not correspond to any difference in the engagement. The publication of RevR6 (2006) aligned the terminology used with developments in the international regulatory space. In light of the clarity and precision of reporting guidance, all the reports in 2006 were titled "assurance or independent assurance report".

8.2.3.2 Reported objective and scope

Prior to the introduction of RevR6(2004), objectives of the engagements were not explicitly stated but implied in statements describing the engagement or the responsibility of the provider. Findings reflect that assurance engagements on sustainability reports took various forms. For example, it has been referred to as a "review" conducted based on agreed on procedures...to examine the reporting of the information and data from SCA"; "an audit" to "examine the contents of SAS' environmental report" and hints of advisory services "Since there are no generally accepted standards for the contents and structure of an environmental report, neither in Scandinavia nor internationally, SAS has continuously discussed with us which information should be disclosed. As a basis for making this selection we have used Deloitte & Touche's Checklist..."where the constant discussion between the reporting company and the assurance provider brings to question the input of assurance providers in the reporting company's decision as to the contents of the report in particular when the provider's checklist has been used as the reporting criteria.

The introduction of Rev6 led assurors to word their objective in a standardised manner in line with ISAE3000 "to express whether there are any indications that the sustainability report is not, in all material respects, drawn up in accordance with the criteria stated" regardless of the scope. The only variation of reports, giving a hint of the scope of engagement, lies in the difference in subject matter and criteria of evaluation each engagement is based.

8.3.2.3 Reporting of work performed

Procedures taken for each assurance engagement had been summarised in bullet points even prior to the publication of RevR6. Work performed include interviews with management of the reporting company, reviews on systems, documentation, reporting principles, scope limitations of contents etc.. These areas have been drafted into the RevR6. Work performed described in assurance reports contains the terms limited review, review and overall review. For example, review of systems and routines was carried out in SCA's engagement in 2000 and an overall review in 2001. Limited review of systems and routines was carried out in SKF's engagement in 2002 and a sample based testing of the reliability of the system was conducted in 2005 by another provider.

Assurance providers continue to exercise discretion on terminology adopted prior to the publication of the recommendation with no explanation on the differences in the extent or scope of work performed.

8.3.2.4 Reporting assurance providers conclusions

Prior to the publication of RevR6, engagements conducted on environmental and sustainability reports include audits and reviews. For example, assurance providers of

Deloitte conducted an audit on SAS environmental report from 1997-2000[768], conclusions were therefore positively phrased. For example in SAS' 2000 engagement, Deloitte concluded:

"Based on the above reviews, it is our opinion that the data and the information in the environmental reporting is supported by data obtained with due care from the operating units, and that the reports on environmental conditions and goal fulfilment relative to the established action places provide an in all material aspects true and fair view of the reported parts of the company's operations".

The other two companies in group 1 engaged assurance providers to perform reviews on their reports resulting in a negatively phrased conclusion. For example in SKF's engagement in 2002, Deloitte concluded:

"Based on our limited review, nothing has come to our attention which indicates that information in the sustainability reports is not supported by data obtained with due care from the company's units. Also nothing has come to our attention, which indicates that the sustainability report does not give a reasonable view of those aspects that have been reported of the company's operations that related to sustainability."[769]

With the publication of RevR6, only review engagements were conducted. An example of a negative conclusion was provided in §5.6 of the RevR6 "in our review, nothing has come to our attention indicating that XX's sustainability report 200x has not been materially drawn up in accordance with the criteria stated". Assurance providers have since phrased conclusions in a similar form relating the subject matter to the objective of the engagement (whether a report has been prepared according to stated criteria). This includes engagements conducted using AA1000AS(2003) jointly.

From the above, it is evident that assurance providers have made certain choices to standardise the format, title, objectives and conclusion of the assurance report. Findings also show assurance providers retaining discretion on the use of certain terminology pertaining to work performed. In engagements where both RevR6 and AA1000AS (2003) are applied, assurance providers have exercised discretion not to further comment in the assurance report on materiality, completeness and responsiveness or provide additional commentary on the progress of reporting and assurance since last report or make suggestions for improvement in the reporting and processes, systems and competencies for the next cycle as recommended by AA1000AS(2003). Instead, all issues and recommendations are provided in an

---

[768] Although the time frame lies outside the scope of this analysis, it is an important point to highlight

[769] The engagement was conducted using as a guide FEE Guidelines "providing assurance on sustainability reports"

internal report to the reporting organisation which most assurance providers make clear in the assurance report.

8.3 How does RevR6(2006) work from a social science perspective?

This part of the analysis has adopted a social science perspective to analyse the exercise of discretion by assurance providers in the context they operate. It analyses how RevR6(2006) had influenced the choices made by assurance providers.

The findings from chapters 7 and section 8.2 of this thesis reflect that RevR6 serves as a site to bring into Swedish practice the developments in the international regulatory space. The standard legitimised the "incremental approach" adopted by the GRI in reporting and supported by the IAASB task force that commented on NIVRA's ED3410 in 2005. According to the GRI, the incremental approach begins with partial adherence to the reporting principles and/or report content in the Guidelines and gradually moves to fuller adoption. RevR6's provision on scope limitations allows this approach as reflected in the findings the differences in the scope and scale of assurance engagements conducted in Sweden.

Analysis of the subject matter of assurance engagements over time shows that reporting companies have increased the scope of the environment or sustainability report subject to assurance. The assurance 'stamp' in these cases cannot be conceived as putting an end of more transparent disclosure.

Company	Scope of first engagement	Years	Development over years
SAS	Environment report	11	Sectional increase in scope and breath: environment to sustainability to GRI Index and stakeholder dialogue reported at website
SKF	Sustainability report	6	Sustainability and Compliance with GRI (All information)
SCA	Information provided by the resource management system	9/2[770]	Increase to include tables, diagrams and graphs under 2 "social" headings (employee and H&S) by another assurance provider. 2007, 14/70 pages)
Trelleborg	3 indicators (2 / 42 pages)	2	Increase in number of indicators to 6 (6/41 pages)
Vin & Spirit	Information presented in tables diagrams and graphs within 9 pages of responsibility section of annual report	2	Continue to be restricted to tables diagrams and graphs within 11 pages of responsibility section of annual report

Figure 8.6 Development in the scope of sustainability assurance engagements

The above shows that assurance providers were given wide discretion to determine the scope limitations with regard to subject matter. There is no indication that accepting assurance engagements which are narrower than the scope of the sustainability report would discourage reporting companies from increasing the scope of subject matter to be assured in subsequent years.

The analysis of the contents of standards in chapter 7 has established that the provisions of RevR6 replicate the ISAE in many areas. The choice to exercise little

[770] From 2006, SCA had two providers, Provider 1 on Environmental report, PROVIDER 2on HR section

discretion and granting it down to practitioners performing the engagement strongly indicate that Far does not have the ambition to tackle issues of assurance debated on, in the international regulatory space. Expert interviews reveal that "stakeholder methodology... not easy for accountancy profession to take on board. (There will be) a lot of development on standard and discussion for the years to come".

RevR6 provided guidance on the GRI and stakeholder issues as examples and vague general provisions. This approach created awareness of issues relevant to assuring sustainability reports and put the onus on users of the standard to keep abreast the development of these issues. This flexible approach also saved Far from committing to develop its own conceptual understanding on terms such as materiality in sustainability assurance and solving the complex problems raised in the sustainability assurance regulatory space. The introduction of RevR6, "gathered" the assurance engagements on sustainability reports (which took the form of audits, reviews and agreed upon procedures prior to 2004) into a common pool that was in 2006 labelled as assurance services.

8.3.1 Effect of RevR6 on different types of engagement

The discretion allowed in RevR6 has granted assurance providers the choice to perform engagements to serve different needs of the reporting organisations. The representatives of all companies interviewed state that the purpose of engaging assurance is to improve reporting and the credibility of their sustainability reports. Interviews with companies further reveal the range of engagement performed under RevR6 and the "role of providers".

8.3.1.1 Narrow scope engagements

Interviews with representatives of two companies with assurance engagements limited in scope reflect the assurance engagement's role as a management tool. The representative from Trelleborg attributed the company's first assurance effort to internal driving forces and described the engagement as more a self-assessment learning exercise which resulted in an assurance report published to readers. The representative of Vin & Sprit also mentions that "*having an external party pointing out and putting words to weaknesses is a great tool for getting approval for further focus on CR issues*" reflecting the engagement's usefulness to enhance the legitimacy of CSR issues within the organisation. In the two above cases, assurance was commissioned late in the process with scope limited to a small part of the report. Both reporting organisations perceived the focus of assurance work to be centred on verifying data, clarifying definitions and tracing data to source documents and making recommendation within those relevant areas. The management also perceived their involvement to be minimal, to supply assurance providers the information needed to conduct the assurance.

8.3.1.2 Wider scope engagements

For companies that have commissioned assurance for a number of years, the scope of the engagement and results from the interviews suggest a broader purpose.

SAS' relationship with Deloitte in the area of environmental reporting and assurance had been established since 1995 when Deloitte was hired as SAS consultants. SAS' 1997-2000 environmental report had been audited by Deloitte at a reasonable level of assurance. An interview with the representative of SAS suggests that assurance standards had "minimum impact on the substance of engagement"[771]. Engagements were a continuation of "getting facts right" and "fine tuning of internal processes to gain greater credibility". The representative of SAS interviewed spoke of the companies role not only as a supplier of required information but also having a part in identifying issues and sites targeted for improvement in the assurance engagement.

SCA has adopted several ways to improve the credibility of its environmental reports. In 1998, independent opinions were voiced by experts from Sustainability, the European Commission, ACCA, FEE and WWF in 1999 an environmental verifier was commissioned. Since 2000, Deloitte has been commissioned to conduct a review on SCA's Environmental report. Interview with the company highlight that unlike previous engagements, Deloitte met the company's needs in improving and fine tuning SCA's system. In addition, "specialised studies" were conducted to focus on credibility of data and measures to improve the system. The representative interviewed informed that management has some influence over the engagement through joint decisions on sites chosen and setting the criteria of "places that are not excellent". From the above examples, one could reflect that although the engagements provide a quality label to the general reader, it has the purpose of instituting improvements within the organisation through advice and recommendations from the assurance provider. For both SAS and SCA, the competence and tools of the assurance provider were the key factors affecting their choice of engaging an accountancy provider.

8.3.1.3 Engagements performed with RevR6 and AA1000(AS)

SKF engaged Deloitte from 2002-2004 and KPMG from 2005 and onwards. SKF has also commissioned with LRQA and later DNV (non-accountancy providers) to

---

[771] Far standard drafted based on best practice of auditors. Auditors were leading practitioners in the field and significant input in drafting of standards

audit SKF's management systems (ISO14001 and OHSAS18001) while commissioning KPMG (previously Deloitte) to audit the sustainability report. The change in assurance provider from Deloitte to KPMG was tied with the change in the financial auditor as the representative interviewed believed that "an integrated report requires integrated verification". The purpose to commission assurance and decision to bring in AA1000AS was according to the representative of SFK to "enhance the level of verification and improve effectiveness and validity of the report for stakeholders". In Sandviks's case, the engagement was commissioned to "enhance the credibility of the report to the readers and to supplement the internal audit". AA1000AS was regarded as a tool used by the auditors.

According to the representatives of both companies, the management was perceived to have some influence on areas that should be the focal point of the assurance engagement. With reference to the work carried out by the assurance provider, the liaison role played by assurance providers between management and stakeholders pertaining to stakeholders' information requirements is suggested.

In the engagements conducted according to RevR6(2006), findings suggest a remote link between current planning procedures and the user's perspective even when stakeholder dialogue is in place, assurance providers exercise their discretion according to the needs of the reporting organisation. The finding is supported by expert interview where *the expert stressed* the importance of "value added" in establishing the business case in Sweden. "This is voluntary, we are there to help (the) client reach application of GRI format as high as they can and develop quality in reporting".[772] Through the interviews, it is evident that the companies put substantial weight on the benefits of the recommendations of the assurance provider.

[772] Interview with expert

The following table presents a summary of the above findings.

Gp	Co.	Scope	Purpose	Focus of work	Duration	Management's perceived Involvement
2	Trelleborg	Limited using RevR6	Verification Self –assessment	Focus on indicators: Verify data, tracing data to source, definitions of terms	About a month	Information provider Minimum involvement
2	Vin & Sprit	Limited using RevR6	Verification, improve reporting	Verify data presented, tracing data to source	2-3 weeks	
1	SCA	Limited using RevR6	Credibility through internal improvement	Specialized studies, verify information	Year round	Information provider Joint decision maker in site selection
1	SAS	Full using RevR6	Credibility through internal improvement and better reporting	Focus on problem sites & verify information, scope of contents of report	Year round	
1	SKF	Full using RevR6 & AA1000AS	The above and stakeholder transparency	Focus on problem areas and interview stakeholders feedback on company's reporting	9months	Information provider: Point out areas for auditors to focus on
2	Sandvik	Full using RevR6 and AA1000AS	Credibility and supplementing internal audit	Focus on problem areas and interview stakeholders feedback on company's reporting	10 months	Information provider: Point out areas for auditors to focus on

Figure 8.7 The nature of scope of sustainability assurance engagements in Sweden.

8.4 The interplay of RevR6 with hard law

This section presents the second part of the analysis on how standards work from a legal perspective. Far is a non-state standard setter; unless granted by hard law, its standards have no legal status. Assurance providers adopting RevR6 however, operate within the Swedish jurisdiction. It is therefore crucial to determine whether sustainability assurance already falls within the scope of hard law in Sweden. This has implications on the character and role of RevR6 which will determine the interplay of RevR6 with hard law to regulate sustainability assurance. The next section presents a brief background to highlight relevant developments in audit legislation in Sweden. A legal lens to perform the analysis will be adopted in section 8.4.2 of the thesis.

8.4.1 Background

In Sweden, the custom of audits existed prior to legislative efforts to make it a mandatory requirement. The 1895 Companies Act stipulated that companies limited by shares should have one or more auditors elected at the shareholders meeting. The statute required auditors to examine and report on the accounts of the company and on the administration of the company's affairs. [773] The Kreuger fraud unrevealed in 1932 had a significant effect on audit regulation, and had an impact on the 1944 Companies Act.[774] The 1944 Companies Act required statutory audit to be performed by a qualified auditor (authorized or approved public accountant) on companies registered on the stock exchange. Provisions were also developed pertaining to the scope of the audit and contents of the audit report.[775] The role of auditor was articulated not only in relation to the investors but the public and the state.[776]

The 1975 Companies Act (1975:1385) introduced a requirement for auditors to comply with "god revisionssed" (generally accepted auditing standards (in Sweden)). This requirement stands also in the 2005 Companies Act (2005:551) (referred to thereafter as the Companies Act). According to the proposition to the Companies Act, 1997/98:99 *Aktiebolagets organisation*, the general principle 'generally accepted auditing standards' was structured in the Act as legislators recognise that rules to

---

[773] Wallerstedt, E. (2001), "The Emergence of the Big Five in Sweden" European Accounting Review, 10(4), p. 847.

[774] Ibid, p.853. See also , SOU (1941:9) *Lagberedningens förslag till Lag om Aktiebolag.*

[775] Larsson, B. (2005) "Auditor regulation and economic crime policy in Sweden, 1965–2000". Accounting, Organizations and Society 30, p. 130-131.

[776] ibid.

regulate the audit cannot be made exhaustive.[777] The reliance on them therefore could compromise or limit an auditor's work and as a result undermine the objective of regulation.

Harmonization efforts by the EU had not led to major changes with regard to the function of the statutory audit in the Companies Act.[778] The audit of the administration of the company by the board of directors or managing director continues to be within the scope of the statutory audit in Sweden. The Fourth and Seventh Company Law Directives and subsequently the Accounts Modernisation Directive were transposed into national law mainly through the Annual Accounts Act[779], and the Eighth Directive mainly through the Auditors Act.[780]

The Supervisory Board of Public Accountants[781] (Revisorsnämden; thereafter RN) was established in 1995 when the Auditors Act came into force and operates under the Department of Justice. It serves the role of as a public oversight body. Its function and authority are stipulated in the Auditors Act. The RN is financed by application and annual fees from qualified auditors and, registration and renewal fees from audit firms ordained by the government.[782]

---

777 Prop. 1997/98:99, p.147.

778 Eighth Council Directive 84/253/EEC of 10 April 1984 based on Article 54 (3) (g) of the Treaty on the approval of persons responsible for carrying out the statutory audits of accounting documents, and the new Directive 2006/43/EC on statutory audit of annual accounts and consolidated accounts and amending Council Directives 78/660/EEC and 83/349/EEC.

779 Fourth Council Directive 78/660/EEC of 25 July 1978 based on Article 54(3)(g) of the Treaty on the annual accounts of certain types of companies. Seventh Council Directive 83/349/EEC of 13 June 1983 based on Article 54(3)(g) of the Treaty on consolidated accounts. Directive 2003/51/EC of the European Parliament and of the Council of 18 June 2003 amending Directives 78/660/EEC, 83/349/EEC, 86/635/EEC and 91/674/EEC on the annual and consolidated accounts of certain types of companies, banks and other financial institutions and insurance undertakings Prop. 1997/98:99, p. 147.

780 Refer to footnote 6.

781 Within the RN is a board that may set rules or standards and take decisions in disciplinary hearings. The board consists of nine members appointed by the government: a chairman and eight other members. The ordinance instructing the RN provides that the Chairman or Vice Chairman should be a lawyer or has experience as a judge. Two members should be authorized or approved public accountants and other members of the board should have worked with or been involved in activities that deal with auditing.[781]

782 http://www.revisorsnamnden.se/rn/english/supervision.html.

Relevant this thesis are Chapter 9 of the Companies Act governs the statutory audit in Sweden for companies limited by shares, the Auditors Act and the Auditors Ordinance (1995:665).

8.4.2 Is sustainability assurance regulated by hard law?

Prior to the analysis of the relevant statutes, it is important to point out that the Companies Act and the Auditors Act regulates the work of qualified auditors. In addition to performing the audit, auditors may take on other engagements. Sweden has a two-tier qualification system for auditors. Both authorized and approved public accountants are recognized as qualified auditors. The requirements of qualification are specified in sections 4 and 5 of the Auditors Act.

Chapter 9, section 11 of the Companies Act, stipulate that only authorized or approved accountants may be appointed the auditor of a company.[783] This provision however, according to chapter 9, section 19 of the Act, does not preclude registered accounting firms to be appointed as auditors. In such cases, provisions such as those regarding conflicts of interest, retention of assistants, the right to be present at the shareholders meeting, are applicable to the auditor primarily responsible for the engagement.

In the case of sustainability assurance, assurance providers may refer to the registered accounting firm, the qualified auditor or the specialist member of Far (sustainability experts). It is only the former two that the stipulations of the Companies Act and Auditors Act apply.

---

[783] In general listed companies, companies of a certain size defined by the number of employees and new value of its assets as well as companies in specific industries such as banking are to be audited by authorized public accountants or approved accountant that has met additional qualification requirements. An auditors term of office is for 4 financial years ending after the AGM where accounts are presented.[783]

Relevant provisions of the Auditor's Act (2001:883)

The Auditors Act regulates among other issues, the role of RN, registration and qualification requirements of auditors, the registration of public accounting firms and the professional obligations of auditors. The Auditors Ordinance supplements the Auditors Act with more specific rules in particular areas. For example, the ordinance specifics details of education and practical training requirements to qualify as auditors.

The Auditors Act grants the RN authority of supervise audit business; auditors (approved or authorised public accountants, qualified auditors) and registered public accounting firms. RN also has the role to ensure that generally accepted auditing standards are developed in an appropriate way. [784]

Section 2(8) of the Auditors Act defines audit business as:

"a. business that consists of such examination of administration or financial information as follows from statute, articles of association, by-law, or contract and which results in a report or some other document that is intended to constitute a basis for assessment also for a person other than the client; and

b. advice or other assistance occasion by observations during the examination in accordance with a"

The aforementioned provisions highlight two unique features pertaining to the supervisory powers of RN. First, the supervision of auditors is not confined to the statutory audit but engagements that fall under the definition of 'audit business'. Second, the supervision does not only apply to qualified auditors but also the registered public accounting firms. It is therefore important to examine the legal concept of 'audit business' (revisionverksamhet). If the sustainability assurance engagement falls within the scope of 'audit business'; the engagement is regulated by law. It will fall under the supervision of RN and qualified auditors are required to comply with rules such as those regarding auditor independence and the acceptance of extraneous business that could undermine confidence of the audit. [785]

[784] Auditors Act, sec. 3(2) and 3(4).

[785] Auditors Act, sec. 20-21 and 25.

8.4.3 Can sustainability assurance be an 'audit business' according to the Auditors Act?

The provision in Section 2(8) of the Auditors Act, can be divided into three parts to evaluate whether sustainability assurance falls within the definition of the 'audit business'. The first part of the Act states that the examination entails the administration and financial information. In this case, evaluation has to be made on whether reporting of sustainability information or sustainability information reported fall with the scope of administration or financial information. The second part states that assurance engagement should be required by legislation, articles of association, by-law, or contract. This entails an examination as to whether sustainability assurance has been commissioned by virtue of the above requirements. The third part states that the engagement should result in a report or other document that is relied upon by third parties constituting a basis to make assessments, requiring an assessment of whether there is a third party dimension.

Administration or financial information

There is no clear specific guidance defining administration or financial information in both the Companies Act and the Auditors Act. The distinction between Administration and financial information is not crucial for this analysis as both constitutes the subject matter of 'audit business'. Drawing from the reporting requirements of auditors from the Companies Act, the auditor is obliged not only to report his or her conclusions on the financial report of the company, but also other matters. These include the management's performance or non-performance of duty that may give rise to liability of the firm, and any decisions that are contrary to the articles of association, accounting, taxation and Company Law requirements.[786]

Section 6.2.1 of this thesis highlighted recent developments in reporting that strongly associates the reporting of environmental and social matters with financial information and the administration of the company. First, the 1999 amendment to the Annual Accounts Act, required environmental disclosures that are important for the appraisal of the corporation's status, results and future performance, to be included in the Board of Directors' report.[787] It was expressed in Proposition 1996/97:167 *Miljöredovisning och miljöinformation i näringslivet,* that environmental disclosures on the impact of a corporation's activities on the natural environment could improve the financial information issued by the reporting organisation.[788]

---

786 Companies Act, ch. 9 sec. 33.

787 Annual Accounts Act, ch. 6 sec. 1.

788 Prop. 1996/97:167. p. 13 and 20.

Second, in the transposition of the Accounts Modernisation Directive, the Annual Accounts Act imposed relevant disclosure requirements of non-financial information in the Board of Directors' reports. Specifically mentioned were environment and employee information, to the extent needed to understand the development and performance of the reporting company.[789]

Although this is not a source of law, it is important to point out that in the external reporting guidelines issued to state-owned companies, the guidelines stated that reporting (including sustainability reporting) provides a basis for feedback and assessment of the activities, targets and objectives set by the company. The guidelines further state that *"the boards of the state-owned companies are responsible for the companies' accounting and reporting complying with these adopted guidelines"*[790] and that reports are required by the owners and other stakeholders to assess the company's sustainability performance.[791] This establishes a link between sustainability reporting and the administration of the company.

The above developments in the legal reporting framework strongly support that sustainability information that affect a company's performance, development and position, qualifies as administration and financial information.

The type of engagement and third party dimension

The legal concept of 'audit business' was developed using the statutory audit as a focal point or point of departure.[792] From the statutory audit, the purpose, scope and elements of audit is derived. From a legal view point, the purpose of auditing is to ensure the accuracy of and confidence in reported financial information[793] and administration of the company by the board of directors[794] through engagements where auditor independence and competence are guaranteed.[795] The audit serves not only to protect shareholders but the interest of other parties in society.[796] Within its scope, 'audit business' includes the examination of the financial report and

---

[789] Annual Accounts Act, ch. 6 sec. 1.

[790] Swedish Ministry of Enterprise, Energy and Communications "Guidelines for external reporting by state-owned companies", 29th November 2007, p. 1 and 2.

[791] Ibid.

[792] Prop. 2000/01:146 p. 39-40.

[793] Prop. 2001/01:146 p. 40.

[794] Ibid.

[795] Prop. 2001/01:146 p. 40.

[796] Prop. 1975:103, p. 244, prop 1997/98:99 p. 132.

administration of the company. It provides that auditors have the duty to provide management with criticism and comments required by the generally accepted accounting principles.[797] The recommendations provided by auditors as a result of the engagement, have been regarded in other preparatory works as necessary and desirable to the company.[798]

The proposition to the Auditors Act states that an 'audit business' is not confined to the statutory audit, but can include three other types of audit-like engagements under specific conditions. The three types of engagements include: other audits that are required by law (e.g., audit of control balance sheet, issue prospectus, etc.); other reviews than audit by statute and engagements commissioned by virtue of the reporting companies contract with third parties (e.g. Stockholm Stock exchange quotation contract requiring interim financial statements be audited).[799]

With reference to section 8.3 of this thesis, sustainability information is reported by companies the following ways:

In the board of directors' report;

In a sustainability report attached to the end of the annual report; or

In a separate sustainability report.

Two questions can be raised, given the above facts, to analyse if sustainability assurance is regulated by hard law.

Can assurance engagements provided on the sustainability information, reported in the three above-mentioned reports, fall under the scope of the statutory audit?

For sustainability assurance engagements that do not fall within the scope of the statutory audit, can they fall within the scope of 'audit business'?

Question (i) - Sustainability assurance - audit by statue?

---

[797] Companies Act, Ch 9. sec. 4, 6, 27-34.

[798] Prop. 2001/01:146 p. 40-41.

[799] Prop. 2001/01:146, p. 40-42

The statutory audit in the Companies Act will be used as an example to discuss audit by statute.[800] Chapter 9 of the Companies Act contains provisions relevant to the audit of a company limited by shares. Chapter 9 section 1 of the Act, states that a public company shall have at least one qualified auditor elected by the general meeting of shareholders. As mentioned in section 8.4.2 of the thesis, registered accounting firms can also be appointed as auditors. The auditor, unless otherwise stated refer to the financial auditor in this discussion.

Chapter 9 section 3 of the Companies Act state:

"An auditor shall examine the company's annual report and financial statements, as well as the management of the company by the board of directors and the managing director. The audit shall be as detailed and extensive as required by generally accepted auditing principles"

The subject matter that falls under the definition of the statutory audit includes the financial accounts, the annual report and the administration of the company. In light of the developments discussed in section 8.4.2.2a of the thesis, in a scenario where assurance services is performed on sustainability information reported in the Board of Directors report, with the conclusions of the auditor expressed in the audit report, it is likely that such engagements would fall within the scope of an audit by statute.

When sustainability information is reported in a separate report, with the conclusions of the auditor expressed in an assurance report separate from the audit report; the situation becomes more ambiguous. As the nature of the engagement and terms of agreement between the auditor and reporting organisation vary, the evaluation of whether it can fall within the scope of the statutory audit has to be performed on a case by case basis. The following evaluates possible two scenarios. Chapter 9 section 4 of the Companies Act, requires auditors to comply with instructions issued at the shareholders' meeting, where such request do not contravene applicable law, the articles of association or general generally accepted accounting principles. In a scenario whereby the auditor has received instructions at the shareholders meeting to perform assurance services on the sustainability report, it is likely that the assurance engagement on the sustainability report falls under the scope of the statutory audits.[801] Even if there are "no instructions from the shareholders meeting", it is still possible to argue that assurance of a sustainability report by the financial auditor of a

---

[800] In Sweden, the statutory audit can refer to the audit required by the Companies Act, but also other Acts such Associations law.

[801] Diamant, A. (2004) "Revisors Oberoende" p. 158-161.

company can be an audit by statute. The judgment by the Supreme Court, in the case NJA1996 p.224 can be used to support this standpoint.

Case summary

Scandinavian Clinics AB (SCAB) appointed BJ as their auditor during the fall in 1985 to 86, the period where the company had financial difficulties. Scandinavian Clinics Försäljnings AB (SCFAB) was established from October 85 to autumn 86 which carried out sales activities. BJ established a preliminary balance sheet showing SCAB's assets to be worth 2 462 006 SEK which actual worth only amounted to 350 000 SEK. The assets were transferred to SCFAB for 2 530 000 SEK. Första Sparbaken granted SCFAB a loan of 1 200 000 SEK made use of preliminary balance sheet by BJ. The company was unable to repay the loan which led to the legal action taken by the bank against BJ.

The Supreme Court ruled that:

"Even if the contribution made by the auditor in respect of the re-construction of SCAB and the transfer of its assets to some degree went beyond his strict obligations as auditor, it was in his capacity as auditor that the company management turned to him for assistance in the reconstruction of the company. His action must therefore to classified among such measures, relating to management, as are referred to in chapter 10 section7 (now chapter 9 section 3) of the Swedish Companies Act. To the extent that the auditor has acted contrary to generally accepted auditing standards, he has therefore also violated this section of the law."[802]

The auditor of the company may be commissioned to perform sustainability assurance engagements by virtue of his or her capacity as an auditor for the company. In such a scenario, with reference to the Supreme Court's reasoning, it can be argued that there is a presumption that sustainability assurance engagements can fall under the definition of the statutory audit by virtue of the Companies Act.

Question (ii) Sustainability assurance – audit business?

In this section, an attempt is made to evaluate whether sustainability assurance engagements that are not statutory audits can fall within the scope of 'audit business' as 'other review that audit by statute'. In the scenarios evaluated above, the financial auditors of the company had been commissioned to perform the sustainability assurance services. In a scenario where the sustainability information is reported in a

[802] NJA 1996 p.224, translated by Moberg. K.(2004) in "Auditors' Liability for Damages" Scandinavian Studies in Law, 45, Stockholm University Law Faculty p.229-230.

separate sustainability report accompanied by an assurance report, and the qualified auditor commissioned to perform the engagement is not the financial auditor; the engagement is likely to falls outside the scope of the statutory audit. The proposition to the Auditors Act 2001/01:146 *Oberoende, ägande och tillsyn i revisionsverksamhet*, sets forth three criteria that guide the interpretation of an 'audit business'.[803]

First, an evaluation has to be made on whether the engagement resembles an audit or requires the auditor to apply various auditing techniques. In the preparatory works, the review of the control balance sheet, issue of prospectus, auditing of the accounts and administration of non-profit organizations were given as examples of activities that within the scope of audit business. The subject matter of these examples is financial in nature, audited or reviewed against established criteria. This raises the question if assurance on sustainability information can be regarded as an activity resembling an audit? Although the methodology had been developed drawing from the financial audit, the nature and characteristics of sustainability information is different from financial information. Sustainability information unlike financial information does not have a common unit of measurement or self-balancing double entry system. Sustainability reports contain more qualitative disclosures requiring a greater reliance on subjective assessments of the assurance provider. In addition, the internal control systems cannot be relied on for all sustainability information as they may not be available or sufficiently mature. As such, even though the broad purpose of the financial audit and sustainability assurance is to improve the credibility of information, the specific techniques which assessments are made differ. Furthermore, reporting standards making up the criteria of sustainability assurance are not as mature as the financial audit and is undergoing rapid development.

Second, consideration is given on whether sustainability assurance engagements results in a report intended for third parties other than the client[804]. The proposition to the Auditors Act 2001/01 specifies that engagements that have no resemblance to audit e.g. tax advisory services and engagements that are not for third parties e.g., agreed-upon procedures are not 'audit business'.[805] With reference to RN case Ö11, the RN further guides that without the third party intention, the fact that the client encloses the report to a third party does not affect bring the engagement within the legal scope of the audit.[806] A company may engage an audit firm to provide advisory services to the management, if the report is solely for the use of management, the engagement fall outside the scope of an audit business. Assurance engagements from

---

[803] Prop. 2001/01:146, p.41.

[804] Prop. 2001/01. p. 41.

[805] Ibid,, p. 42.

[806] RN Ö11, sec 3.1.5.

the audit profession's perspective (following the assurance framework), results in the issue of a report addressed to readers that expresses his or her conclusion and conveys a level of assurance based on the work performed. This raises the questions if it fulfils this criterion if the assurance provider has been commissioned by the management; and whether it makes a difference if assurance providers are appointed at the annual general meeting by the shareholders as in the case of financial audits. Some uncertainty prevails, but the judgment NJA 1996:224 from the Supreme Court in Sweden can be an argument to support that an agreement with the management is sufficient to regard the engagement as 'other review than audit by statute'.

Third, consideration is given to whether assurance providers are engaged, or if their reports have been relied on due to the competence and qualifications as approved or authorised public accountants.[807] Sustainability assurance engagements carried out in multidisciplinary teams and co-signed or in a particular case signed by specialist members of Far, reflecting the extent an engagement rely on skills other than that of the traditional audit. From one perspective, statutory audits in Sweden require financial auditors to audit the administration of the company; auditors are familiar with broader issues environmental and social issues. On the other hand, to perform sustainability assurance engagements require the assurance provider to have knowledge and competence to *assess sustainability and associated technical risk*, national and/or international agreements, rules, conventions related to sustainability and knowledge of standards, such as ISO 14001 and SA 8000; GRI's "Sustainability Reporting Guidelines". These areas are not given any mention in the list of theoretical courses of study specified in the Auditors Ordinance.[808] Although interested candidates can choose to study these areas, it is to a large extent optional. This raises the question if the auditor is commissioned by virtue of his qualifications as an auditor or is the audit firm engaged to perform such services as it has the resources to assemble multidisciplinary teams to undertake such engagements?

---

[807] Prop.2000/01:146 p. 41.

[808] Auditors Ordinance, RNFS 1996:1 60 points in business administration which accounting, auditing, financial, financial control given priority, 30points in law 20points in tax, 20 in either computer science, economics, statistics or mathematics and 10 points where the undergraduate can make a choice.

8.4.4 The interplay of RevR6 with hard law to regulate sustainability assurance

The above analysis highlights that the legal concept of an audit defined by the Auditor's Act as 'audit business' includes within its scope not only the statutory audit, but for example other reviews than audit by statute. The fact that the legal concept of an audit is built on the statutory audit creates a link and interplay between the two statutes in the regulation of audit business by legal institutions.

Chapter 9 section 3 of the Companies Act require the statutory audit to be performed in accordance with generally accepted auditing standards. Likewise, section 3 of the Auditors Act grants RN supervisory authority over 'audit business', auditors and registered public accounting firms and public oversight over the development according to generally accepted auditing standards. The supervision of 'audit business' by RN according to the proposition to the Auditors Act 2001/01:146 should include the quality of work performed by auditors defined by generally accepted auditing practices and the professional code of ethics.[809] How are standards of the profession linked to this legal concept?

Doctrinal writings characterises requirements such as "generally accepted auditing standards" as an "incomplete rule". [810] Generally accepted auditing standards is therefore a legal concept where material content is not provided by statues but will be specified and evolve through standards and practice outside the legal system.[811] This legislative technique awards flexibility for generally accepted auditing practices to continually adapt to the dynamic conditions in practice.[812] According to preparatory works, proposition 1997/98:99 *Aktiebolagets organisation*, standards serve as guidance.[813] The standards of the profession have no legal status but can be used as guidance to determine generally accepted auditing standards in Sweden. It also makes explicit that the final interpretation of the generally accepted auditing practices lies with the courts and not the standard setting organisations of the audit profession.[814]

---

809 Prop. 2001/01:146, p. 43-44.

810 Thorell, P. "Redovisning rättsliga ställning" Thorell refers to "Ofullständigt regler" translated by the author as "incomplete rule". p.115

811 Ibid., p.115-116

812 SOU 1971:15 p. 266.

813 Prop. 1997/98:99 p. 147.

814 Ibid.

This raises the question if there is a statute requiring RevR6 provide guidance to fill the gap?

Three scenarios can be drawn from the above analysis. First, if sustainability assurance is considered to lie within the scope of the statutory audit, the engagement is to be conducted and undergoes quality supervision by RN according to generally accepted auditing standards as stipulated by the Companies Act and Auditors Act respectively. In the performance of an engagement, generally accepted auditing standards is complied with presumable through following standards set by Far in this case RevR6. RN would use RevR6 as a guideline in supervising the quality of sustainability assurance engagements. Though RevR6 has no legal status, it nevertheless has some effect as it is used as a tool to define quality. Although the RN relies to a large extent on standards set by Far, they retain the preferential right of interpretation in cases of disagreements.

Second, if sustainability assurance falls outside the scope of the statutory audit but within the scope of the 'audit business' as other review than audit by statue; the engagement falls within the scope of RN supervision according to the Auditors Act. Third, if sustainability assurance falls outside the scope of 'audit business', the RN has no statutory rights to challenge the quality of those services, RN's scope of supervision for activities is confined to whether the engagement has been performed abiding by the professional code of ethics.[815]. RevR6 will be used by Far members by virtue of their Far membership there is no direct interplay with hard law.

The developments in sustainability assurance in practice have been heavily influenced by developments in sustainability reporting and international trends in this area. The diversity in reporting practices and terms of assurance engagement and dealing with interpreting existing statutes' in the regulation of a new phenomenon has raised legal issues that could be interesting for research in the future.

In this section, an analysis has been performed to determine the larger regulatory space which RevR6 works with the objective of analysing the potential interplay of RevR6 with hard law in regulating sustainability assurance performed by the accountancy profession in Sweden. Building the analysis from the relevant legal sources and raising issues particular to sustainability assurance allows the conclusion that sustainability engagement can fall under the definition of 'audit business'.

---

[815] Prop 2001/01:146 p. 43-45

8.5 Discussion – A decentred view of regulation

The legal analysis from section 8.4 concludes that RevR6 had been developed "outside" the existing legal institutions regulating auditors in Sweden. It is a standard by a private organisation. From a legal point of view, the RevR6 is not a source of law in itself. Its role in the legal system, if any is dependent on its status granted by hard law. A legal analysis of the regulatory framework regulating auditors indicates a potential interplay between RevR6 and hard law by virtue of the Companies Act and Auditors Act. If sustainability assurance is considered 'audit business' RevR6 could serve as guidance to both RN when it conducts its supervisory role and the court. As it stands today (in 2011), until jurisdiction is exercised by a legal authority, this fact remains unclear.

The decentred view of regulation takes us away from this traditional mode of thought. It forces an examination of "what is it we are looking at, and looking for, when we seek to analyse regulation."[816] Decentred regulation involves not only a shift in regulatory locus of from the state to potentially multiple sites but also the adoption on the state's part, of particular strategies of regulation.[817] This theoretical framework recognizes the initiatives that can come from different non-state actors. These actors have different philosophical beliefs, different legitimate communities of account to and operate in different context. Without a hierarchical structure or a 'central co-ordinator', a regulated community can therefore be subjected to different courses of actions, prescribed by different actors.

The expansion of business activities across borders in the era of globalization has driven professional accounting firms to grow out of their national regulatory structure and engage in other services than the traditional audit. In the light of the difficulties faced by legislators, one strategy adopted by legislators when drafting the 1975 Swedish Companies Act was to impose a general requirement for auditors to perform the audit according to "generally accepted auditing standards". With this provision, the locus of regulating the profession shifts from the state to the standard setting bodies of the profession. This has been based on the belief that these standard setters are more equipped with the resources and capacity to keep up with the complex dynamic environment in which their members operate. Even if the courts remain the final interpreter of generally accepted auditing standards, the shift in the regulatory locus is accompanied by systemic challenges as the profession has different interests and ideas with regards to what is regulation and they regulatory goals.

---

[816] Black, J. (2002), p. 27.

[817] Ibid.

Findings from chapters 5 and 6 suggest that the profession's logic for regulating assurance or sustainability assurance differs from the state. The profession was primarily interested in establishing standards to maintain the quality of services and the "business case" to establish the profession as primary providers of the service. In the light of the financial scandals and IFAC's goal to restore confidence in the audit, the profession also made attempts through the structure of its standards to separate the core financial audit from other assurance services. The analysis of the contents of RevR6 showed that although the standard had been drafted to guide its members to perform engagements on assuring sustainability reports (a specific subject matter), the provisions of IFAE and ISAE3000 were in many key areas replicated. The choice of Far to limit its own discretion indicate its intention of granting assurance providers a in performing assurance engagements in Sweden. Analysis of the linguistic structure showed that RevR6(2006) was packed with vague, evaluative and technical terms and potentially over-inclusive provisions. The empirical evidence up to 2008 illustrated the amorphousness of RevR6. The construction of the provisions of the standard has allowed assurance providers the discretionary space to exercise discretion in a way that is compatible with existing practices catering to the needs of reporting organisations and resources allocated to assurance activities. Providers have accepted a range of engagements of varying scope and purpose and applied other standards jointly with it in performing an engagement.

It is important to point out that the supervisory role that RN might have over sustainability assurance was not designed specifically to supervise these engagements but could apply by default due to the arguable over-inclusive definition of audit business in the Auditors Act. This leads to the question if there are clashes in regulatory goals and logic between the standard setter (the profession) and the body potentially supervising sustainability assurance engagements (the RN)? From the documentation of the profession and legal sources of law, both parties have a common objective of improving audit or assurance quality. However, differences are prevalent in the way the role of the auditor (assurance provider) and the concept of audit are defined. The commercial logic of the profession has been evident from the emphasis on the business case to encourage the commission of sustainability assurance.[818] According to the legal preparatory works, auditors has the role of a professional trustee ensuring quality to protect the various communities that place reliance on the assurance providers report pertaining to the information provided by the firm.[819] The dominant logic of regulation could be dependent on whether the state exercises jurisdiction in the field and the extent enforcement bodies perceives the risk and significance of the activity and exercise their supervisory powers.

---

818 Refer to chapter 4 and 5 of the thesis

819 Prop 2000/01:146

In light of the above findings, RevR6 could be utilized by RN as a guide determining general accepted auditing practice in performing its supervisory role. This raises the questions as to whether the quality of supervision will be bounded by the quality of standards and what the potential impact of RN supervision would bring to the quality of assurance on sustainability reports. On the other hand, the RevR6 had kept pace with developments in international sustainability reporting and assurance exposing a rather passive Swedish market to international best practices in this field, the very reason why legislators left standards to the profession. These dilemmas can perhaps be better dealt with through regulatory conversations, or other means to better coordinate or steer the emergence of regulatory initiatives in these new areas of "audit like" services and alleviate systemic problems of decentred regulation.

8.6 A note on RevR6(2008)

RevR6(2008) is a translation of NIVRA's 3410N Assurance on sustainability reports with very few adjustments relevant to Sweden.[820] The standard had been introduced in Chapter 5, section 5.3.2.3(b), (c) and (d). The Dutch exposure draft 3410 (ED3410) consisted 141 sections 42 pages long accompanied by 32 pages of appendix. With extensive comments by the IAASB task force (with input from SEAP), FEE and other interested parties, ED3410 was revised adopting the "clarity style" of drafting in line with the drafting conventions adopted by IAASB in the clarity project. To recapitulate, the clarity project purports principles based standards. The users of the standard are obliged to follow the requirements and exercise discretion in line with the objectives the standard pursues. Guidance notes accompanying the standard serve only to aid users to interpret the recommendation but are not requirements.

The finalised NIVRA standard 3410N is divided in sections: introduction, objective, definitions and requirements consisting 39 sections, accompanied by 67 points of application and other explanatory material or guidance notes. In the Swedish translation, four sections and two examples of assurance reports were added. They pertained to the applicability of IFAC's code of ethics for professional accountants and ISQC; the need for professional scepticism when conducting the engagement; the treatment of after closing events and other reporting responsibilities to further align

[820] E.g., it has also been made explicit RevR 6(2008) the definition of "internal specialist" referring to experts that are employed by the firm providing assurance services; sustainability standards used by Swedish reporters; how internal control can be used in limited assurance engagements etc.

the standard to ISAE3000. The latter two sections were also part of RevR6(2006). The two examples include a hybrid and a limited assurance engagement report.

The following does not attempt to make a comparative analysis of the standard by highlight two important features introduced by RevR6(2008): conditions of acceptance; levels of assurance and work performed that could affect how sustainability assurance is performed in Sweden.

8.6.1 Issues pertaining to conditions of acceptance

Analysis of the Swedish case indicate that assurance providers have exercised discretion granted by RevR6(2006) and accepted a wide range of engagements. Additional conditions of acceptance have been imposed in RevR6(2008). First, there has to be a rational purpose for the assurance engagement and second, a justifiable assumption that management of the reporting company is acting in good faith.[821] In addition, RevR6(2008), however contains two provisions on evaluating the reporting criteria adopted by the reporting organisation. Section 12 is a general provision, similar to RevR6 requiring assurance providers to evaluate suitable criteria and specifies the courses of action to pursue if the criteria are unsuitable. Section 13 introduces guidance specific to assurance on sustainability reports. The provision requires the assurance provider to assess the decision making process of management pertaining to the choice of issues and depth of issues published in the sustainability report. It imposes that at minimum the assurance provider should verify that the report is not misleading such that it satisfies the legitimate information needs of the report's intended users. Legitimate information needs according to the standard refer to information needs an "average representative" of the parties interested in the conduct of the organization or content of the organization's report *can demand based on legislation, general accepted reporting standards, case law or agreements between the reporting organization and these intended users or user group.* [822]

The guidance notes provide more specific guidelines on evaluating the reporting criteria. It states the five characteristics of suitable criteria and builds on the list of examples of available sustainability reporting guidance from RevR6(2006).[823] It

[821] RevR8(2006), sec. 6

[822] RevR6(2008) sec 4, 2010 version sec 5

[823] RevR6(2008) T18, 19, 20. RevR6(2006) mentions the GRI, RevR6(2008) provides two other examples, report from the World Business Council for Sustainability Development and AccountAbility standards.

further provides guidance in cases where the GRI is not the reporting criteria. In such cases, auditors are recommended to evaluate the reporting criteria according to 2 sections of the GRI G3 guidelines on defining the report content, quality and boundary of the report and standard disclosures and other sources such as industry wide agreements. In all cases, the reporting criteria adopted by the firm are to be described in the companies reporting policy. The guidance notes also recommend a test of completeness to be performed that to assess if legitimate information needs of the intended users are met. The completeness test guides the assurance provider to consider if the sustainability report contains relevant topics in sufficient depth for the six groups of intended users. The intended users include: suppliers, employees, customers, stakeholders in the context of the reporting organizations impact on social aspects of society and / environment and financial and economic stakeholders.

In comparison to RevR6(2006) that refers to stakeholders and readers in a broad manner, RevR6(2008) adopts the term "intended user" and specifies the information needs of an "average representative of a user group"[824]. Although RevR6(2008) introduced more specific requirements on the evaluation of suitability of reporting criteria and makes clear that users refer to intended group, the new provisions also raises new questions. One particular question highlighted by IAASB's task force was how the average representatives of each group could be identified? Others include how rational purpose can be determined, what is justifiable assumption and how legitimate information needs of intended users when the scope of the assurance engagement is more limited than the scope of reporting?

If the guidance notes were obligatory, RevR6(2008) could have an impact on the sustainability assurance engagements that can be accepted in Sweden. The guidance notes however only provide and aid to interpreting the standard and as such, there is no obligation for users to adopt those recommendations. This leads to the conclusion that given the broad evaluative conditions in the standard along with the provision for possible scope limitations in accuracy and completeness (as long as these limitations are clearly reported) are likely to have a minimal effect on the wide range of engagements accepted by current assurance providers in Sweden.[825]

It is interesting to note that RevR6(2006) provided that the assurance engagement has to be conducted with the auditors abiding by ethical requirements in the Auditors Act (2001:883), Auditors Ordinance (1995:665) and rules of RN (Revisorsnämdens föreskrifter). Both auditors and other assurance providers are obliged to follow the

---

[824] The distinction between stakeholders and intended users was a point raised in IAASB's task force comments to NIVRA in 2005 The IAASB task force pointed out in its comments on ED3410 that stakeholder far exceed the intended user (those expected to read the report).

[825] RevR 6(2008) sec. 16

Professional code of ethics and quality control standards by Far. RevR6(2008) however require assurance services in accordance to IFAC Code of Ethics for Professional Accountants and ISQC1. This provision does not have legal effect as hard law is legally binding.

8.6.2 Issues pertaining to planning and performing an engagement

RevR6(2008) provides guidance for assurance performed at a reasonable, limited level as well as a hybrid of the two types of engagement. It distinguishes the work to be performed for assurance providers to conclude with a reasonable level of assurance with that of limited assurance. According to RevR6(2008), reasonable assurance engagements require assurance providers to make inquiries, analytical review and test for relationships, verification of information against source documents and test for details whereas the main procedures of a limited assurance engagement only involves the first three aforementioned procedures. Examples given in the guiding notes on test for details include investigating the correct use of established methods or formulas, investigation of timeliness of entering key data into systems and recalculating and re-measuring significant emissions or waste production which are not required in limited assurance engagements.[826]

RevR6(2006) only provides guidance for limited assurance engagements. From the analysis in chapter 7 and section 8.2, RevR6(2006) does not impose an upper limit on the work performed for the auditor for limited assurance engagements but state that *"the review procedures should be manifested in a balanced combination of review of internal control of the company's management systems and substantive review procedures, both analyses and detailed testing"*. When performing the engagement, practitioners have exercised wide discretion of work to be carried out according to needs of each engagement. The introduction of reasonable level assurance engagements and hybrid engagements makes it important to make a distinction on work performed for reasonable and limited assurance engagements. Evidence reveal that limited assurance engagements performed in 2006 in Sweden incorporate a substantial level of detailed testing, which raises questions of how this can be dealt with when RevR6(2008) is applied. Would hybrid engagements solve the problem and if the reporting company commission a limited assurance engagement, should detailed testing be "unnecessary" as a source of evidence for assurance providers to make the same conclusions?

---

[826] RevR 6(2008)T46

8.6.3 Issues relating to RevR6(2008)

The members of Far have been active in the international sustainability assurance regulatory space and have taken a proactive approach in the development of standards in this field. The IAASB is currently conducting an issue study and making revisions to ISAE3000 and will look into NIVRA standard 3410N in the near future. The adoption of NIVRA's standard raises the question if Far had made a commitment too early in the process. NIVRA had taken into accounts comments made during ED3410's exposure period in particular the comments made by the IAASB task force, there are however issues where debates persist.

For example, RevR6(2008) adopts NIVRA's provision to distinguish the level of assurance by the extent of evidence gathering procedures.[827] The ISAE3000 on the other hand does not "fix" the nature, timing and extent of evidence gathering procedures. Procedures for limited assurance engagements are "deliberately limited" relative to reasonable assurance engagements. The IAASB task force expressed doubts that fixing the extent of evidence gathering procedures was appropriate at "this early stage of evolution".[828] RevR6(2008) has also readopted the terms "audit" and "review". The terms are not in line with ISAE3000 that adopt the terms reasonable and limited assurance.

These observations contribute to our understanding of the current issues of sustainability assurance in Sweden.

---

[827] This has been adopted in line with guidance in NIVRA's 3410N with a few differences e.g. how internal control is regarded in limited assurance engagements.

[828] Comments of the IAASB task force to NIVRA

9. Summary and Final remarks

9.1 Introduction

This thesis has been motivated by the challenge to understand regulation in contemporary society where regulatory gaps are not always filled by state initiatives but rather by that of non-state actors. The purpose is to contribute to the understanding of regulation by non-state standard setters through analysing the dynamics of regulating sustainability assurance through standards. In the first chapter, the notion of decentring motivated a revisit of four fundamental questions on regulation. These questions: "Who is/are regulating? What is the character of standards set by non-state standard setters? How are they interpreted? How are they enforced?" provided the building blocks from which the research questions of this thesis were formulated.

The case of sustainability assurance was adopted to analyse regulation through standards by the accountancy profession (a non-state actor) in an area outside its core competence. The analysis from 1997 to 2010 allowed the capture of the context within which sustainability assurance standards had been drafted and applied, the developments in the contents of standards, and the dynamic interaction between them in regulating sustainability assurance.

This thesis has drawn on the applied socio-legal perspective, which enables looking beyond the ontologically rule model used in traditional legal research. This approach draws our focus on issues reflecting the concerns of law, regulators and policy makers. The analysis had been conducted using a series of different paradigms to better understand how standards develop and how standards work to regulate sustainability assurance. Standards by non-state actors are developed in a regulatory space outside of a state's law making institutions. The context within which standards develop has therefore been analysed using a social science interpretive paradigm. The analysis of how standards work has been conducted in two steps. First, the scope and linguistic structure of the provisions of standards were analysed to determine the discretionary space granted to the regulated parties. This was carried out using socio-legal theories which had been built from linguistic and jurisprudence scholarship. This was followed by an analysis performed from a social science perspective, on how users of the standard exercise discretion granted to them in the context within which they operate. In parallel with this analysis, a legal analysis has been conducted as to whether sustainability assurance is regulated by existing legislation in Sweden. The legal analysis defines the potential interplay between standards by the accountancy professionand law from a legal perspective, The findings using this multi-paradigm analysis provides for a more complete understanding of how standards work.

Detailed discussions of the findings have been presented at the end of each chapter. The purpose of this chapter is to re-summarise the important findings and discuss the contributions of this thesis.

9.2 Dynamics of regulating sustainability assurance

9.2.1 The development of standards

The thesis started with raising the question on how the assurance concept has been framed and codified by the IAASB. The understanding of the problems and issues that emerged in the course of developing the IFAE and ISAE3000, the way the assurance concept has been codified into standards and the structure of IAASB's pronouncements provided an explanatory source for the findings of the other research questions.

The presence of different assurance providers offering a myriad of services in the name of "sustainability assurance" led to concerns about their quality and comparability. The regulatory gap pertinent to the quality of engagements and need for harmonisation led to different initiatives from non-state actors to call for or /and set standards.

The second question focuses research on how the perceived regulatory gaps concerning sustainability assurance have been filled by standards. The choice of 'how regulatory gaps are filled' rather than 'who is regulating' focused the inquiry on the activities actors engage in, the interactions and patterns of influence that occur in the regulatory space. The findings have enhanced our understanding of the context of standard setting.

In most jurisdictions, the role of legislators and due process for law making are provided in statutes. In the regulation of sustainability assurance however, the development of standards had been a result of the dynamic interactions between: the accountancy and non-accountancy actors; and between standard setters and organisations that do not set standards but have an interest in defining regulatory issues in the regulatory space. Interestingly, the IAASB, the standard setter of the accountancy profession constructed for itself a very limited role in the sustainability assurance regulatory space. AccountAbility (a non-accountancy standard setter) and FEE (a representative organisation for the accountancy profession in Europe) had been most active in the regulatory space in the period of study. Both organisations consequently played significant parts in establishing and shaping the regulatory issues that influenced the development of standards.

Findings support the literature reviewed in chapter 5 that the entry, exit and level of participation of non-state actors is primarily driven by the how they legitimise their role to relevant communities in society. An interesting observation made in this thesis is that the FEE has been consistently enthusiastic about sustainability assurance standards. The fact that FEE is a non-standard setter that represents the accountancy profession in Europe suggests that the call for standards have been driven by professional associations at the national level. This counters the conventional "top-down" trends of standard setting. The IAASB appears to have a passive role in this development. This could be due to its priority to engage resources to complete the clarity project, its earlier investment in the drafting of ISAE3000 between 1997 and 2003 and the availability of this generic assurance standard to provide guidance. But, on the other hand, it is important not to understate its influence. IAASB formed the SEAP and a task force to comment (in its own capacity) on NIVRA's ED3410. The extensive comment letter to NIVRA highlighted the issues pertaining to sustainability assurance and drafting conventions, which today continue to give direction and to shape the development of standards.

It has also been observed that patterns of influence between actors change over time. Changes had been driven by interactions between actors; practices in the assurance market and cross representation of members between actor groups. Three phases have so far been identified. In the first phase, actors attempted to establish the regulatory space through constructing problems, identifying issues and inviting, or excluding others into this space (1997-2002), this was followed by a phase of learning and cooperation between the actors (2003-2005). In the third phase (2006-2010), actors reduced their cooperative efforts. Instead they refocused on their own institutional objectives and engaged in interactions, to protect their own discourse and terminology.

With no legal authority having been granted, non-state actors engage in different legitimising strategies to stake claims within the regulatory space. To legitimise its entry into the space and its standards, AccountAbility engaged in input and throughput legitimacy strategies. Standards were set through multi-stakeholder participation appealing to the democratic ideals of standard setting in contemporary society. It further engaged in output legitimacy strategies through publishing papers on the applicability and usefulness of its standards when they are applied with other reporting and assurance standards.

In comparison, the accountancy profession has capitalised on their institutionalised role in society and their expertise by virtue of the financial audit, the way it which it legitimise its entry in the regulatory space. Analysis also revealed a shift in the way the accountancy profession constructed legitimacy over time. Their initial efforts were complemented by collaborative efforts with other standard setters such as AccountAbility, and evoking legitimacy claims through the support of third party organisations, particularly after AccountAbility staked its claim in the regulatory space.

The patterns of influence in the regulatory space highlighted the normative challenges of decentred regulation. The polymorphous actors grounded by different philosophical pursuits brought competing concepts and competing regulatory logic into the space. Problems in the regulatory space had been constructed, debated and negotiated, without the presence of an authority or central co-ordinating body making the final decisions. This is a role the state normally assumes in the development of state-centred regulation. What should be the objective of regulation, what issues and how they should be regulated depended on the various communities the non-state actors' legitimacy strategies are targeted towards. Fragmentation and regulatory overlap were evident particularly during the second and third phases identified above.

At the international level, the AA1000AS by AccountAbility remain the only international standard drafted specifically to guide sustainability assurance. Providers from the accountancy profession continue to apply ISAE3000 obligated by their membership to IFAC. NIVRA's 3410N has a "pseudo-international" status by virtue of IAASB's intention of using it as a point of departure to develop standards for assurance on sustainability reports. This project has however been postponed due to problems related to the criteria, the development of the International Integrated Reporting Framework (IIRF) and different demands on assurance by different jurisdictions.[829]

In the Swedish case, efforts to draft and revise standards on sustainability assurance had been voluntarily taken by Far. It is neither required by law or membership to IFAC to take this initiative. RevR6 had been developed "outside of" existing legal institutions drawing heavily on international standards. The development of RevR6 had also been dependent on sustainability reporting practices and the maturity of reporting systems of the companies that publish sustainability reports in the Swedish market.

9.2.2 How standards work

The third question of the thesis focuses on how standards work. An analysis has been carried out on how sustainability assurance standards work in Sweden. The empirical

---

[829] IAASB Consultation Paper, Proposed IAASB Strategy and Work program for 2012-14, 2/1/2011, p. 36 – 37.

observations of standards in action provided better understanding of the regulatory instrument, its properties and inherent limitations. Findings show that the scope, linguistic structure and character of standards affect how and who interprets the standard.

To recapitulate, the IFAE provides a framework defining the elements of assurance; the ISAE3000 provides the general principles to guide assurance on non-historical financial information whereas RevR6, in Sweden provide guidance on assurance engagements on sustainability reports based on ISAE3000. In this case, discretion had been exercised at two levels: the IAASB in constructing the contents of ISAE3000 and Far in aligning contents of RevR6(2006) to ISAE3000. The analysis shows the presence of replications of IFAE in ISAE3000; likewise, replications of IFAE and ISAE3000 in RevR6. This led to the conclusion that the extent which discretion is distributed in a multi-levelled regulatory regime depends on the choices exercised by standard setters at each level. Discretion is granted to standard setters at the next level who decides the extent of discretion it intends to exercise; discretion that is not exercised is left to the end user.

In this case, IAASB's objective to create a generic assurance standard on non-historical financial assurance engagements has limited its own discretion in defining the general principles of assurance. The over-inclusive nature of the standard, the lack of agreement on the determinants of assurance and the linguistic structure of the provisions in ISAE3000 granted Far wide discretion on setting sustainability assurance standards. The analysis of RevR6(2006) indicated however, that Far had chosen to delegate most of that discretion to assurance providers as both standards (reporting and assurance) and the market are developing. Most importantly, issues specific to sustainability assurance e.g., the criteria remain to be defined. Findings show that the assurance providers exercised wide discretion on the elements of assurance in engagements performed according to RevR6(2006). The amorphousness of RevR6(2006) allowed their necessary adaptation to suit practical circumstances and assurance providers to customise the scope of the assurance to each engagement. The wide discretion has also allowed users of standards to cherry pick or jointly-use different standards, performing reconciliations of any potential conflicts through interpretation. This was evident in engagements where assurance providers adopted the AA1000AS in conjunction with RevR6.

Black puts forth "The greater the shared understanding of the rule and practices it is addressing, the more the rule maker can rely on tacit understanding as to the aim of the rule and the context in which it operates, the less the need for explicitness, and the greater the degree to which simple, vague rules can be used"[830] Can we look to shared

[830] Black, J (1997), p. 31.

understanding of interpretive community, in this case assurance providers in Sweden to ameliorate rules' limitations? In the Swedish case, findings suggest that assurance providers share in common the importance of the "business case". Sustainability assurance engagements have been put forth by the profession to "make reports better" with emphasis placed on the quality of reports and its internal benefits to the reporting organisation. This raises the question as to what can we reasonably expect from RevR6(2006)? Perhaps the control of discretion is less pivotal if one accepts that standards are a legitimacy tool, a means to keep other assurance providers out of the market or for Swedish practice to keep pace with international trends?

Although RevR6 is developed by a Far and used by the members of Far; the fact that the users of the standard operate under state jurisdiction hints that an understanding of how it works cannot preclude a legal analysis of whether there is interplay between RevR6 and hard law. Chapter 8 highlights that there is potential interplay between RevR6, the Companies Act and the Auditors Act. By virtue of the requirement "god revisionssed", legislators have delegated to standard setters to give guidance through their standards which appertains to the concept of 'generally accepted auditing standards. With reference to Chapter 9 Section 3 of the Companies Act and sections 2(8a) and 3 of the Auditors Act; in scenarios where sustainability assurance engagements fall within the scope of the "statutory audit" or "audit business", RevR6 will be utilised as guiding material by both RN and the courts. On the other hand, if sustainability assurance falls outside the scope; there will be no direct interplay with hard law. The final interpretation of generally accepted auditing standard however lies with the courts. Findings lead to the conclusion that from a legal perspective, the potential role of RevR6 has been provided for in the legal system. The development of RevR6 is dependent on whether sustainability assurance falls under the scope of an "audit business". If so, the RN will have oversight over the development of the standard.

To conclude, sustainability assurance engagements performed by the accountancy profession in Sweden are guided by RevR6. RevR6(2006) was aligned with ISAE3000 that represents the accountancy profession's (audit profession's in the case of Sweden) conceptual understanding of assurance. The guidance in RevR6 specific to sustainability reports drew on international standards based on exchange of expertise by virtue of power of non-state actors. It however is potentially supervised by RN, based on the legal concept of an audit.

The scenario above shows that regulation by the profession in this area exists and works alongside state-centred regulation. The fragmentation that arises due to conflicting interpretations of the concept, role of the audit as well as the regulatory objectives between the profession and the state brings about systemic challenges. The incoherent regulatory goals within a regulatory system create uncertainty which potentially compromises the effectiveness of regulation. A clear regulatory overlap and conflict between legally and non legally binding standards are resolved in favour of the former. However, when regulatory overlaps are unclear, regulated parties operate in grey zones awaiting legal institutions' clarification of the law. The state has the ability to re-establish the public and private dichotomy through exercise its coercive powers which could determine the role, development and provisions of standards set by non-state actors. Until this power is asserted, the dominant logic of regulating sustainability assurance engagements in my opinion lies with the profession.

9.2.3 Limitations of this study

This thesis has several weaknesses. First, the research design to analyse how discretion is exercised by assurance providers could be more comprehensive. The design did not differentiate important subgroups within the assurance provider. The qualified financial accountant, the risk management department, the sustainability specialist have been regarded as one group. In addition, the accounting firms' strategies, interpretations of standards, methodologies and working papers did not form part of the corpus of documents necessary for a comprehensive understanding of how choices are made and who makes those choices within the accounting firms.

Second, it may be questioned if the exercise of discretion by the assurance provider be analysed without the assurance provider? The limited access to the "profession at work" has been noted in chapter 8 of this thesis. The study sought evidence around the assurance providers exercising choices through analysing the output of the assurance providers and the representatives of the reporting companies who dealt with the assurance providers. Furthermore, experts who are also practitioners have been interviewed, to increase the reliability of the inferences made. Still, they are no substitutes.

Third, the study was focused on how ISAE3000 worked as a standard to guide Far in the drafting of sustainability assurance standards. As a result, RevR6(2006) was chosen as a focal point to analyse how standards work. This choice has contributed to this thesis being somewhat "outdated" as RevR6(2006) is no longer in use and much has changed in today's (2011) practice. This poses the question as to whether this thesis is of any value. This thesis does not claim to study today's practices or the value of sustainability assurance to companies, the purpose is to enhance our understanding of how standards work in the multi-levelled regulatory regime of the accountancy profession. Findings contribute to the understanding and awareness of how discretion can be distributed by the structure and provisions of standards. The issue of "time" is important but perhaps, secondary in this case.

9.3 Contributions of this study

This thesis has attempted to bridge between the legal and social science disciplines to understand the dynamics of regulating sustainability assurance through standards. It introduces an approach that considers the context, contents and the interaction between the two. This approach takes into account the multi-faceted nature of regulation which explains how standards are developed and how they work in greater depth.

Social science research on sustainability assurance focuses primarily on the context which sustainability assurance providers operate. Stakeholder and legitimacy theories have been used to explain the various different interests driving assurance practices and assurance providers engaging in various different strategies to gain legitimacy from relevant communities. This thesis contributes to social science research by enhancing our understanding of the role played by standard setters in legitimising sustainability assurance and their contribution to the development of sustainability assurance practice both at the international level and in Sweden. Second, it brings to the social science field a socio-legal perspective. The study has drawn on the decentred view of regulation and the dimensions of rules which enhance our understanding of how standards as a regulatory tool function, its inherent characteristics and its construction which affects the distribution of discretion. An understanding of how standards work highlights what we can realistically expect from standards.

Social science research on audit regulation tends to place focus on the context and sites where audit regulation develops and the impact of standards on behaviour of regulated parties. The regulatory tool is often neglected. This study has shown the importance of understanding the structure and contents of standards in order to determine how discretion has been distributed. It also provides a complementary source of data to other qualitative techniques to understand the regulatory intentions of standard setters at each level. The findings of this study support literature reviewed that "principles" and "rules" are not meaningful descriptions of assurance standards but a false theoretical divide. An analysis of the provisions of standards in light of the scope, linguistic structure and character of standards allows a more thorough understanding of controls (or the lack of it) imposed by standards.

This study also brings attention to the legal lens as it determines the character of standards and the potential interplay between standards set by non-state actors and law. Standards developed external to existing legal institutions should not be prematurely eliminated to have no legal status. The fragmentation of contemporary regulation suggests the possible overlap given that there is no centre of authority coordinating the development of standards in a decentred regime. The Swedish case serves as a reminder that the regulation of the "global audit profession" and standard setting at an international level has not displaced the state and the interplay, development of standards and use of standards is dependent on whether the state had claimed jurisdiction in a particular area.

This thesis contributes to socio-legal and social science literature providing an empirical case of how the accountancy profession is regulated outside its traditional area of competence. In performing the research, it has been observed that the accountancy profession is bounded by the discourses of the financial audit, their role as the "audit profession" and existing institutional structures at both the international and national level regulating the financial audit. Findings however are case specific and generalisations cannot be drawn to apply to regulation of non-state actors in general. Explanations derived from this particular case study however can be extended to other cases which have similar theoretical characteristics. For example, the framework applied can be used to study regulation of other types of non-historical financial information which the accountancy profession are involved in such as assurance of greenhouse gas statements.

From a policy perspective, this thesis has highlighted the importance of the socio-political environment external to the legal structures which non-state standards are set. It draws attention to the importance of regulators to monitor the developments of non-state standards as these standards have a life outside the existing legal institutions in control behaviour of actors. Knowledge of the context which standards are set and the regulatory objectives of non-state standard setters, stand legislators in better stead to make decision on whether to let these regimes exist alongside independent of state regulation; when to step in; whether there is potential interplay and how to use these alternative regimes achieve a policy objectives. It is well accepted that the stability and predictability of legal institutions creates stability in the environment for economic entities to operate. These entities however operate in and have to adapt to an increasingly dynamic social, commercial and technological environment. With more policy decisions being made at a global level and the emergence of non-state standard setters filling the regulatory gap; from time to time, these legal institutions inevitable have to be reviewed to keep up with the dynamic environment which they preside. This allows legislators to keep track of the shifting sites of non-state regulation and their dominant regulatory logic. It has been shown that without the state as a central coordinating mechanism, uncertainty can be a problem. Uncertainty resulting from a potential regulatory overlap with existing legal structures can occur. Whether these standards are a problem or solution may be dependent on a state's legal institutions' vigilance in monitoring its interplay or possible interplay with hard law.

9.4 Future research

As sustainability assurance is a developing phenomenon, there are many potential areas of research that can follow this initial exploratory study. The following mentions a few areas that are important in light of the recent developments in this field.

When the research project commenced at the end of 2004, Far had just published the first version of RevR6(2004). Since then, RevR6 had gone through two revisions, AA1000AS(2003) had been revised; the ISAE3000 was made obligatory and is undergoing its first revision, NIVRA's 3410N has been adopted by Far; the Swedish government has made assurance on sustainability reports mandatory for state-owned companies. These are among a few of the more important examples of developments in the regulatory landscape on sustainability assurance. The dynamic environment in which sustainability assurance standards are developing has made the journey to open the standards black box a challenging one. As the regulation of sustainability assurance is a relatively new phenomenon, it was necessary to describe what has happened and what is happening prior to performing the analysis. In light of the recent development, each of the empirical chapters can be updated and dealt with in greater scope and depth. Studies on other countries as well as comparative studies between countries can be conducted. For example, a comparative study between Sweden, the UK and the Netherlands would be interesting. In the UK, assurance providers from the accountancy profession do not dominate the market for sustainability assurance. Assurance providers are applying ISAE3000 to carry out sustainability assurance engagements in the absence of guidance from the APB in this area. Assurance engagements in Sweden are guided by RevR6(2008) which is a standard that is a close replica of NIVRA's 3410N adopted by assurance providers in the Netherlands. A comparison of the practices between the three countries would enhance our understanding of how sustainability assurance standards work in different institutional contexts.

The IAASB embarked on a project to revise existing ISAs according to the clarity drafting convention. NIVRA's 3410N has been drafted in clarity style where most of the detailed guidance in its original exposure draft had been reclassified as 'application and other explanatory material'. This enables NIVRA's 3410N to be more "transplantable" to other countries. This raises a number of questions: to what extent will harmonisation be achieved in the practice of sustainability assurance? How would different jurisdictions regard application and other explanatory material and would it be filled with such details that could limit the flexibility of an engagement (the very benefit that principles-based regulation claims) leading to a paradox?

Some observations were made in the course of collecting empirical evidence. A small group of individuals from the large accounting firms have been active in the constructing exchange, support and collaboration networks, driving sustainability assurance and the development of standards both at the international and national levels. An analysis of how epistemic communities emerge, provide knowledge and frame issues for politicians and key decision-makers in this field would contribute to a more in-depth understanding of the socio-political context which standards develop.

In 2009, the Prince of Wales convened a high level meeting of investors, standard setters, companies, accounting bodies and UN representatives. The meeting brought together the The Prince's Accounting for Sustainability Project and the GRI to oversee the creation of a generally accepted integrated reporting framework. The International Integrated Reporting Committee (IIRC), has within the committee, "powerful, international cross section of leaders from the corporate, investment, accounting, securities, regulatory, academic and standard-setting sectors as well as civil society."[831] It published a discussion paper in September 2011 "Towards Integrated Reporting – Communicating Value in the 21st Century" inviting input on the purpose of integrated reporting and proposals with regard to the development of an IIRF. The framework has the objective to provide guidance to organizations to report in a clear, concise, connected and comparable manner on a broad set of information needed by investors and other stakeholders to assess the organization's long-term prospects. A pilot program to test the framework was launched in October 2011. This initiative has gained wide support including IFAC president who is a member of IIRC.

"Financial information is not sufficient. We have to provide information on sustainability, on social issues and environment, and it has to be done in an integrated way with a financial report."

Goran Tidstrom, President of the International Federation of Accountants and member of the IIRC[832]

The developments in the reporting landscape have the potential to shape the assurance market, as well as the role and obligations of financial auditors. The development of the IIRF raises many questions and potential areas of research. Some potential research topics building on this thesis are: who are the actors and how are they involved in IIRF? How would the accountancy profession construct the audit and assurance into the regulatory space? What are the implications on the scope of the audit, qualifications and responsibilities of auditors? How would the integration of the financial with non-financial reporting affect the current sustainability assurance market? Will consultancy providers have a role in future?

It also raises many legal issues pertaining to the Swedish case. In particular, would audit of the full integrated report be considered an audit by statute? How would it affect the interpretation of "audit business"? How would it affect the supervision of these activities by the RN and RevR6(2008)'s interplay with the law?

---

[831] http://www.theiirc.org/about/making-happen/ 14th October 2011.

[832] http://www.theiirc.org/about/ 14th October 2011.

9.5 Final comments

Regulation in contemporary society can be characterised by state regulation; regulation by non-state actors and hybrid forms of co-regulation at international regional and transnational levels using a myriad of regulatory tools of different nature and character. The traditional legal perspective focuses state-centred regulation, the multifaceted world of regulation through non-state actors influencing the behaviour of individuals in society have been largely precluded from the authority paradigm. Although deregulation policies pursued by states in the 1980s and 90s have shifted sites of regulation from the state, there is also a contra-trend of standards by non-state standard setters becoming hard law.

Interdisciplinary efforts are needed to understand the dynamics and challenges of regulation in contemporary society. This not only contributes to the understanding of regulation but the changing role of the state and legislators. The importance of understanding the social, political, market and institutional forces shaping the development of standards, and the nature and character of regulatory tools by non-state standard setters cannot be neglected. This knowledge and understanding sets foundation for new relationships to emerging between the state and other actors involved in regulation such that these fragmented regimes can work together, in a more coherent manner to achieve policy objectives.

Appendix 1 Key issues raised and discussed by FEE in 1999

Issues	Assurance and the level of assurance	Approach and responsibility of assurance	Engagement acceptance	Compliance with environmental law	Risk and internal control	Materiality	Reporting
General Consensus	Use of the term assurance Lack of a reporting framework	Flexible approach using multidiscipline-ry teams Preference for one report	Scope limitations on subject matter should not deter acceptance of engagements Importance of reporting clarity Importance of developing reporting standards (criteria for engagement) Involvement of stakeholders in determining criteria	Compliance with the environmental law is the responsibility of the reporting organisation's management	Assurance providers should only evaluate those procedures within the EIS or EMS relevant to the engagement objectives be stated it clearly in the assurance report	Importance of materiality	Clarity fundamental to close expectation gap

Issues to be further debated	Reporting framework Nature of subject matter Lack reliable appropriate evidence Conceptual underpinnings of assurance Assurance providers' reporting relationship with the reporting organisation	Who should be the legitimate coordinator? Applicability of ISA620?	Scope limitations on obtaining evidence? Applicability of ISA720 How to deal with information in the report that is not part of the assurance engagement? How stakeholders can be involved? Role of audit committee?		Suitability of the traditional audit risk model?	What constitutes materiality? How materiality should be considered for assurance on environmental reports? How to address completeness	How to report?

Appendix 2 State-owned companies reporting principles

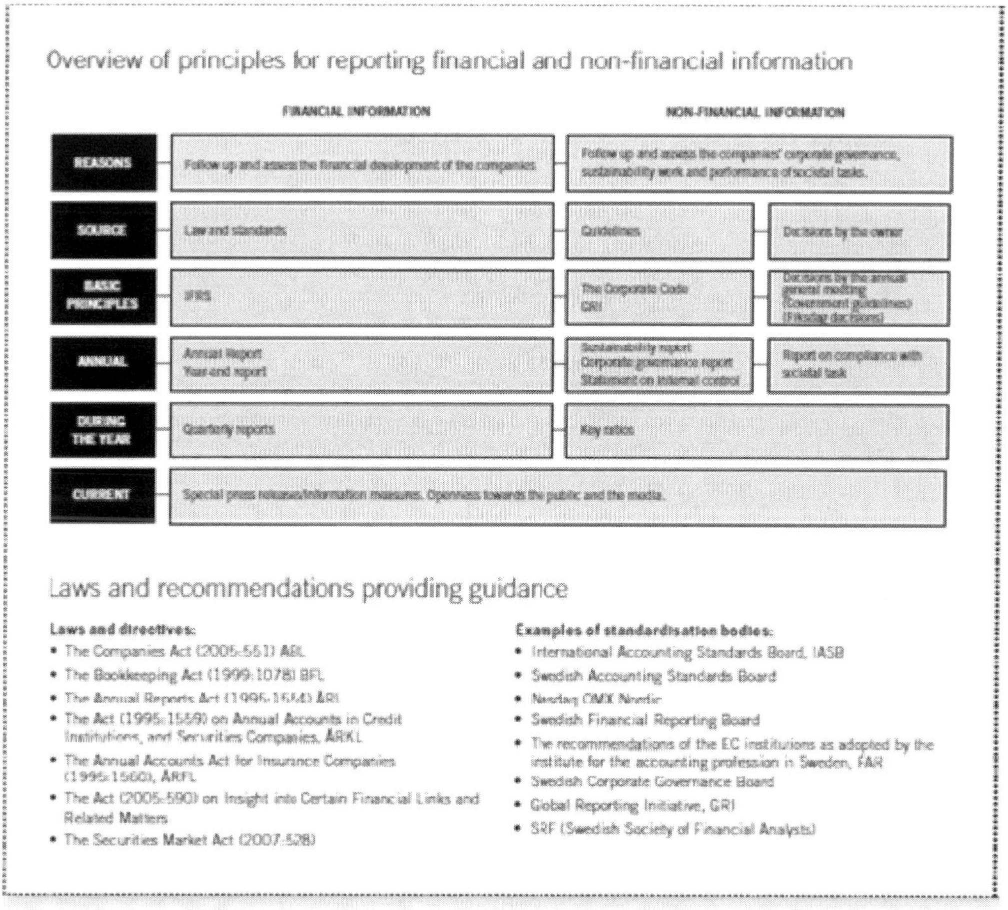

Overview of principles for reporting financial and non-financial information

	FINANCIAL INFORMATION	NON-FINANCIAL INFORMATION	
REASONS	Follow up and assess the financial development of the companies	Follow up and assess the companies' corporate governance, sustainability work and performance of societal tasks.	
SOURCE	Law and standards	Guidelines	Decisions by the owner
BASIC PRINCIPLES	IFRS	The Corporate Code GRI	Decisions by the annual general meeting (Government guidelines) (Riksdag decisions)
ANNUAL	Annual Report Year-end report	Sustainability report Corporate governance report Statement on internal control	Report on compliance with societal task
DURING THE YEAR	Quarterly reports	Key ratios	
CURRENT	Special press releases/information measures. Openness towards the public and the media.		

Laws and recommendations providing guidance

Laws and directives:

- The Companies Act (2005:551) ABL
- The Bookkeeping Act (1999:1078) BFL
- The Annual Reports Act (1995:1554) ÅRL
- The Act (1995:1559) on Annual Accounts in Credit Institutions, and Securities Companies, ÅRKL
- The Annual Accounts Act for Insurance Companies (1995:1560), ÅRFL
- The Act (2005:590) on Insight into Certain Financial Links and Related Matters
- The Securities Market Act (2007:528)

Examples of standardisation bodies:

- International Accounting Standards Board, IASB
- Swedish Accounting Standards Board
- Nasdaq OMX Nordic
- Swedish Financial Reporting Board
- The recommendations of the EC institutions as adopted by the institute for the accounting profession in Sweden, FAR
- Swedish Corporate Governance Board
- Global Reporting Initiative, GRI
- SRF (Swedish Society of Financial Analysts)

Source: Annual report state –owned companies 2010

Appendix 3 Interview questions and profile of company representatives

1a. What are the driving forces and purpose of sustainability reporting and third party assurance for Company X. How has this changed over time?

b. At which level or department of Company X, are sustainability reporting and assurance decisions taken? Which party serves as a point of contact with assurance providers?

Guiding questions

- Whether to report?
- The adoption of GRI guidelines?
- Should reports be subjected to external verification?
- Which external assuror to be commissioned?
- Did Company X have consultants with regards to sustainability reporting?

2a Can you please describe how the assurance process was carried out and give examples at each stage of the assurance process, the key activities performed by the assurance providers and the type of issues that you and your assurance provider will discuss.

Guiding questions

- How many auditors have conducted the engagement, their team size and composition?
- What was the duration of work performed by assurance providers on information presented by Company X
- What is your role or the role of other key personnel of the company in the assurance process?

3a. In commissioning third party assurance

- What did Company X hope to achieve?
- How has Company X benefited?
- Was there any reaction from stakeholders?

b. Why did Company X choose Audit firm Z? Were other providers considered?

c. What is the percentage of fee charged by assurance provider on sustainability issues as compared to a financial audit?

Company	Duration of interview	Interviewee profile
SAS	62 minutes	Group Environmental Director
SCA	57 minutes 20 minutes	Vice President Environmental Affairs Senior Vice President, Sustainability
Trelleborg	48 minutes	PR Manager, Corporate Communications
Vin & Sprit	30 minutes*	Corporate Responsibility Manager
Sandvik	42 minutes	Group Assurance, in charge of the company's sustainability work
SKF	20 minutes* 23 minutes	Representative on sustainability issues

*The interview questions were also answered via email.

Appendix 4 Issues raised on ED3410

Major Issues	Issues identified by the task force and SEAP in March 2005	Major issues raised in meetings[833] and by Board members	Comments of the task force to NIVRA
Scope of Guidance (ED3410 par 7)	Definition of sustainability report Does guidance apply to reports that cover only one or two of the components? Must the report be stand alone? Does the engagement cover assertion based or direct reporting?		The task force considered important to make clear its applicability with regard to reporting form and adopts the term "performance" and recommends the "incremental approach" in line with GRI. It suggest that the report is product than the function it is intended to serve and found explicit reference to relevant target group of stakeholders not necessary in the definition of sustainability reporting
Suitable	The practitioners'	Raised in March	IFAE: 35 suitable

[833] Audio recording and minutes

Criteria (ED3410 par 9, 46-50)	involvement to supplement criteria by setting minimum requirements for the content of sustainability report, is it consistent with role of practitioners from IAASB's perspective Should guidance be linked to a particular criteria e.g. GRI	meeting: Importance of public availability of criteria Concerns raised on "supplement criteria"	criteria are context sensitive. Allowing for different criteria is in line with the IFAE. Made available to intended users The assurance providers supplementary criteria is not consistent with IFAE and ISAE w.r.t role of criteria, assurance standards and assurance practitioners (B3)
Scope of engagement (ED3410 17-20)	Does not appear to envisage possibility of extending beyond reporting entity For limited scope engagement is there sufficient guidance on engagement acceptance and reporting? Practitioner's responsibility with regard to other information?		Better consistency in the document suggesting GRI Boundary Protocol as a reference Engagement consisting of audit and review elements is consistent with IFAE and ISAE as long as there is a rationale distinction between elements audited and reviewed Difficulty expressed in assuring completeness of report when there are scope limitations
Stakeholder Engagement ED3410 par	Stakeholders and intended users are used synonymously in practice they may	Raised in March meeting by board member:	Suggest importance of clearly specifying the responsibilities of management and

7, 78	not be the same? How is "average representative defined"? Implications of minor stakeholders? Extent of practitioner involvement in the stakeholder engagement to meet requirements specified ascertain most relevant stakeholders are in the target group and whether justified information requirement of the average representatives of each targeted group of stakeholders have been met	Identification of stakeholders	practitioner in stakeholder dialogues
Materiality and completeness (ED3410 par 25-36; 10, 17, 20, 117)	Does guidance take into account qualitative nature of sustainability information Definitions of materiality in relation to significance and relevance? Definition of Completeness?	June meeting: Completeness Was regarded to be broader than a cost benefits consideration. Concerns were raised as to whether completeness made it impractical to conclude reasonable assurance. It was noted that completeness consists separate	Whether a positive conclusion can be expressed given issues related to completeness: Nature of subject matter Potential stakeholders that may be relevant Time and cost constraints reasonably imposed on assurance

		components: identifying relevant information which can be performed through stakeholder engagement and full disclosure those area. The latter was identified as a significant area of concern.[834]	engagements Puts in perspective distinction between materiality and relevance
Joint Responsibility (ED3010)	Is joint responsibility allowed? ED does not address in detail issues that will arise when relying on the work of others e.g, if the financial auditor was not from the same audit firm as the sustainability assurance provider		IFAE and ISAE touches on joint responsibility but does not provide guidance
Reasonable and limited assurance (ED3410 par7, 105-108)	Are audit and review consistent with reasonable assurance and limited assurance engagements in the framework? Is evidence gathering procedures adequate to ensure consistency? Are specifications of	Discussed in the June meeting in relation to completeness	The terms audit and review are not consistent with IFAE Inconsistencies detected with regard to evidence gathering procedures E.g. review engagements according to ED3410 does not have to test

[834] Minutes June meeting par 9 agenda item 1A

	procedures to determine nature timing and extent of work for an engagement appropriate given the subject matter?		effectiveness of internal controls; ISAE3000 allows a more flexible approach to determine nature, timing and extent of procedures The possibility to issue a qualified conclusion accompanied by negative assurance is not consistent with IFAE
Other matters	Documentation; fraud, quality control; drafting style, consistency of terminology with IFAE and ISAE3000		Important to ensure practitioner's responsibility are clearly described in issues pertaining to fraud.

Bibliography

Abbott, A. (1988) The System of Professions: an Essay on the Division of Expert Labour, Chicago University Press

Alford, S. (1999) "A Critical Appraisal of Assurance Services Development in Australia", Australian Accounting Review, 9(18), p. 11-16

Adams, C. (2004) "The Ethical, Social and Environmental Reporting-Performance Portrayal Gap", Accounting, Auditing & Accountability, 17(5), p. 731-757

Adams, C.; Evans R. (2004) "Accountability, Completeness, Credibility and the Audit Expectations Gap", Journal of Corporate Citizenship, 14, p. 97-115

Amdou, T. (2008) "For the record? Understanding Regulatory Processes through Archival Materials", Cardiff University Law School, p.183-200

Arena, M.; Jeppesen, K. (2010) "The Jurisdiction of Internal Auditing and the Quest for Professionalization: The Danish Case", International Journal of Auditing, 14(2), p. 111-129

Baldwin, R.; Cave, M. (1998) Understanding Regulation, Oxford University Press

Balkin, J. (1996) Interdisciplinary as Colonization,

http://www.yale.edu/lawweb/jbalkin/writings.htm

Ball, A.; Owen, D.; Gray, R. (2000) "External Transparency or Internal Capture? The Role of Third-Party Statements in Adding Value to Corporate Environmental Reports", Business Strategy and the Environment, 9(1), p. 1-23

Bogner, A. and Menz, W. (2010) "The Theory Generating Expert Interview: Epistemological Interest, Forms of Knowledge and Interaction", in Bogner, A.; et al., Interviewing Experts, PalgraveConnect, p. 43-72

Banakar, R. (2009) "Law Through Sociology's Looking Glass: Conflict and Competition in Sociological Studies of Law" in Denis, A.; Kalekin-Fishman, D., The New ISA Handbook in Contemporary International Sociology: Conflict Competition and Cooperation, Sage

Banakar, R.; Travers, M. (2005) Theory and Method in Socio-legal Research, Hart Publishing

Benett, B; Bradbury, M.; Prangnell, H. (2006) "Rules, Principles and Judgement in Accounting Standards", Abacus, 42(2), p. 189-203

Black, J. (2009) "Legitimacy, Accountability and Polycentric Regulation: Dilemmas, Trilemmas and Organisational Response" in Peters, A., Koechlin, L., Förster, T. & Zinkernagel, G. ,Non-state actors as standard setters, Cambridge University Press, p. 241-269

Black, J. (2008) "Forms and Paradoxes of Principles-based regulation", LSE Law, Society and Economics Working Papers, 13-2008. Department of Law, London School of Economics and Political Science, London, UK, 3(4), p. 425.

Black, J. (2007) "Making a Success of Principles-based Regulation", Law and Financial Market Review, p. 193-196.

Black, J. (2002) "Critical Reflections on Regulation" CARR; London School of Economics

Black, J. (2001) "Decentring Regulation: Understanding the Role of Regulation and Self-Regulation in a 'Post-Regulatory' World", Current Legal Problems, 54(1), p. 103-146

Black, J. (1997) Rules and Regulators, Clarendon Press

Blokdijk, J.; Drieenhuizen, F. (1992) "The Environment and the Audit Profession", European Accounting Review, 1(2), p. 437-443.

Bogner, A.; Menz, C. (2010) "The Theory-Generating Expert Interview: Epistemological Interest, Forms of Knowledge, Interaction" in Bogner, A.; Liittg, B.; Menz, C. Interviewing Experts, PalgraveConnect, p. 43-73

Boli, J. (1999), "Conclusion: World Authority Structures and Legitimations" in Constructing World Culture International Non-governmental Organisations. Stanford University Press

Braithwaite, J. and Drahos, P. (2000) Global Business Regulation, Cambridge University Press

Brunsson, N.; Jacobsson, B. (2000) A World of Standards, Oxford University Press

Bryman, A.; Bell, E. (2007) Business Research Methods Second Edition, Oxford University Press

Caramanis, C. (2002) "The Interplay between Professional Groups, the State and Supranational Agents: Pax American in the age of Globalisation" Accounting, Organisations and Society, 27, p. 379-408.

Collison, D.; and Slomp, S. (2000) "Environmental Accounting, Auditing and Reporting in Europe: the Role of FEE" European Accounting Review, 9(1), p. 111-129

Cortell, A. and Davis, J.(1996) "How do International Institutions Matter? The Domestic Impact of International Rules and Norms", International Studies Quarterly, 40(4) p 451-478

Courville, S. et al (2003) "Introduction: Auditing in Regulatory Perspective" Law and Policy, Vol. 25, No. 3 July 2003 p. 179-184

Cowan, D.; Dopart, P.; et al. (2010) "A Cross-Sectional Analysis of Reported Corporate Environmental practices" Regulatory Toxicology and Pharmacology, p. 524-538.

Dando, N.; Swift, T.(2003) "Transparency and Assurance: Minding the Credibility Gap" Journal of Business Ethics 44(2-3), p.195-200

De Bellis, M. (2006) "Global Standards for Domestic Financial regulators" Global Jurist Advances, 6 (3), p. 1-24.

De Cruz, P. (1999) "Comparative Law in a Changing World", Second edition, London: Cavendish Publishing Limited

Deegan, C. Cooper, B & Shelly, M. (2005) "An Investigation of TBL Report Assurance Statements: International Evidence" European Accounting Association Conference Paper.

Dennis, I. (2010) "Clarity Begins at Home: An Examination of the Conceptual Underpinnings of the IAASB's Clarity Project" International Journal of Auditing, 14, p.298-316.

Djelic, M.; Sahlin-Andersson, K. (2007) Transnational Governance, Cambridge University Press

Diamant, A. (2004) Revisors Oberoende, Iustus Förlag

Dworkin, R. (1967) The Model of Rules University of Chicago Law Review

Elliot, R. (1994) "Confronting the future: Choices for the attest function" Accounting Horizons, 8(3), p. 107.

Elliott, R., Pallais. D (1997) "Are You Ready for New Assurance Services?" Journal of Accountancy, June, p. 47–51.

Elliott, R. (1997) "Assurance Services Opportunities: Implications for Academia", Accounting Horizons 11(4), Elliot, R., Pallais. D (1997) "First know your market" Journal of Accountancy, online issues, July. www.aicpa.org/PUBS/JOFA/july977elliot.htm.

Elliot, R. (1994) "Confronting the future: Choices for the attest function" Accounting Horizons, 8(3), p. 107.

Fish, S. (1984) "Fish v. Fiss" Standford Law Review, 36(6) p. 1325-1347

Foucault (1991) "Governmentality" in Burchell, G., Gordon, C. and Miller, P. The Foucault effect: Studies in Governmentality, University of Chicago.

Galligan, D. (2007) Law in Modern Society, Oxford University Press.p. 34-38.

Gioia, D.; Pitre, E. (1990) "Multiparadigm Perspectives on Theory Building" The Academy of Management review; 15(4), p.584-602

Gray, R. (2010) "Is Accounting for Sustainability Actually Accounting for Sustainability and How Would We Know" An Exploration of Narratives of Organizations and the planet, Accounting, Organizations and Society, 35(1) p. 4-61

Gray, R.; Milne, M. (2002), Sustainability reporting: Who's Kidding Whom? Chartered Accountants Journal of New Zealand, 81(6), p. 66–70

Gray, R. (2000) "Current Developments and Trends in Social and Environmental Auditing, Reporting and Attestation: A Review and Comment, International Journal of Auditing, 4(3), 247-268

Georgosouli, A. "Regulatory Interpretation: Conversational or Constructive? Oxford Journal of Legal Studies, Vol. 30, No. 2 (2010), pp. 361–384.

Gunningham, N. and Rees, J. (1997) "Industry self-regulation: An institutional perspective", Law and Policy, 19(4), p. 364-366.

Hart, H (1961) The Concept of Law, Oxford

Hancher, L.; Moran, M. (1989) "Organizing Regulatory Space" in Hancher, L.; Moran, M.(1989) Capitalism, Culture, and Economic Regulation, Clarendon Press

Hasan M.; Roebuck P.; Simnett; R. (2003) "An Investigation of Alternative Report Formats for Communicating Moderate Levels of Assurance" Auditing 22(2): 171-187

Hawkins, K (1992a), The Uses of Discretion , Oxford University Press

Hawkins, K. (1992b), 'The Use of Legal Discretion: Perspectives from Law and Social Science' in Hawkins, K.,The Uses of Discretion, Oxford University Press

Harris, D (1983) "The development of socio-legal studies in the United Kingdom" Legal studies 2, p315

Hedberg, C. and von Malmborg, F. (2003) "The Global Reporting Initiative and corporate sustainability reporting in Swedish companies" Corporate Social Responsibility and Environmental Management, 10(3), p. 153-164

Henning, R. (2000) "Selling Standards" in Brunsson, N.; Jacobsson, B. (2000) A World of Standards, Oxford University Press

Hofmann (1999) "Non state actors as new subjects of International Law: from the traditional state order towards the law of the global community

Humphrey, C.; Loft, A. and Woods, M. (2009) "The global audit profession and the international financial architecture: Understanding regulatory relationships at a time of financial crisis" Accounting, Organizations and Society, 34(6-7), figure 1.

Humphrey, C. and Loft, A. (2011 forthcoming) "Standardizing global financial reporting: Global governance without global government"

Hutter, B. (2006) "The Role of Non-state Actors in Regulation", in: Schuppert, Gunnar Folke, Global Governance and the Role of Non-state Actors, Nomos Publishers, Berlin

Kamp-Roelands, N. (2002) Towards a Framework for Auditing Environmental Reports, Tillburg University.

Kolk, Ans "Trends in Sustainability reporting by the Fortune Global 250" Business Strategy and the Environment, year: 2003, volume: 12, issue: 5, pages: 279-291

Keohane, R., Nye, J. (1989) Power and Interdependence: World Politics in Transition, Second edition, Boston, Little, Brown

Kerwer, D. (2005) "Rules that Many Use: Standards and Global regulation" Governance: An International Journal of Policy, Administration, and Institutions, 18(4), p.611-632

KolK, A..; Perego, P (2010) "Determines of the Adoption of Sustainability Assurance Statements: an International Investigation 19, p. 182-198

Larsson, B. (2005) "Auditor regulation and economic crime policy in Sweden, 1965–2000". Accounting, Organizations and Society 30, p. 127-144

Lewis, M.; Grimes, A. (1999) "Metatriangulation: Building Theory from Multiple Paradigms" Academy of Management Review, 24(4), p.672-690

Loft, A.; Humphrey, C.; Turley, S.(2006), "In pursuit of global regulation: Changing governance and accountability structures at the International Federation of Accountants (IFAC)" Accounting, Auditing & Accountability, 19(3), p. 428-451.

McBarnet, D.; Whelan, A. (1991) "The Elusive Spirit of the Law: Formalism and the struggle for legal control, Modern Law Review, 54, p.848-873

MacDonald, L.; Richardson, A. (2004) "Identity, appropriateness and the construction of regulatory space: the formation of the Public Accountant's Council of Ontario "Accounting, Organizations and Society, 29(5-6), p. 489-494

Moerman, L.; Van Der Laan,S. (2005) "Social reporting in the tobacco industry: all smoke and mirrors?", Accounting, Auditing & Accountability, 18(3), 374-389

Moberg, K.(2004) in "Auditors' Liability for Damages" Scandinavian Studies in Law, 45, Stockholm University Law Faculty

Morgan, B.; Yeung, K. (2007) "An Introduction to Law and Regulation" Cambridge University Press

Mörth, U. (2004) Soft law in Governance and Regulation: An Interdisciplinary Analysis, Cheltenham: Edward Elgar

Miller, G.; Dingall, R.(1997) Context and Method in Qualitative Research, Sage Publications

Meuser, M.; Nagel, U. (2010) "The Expert Interview and Changes in Knowledge Production", in Bogner,A.; Liittg, B.; Menz, C. Interviewing Experts, p. 17-36..

Nassani, M. (1995) "Fruits, Salads and Smoothies: A working definition of interdisciplinarity" Journal of Educational thought, p. 119-126.

Nugent, M. (1999) "The Australian Accounting Role in Developing Assurance Services", Australian Accounting Review, 9(2), p. 3-10.

Nergelius, J. (2000) "Constitutional Law" in Bogdan, M. Swedish Law in the New Millennium", Norstedt Juridik , p. 65-88.

O'Dwyer, B., Owen, D.,Unerman, J. (2011) "Seeking legitimacy for new assurance forms: The case of sustainability assurance", Accounting, Organizations and Society , 36, (1), 31-52.

O'Dywer, B.(2011 forthcoming). The Case of Sustainability Assurance: Constructing a New Assurance Service. Contemporary Accounting Research

O'Dwyer, B.; Owen, D. (2005) "Assurance Statement Practice in Environmental, Social and Sustainability Reporting: a Critical Evaluation" The British Accounting Review, 37(2), p. 205-229

O'Dwyer, B. (2003) "Conceptions of Corporate Social Responsibility: The Nature of Management Capture" Accounting, Auditing and Accountability Journal, year: 2003, 16(4) p. 523-557.

Owen, D., Swift, T., Humphrey, C., Bowerman, M.(2000) "The New Social Audits: Accountability Managerial Capture or the Agenda of Social Champions?. European Accounting Review 9 (1), 81–98.

Owen, D. and O'Dwyer, B. (2007) "Corporate Social Reporting and Stakeholder Accountability: The missing link" Accounting, Organizations and Society , 32(7-8), p. 649-667

Pallais, D. (1996) "Assurance Services: Where are We Going: Where We're Going". Journal of Accountancy, 182(3), p.16-17

Park, J. and Broson, T. (2005), "Experiences of and views on third-party assurance of corporate environmental and sustainability reports" Journal of Cleaner Production, 13(10-11). p. 1095-1106

Peters, A.; Koechlin, L.; Förster, T.; Zinkernagel, G. (2009) Non-state Actors as Standard Setters, Cambridge University Press

Picciotto, S. (2002) "Introduction: Reconceptualization Regulation in the Era of Globalization", Journal of Law and Society, 29(1), 1-11

Power, M. (1994) The Audit Explosion. London: Demos.

Power, M. (1996) "Making Things Auditable" Accounting, Organizations and Society, 21(2-3), p.289-315;

Power, M. (1997b) "Expertise and the Construction of Relevance: Accountants and Environmental Audit" Accounting, Organizations and Society, 22(2), p. 123-146

Quack, S. (2010) "Law, Expertise and Legitimacy in Transnational Economic Governance: an Introduction" Socio-Economic Review, 8, p. 3-16

Raz, J. (1972), "Legal Principles and the Limits of Law" Yale Law Journal, p. 823-854

Rose, N. and Miller, P. (1992) "Political Power beyond the State: Problematics of Government" British Journal of Sociology 43(23) p.173-205

Rosneau, J. (2006) "The Study of World Politics : Globalisation and Governance" London Routledge

Saunders, M.; Lewis, P.; Thornhill, A. (2009) Research Methods for Business Students, Fifth Edition, Prentice Hall

Samuel, G. (2009) "Interdisciplinarity and the Authority Paradigm: Should Law Be Taken Seriously by Scientists and Social Scientists?" Journal of Law and Society, 36(4), p.431-459

Scapens, R. (2007) "Using case studies in Accounting Research: The Role of Theory. Lund University workshop

Schlag (1985) Rules and Standards UCLA Law Review, 33, p. 379-430

Schauer (1997) "Prescriptions in Three Dimensions" Iowa Law Review, p. 911-922

Schneider, C (1992) "Discretion and Rules: A Lawyer's View" in Hawkins, K. The Uses of Discretion, Oxford University Press, p. 47-88.

Scott, P., Jackson, R.(2002) "Environmental, Social and Sustainability Reporting on the Web: Best Practices" Corporate Environmental Strategy, 9(2), p. 193-202.

Schipper, K. (2003) "Principles based Accounting Standards", Accounting Horizons, 16(3), p. 219-232

Simnett, A. ;Vanstralen; Chua,W.(2009) "Assurance of Sustainability Reports' The Accounting Review, 84 (3), p. 937-967

Sinclair, D. (1997) "Self-regulation versus Command and Control? Beyond False Dichotomies" Law and Policy", 19(4), p. 529-541

Smith, K. (2002) "What is Regulation: A Reply to Julia Black" Australian Journal of Legal Philosophy, 27, p. 37-46.

Suddaby, R.; Cooper, D.; and Greenwood, R. (2007) "Transnational regulation of professional services: Governance dynamics of field level organizational change" Accounting, Organizations and Society, 32(4-5), p. 333-362.

Swift, T. and Dando, N., "From Methods to Ideologies: Closing the Assurance Expectations Gap in Social and Ethical Accounting Auditing and Reporting" Journal of Corporate Citizenship, 8. p. 81–90.

Suchman, M. (1995) "Managing Legitimacy: Strategic and Institutional Approaches" Academy of Management Review 20(3), p. 571-610.

Tamm-Hallström, K. (2004) Organizing international standardization, Cheltenham: Edward Elgar

Thorell, P. (2008) Redovisning och juridik, Iustus Förlag

Thomas, P. (1997) 'Socio-Legal Studies: The Case of Disappearing Fleas and Bustards', in Thomas, P., Socio-Legal Studies. Aldershot: Dartmouth. p. 1–22.

Travers, M. (2001) "Sociology of Law in Britain", American Sociologist, 32, p.26-40

Tomlins, C. (2007) "How Autonomous is Law?", Annual Review of Law and Social Science, 3, p. 46-49

Twining, E. and Miers, D.(1985), How to Do Things With Rules, Second edition, Butterworths

Tweedie, D. (2007) 'Take it from the Top', 'Can Global Standards be Principle Based?'Journal of Applied Research in Accounting and Finance,

Tomkins, C. (2001) "Interdependencies, Trust and Information in Relationships, Alliances and Networks", Accounting, Organizations and Society, 26(2), p.161-191

Unerman, J.; Bennett, M. (2004) "Increased Stakeholder Dialogue and the Internet: Towards Greater Corporate Accountability or Reinforcing Capitalist Hegemony?" Accounting, Organizations and Society 29(7) p. 685-707

Vick, D. (2004), "Interdisciplinarity and the Discipline of Law", Journal of Law and Society, 31(2), p.1163-193

Vogel, H. (2000) "Sources of Swedish Law" in Bogdan, M. "Swedish Law in the New Millennium", Norstedt Juridik p. 48-64

Wallage, P. (2000) "Assurance on sustainability reporting: An auditor's view" Auditing, p, 53-65.

Wallerstedt, E. (2001), "The Emergence of the Big Five in Sweden", European Accounting Review, 10(4), p.843-867

Young, J. (1994) "Outlining Regulatory Space: Agenda Issues and the FASB", Accounting, Organizations and Society, 19(1), p. 83-109

Official Documents

Legal Sources

Sweden

LEGISLATIONS

Aktiebolagslag (2005:551)

Revisorslag (2001:883)

Förordning om revisorer (1995:665)

Årsredovisningslag (1995:1554)

Miljöbalk (1998:808)

Revisorsnämndens föreskrifter (1996:1)

Årsredovisningslag (1995:1554)

PREPARATORY WORKS

SOU 1941:9 Lagberedningens förslag till Lag om Aktiebolag.

PROPOSITIONS

Prop. 1975:103	Regeringens proposition med förslag till ny aktiebolagslag, m.m
Prop. 1996/97:167	Miljöredovisning och miljöinformation i näringslivet
Prop. 1997/98:99	Aktiebolagets organisation
Prop. 2000/01:146	Oberoende, ägande och tillsyn i revisionsverksamhet

CASE

NJA 1996 s.224

European Union

Fourth Council Directive 78/660/EEC of 25 July 1978 on the Annual Accounts of Certain Types of Companies

Seventh Council Directive 83/349/EEC of 13 June 1983 on Consolidated Accounts

Eighth Council Directive 84/253/EEC of 10 April 1984 based on Article 54(3)(g) of the Treaty on the approval of persons responsible for carrying out the statutory audits of accounting documents

Directive 2003/51/EC of the European Parliament and of the Council of 18 June 2003 amending Directives 78/660/EEC, 83/349/EEC, 86/635/EEC and 91/674/EEC on the annual and consolidated accounts of certain types of companies, banks and other financial institutions and insurance undertakings

REGULATION (EC) No 1221/2009 OF THE EUROPEAN PARLIAMENT AND OF THE COUNCIL of 25 November 2009 on the voluntary participation by organisations in a Community eco-management and audit scheme (EMAS), repealing Regulation (EC) No 761/2001 and Commission Decisions 2001/681/EC and 2006/193/EC

Other Official Documents

AccountAbility

AA 1000 Framework 1999

AA1000AS (2003)

AA1000AS(2008)

AA1000APS(2008)

AA1000 Guidance Note on the Principles of Materiality, Completeness and Responsiveness as they Relate to the AA1000 Assurance Standard 2006

The State of Sustainability Assurance 2003

The Materiality Report: Aligning Strategy, Performance and Reporting 2006

Zadek, S & Merme, M "Redefining Materiality: Practice and Public Policy for Effective Corporate Reporting" July 2003b

Zadek, S. & Raynard, "The future of sustainability assurance" ACCA Research Report No, 86 by ACCA working with AccountAbility 2004

Zadek, S., Raynard, P. & Forstater, M, "What Assures: Listening to the words of assurance" 2006

AccountAbility Research Report "The State of Responsible Competitiveness 2007: making sustainable development count in global markets" July 2007, p. 22.

AICPA

AICPA (1997) "Report of the Special Committee on Assurance Services".

AICPA Thomas Ray: international assurance standards exposed for comment. October 1997

European Union

Commission Recommendation of 30 May 2001 on the recognition, measurement and disclosure of environmental issues in the annual accounts and annual reports of companies (notified under document number C(2001) 1495)

European Commission "Promoting a European Framework for Corporate Social Responsibility" Green Paper July 2001

Far

FARS SAMLINGS VOLYM 2004 DEL II

Förslag till rekommendation oberoende översiktlig granskning av frivillig separat hållbarhetsredovisning feb 2004

RevU5 Revisorns beaktande av miljöfrågor vid granskning av årsredovisningen

FARS SAMLINGS VOLYM 2007 DEL II

RevR 6 Förslag till rekommendation RevR6 Oberoende granskning av frivillig separat hållbarhetsredovisning

FARS SAMLINGS VOLYM 2008 DEL II

RevR 6 Oberoende granskning av frivillig separat hållbarhetsredovisning

FARS SAMLINGS VOLYM 2009 DEL II

RevR 6 Bestyrkande av hållbarhetsredovisning

FARS SAMLINGS VOLYM 2011 DEL II

RevR 6 Bestyrkande av hållbarhetsredovisning

RevU5 Revisorns beaktande av icke-finansiella upplysningar vid granskning av årsredovisningen

Far Articles of Association

Far bidrar till hållbar utveckling" Balans nr 2 2004.

European Sustainability Reporting Association report for Sweden 2006, 2007, 2008.

Far Engelska Ordbok.

FAR Press release, 11th March 2004.

FEE

FEE (1995) Environmental Accounting, Reporting and Auditing: Survey of Current Activities and Developments within the Accountancy Profession.

FEE (1996) Research Paper: Expert statements in environmental reports.

FEE (1999) Discussion Paper "Providing assurance on environmental reports"

FEE(1999)Position Paper "Towards a Generally Accepted Framework for Environmental Reporting

FEE(2000) Analysis of responses to FEE Discussion Paper

FEE (2002)Discussion Paper Providing Assurance on Sustainability Reports

FEE(2003) Benefits of Sustainability Assurance

FEE(2003b) Issues Paper - Principles of Assurance: Fundamental Theoretical Issues with respect to Assurance in Assurance Engagements - Full-Length Version

FEE (2004) Call for Action: Assurance for Sustainability (2004)

FEE (2006) Discussion Paper Key Issues in Sustainability Assurance-An Overview

FEE Comment Letter on Exposure Draft of Guidance for the Use of AA 1000 AS (2008)

FEE Comment Letter to AccountAbility 18/3/2009, Letter to Mr. Alan Knight, Head of Standards and Related Services, AccountAbility on Consultation on AA1000 Assurance Standard Revision Draft issued on 17/7/2008

FEE Comment letter to AccountAbility, 22/9/2008

FEE Comment Letter to Mr. Richard Howitt, MEP Assurance on Sustainability Reports 6/12/2006

FEE Comment Letter to AccountAbility on Consultation on Guidance Note on Assurance Levels & Statements as they relate to AA1000 Assurance Standard 26/7/2007

FEE Comment letter to AccountAbility, 22/9/2008

FEE correspondence with IFAC ref IAASB consultation paper on assurance aspects of G3 23/5/2006

GRI

GRI (2001), Overarching Principles for Providing Independent Assurance on Sustainability Reports,

Global Reporting Initiative Verification Working Group, Boston, MA, .

Global reporting Standards 2002, G2

Global reporting Standards 2006, G3

http://web.archive.org/web/20010710133538/globalreporting.org/News/June1999Update.htm downloaded 2/2011

IFAC

IFAC Code of Ethics for Professional Accountants, Parts A and B.

International Standards on Quality Control 1

ED 1 Reporting on the Credibility of Information 1997

ED 2 Assurance Engagements 1999

International standards on assurance engagements 100

International standards on assurance engagements 2000

International standards on assurance engagements 3000

International Framework for Assurance Engagements (issued with ISAE100, ISAE2000 and ISAE3000)

IAASB Consultation Paper, Proposed IAASB Strategy and Work program for 2009-2011

IAASB Consultation Paper, Proposed IAASB Strategy and Work program for 2012-14

Proposed Terms of Reference, Preface to the International Standards on Quality Control, Auditing, Assurance and Related Services and Operations Policy No 1 – Bold Type Lettering,

IFAC (2001) Rebuilding Confidence in Financial Reporting

IFAC policy position 1: IFAC's position on professional regulation, December 2007

IFAC(2010) Handbook of International Quality Control, Auditing, Review, Other Assurance, and Related Services Pronouncements.

Feedback to NIVRA 2005 Meeting Agenda Item 3A and 7

IAASB Annual reports

IAASB Meetings 2002, 2003, 2005